AF772675

Nuclear Analytical Methods in the Life Sciences 1994

Nuclear Analytical Methods in the Life Sciences 1994

Edited by

Jan Kučera

Nuclear Physics Institute,
Řež near Prague, Czech Republic

Ivan Obrusník

Czech Hydrometeorological Institute,
Prague, Czech Republic

and

Enrico Sabbioni

CEC, Environmental Institute, Joint Research Centre,
Ispra, Italy

Humana Press, Totowa, New Jersey

This work also appears as volumes 43–45 of the Humana Press journal,
Biological Trace Element Research, Editor-in-Chief: Gerhard N. Schrauzer.

ISBN 0-89603-300-7

© 1994 Humana Press Inc.
999 Riverview Drive, Suite 208
Totowa, New Jersey 07512

All rights reserved.

No part of this book may be reproduced, stored in a retrieval system, or transmitted in any form or by any means, electronic, mechanical, photocopying, microfilming, recording, or otherwise without written permission from the Publisher.

This publication is printed on acid-free paper. ∞
ANSI Z39.48-1984 (American National Standards Institute) Permanence of Paper for Printed Library Materials.

Photocopy Authorization Policy:
Authorization to photocopy items for internal or personal use, or the internal or personal use of specific clients, is granted by Humana Press Inc., provided that the base fee of US $4.00 per copy, plus US $00.20 per page is paid directly to the Copyright Clearance Center at 222 Rosewood Dr., Danvers, MA 01923. For those organizations that have been granted a photocopy license from the CCC, a separate system of payment has been arranged and is acceptable to Humana Press Inc. The fee code for users of the Transactional Reporting Service is: [0-89603-300-7/94 $4.00 + $00.20].

Printed in the United States of America. 9 8 7 6 5 4 3 2 1

CONTENTS

---

*For papers with multiple authorship, the asterisk identifies the author to whom correspondence and reprint requests should be addressed.

Preface

The 1993 International Conference on Nuclear Analytical Methods in the Life Sciences (NAMLS) was a continuation of a series of conferences held by the International Atomic Energy Agency (in Amsterdam in 1967, in Bled in 1972, in Vienna in 1978) and by the National Institute of Standards and Technology in Gaithersburg in 1989. The continuous aim of these conferences has been to bring together those who are working as environmentalists, toxicologists, and physicians with those who are working in the fields of development and application of nuclear analytical methods.

The NAMLS–93 Conference was held at the University of Agriculture in Prague. Traditionally, results from multi-disciplinary research on both the methodological aspects and the development of nuclear analytical methods and their application to various field of the life sciences, quality assurance, and comparison with other methods for trace element analysis and speciation were presented. A special emphasis was given to the problems of analysis of biological materials.

Since the last conference, nuclear analytical methods have passed through a major developmental phase toward refinement and optimization of the procedures applied to biological and environmental samples and toward the improvement of accuracy of results. The inherent potential of these methods for the accurate determination of trace and ultrace element concentrations, especially in the case of neutron activation analysis, and the multielemental character of these methods seem to promise a good future in a strong competition with other rapidly developing nonnuclear methods. These features along with more extensive applications of nuclear analytical methods in the life sciences, can also help in keeping large nuclear facilities (reactors, accelerators) in operation, which is becoming a problem in various countries. From the list of the conference participants can also be seen that nuclear analytical methods are utilized in the life sciences not only in countries with the long tradition in this field, but also in many developing countries, and this is also promising for the future.

The NAMLS–93 Conference was organized by the Ioannus Marcus Marci Spectroscopic Society (formerly Czechoslovak Spectroscopic Society) in cooperation with the University of Agriculture, Prague, Joint Research Centre of the Commission of the European Communities, the Environment Institute, Ispra, Italy, and the International Atomic Energy Agency. The two latter institutions supplied both financial support and scientific expertise. The meeting was cosponsored by the Sigma-Aldrich Ltd., Prague and Junkers Software Production Ltd., Prague. We are also indebted to many people and organizations for their help in organizing the meeting. Special thanks are due to J. Faltejsek of the Czech Ecological Institute, Prague for his assistance in preparing the manuscripts for publication.

We hope that the Conference helped in exchanging ideas and stimulating further collaborative actions between scientists working in the above-mentioned fields. We also hope that the results of the meeting presented in this special issue will encourage the readers to participate in the next meeting, presumably within four years.

Jan Kučera
Ivan Obrusník
Enrico Sabbioni

History of Development and Applications of Nuclear Analytical Methods in the Czech Republic

MILOSLAV VOBECKÝ

Academy of Sciences of Czech Republic, Institute of Analytical Chemistry, 142 20 Praha 4, Czech Republic

ABSTRACT

This article presents the evolution of the field of nuclear analytical methods, as well as its inventions and applications from the 1960s to the present in the territory of the former Czechoslovakia.

Index Entries: Reactor neutron activation analysis; radiochemical neutron activation analysis; substoichiometric separation; replacement substoichiometric separation; instrumental neutron activation analysis; high-energy neutron capture γ-ray analysis; γ activation analysis; neutron depth profiling; Rutherford back scattering spectrometry; proton-induced x-ray emission; applications of radioanalytical methods.

The beginning of the field called Radiochemistry is tightly connected to the occurrence of uranium ore in Bohemia and its subsequent processing on uranium pigments. Thanks to a successful experiment of metallurgist A. Patera, the processing of uranium ore (pitchblende) on uranium pigments for the glass industry commenced at Jáchymov in 1856. The wastes after chemical treatment enriched with radium were used by M. and P. Curie in their experiments leading to the discovery of the first radioactive elements. Since 1906, radium salts were produced at Jáchymov as well *(1)*. The discovery of induced uranium fission unfortunately caused the total exploitation of this rare locality for nonacceptable purposes.

Methodological development in the field of nuclear analytical methods has been very closely connected with the progress in nuclear physics, nuclear chemistry, and electronics.

Good conditions for development of the entire nuclear research and its applications in Czechoslovakia became available only in the second

Biological Trace Element Research Editor: G. N. Schrauzer © 1994 by Humana Press Inc.

half of the 1950s. The first step was the establishment of the Institute of Nuclear Physics and Faculty of Technical and Nuclear Physics at the Charles University in Prague in 1955. In the structure of the above mentioned faculty that was later transformed to the Faculty of Nuclear Science and Physical Engineering (FNSPE) of the Czech Technical University (CTU), we could find the Department of Nuclear Chemistry, which was built up and directed by Vladimír Majer. He is, by the way, the author of a textbook titled *Radiochemistry*, which was already issued in 1942 as the first Czech monograph in this field *(2)*. He lectured radiochemistry at the Faculty of Chemical Engineering of the CTU in Prague since 1946.

REACTOR NEUTRON ACTIVATION ANALYSIS

A milestone in the development of the nuclear analytical methods was the beginning of routine operation of our first experimental nuclear reactor VVR-S at the Nuclear Research Institute of the Czechoslovak Academy of Sciences in the early 1960s. The first report concerning application of radiochemical neutron activation analysis was published by Šimková, Kukula, and Slunečko in 1962. At that time, a team of pioneers within the Radiochemical Division of the Nuclear Research Institute applied neutron activation analysis (NAA) method to the analysis of pure materials (aluminum, silicon) *(3)*, alloys *(4)*, organic polymers *(5)*, and many different matrices *(6)*.

In this period, Rakovič from the Biophysical Institute of the Charles University concentrated attention on application of NAA to the analysis of biological materials *(7)*. He is also the author of the first Czechoslovak monograph dedicated to activation analysis issued in 1970 *(8)*. Determination of zinc in different biological materials using radiochemical NAA was also published by Babický from the Isotope Laboratory of the Academy in 1966 *(9,10)*.

A smart principle of substoichiometric separation can be considered as a breakthrough in the methodology of the radiochemical activation analysis. The scientific team of Růžička, Starý, and Zeman of the Department of Nuclear Chemistry of the FNSPE developed the methodology and applied this principle in various fields *(11–13)*. Furthermore, many methods for determination of trace concentrations of elements based on isotope exchange were also developed at this department, e.g., determination of phosphorus *(14)*, organomercury compounds *(15,16)*, and many others.

Obrusník and Adámek developed replacement (displacement) substoichiometry *(17–19)* for separation of elements with low extraction constants of their chelates, e.g., for the separation of indium, tin, thallium, and so on.

Tavendale's discovery of a semiconductor detector of γ rays *(20)* has opened unprecedented possibilities for radioanalytical applications. In Czechoslovakia, the first detectors were developed and prepared by Trousil at the Institute of Solid-State Physics of the Czechoslovak Academy of Sciences in 1965. Vobecký started using high-resolution γ-ray spectrometry for NAA already in 1966 *(21)*. Later on, this technique was exploited for element determination using prompt rays from radiation capture of neutrons by atomic nuclei *(22,23)* for nondestructive burn-up determination of nuclear fuel and for determination of isotopic composition of uranium *(24)*. In this period, attention was paid especially to the materials of mineral origin, e.g., tektites *(25)*, meteorites *(26)*, minerals, rocks, and lunar samples (Apollo 11, Apollo 12) *(27)*.

The introduction of instrumental neutron activation analysis (INAA) increased the concern about possible automation of activation analysis. Many of our laboratories have been consecutively equipped with necessary spectrometric systems. In addition to the establishment of the NAA group in the Nuclear Research Institute, the NAA method was also introduced in the fields of geology, material research, and the uranium industry, as well as biological and medical research. The NAA laboratory in Řež was oriented especially toward studying environmental problems of analysis of coal, fly ash, atmospheric aerosol, and hair samples *(28–30)*. Optimization of the analytical process and quality assurance of analysis were systematically studied in this laboratory by Obrusník *(31)*. In spite of the significant contribution of INAA to the instrumentation of the analytical process, some analytical tasks could only be solved using radiochemical NAA. From the beginning of the 1970s, Kučera *(32,33)* concentrated his efforts on reliable determination of very low concentration of toxic and/or essential elements (As, Cd, Cu, Hg, Mn, Mo, Ni, Pt, Sb, Se, V, and so forth) in biological and environmental materials, and regularly took part in the preparation of Czechoslovak and international reference materials *(34)*. Presently, he serves as a cooperating analyst for the US NIST *(35)* and is involved in vanadium determination in human tissues *(36,37)*.

FAST NEUTRON ACTIVATION ANALYSIS (FNAA)

The installation of several neutron generators in the 1970s has allowed the application of FNAA to the analysis of mineral raw materials, and alloys, and to the development of a continuous NAA *(38,39)*. A neutron generator constructed at the Physical Institute of Slovak Academy of Sciences in Bratislava was applied for this purpose. Bartošek and Mašek from the Institute of Geophysics in Brno contributed to a solution of the dead time correction problems and developed one of the first NIM modules for this purpose *(40,41)*.

HIGH-ENERGY NEUTRON CAPTURE γ-RAY ANALYSIS

Prompt high-energy γ radiation followed radiation capture of neutron during irradiation of samples by means of isotope neutron sources ^{241}Am-Be and ^{252}Cf was suggested to be used for the control of technological processes, for instance, in power plants. It was used for determination of silicon, sulfur, and iron in coal *(42,43)*, nickel, and chromium in ores *(44)*, and so on. This work was carried out at the Department of Physics of the FNSPE in cooperation with the Institute of Nuclear Biology and Radiochemistry.

γ ACTIVATION ANALYSIS (GAA)

Methodology investigations related to the application of GAA especially for the analysis of mineral raw materials were systematically performed by Řanda at the Institute of Raw Materials in Kutná Hora *(45,46)*. The first experiments were performed by using microtrons of the Institute of Physical Problems by Kapitsa in Moscow and also in the Joint Institute for Nuclear Research in Dubna. Since 1981, the microtron of the FNSPE in Prague constructed by Šimáně and Vognar has been utilized for this purpose. Since then, another microtron has been installed at the Institute of Raw Materials in Kutná Hora.

NEUTRON DEPTH PROFILING, RUTHERFORD BACK SCATTERING SPECTROMETRY, AND PIXE

Another important group of nuclear analytical methods that enables nondestructive determination of spatial distribution of elements with good depth resolution is presently used at the Institute of Nuclear Physics, the Academy of Sciences of Czech Republic. The methods of neutron depth profiling (NDP), Rutherford back scattering spectrometry (RBS), and particle-induced x-ray emission (PIXE) are utilized for solution of both the scientific and technological problems related to solid-state physics *(47–50)*. A Van de Graaff accelerator located there serves as a source of beams of protons, deuterons, and α particles with the energies in the range of 1 to 3 MeV, and the beam currents up to 5 μA for RBS and PIXE methods. The RBS method was applied here for the first time by Habanec *(51)* at the end of the 1960s. In this laboratory, nuclear reactions (n,γ) and (n,α) on nuclei of light elements induced by thermal neutron extracted from the nuclear reactor have also been exploited *(52–54)*.

TASK GROUP OF INSTRUMENTAL RADIOANALYTICAL METHODS

Increasing interest in application of the high-resolution γ-ray spectrometry stimulated the establishment of the Task Group of Instrumental Radioanalytical Methods of the Czechoslovak Spectroscopic Society, now the Ioannes Marcus Marci Spectroscopic Society. The group has been established in 1971, when the main activity was concentrated on INAA methodology and automated processing of γ-ray spectra measured by semiconductor detectors. The group attracted the majority of scientists from laboratories of the Academy of Sciences, universities, and research institutes as well as from industry. In addition to seminars and courses, national conferences on Instrumental Activation Analysis (IAA) have been held every year until now. A review dealing with the development of activation analysis in Czechoslovakia until 1977 was presented at the Conference on Instrumental Activation Analysis—IAA 77 *(55)*.

In the present period of structural changes, many radioanalytical laboratories are facing problems of competition with other analytical techniques. It difficult to say now in which direction the development and extent of application of radioanalytical methods in the Czech Republic will go, although we are sure that in some fields and applications, these methods are indispensable.

REFERENCES

1. V. Majer, *Základy jaderné chemie* (in Czech), SNTL, Prague, 1961. V. Majer, *Grundlagen der Kernchemie*, 2nd ed. (in German), Carl Hauser Verlag, München, Wien, 1982.
2. V. Majer, *Radiochemie* (in Czech), The Union of Czech Mathematician and Physicists, Prague, 1942.
3. F. Kukula, J. Slunečko, and M. Šimková, *Determination of Copper in Aluminum* (in Czech), Report UJV 672, Nucl. Res. Inst. Czechoslovak Acad. Sci. Řež, 1962.
4. M. Šimková, *Hutnické listy* (in Czech), **18,** 357 (1963).
5. M. Šimková, F. Kukula, and J. Slunečko, *Chem. Zvesti* (in Czech), **19,** 115 (1965).
6. M. Vobecký and O. Knotek, *Chem. listy* (in Czech), **58,** 15 (1964).
7. M. Rakovič and H. Talpová *Čas. Lék. čes.* (in Czech), **103,** 632 (1964).
8. M. Rakovič, *Activation Analysis*, Academia, Prague, 1970.
9. A. Babický and D. M. Taylor, *Českoslov. stomatologie* (in Czech), **66,** 167 (1966).
10. J. Pařízek, J. C. Boursnell, Mary F. Hay, A. Babický, and D. M. Taylor, *J. Reprod. Fert.* **12,** 501 (1966).
11. J. Růžička, J. Starý, and A. Zeman, *Talanta* **10,** 905 (1963).
12. A. Zeman, J. Růžička, and J. Starý, *Talanta* **10,** 685 (1963).
13. J. Růžička and J. Starý, *Substoichiometry in Radiochemical Analysis*, Pergamon, Oxford, 1968.
14. A. Zeman and K. Kratzer, *Radiochem. Radioanal. Lett.* **27,** 217 (1976).
15. J. Starý and J. Prášilová, *Radiochem. Radioanal. Lett.* **24,** 143 (1976).
16. J. Starý and J. Prášilová, *Radiochem. Radioanal. Lett.* **26,** 193 (1976).
17. I. Obrusník and A. Adámek, *Substoichiometric Replacement in Activation Analysis*, Report UJV 1970, Nucl. Res. Inst., Czechoslovak Acad. Sci., Řež, 1967.
18. I. Obrusník and A. Adámek, *Talanta* **15,** 433 (1968).
19. I. Obrusník, *Talanta* **16,** 563 (1969).
20. A. J. Tavendale, *IEEE Trans. Nucl. Sci.* ***NS*** **11,** 191 (1964).
21. M. Vobecký and F. Petrů, *Collection Czechoslov. Chem. Commun.* (in German), **33,** 3903 (1968).
22. J. Juna, K. Konečný, and M. Vobecký, *Collection Czechoslov. Chem. Commun.* **34,** 1605 (1969).

23. K. Konečný, M. Vobecký, and J. Juna, *Jaderná Energie* **15,** 128 (1969).
24. J. Frána, M. Vobecký, and J. Krišťák, *Radiochem. Radioanal. Lett.* **1,** 41 (1969).
25. J. Kuncíř, J. Benada, Z. Řanda, and M. Vobecký, *J. Radioanal. Chem.* **5,** 369 (1970).
26. M. Vobecký, J. Frána, Z. Řanda, J. Benada, and J. Kuncíř, *Radiochem. Radioanal. Lett.* **6,** 237 (1971).
27. M. Vobecký, J. Frána, J. Bauer, Z. Řanda, J. Benada, and J. Kuncíř, *Proceedings of the Second Lunar Science Conference*, vol. 2, The M.I.T. Press, 1971, *Geochim. Cosmochim. Acta*, **Supplement 2** (1971).
28. I. Obrusník, *Chem. Anal.* **35,** 221 (1990).
29. I. Obrusník, B. Stárková, J. Blažek, and V. Bencko, *J. Radioanal. Chem.* **54,** 311 (1979).
30. I. Obrusník, *J. Hyg. Epid. Microbiol. Immun.* **30,** 11 (1986).
31. I. Obrusník and K. Eckschlager, *J. Radioanal. Nucl. Chem., Articles* **112,** 233 (1987).
32. J. Kučera, *Radiochem. Radioanal. Lett.* **24,** 215 (1976).
33. J. Kučera and L. Soukal, *J. Radioanal. Nucl. Chem., Articles* **168,** 185 (1993).
34. J. J. M. De Goeij, L. Kosta, A. R. Byrne, and J. Kučera, *Anal. Chim. Acta* **146,** 161 (1983).
35. J. Kučera, L. Soukal, and J. Horáková, *Fresenius J. Anal. Chem.* **345,** 188 (1993).
36. J. Kučera, A. R. Byrne, A. Mravcová, and J. Lener, *Sci. Total Environ.* **15,** 191 (1992).
37. J. Kučera, J. Lener, and J. Mňuková, this vol.
38. V. Kliment and J. Tölgyessy, *Isotopenpraxis* (in German) **7,** 446 (1971).
39. V. Kliment and J. Tölgyessy, *Isotopenpraxis* (in German), **8,** 211 (1972).
40. J. Bartošek, F. Adams, and J. Hoste, *Nucl. Instrum. Methods* **103,** 45 (1972).
41. J. Bartošek, J. Mašek, F. Adams, and J. Hoste, *Nucl. Instrum. Methods* **104,** 221 (1972).
42. M. Vobecký, Z. Janout, J. Koníček, and S. Pospíšil, *J. Radioanal. Nucl. Chem. Lett.* **136,** 411 (1989).
43. Z. Janout, J. Koníček, S. Pospíšil, M. Vobecký, and R. Drahoňovský, *J. Radioanal. Nucl. Chem. Lett.* **136,** 423 (1989).
44. M. Vobecký, Z. Janout, S. Pospíšil, and J. Dáňa, *J. Radioanal. Nucl. Chem. Articles* **83,** 123 (1984).
45. Z. Řanda, B. Špaček, J. Kuncíř, and J. Benada, *Nondestructive Gamma Activation Analysis of Mineral Materials*, Nuclear Information Centre, Prague, 1981.
46. Z. Řanda and F. Kreisinger, *J. Radioanal. Chem.* **77,** 279 (1983).
47. V. Hnatowicz, V. Havránek, and J. Kvítek, *Nucl. Instrum. Methods Phys. Res.* **B 53,** 337 (1991).
48. V. Hnatowicz, V. Havránek, J. Kvítek, and V. Peřina, *Nucl. Instrum. Methods Phys. Res.* **B 72,** 224 (1992).
49. V. Hnatowicz, V. Havránek, J. Kvítek, V. Peřina, V. Švorčík, and V. Rybka, *Nucl. Instrum. Methods Phys. Res.* **B 80/81,** 1059 (1993).
50. V. Hnatowicz, V. Havránek, J. Kvítek, and I. Obrusník, *PIXE-INP Computer Code for PIXE Analyses of Thin and Thick Samples*, Int. Conf. on Ion Beam Analysis, IBA-11, July 11, 1993, Balatonfüred, Hungary.
51. J. Habanec, M. Zárobský, A. Petřina, and Z. Janoušek, *J. Radioanal. Chem.* **13,** 213 (1973).
52. J. Červená, V. Hnatowicz, J. Hoffmann, Z. Kosina, J. Kvítek, and P. Onheiser, *Nucl. Instrum. Methods* **188,** 185 (1981).
53. V. Havránek, V. Hnatowicz, J. Kvítek, J. Vacík, and J. Hoffmann, *Nucl. Instrum. Methods Phys. Res.* **B 73,** 523 (1993).
54. J. Červená, J. Kvítek, J. Honzátko, and K. Konečný, *Radiochem. Radioanal. Lett.* **38,** 321 (1979).
55. P. Kotas, *Radiochem. Radioanal. Lett.* **32,** 209 (1978).

METHODOLOGY OF NUCLEAR ANALYTICAL METHODS

Nuclear Analytical Methods in the Life Sciences

J. J. M. de Goeij

Interfaculty Reactor Institute, Delft University of Technology, Mekelweg 15, 2629 JB Delft, The Netherlands

ABSTRACT

A survey is given of various nuclear analytical methods. The type of analytical information obtainable and advantageous features for application in the life sciences are briefly indicated. These features are: physically different basis of the analytical method, isotopic rather than elemental determination, no interfering effect of electrons and molecular structure, and penetrating character of nuclear radiation. Suggestions are given for exploitation of the sometimes unique potentials of nuclear analytical methods, particularly when requiring considerable investment for equipment, supporting facilities, and specialized staff.

Index Entries: Nuclear analytical methods; nuclear analytical chemistry; nuclear analytical physics; nuclear analytical techniques; radioanalytical chemistry; radioanalytical physics; life sciences.

INTRODUCTION

Before discussing nuclear analytical methods, a definition should be given of the term "analytical." Traditionally, a distinction is made between a qualitative analysis (which analytes are present?) and a quantitative analysis (how much of them are present?). Quite often, this information alone is insufficient, and other information on the analyte is required as to chemical form, distribution, structure, kinetics, and/or (reaction) mechanism. Within the frame of this article, obtaining this information will also be considered as analysis.

Since the discovery of the phenomenon of radioactivity about a century ago and particularly during the last half century when nuclear sciences and instrumentation were developed, nuclear analytical methods have provided an increasing potential for applications in various disciplines, not the least of which is in the life sciences. Nuclear analytical methods—from now on abbreviated NAMs—are based on properties of the nucleus of the atom, in contrast to nonnuclear methods, which are based on properties of the atom or molecule, where actually the electrons play the decisive role. In NAMs, the nuclei of isotopes (belonging to a given element) are distinguishable, and thus, nuclear methods are determining isotopes, rather than elements as a combination of isotopes.

It should be mentioned that—either fundamentally or practically—no sharp borderline can be drawn between nuclear and nonnuclear analytical methods. For instance, mass spectrometry is actually a nonnuclear method, since it deals with the atom, rather than with the nucleus, but it is also an isotopic method because of the fact that the atomic mass (actually depending on the nucleus mass) gives the characteristic signal. Thus, mass spectrometry is most times classified as a nonnuclear method, as will be done here, too. However, the use of isotopically enriched material in conjunction with mass spectrometry will be considered as an NAM. Analysis of materials via γ-ray or β-ray absorption is not considered here to be NAMs, since the relevant interaction of nuclear radiation with matter is predominantly an electronic process.

On the other hand, particle-induced X-ray emission as one of the varieties of ion beam analysis deals with electron shell ionization, and thus, is a nonnuclear method. However, since particle-induced X-ray emission requires almost the same nuclear equipment as other varieties of ion beam analysis, particle-induced X-ray analysis is practically incorporated as an NAM in the domain of nuclear physicists.

Quite often, the nucleus may be treated separately from the corresponding electrons, except that the nucleus structures the electron shells and gives mass to the atom. However, there are specific interactions (coupling) between the energy levels of electrons and nuclei, and vice versa. Sometimes such interactions provide analytically interesting possibilities.

SURVEY OF NAMS

Nuclear parameters that may serve as a base for NAMs are: spin and magnetic moment, mass, excited states and associated parameters, probability for a given nuclear reaction, and—only for radioactive nuclei—half-life, types and energies of the emitted radiation. Table 1 lists a survey of NAMs grouped into seven categories. Also, the typical analytical information obtainable is given. It should be noted that some groups overlap, and some of the modes listed may also be classified into another category, e.g., particle-induced γ emission, now listed as the ion beam method, may also be considered as a mode of activation analysis. As may be seen for activation analysis, sometimes a wide variety of modes exist.

The survey points at NAMs, which have been proven to be useful in the field of life sciences. Other nuclear methods, which have not or only scarcely been used as an analytical tool in the life sciences, are considered as "non-established" NAMs. These methods comprise: analysis via

Table 1
Survey of Nuclear Analytical Methods

Ion beam analysis (IBA)	
modes	elastic or inelastic scattering methods
	with or without excitation of the nucleus
	with or without the use of a microbeam
examples	Rutherford backscattering (RBS)
	elastic recoil detection analysis (ERDA)
	particle-induced X-ray emission (PIXE)
	nuclear reaction analysis (NRA)
	particle-induced gamma ray emission (PIGE)
information	(semi)quantitative, qualitative, and/or distribution
Nuclear magnetic resonance spectroscopy (NMR)	
modes	with or without pulse excitation
	with or without isotopic substitution
	chemical analysis or imaging
	analysis of liquids or analysis of solids
examples	magic angle spinning NMR (MAS-NMR)
	magnetic resonance imaging (MRI)
information	chemical form, structural, quantitative, and/or distribution
Neutron scattering and diffraction	
modes	inelastic scattering or diffraction mode
	with or without polarized neutrons
	with or without isotopic substitution
examples	small angle neutron scattering (SANS)
	quasielastic and inelastic neutron scattering
information	structural, chemical form, quantitative, and/or kinetics (more properly: dynamics)
Mössbauer spectroscopy	
modes	emission or absorption mode
	with or without isotopic enrichment
	with or without low temperature
information	structure, chemical form, and/or quantitative
Isotope dilution analysis (IDA) and related methods	
modes	with stable or radioactive isotopes
	direct or reverse mode
	with or without substoichiometric amounts of reagent
examples	isotope dilution mass spectrometry (IDMS)
	various biochemical assays, e.g., radioimmunoassay (RIA)
	isotopic exchange methods
information	quantitative
Tracer studies	
modes	with stable or radioactive isotopes
	natural or artificial radioactivity
	chemical analysis or imaging
examples	positron emission tomography (PET)
	single photon emission computer tomography (SPECT)
information	kinetics, distribution, (reaction) mechanisms, and/or quantitative

Table 1 *(continued)*

Activation analysis	
modes	prompt or delayed mode
	neutron, charged-particle or γ-photon activation
	instrumental or radiochemical mode
examples	instrumental neutron activation analysis (INAA)
	radiochemical neutron activation analysis (RNAA)
	charged particle activation analysis (CPAA)
	fast neutron activation analysis (FNAA)
	photon activation analysis (PAA)
	cyclic instrumental neutron activation analysis (CINAA)
	epithermal neutron activation analysis (ENAA)
	prompt γ neutron activation analysis (PGNAA)
	neutron depth profiling (NDP)
information	quantitative and/or qualitative

perturbed angular correlations, analysis via positron annihilation (angular and time effects) and positronium chemistry, analysis via alterations in half-life of electron capture radionuclides, and analysis via muon spin resonance. For the underlying nuclear and other physical principles of the various "established" NAMs listed in the table, the reader is referred to textbooks and review articles.

FEATURES OF NAMS

Typical characteristic features of NAMs are grouped into four categories. However, it should be stressed that these features can be distinguished, but not always separated or considered isolated. In principle, each of the nuclear methods listed in Table 1 displays one or more of these features. The four categories of features are given below.

Different Basis of NAMs Compared to Nonnuclear Analytical Methods

Compared to nonnuclear analytical methods, NAMs have a totally different physical basis. Moreover, the nuclear parameters vary from element to element in a different way compared to those for atomic—quite often purely chemical—parameters of the various nonnuclear analytical methods. This implies that NAMs create an interesting supplementation of the nonnuclear analytical possibilities, particularly since the strong and weak points of NAMs are distributed over the Periodical Table of Elements in a different way compared to those of the nonnuclear analytical methods. About half of the atoms in the living cells are hydrogen, being invisible in X-ray structural analysis, but *par excellence* accessible for neutron scattering/diffraction and nuclear magnetic resonance. Examples of neutron scattering/diffraction from molecular biology are: analysis of protein crystals, lipid layers of biological membranes, and dynamics of hydration water in proteins.

In cases where NAMs and nonnuclear analytical methods provide in principle the same information, the difference in physical base between both types of methods makes NAMs well suited as independent methods of reference, particularly taking into account also the advantages listed under Isotopic Analysis Rather than Elemental Analysis and No Interfering Effect of Electronic and Molecular Structure. For instance, in most certification programs of trace elements in biological reference materials, isotope dilution and activation analysis have become indispensable tools.

Finally, NAMs may also act as "pioneers" for nonnuclear methods. NAMs disclose analytically new fields of interest. Once the importance of the new fields is recognized, efforts will be spent to gain access to the new field with nonnuclear methods as well. A striking example is the present set of immunological assays, which began at the time with the radioimmunological assay of iodine-containing hormones. Now there are several competing immunological methods without radioactive label, e.g., enzyme-linked immunosubstrate analysis. Radiochemical neutron activation analysis is for many elements at trace and ultratrace levels a reliable method. The results obtained in the life sciences with this method have stimulated the improvement of nonnuclear methods, e.g., for the reliable determination of selected trace elements, including chromium and mercury.

Isotopic Analysis Rather than Elemental Analysis

Most elements cover two or more stable isotopes and several radioactive isotopes. Chemically, all these isotopes behave identically. However, the nuclear parameters of the nuclei in the various isotopes of a given element differ and, thus, their responses in NAMs. These characteristics enable tracer experiments, where the tracer and tracee (the compound to be traced) behave chemically identically, but physically differently, so that tracer and tracee can be distinguished.

This principle is the basis of various NAMs, e.g., isotope dilution analysis and radioimmunoassay. It is also the basis for tracer studies with radioisotopes or with isotopically enriched stable tracers to obtain information on spatial, temporal, kinetic, and (reaction) mechanistic aspects, all four aspects being rather important in the field of the life sciences. Tracers (labeled compounds) are used in the study of many chemical, biochemical, and biological processes and methods, giving a substantial extension of the nonnuclear analytical methods. For instance, the use of ^{14}C and ^{3}H (measurement via their radioactivity), ^{15}N, ^{18}O, and ^{2}H (measurement via mass spectrometry), and ^{13}C, ^{17}O, and ^{2}H (measurement via nuclear magnetic resonance) as a label in a variety of organic compounds has contributed substantially to the increase of knowledge concerning (reaction) mechanisms and kinetics in various branches of the life sciences. It has also contributed to the development of new pharmaceuticals and other important organic compounds.

Caveat: With nuclear methods, isotopes rather than elements are determined. Using radiotracers, the radioactivity signal is only indicative for the tracee, when the specific activity (the ratio between radiotracer and tracee) is constant and uniform. However, this ratio can easily change because of exchange with nonlabeled pools of the tracee in the system under study. Determination of specific activities and compartmental modeling may be helpful tools in overcoming this problem. A comparable warning and solution apply to the use of stable enriched isotopes as a tracer.

Dealing with light elements and sometimes complicated and multistep biological processes, slight, but measurable isotopic fractionation may occur. The occurrence of these fractionations has been used fruitfully for the study of reaction mechanisms and kinetics in some branches of the life sciences.

No Interfering Effect of Electronic and Molecular Structure

In many nonnuclear analytical methods, the signal relevant to detection depends on the chemical state of the analyte, and consequently, the occurrence of an analyte in more than one chemical state may act as an interference, leading to erroneous results. Since the nucleus is rather insensitive to such effects, NAMs are in principle not sensitive either. This contributes—in addition to the nuclear-physically different base—to possibilities of, e.g., activation analysis as a method of reference. On the other hand, this insensitivity is also a drawback, particularly in the life sciences, since quite often information on a particular form of an element is required, e.g., methylmercury rather than total mercury.

As mentioned above, there are specific interactions (coupling) between the energy levels of electrons and nuclei, and vice versa. Although such interactions generally are rather weak, they may occasionally provide interesting possibilities to give information about electronic and molecular structures. This is the case for analysis via the Mössbauer effect and via nuclear magnetic resonance spectroscopy. In the life sciences, the application of the Mössbauer effect is practically limited to a very few elements, in particular, to iron in compounds of hematological interest. Nuclear magnetic resonance has gained substantial importance in the life sciences for structural analysis, since some of the constituting elements in organic materials have suitable nuclei, e.g., ^{1}H, ^{13}C, and ^{31}P.

In radiotracer studies as well as in isotope dilution analysis and in related methods, the signal itself is not influenced by the chemical form of the radioactive atom. However, these methods are actually hybrid analytical methods, where generally the chemical (c.q. biological, physiological, biochemical) part ensures the selectivity, and the nuclear part ensures the sensitivity. In biochemistry and clinical chemistry, many enzymatic and immunological analytical methods are based on this approach. A recent example is DNA "fingerprinting," where the biochemical treatment ensures a very high selectivity (down to a single person), but the radioactive ^{32}P in the probe enables the required high sensitivity, so that samples with even minute amounts of DNA can be used.

Penetrating Character of Nuclear Radiation

In most NAMs, the excitation and de-excitation signals penetrate through matter, thereby enabling noninvasive measurements without disturbing the processes to be studied. This is the case where radio waves, a magnetic field (nuclear magnetic resonance), neutrons (neutron scattering/diffraction, neutron activation analysis), and γ radiation (activation analysis, Mössbauer spectroscopy, radiotracer investigations, and isotope dilution analysis with radioactive isotopes) are used.

Important examples of noninvasive measurements for humans are: magnetic resonance imaging, single photon emission computerized tomography, in vivo neutron activation analysis, nuclear magnetic resonance localized spectroscopy, and positron emission tomography. The first two methods have proven to be very useful for routine application in medical diagnosis. The third method is also applied in diagnosis, but much less frequently. The last two methods have been shown to be extremely useful in physiological studies. For instance, positron emission tomography with ^{11}C- or ^{18}F-labeled compounds has provided much insight in various metabolic and pharmacologic processes, particularly in the otherwise rather inaccessible brain; nuclear magnetic resonance localized spectroscopy using natural ^{31}P levels is able to follow in vivo the phosphorus metabolism inside muscle tissue, e.g., the breakdown of ATP and creatinine phosphate on physical exercise.

On the other hand, ion beam analysis scattering methods and charged particle activation analysis deal with the aspect of the relatively short range of charged particles in matter, which implies relatively small samples and analyses of surface and near-surface layers only. An interesting example of ion beam analysis is the depth profiling of fluoride in teeth using nuclear reaction analysis with resonance photons.

PROMOTION OF APPLICATION OF NAMS IN THE LIFE SCIENCES

Generally, NAMs require substantial investments. Although instruments for nuclear magnetic resonance and mass spectrometry (for isotope dilution mass spectrometry and tracer studies with stable isotopes) are rather expensive, they have found wide application in the life sciences. This is because of the fact that these instruments are built as dedicated to the specific purpose, the associated costs can be shared by groups from different organizations, and the information obtainable is valuable, if not, indispensable.

Some other NAMs also require expensive instrumentation, e.g., ion beam analysis, neutron scattering/diffraction, and activation analysis. Particle accelerators and nuclear research reactors play a dominant role here. Generally, these machines are not built and operated specifically for NAMs, but for a variety of purposes. To perform NAMs, specific auxiliary equipment connected to these machines has to be developed, assembled, and/or tested by the researchers themselves. Moreover, the experimental procedures and complexity of the results often require substantial support from the technical and scientific staff of the nuclear laboratory. An additional constraint is that these machines and auxiliary equipment deal with ionizing radiation. This implies the necessity of a health or safety department, special measures to deal with ionizing radiation, and training of the researchers involved, all adding up to costs and efforts. To a certain extent, this also applies to radiotracer studies, radioisotope dilution analysis, and Mössbauer spectroscopy, which do not require operation at the site of the nuclear machines.

An additional problem is that public attitude toward radioactivity and radiation has become rather critical, and this has also affected the universities and other scientific institutions. For instance, professorial chairs in the nuclear field that became empty were quite often discontinued, whereas scientists well trained in the nuclear field had to move to other fields. Some nuclear reactors and particle accelerators and associated departments were closed, whereas others could not function at the state-of-the-art level because of a shortage of funds.

Against this background, it is believed that full exploitation of the possibilities of NAMs requires the following aspects:

1. A policy keeping regionally nuclear machines (nuclear reactors and particle accelerators) up to date and accessible for students and researchers;
2. Performing NAMs at the state-of-the-art level;
3. Performing NAMs in an interdisciplinary approach;
4. Using NAMs well directed to a meaningful problem; and
5. Dissemination of information to relevant groups.

These aspects are explained below.

A policy to keep nuclear machines available and accessible for students and researchers implies a shared responsibility by and cooperation between universities and associated institutions. Machines should be fully exploited by using them for a variety of purposes, especially for practical problems. Multidisciplinary exploitation and thus reduction of costs per investigation as well as solving socially important problems (including many in the field of the life sciences) have been shown to be helpful in creating support from the general public and/or government. In this frame also, partial commercial exploitation is useful, e.g., silicon transmutation doping, instrumental neutron activation analysis, and radionuclide production.

NAMs should be performed at the state-of-the-art level, since many applications and methods presented years ago can obtain a new life, when modern types of radiation sources, detectors, chemical separations, and/or software are used. Although NAMs have good prospects for quality, this must be actually demonstrated. To be successful, NAMs should not be taken as they are available from the literature, but always adapted to the specific problem at hand, requiring that the experts in NAMs and the researchers in the life sciences learn something about each

other's field. Development and application of NAMs should occur in scientific and interdisciplinary cooperation. Experience shows that for most students in the nuclear field, it is an attractive feature to implement their activities in other disciplines.

Furthermore, adequate attention should be paid to education and information, being clear, realistic, and fair as to other nonnuclear methods. Thus, there should be no advertising of nuclear methods as a panacea for problems or neglect of practical problems (e.g., licensing, radiation protection, and radioactive waste), as unfortunately has been done in the past. The analytical tasks in the life sciences cover such a wide range that the entire spectrum of both nonnuclear and nuclear analytical possibilities is badly needed.

Finally, a continuous exchange has to occur between the researchers in the life sciences and experts in NAMs. This "crossfertilization" should occur at all possible levels, e.g., in scientific and professional training, in practice of science, and in publishing. Once the typical advantages of NAMs are known—or even better—experienced, the researchers in the life science will be motivated to cope with some "inconveniences," e.g., finding partners for cooperation, acquiring funds for travel, sharing authorships, learning more about the specific aspects of the field of the other partner, and becoming familiar with radiation and radiation protection.

Use of k_0-NAA in the Life Sciences

F. De Corte,*[,1] A. Simonits,[2] Lin Xilei,[3] and M. C. Freitas[4]

[1]*Laboratory of Analytical Chemistry, Institute for Nuclear Sciences (INW), University of Gent, Proeftuinstraat 86, B-9000 Gent, Belgium;* [2]*KFKI Atomic Energy Research Institute, Nucl. Chem. Dept., Budapest, Hungary;* [3]*Institut für Radiochemie der Technischen Universität München, Garching, Germany; and* [4]*Instituto Nacional de Engenharia e Tecnologia Industrial (INETI), Instituto de Ciências e Engenharia Nucleares, Departamento de Química, Sacavém, Portugal*

ABSTRACT

The applicability and usefulness of k_0-based reactor neutron activation analysis (NAA) in the life sciences is evaluated from the following examples: 1. Instrumental NAA of NIST SRM 1633a coal fly ash, as a quality assessment; 2. Radiochemical NAA of Versieck's reference human serum, and—herewith associated—the development of practical correction procedures for neutron-induced reaction interferences and of improved methods to evaluate the detection efficiency and the correction for true coincidence; and 3. Determination of the lanthanides in plant leaves and lichens near a Portuguese coal-fired power station, which led to the introduction of the Westcott formalism and to the use of a low-energy photon detector. As concluded, k_0-based NAA is at present capable of tackling a large variety of analytical problems when it comes to the multielement determination in environmental and biological matrices.

Index Entries: Neutron activation analysis; k_0 standardization; life sciences; biological materials; environmental materials; coal fly ash; human serum; lanthanides; plant leaves; lichens.

INTRODUCTION

The concept of the k_0 standardization of (n, γ) reactor neutron activation analysis (NAA) was launched in 1975 *(1)* with the aim to eliminate:

*Author to whom all correspondence and reprint requests should be addressed.

1. The tedious experimental work of the relative method (especially in multielement analysis), which requires the preparation of standards, their counting, and the evaluation of their γ spectra;
2. The inflexibility of the single-comparator method, which is strictly bound to "personal" experimental conditions with respect to irradiation and counting; and
3. The inaccuracy of the absolute method, which is originating from errors and inconsistencies in published data for cross-sections and γ intensities (and occasionally also isotopic abundances): these are replaced by compound nuclear data, k_0 factors, which are tabulated as highly precise and accurate experimentally determined values.

In short, the ultimate goal envisaged was to increase the manageability (experimental simplicity, versatility, accuracy) of NAA, making it less prone to the risk of human errors and more suitable for computerization and automation.

According to the k_0-standardization method, the concentration ρ of an element x is obtained from coirradiation of the sample with a comparator in a calibrated reactor position, followed by γ spectrometry with a calibrated detector:

$$\rho\ (x) = \{(N_p/t_m/SDC)_x/(N_p/t_m/SDC)_{\text{comp}}\} \cdot (w_{\text{comp}}/w_{\text{samp}}) \cdot [1/k_{0,\text{comp}}\ (x)] \cdot \{[f + Q_{0,\text{comp}}(\alpha)]/[f + Q_{0,x}(\alpha)]\} \cdot (\varepsilon_{p,\text{comp}}/\varepsilon_{p,x}) \quad (1)$$

In Eq. (1): N_p = full-energy peak area, corrected for losses (dead time, random and true coincidence, burn up); $S = 1 - \exp(-\lambda t_{\text{irr}})$, with t_{irr} the irradiation time and $T = (\ln 2)/\lambda$ the half-life; $D = \exp(-\lambda t_d)$, with t_d the decay time; $C = [1 - \exp(-\lambda t_m)]/\lambda t_m$, with t_m the measuring time; w = mass of sample or comparator; $k_0 = k_{0,\text{comp}}(x) = M_{\text{comp}}\ (\theta\ \gamma\ \sigma_0)_x\ /\ M_x\ (\theta\ \gamma\ \sigma_0)_{\text{comp}}$, with M the molar mass, θ the isotopic abundance, γ the absolute gamma intensity, and σ_0 the (n, γ) cross-section at 2200 ms^{-1} neutron velocity; f = ratio of the thermal (i.e., subcadmium) to epithermal neutron fluence rate (in the Høgdahl convention [for the Westcott formalism: *see* infra]; $Q_0(\alpha) = (Q_0 - 0.429)\bar{E}_r^{-\alpha} + 0.429\ /\ [(2\alpha + 1)(0.55)^\alpha]$, with $Q_0 = I_0/\sigma_0$ the ratio of the resonance integral to the 2200 ms^{-1} cross-section, $\bar{E}_r$ the effective resonance energy, and α a measure for the deviation of the epithermal neutron fluence rate distribution ($\propto 1/E^{1+\alpha}$) from the ideal 1/E behavior (in the Høgdahl convention [for the Westcott formalism: *see* infra]); ε_p = full-energy peak detection efficiency, taking into account γ attenuation, for the actual source-detector configuration.

In former publications, more detailed information can be found on the fundamentals of the method and specific sources of errors *(2,3)*, the nuclear data library (k_0, Q_0, $\bar{E}_r$, T, and so forth), and the modification of Eq. (1) in case of complex activation and/or decay *(4,5)*, the evaluation of the neutron spectrum parameters f and α *(6–8)*, the determination of the Ge-detector characteristics ε_p and P/T (the peak-to-total ratio, yielding $\varepsilon_t = \varepsilon_p/[P/T]$ for true-coincidence correction) *(9–11)*, the computerization and automation of the method *(5,12,13)*, and the recent introduction of some extensions and refinements with respect to the conventional description of the neutron spectrum and to the source-detector configuration—aiming at a more general applicability *(14–17)*. Let it only be recalled here that at present accurately measured k_0 factors and related nuclear data are available *(4,5)* for the most prominent γ rays of 122 analytically interesting radionuclides. For convenience to condense published data, these "$k_{0,\mathrm{Au}}$" factors are expressed vs $^{197}\mathrm{Au}(n,\gamma)^{198}\mathrm{Au}$ with E_γ = 411.8 keV, but in practice, any suitable comparator can be used, since:

$$k_{0,\mathrm{comp}}\,(x) = k_{0,\mathrm{Au}}(x)/k_{0,\mathrm{Au}}(\mathrm{comp}) \qquad (2)$$

Nevertheless, it should be remarked that in order to improve the accuracy and traceability of the k_0 standardization, the IRMM (former CBNM), Geel/Belgium issued an Al-0.1% Au certified reference material [(IRMM-530)] *(18)*.

After its introduction in 1975, the NAA k_0-standardization method has grown into an operational tool, as it became clear at the (first) International k_0 Users Workshop—Gent held at Astene/Belgium, September 30–October 2, 1992 *(13)*. Another paper illustrative in this respect, showing at the same time a broad international cooperation *(5)*, was presented at the recent MTAA-8 Conference (Vienna, 1991), and gave a survey of a large variety of scientific and industrial applications, including the analysis of biological, geological, and geochronological samples, various kinds of reference materials, high-purity products and ceramics, and a number of environmental pollution indicators. In the present article, a more detailed evaluation is made of the applicability, together with the specific problems and their solutions, of k_0-based NAA in the analysis of biological and environmental samples. The discussion is based on the following examples:

1. Instrumental multielement NAA or NIST SRM 1633a coal fly ash;
2. Radiochemical NAA of Versieck's reference human serum; and
3. Determination of the lanthanides in environmental materials (plant leaves, lichens) near a Portuguese coal-fired power station.

ACCURACY ASSESSMENT: INAA OF NIST SRM 1633A COAL FLY ASH

Instrumental NAA (INAA) of NIST SRM 1633a coal fly ash was one of the elaborate controls performed in the early 1980s to assess the accuracy of the k_0-standardization method *(19)*. Therefore, independent analysis results were obtained at the INW (Gent) and the KFKI (Budapest). This was achieved by using different experimental procedures with respect to sample preparation, irradiation, counting, peak

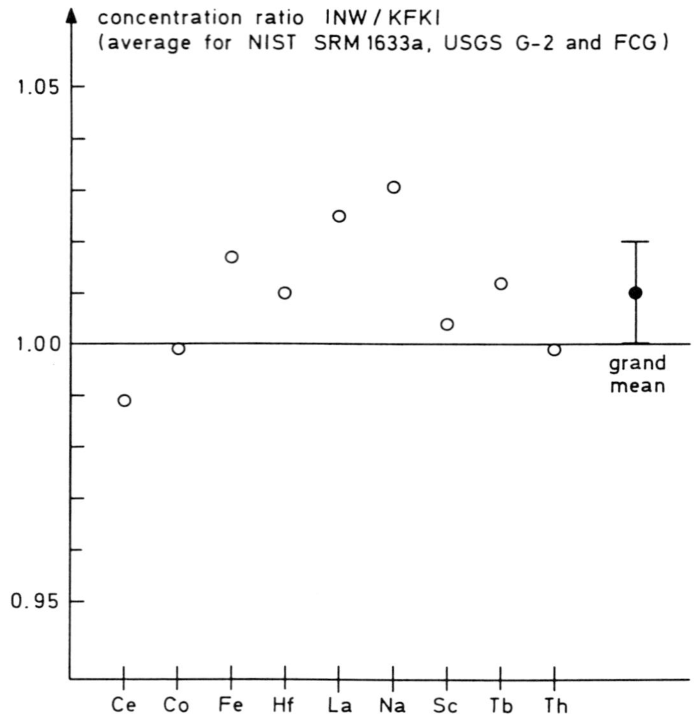

Fig. 1. Average ratios of the concentrations of "good counting statistics"—elements determined at the INW (Gent) and the KFKI (Budapest) via k_0-based INAA in the reference materials NIST SRM 1633a coal fly ash, USGS G-2 granite, and FCG fired clay of Ghent *(19)*.

integration, and other calculational methods. For instance, whereas at the INW, considerable conversions and corrections had to be introduced for geometry and true coincidence—since voluminous samples were measured on top of the Ge(Li) detector—this was not the case at the KFKI, where quasi-point samples were positioned at ~15 cm distance from the detector. On the other hand, the results of both institutes evidently were based on the same nuclear data library (k_0, and so on).

The correspondence between the INW and the KFKI analysis results is shown in Fig. 1 as an average for the elements determined with good counting statistics in NIST SRM 1633a, granite USGS G-2, and fired clay of Ghent FCG *(19)*. As seen, the grand mean of the INW/KFKI concentration ratio amounts to 1.010 ± 0.010 (2σ SE).

The consistency between the overall INW-KFKI results and the concentrations specified by NIST is demonstrated in Table 1 for the elements in SRM 1633a with certified values. It should be noted that, as compared to the data presented in *(19)*, additional certified concentration values could not be included for Al, Mn, Sb, and V, whereas some INW-KFKI results were slightly revised (maximum 1%) in accordance with the updated nuclear data library (in *4,5*) superseding the one in use formerly

Table 1
NIST SRM 1633a Coal Fly Ash[a]

	Concentration		
Element[b]	Overall INW - KFKI result	Certified NIST value	Ratio (INW - KFKI/NIST)
Al, %	14.0 ± 0.3[c]	14.3 ± 1.0[c]	0.979 ± 0.071[c]
As	151 ± 5	145 ± 15	1.041 ± 0.113
Ca, %	1.16 ± 0.11	1.11 ± 0.01	1.045 ± 0.096
Cr	197 ± 6	196 ± 6	1.005 ± 0.042
Fe, %	9.47 ± 0.42	9.4 ± 0.1	1.007 ± 0.046
K, %	1.83 ± 0.09	1.88 ± 0.06	0.973 ± 0.058
Mg, %	0.52 ± 0.09	0.455 ± 0.010	1.143 ± 0.194
Mn	180 ± 2	179 ± 8	1.006 ± 0.046
Na, %	0.175 ± 0.003	0.17 ± 0.01	1.029 ± 0.063
Rb	138 ± 9	131 ± 2	1.053 ± 0.075
Sb	6.7 ± 0.25	6.8 ± 0.4	0.985 ± 0.069
Se	10.3 ± 0.25	10.3 ± 0.6	1.000 ± 0.063
Sr	857 ± 29	830 ± 30	1.033 ± 0.051
Th	24.4 ± 0.4	24.7 ± 0.3	0.988 ± 0.022
U	10.2 ± 0.25	10.2 ± 0.1	1.000 ± 0.026
V	302 ± 14	297 ± 6	1.017 ± 0.053
Zn	234 ± 15	220 ± 10	1.064 ± 0.085
	Weighted mean		1.003
	External (observed) error[c]		± 0.010_5
	Internal (expected) error[c]		± 0.011_5
	Reduced chi-square	χ^2_R	= 0.83

[a]Comparison of the overall INW-KFKI results (obtained via k_0-standardized INAA) with certified values (in ref. *19* the NIST values for Ce and Co were erroneously considered as being certified).

[b]Values given in μg/g unless otherwise indicated.

[c]2σ SE.

(in *20*). It should also be remarked that, in addition to the certified elements mentioned in Table 1, INW-KFKI results were obtained for Ba, Ce, Co, Cs, Eu, Ga, Hf, Sc, and Ti. Since for standardization only an Al-0.1% Au wire and a Zr-foil [for f and α monitoring *(6–8)*] were coirradiated with the sample, this underlines the k_0 method's capability of multielement and even panoramic (survey) analysis, where concentrations or detection limits can be determined for all elements considered to be relevant *(21)*.

RADIOCHEMICAL NAA OF VERSIECK'S REFERENCE HUMAN SERUM

INAA for trace and ultratrace elements is not possible in quite a few environmental and especially biological samples (but also in others like geological materials) that have complex matrices giving rise to major radioactivities from many elements. Sometimes this problem can be solved by irradiating the samples under a Cd cover (epicadmium NAA [ENAA]), viz. when it concerns the determination of elements (such as Mo, Ag, Ta, U, and so forth) via an (n,γ) reaction with a high Q_0 ratio and the suppression of interfering activities induced by low-Q_0 reactions (as for ^{24}Na, ^{42}K, ^{32}P, and so on). The possibilities of k_0 standardization of ENAA were studied in detail *(22)*, and as concluded, it is sufficient to put $f = 0$ in Eq. (1) and to multiply each Q_0 by the corresponding F_{Cd}, the Cd-transmission factor for epithermal neutrons (usually = 1), as tabulated in *(4)*. It is clear that in this case, the quality of the analysis results will be somewhat affected by the uncertainty on the Q_0 values, and also the uncertainty on the α-factor will play a larger role.

A more general way out for the above mentioned sensitivity problem is offered by radiochemical neutron activation (RNAA). In the mid-1980s, the performance of k_0-based RNAA was investigated *(23)* for the determination of 13 trace elements (Ag, Co, Cr, Cs, Fe, Hg, Rb, Sb, Sc, Se, Sn, Sr, and Zn) in Versieck's "second-generation" biological reference mate-rial (freeze-dried human serum) *(24)*. Lyophilized and ashed serum samples were irradiated for 13 d at neutron fluence rates of 10^{17} m^{-2} s^{-1} and 10^{18} m^{-2} s^{-1}, respectively. For the k_0 standardization, an Al-0.1% Au wire (1 mm diameter) and a Zr foil (25-μm thickness) were coirradiated with the samples. As a check, the standardization was also done in a relative way, based on the simultaneous irradiation of standards that next underwent the same procedure as the samples. The radiochemical procedure applied was based on inorganic ion retention and was followed by γ spectrometry, on two Ge detectors, of the four fractions obtained: the inorganic ion exchangers AAO (acid aluminum oxide), HAP (hydrated antimony pentoxide) and HMD (hydrated manganese dioxide), and the eluant (3.3 mol/L $HClO_4$). A comparison of the results is shown in Fig. 2. It can be seen that the concentrations obtained by the k_0 method and the relative method are consistent, and are also in good agreement with the certified values (where available). Note that Sn was an unexpectedly detected element for which no standard was foreseen, so that quantitative determination was only possible according to the k_0 method.

In addition to the comparison of concentrations, the above-described analysis led to a number of interesting observations and developments. On the one hand, these were related to the correction for neutron-induced reaction interferences, such as—in the present case—^{54}Fe$(n,\alpha)^{51}$Cr and ^{44}Ca$(n,\gamma;\ \beta^-;\ n,\gamma)\ ^{46}$Sc in the determination of Cr and Sc via ^{50}Cr(n,γ) ^{51}Cr and ^{45}Sc(n,γ) ^{46}Sc, respectively. A detailed study *(25)* revealed that the correction for these interferences can be calculated in a sufficiently accurate way by employing the usual neutron fluence rate monitors in the k_0 method, namely a Zr-foil and a dilute Al-Au alloyed wire. It should be emphasized that this holds as well for the determination of the fast neutron fluence rate (needed in the correction for the $[n,\alpha]$ induced ^{51}Cr interference) via the threshold reaction ^{90}Zr $(n,2n)$ ^{89}Zr (E_γ = 909.1 keV]. On the other hand, the comments made in ref. 23 with respect to the counting of the AAO, HAP, HMD, and eluate fractions are

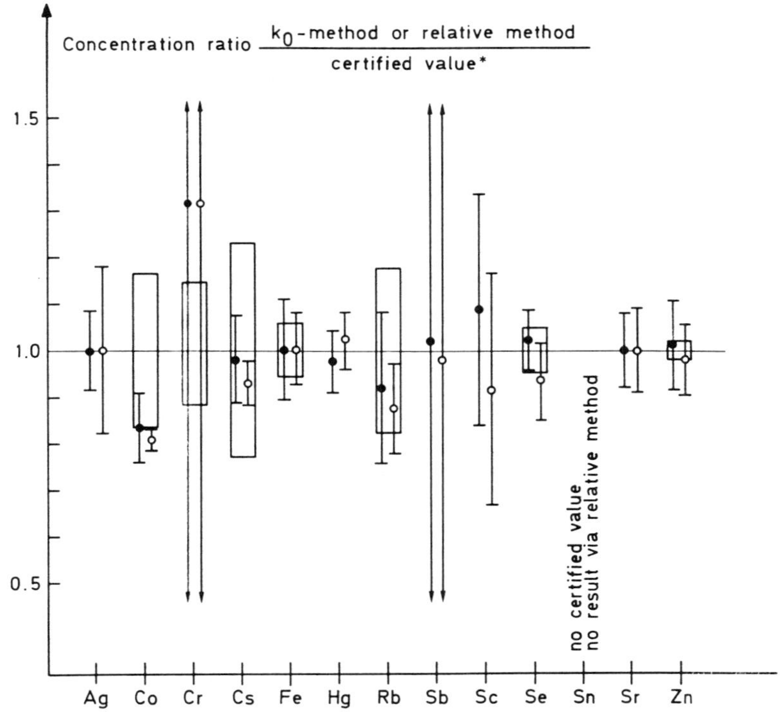

Fig. 2. Ratio of elemental concentrations in Versieck's reference human serum obtained via the k_0 method (●) and the relative method (○) to the certified values (* or in case no certified value is available to the average of the "k_0" and the "relative" results). The error bars represent the 2σ SD. The height of the rectangles indicates the 95% confidence limits of the certified values (*see* refs. *23,24*).

indicative of a more often encountered situation, where the Ge detector's detection efficiency and corrections for true coincidence have to be evaluated for voluminous samples with a not so accurately known composition. Whereas formerly—as mentioned in *(23)*—the efficiency calculation via computer code SOLANG *(9)* was restricted to small-diameter sources, recently algorithms were developed and introduced in the program EXTSANGLE *(17)* so that sources with a diameter larger than that of the detector can also be dealt with. As a further extension, this was followed by the handling of still other interesting counting geometries, such as Marinelli beakers and well-type detectors *(26,27)*. In the work described in *(23)*, the height of the measured AAO, HAP, HMD, and eluate fractions was kept as small as possible in order to minimize errors in the efficiency calculation (in fact, in the correction for γ attenuation) caused by inaccurate knowledge of composition and packing density. This layer-like counting geometry is not always practical or feasible to realize, however, and then it may be interesting to make use of linear attenuation coefficients μ (to be given as input data in SOLANG or EXTSANGLE),

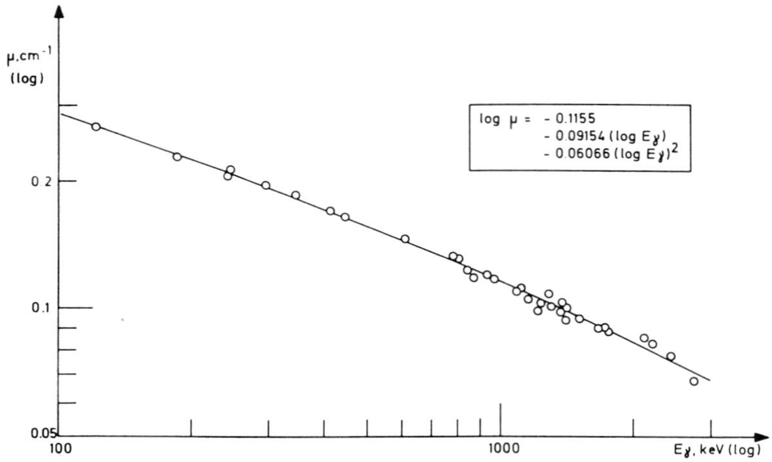

Fig. 3. Linear attenuation coefficient μ experimentally determined for a soil sample according to Eq. (3).

which are obtained from direct experimental determination instead of calculation. This can be achieved via the counting of γs emitted by some suitable point sources, positioned at a large distance from the detector, without and with an interposed layer with thickness, *d*, of the material investigated. With the observed count rates denoted by N_0 and N, respectively, one has that:

$$\mu \text{ (in cm}^{-1}) = (N/N_0)/d \text{ (in cm)} \tag{3}$$

An example of a μ vs E_γ curve registered for a soil sample with the aid of ^{152}Eu, ^{226}Ra, and ^{24}Na point sources is shown in Fig. 3. Another problem related to the counting of samples with large height is the computation of the correction factor (COI) for true-coincidence effects. As outlined in *(28)*, this correction can no longer be calculated accurately with the detection efficiency for the whole source as an input, but rather should be based on the integration of point elements in the source. In the present work, for a cylindrical sample with height, *H*, it is proposed to use the following numerical integration procedure:

$$\text{COI (cyl)} = \int_0^H \text{COI}(h) \cdot \bar{\Omega}(h) \cdot dh/(\bar{\Omega}\text{ (cyl)} \cdot H) \tag{4}$$

where $\bar{\Omega}$ (cyl) is the effective solid angle for the whole cylinder, as obtained via SOLANG or EXTSANGLE, whereas $\bar{\Omega}$ (*h*) and the corresponding COI(h) refer to values calculated for zero-height circular sources at a distance, *h*, from the base of the cylindrical sample—taking into account the attenuation layer with thickness h. Figure 4 shows, for an HPGe detector with 38% relative efficiency in use at the INW, a comparison of "erroneous" (based on the source as a whole) and "correct" (Eq. [4]) COIs in case of the 657.8 and 1173.2 keV lines of ^{110m}Ag and ^{60}Co, respectively, in soil cylinders (on top of the detector) with 5-cm diameter and heights, *H*, going up to 10 cm. As seen, the discrepancies exceed 1% from H ≃ 1 cm on for the 657.8-keV line and from *H* ≃ 2 cm

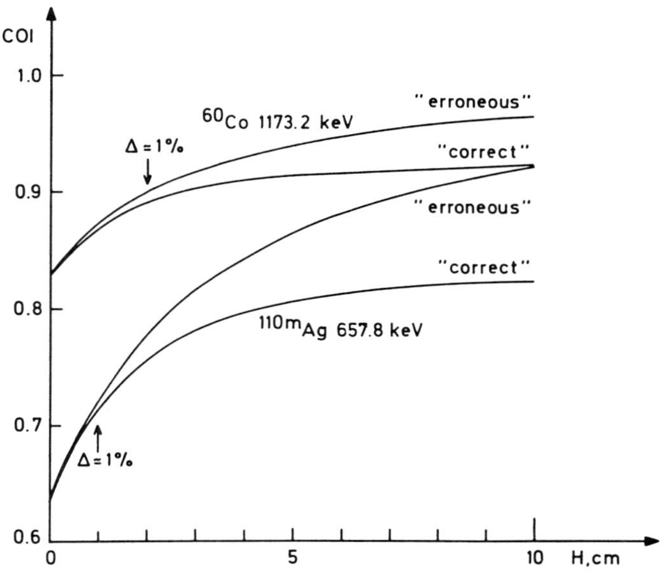

Fig. 4. Comparison of "erroneous" (considering the source as a whole) and "correct" (Eq. [4]) COI factors for the ^{60}Co 1173.2 and the ^{110m}Ag 657.8 keV lines when measuring soil cylinders with height, H, on top of a 38% efficiency Ge detector.

on for the 1173.2-keV line. Full details and results of the here proposed method will be published in due course.

DETERMINATION OF THE LANTHANIDES IN PLANT LEAVES AND LICHENS NEAR A PORTUGUESE COAL-FIRED POWER STATION

As part of an extensive study made with the aim of investigating the effect of the operation of a Portuguese coal-fired power station (C.T. Sines) on the environment, the determination of the lanthanides in plant leaves (*Cistus salvifolius* and *Inula viscosa*) and lichen (*Parmelia sulcata*) via k_0-based INAA was undertaken at the INETI, with the cooperation of the INW. In order to accomplish this successfully, some extensions and modifications had to be introduced in the k_0 method. These were related to (1) the fact that many of the lanthanides are determined with better accuracy and sensitivity via measurement of low-energy γ lines, for which a large Ge-detector is not suited, and (2) the non-1/v character of the thermal neutron cross-section $\sigma(v)$ (v = neutron velocity) for $^{151}Eu(n,\gamma)$, $^{164}Dy(n,\gamma)$, $^{168}Yb(n,\gamma)$, and $^{176}Lu(n,\gamma)$, which rules out the use of Eq. (1) written in the Høgdahl convention requiring $\sigma(v) \propto 1/v$ up to ~ 1.5 eV.

As to the measurement of low-energy γs, extensive work was done leading to the introduction of a low-energy photon detector (LEPD) in k_0-

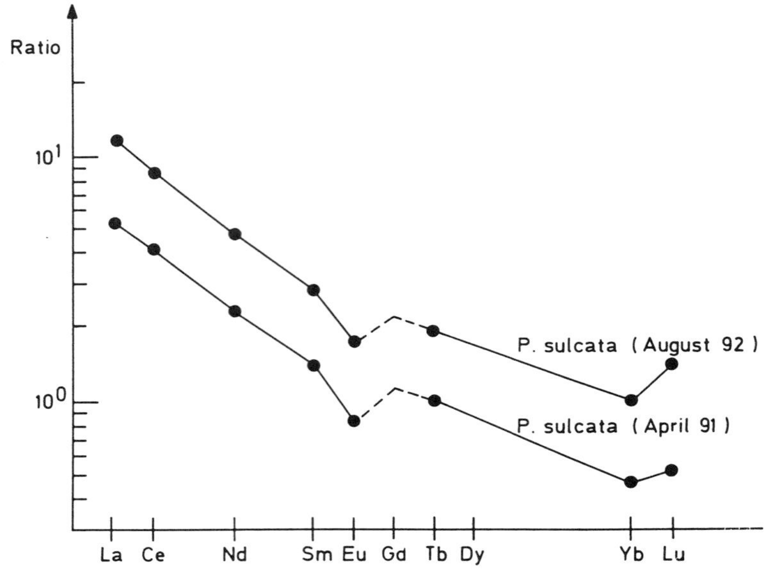

Fig. 5. Chondrite normalization for the lanthanide contents obtained via k_0-based INAA in *Parmelia sulcata* sampled south of the coal-fired power station at Sines.

based NAA: practical procedures were elaborated for calibrating the LEPD with respect to its detection efficiency and peak-to-total ratio *(15)*; algorithms and computer codes with up-to-date nuclear data files were developed for the calculation of coincidence correction factors, taking into account γ-γ, γ-KX and γ-LX coincidences (with KX and LX originating from internal conversion and/or orbital electron capture) *(16,29)*; accurate k_0 factors were experimentally determined *(5)* and $^{99}Mo/^{99m}Tc$ ($E\gamma$ = 140.5 keV) was selected as the most suitable comparator for which the effect of epithermal neutron self-shielding was studied in detail *(30)*.

Concerning the handling of "non-1/v" (n,γ) reactions, a changeover from the Høgdahl convention to the Westcott formalism was proposed *(14,31)*. This leads in Eq. (1) to the replacement of the terms $f + Q_0(\alpha)$ (in nominator and denominator) by $g(T_n) + r(\alpha)\sqrt{T_n/T_0} \cdot s_0(\alpha)$, where $g(T_n)$ is the Westcott g-factor at neutron temperature T_n (equal to 1 for $1/v$ cases), $r(\alpha)\sqrt{T_n/T_0}$ is the modified spectral index, and $s_0(\alpha)$ is a measure for the reduced resonance integral to 2200 ms^{-1} cross-section ratio *(14)*. User-friendly methods were worked out for the monitoring of T_n and $r(\alpha)\sqrt{T_n/T_0}$ (including the ongoing manufacturing and certification of a Zr-Au-Lu alloy by the IRMM), and experimentally determined k_0 factors together with re-evaluated $g(T_n)$ and s_0 values were added to the nuclear data library *(5,14)*.

As usual in the determination of the lanthanides, one of the quality controls that can be performed is the chondrite normalization. That this is reassuring for the results of k_0-based INAA, after introducing the Westcott formalism and the LEPD, is shown in Fig. 5 for the lichen *Parmelia sulcata* sampled in two subsequent years south of the Sines power sta-

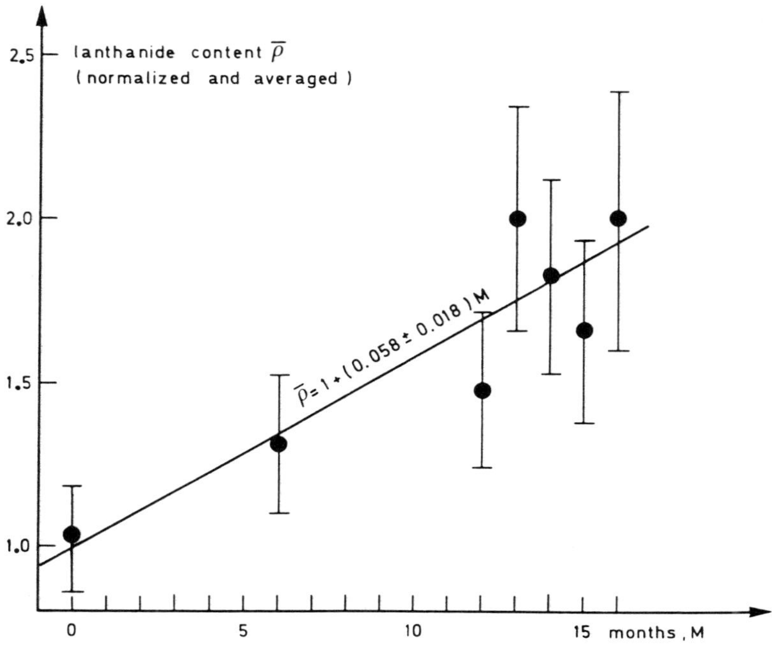

Fig. 6. Increase with time of the lanthanide content (normalized and averaged over all lanthanides) of *Parmelia sulcata* sampled south of the coal-fired power station at Sines (time scale: from April 1991 to August 1992).

tion. Figure 6 shows one of the practical results of this environmental study, indicating a significant increase with time of the lanthanide content in *Parmelia sulcata*. This effect may originate from particulates of the coal and ash deposits being transported through the air and falling out on the lichen (as observed by scanning electron micrography) where absorption occurs *(32)*.

CONCLUSION

From the developments and results reported in this article, it is concluded that at present, k_0-based NAA is capable of tackling a large variety of analytical problems when it comes to the multielement determination in environmental and biological matrices. Although not discussed here, other interesting applications take shape at the horizon, the most challenging one being undoubtedly the use of the k_0 method in prompt-γ cold-neutron activation analysis *(33–35)*.

ACKNOWLEDGMENTS

Thanks are due to A. De Wispelaere for technical assistance. The financial support of the National Fund for Scientific Research, the Inter-University Institute for Nuclear Sciences (F.D.C.), and the Hungarian

National Research Fund/OTKA Contract No. 1838 (A.S.) is gratefully acknowledged. F.D.C. is Research Director of the Belgian National Fund for Scientific Research.

REFERENCES

1. A. Simonits, F. De Corte, and J. Hoste, *J. Radioanal. Chem.* **24,** 31–46 (1975).
2. F. De. Corte, A. Simonits, A. De Wispelaere, and J. Hoste, *J. Radioanal. Nucl. Chem., Articles* **113,** 145–161 (1987).
3. F. De Corte, A. Simonits, A. De Wispelaere, and A. Elek, *J. Radioanal. Nucl. Chem., Articles* **133,** 3–41 (1989).
4. F. De Corte and A. Simonits, *J. Radioanal. Nucl. Chem., Articles* **133,** 43–130 (1989).
5. F. De Corte, A. Simonits, F. Bellemans, M. C. Freitas, S. Jovanović, B. Smodiš, G. Erdtmann, H. Petri, and A. De Wispelaere, *J. Radioanal. Nucl. Chem., Articles* **169,** 125–158 (1993).
6. A. Simonits, F. De Corte, and J. Hoste, *J. Radioanal. Chem.* **31,** 467–486 (1976).
7. F. De Corte, K. Sordo-El Hammami, L. Moens, A. Simonits, A. De Wispelaere, and J. Hoste, *J. Radioanal. Chem.* **62,** 209–255 (1981).
8. F. De Corte, S. Jovanović, A. Simonits, L. Moens, and J. Hoste, *Atomkernenergie-Kerntechnik,* **suppl. to 44,** 641–647 (1984).
9. L. Moens, J. De Donder, Lin Xilei, F. De Corte, A. De. Wispelaere, A. Simonits, and J. Hoste, *Nucl. Instrum. Methods* **187,** 451–472 (1981).
10. L. Moens and J. Hoste, *Int. J. Appl. Radiat. Isotop.* **34,** 1085–1095 (1983).
11. F. Bellemans, J. Faltejsek, A. De Wispelaere, and F. De Corte, in *Proceedings of the International k_0 Users Workshop—Gent,* (F. De Corte, ed., Lab. Anal. Chem., INW, RUG, 1992, pp. 69–72.
12. Lin Xilei, F. De Corte, L. Moens, A. Simonits, and J. Hoste, *J. Radioanal. Nucl. Chem., Articles* **81,** 333–343 (1984).
13. F. De. Corte, ed., *Proceedings of the International k_0 Users Workshop—Gent* Lab. Anal. Chem., Inst. Nucl. Sci., Univ. Gent, 1992.
14. F. De Corte, F. Bellemans, P. De Neve, and A. Simonits, *J. Radioanal. Nucl. Chem., Articles* **179,** 93–103 (1994).
15. F. De Corte, M. C. Freitas, and A. De Wispelaere, *Nucl. Instr. and Methods* **A299,** 335–343 (1990).
16. F. De Corte and M. C. Freitas, *J. Radioanal. Nucl. Chem., Articles* **160,** 253–267 (1992).
17. N. Mihaljević, S. Jovanović, F. De Corte, B. Smodiš, R. Jaćimović, G. Medin, A. De Wispelaere, P. Vukotić, and P. Stegnar, *J. Radioanal. Nucl. Chem., Articles* **169,** 209–218 (1993).
18. C. Ingelbrecht, F. Peetermans, F. De Corte, A. De Wispelaere, C. Vandecasteele, E. Courtijn, and P. D'Hondt, *Nucl. Instrum. and Methods* **A303**, 119–122 (1991).
19. F. De Corte, A. Demeter, Lin Xilei, L. Moens, A. Simonits, A. De Wispelaere, and J. Hoste, *Isotopenpraxis* **20,** 223–226 (1984).
20. L. Moens, F. De Corte, A. De Wispelaere, J. Hoste, A. Simonits, A. Elek, and E. Szabó, *J. Radioanal. Nucl. Chem. Data Section* **82,** 385–452 (1984).
21. F. De Corte, A. De Wispelaere, M. Van den Boer, D. Bossus, and R. van Sluijs, *Anal. Chim. Acta* **254,** 127–143 (1991).
22. T. El Nimr, F. De Corte, L. Moens, A. Simonits, and J. Hoste, *J. Radioanal. Chem.* **67,** 421–435 (1981).
23. Lin Xilei, D. Van Renterghem, R. Cornelis, and L. Mees, *Anal. Chim. Acta* **211,** 231–241 (1988).

24. J. Versieck, L. Vanballenberghe, A. De Kesel, J. Hoste, B. Wallaeys, J. Van den Haute, N. Baeck, H. Steyaert, A. R. Byrne, and F. W. Sunderman, *Anal. Chim. Acta* **204,** 63–75 (1988).
25. Lin Xilei, D. Van Renterghem, F. De Corte, and R. Cornelis, *J. Radioanal. Nucl. Chem. Articles* **133,** 153–165 (1989).
26. N. Mihaljević, S. Jovanović, F. De Corte, P. Vukotić, J. Faltejsek, J. Kučera, D. Piccot, and A. De Wispelaere, *Proceedings of the International k_0 Users Workshop—Gent,* F. De Corte, ed., Lab. Anal. Chem., INW, RUG, 1992, pp. 53–57.
27. S. Jovanović, N. Mihaljević, F. De Corte, L. Vancraeynest, J. Faltejsek, P. Vukotić, S. Dapčević, J. Kučera, A. De Wispelaere, F. Carrot, C. Dardenne-Deschamps, and N. Deschamps, *Proceedings of the International k_0 Users Workshop—Gent,* F. De Corte, ed., Lab. Anal. Chem., INW, RUG, 1992, pp. 59–62.
28. K. Debertin and R. G. Helmer, in *Gamma- and X-ray Spectrometry with Semiconductor Detectors,* North-Holland, Amsterdam, 1988, pp. 264–265.
29. M. C. Freitas and F. De Corte, *Proceedings of the International k_0 Users Workshop—Gent,* F. De Corte, ed., Lab. Anal. Chem. INW, RUG, 1992, pp. 63–67.
30. M. C. Freitas and F. De Corte, *Proceedings of the International k_0 Users Workshop—Gent,* F. De Corte, ed., Lab. Anal. Chem., INW, RUG, 1992, pp. 115–119.
31. F. De Corte, P. De Neve, F. Bellemans, and A. Simonits, *Proceedings of the International k_0 Users Workshop—Gent,* F. De Corte, ed., Lab. Anal. Chem., INW, RUG, 1992, pp. 163–166.
32. M. C. Freitas, Ph.D. Thesis, University of Gent (1993).
33. R. M. Lindstrom, R. F. Fleming, R. L. Paul, and E. A. Mackey, *Proceedings of the International k_0 Users Workshop—Gent,* F. De Corte, ed., Lab. Anal. Chem., INW, RUG, 1992, pp. 121–124.
34. M. Rossbach and F. De Corte, *Proceedings of the International k_0 Users Workshop—Gent,* F. De Corte, ed., Lab. Anal. Chem., INW, RUG, 1992, pp. 125–129.
35. A. Simonits and G. Molnár, private communication.

Application of Short-Lived Radionuclides in Neutron Activation Analysis of Biological and Environmental Samples

F. GRASS,*[,1] M. BICHLER,[1] J. DORNER,[1] H. HOLZNER,[1] A. RITSCHEL,[1] A. RAMADAN,[2] G. P. WESTPHAL,[1] AND R. GWOZDZ[3]

[1]*Atominstitut, Vienna, Austria;* [2]*Atomic Energy Authority, Cairo, Egypt, and Tracechem, Copenhagen, Denmark*

ABSTRACT

The application of short-lived nuclides, especially in connection with the ^{6}LiD-converter, in biological and environmental samples is demonstrated on I and Br determination in human urine, on I in pet food, and on the analysis of all the halogens in volcanic gases in a single activation. Trace element determination in lichens indicates polluted and unpolluted areas. The use of the .74-s ^{38m}Cl enables the rapid screening of great number of samples.

Index Entries: Activation analysis; short-lived nuclides; trace elements; lichen; urine; pollution, ^{6}LiD-converter; halogens; DOWEX 1.

INTRODUCTION

The use of short-lived nuclides in activation analysis has the advantage of immediately available results. However, some special equipment is needed: A fast-transfer system and a high-rate measurement system. Since not only short-lived nuclides are activated in a short activation time, but longer-lived nuclides are activated simultaneously, we should rather use the term short-time activation analysis. There are a number of elements that form only short-lived nuclides by neutron activation: B, the

*Author to whom all correspondence and reprint requests should be addressed.

20-ms ^{12}B; Li, the 840-ms ^{8}Li; F, the 11.3-s ^{20}F; Y, the 16-s ^{89m}Y; and Pb, the 800-ms ^{207m}Pb. Of these nuclides, B, Y, and Pb do not have the necessary sensitivity for analysis of biological samples activated in the usual research reactors and, therefore, need enrichment procedures, but may be used for analyses in environmental samples containing higher concentrations of these elements, such as Pb in road dust *(1)*, in engine exhaust if leaded fuel is used *(2)*, or in the fumes of Pb smelters. The enrichment of B from oil well water was performed by its BF_4^- complex on a Dowex 1X8 anion exchanger, which was then activated. The energetic β-particles were counted by a Cerenkov counter, and the decay curves analyzed *(3)*.

There are some elements having longer-lived activation products that may be analyzed more rapidly by their short-lived states, e.g., for screening of a large number of samples. Examples are 740-ms ^{38m}Cl, 17.5-s ^{77m}Se, 4.9-s ^{79m}Br, and the 18.7-s ^{179m}Hf to mention only the nuclides with half-lives of <1 min. Chlorine screening with ^{38m}Cl was performed on iron concrete after PVC fires in which HCl was emitted and may have influenced the stability of the iron construction *(4)*. The screening of organic chlorine containing sewage oil was proposed by Zeisler (IAEA, personal communication, 1993) and is obviously a possible application, since the increased sensitivity to environmental problems makes analysis compulsory and the throughput in LC-MS analysis is somewhat restricted. The increasing attention that Se analyses are receiving, as seen in this conference, would make it seem advantageous to use the 17.5-s ^{77m}Se instead of the ^{75}Se where highest sensitivity is not necessary. The 18.7-s ^{46m}Sc may substitute the long-lived ^{46}Sc as a measure for the crust fraction of environmental samples. Fluorine analyses using ^{20}F have been widely used for determination of F in bones, in a bone biopsy, and in environmental samples *(5)*. We will not include already published applications, but limit ourselves to new ones.

METHODS

Irradiation

Most of the irradiations were performed in the Vertical Fast Irradiation and Measurement System (VFIMS) of the institute's 250-kW TRIGA Mark II reactor, without and with the ^{6}LiD converter *(6)*. The inserted converter doubles the fast flux, but even more pronounced is the reduction of the thermal flux. This enables a reduction of most matrix activities as shown in Fig. 1, where the activity ratios of some elements activated without and with the converter *(7)* are plotted vs the ratios of the thermal activation cross-section and the resonance integral *(8)*. Therefore, an irradiation time of minutes is possible without producing an undue matrix activity and enables the determination of elements with minute nuclides. Irradiation times without the ^{6}LiD converter are in the range of 5–30 s at a flux of 2.5×10^{12} cm^{-2} s^{-1} and a flux ratio of 12, and with the ^{6}LiD converter 300–600 s *(9)*.

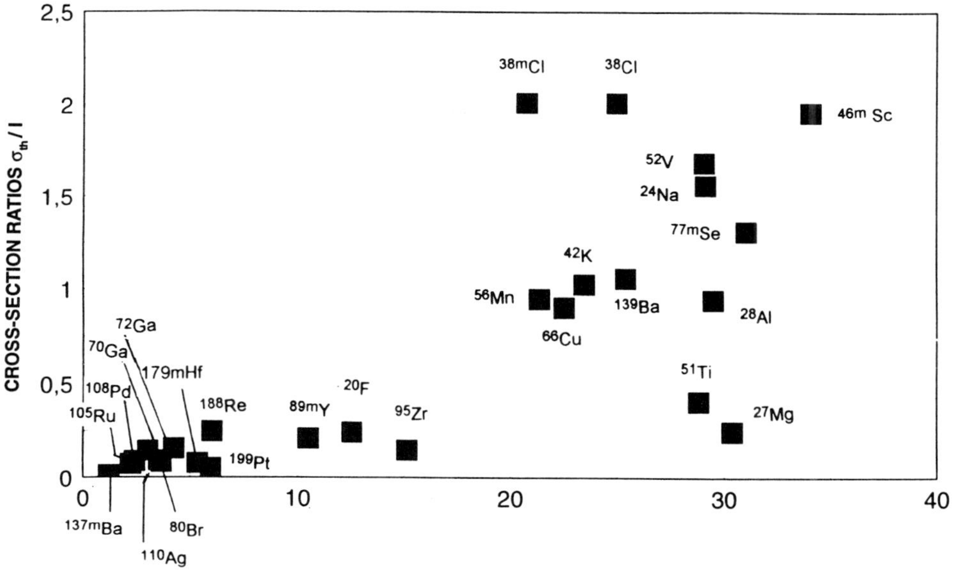

Fig. 1. Activity ratios of activation without and with the ^{6}LiD converter vs the cross-section ratios σ_{th}/I.

Spectrometry Systems

The γ-spectrometry systems at the institute's fast-transfer systems are optimized for resolution and throughput rate. For the best performance at high counting rates, *n*-type high-purity germanium detectors of 15 and 20% relative efficiency* are used together with the Pre-Loaded Filter (PLF) pulse processing systems** rendering a typical low-rate resolution of 1.8 keV at 1332 keV. The maximum throughput rate is 110 kc/s in the high-throughput mode, which gives the worst-case resolution of 3.3 keV. In the lower throughput mode, which is used in this article, the maximum throughput rate is 80 kc/s with the worst-case resolution of 2.6 keV. To enable quantitative measurements of spectra with rapidly varying intensity and sample composition, the system is equipped with real-time loss correction modules ND 599 LFC based on the Virtual Pulse Generator method of counting loss correction *(11)*. The pulses are fed into a Canberra-ND ACCUSPEC multichannel analyzer installed in a 486 or in a 386 PC. The performance of the pulse processing system is shown in Figs. 2 and 3.

*Relative to a 3" × 3" NA detector.

**The PLF amplifier was substantially improved by simplifying the original design *(10)* to maximize the throughput and produce an exponential pulse-interval distribution to enable conventional correction of ADC dead-time losses. ND 599 LFC counting loss correctors could then be used without modification.

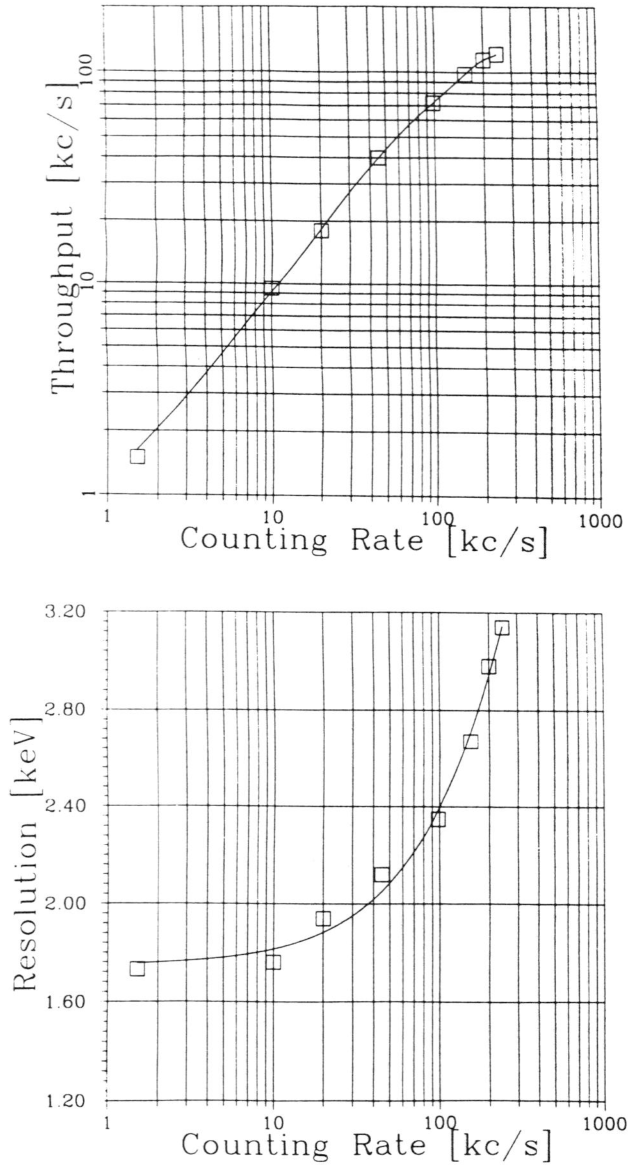

Fig. 2. Resolution and throughput of the preloaded filter system vs counting rate.

Data Processing

The spectra recorded with the multichannel analyzer were processed in three ways according to needs: for a proper assessment of the counting statistics, two spectra are recorded simultaneously—a dead-time and pile-up corrected spectrum and an uncorrected spectrum. The peak positions of the latter are selected, and the errors of the peak areas are calculated. The true peak areas are calculated from a comparison of

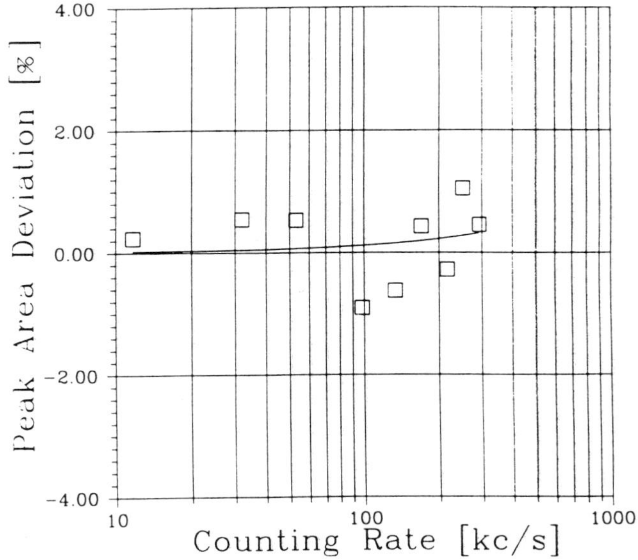

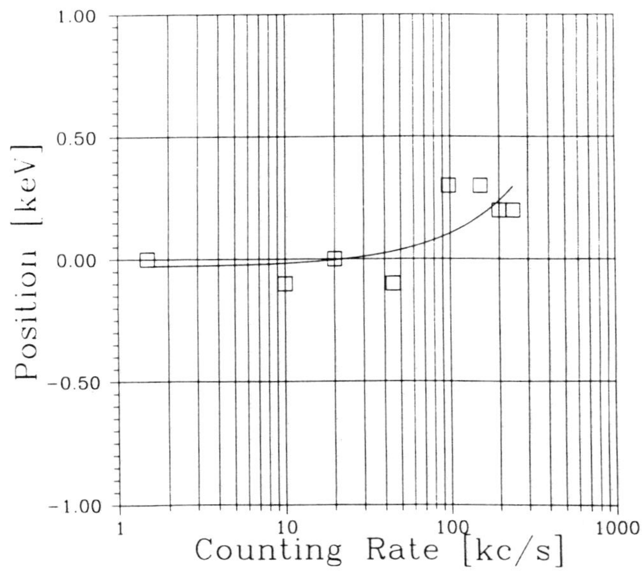

Fig. 3. Peak area deviation and position shift vs counting rate.

corrected and uncorrected peak areas using a special peak evaluation program (*12*). The data are printed immediately after the measurement. This enables the best evaluation of areas and their errors, but it is time-consuming.

The second evaluation method, FAST, needs only the corrected spectra, and determines peak areas and concentrations from a data string after all the measurements are completed *(13)*. If only a few nuclides are to be analyzed, the printout of the regions of interest (ROI) of the corrected spectrum is sufficient. All three methods were applied in this work.

Sample Preparation

The samples for short-time activation analysis are prepared by the usual procedure: after homogenizing the samples in an agate mill, they are weighed into short polyethylene tubes plugged at one end, then plugged at the other end, and heat-sealed. Lichen samples were dried at 40–50°C to enable the necessary homogenization. That was performed in an agate mill if the sample was large enough. Small samples were crushed in an agate mortar. Liquid samples and the exchanger slurries were injected with a syringe into a preformed irradiation container and heat-sealed.

APPLICATIONS

Having determined B in oil-well water and sea water in the form of its BF_4^- complex by the 20-ms ^{12}B after enrichment on Dowex 1X8 *(3)* Sabbioni (Ispra, personal communication, 1992) suggested determining B in urine by the same method. So we prepared small polyethylene columns containing 0.6 mL of a Dowex 1X8 of the 100–200 mesh fraction in its F form and pumped up to 700 mL urine through this column after having added 2 mL of 40% HF. Unfortunately, our registration system in VFIMS broke down and could not be replaced in time so this analysis had to be postponed. However, we were left with a number of prepared samples and decided to find out whether other elements were also enriched on the anion exchanger. In this pilot experiment, we found that I and Br were fully absorbed (*see* Fig. 4) and could be determined quantitatively, a fact that was confirmed by analyzing one sample in two fractions of greatly differing volumes in two columns: the respective iodine and bromine concentrations were found to be practically the same in both cases. A few results are shown in Table 1. Though the method was tested only on a restricted number of samples and needs more detailed studies (e.g., to decrease the I and Br content of the ion exchanger), various aspects of this method seem promising. The irradiation of the VFIMS was performed using the ^{6}LiD converter with an analysis scheme of 600-0-60, 600-60-300, and 600-360-600 s, denoting irradiation time - decay time - measurement time. Besides the halogens, Mg and Ca were found in the exchanger, possibly as deposits or as complexes attached to the exchanger. A separate study would be required to confirm this assumption.

Another application of VFIMS with the converter is the analysis of I in pet food. A 300-30-600 analysis scheme is sufficiently sensitive for this determination in the 0.5–50 µg/g range. If larger amounts of sodium are present, a longer decay time is required to let the 39-s ^{23}Ne (439 keV) decay, since it nearly coincides with the 443 keV line of ^{128}I.

A further application is the determination of all the halogens in the condensates of volcanic gases. Bichler and his team took samples from

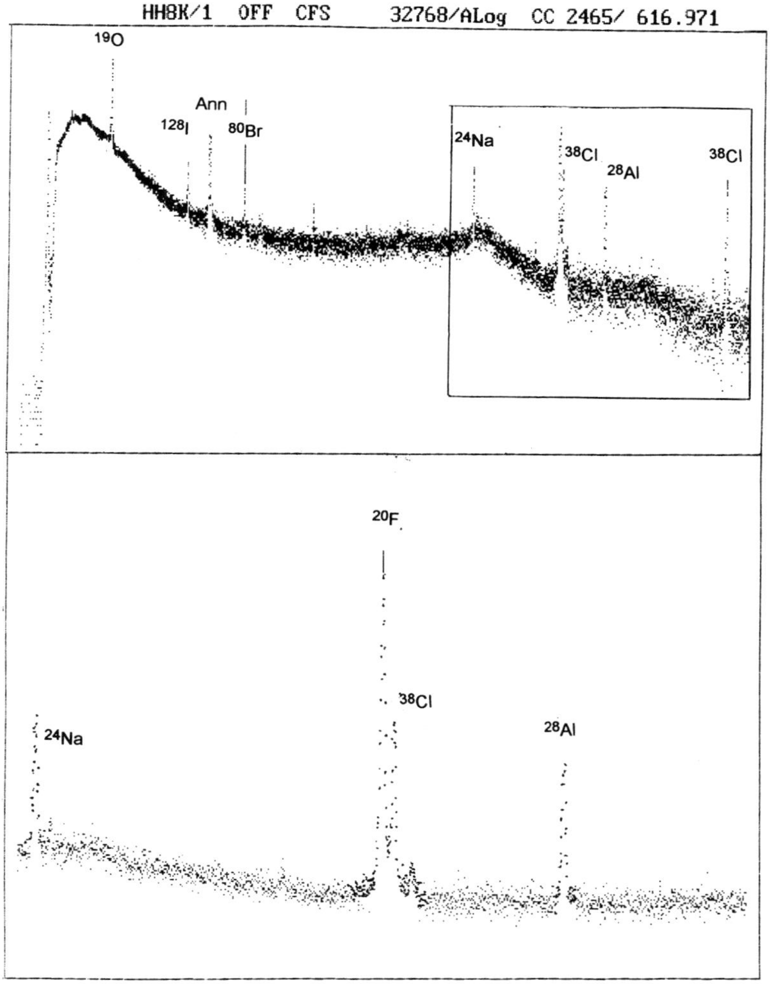

Fig. 4. Anions absorbed from 400 mL morning urine by a 0.6-mL Dowex 1X8 column, ^{6}LiD converter. t_i = 600, t_d = 0, t_m = 60 (s).

Table 1
Anions Absorbed on a 0.6-mL Dowex 1X8 Column from Human Urine

Test person	Sample		μgI	μg Br	μgI/mL	μg Br/mL
female 63 yr	morning urine	400 mL	6.22	15.0	0.016	0.0375
male 66 yr	morning urine	400 mL	7.66	44.6	0.019	0.112
male 66 yr	urine	280 mL	4.83	26.6	0.017	0.095
male 66 yr	urine	700 mL	10.34	80.9	0.015	0.116
column blank	Dowex 1X8	0.6 mL	0.43	0.32		

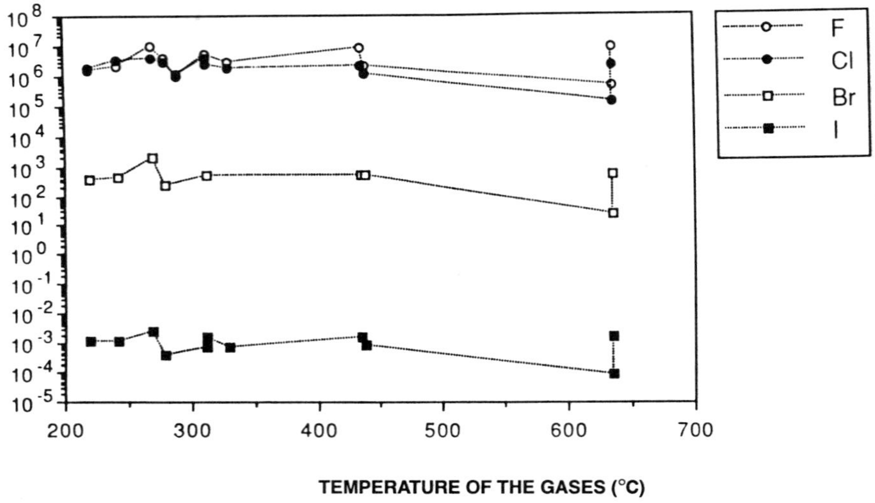

Fig. 5. Halogen contents in condensates of volcanic gases (μg/m^3) sampled at Fossa, Vulcano (Eolian Islands) vs temperature of the gases (°C).

fumaroles of Vulcano. The details of sampling (which is the most difficult procedure) will be presented elsewhere. For the short-time activation analysis, 0.5 mL of the condensate water were heat-sealed in the polyethylene containers and analyzed in VFIMS without the converter for F and Cl according to the scheme 22-0-22 s. Though the 443 keV peak of ^{128}I and the 617 keV peak of ^{80}Br were found, the peak areas were small, so that only an evaluation with a large error was possible. However, using the scheme 300-300-600 s with the converter, all the halogens were analyzed in a single irradiation. The aim of these measurements was to prove whether the composition of the condensates is dependent on the temperature of the fumaroles. Figure 5 shows the results of these measurements. There is a difference of nine orders of magnitude in the concentrations, but the ratios between the halogens remain nearly constant.

Moss and lichen analyses have attracted much attention as biomonitors in recent years *(14,15)*, as is also seen from this conference program. For an application of short-time activation analysis, we obtained from Steinnes (Univ. Trondheim, Norway) a few moss samples on which different techniques of analysis were performed, to find out which technique is best suited for this purpose. We used short-term activation analysis (AA) without and with the ^{6}LiD converter, AA with middle and long-lived nuclides, and TXRFA on the same sample. A series of analyses was performed on a test sample that had already been homogenized. However, it turned out that the last portions of the sample had a different composition, showing that segregation of heavier particles is a critical point in analyses of biomonitors. The results of these measurements are shown in Table 2.

Table 2
Moss DK1 Norway[a]

Element, nuclide	VFIMS	VFMIS + CONV	AA long	TXRF
Na ^{24}Na	—	446 (29)	435 (20)	—
Mg ^{27}Mg	3430 (215)	3060 (325)	—	—
Al ^{28}Al	893 (27)	890 (50)	—	—
Cl ^{38}Cl	448 (10)	460 (20)	—	—
K ^{42}K	—	—	6400 (800)	7015 (540)
Ca	—	—	—	3518 (230)
Sc ^{46m}Sc	89 (5)[b]	76 (10)[b]	—	—
^{46}Sc	—	—	92 (2)[b]	—
Ti	—	—	—	87 (7)
V ^{52}V	4.20 (0.05)	4.15 (0.1)	—	—
Cr ^{51}Cr	—	—	1.20 (0.15)	—
Mn ^{56}Mn	116 (1)	95 (8)	—	125 (8)
Fe ^{59}Fe	—	—	630 (70)	644 (40)
Co ^{60}Co	—	—	0.10 (0.05)	—
Ni	—	—	—	3.4 (0.8)[c]
Cu ^{66}Cu	—	80 (10)	—	114 (9)
Zn ^{65}Zn	—	—	49 (2)	48 (4)
As ^{76}As	—	0.66 (0.3)	—	—
Se ^{77m}Se[c]	0.70 (0.07)	—	—	0.38[d]
^{75}Se	—	—	0.68 (0.03)	—
Br ^{80}Br	28 (4)	32 (6)	—	—
Rb ^{86}Rb	—	—	14 (3)	17 (1)
Sr ^{85}Sr	—	—	25 (8)	15 (1)
Sb ^{124}Sb	—	—	0.36 (0.02)	—
I ^{128}I	—	4.2 (0.3)	—	—
Cs ^{137}Cs	—	—	250 (25)[b]	—
Ba ^{131}Ba	—	—	5.9 (0.2)	—
^{137m}Ba	—	5.6 (0.9)	—	—
La ^{140}La	—	—	0.38 (0.05)	—
Eu ^{152}Eu	—	—	40 (3)[b]	—
Au ^{198}Au	—	—	1.1 (0.2)[b]	—
Hg ^{203}Hg	—	—	100 (10)[b]	—
Pb	—	—	—	25 (1)

[a]All data in mg/kg, if not otherwise stated, assumed error in ( ).
[b]µg/kg.
[c]extraction.
[d]extraction, overlaid by Pb.

These data show that each technique has some advantage. However, most of the data were obtained by the AA with long-lived nuclides. The time of analysis by these methods varies widely. For AA in VFIMS the scheme 30-0-30,300 is used; in VFIMS with converter 600-0-300,600, 1200; AA with long-lived nuclides are performed with a scheme 12 h-200 h-4 h and are activated at a flux of 8×10^{13} cm^{-2} s^{-1} in the ASTRA reactor; with TXRF, the samples are analyzed in a 1200-s cycle.

Influenced by this study, we collected lichens (most of the samples were *Pseudevernia furfuracea*) from the Mur-Mürz valley. This is geologically situated on the border of a paleozoic schist region bounded on both sides by slate, phyllite, and paragneiss mountains. The valley bottom is the traditional region of the Austrian steel industry. This concentration of heavy industry in a narrow valley renders it interesting for environmental studies. In most cases the lichens were sampled from larch trees, in some cases from spruce trees, and only few samples from deciduous trees. To ensure that no lichen was analyzed that was too young to have collected dust for more than 7 yr, we measured the Cs activities when the age of the lichens seemed doubtful. These activities are a rough measure for the age of the lichen: If there is a ^{137}Cs surplus greater than the ^{137}Cs/^{134}Cs ratio corresponding to the Chernobyl accident, the lichen is assumed to contain ^{137}Cs from the nuclear weapon tests in the atmosphere. The Chinese weapon tests were the last atmospheric tests in 1976–1980, so a minimum age of 12 yr may be assumed.

Other samples to test the background level for some elements were taken from a recreational area without any industry near Schladming from a region of dachstein limestone at Ramsau and from the gneisses of Schladminger Tauern at the Reiteralm. Only small samples were available from this region, so only the ^{137}Cs activity of these samples is listed together with the analytical data in Table 3.

This table shows that the Mn content is enhanced in the gneiss region. The low Al values show that very little dust was present when the lichen was sampled in the wintertime. The V value in FL6 must be an outlier. The ^{137}Cs activities were higher than those in the Mur-Mürztal region.

Figure 6 shows a series of measurements in a profile NNE to SSW from Maria Schutz to Stuhlegg on a ridge parallel to the valley bottom of the Mur-Mürz valley. The Mn and the V data are comparable to the Dachstein data. A telecommunication tower and a larch tree are indicated in this figure. This larch was uprooted so that a dependence of some trace elements on the sampling height could be estimated; *see* Fig 7. Generally, the lichens were sampled 1–2 m from the ground. We can conclude from the results that the Mn content of the lichens near the steel industry and in the valleys near the SSW region is strongly enriched compared to the NNW region. The V content in lichens is more widely distributed, since apart from the steel industry, there is an additional source through the burning of cheap heavy oil with a high V content by other larger consumers.

The geographic distribution of the iodine content in lichens in Styria deserves special interest, since in this region goiter was endemic in former days. Now the iodine demand in human nutrition is covered by addition of iodine to cooking salt. From Fig. 8 it is clearly seen that iodine is enriched in the elevated regions and near the industrialized

Table 3
Trace Elements in Lichen of a Recreation Area Near Schladming

Species	height, m	Mn[a]	V	Al	Cl	Br	^{137}Cs Bq/g
REITERALM, gneiss							
FL 2 Pseudever. furf.	1900	77	3.4	474	1268	12	16.6
FL 3 "	1500	64	3.9	526	343	11	14.6
RAMSAU dachstein limestone							
FL 6 Pseudever. furf.	1250	27	9.0	442	1424	13	6.9
FL 5 "	1400	24	3.0	240	1027	8.8	11.3
FL 4 Usnea hirta + Evernia divaricata	1400	28	1.9	193	1093	7.4	7.9
FL 7 Pseudever. furf. very exposed	1800	82	2.8	231	1100	9.3	16.6

[a]Data for Mn, V, Al, Cl, and Br given in mg/kg.

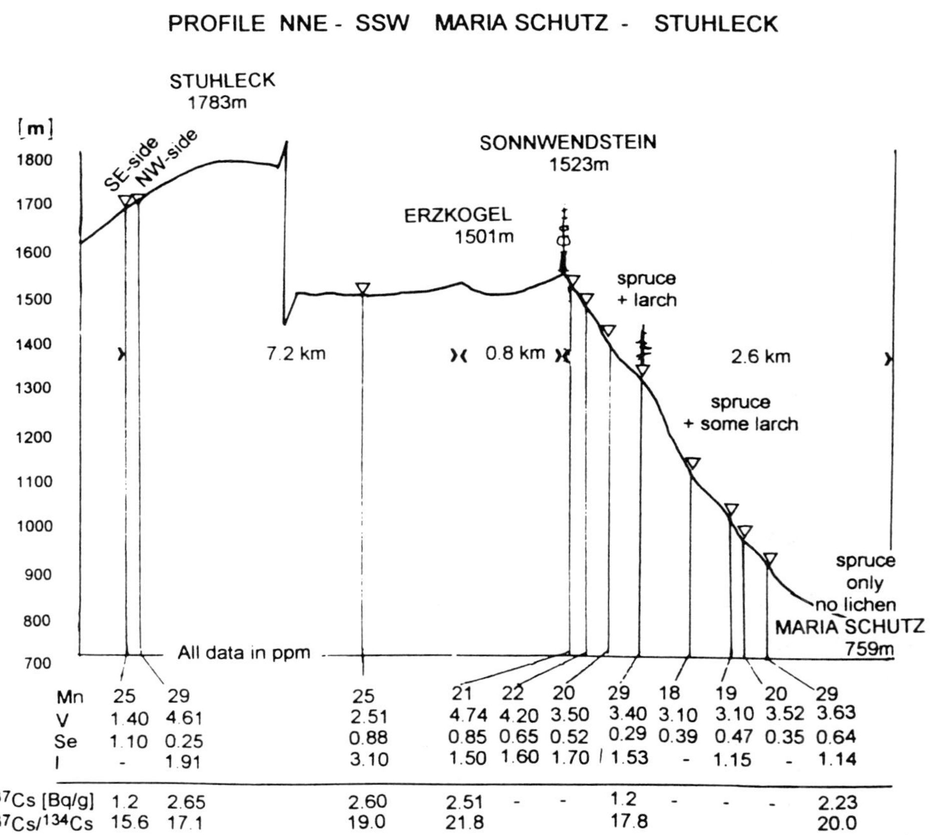

Fig. 6. Trace elements in lichens on an exposed mountain ridge.

ANALYSIS OF LICHEN (PSEUDEVERNIA FURFURACEA) FROM AN UPROOTED LARCH, SONNWENDSTEIN SOUTH SLOPE 1300m (all data in mg/kg)

^{137}Cs Bq/g	^{137}Cs / ^{134}Cs	HEIGHT	Mn	V	Se	Al	Cl	Br	Mg	Ca
2.0	15.6	25m	19	3.4	.34	920	1702	12	6.5	1240
2.9	19.5	23m	20	4.6	.59	1250	2180	19	7.8	710
2.7	19.5	17m	20	4.6	.45	1211	1900	20	7.2	760
2.4	20.1	9m	19	3.7	.37	977	1970	17	7.8	222
1.2	17.8	1m	29	3.4	.29	756	1215	10	11	238

Fig. 7. Analysis of lichen from an uprooted larch: trace element distribution as function of the height.

zones, and the figure shows low values in the isolated valleys. A cluster analysis with the full set of data is under way and will be published elsewhere.

To illustrate the advantage of short-time activation analysis for screening of Cl, we compared the Cl content of lichen measured by the schemes 5-0-5 s and 30-30-300 s. Although a ^{38m}Cl measurement including the manual handling takes 30 s, the ^{38}Cl measurement takes 12 times longer. This demonstrates impressively the applicability of organic Cl determinations by short-time activation analysis in different samples, including water samples after a short separation procedure.

SUMMARY

The advantages of activation with and without the ^{6}LiD-converter are demonstrated by the determination of iodine and bromine in human urine and all the halogens after a single activation in volcanic gases. Lichen analysis using these and other methods allows the choice of the most convenient one for the purpose. The distribution of Mn, V, and I in lichen from an industrialized region in Styria indicates unpolluted, polluted, and I-deficient areas. The application of the .74-s ^{38m}Cl allows the fast screening of a great number of samples.

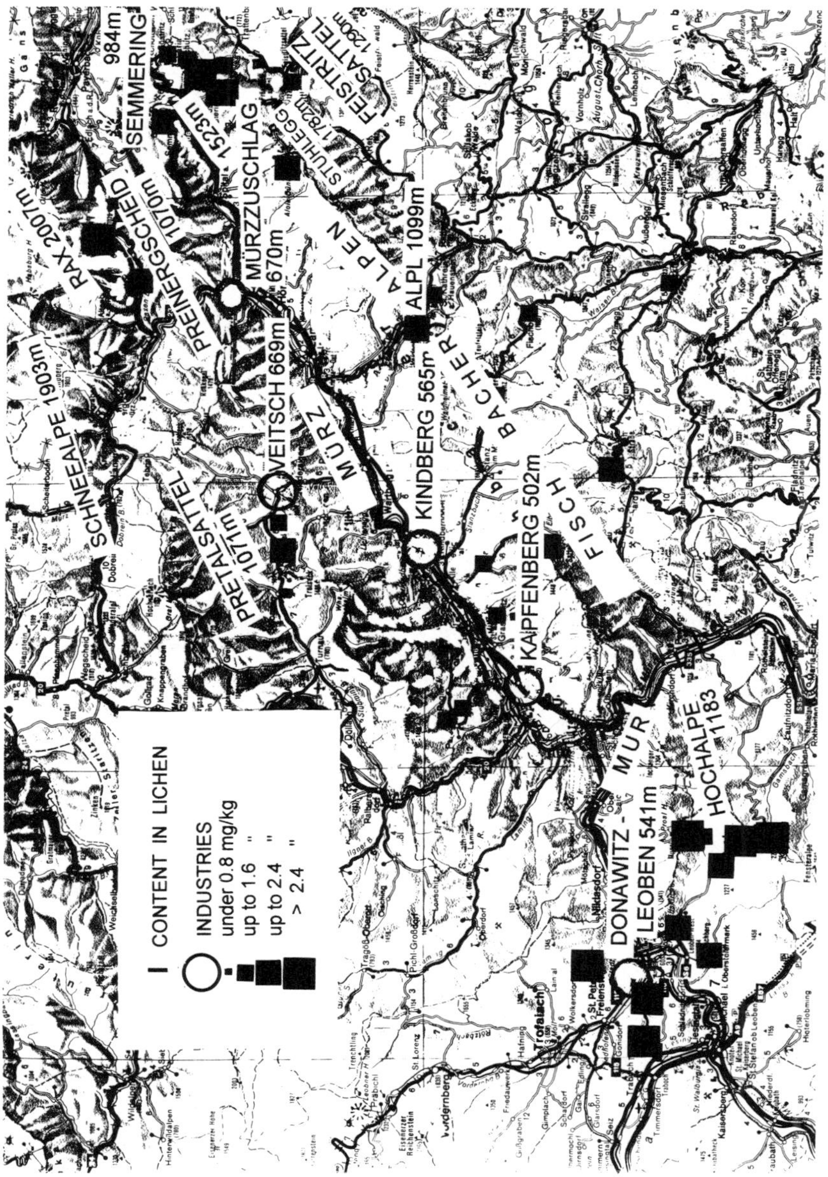

Fig. 8. Iodine content in lichen from a highly industrialized region of Styria, Austria.

REFERENCES

1. S. S. Ismail, F. Grass, P. Schindler, and J. O. Schmidt, Proc. 5th Intern. Conf. of Nucl. Methods in Environmental and Energy Research, DOE, CONF-840408 (1984) 49.
2. F. Grass and M. Bichler, Activation Analysis in Environmental Protection JINR DUBNA 1993.
3. F. Grass, *J. Radioanal. Nucl. Chem.* **112,** 347 (1987).
4. F. Grass, Nuclear Methods in Environment and Energy Res. 4th Intern. Conf. April 1980, Columbia DOE-Conf. 800433.
5. R. Gwozdz, F. Grass, and J. Dorner, *J. Radioanal. Nucl. Chem.* **169,** 52 (1993).
6. A. Salahi, F. Grass, F. Bensch, T. Kasa, E. Seidl, and S. Roth, *J. Trace & Microprobe Techn.* **229,** (1988).
7. H. Holzner, Dipl. work TU Wien, 1993.
8. G. Erdtmann, *Neutron Activation Tables*, Verlag Chemie, 1976.
9. F. Grass, M. Bichler, J. Dorner, H. Holzner, S. S. Ismail, B. Lipp, and G. Wiesinger, IAEA-SM-325/175 Karlsruhe, Application of Isotopes and Radiation in Conservation of the Environment, 571 (1992).
10. G. P. Westphal, *Nucl. Instr. Methods* **A299,** 261 (1990).
11. G. P. Westphal, *J. Radioanal. Chem.* **70,** 387 (1982), US Pat. 4476384.
12. G. P. Westphal and B. Lipp, Modified software in use since 1991.
13. R. Gwozdz, Peak evaluation program FAST in use since 1992.
14. M. C. Freitas, F. de Corte, M. A. Gouveia, M. I. Prudenico, and J. M. P. Cabral *J. Radioanal. Nucl. Chem.* **169,** 219 (1993).
15. M. C. Freitas, F. M. Catarino, C. Branquinho, and C. Magvas, *J. Radioanal. Nucl. Chem.* **167,** 309 (1993).

Analytical Applications of Guided Neutron Beams

RICHARD M. LINDSTROM,* ELIZABETH A. MACKEY, AND RICK L. PAUL

Inorganic Analytical Research Division, National Institute of Standards and Technology, Gaithersburg, MD 20899

ABSTRACT

Guided beams of thermal and cold neutrons have become available to analysts at several reactors during the past decade. The very pure beams from these guides have led to lower backgrounds and higher sensitivities for prompt-γ activation analysis (PGAA), and thus to new applications for this technique. For analytical accuracy, the details of neutron scattering within the sample need to be taken into account; this consideration is especially important for most materials of biological origin.

Index Entries: Cold neutrons; neutron guide; prompt-γ activation.

INTRODUCTION

A neutron guide is a reflecting channel that efficiently transports a beam of slow neutrons from the source to a distant instrument, such as a prompt-γ activation analysis (PGAA) spectrometer. The guide serves as a filter to increase the population of desirable slow neutrons compared to fast neutrons and γ-rays, which diminish with the square of the distance from the neutron source. The transmission efficiency is inversely proportional to the temperature of the neutron source, so that most guided beams are associated with reactor cold sources. The term "cold" usually is applied to neutrons with a wavelength > 4 Å, the most probable wavelength of a 30 K Maxwellian distribution. This corresponds to an energy of 0.005 eV and a velocity of 1000 m/s.

*Author to whom all correspondence and reprint requests should be addressed.

Biological Trace Element Research Editor: G. N. Schrauzer © 1994 by Humana Press Inc.

The analytical advantages of cold neutrons have been well documented *(1,2)*. Since capture cross-sections are generally inversely proportional to the velocity, neutrons from a moderator at 30 K give a threefold higher reaction rate per target atom than those from a source at 300 K. By bending the guide or by inserting a filter, the beam can be made free of fast neutrons and γ-rays; as a result, the background for PGAA is usually lower than with simple collimated beams of thermal neutrons, especially at low γ-ray energies. Moreover, because the shielding for guided beams can be made compact, the induced radiation from the sample can be detected with much better geometric efficiency than with traditional beams, and thus the sensitivity for PGAA and related techniques is higher. Because of improved geometry, an optimized PGAA system using a guided thermal beam *(3)* is competitive with systems using conventional reactor beams with flux density two orders of magnitude higher *(4,5)*. The advantage of higher efficiency, however, carries with it the need for increased alertness to the control of sample-detector geometry *(6)*.

A thermal guide has been in use for a decade at Kyoto for measurement of boron in tissues in support of neutron-capture cancer therapy *(3)*. The first use of cold neutrons for PGAA was reported in 1973 *(7)*; in 1986, large Ge detectors were used with a guided thermal beam at ILL Grenoble *(8)* and the first permanent cold-neutron instrument was installed at KFA Jülich *(9,10)*. More recently, cold-neutron PGAA has become established at NIST *(11)* in the USA, and on cold and thermal guides at the JRR-3M reactor in Japan *(12,13)*. Other instruments are under construction for cold sources at Cornell *(14)*, the Budapest Neutron Centre *(15)*, and the University of Texas *(16)*. There will also be a station at the Dhruva reactor at BARC in India *(17)*.

APPLICATIONS

Prompt γ activation analysis has been used to determine the composition of a wide variety of materials, from standard rocks *(8,18–21)* to foods *(22–24)*. PGAA is most useful in conjunction with instrumental neutron activation analysis (INAA). The two techniques both use reactor neutrons and similar instrumentation, and measure complementary suites of elements. Most importantly, a number of the low-Z elements that form the major part of tissue and rocks (e.g., H, B, C, N, Si, and S) do not give γ-emitting radioactive products on neutron capture, but do emit prompt capture γ-rays. By contrast, INAA is most sensitive to trace elements, not the major elements in common materials. This complementarity of the two techniques has been exploited for the analysis of NBS Coal Standard Reference Materials 1632a and 1635 *(25)*, in which 16–17 elements were determined by PGAA and 36–40 by INAA; 8–10 elements were measured by both techniques. The combination has been applied effectively to the analysis of marine bivalves *(26)*.

Recently, the advantages of guided neutron beams in PGAA are being exploited in numerous areas of research. Improved sensitivity is leading to new applications, especially for the nondestructive determination of small amounts of hydrogen *(27)*. No other method involves such a small number of physical and chemical assumptions: ^{1}H gives a unique capture γ-ray at 2223.23 keV, and the energy and capture probability are independent of the chemical state of hydrogen. With cold neutrons at JRR-3M, a 1-μg detection limit for hydrogen has been attained, and boron has been measured in reactor-grade graphite and beryllium with a detection limit of 4 ng for a 1-g sample *(13)*. Numerous measurements of hydrogen have been performed at NIST (mostly unpublished) in samples of metal standards, fullerenes and their derivatives, potassium titanophosphate, lithium niobate, mesoporous silicon, sputtered amorphous carbon, thin diamond films, hydrothermally grown quartz, silicon, electrothinned germanium, zeolites and sodalites, and titanium jet engine compressor blades.

Stoichiometry measurements (elemental ratios) have been made on samples of various substituted fullerenes, an NH_4·Rh·Cl complex, Y·Ca·Ba nickelates, and tungstophosphoric acids. Chlorine, boron, and other elements have been measured in suites of basalts and meteorites. Hydrogen and seven major elements were measured in glass fibers used for neutron focusing lenses. These measurements were performed to provide compositional information to basic and applied research programs at NIST and elsewhere, having widely varying applications in science and technology. Some alkali-metal fullerene derivatives, for example, are superconductors, and other derivatives *(28)* have been shown to inhibit the viruses that cause AIDS.

The greatly improved γ-ray background at low energy afforded by guided beams compared with conventional thermal beams makes possible the use of low-energy capture γ-rays *(29,30)* for analytical purposes. The rare earth elements are especially rich in this energy range; these elements have been shown to be useful in environmental studies.

THE SPECIAL CASE OF BIOLOGICAL MATERIALS

The large amount of hydrogen in most tissues and other materials of biological origin cause a special problem for neutron beam techniques: neutron scattering in samples that are not infinitely thin. The thermal scattering cross-section of a free hydrogen atom is 20.5 b; if the hydrogen is bound into a heavy molecule, the cross-section is fourfold higher, and for cold neutrons both absorption and scattering cross-sections increase without limit as the neutron energy approaches zero *(31)*. As a result, cold neutrons have a probability of scattering of about 50% after penetrating <1 mm of carbohydrate, protein, or water. The combination of hydrogenous samples and cold neutron beams with shallow penetration depth introduces new considerations for PGAA of biological materials, as well as for other applications, such as boron neutron capture cancer therapy *(32)* and autoradiography of paintings *(33)*.

Workers with tissues and other hydrogenous matrices have known for many years *(34)* that the analytical sensitivity (counts/s·g) of PGAA depends on the hydrogen content of the sample *(35)*. The potential bias in the analytical results has been controlled in practice by matching standards to the samples *(24)*. The effect is caused by a change in the neutron's effective path length in passing through a scattering material, and hence a change in the probability of its being captured, which depends on the size, shape, and composition of the analytical specimen *(36)*. For a purely scattering sample, unlike an absorber, the bias vanishes for spherical targets *(37,38)*, a corollary of the fact that there is no bias for a scattering sphere in an isotropic flux *(39)*. Monte Carlo calculations and experiments agree quantitatively for a commonly used sample geometry, a scattering and absorbing disk inclined 45° to the beam *(40)*. An attempt has been made to minimize scattering and absorption by irradiating samples as thin, large-area slabs *(41)*, but forming biological samples into uniform sheets without altering their composition is difficult and the advantage of close counting geometry is partly lost. This stratagem requires further study.

In addition, the background resulting from neutron capture in the surrounding shielding and support structure increases when neutrons are scattered (and for crystalline materials diffracted) from the sample *(9,42)*. Because of the larger cross-section at low neutron velocity, the effects are more severe for cold neutrons *(43)*, and there is still another phenomenon to be considered: When cold neutrons scatter in a room-temperature sample, the average neutron energy increases, and the capture cross-section and analytical sensitivity thereby decrease. Preliminary work has shown that decreasing the sample temperature alleviates this effect. This topic is discussed in another contribution to this conference *(44)*.

These difficulties (except background enhancement) can be avoided by measuring element ratios *(3,10,45,46)*. The approach is similar to the classical k_0-standardization method of neutron activation analysis, with the important proviso that the element chosen for the standard must be an intimate part of the sample *(47)*. This dependence on the nature of the sample is inconvenient for analysts (and their customers) who are accustomed to reporting measurements in units of g of analyte/g of sample, but in many cases, an elemental ratio is in fact the information that is desired. Recent examples from work at NIST include the determination of trace hydrogen in intact gas-turbine compressor blades where the sample mass was ill-defined *(48)* and the analysis of major components in substituted C_{60} compounds where the stoichiometry was to be determined *(49)*.

FUTURE DEVELOPMENTS

Work in progress at several laboratories on multilayer supermirrors suggests that the collection efficiency of neutron guides may be improved by as much as a factor of ten *(50)*. Recent advances in focusing neutron optics will lead to greatly increased flux density in neutron beams *(51)*. Capillary lenses, made up of hundreds of thousands of microguides, are especially promising. A gain of nearly an order of magnitude in flux density, condensed to a spot <1 mm in diameter, has been demonstrated with a first-generation lens *(52)*; the theoretical limit in gain and spot size is much better. A neutron microprobe appears within reach in the near future. Bundles of capillaries can also be used, with or without focusing, to bend beams of thermal or cold neutrons away from interfering radiations *(53)*; this technology may be much less expensive than traditional neutron macroguides and, thus, may open the use of clean beams to smaller reactors.

In addition to cold-neutron PGAA beams in existence and under construction, new facilities will greatly expand access to cold neutrons for researchers. A committee of the US National Academy of Sciences *(54)* recommended in 1984 that a major laboratory be constructed around the cold source at the NIST reactor to serve the American materials science community for the next decade; this recommendation has led to the establishment of the present Cold Neutron Research Facility *(55)*. The committee further recommended that a new-generation Advanced Neutron Source (ANS) reactor be built for the long term; a recent study has reinforced that recommendation *(56)*. The ANS will be a 330 MW heavy-water reactor at Oak Ridge National Laboratory, with a peak thermal flux density of $\phi = 7.5 \times 10^{15}$ n/cm^2s, and a D_2O reflector with over 30 m^3 of volume having $\phi > 10^{14}$ n/cm^2s *(57)*. On completion in 2002, the ANS will have 29 thermal, hot, cold, and very cold neutron beams, serving 48 neutron-scattering, fundamental physics, and analytical chemistry instruments *(58)*. The latter comprise two low-background cold-neutron PGAA stations and two neutron-depth profiling stations, each with $\phi > 10^{10}$ n/cm^2s, seven pneumatic tubes, and associated laboratories to accommodate visiting users *(59)*. The first ANS users' meeting will be held in November 1993.

Analytical applications of guided neutron beams seem sure to increase in the future. The growing number of these beams at research reactors are making them more widely available for analytical use in the life sciences and in other fields. Their lack of fast neutrons and γ-rays compared to conventional collimated beams, and the improved analytical sensitivity, which is attainable as a consequence, are leading to broader applicability to chemical analysis.

ACKNOWLEDGMENTS

We thank D. L. Anderson, R. R. Greenberg, D. F. R. Mildner, and C. P. Zeissler for useful discussions. Contributions of the National Institute of Standards and Technology are not subject to copyright.

REFERENCES

1. R. M. Lindstrom and D. L. Anderson, in *Capture Gamma-Ray Spectroscopy and Related Topics—1984 (AIP Conf. Proc. 125,* American Institute of Physics, New York, 1985, p. 810–819).

2. R. M. Lindstrom and C. Yonezawa, in *Prompt Gamma Neutron Activation Analysis*, Z. B. Alfassi and C. Chung, eds., CRC Press, Boca Raton (to be published in 1994).
3. T. Kobayashi and K. Kanda, *Nucl. Instrum. Methods* **204,** 525–531 (1983).
4. M. P. Failey, D. L. Anderson, W. H. Zoller, G. E. Gordon, and R. M. Lindstrom, *Anal. Chem.* **51,** 2209–2221 (1979).
5. A. G. Hanna, R. M. Brugger, and M. D. Glascock, *Nucl. Instrum. Methods* **188,** 619–627 (1981).
6. G. E. Nicolaou, Y. S. Khrbish, and N. M. Spyrou, *Appl. Radiat. Isot.* **37,** 1219–1224 (1986).
7. R. Henkelmann and H. J. Born, *J. Radioanal. Chem.* **16,** 473–481 (1973).
8. S. A. Kerr, R. A. Oliver, P. Vittoz, G. Vivier, F. Hoyler, T. D. MacMahon, and N. I. Ward, *J. Radioanal. Nucl. Chem.* **113,** 249–258 (1987).
9. R. M. Lindstrom, R. Zeisler, and M. Rossbach, *J. Radioanal. Nucl. Chem.* **112,** 321–330 (1987).
10. M. Rossbach and N. T. Hiep, *Fresen. J. Anal. Chem.* **344,** 59–62 (1992).
11. R. M. Lindstrom, R. Zeisler, D. H. Vincent, R. R. Greenberg, C. A. Stone, D. L. Anderson, D. D. Clark, and E. A. Mackey, *J. Radioanal. Nucl. Chem.* **167,** 121–126 (1993).
12. C. Yonezawa, A. K. H. Wood, M. Hoshi, Y. Ito, and E. Tachikawa, *Nucl. Instrum. Methods* **A329,** 207–216 (1993).
13. C. Yonezawa, *Anal. Sci.* **9,** 185–193 (1993).
14. D. D. Clark, C. G. Oullet, and J. S. Berg, *Nucl. Sci. Eng.* **110,** 445–454 (1992).
15. G. Molnár, Z. Révay, Á. Veres, A. Simonits, and H. Rausch, *J. Radioanal. Nucl. Chem.* **167,** 133–137 (1993).
16. B. W. Wehring, K, Ünlü, J.-Y. Kim, and C. Rios-Martinez, *Trans. Am. Nucl. Soc.* **68,** 161–162 (1993).
17. S. Gangadharan, personal communication (1988).
18. D. B. Curtis, E. S. Gladney, and E. T. Jurney, *Geochim. Cosmochim. Acta* **44,** 1945–1953 (1980).
19. M. D. Higgins, *Geostand. Newslett.* **8,** 31–34 (1984).
20. D. L. Anderson, Y. Sun, M. P. Failey, and W. H. Zoller, *Geostand. Newslett.* **9,** 219–228 (1985).
21. W. P. Leeman, *Geostand. Newslet.* **12,** 61–62 (1988).
22. D. L. Anderson, W. C. Cunningham, and E. A. Mackey, *Fresen. J. Anal. Chem.* **338,** 554–558 (1990).
23. D. L. Anderson, W. A. Cunningham, and E. A. Mackey, *Biol. Trace Element Res.* **27,** 613–622 (1990).
24. D. L. Anderson, W. C. Cunningham, and G. H. Alvarez, *J. Radioanal. Nucl. Chem.* **167,** 139–144 (1993).
25. M. S. Germani, I. Gokmen, A. C. Sigleo, G. S. Kowalczyk, I. Olmez, A. M. Small, D. L. Anderson, M. P. Failey, M. C. Gulovali, C. E. Choquette, E. A. Lepel, G. E. Gordon, and W. H. Zoller, *Anal. Chem.* **52,** 240–245 (1980).
26. R. Zeisler, S. F. Stone, and R. W. Sanders, *Anal. Chem.* **60,** 2760–2765 (1988).
27. R. M. Lindstrom, R. L. Paul, D. H. Vincent, and R. R. Greenberg, *J. Radioanal. Nucl. Chem.* **180** (in press. 1994).
28. R. Baum, *Chem. Eng. News* **2 Aug. 1993,** 3.
29. P. Durner, T. von Egidy, and F. J. Hartmann, *Nucl. Instrum. Methods* **A278,** 484–490 (1989).
30. R. Henkelmann, *Z. Physik* **258,** 315–320 (1973).
31. J. R. Lamarsh, *Introduction to Nuclear Reactor Theory*, Addison-Wesley, Reading, MA, 1966.
32. M. Papaspyrou and L. E. Feinendegen, *Nucl. Sci. Eng.* **110,** 349–354 (1992).

33. C. O. Fischer, J. Kelch, C. Laurenze, W. Leuther, and K. Slusallek, *Kerntechnik* **51,** 9–13 (1987).
34. D. L. Anderson, G. E. Gordon, W. B. Walters, W. H. Zoller, M. E. Kitto, and R. M. Lindstrom, in *NBS Techn. Note 1190*, 1984, p. 148–155.
35. D. Trubert, G. Duplâtre, and J. C. Abbé, *Appl. Radiat. Isot.* **42,** 699–705 (1991).
36. J. R. D. Copley and C. A. Stone, *Nucl. Instrum. Methods* **A281,** 593–604 (1989).
37. E. A. Mackey, G. E. Gordon, R. M. Lindstrom, and D. L. Anderson, *Anal. Chem.* **63,** 288–292 (1991).
38. E. A. Mackey, G. E. Gordon, R. M. Lindstrom, and D. L. Anderson, *Anal. Chem.* **64,** 2366–2371 (1992).
39. J. R. D. Copley, *Nucl. Instrum. Methods* **A307,** 389–397 (1991).
40. E. A. Mackey and J. R. D. Copley, *J. Radioanal. Nucl. Chem.* **167,** 127–132 (1993).
41. M. Rossbach, *Anal. Chem.* **63,** 2156–2162 (1991).
42. D. L. Anderson and E. A. Mackey, *J. Radioanal. Nucl. Chem.* **167,** 145–151 (1993).
43. R. L. Paul and E. A. Mackey, *J. Radioanal. Nucl. Chem.* **181** (in press 1994).
44. E. A. Mackey, this conference.
45. C. A. Stone, S. F. Stone, M. J. Blackman, J. S. Olin, E. Myers, and D. D. Clark, *Trans. Am. Nucl. Soc.* **62,** 224–225 (1990).
46. R. M. Lindstrom, R. F. Fleming, R. L. Paul, and E. A. Mackey, in *Proceedings of the International* k_0 *Users Workshop*, Universiteit Gent, Belgium, 1992, p. 121–124.
47. M. Rossbach and F. De Corte, in *Proceedings of the International* k_0 *Users Workshop*, Universiteit Gent, Belgium, 1992, p. 125–129.
48. R. L. Paul and R. M. Lindstrom, *Rev. Progr. Quant. Nondest. Eval.* **13,** (in press, 1994).
49. D. A. Neumann, J. R. D. Copley, D. Resnik, W. A. Kamitakahara, J. J. Rush, R. L. Paul, and R. M. Lindstrom, *J. Phys. Chem. Solids* **54,** 1699–1712 (1993).
50. C. F. Majkrzak and J. L. Woods, eds., *Neutron Optical Devices and Applications (SPIE-1738)*, Soc. Photo-Opt Instrum. Eng., Bellingham, WA, 1992.
51. D. Mildner, H. Chen, G. Downing, and V. Sharov, *J. Neutron. Res.* **1,** 1–11 (1993).
52. M. A. Kumakhov and V. A. Sharov, *Nature* **357,** 390–391 (1992).
53. Q. F. Xiao, H. Chen, D. F. R. Mildner, R. G. Downing, and R. E. Benenson, *Rev. Sci. Instr.* **64,** 3252–3257 (1993).
54. F. Seitz and D. E. Eastman, *Major Facilities for Materials Research and Related Disciplines*, National Academy of Science, Washington, DC (1984).
55. H. J. Prask, J. M. Rowe, J. J. Rush, and I. G. Schröder, J. *Res. Nat. Inst. Stds. and Technol.* **98,** 1–14 (1993).
56. Panel on Neutron Sources, DOE/ER-0576P (1993).
57. F. J. Peretz, ORNL-TM-12184 (1992).
58. D. L. Selby, R. M. Harrington, and P. B. Thompson, ORNL-6273 (1993).
59. L. Robinson and F. F. Dyer, *J. Radioanal. Nucl. Chem.* **167,** 7–10 (1993).

Developments in Tomographic Methods for Biological Trace Element Research

N. M. SPYROU,*[,1] J. M. SHARAF,[1] AND S. RAJESWARAN[1,2]

[1]*Department of Physics, University of Surrey, Guildford, Surrey, GU25XH;* [2]*MRC, Cyclotron Unit, Hammersmith Hospital, London, UK*

ABSTRACT

Neutron-induced γ-ray emission tomography for quantitative determination of the concentration and distribution of elements in a selected plane through a biological specimen is briefly explained and applied by way of illustration to the analysis of gallstones. A system capable of carrying out studies of the binding site of ^{75}Se in different matrices using time differential perturbed angular correlation spectroscopy is also briefly described. Developments in the detector technology of positron emission tomography have allowed small-diameter imaging devices to be built for in vivo preclinical evaluation of new tracers in small animals and are discussed in the context of a proposed experiment combining the techniques mentioned above.

Index Entries: Perturbed angular correlation; positron emission tomography; selenium; gallstones; elemental analysis; instrumental neutron activation analysis; neutron-induced γ-ray emission tomography.

INTRODUCTION

The techniques of X-ray transmission tomography and γ-ray emission tomography in clinical applications are well established and have reached maturity since the introduction by Hounsfield in 1972 of the first commercial CT-scanner produced by EMI. We introduced the concepts of

*Author to whom all correspondence and reprint requests should be addressed.

 © 1994 by Humana Press Inc.

neutron-induced γ-ray emission tomography (NIGET) in delayed γ-ray mode *(1)* and in prompt γ-ray mode *(2)* by combining the principles of tomography with the sensitivity of instrumental neutron activation analysis (INAA) in order to obtain, nondestructively, not only the concentration of elements, but also their distribution in a section through the sample under investigation. Application of the technique is considered to be useful in extracting this additional information about elemental distribution, provided that the intensity and the half-lives of the induced activity of the nuclides of interest allow a statistically significant number of counts to be collected during the period when projection data of the radionuclide profiles, for a predetermined number of angles through 360° around the sample, are acquired, in order to reconstruct a tomographic map of elemental concentration. The technique can be multielemental since high-resolution γ-ray spectrometry is usually employed and spatial resolution can be <1 mm *(2,3)*. However, the use of more than one detector in the scanning rig would make better use of the time available for data acquisition and improve detection limits. Although multiple-detector arrays of good spatial resolution are becoming available through developments in positron emission tomography (PET), these are based on scintillation detectors and therefore suffer in comparison with high-energy resolution semiconductor detectors. Systems that incorporate arrays of high-energy resolution semiconductor detectors specifically designed for research in trace element analysis with tomography would, at present, be prohibitively expensive.

Trace element research, however, is not confined to only determining how much of an element is present (its concentration) and where it is to be found in a biological system (its location and distribution), but also includes what it is attached to and what its "chemical environment" (its speciation) is. The latter is becoming increasingly important as information about the nature of the microenvironment of the specimen, whether this is expressed in terms of its chemical form or any other property, is being sought. This is where perturbed angular correlation (PAC) is playing an useful role. Since the surroundings, chemical or physical, of a radioactive nucleus may influence its decay in terms of the spin direction and, thus, the emission produced, one can study the radiation emitted and reach conclusions about the surroundings in question. For a radiative cascade, where two γ-rays are emitted, PAC coincident spectroscopy is the tool used to study the disturbed angular correlation between the two photons. It has two important features for a speciation method: it is noninvasive, and it is sensitive enough to be applied to elements, present at trace levels, which is the case when dealing with very dilute systems like the ones frequently encountered in biomolecular investigations.

Early work with PAC concerned the experimental determination of the g-factors of excited nuclear states for certain elements. Applications were extended to the field of material science and particularly of various crystals, since a PAC study can provide a measure of the nuclear relaxation time and thereby yield the rotational correlation time of a molecule to which the radionuclide is bound, under the influence of various factors, such as temperature and pressure *(4)*. Much work followed in the important field of study of impurity defects in metals, semiconductors, insulators, and intermetallic compounds *(5)*. PAC studies in the biological and biomedical field are more recent, and Bauer *(6)* gives a review of the PAC method in protein identification, conformation, and dynamics. Various applications are described, such as the study of local dynamical features of proteins using ^{197}Hg labeling, the identification of the binding sites of metal in living cells, and the complexity between small molecules and biological macromolecules *in situ* and in vivo *(7–9)*. The research encompasses the study of enzymes and, interestingly, the binding of $^{111}Cd^{m}$ to DNA molecules *(10)*. A brief description of the developments and applications of these methods in our laboratory follows.

METHODS AND DISCUSSION

Neutron-Induced γ-Ray Emission Tomography (NIGET)

Determination of the concentration and distribution of elements in biological systems using NIGET in the study of underlying mechanisms for the formation of stones (salivary gland [11], kidney, and gallbladder) continues, and more recently is being applied to the analysis of normal and pathological human breast tissues. Figure 1 describes the procedure for obtaining the distribution of the concentration of elements in a selected "slice" through the sample, after neutron irradiation when the intensity of delayed γ-rays is measured. For γ-ray emission tomography to yield quantitative results, attenuation of the photons emitted by the sample must be taken into account; this is done by acquiring and reconstructing a transmission tomographic image that represents the distribution of the photon linear attenuation coefficient (a function of density and atomic number) in the same "slice" and correcting the emission data. The transmission image is also useful because it provides information about the structure of the sample. Superposition, therefore, of the emission image on that obtained by transmission tomography provides an accurate distribution of the elemental concentration and its exact location in the sample (fused image).

Figure 2(A) shows one of two gallstones removed during surgery from an elderly male patient; it has an irregular shape with major and minor diameters 7 ± 0.5 and 6 ± 0.5 mm, respectively. Figure 2 (B) is an emission tomography image representing the distribution of the concentration of Na, via the 1.36 MeV γ-ray energy of ^{24}Na, in a "slice" through the middle of the gallstone. Data were acquired after irradiation for 8 h in a thermal neutron flux of 1.2×10^{12} $ncm^{-2}s^{-1}$, at the Imperial College Reactor Centre, and a waiting time of about 26 h. Each pixel is 1 × 1 mm, and the thickness of the "slice" is also 1 mm; thus, this represents the concentration of the element per unit volume (1 mm^3). As can be seen, Na concentrates toward the center of the gallstone. Other radionuclides were detected in the stone (e.g., ^{42}K, ^{82}Br, and ^{198}Au), but not ^{47}Ca, con-

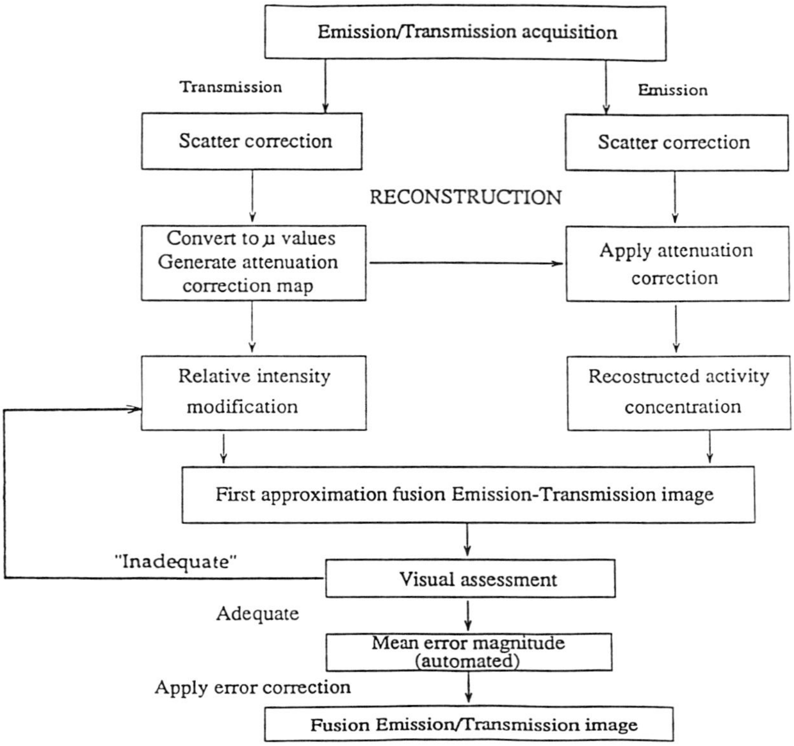

Fig. 1. The procedure followed for accurate determination of the distribution of radionuclide concentration in emission tomography using transmission tomography for attenuation correction and correct localization.

trary to expectations *(12)*. If insufficient bile salts or lecithin is present in bile, or if there is excessive cholesterol, the cholesterol precipitates out of solution and crystallizes to form cholesterol microcrystals. The fusion of single microcrystals is the beginning of 95% of all gallstones *(13)*. However, the process by which cholesterol microcrystals unite and grow is unknown *(14)*, and factors, such as inflammation, stasis, mucus content, bile pigment, pH, and so forth, which do not appear to be significant in the precipitation of cholesterol microcrystals, may be important in the process by which macroscopic cholesterol gallstones are formed. Since more than 10% of the adult population suffers from this disease, the problem justifies more detailed study *(12)*. Rajagopalan and Lindenbaum *(15)* claimed that the presence of Na and K in bile causes the precipitation of small particles of Ca bound to bile salt micelles, which may form the nucleus for cholesterol precipitation.

Perturbed Angular Correlation (PAC)

We started to design and construct a system capable of time differential perturbed angular correlation (TDPAC) measurements for ^{75}Se a few years ago *(16)*, and an instrument has now been built meeting most

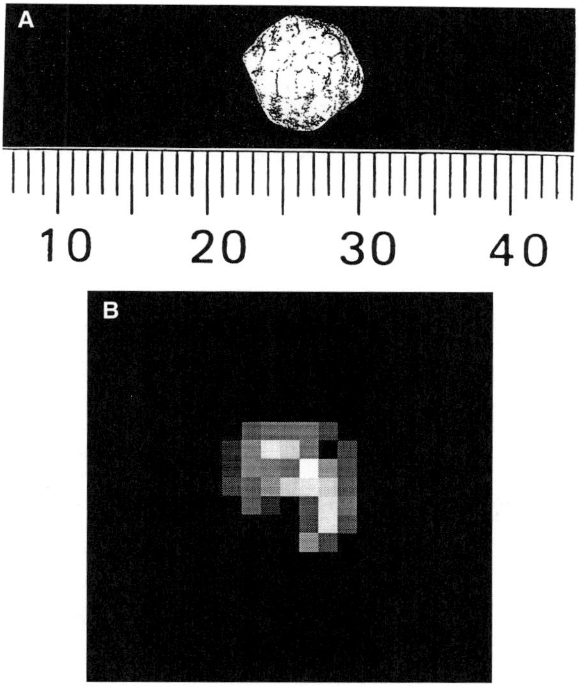

Fig. 2. (A) Physical size and structure of the gallbladder stone. (B) The distribution of Na in the stone through a central plane; the concentration of Na is shown on a gray level scale with white indicating the highest value and each pixel representing 1 × 1 mm area of the plane.

of the requirements. It is based on BaF_2 fast scintillator detectors and has a time resolution (fwhm) of the order of 350 ps with scope for significant improvement. This should now allow us to study matrices to which ^{75}Se is bound, since the challenge so far for TDPAC spectroscopy of this particular radionuclide has been the relatively short intermediate lifetime (280 ps) state between the two cascading γ-rays of interest. Table 1 lists the radionuclides that can be used for PAC studies; note that time-integrated perturbed angular correlation (TIPAC) spectroscopy does not impose the stringent time-resolution performance required in applications of TDPAC.

Positron Emission Tomography (PET)

Positron emission tomography is now an established method for the in vivo measurement of the regional distributions of tracer compounds labeled with cyclotron-produced positron emitting radionuclides (^{11}C, ^{13}N, ^{15}O, and ^{18}F). The distribution of activity, a function of position and time, is obtained from external coincidence detection of the two almost antiparallel photons that are emitted as a result of the annihilation of a

Table 1
Radionuclides Suitable for Perturbed Angular Correlation (PAC) Studies

Nuclide	Half-life	Lifetime of intermediate level (ns)	Gamma-ray energies of interest (keV)	Relative intensities (%) I_β^* $I\gamma_1$ $I\gamma_2$
^{181}Hf	42.5 days	10.6	133 - 482	93 - 99.5 - 85
^{111m}Cd	48.6 minutes	84	150 - 247	99 - 100 - 100
and ^{111}In **	48.6 minutes	84	150 - 247	99 - 100 - 100
^{75}Se	120.4 days	0.28	121 - 280	94 - 64 - 84
^{60}Co	5.26 years	0.7 (ps)	1173 - 1332	99 - 100 - 100
^{133}Ba	7.2 years	6.3	356 - 81	78 - 91 - 100
^{203}Pb	52.1 hours	0.28	401 - 279	5 - 85 - 100
^{199}Hg	65 hours	1.9	191 - 77	2 - 93 - 100
^{154}Eu	16 years	1.18	1278 - 123	37 - 100 - 100
^{77}Br	57 hours	9.3	58 - 239	19 - 35 - 100
^{57}Co	270 days	98	122 - 14	99 - 89 - 100
^{244}Am	10.1 hours	0.1	99 - 43	72 - 100 - 100
^{48}Cr	23 hours	10.9	116 - 310	100 - 100 - 100
^{120}Sb	5.8 days	5.5	200 - 90	100 - 100 - 99
^{237}U	6.8 days	63	208 - 59	43 - 89 - 94
^{204}Bi	11.2 hours	260	1203 - 375	85 - 99 - 100
^{161}Tb	6.9 days	29	49 - 26	64 - 62 - 100
^{44}Ti	48 years	153	78 - 68	100 - 100 - 100
^{83}Rb	83 days	150	32 - 10	23 - 100 - 100
^{168}Tm	85 days	110	448 - 198	40 - 100 - 100

*Iβ is the relative intensity of the preceding β-decay or electron capture.
**Indium and Cadmium result in the same cascade.

positron with an electron. Clinical imaging systems involve either circular geometry with rings or polygons around the patient or position-sensitive detectors, such as γ cameras, multiwire proportional chambers, or arrays of detectors on either side *(17)*. PET's ability to generate quantitative, dynamic, functional images provides unique insights into tissue physiology, biochemistry, and drug kinetics. Tracers currently in use in humans enable the measurement of blood flow and metabolism and the characterization of some neurotransmitter systems *(18)*. Its success has meant that much effort has been put toward improving the spatial resolution of PET systems, nearly up to the limiting intrinsic resolution of 2–3 mm dictated by the range of the positrons in tissues, i.e., from the point of emission of the positron by the nucleus to the end of its path when it annihilates.

There is a need for increased sensitivity because patient dose considerations and coincident counting make PET photon limited. This is achieved in some designs by removing the septa separating the rings of detectors and allowing coincidence to be recorded between detectors in different ring planes. Interesting developments, not the least of which is the possibility of carrying out truly 3-D imaging and the evolution of a small-diameter or mini-PET system, based on the latest generation of multicrystal, high spatial resolution PET detectors, have taken place. The impetus for a system to carry out in vivo PET studies in small animals to aid the preclinical evaluation of new tracers and the interpretation of signals obtained with existing clinical tracers has been given because data sets from rodents have generally been obtained postmortem, for example, by measuring the radioactivity in brain regions at selected times after intravenous injection of the labeled compound. Apart from being an animal-intensive and time-consuming procedure, errors arise from this because of interanimal variation. The potential therefore exists for the noninvasive acquisition of regional, differential tracer kinetics from individual animals *(19)*.

A small-diameter system by CTI/PET Systems, Inc, Knoxville, TN (the RAT PET) was delivered to the MRC Cyclotron Unit at Hammersmith Hospital in May 1992 consisting of the detector blocks (the scintillation crystal is bismuth germanate, BGO) arranged in a ring of 115-mm diameter. Each detector block is a $50 \times 23 \times 30$ mm^3 vol of BGO, which is segmented into an array of 7×8 crystals, and as with earlier designs, cuts of about 0.3 mm of varying length, coated with a reflecting material, have been made along the block to act as light guides within the BGO matrix and transmit the scintillation light to the photomultiplier tubes. There are therefore eight detector rings with an axial field of view of 50 mm, with no septa present between the rings; *see* Fig. 3. The absolute 3-D detection efficiency of the scanner for a point source at the center was found to be 8% when using a 250–850 keV window. This is lower than the theoretical value of about 13% estimated solely on the solid angle, because of the gaps between the blocks. The transaxial spatial resolution (fwhm) using ^{18}F is ~3 mm at 5 mm from the center of the field of view, degrading rapidly to ~6 mm at 40-mm radius; the axial resolu-

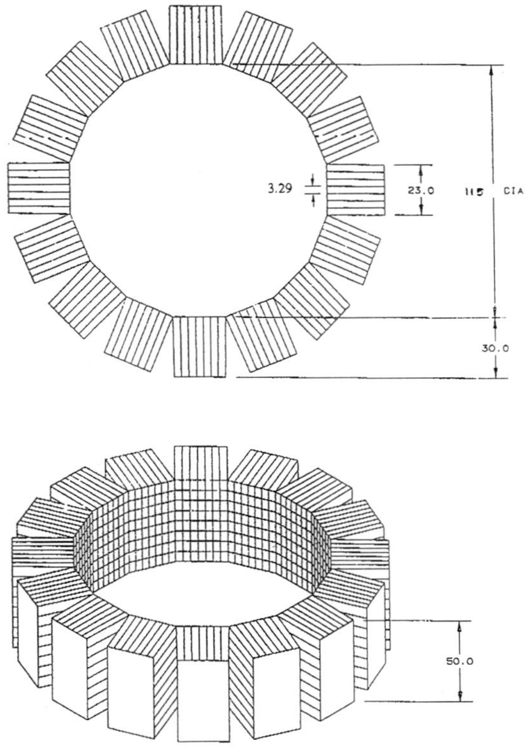

Fig. 3. The dimensions of the small-diameter PET scanner are given, showing the 16 detector block arrangement (a total of 896 detectors of BGO) and the eight rings of detectors.

tion is ~5 mm *(20)*. Two energy windows can be set for each detector, one for the full-energy photopeak and the other for estimation of scattering.

We have proposed an experiment this summer to make use of the above system in determining the activity distribution of a suitable radionuclide that emits two γ-rays in cascade by employing the two energy windows. From the data acquired, it should also be possible to obtain the angular correlation between the two γ-rays emitted, therefore combining coincident γ-ray emission tomography with perturbed angular correlation spectroscopy. If successful, the method should yield both the distribution of radionuclide concentration in a plane or slice (tomography), and provide information about its "binding" site (speciation). It has been proposed to call this method "angular correlation coincident tomography" or ACCT.

ACKNOWLEDGMENTS

N.M.S acknowledges that this article is in celebration of the 80th birthday of his father, Marc Nicholas Spyrou. S. D. R. is supported by an

MRC research studentship, and J. M. S. is on research study leave from the University of Jordan supported by the Jordanian government.

REFERENCES

1. N. M. Spyrou, *J. Radioanal. Nucl. Chem.* **110,** 2, 641 (1987).
2. N. M. Spyrou, Kusminarto, and G. E. Nicolaou, *J. Radioanal. Nucl. Chem.* **112,** 1, 57 (1987).
3. N. M. Spyrou, F. A. Balogun, and G. Davies, *J. Radioanal Nucl. Chem.* **113,** 2, 417 (1987).
4. E. Karlsson, E. Matthias, and K. Siegbahn, eds., *Perturbed Angular Correlations*, Elsevier, North Holland, Amsterdam, 1964.
5. E. Reckagel and J. C. Soares, eds., *Nuclear Physics Applications on Material Science*, Kluwer Academic Publishers, Dordrecht/Boston/London, 1988.
6. R. Bauer, *Quart. Rev. Biophys.* **18,** 1, 1764 (1985).
7. T. K. Liepert, J. D. Balderschweiler, and D. A. Shirley, *Nature* 220, 907 (1968).
8. H. Singh, H. S. Binarch, S. S. Ghumman, and H. S. Sahota, *Int. J. Rad. Appl. Instrum.* **A41,** 9, 797 (1990).
9. G. Graf, J. C. Glass, and L. L. Richer, *Adv. Exp. Med. Biol.* **48,** 639 (1974).
10. R. D. Viss, A. J. J. Boss, S. Idjenga, and H. Verheul, *Nucl. Inst. Methods* **163,** 265 1979).
11. N. M. Spyrou, J. M. Sharaf, and S. Sarkar, *J. Radioanal. Nucl. Chem.* **167,** 2, 453 (1992).
12. A. T. Al-Kinana, I. A. Harris, and D. E. Watt, *Phys. Med. Biol.* **29,** 175 (1984).
13. G. J. Tortora and N. P. Anagnostakos, *Principles of Anatomy and Physiology*, 5th ed., Harper and Row, New York, 1987.
14. W. W. LaMorte, D. J. Schoetz Jr., D. H. Birkett, and L. F. Williams Jr., *Gastroenterology* **77,** 580 (1979).
15. N. Rajagopalan and S. Lindebaum, *Biochem. Biophys., Acta,* **711,** 66 (1982).
16. S. N. Mallion and N. M. Spyrou, *Nucl. Inst. Meth. Phys. Res.* **A299,** 610 (1990).
17. K. Kouris, N. M. Spyrou, and D. F. Jackson, *Imaging with Ionising Radiation*, vol. 1, Surrey University Press/Blackie and Son, Glasgow, 1982.
18. L. Eriksson, M. Dahlbomand, and L. Widen, *J. Microscopy* **157,** 305 (1990).
19. S. Rajeswaran, S. P. Hume, J. E. Cremer, J. Young, D. L. Bailey, J. Ashburner, S. K. Luthra, A. K. P. Jones, and T. Jones, *J. Neurosci. Methods* **40,** 223 (1991).
20. S. Rajeswaran, W. F. Jones, L. G. Byers, J. Young, C. W. Williams, M. Andreasco, S. P. Hume, A. A. Lammerstsma, and T. Jones, in *IEEE Medical Imaging Conference Record*, 1992, pp. 985–88.

Use of Total Reflection X-Ray Fluorescence Analysis in the Life Sciences

PETER WOBRAUSCHEK

Atominstitute of the Austrian Universities, A-1020 Vienna, Schüttelstr. 115, Austria

ABSTRACT

Total Reflection X-ray Fluorescence Analysis (TXRF) is accepted as a powerful analytical tool. TXRF is basically an energy-dispersive technique with the sample excitated in total reflection geometry. Simultaneous detection of almost all chemical elements and lower limits of detection for medium Z elements in the range of a few hundred fg are achievable at optimized excitation conditions. Aqueous and acidic solutions are the preferred sample types, so a lot of applications in the life sciences are given. For solids and other samples, special preparation techniques are required.

Index Entries: X-ray fluorescence analysis (XRF); energy-dispersive XRF (EDXRF); total reflection XRF (TXRF).

INTRODUCTION

The basic principles of TXRF were published more than 20 years ago (1–4). TXRF combines the physical phenomenon of X-ray total reflection on the plane polished surface of a reflector carrying the sample in its center and energy-dispersive measurement (ED) of the emitted fluorescence radiation. Therefore, all types of available detectors (Si(Li), HP-Ge) and appropriate electronics can be used. In Fig. 1, the scheme of conventional EDXRF is presented. The result is the spectrum of the characteristic lines superimposed on a background. The background is caused by elastic and inelastic scattering of primary photons on sample and substrate. This

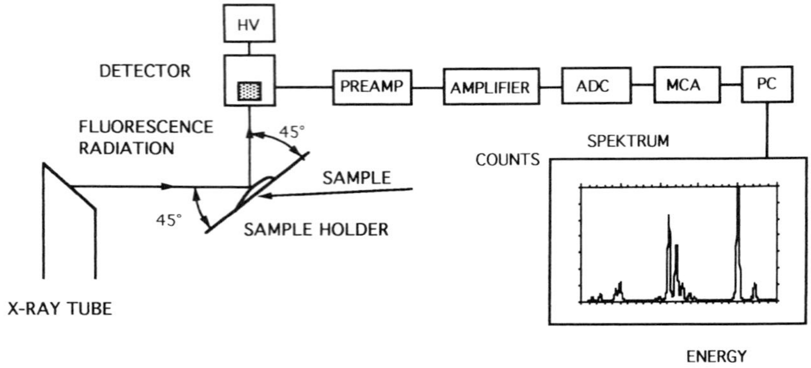

Fig. 1. Scheme of the conventional EDXRF setup and instrumentation.

background is the limiting factor for the determination of low concentrations. The definition of the lower limits of detection (LLD) is:

$$LLD = (3/S) \cdot (I_B/t)^{1/2} \quad (1)$$

where *LLD* is the detection limit (mass or concentration), *S* is the sensitivity (counts per second per mass or concentration), and *t* is the counting time (seconds). This implies that an increase of *S* and *t* and a decrease of I_B improves the *LLD*. TXRF offers the advantage to increase the sensitivity and to reduce the background inherently. In Fig. 2, the schematic setup of TXRF can be seen offering the following features:

1. Efficient excitation of the sample by both the primary and the reflected beam yielding a doubling of the fluorescence signal;
2. Reduction of the background because the X-rays scarcely penetrate into the reflector (penetration depth a few nanometers);
3. Efficient detection geometry owing to the large solid angle (detector can be positioned as close as 1 mm to the sample); and
4. The samples form a thin layer of residues on the reflector; thus, thin film approximation is applicable for the quantitative analysis.

INSTRUMENTATION

In the standard case for TXRF, the line focus of a high power X-ray tube is used (2–3 kW, 8 mm × 0.04 mm, anodes: Cr, Cu, Mo, W). The increase of the power (P = 18 kW, same focus) if rotating anodes are used results in an improvement of the *LLD* according to the square root of P_1/P_2. Synchrotron radiation has also been successfully used. In this case, the outstanding properties of high brilliance, linear polarization,

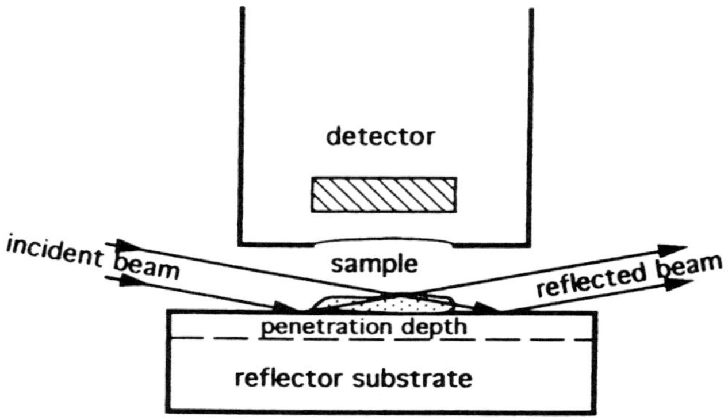

Fig. 2. Scheme of the setup for TXRF.

and tunability of the excitation energy led to further improvements of the *LLD*. The importance of synchrotron radiation has been shown in the case of samples, where the matrix cannot be removed *(5)*.

Reflector materials like quartz, Synsil®, Suprasil®, hyper-pure Silicon, hyper-pure Ge, Boron Nitride, and Perspex have been tested. The only requirement is that the necessary surface quality for Total Reflection (flatness $\lambda/20$, mean roughness 1–10 nm) have to be fulfilled.

Regarding the detection of the fluorescence radiation, all solid-state detectors can be used. For light elements, an ultrathin entrance window is necessary, and special qualities for the crystal and the electronics are required. For medium Z elements, an Si(Li) with a 7-µm Be-window is standard. For K-shell excitation of higher Z elements, only the HP-Ge has sufficient detection efficiency.

On the market, there are commercially available units including sample changer and multiple-tube excitation (Mo and W) (6–8). An attachment TXRF module is available *(9)*.

Insertion devices like cutoff reflector *(10)* and multilayer structure *(11)* in the primary beam path led to alterations of the spectral distribution. This results in further improved background conditions, and if monoenergetic excitation is chosen, the mathematical procedure of converting intensities in concentrations is simplified.

SAMPLE PREPARATION

The samples are preferably aqueous or acidic solutions where the liquid matrix is evaporated. The residue remains on the surface of the reflector. Up to 100 µL are pipeted in the center of the reflector and dried. Sample preparation techniques range from simple direct pipeting to sophisticated chemical procedures converting the sample state into the liquid form or preconcentration–separation methods (12–14). Figure 3 shows a flow chart. In all preparations finally, an internal standard element of known concentration must be added.

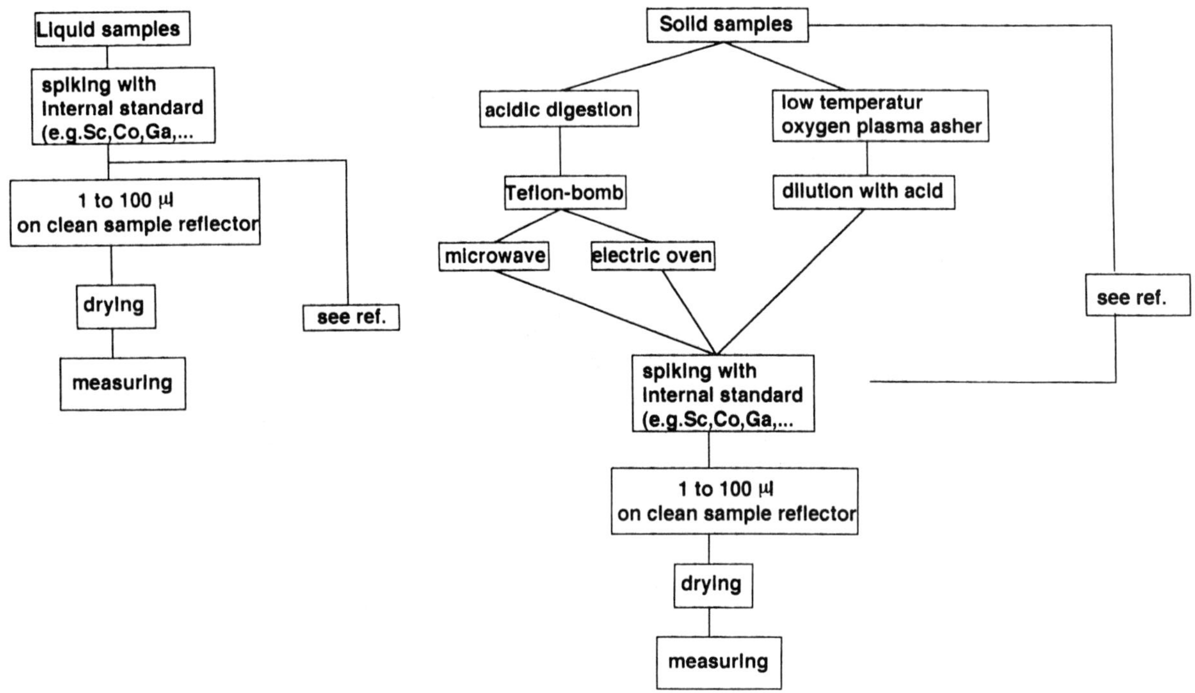

Fig. 3. A flow chart showing different ways of sample preparation.

Table 1
Applications of TXRF

Environment	
Water	drinking, waste, river, sea, rain, fog condensate
Air	airborne particles, aerosols, dust, car-exhaust gases, coal fly ash
Soil	sediments
Biosphere	leaves, grass, algae, plants, vegetables, mushrooms, roots, needles, lichens, mosses
Medicine	
Body fluids	whole blood, serum, cell fluids, urine, sweat
Tissue	lung, liver, hair, bone, nail
Technics	thin films, coatings, thin layers, surface contamination on Si wafer, tar condensates, oil, grease, fuel
Archaelogy	pottery
Fine Art	paintings
Forensic	residues, fingerprints

APPLICATIONS

The wide field of applications in various scientific disciplines can be seen in Table 1.

DETECTION LIMITS

Detection limits that can be obtained under optimized conditions are in the range of several femtograms. Using standard TXRF equipment, detection limits in the picogram range are achieved. As a general remark, minimum detectable amounts depend on the finally remaining residue on the sample reflector, which is determined by the used sample preparation technique *(15)*. As an example, ppt levels in sea water can only be detected after removal of NaCl and special preconcentration *(16)*. Figure 4 shows achievable *LLD* for optimized conditions performed by a rotating anode system with a TXRF spectrometer in a vacuum chamber.

QUANTIFICATION

TXRF is sensitive to changes in the geometry of the experimental setup. This regards not only the mechanical reproducibility after a sample change, but also the sample distribution on the reflector. Therefore, the samples can be reasonably quantified only after addition of an internal standard element of known concentration to the sample. The choice of that internal standard should be made according to the sample. Only intensity ratios with reference to the internal standard are used to estab-

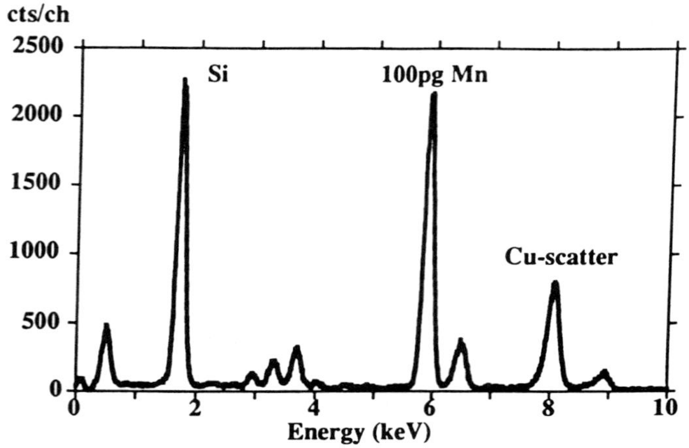

Fig. 4. Spectrum obtained with a rotating anode tube. Operating conditions: Cu 40 kV, 190 mA, 500 s. Extrapolated *LLD* for Mn: 170 fg.

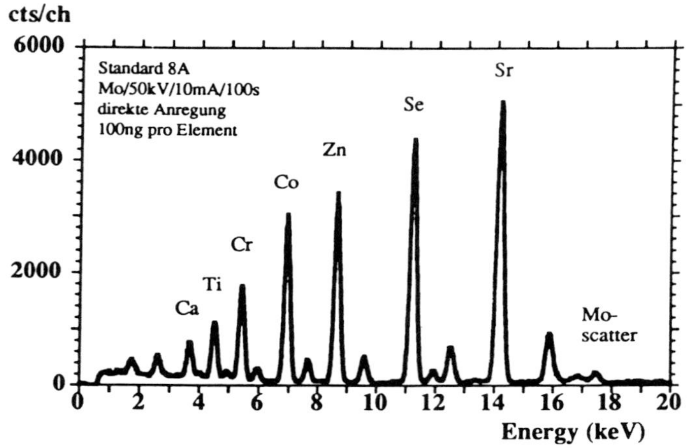

Fig. 5. Spectrum of a calibration standard, 100 ng of each element. Operating conditions: Mo 50 kV, 10 mA, 100 s.

lish calibration curves and corresponding relative sensitivities. These data can be achieved experimentally or theoretically using fundamental parameters. Figure 5 shows a spectrum of a multielement standards used for calibration.

CONCLUSIONS

TXRF is capable of determining elements at ultratrace levels in various matrices. In many cases, the sample can be directly analyzed. In all other cases, the sample preparation does not require more effort than used for other analytical techniques (AAS, ICP-MS, RNAA, and so on). The sample volume required is in the microliter range, which is advantageous when only small sample volume is available. TXRF can be considered as laboratory scale where the instrumentation is a compact unit. With automatic sample changer and an appropriate software, a high throughput of samples is possible. Because of the analytical features, TXRF is a reliable "work horse" for routine analysis in the ultratrace region and is a perfect complementary tool to INAA or RNAA.

ACKNOWLEDGMENTS

The author would like to express his gratitude to the team of the Atominstitut, R. Görgl, P. Kregsamer, W. Ladisich, R. Rieder, and C. Streli for their valuable contributions and discussion in experiment and theory of TXRF.

REFERENCES

1. Y. Yoneda and T. Horiuchi, *Rev. Sci. Inst.* **42,** 1069 (1971).
2. H. Aiginger and P. Wobrauschek, *Nucl. Instr. Methods* **114,** 157 (1974).
3. P. Wobrauschek and H. Aiginger, *Anal. Chem.* **47,** 852 (1975).
4. J. Knoth and H. Schwenke, *Fresenius Z Anal Chem* **291,** 200 (1978).
5. F. Hegedüs, P. Wobrauschek, W. Sommer, R. Ryon, C. Streli, P. Winkler, P. Ferguson, P. Kregsamer, R. Rieder, M. Victoria and A. Horsewell, *X-ray spectrometry* **22,** 277 (1993).
6. EXTRA II, Atomica, Oberschleiβheim, Germany.
7. Rigaku Corporation, 4-15-3, Nishi-Shinjuku-ku, Tokyo 190, Japan.
8. Technos, Neyagawa City, Osaka 572, Japan.
9. P. Wobrauschek, Atominstitut, Vienna.
10. H. Aiginger and P. Wobrauschek, *Adv. X-ray Anal.* **28,** 1 (1985).
11. W. Ladisich, R. Rieder, P. Wobrauschek and H. Aiginger, *Nucl. Instr. Methods* **A330,** 501 (1993).
12. A. Prange, *Spectrochim. Acta* **44B,** 437 (1989).
13. A. von Bohlen, R. Eller, R. Klockenkämper and G. Tölg, *Anal. Chem.* **59,** 2251 (1987).
14. R. Pepelnik, B. Erbslöh, W. Michaelis, and A. Prange, *Spectrochim. Acta* **48B,** 223 (1993).
15. R. Klockenkämper and A. VonBohlen, *J. Anal. Atom. Spectrom.* **7,** 273 (1992).
16. A. Prange, A. Knöchel and W. Michaelis, *Anal. Chim. Acta* **172,** 79 (1985).
17. W. Ladisich, doctoral thesis, TU-Vienna 1993.

Applications of PIXE in the Life Sciences

ZOLTÁN SZÖKEFALVI-NAGY

KFKI Research Institute for Particle and Nuclear Physics, Hungarian Academy of Sciences, P.O. Box 49, Budapest 114, H-1525 Hungary

ABSTRACT

During the last decade, particle-induced X-ray emission spectrometry (PIXE) has been accepted by the analytical chemistry community as a standard method. Instead of routine bulk analysis of biomedical samples where several competing analytical techniques are available, the full strength of PIXE could be exploited in special applications where also location of the detected ions is desirable. In this article, two different approaches for this purpose are discussed and illustrated: (1) nuclear microscopy, where micro-PIXE utilizing a proton beam spot of approx 1 μm^2 or less promises unique perspectives to obtain elemental maps at the cellular level, and (2) *in situ* PIXE analysis of the minute metalloprotein volumes concentrated in thin-layer electrophoretograms.

Index Entries: PIXE; particle-induced X-ray emission; location of trace elements; nuclear microscopy; metalloproteins; gel electrophoresis; biomedical research.

INTRODUCTION

The large majority of papers claiming to use nuclear analytical methods in the life sciences report trace analyses carried out by neutron activation analysis. Other techniques, like charged particle activation (CPA), prompt nuclear reaction analysis (NRA), or particle-induced X-ray emission spectroscopy (PIXE), can only gradually "intrude" into the field *(1)*. Among these ion-beam techniques, PIXE is certainly playing the leading role. One reason is that during the last decade, PIXE has been accepted by the analytical chemistry community as a standard method for quantitative elemental analysis *(2)*. Its multielemental character and the generally accepted fact that PIXE is most suitable for analysis of trace elements of medium and high atomic number in a matrix composed of light elements make PIXE the most useful technique in environmental, biological, and medical applications *(3)*. All of these subjects belong, in a general sense, to life sciences. Taking into account, however, that the method requires access to an accelerator operated by specially experienced staff, routine bulk analysis of a great many samples is generally not the case where PIXE would be the first choice. Analysis of environmental samples, especially that of thin air filters, is an important exception in this respect, mainly as a consequence of the relative simplicity of both sampling and measurement automation. This article, however, is confined to dealing with biomedical applications only.

The papers presented at three recent conferences, including the first dedicated Bio-PIXE Symposium, demonstrate how wide-ranging the problems are where the conventional macro-PIXE technique with particle beam diameters in the millimeter range or micro-PIXE utilizing carefully focused beams down to even submicrometer sizes is successfully applied *(4–6)*. Without going into details, a few randomly selected examples can illustrate the situation. Sha et al. *(7)* have tried to find a correlation between trace element contents in whole blood and acute leukemia, whereas the paper by Pinheiro et al. *(8)* reports on measurements of alterations of trace element concentrations associated with myocardial infarction in serum and packed blood cells. In order to find a good bioindicator of environmental pollution, trace element distributions in nontreated intact ants collected in different countries were analyzed by scanning PIXE *(9)*. Nielsen et al. *(10)*, on the other hand, analyzed fossilized dinosaur bone and surrounding rock to investigate the relationship between the bone and rock.

Precise and accurate determination of bulk trace element concentrations in biomedical samples is certainly very important, but the usefulness of the results in exploring the structural and functional role of the elements of interest, which should be the ultimate purpose of such studies, is rather uncertain. Effective differentiation between inherently bound and possibly adsorbed ions, as well as location of the bound ions, is obviously necessary. The versatility of the PIXE technique allows different approaches to try to solve these problems. Two of them will be described in detail in this article. The application of the micro-PIXE technique utilizing proton beam spot of approx 1 μm^2 or less promises unique perspectives for obtaining elemental maps at the cellular level. A completely different strategy to speciate the trace metals of interest could be the combination of PIXE with appropriate biochemical separation techniques, that is, keeping track of both the appearance and quantity of particular trace ions during separation. In this article, these two research strategies will be illustrated by selected micro-PIXE applications and by reviewing a novel ion-beam evaluation technique of thin-layer electrophoretograms.

SPATIAL LOCATION OF IONS AT THE CELLULAR LEVEL

The important concentration differences of certain ions as Na, K or Ca maintained between the extra- and intracellular regions and the ionic transport across the cell membrane play a crucial role in basic physiological processes. Abnormal local accumulation of toxic elements or deficiency in essential ones can be the result or even the initiator of malfunctions in basic physiological processes, often culminating in serious diseases of the whole living organism. In biomedical research and also as part of everyday medical treatments, specially selected elements are intentionally dosed in harmless amounts as tracers. Therefore, there is a very natural demand for suitable techniques to measure directly the concentrations and distributions of trace and major elements in cellular dimensions. As a result of the continuous development of the strong focusing electromagnetic lens systems *(11)*, nuclear microscopy involving micro-PIXE for elemental mapping and scanning ion transmission microscopy (STIM) for the corresponding imaging gradually became a valuable tool for biomedical research. The forthcoming brief summary of three selected experiments is intended to illustrate the present power of micro-PIXE in the topics mentioned at the beginning of this section.

Moretto et al. *(12)* performed quantitative mapping of intracellular cations in the human amniotic membrane. The PIXE microbeam in Bordeaux was used to measure the effect of Mg and taurine on the cell membrane permeability to monovalent cations of Na, Mg, P, S, Cl, and K. The proton beam of 1 MeV energy was focused down to a 2-μm spot diameter. The scanned area was smaller than $100 \times 100\ \mu m^2$. Because of the unexpected large-scale modifications in the electrolyte concentrations, it was not possible to point out the expected protective effect of Mg, but important observations were made concerning the role of the amnion compact lamina, acting as buffer for fixing minerals. The consequences of this role have to be taken into account in the interpretation of electrophysiological experimentations.

Watt and Landsberg carefully analyzed senile plaques from five known cases of Alzheimer's disease using the Oxford Scanning Microprobe *(13)*. Less than 10% of the immunohistochemically stained plaque cores contained Al deposits, usually colocalized with Si and frequently with Ca, Fe, and Ti. Additional analysis of a wide selection of reagents used in staining supports the suspicion that many commercially available reagents also contain alumino-silicates. In order to eliminate probable errors caused by the contamination, a PIXE analysis was performed on unstained, untreated tissues, too. In these measurements, STIM imaging was used to identify and localize senile plaques and cores. The results show no evidence of Al in the plaque cores at the detection limit of 15 ppm. These results are claimed to resolve a current controversy concerning the presence or absence of Al within the plaque cores.

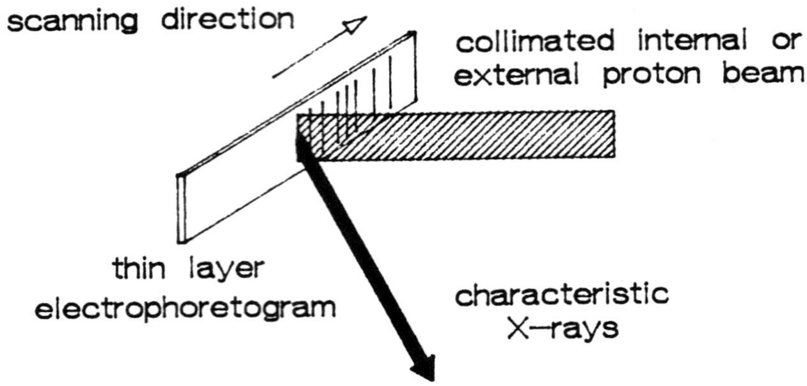

Fig. 1. Basic scheme of the ion-beam evaluation of a thin-layer electrophoretogram.

Cholewa et al. *(14)* applied the Melbourne microprobe to trace synthesized organometallic and inorganic drugs designed for antiviral activity against HIV viruses in white blood cells. The idea of tracing is based on the PIXE measurements of the inherent heavy metal content (Co and W) of the drugs. In spite of the serious difficulties in preparing suitable control samples and the unusual beam sensitivity of the lymphocyte cells, the heavy metal content of the drugs of interest have proven to be adequate for the identification and quantitative mapping of the drug distributions within their target cells.

LOCATION OF PROTEINS IN THIN-LAYER ELECTROPHORETOGRAMS

Isolation, purification, and separation of proteins or other macromolecules are essential and mandatory steps in the majority of biochemical experiments. Different versions of electrophoresis belong to the most widely used biochemical separation techniques. Molecules having different molecular weights migrate with different velocities, finally forming isolated narrow bands. The most common standard technique used to detect these bands involves the use of stains and optical densitometry, but only the molecular weights of the proteins present can be determined in these measurements. More tedious and time-consuming methods utilizing autoradiography of radiolabeled analytes or immunoreaction with specially prepared antisera could provide some structural and functional information, but no routinely usable technique was known to identify or even quantify the possibly protein-bound inherent metal ions playing an essential role in the function of the so-called metalloproteins. In our laboratory, the direct PIXE scanning of the electrophoretograms was proposed *(15)*. The basic idea is illustrated in Fig. 1. The self-supporting sample (at first a specially dried polyacrilamide gel [PAGE] section) was directly bombarded by a properly collimated proton beam of 2–3 MeV energy and about 10 nA intensity. The characteristic X-rays produced

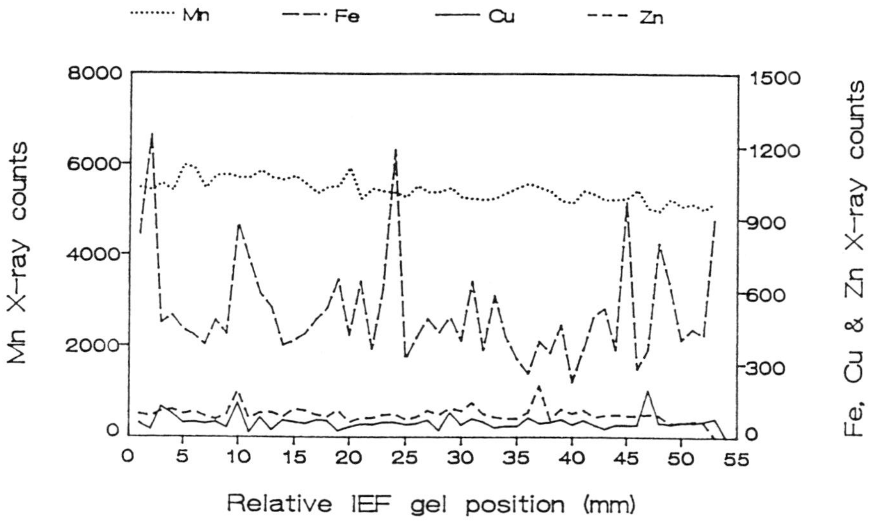

Fig. 2. Mn, Fe, Cu, and Zn distribution along an isoelectric focusing electrophoretogram containing pork meat protein bands measured by PIXE scanning. (The Mn signals are originating from the supporting Gelbond sheet.)

were detected with an energy-dispersive X-ray spectrometer. In the case of a decomposed bacterial hydrogenase, not only were the metal-containing subunits easily identified, but very important structural information was also obtained. In addition to the unambiguous confirmation that the native enzyme contains both Fe and Ni ions, it has also turned out that after decomposition, the Fe ions remained bound to the larger subunit of the enzyme, whereas the smaller one remained bound to the Ni ions *(16)*. The quantitative capabilities of this so-called PIXE-PAGE method were demonstrated by a series of measurements on PAGE samples containing given amounts of HiPIP, an iron-containing protein *(15)*. The intensities of the Fe X-rays detected were found to be proportional to the amounts of proteins supplied for the electrophoresis, and bands containing <1 μg protein were detected by measuring their Fe content. Recently, the technique was successfully applied to measuring the distribution of essential trace metal ions among pork meat proteins separated by isoelectric focusing electrophoresis *(17)*. In this technique, the thin agarose substrate in which the electrophoresis takes place is layered on a thick plastic sheet called Gelbond. This material contains a considerable amount of Mn upsetting the detection of this metal in the proteins, but offering, on the other hand, an "internal standard" for easy normalization. In Fig. 2, results of a PIXE scan are displayed, apparently indicating the presence of several metalloproteins.

Similar to nuclear microscopy, where the micro-PIXE scans were completed with simultaneous STIM measurement providing density information of the sample, the simultaneous monitoring of the protein content of the bands is possible by the simultaneous particle-induced γ-ray emission (PIGE) measurement utilizing an appropriate nuclear reaction on nitrogen nuclei. The sharp resonance in the ^{14}N (p,p′γ) ^{14}N nuclear reaction of about 4 MeV proton energy can be a promising choice for this purpose. Because of the large inherent nitrogen content of polyacrylamide substrate, this idea obviously does not work with PAGE, but in the case of nitrogen-free cellulose acetate electrophoretograms (CAE), successful exploratory measurements were made *(18,19)*. This combined PIXE-PIGE-CAE technique can also be used to study the details of the protein-staining procedures, including the possible loss of the weakly bound metal ions during the staining process.

ACKNOWLEDGMENTS

Dedicated to my father, Professor Béla Szökefalvi-Nagy, on his 80th birthday. This work was supported in part by the Hungarian National Science Foundation (OTKA I/3) under Research Contract No. 1819.

REFERENCES

1. R. Zeisler and V. P. Guinn, eds. *Nuclear Analytical Methods in the Life Sciences,* Humana, Clifton, NJ, 1990.
2. S. A. E. Johansson, *Nucl. Instrum. Methods Phys. Res.* **B75,** 589 (1993).
3. S. Morita, *Int. J. PIXE* **2,** No. 1, ix–xi (1992).
4. U. Lindh, ed., *Nucl. Instrum. Methods Phys. Res.* **B77** (1993).
5. K. Ishii, H. Orihara, and S. Morita, eds. *Int. J. PIXE* **2,** No. 3–4 (1992).
6. M. Uda, ed., *Nucl. Instrum. Methods Phys. Res.* **B75** (1993).
7. Y. Sha, P. Liu, P. Zhang, G. Liu, H. Lin, B. Yang, and L. Qian, *Int. J. PIXE* **2,** 569 (1992).
8. T. Pinheiro, W. Maenhaut, U. Wätjen, and M. J. Halpern, *Int. J. PIXE* **2,** 469 (1992).
9. K. Kitao, *Int. J. PIXE* **2,** 405 (1992).
10. D. W. Nielsen, L. B. Rees, N. F. Mangelson, and M. W. Hill, *Int. J. PIXE* **2,** 413 (1992).
11. F. Watt and G. W. Grime, eds. *Principles and Applications of High-Energy Ion Microbeams,* Adam Hilger, Bristol, 1987.
12. Ph. Moretto, Y. Llabador, M. Simonoff, L. Razafindrabe, M. Bara, and A. Guiet-Bara, *Nucl. Instrum. Methods Phys. Res.* **B77,** 275 (1993).
13. F. Watt and J. P. Landsberg, *Nucl. Instrum. Methods Phys. Res.* **B77,** 249 (1993); J. P. Landsberg, B. Mcdonald, and F. Watt, *Nature* **360,** 65 (1992).
14. M. Cholewa, G. J. F. Legge, H. Weigold, G. Holan, and C. Birch, *Nucl. Instrum. Methods Phys. Res.* **B77,** 282 (1993).
15. Z. Szökefalvi-Nagy, I. Demeter, Cs. Bagyinka, and K. L. Kovács, *Nucl. Instrum. Methods Phys. Res.* **B22,** 156 (1987).
16. Z. Szökefalvi-Nagy, Cs. Bagyinka, I. Demeter, K. L. Kovács, and L. H. Quynh, *Biol. Trace Element Res.* **26–27,** 93 (1990).
17. E. Dworschák, Gy. Hajós, Z. Szökefalvi-Nagy, H. J. Woo, D. K. Kim, and H. W. Choi, *Meat Sci.* (to be published).
18. Z. Szökefalvi-Nagy and J. Räisänen, *Nucl. Instrum. Methods Phys. Res.* **B66,** 258 (1992).
19. Z. Szökefalvi-Nagy, I. Demeter, K. Hollós-Nagy, and J. Räisänen, *Nucl. Instrum. Methods Phys. Res.* **B75,** 165 (1993).

Radiotracer Method in the Study of Environmental Speciation and Migration of Contaminants

P. Beneš

Department of Nuclear Chemistry, Czech Technical University, Břehová 7, 115 19 Praha 1, Czech Republic

ABSTRACT

Principles, advantages, and limitations of the use of radiotracer method for the analysis of speciation and migration of contaminants in the environment are briefly discussed. Several recent examples of use in the author's laboratory are given: development of the separation method for methylmercury and inorganic mercury in hair, analysis of the speciation of cadmium in soil solutions, study of the interaction of ^{137}Cs and ^{58}Co with suspended sediments in river water, and determination of input data for mathematical modeling of radiocesium migration in a small river.

Index Entries: Radiotracer method; speciation; migration; mercury; cadmium; ^{137}Cs; ^{58}Co; hair; soils; surface waters.

INTRODUCTION

The toxicity and other environmental effects of contaminants released by humans strongly depend on the physicochemical forms (speciation) of contaminants and on their migration (transport) in the environment. Therefore, the speciation and migration have been extensively studied in the past two decades. Considerable difficulties have been encountered in these studies owing to the very low environmental concentrations of the contaminants, and to the complexity of environmental processes and systems (1–4). Some of these difficulties can be overcome using the radiotracer method either in the laboratory or *in situ*. In this article, the principles of the use of the radiotracer method for the given purposes are briefly outlined and the use is illustrated with several recent studies.

 © 1994 by Humana Press Inc.

SPECIATION STUDIES

Ions, molecules, colloidal or particulate forms, phase associations, or solid-phase bonds of contaminants in environmental media and their relative abundances can be determined or characterized by a number of methods falling into three broad categories: thermodynamic and/or kinetic calculations, direct analysis of environmental samples, and model experiments *(5)*. The calculation methods are of limited value, since their use is often hampered by the lack of reliable input data. The other methods suffer from the low sensitivity, accuracy, or selectivity of experimental procedures employed for the determination or separation of individual forms (species) of contaminants *(3,6)*. Among the most frequent reasons for errors in the speciation analysis is the shift of equilibria in the analyzed sample induced by the analytical or separation procedure itself. Therefore, new, noninvasive procedures are much needed.

The experimental analysis of the speciation of many contaminants can be greatly facilitated using the radiotracer method *(5–9)*. The method consists of labeling one or more forms of a contaminant with a suitable radionuclide (isotopic or nonisotopic with the contaminant), and tracing the form(s) during a separation or another physicochemical process. Its principal advantages are ease and sensitivity of the tracing.

Development of Speciation Procedures

Radioactive labeling helps in the development and checking of methods used for the storage or pretreatment of environmental samples, and for separation of individual forms of contaminants. Basic conditions for the application of the radiotracer method for this purpose are the negligible isotope exchange between contaminant forms to be separated or investigated, and the identical behavior of the label and the labeled form in the analyzed process. The latter condition cannot always be directly checked, and then its validity should be proved indirectly. Such an application can be illustrated by the development of speciation procedures for mercury in hair.

Limited knowledge existed concerning the behavior of mercury species in the dissolution or leaching of hair, and concerning selectivity of separation of inorganic and organic forms of mercury from hair. Therefore, two procedures suitable for the separation were developed and tested: solvent extraction of dissolved hair *(10)* and acid leaching of hair *(11)*. At first, the procedure for preparation of methylmercury labeled with ^{203}Hg had to be proposed. The procedure employed isotope exchange in the aqueous solution of nitric acid containing silver nitrate. Then suitable conditions for labeling human hair with radioactive methylmercury (MeHg) or inorganic mercury (Hg_{in}) were found. The efficiency and homogeneity of the labeling as well as the long-term stability of the labeled species in hair were checked radiometrically and were very good. The procedure is also useful for preparation of standard samples of spiked hair.

Various possibilities of dissolution of labeled hair were tested for their efficiency, loss of mercury by volatilization, and for a change in speciation of mercury label in hair. A procedure using hot 10 mol/L NaOH proved to be satisfactory. Then, optimum conditions for the separation of MeHg from Hg_{in} from the dissolved hair were investigated using ion-exchange and solvent extraction. The best separation was achieved by the extraction of MeHg with benzene at pH 0.5–1 (H_2SO_4) in the presence of 0.5 mol/L KI. Measurement of activity of the extract proved that the extraction was very selective, and the separation factor was close to 10^4 *(10)*.

Leaching of hair labeled with ^{203}MeHg or $^{203}Hg_{in}$ using 0.1–5 mol/L HCl proved that MeHg is leached to a much larger extent than Hg_{in}. On the basis of these experiments, the procedure was elaborated for selective separation of these mercury species from hair by leaching with 2 mol/L HCl for 4 h at room temperature. The procedure was used for the determination of mercury species in a sample of unlabeled human hair in combination with cold vapor AAS *(11)*. A very good reproducibility was obtained, and the average result agreed well with the result of determination of methylmercury in the same hair using solvent extraction. The good agreement of the data obtained on the speciation of natural mercury in hair using two very different separation methods suggests that natural mercury species from hair behaved in the separation similarly to the labeled species used in the development of both the separation methods. This confirms the validity of basic assumptions of radiotracer methods applied in the development.

Determination of Speciation in Environmental Samples

The radiotracer method can also be used for the direct or indirect determination of species of isotopically labeled contaminant. Then either full isotopic equilibrium of the label with all contaminant forms present is achieved and the forms are determined from the behavior of the label in a suitable process, or the rate of the isotopic exchange is measured and used for characterization of the forms. In the former case, common methods of speciation analysis are employed. Here the radiotracer method enables or facilitates the use of noninvasive methods, such as self-diffusion *(12)* or free-liquid electrophoresis *(13)*. The crucial condition in this case is the attainment of full isotopic equilibrium, and thus, a correspondence of physicochemical states between the radiotracer and its analyzed counterpart *(14)*. The attainment should be checked at least by following the time dependence of the speciation of the label or parameter characterizing it *(15)*.

The advantages and limitations of the method can be demonstrated in the study of speciation of cadmium in soil solutions. The study was carried out by model experiments where carrier-free ^{109}Cd in cationic form was added to prefiltered model solution obtained by leaching of forest soil samples with distilled water *(16)*. Speciation of the added ^{109}Cd was studied using electrophoresis, centrifugation, and dialysis as a function of time elapsed from the labeling. Simultaneously, the speciation of stable cadmium present in or added to the same solutions was determined by anodic stripping voltammetry *(17)*. The very small effect of the elapsed time on the results indicated rapid establishment of isotopic equilibrium between the added ^{109}Cd and stable cadmium present in the solution in excess over ^{109}Cd. Virtually no removal of ^{109}Cd by centrifugation from the labeled soil leachates indicated that ^{109}Cd was not present in the form of particles with diameters larger than 26 nm and densities of 1.5 g·cm^{-3} or more. However, the electrophoretic mobility of ^{109}Cd was $<3 \times 10^{-4}$ cm^2 s^{-1} V^{-1} or negative (i.e., toward anode) as compared with the positive electrophoretic mobility of Cd^{2+} ion equal to 5.7 $\times 10^{-4}$ cm^2 s^{-1} V^{-1}, determined in 0.001 mol/L HNO_3.

These results proved that ^{109}Cd and also stable cadmium were present in the analyzed leachates in the form of neutral or negatively charged complexes or small colloids. Computation of chemical equilibria based on the concentration of inorganic anions in the leachates excluded the presence of significant amounts of neutral or anionic inorganic complexes of cadmium. Consequently, cadmium was associated with organic substances of molecular or colloidal dimensions, probably with fulvic acids whose presence in the leachates was proved by absorption spectrophotometry. The results of electrophoresis enabled the characterization, at least semiquantitatively, of the abundance and electrophoretic mobility of the associates formed.

The existence of organic associations of cadmium in the same leachates was also indicated by anodic stripping voltammetry (ASV), but the abundance of the (nonlabile) forms determined by ASV was much lower than that estimated from the electrophoresis. This fact is the result of the dissociation of the associates during electrodeposition of cadmium (the first step of ASV), whereas the noninvasive character of free-liquid electrophoresis enabled the determination of electrophoretic mobility unaffected by a shift of equilibria.

The organic complexes of cadmium were further characterized by the dialysis of ^{109}Cd from soil leachates through cellophane membrane. The incomplete dialysis against water proved the presence of organic associations with molecular weights larger than 10^4. The dialysis against the same, but unlabeled leachate was always complete. This finding points to the reversible nature of the organic forms of cadmium.

It cannot be excluded that stable cadmium was present in the soil leachates in a kinetically inert form that underwent negligible or very slow isotopic exchange with cationic ^{109}Cd. Such a form would naturally escape detection by the radiotracer method used, which represents one of the limitations of the method. However, existence of such totally inert forms is improbable, and if these forms exist, they are probably of small environmental importance. By no means can the existence of a kinetically inert form of cadmium explain the different results obtained by the radiotracer method and anodic stripping voltammetry. On the other hand, the absence or slowness of the isotopic exchange can be used for detection of kinetically inert forms of trace elements by the radiotracer method combined with another method *(18)*.

MIGRATION STUDIES

The radiotracer method also represents an excellent tool in the research of migration of contaminants in the atmosphere, hydrosphere, and lithosphere. The method is particularly suitable for the study of interaction of contaminants with solid phases in the environmental media and for obtaining data necessary for modeling the migration.

Interaction of Contaminant with a Solid Phase

The interaction (i.e., uptake and release) of contaminants with environmental solid phases strongly affects the speciation and migration of contaminants in the atmosphere, hydrosphere, and lithosphere. Present knowledge of the mechanisms and physicochemical laws of the interaction are rather imperfect, which causes problems in the description of contaminant transport using mathematical models and makes proper evaluation of field analytical data difficult. The interaction can be most easily studied using laboratory experiments and the radiotracer method. Of special importance is the application of the radiotracer method to the study of uptake and release of artificial radionuclides by suspended solids in surface waters, because the predominant mechanism of these processes is the isotopic exchange with stable isotopes of the radionuclides.

An example of such application is our study of the interaction of ^{137}Cs and ^{58}Co with suspended sediments in a small river receiving waste waters from a nuclear power plant. Laboratory batch experiments were carried out where the distribution of ^{137}Cs and ^{58}Co between the sediments and river water was measured using the radionuclides of a high specific activity *(19)*. The uptake and release of the radionuclides by the sediments were studied as a function of pH and composition of the water, cesium or cobalt concentration, composition of the sediment, contact time, liquid/solid ratio (V/m), and temperature. The results suggest that radiocesium is sorbed on fresh-water solids by isotopic exchange with stable cesium contained on the clay minerals of the solids, and radiocobalt is predominantly exchanged for stable cobalt held on the solids by chemisorption and ion exchange. From the six factors studied, the following can significantly affect the interaction: contact time, liquid/solid ratio, temperature, composition of solid phase, concentration of cesium and potassium (only for radiocesium), and pH (only for radiocobalt). Radiocesium and radiocobalt are efficiently adsorbed on suspended solids present in river water, and their uptake and release by the solid phase in rivers thus occur predominantly in suspension. Therefore, mathematical modeling of their migration in rivers should preferably consider the uptake and release by the suspended solids, as well as the sedimentation of the solids and resuspension of bottom sediments.

The effects of the following parameters on the uptake and release should be directly included in the migration model: contact time (kinetics of the interaction), *V/m*, and temperature. The easiest way is to express the effect of time in the form of a suitable kinetic function and to determine the dependence of kinetic and equilibrium parameters of this function on V/m and temperature. Kinetic functions suitable for the purpose were recently proposed *(20)*, and the dependence of their parameters on V/m was determined for Cs and Co *(21)*. The effect of the other factors can be taken into account in the form of variability of the parameters caused by variation in the pH, and composition of water and of suspended solids in the modeled river. Consequently, the parameters should be sought by laboratory experiments with repeatedly sampled unfiltered water *(19,21)*.

Finding Input Data for Migration Models

Apart from the parameters characterizing the interaction, a number of other input data are needed for modeling the migration of contaminants in the environment. The weak point of applicability of many migration models is the lack of or low accuracy of certain input data for the model. Some of the data can be obtained or made more accurate by means of the radiotracer method. Analysis of the migration of an artificially introduced or naturally present radionuclide in the system can greatly facilitate determination of the dispersion coefficients, sedimentation, resuspension rates, and so forth. Radionuclides can also be used for verification of transport models.

Recently, we described the use of an accidental release of radiocesium into a stream for obtaining input data on the distribution of resuspendable sediments along the stream and on the mean coefficient of exchange of solid phase between suspended solids in water and bottom sediments. A simple migration model was proposed whose applicability for the studied system was corroborated using the data on the time changes of the distribution of radiocesium in sediments of the stream *(22)*.

CONCLUSION

The radiotracer method represents the efficient tool in the investigation of environmental transport and speciation of contaminants. Despite the significant progress in the use of the method for this purpose, there are still many problems connected with it. Consequently, the application of the radiotracer method in environmental research deserves further study.

REFERENCES

1. M. Bernhard, F. E. Brinckman, and P. J. Sadler, eds., *The Importance of Chemical "Speciation" in Environmental Processes,* Springer Verlag, Berlin, 1986.

2. J. R. Kramer and H. A. Allen, eds., *Metal Speciation, Theory, Analysis and Application,* Lewis, 1988.
3. G. A. Batley, ed., *Trace Element Speciation: Analytical Methods and Problems.* CRC Press, Boca Raton, FL, 1989.
4. W. Salomons and U. Foerstner, *Metals in the Hydrocycle,* Springer Verlag, Berlin, 1984.
5. P. Beneš, in *The Environmental Behaviour of Radium,* vol. 1, International Atomic Energy Agency, Vienna, 1990, pp. 273–299.
6. P. Beneš and V. Majer, *Trace Chemistry of Aqueous Solutions,* Elsevier, Amsterdam, 1980.
7. H. J. M. Bowen, E. Page, I. Valente, and R. J. Wade, *J. Radioanal. Nucl. Chem.* **48,** 9 (1979).
8. M. Amdurer, D. Adler, and P. H. Santschi, in *Trace Metals in Sea Water,* C. S. Wong, ed., Plenum, New York, 1983, pp. 537–562.
9. H. A. Das, A. Faanhof, and H. A. van der Sloot, *Radioanalysis in Geochemistry,* Elsevier, Amsterdam, 1989.
10. K. Kratzer, P. Beneš, and V. Spěváčková, Separation of methyl-mercury from human hair by solvent extraction, *Int. J. Environ. Anal. Chem.,* in press.
11. K. Kratzer, D. Kolihová, P. Beneš, V. Spěváčková, and J. Žilková, *J. Anal. At. Spectrom.* **9,** 303 (1994).
12. P. Podhájecký, A. Gosman, and P. Beneš, *J. Radioanal. Chem.* **57,** 253 (1980).
13. P. Beneš and J. Glos, *J. Radioanal. Chem.* **52,** 43 (1979).
14. P. Beneš and E. Steinnes, *Int. J. Environ. Anal. Chem.* **4,** 263 (1976).
15. P. Beneš and M. Obdržálek, in *Proc. Sem. Hydrochémia 80,* Water Research Institute, Bratislava, 1980, pp. 139–156.
16. J. Mizera, Diploma Thesis, Czech Technical University, Prague, 1993.
17. P. Beneš and L. Loub, unpublished results.
18. M. Piro, M. Bernhard, M. Branica, and M. Verzi, in *Radioactive Contamination of the Marine Environment,* International Atomic Energy Agency, Vienna, 1973, pp. 29–44.
19. P. Beneš et al., *J. Radioanal. Nucl. Chem., Articles* **132,** 209, 225 (1989), **133,** 359 (1989), and **159,** 201 (1992).
20. P. Beneš, P. Picat, M. Černík, and J. M. Quinault, *J. Radioanal. Nucl. Chem., Articles* **159,** 175 (1992).
21. P. Beneš and M. Černík, *J. Radioanal. Nucl. Chem., Articles* **159,** 187 (1992).
22. P. Beneš, M. Černík, and O. Slávik, *J. Environ. Radioact.* **22,** 279 (1994).

Comprehensive RNAA of Cadmium, Cobalt, Nickel, and Copper Using ^{109}Cd, ^{57}Co, and Reactor-Produced ^{67}Cu as Radioisotopic Yield Monitors

A. R. BYRNE* AND M. DERMELJ

Department of Nuclear Chemistry, "J. Stefan" Institute, 61000 Ljubljana, Slovenia

ABSTRACT

An existing radiochemical NAA procedure for Cd, Co, and Cu was improved to allow determination of individual radiochemical yields by the radioisotopic tracer technique, thus eliminating errors owing to variable recovery. ^{109}Cd was used as tracer for Cd determination via $^{115}Cd/^{115m}In$, ^{57}Co for Co via ^{60}Co, and potentially for Ni via ^{58}Co, whereas as a novelty ^{67}Cu, produced by reactor irradiation of ZnO of natural isotopic composition (by the ^{67}Zn [*n,p*] ^{67}Cu reaction) was used for Cu via the indicator nuclide ^{64}Cu. The simple production and purification of ^{67}Cu by anion exchange is described. Results for biological RMs are given and discussed.

Index Entries: Radiochemical NAA; copper; cobalt; cadmium; ^{67}Cu; radiochemical yield.

INTRODUCTION

One of the major potential errors in radiochemical neutron activation analysis (RNAA) is variation of the chemical yield of the separation. The advantages of individual yield measurements have been discussed thoroughly by Heydorn *(1)*, who also coined the term "comprehensive NAA" for a procedure in which the yield of the separation is measured. Apart from yield measurements based on chemical analysis of carrier recovery and the reactivation technique, the use of a suitable radioisotopic tracer

*Author to whom all correspondence and reprint requests should be addressed.

to monitor recovery of the induced characteristic nuclide is often the easiest and most accurate technique. This is because the yield measurement takes place simultaneously with the measurement of the indicator activity, and errors resulting from geometry, pulse pile-up, or dead-time losses are compensated for.

In the present work, in addition to using ^{109}Cd ($t_{1/2}$ = 433 d, $E\gamma$ = 88 keV) as isotopic yield tracer for Cd determined via ^{115}Cd(^{115m}In), and ^{57}Co ($t_{1/2}$ = 271 d, $E\gamma$ = 122 keV) as simultaneous yield monitor for both Co via ^{60}Co ($t_{1/2}$ = 5.7 y, $E\gamma$ = 1117, 1332 keV) and potentially for Ni via fast neutron activation by ^{58}Ni(n,p)^{58}Co ($t_{1/2}$ = 71 d, $E\gamma$ = 810 keV), as a novelty, ^{67}Cu ($t_{1/2}$ = 62 h, $E\gamma$ = 184.5 keV) was prepared in the laboratory by neutron activation of zinc oxide of natural isotopic composition, and applied as an isotopic yield monitor in determination of Cu via ^{64}Cu ($t_{1/2}$ = 12.7 h, $E\gamma$ = 511 keV). Thus, for all the elements determined by a simultaneous RNAA procedure based on dithiocarbamate extraction, previously developed in our laboratory *(2)*, the chemical yields were determined for each analysis by the radioisotopic tracer technique. This procedure produced two fractions for γ spectroscopy, one containing Cu and Co(Ni) and the other Cd. The method was then applied to the analysis of standard or certified biological reference materials (SRMs/CRMs).

EXPERIMENT

Preparation of ^{67}Cu Tracer

One gram of ZnO (Merck, p.a. quality) was irradiated for 2–20 h in the "F" ring of our TRIGA Mk II reactor, where the thermal fluence rate ≈ the fast fluence rate ≈ 3×10^{12} n·cm^{-2} s^{-1}.

After cooling for 8 or 9 d to allow ^{69m}Zn, ^{64}Cu, and other short-lived nuclides to decay, the separation of ^{67}Cu from ^{65}Zn was based on a single-step anion-exchange process in 2 mol/L HCl. The ZnO was dissolved in 3 mL 9 mol/L HCl and warmed on a hot plate to reduce the volume to about 0.5 mL. A little Cu^{2+} carrier (microgram amounts) may be added (to help sharp elution of Cu from the column), but this is not necessary, and ^{67}Cu may be prepared in "no carrier added" form. The zinc solution is taken up in 10 mL of 2 mol/L HCl and transferred to the top of an anion-exchange column of Dowex 1X8, 100–200 mesh, Cl^- form, of dimensions 90 × 20 mm ϕ fitted with a stopcock, and previously conditioned with 2 mol/L HCl.

^{67}Cu is eluted at a flow rate of 1–2 mL·min^{-1} with 2 mol/L HCl. The first 15 mL are discarded, and the next 50 mL contain over 98% of the ^{67}Cu and ^{64}Cu impurity. The volume is reduced by heat, and the acidity and volume adjusted as desired.

^{67}Cu and ^{64}Cu contents are measured by γ spectrometry on a calibrated HPGe detector. Calibration for ^{67}Cu can be performed by reference to a uranium solution (^{235}U line at 185.7 keV) or a ^{226}Ra standard at 186.1 keV. Further details of ^{67}Cu production can be found elsewhere *(3)*.

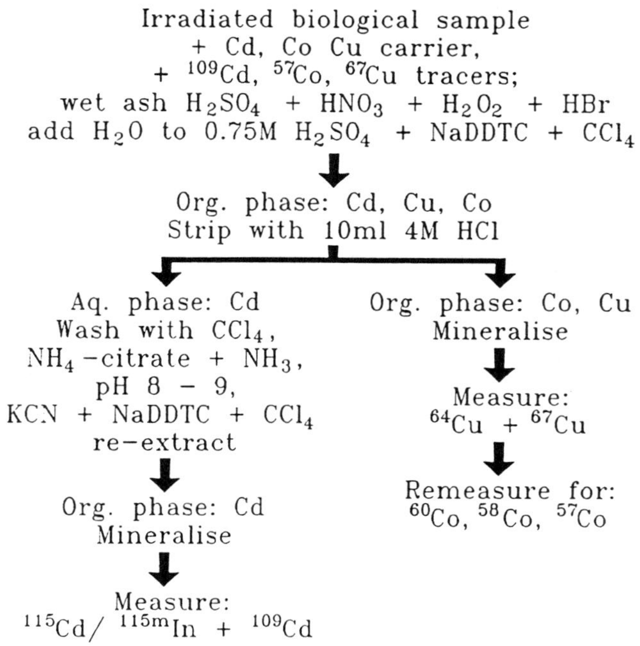

Fig. 1. Radiochemical separation scheme for Cu, Co, and Cd.

Determination of Cu, Cd, and Co (Ni)

Irradiation and Standards

Carrier solutions of 1 $mg{\cdot}g^{-1}$ were diluted to 50 μg Cu and $Cd{\cdot}g^{-1}$ or 300 $\mu g{\cdot}g^{-1}$ for Co, and used as irradiation standards. Samples were packed in cleaned polyethylene ampules and irradiated for about 20 h at a thermal fluence rate of 1.2×10^{12} $n{\cdot}cm^{-2}$ s^{-1} with standards taped alongside.

Procedure

The procedure is basically identical to that given earlier *(2)* and for the sake of completeness is summarized in the scheme shown in Fig. 1. All three elements were coextracted from dilute sulfuric acid after wet ashing and Cd stripped with 4 mol/L HCl. It was purified by re-extraction at pH 8–9 in the presence of citrate and cyanide. Cu and Co(Ni) were measured together after mineralization of the organic phase.

Activity Measurements

Separated fractions were measured in 5-mL vials in a well-type HP Ge detector and Canberra 90 MCA. A weighed aliquot of the radioisotopic tracers was measured after the separated fractions and their standards to obtain the chemical yield from the relative peak heights of the tracer added to the sample and the standard, and their masses. In measuring ^{64}Cu, the ^{67}Cu radioisotopic tracer standard was checked for ^{64}Cu content, and a correction made if necessary (in practice, this was negligible for samples with Cu contents in the $\mu g \cdot g^{-1}$ range).

The measurement of cobalt as ^{60}Co, with added ^{57}Co, was made in a second, longer count after quantifying copper, since the induced ^{60}Co activity is generally very much smaller than that of ^{64}Cu. Ni via ^{58}Co is also measured in this count, if sufficiently activated.

RESULTS AND DISCUSSION

In the original version of the procedure *(2)*, the radiochemical yields for Cd, Co, and Cu were measured spectrophotometrically, which was tedious and time-consuming, or in routine work, an average yield was assumed, which was inaccurate. The present modified procedure using radioisotopic tracers as yield monitors enables the measurement of yields from the γ spectrum of the indicator nuclide, and is thus efficient and accurate. This technique compensates for errors owing to geometry, dead-time, or pulse pile-up with an overall gain in quality. In the present work, in addition to the use of ^{109}Cd and ^{57}Co, which have previously been used for yield determinations in RNAA of Cd and Co(Ni), home-produced ^{67}Cu was applied as an original solution to the determination of Cu yields in RNAA. ^{67}Cu in useful tracer quantities (0.1–1 μCi for 2–20 h irradiation) could be produced from 1 g ZnO of natural isotopic composition, with <2% ^{64}Cu content 10 d after irradiation.

Mushtaq et al. *(4)* described production of ^{67}Cu in larger quantities using expensive ZnO isotopically highly enriched in ^{67}Zn. These authors also used 0.5 mol/L HCl for elution of Cu from Zn, whereas the maximum retention of Zn on the anion exchanger is close to 1.8 mol/L HCl *(5)*.

The γ spectrum of ^{67}Cu produced in the way described is shown in Fig. 2, and a spectrum of the copper and cobalt fraction isolated by the procedure from NBS spinach is shown in Fig. 3.

Results obtained by the method developed are shown in Tables 1 and 2, including for comparison results obtained (by operator 2) for spinach without measurement of the chemical yield. It may be seen that the mean chemical recoveries (obtained by operator 1 for Cd, Co, and Cu were 74.6 ± 17.8, 90.3 ± 4.2, and 92.4 ± 5.5%, respectively. Evidently for Cd, the yields are sometimes rather low; this is even more marked for the results obtained (Table 2) by operator 2, and this is believed to represent poor control of the acidity in the first extraction. Operator 2 also obtained some low yields for Co (Table 2). Several conclusions can be drawn from these results:

1. Good agreement was obtained with the certified values when individual chemical yields were measured;
2. The application of an individual chemical yield produces a good result, even when the yield is very low;

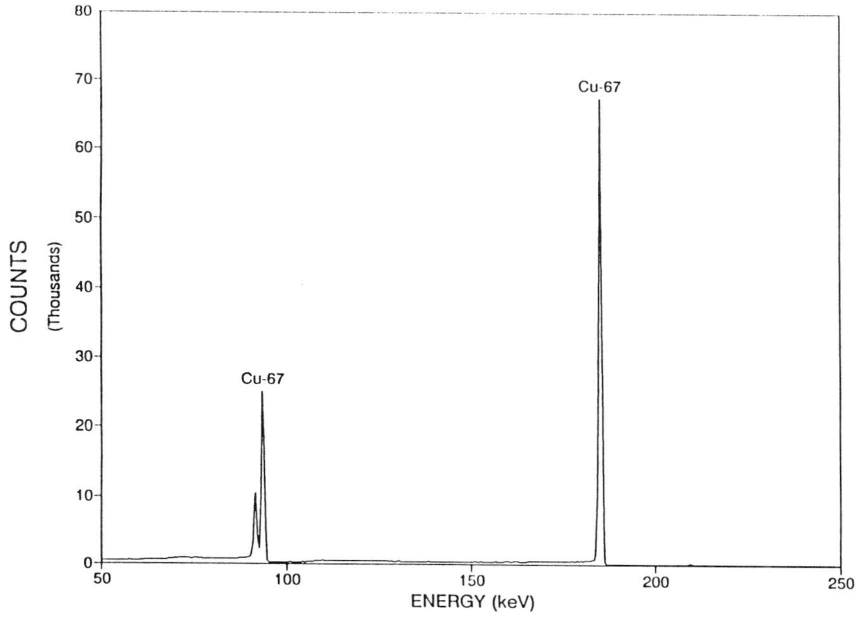

Fig. 2. γ Spectrum of ^{67}Cu prepared from reactor-irradiated ZnO of natural isotopic composition.

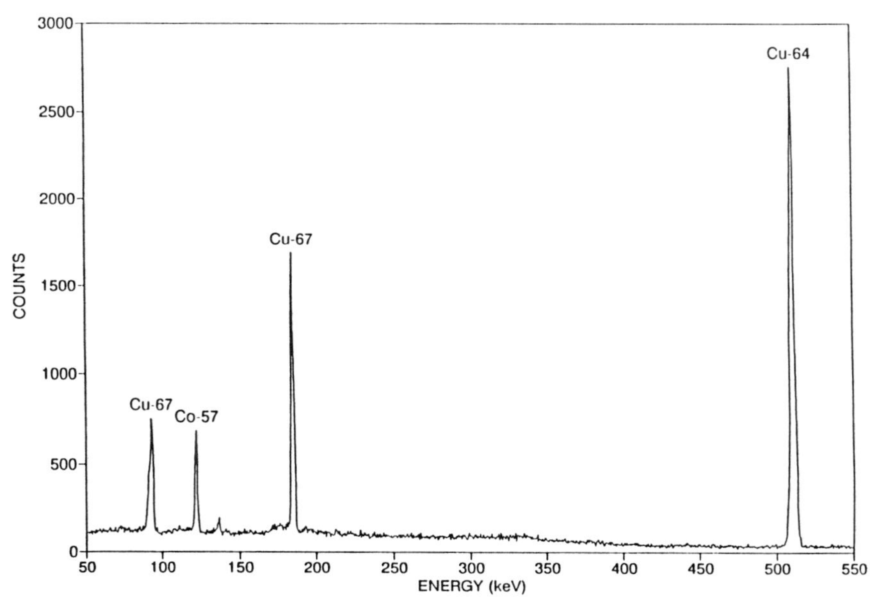

Fig. 3. γ Spectrum of copper and cobalt fraction, isolated from NBS spinach, with ^{67}Cu and ^{57}Co tracers.

Table 1
Results for NBS Oyster Tissue SRM-1566 in mg/kg (dw)

Cd		Co		Cu	
% yield	result	% yield	result	% yield	result
Operator 1:					
59.5	3.26	90.4	0.344	96.8	57.6
94.9	3.50	90.9	0.382	97.4	57.4
54.7	3.28	89.5	0.343	100.7	57.9
78.3	3.68	80.2	0.359	81.8	69.3
66.4	3.85	87.2	0.308	90.1	67.2
93.2	3.60	91.2	0.343	97.5	64.9
82.2	4.01	93.7	0.359	93.3	69.6
89.9	3.76	90.8	0.342	92.0	64.1
88.4	3.62	93.7	0.338	96.0	61.3
93.8	3.41	94.9	0.308	90.5	60.4
39.4	3.32	90.5	0.349	92.3	58.6
52.4	3.57	93.4	0.341	91.0	64.4
82.3	3.62	84.2	0.359	92.9	69.8
68.7	3.58	85.4	0.363	81.5	69.1
means:					
74.6 ± 17.8	3.58 ± 0.21	89.7 ± 4.2	0.346±0.020	92.4 ± 5.5	63.7 ± 4.8
Operator 2:					
- *	2.77	-	0.25	-	47.0
-	2.36	-	0.26	-	40.9
-	2.43	-	0.24	-	46.3
-	2.79	-	0.23	-	48.0
-	2.31	-	0.36	-	49.0
-	2.41	-	0.27	-	45.5
				-	52.2
				-	50.1
				-	59.6
				-	54.3
mean:	2.52 ± 0.21		0.27 ± 0.05		49.3 ± 5.2
certified:	3.5 ± 0.4		(0.4)		63.0 ± 3.5

* = not determined

Table 2
Results for NBS Spinach in mg/kg (dw)

Cd		Co		Cu	
% yield	result	% yield	result	% yield	result
Operator 2:					
Old SRM 1570					
75.5	1.48	62.4	1.41	88.7	13.3
90.1	1.39	40.9	1.41	90.4	13.2
7.86	1.46	14.2	1.33	92.6	12.9
means	1.44 ± 0.05		1.38 ± 0.05		13.1 ± 0.2
certified	(1.5)		(1.5)		12 ± 2
New candidate SRM 1570a*					
55.9	2.78	15.8	0.432	91.7	12.5
96.8	2.86	41.2	0.425	97.9	13.0
88.8	2.86	67.8	0.433	90.5	13.1
88.5	2.96	82.9	0.376	91.6	12.6
51.7	2.96	90.1	0.395	100.0	12.9
75.1	2.69	84.1	0.355	98.7	11.9
53.8	2.75	80.4	0.335	91.4	12.2
means					
72.9 ± 19.0 (7)	2.84 ± 0.10		0.393 ± 0.39	93.3 ± 4.0 (10)	12.6 ± 0.4

* not yet certified

3. The standard deviation of the results is smaller when individual chemical yields are measured; and
4. For Oyster Tissue (Table 1), application of a mean yield to the results of operator 2 produces results that are somewhat too low; however, this is evidently a risky proceeding, in view of the variation of the yields for Cd and occasionally Co.

REFERENCES

1. K. Heydorn, *Neutron Activation Analysis for Clinical Trace Element Research*, vol. 1, CRC Press, Boca Raton, FL, 1984.
2. M. Dermelj, A. Vakselj, V. Ravnik, and B. Smosdiš , *Radiochem. Radioanal. Lett.* **41,** 149 (1979).
3. A. R. Byrne, Report IJS DP-6693, J. Stefan Institute, Ljubljana, Slovenia, March 1993.
4. A. Mushtaq, H. M. A. Karim, and M. A. Khan, *J. Radioanal. Nucl. Chem. Articles* **141,** 261 (1993).
5. K. A. Kraus and F. Nelson., Proc. 1st Int. Conf. Peaceful uses of Atomic Energy, Geneva, **7,** 113 (1956).

Enhanced Sensitivity for the Determination of Selenium by INAA

M. Makarewicz and R. Zeisler*

IAEA Laboratories Seibersdorf, International Atomic Energy Agency, POB 100, A-1400 Vienna, Austria

ABSTRACT

A working analytical procedure for selenium determinations by INAA, using γ spectrometry with a germanium well-type detector, is presented. The spectral line at 400.7 keV is employed, enhanced with the effect of energy summing in true coincidence of cascade photons. The main advantage of the method is high efficiency and reliable, interference-free results; potential drawbacks are increases in input count rate and pile-up losses, decrease in resolution, and sensitivity to errors in sample-to-detector geometry. The procedure is applied to certification analysis of reference materials, large-area biomonitoring by oak leave samples, and determination of separated proteins.

Index Entries: Selenium; sensitivity; instrumental neutron activation analysis; reference materials; γ spectrometry; efficiency; well-type detector.

INTRODUCTION

Selenium commonly is assayed through its long-lived ^{75}Se isotope (half-life 120 d) produced in the reaction of neutron capture ^{74}Se (n,γ) ^{75}Se. Low abundance of ^{74}Se (0.9%) is balanced in the reaction rate by a relatively high neutron cross-section of 52 b for thermal neutrons. Hence, Se can be detected by NAA down to the picogram range *(1)*. However, this "interference-free" sensitivity is not reached in INAA. The most intense γ lines of ^{75}Se, those at 136.0 and 264.7 keV often used for Se evaluation, are interfered with by those at 136.2 keV of ^{181}Hf (half-life of 42.4 d) and 264.1 keV of ^{182}Ta ($t_{1/2}$ = 115 d). The line at 279.5 keV coincides with that at 279.2 keV of ^{203}Hg ($t_{1/2}$ = 46.6 d), and the line at 121.1 keV

*Author to whom all correspondence and reprint requests should be addressed.

Biological Trace Element Research Editor: G. N. Schrauzer © 1994 by Humana Press Inc.

with that at 121.8 keV of ^{152}Eu ($t_{1/2}$ = 13.3 yr). The magnitude of these interferences is particularly high in geological and environmental materials, and usually cannot be corrected accurately. Also in biological materials, these interferences may make determination of Se less reliable. The line at 400.7 keV is interference-free; however, in common procedures, its low-emission probability (11.6%) limits the applicability to cases with low Compton background and higher concentrations.

We employ the γ line at 400.7 keV with an enhanced counting rate. The physical phenomenon that the γ decay occurs from the same level mainly via three branches, directly to the ground state, and via lines at 121.1 and 279.5 keV, and 136.0 and 264.7 keV, may contribute to the 400.7-keV line in a true coincidence summing. The rate of the effect depends first of all on the solid angle in a source-detector configuration. We use a well-type germanium detector to increase the rate of the summing effect (large solid angle).

METHOD

Analytical Procedure

In the standard analytical procedure, samples of approx 0.2 g in mass, closed in ampules made of ultrapure quartz, are irradiated with thermal neutrons (fluence rate ~7 × 10^{13} $cm^{-2}s^{-1}$) in the ASTRA reactor (Austrian Research Center Seibersdorf) for 0.5–24 h. After a period of time of 4–6 wk, necessary for decay of short-lived activation products, γ-spectra of the samples are recorded with the spectrometric system. The well detector (HP-Ge crystal with vertical cryostat; 20% relative efficiency, and FWHM of 1.9 keV at 1.3 MeV), manufactured by Canberra (Meriden, USA), is part of the γ spectrometry system consisting of spectroscopy amplifier Tennelec (Oak Ridge, TN) TC 244 (6 μs, PUR OFF), ADC Canberra 8075 (8K), and an acquisition board Canberra S-100. No hardwired live-time correction was established between the amplifier and the ADC. Counting time is 4–8 h. The spectra are evaluated with the program GAMANAL/GANAAS *(2)*. Concentration of elements and their uncertainties are calculated with a PC program written by one of us. Final data processing for presentation is done in a PC spreadsheet.

Correction for Pile-up

The observed peak area N' is corrected for pile-up and other losses with the semi-empirical function of the ADC dead time:

$$N = N'/(1 - a \cdot DT) \quad (1)$$

where N is corrected peak area; DT is the current ADC dead time, in particular, $DT = (T - t)/T$, where T is the true counting time and t is the live time; and a is a constant specific for both detector and nuclear equipment used determined in a series of additional measurements. In particular, a is the slope of a straight line fitted to N'/T vs DT. It follows the method by Wyttenbach *(3)* in which N'/t vs $(T - t)/t$ is fitted. It can be shown that $a = 1 + 2\tau/\vartheta$, where $2\tau/\vartheta$ is the Wyttenbach's fitting parameter.

Correction for Efficiency

Elemental content is determined by comparing the signal of the test sample to that of a standard source:

$$\mu = \mu_S (\dot{N}/\dot{N}_S) (\varepsilon_S/\varepsilon) \quad (2)$$

where μ is content of the element, and $\dot{N}$ is count rate at the γ line of interest, already corrected for pile-up and other losses, decay owing to the fact that the test sample and the standard are measured on different times, and decay during measurement. The subscript s refers to the standard source. The ratio $\varepsilon_S/\varepsilon$ is the correction factor, which takes into account possible difference in detector efficiency for the test sample (ε) and the standard (ε_S). For example, for a sample 4 mm long and a 20-mm standard source, the efficiency correction factor $\varepsilon_S/\varepsilon = \sim 0.75$.

In particular, the detector efficiency ε_h for a volume source of the height h with an activity uniformly distributed over the range from h = 0 to h is given by *(see* Appendix):

$$\varepsilon_h = (1/h) \int_0^h \varepsilon(x)dx \quad (3)$$

where $\varepsilon(h)$ is the efficiency for a point source at the height h in the detector well. The efficiency $\varepsilon_{h,\Delta}$ for an extended source of length Δ, ranging from the height h to $h + \Delta$ is then:

$$\varepsilon_{h,\Delta} = \varepsilon_{h+\Delta} + (h/\Delta) (\varepsilon_{h+\Delta} - \varepsilon_h) \quad (4)$$

We determined the efficiency of the well-type detector vs source position (activated selenium solution closed in a quartz ampule). The experimental results show (Fig. 1) strong dependency of the efficiency $\varepsilon(h)$ for a point source on the source position. The efficiency at the depth of 35 mm (h = 0 in Fig. 1) is over 14 times higher than that at the upper well edge, and it is approx 21 times higher than that at the distance of 5 mm over the detector end cap. The efficiency ε_h for volume sources varies significantly less; however, it should be determined carefully for practical applications. Otherwise the determined concentration is biased with remarkable error (Fig. 2). For instance, underestimation of sample height for 1 mm may result in underestimation of concentration by ~2.5%. The effect is most pronounced for the samples with height between 20 and 30 mm. Overestimation of height of the top and the bottom of a sample will result in an overestimation of concentration. In principle, samples smaller in size are virtually biased with lower uncertainty.

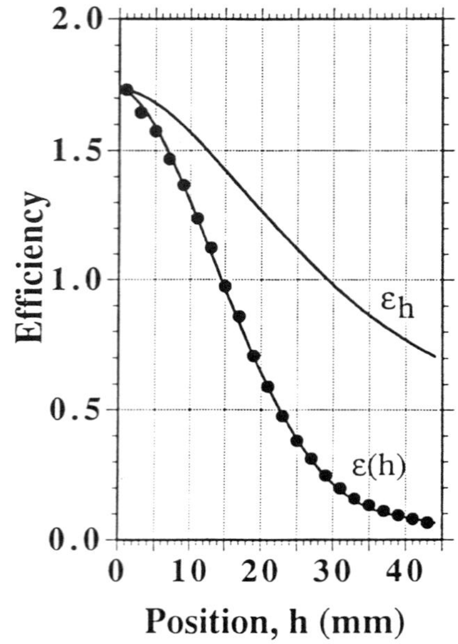

Fig. 1. Efficiency of well-type detector for ^{75}Se peak at 401 kEv vs source position in the well. The well is 40 mm deep; the height $h = 0$ corresponds to the distance of 5 mm above well bottom (the lowest position of a point source, available in our standard counting procedure). The points are experimental efficiency data obtained with finite size standard source, and $\varepsilon(h)$ is the unfolded efficiency for a point source, described by a 4th-order polynomial (Eq. [A1]). Reduced coefficients c_i/c_0 for $i = 0, \ldots 4$: 1, $-4.489 \cdot 10^{-3}$, $-2.917 \cdot 10^{-3}$, $9.647 \cdot 10^{-5}$, $-8.900 \cdot 10^{-7}$, with $c_0 = -4.489 \cdot 10^{-3}$. The efficiency ε_h for volume source of height, h, extending from the height 0 to h, is plotted according to Eq. (A2).

RESULTS AND DISCUSSION

Comparison of Two Standard Analytical Procedures

A standard selenium source 20-mm long was counted in a well-type detector, and with closed-end HP-Ge detector (Ortec GEM, 28% rel. eff., FWHM of 1.8 keV; amp. Ortec 672; ADC Canberra 8075) in a standard geometry (i.e., ampule perpendicular to the detector axis, 4.5 cm above the detector end cap). The enhancement factor, $f = t_c/t_w$, is here defined as a ratio of counting time necessary to obtain results with equal random uncertainties with closed-end detector (t_c) to that of well detector (t_w). The obtained value of $f = 39.2$ means that counting with well-type detector could be approx 40 times shorter in time than that with the closed-

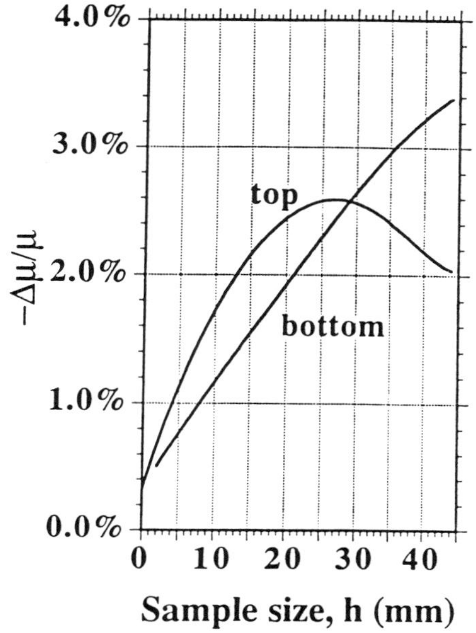

Fig. 2. Uncertainty $\Delta\mu/\mu$ in the result of determination of selenium content (or concentration), as a function of suspected sample height, *h*. The curve top shows relative difference between the concentration that would be obtained and its "correct" value in the case where the sample is 1 mm higher *(h* + 1) than assumed *(h)* in the determination. The curve bottom shows the magnitude of the error when the bottom of the sample is 1 mm higher than assumed. In both cases, $\Delta\mu/\mu$ is negative, and the concentration will be underestimated.

end detector, at the same statistical significance of the results. The following formula for estimation of counting time, *t*, necessary to achieve the net result with required relative uncertainty, s_r, was used (both net peak area, $\dot{N}$, and background, $\dot{B}$, under the peak are in s^{-1}):

$$t = (1/s_r^2)\ (\dot{N} + 2\dot{B})/\dot{N}^2 \qquad (5)$$

Sample Results

Examples of detection limits, calculated according to Currie *(4)*, obtained in routine multielemental analysis of different materials, are given in Table 1. The detection limits are sufficiently low for the purpose. The relatively short irradiation and counting times are factors increasing analysis throughput. Lower detection limits could be achieved with longer irradiation and counting times. The values obtained for certified reference materials show no significant bias despite greatly different matrix composition. In addition, the statistical evaluation of results for 34

Table 1
Limit of Detection (LOD) of Selenium Determined in Routine Multielemental Analysis (INAA) of Different Test and Control Materials, at the γ Line at 401 keV

Material Code	Material Description	Mass (g)	Counting time (h)	Result ± unc. (ng/g)	Certified ± unc. (ng/g)	LOD (ng/g)
		t_i= 0.5 h,	*t_d= 17 weeks*			
CTA-FFA-1	Fine Fly Ash	0.1	8	3800 ± 230	(4600)	600
SRM 1633a	Coal Fly Ash	0.1	17	9220 ± 270	10300 ± 600	500
ZUK-1	Coal Ash (Kansk)	0.1	20	701 ± 35	(300 - 3000)	200
		t_i= 0.5 h,	*t_d= 16 weeks*			
SRM 1941a	Coastal Sediment	0.1	22	4460 ± 400		400
BIL-1	Bottom Silt (Baikal)	0.1	24	974 ± 39		200
SRM 1645	River Sediment	0.1	24	1537 ± 65	(1500)	200
		t_i= 24 h,	*t_d= 11 weeks*			
SRM 1577a	Bovine Liver	0.2	5	742 ± 24	710 ± 70	9
	Oak Leaves (Italy)	0.2	7	52 ± 29		5
SRM 1515	Apple Leaves	0.2	17	48.8 ± 2.6	50 ± 9	7
SRM 1572	Citrus Leaves	0.2	7	26.2 ± 1.3	(25)	5
		t_i= 4 h,	*t_d= 4 weeks*			
	Peroxidase[a]	10 μg[b]	0.8	(32.00 ± 0.45) ng		0.7 ng
			1.7			0.5 ng
			24			0.1 ng[c]

[a]glutathione-peroxidase, separated seleno-protein.
[b]10 μg Protein in 100 mg polyacrylamide.
[c]Projected to 24 h; exactly 0.13 ng.

LOD was calculated according to Currie's (4) definition, based on the spectral interval of double FWHM around the peak in the spectrum. Samples were irradiated at thermal neutron fluence rate of $7{\cdot}10^{13}$ $cm^{-2}s^{-1}$. The symbol t_i means irradiation time, and t_d means approximate cooling time.

samples of oak leaves showed that the line at 401 keV seems to be interference-free. The results based on the other lines are subject to significant bias: +4330% at 121 keV, +40% at 136 keV, +70% at 264 keV, and +1350% at 279 keV. The strong interference to 279 keV is due to ^{203}Hg. Elements belonging to components of geological origin probably interfer with the other lines.

The sensitivity increased by using the 401-keV line has made it possible to instrumentally determine selenium in coal ash material (Fig. 3). This material originated from the Kansk-Achinsk basin, code ZUK-1 *(5)* and is distributed as a test material for intercomparison. The concentration of selenium is ~700 ng/g *(see* Table 1).

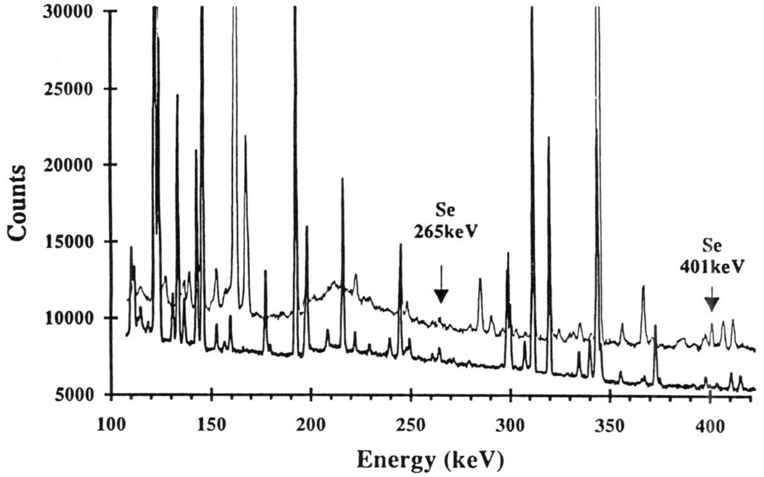

Fig. 3. γ-Spectra of 0.1 g sample made from brown coal ash from Kansk-Achinsk basin, recorded with two different HP-Ge detectors within 70,000 s of counting time. The lower spectrum was collected with a closed-end detector (28% relative efficiency, FWHM of 1.8 keV at 1.3 MeV); the upper spectrum was collected 4 mo later with a well-type detector (20%, 1.9 keV). Selenium can be quantified only in the latter measurement.

SUMMARY

The main advantages of the method are reliable interference-free results and high efficiency for the detection of the analytical signal (401-keV line). The latter permits a significant increase in throughput and/or lower detection limit. Precautions can be taken to minimize adverse effects resulting from high input count rate and sensitivity to counting geometry. The evaluation has shown that errors owing to geometry can be limited to a few percent relative. Also, we have used high-quality conventional amplifiers with proper calibration for pile-up/live-time correction up to about 20,000 counts/s detector input count rate. The preloaded filter amplifier technology developed by Westphal *(6)* would significantly increase throughput, i.e., minimize pile-up and maintain good detector resolution at high count rate. Its implementation will further improve this efficient and accurate INAA procedure for the determination of selenium.

ACKNOWLEDGMENTS

We are grateful to R. Ogris for preparing samples for analysis, and to the Austrian Research Center Seibersdorf for making available to us its irradiation facility at the ASTRA reactor.

APPENDIX

Assuming a point source efficiency, $\varepsilon(h)$, expressed by a nth order polynomial:

$$\varepsilon(h) = \sum_{i=0}^{n} c_i h^i \tag{A1}$$

Eq. (3) takes the following form:

$$\varepsilon_h = \sum_{i=0}^{n} \frac{c_i}{i+1} h^i \tag{A2}$$

The efficiency, $\varepsilon_{h,\Delta}$ (See Eq.[4]), for a source extending from height, h, to $h + \Delta$ is then:

$$\varepsilon_{h,\Delta} = \sum_{i=0}^{n} \frac{c_i}{i+1} \frac{1}{\Delta}\left[(h+\Delta)^{i+1} - h^{i+1}\right] = \sum_{i=0}^{n} \frac{c_i}{i+1} \alpha_i \tag{A3}$$

where:

$$\alpha_i = \frac{1}{\Delta}\left[(h+\Delta)^{i+1} - h^{i+1}\right] = \sum_{k=0}^{i} (h+\Delta)^{i-k} h^k \tag{A4}$$

calculated with the following recurrence formula:

$$\alpha_i = \alpha_{i-1}(h+\Delta) + h^i \tag{A5}$$

for $i = 1, 2, \ldots n$, with $\alpha_0 = 1$.

The polynomial expansion coefficients, c_i, were found by solving a system of linear equations of the type of Eq. (A3) by least-squares method, where the values of $\varepsilon_{h,\Delta}$ at different heights, h, were obtained from experiment.

REFERENCES

1. W. M. Ching, B. deWeerd, and T. C. Stadtman, *Proc. Natl. Acad. Sci. USA* **82,** 347–350 (1985).
2. GANAAS, Computer Manual Series **3,** IAEA, Vienna (1991), 106 pp.
3. A. Wyttenbach, *J. Radioanal. Chem.* **8,** 335–343 (1971).
4. L. A. Currie, *Anal. Chem.* **40,** 586–593 (1968).
5. L. L. Petrov, Yu. N. Kornakov, L. A. Persikova, A. V. Vlasov, V. V. Malykh, and A. Z. Prokopjeva, *Geostandards Newsletter* **1,** 135–137 (1991).
6. G. P. Westphal, K. Jöstl, B. Lipp, and P. Schröder, *J. Radioanal. Nucl. Chem.* **160,** 395–405 (1992).

Effects of Target Temperature on Analytical Sensitivities of Cold-Neutron Capture Prompt γ-Ray Activation Analysis*

ELIZABETH A. MACKEY

Technology Administration, National Institute of Standards and Technology, Inorganic Analytical Research Division, Gaithersburg, MD 20899

ABSTRACT

Cold-neutron prompt γ-ray activation analysis sensitivities are often decreased because of an increase in the average neutron energy on scattering within room temperature targets. Experiments were performed to determine whether target cooling would alleviate these effects. Cooling the targets to 77 K increased hydrogen sensitivity by as much as 25%. Target cooling decreases those effects of neutron scattering on CNPGAA sensitivities that are the result of an increased average neutron energy. However, cold-neutron scattering may also change the average path length traveled, and this effect on sensitivity is not alleviated by controlling temperature.

Index Entries: Prompt γ-ray neutron activation analysis; cold neutrons; neutron scattering; target temperature and sensitivity.

INTRODUCTION

Neutron scattering by hydrogen causes changes in the analytical sensitivities (cps/mg) of both thermal and cold-neutron capture prompt γ-ray activation analysis (PGAA). These changes in sensitivity are mea-

*Contributions of the NIST are not subject to copyright. Certain commercial equipment, instruments, and materials are identified in this article for purpose of clarity; such identification does not imply endorsement by the NIST, nor does it imply that these are necessarily the best available for the purpose.

surable at H concentrations of about one weight percent of H (and greater), so that for accuracy in the analysis of biological or other hydrogenous materials, these effects must be characterized. For thermal neutron (TN) PGAA, these effects have been studied *(1–4)*. The magnitude of the effect of scattering on element sensitivity is a function of the size, shape, and scattering power of the target. The effects of thermal neutron scattering on analytical sensitivities can be accounted for quantitatively by elastic scattering, i.e., scattering in which the energy of the neutron does not change. Results of Monte Carlo simulations performed using an elastic scattering model agree well with experimental results *(4)*. The observed effects on sensitivity are largely the result of elastic scattering, which changes the mean free path of the neutron within the target and, therefore, changes the probability for neutron absorption.

Effects of elastic scattering are minimal for a sphere. The spherical symmetry causes the average neutron path length not to change as much on scattering as it might for a slab or a disk, depending on its size and shape. When there is no net change in the energy of the neutrons, spherical targets are preferred because sensitivities are independent of target size. Thermal neutron PGAA of spheres of urea (6.71% H), of tris-hydroxymethyl-aminomethane (THAM, 9.15% H), and of paraffin (15% H) yielded constant sensitivity as a function of amount of scatterer (or quantity of H) and as a function of size for spheres with diameters from approx 2 to 12 mm *(3)*.

Effects of neutron scattering on analytical sensitivities of cold-neutron (CN) PGAA are different from those of TNPGAA. The effects of cold-neutron scattering, as for thermal neutron scattering, are dependent on the size, shape, and scattering power of the target. However, for several cases in which TNPGAA sensitivities were enhanced, CNPGAA sensitivities for the same targets were decreased *(5)*. For example, for a series of relatively thin targets (thicknesses of approx 2 mm), TNPGAA sensitivities increase with increasing H concentration, whereas CNPGAA sensitivities for the same targets decrease with increasing H content. The effects of neutron scattering on CNPGAA and TNPGAA sensitivities as a function of target geometry also differ. For example, TNPGAA sensitivities for spherical hydrogenous targets remain constant as a function of sphere size, whereas CNPGAA sensitivities for the same targets decrease with increasing sphere size.

Results of initial studies may be explained by cold neutrons gaining energy, on average (neutron warming), on scattering within a room-temperature target *(5)*. Sensitivities for elements for which absorption cross-sections vary inversely with neutron velocity ("$1/v$" elements) and for elements for which there are large resonances were measured within scattering matrices. Sensitivities for the "$1/v$" elements decreased by the same relative amounts, whereas sensitivities for the resonance elements differed significantly from the "$1/v$" nuclides, indicating that scattering causes a net change in the average neutron energy. For most targets studied, sensitivities decreased with increasing amount of scatterer. These results are consistent with the hypothesis that cold-neutron scattering within a room-temperature target results in an increase in the average energy of the neutrons, corresponding to a decrease in absorption cross-sections and, therefore, decreased sensitivities.

In this study, experiments have been performed to determine whether cooling the target reduces that portion of the CNPGAA sensitivity decrease that is caused by neutron warming. If the temperature of the target and the "effective temperature" of the neutron beam were similar, effects of up-scattering and down-scattering would result in little or no change in the average neutron energy. Sensitivities have been measured for a series of hydrogenous disks of different thicknesses and spheres of different diameters at target temperatures of 300 and 77 K.

METHODS

Spheres and disks of reagent-grade THAM were prepared using a procedure described previously *(3)*. All targets were sealed in bags formed from Teflon™ film. A cryostat with a copper cold finger and aluminum cap was used as a sample chamber. A copper foil was attached to the Cu cold finger by a ring clamp, and the targets were held in place on the foil using Teflon™ tape. This chamber fits into a Dewar and the entire apparatus was mounted in a frame designed to facilitate target positioning. *See* Fig. 1. Target positioning was reproducible within 1–4%, depending on the size of the target. Each target was held in position, under vacuum, and cooled to liquid nitrogen temperatures. Hydrogen sensitivities were measured at the time the target was placed in the beam under vacuum, again immediately after the addition of liquid nitrogen, and every 2 min as the target cooled. The sensitivities increased on cooling for all targets measured, and for most of these, constant sensitivity was reached in about 20 min. After constant sensitivity was reached, more liquid nitrogen was added to the Dewar, and the sensitivity was measured every 2 min again for an additional 5–10 min.

All irradiations were done using the CNPGAA apparatus at the NIST's Cold Neutron Research Facility at a reactor power of 15 MW, corresponding to a fluence rate at the target position that is equivalent to about 10^8 thermal neutrons/$cm^2 \cdot s$. γ-Radiations were collected over the range of 150 to about 8000 keV using an intrinsic germanium detector (of 25% efficiency relative to a 3″ × 3″ NaI crystal, with a full width half maximum of 2.2 keV for the 1333 keV line of ^{60}Co), connected to an analog-to-digital converter, which is in turn coupled to a multichannel pulse height analyzer. Spectral data were acquired using a Nuclear Data (ND) acquisition interface module connected to an ND9900 work station. Data reduction was accomplished using ND software programs PEAK and PILEUP.

Fig. 1. Target chamber consisting of cryostat with copper cold finger extended with copper foil, an aluminum cap (shown on the side), a Dewar for liquid nitrogen, and a frame for secure positioning on the optical guide rail below the neutron beam.

RESULTS AND DISCUSSION

Cooling the targets results in an increase in sensitivity for all targets measured. In Fig. 2, sensitivities for H in THAM disks and spheres are plotted as a function of sample mass for both room temperature and cooled targets. For disks, sensitivity increases range from 10 to 25% and for spheres, from 4 to 25%. Sensitivities for disks show the same qualitative trends for targets at 300 and 77 K, but with a smaller overall sensitivity decrease as a function of disk thickness for the 77 K disks. Sensitivities for H in the 300 K spheres decrease with sphere size but those at 77 K are constant within the uncertainties associated with counting statistics (0.5–2.5%) and target positioning in the beam.

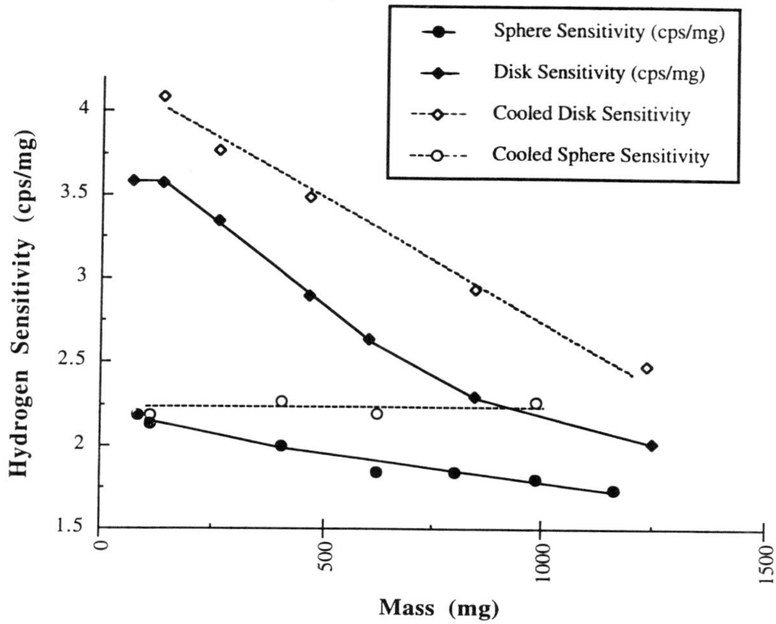

Fig. 2. Cold-neutron prompt γ-ray activation analysis sensitivities for H measured in disks (◇) and spheres (○) of THAM. Open symbols represent 77 K targets and filled symbols, 300 K.

These results are consistent with the hypothesis that cooling the targets eliminates the effects of neutron warming. Constant sensitivity for 77 K spheres of varying diameters indicates that there is no measurable change in the average energy of the neutrons, because spheres should be least affected by elastic scattering. Sensitivities for 77 K disks decrease with increasing thickness, but not to the extent that those for the 300 K disks decrease. Cold neutrons interacting within a cold target show results that are similar to results for thermal neutrons interacting within a 300 K target (described previously). The entire effect for disks is not eliminated by controlling the target temperature because neutron scattering, even without energy change, alters the mean free path of the neutron within the target and, therefore, alters the probability for neutron absorption. However, that part of the sensitivity decrease that is caused by neutron warming is at the very least mitigated, and the remainder of the effect that is caused by scattering in which the average neutron energy is not affected can be studied using Monte Carlo methods.

Future research will include the study of the temperature dependence for other hydrogenous materials, the study of the temperature dependence for nonhydrogenous materials, and the use of a cryostat with temperature control so that other constant temperatures can be achieved for the determination of an optimum temperature.

ACKNOWLEDGMENTS

I gratefully acknowledge the cooperation of the NIST reactor staff during these experiments. I also thank George Lamaze for his advice and interest in this work, and for providing the materials necessary for the target chamber.

REFERENCES

1. J. R. D. Copley and C. A. Stone, *Nucl. Instrum. Methods* **A241,** 593 (1989).
2. E. A. Mackey, D. L. Anderson, R. M. Lindstrom, and G. E. Gordon, *Anal. Chem.* **63,** 288 (1991).
3. E. A. Mackey, G. E. Gordon, R. M. Lindstrom, and D. L. Anderson., *Anal. Chem.* **64,** 2366 (1992).
4. E. A. Mackey and J. R. D. Copley, *J. Radioanal. Nucl. Chem. Articles* **167,** 127 (1993).
5. R. L. Paul and E. A. Mackey, submitted to *J. Radioanal. Nucl. Chem. Articles* Sept. 1993.

Use of ^{191}Pt Radiotracer for the Development of Enrichment Procedures to Detect Natural Levels of Platinum in Biological and Environmental Materials

M. PARENT,*[,1] R. CORNELIS,[1] F. ALT,[2] K. STRIJCKMANS,[1] AND R. DAMS[1]

[1]*Laboratory of Analytical Chemistry, Institute for Nuclear Sciences, Universiteit Gent, Proeftuinstraat 86, B-9000 Gent, Belgium; and* [2]*Institut für Spektrochemie und Angewandte Spektroskopie, Dortmund, Germany*

ABSTRACT

The ^{191}Pt-radiotracer is a powerful tool to develop separation and preconcentration methods for Pt. The radiotracer was produced either through ^{190}Pt (n,γ) ^{191}Pt reaction on Pt enriched in ^{190}Pt, or through ^{191}Ir (p,n) ^{191}Pt reaction on natural Ir, followed by Pt/Ir separation. The solvent extraction behavior of ^{191}Pt(IV) from HCl with tributylphosphate, rubeanic acid in tributylphosphate and thenoyltrifluoroacetone in *n*-butylalcohol/acetophenone, followed by back-extraction with 2 mol/L NH_4OH was studied. The overall recovery from a biological matrix is up to 80%. Trace amounts of ^{191}Pt were also preconcentrated as *bis*-(carboxymethyl) dithiocarbamate chelates on an XAD-4 microcolumn and eluted with EtOH/1 mol/L HNO_3. More than 80% is recovered in 2 mL or about 70% in 300 μL.

Index Entries: Platinum; isolation; production of ^{191}Pt; Pt extraction; biological samples.

*Author to whom all correspondence and reprint requests should be addressed.

 © 1994 by Humana Press Inc.

INTRODUCTION

Since the introduction of automobile exhaust catalytic cleaning, platinum has become a new environmental pollutant *(1)*. This potentially new threat to human health has initiated research to develop analytical methods for Pt detection at very low levels (pg/g). Since the usual analytical techniques, such as graphite furnace atomic absorption spectrometry (GFAAS) *(2,3)* and inductively coupled plasma-mass spectrometry (ICP-MS) *(3,4)* do not allow such low detection limits, a preconcentration of Pt is mandatory. To develop preconcentration procedures, radiotracer studies are very helpful because of the ease of detection.

PRODUCTION OF ^{191}PT RADIOTRACER *(5)*

The choice of an adequate Pt radiotracer depends on requirements concerning suitable half-life and γ-energy. ^{191}Pt (with $t_{1/2}$ = 2.80 d and E_γ = 539 keV) is considered the best choice.

Two methods are used for the production of ^{191}Pt with high specific activity and radionuclidic purity: first via the ^{190}Pt (n,γ) ^{191}Pt reaction by thermal neutron irradiation of a Pt target, enriched in ^{190}Pt (4.19%), in a nuclear reactor with high neutron fluence rate (2.10^{14} $cm^{-2}s^{-1}$) during 4 d, and second via the ^{191}Ir (p,n) ^{191}Pt reaction by proton irradiation of natural Ir with a cyclotron. About 220 mg of $(NH_4)_2IrCl_6$ (<1 µg/g Pt) were irradiated with 14-MeV protons for 1 h with a 0.5-µA beam. In the latter case, a time-consuming liquid–liquid extraction with dithizone in CCl_4 is applied to separate Pt from the Ir matrix. The proton irradiation of Ir provides optimum radionuclidic purity. With a suitable proton energy, the impurities ^{190}Ir and ^{192}Ir can be suppressed, and only ^{191}Pt activity is observed with HPGe γ-spectrometry. The chemical purity of the ^{191}Pt tracer through neutron irradiation of Pt is, however, far superior. The radionuclide isolated from the $(NH_4)_2IrCl_6$ is contaminated with Ir (from target material) and Sn (reducing agent for extraction). The experimentally determined specific activity is comparable for both procedures (±1 MBq/µg Pt), despite losses during the Pt/Ir separation. For practical reasons, the proton production was chosen.

ENRICHMENT PROCEDURES

Extraction of Pt(IV) (6)

After digestion of biological materials and subsequent treatment with aqua regia and HCl, Pt is present as Pt(IV) (H_2PtCl_6). It is preferable to extract Pt(IV), without addition of reducing agents, which are liable to coextract and would interfere later with the GFAAS or ICP-MS measurement of Pt. Three media were investigated: rubeanic acid (systematic name: dithiooxamide) in tributylphosphate (TBP) *(7)*, tributylphosphate *(8)*, and thenoyltrifluoroacetone (TTA) *(9)* in *n*-butylalcohol/acetophenone, all of them out of HCl. Each experiment, consisting of 10 mL of both the aqueous and organic phase, was performed at least in threefold, and included the ^{191}Pt measurements of both the organic and aqueous phases to calculate the total recovery or mass balance. The influence of the acidity of the aqueous phase and shaking time were investigated for the three media. For TBP and TBP + rubeanic acid, 3 mol/L HCl are considered the optimal acidity because of the high Pt extraction yield, whereas 4 mol/L HCl give the best Pt recovery for TTA extraction. Variation of the shaking time from 5 min to 1 h showed no influence on the ^{191}Pt extraction yields (90% for TBP, 85% for TBP + rubeanic acid, and 75% for TTA). In all cases, the total recovery of ^{191}Pt over both phases appears to be about 100%. The optimum TTA/acetophenone ratio was determined to be 3/7. The influence of the Pt-carrier concentration (0.01, 0.1, and 1 μg/L) on the extraction yield was also investigated. The yields varied between comparable boundaries 90–95% for TBP extraction and 78–88% for TTA.

Since the variety of analytical detection methods requires aqueous solutions (e.g., ICP-MS), Pt must be "stripped" from the organic phase. First back-extraction with oxidizing acids in order to decompose the complex was tried. A 1.4 mol/L HNO_3 and a 2% (v/v) $HClO_4$ solution were used. Since the back-extractions were not quantitative, an alkaline medium was chosen. NaOH has given good results, but is not suitable for ICP-MS, which will be the ultimate detection technique, owing to the introduction of unwanted Na^+ ions. NH_4OH, on the contrary, does not introduce elements that interfere with the Pt measurement. The concentration of NH_4OH (1–4 mol/L) does not influence the back-extraction. A quantitative recovery is not mandatory when isotope dilution mass spectrometry (ID-ICP-MS) is applied, since losses of Pt during separation do not change the isotopic ratios.

To check whether the recoveries remain the same in the presence of a matrix, soil, liver, and plant materials were spiked with ^{191}Pt solution, mineralized, and extracted. The overall recovery of Pt after this procedure was examined. The results of the extraction and back-extraction for the three media and the three matrices are given in Table 1. Each extraction and back-extraction consisted of two repetitions. As can be seen from Table 1, the overall recovery of an extraction from TBP and back-extraction with 2 mol/L NH_4OH yielded 75–81%, except for soil. In the case of TTA, the overall recovery is less satisfactory. The recoveries are similar to those obtained without matrix, and although they are not quantitative, these methods are still useful to isolate Pt when subsequent measurement with ID-ICP-MS is performed. Because of the poor back-extraction of Pt from TBP + rubeanic acid, this medium is hardly suitable for the determination of Pt at very low levels. In this extraction, Pt(IV) is reduced to Pt(II) by rubeanic acid, and the much more stable Pt(II) complex is extracted to TBP. The high stability of the Pt(II) complex hinders the back-extraction into NH_4OH.

Table 1
Recovery of ^{191}Pt After Extraction with TBP and TBP + Rubeanic Acid from 3 mol/L HCl, and TTA from 4*M* HCl, Followed by Back-Extraction with 2 mol/L NH_4OH for Different Matrices

matrix	medium	% Pt recovery ± sd		
		extraction	back-extraction	total
liver	TBP	83.6 ± 5.4	96.6 ± 5.4	81.0 ± 9.4
	TTA	69.4 ± 5.6	66.7 ± 1.1	58.0 ± 2.4
	TBP + rub.	86.4 ± 2.1	67.2 ± 4.4	58.0 ± 2.4
plant	TBP	89.7 ± 0.4	83.8 ± 2.2	75.2 ± 1.7
	TTA	87.2 ± 3.2	80.7 ± 1.2	70.4 ± 1.7
	TBP + rub.	86.7 ± 1.3	47.1 ± 7.5	40.7 ± 5.9
soil	TBP	29.9 ± 12.5	87.9 ± 13.3	26.9 ± 13.2
	TTA	70.6 ± 2.2	89.5 ± 4.8	63.2 ± 4.5
	TBP + rub.	92.0 ± 0.8	13.7 ± 5.9	12.3 ± 5.4

Enrichment of Pt with Adsorption Chromatography

Usually anion-exchange chromatography resins do not adsorb negatively charged Pt complexes selectively, whereas some of the complexes are adsorbed too strongly, so elution with strong acids or large volumes are necessary to elute the complex from the resin. In adsorption chromatography, the selectivity can be enhanced by a good choice of chelate. In addition, the complexes are not bound chemically to the resin, so a rapid and complete elution can be expected. Pt is known to form complexes with *S*- or *N*-containing chelates. Many of them, however, are not soluble in an aqueous medium, except for *bis*-(carboxymethyl)dithiocarbamate (CMDTC), which was therefore investigated as complexing agent. The CMDTC was synthesized according to Plantz et al. *(10)*. Pt was preconcentrated as CMDTC-chelate, formed by adding solid CMDTC (1.2% [w/v]) to the sample solution *(11)*. This solution consisted of 1 mol/L H_2SO_4 or 1 mol/L HCl to which 0.3% (w/v) KI and 0.03% (w/v) ascorbic acid, or 0.3% (w/v) $SnCl_2$, respectively, was added as reducing agent. After completion of the reaction, the sample solution was pumped with a flow rate of 0.5 mL/min through a 126-µL microcolumn

(4 mm id, 1 cm length) packed with XAD-4 (50–150 μm) as a hydrophobic sorbent. Preliminary experiments showed that the resin had to be activated with EtOH/1 mol/L HNO_3 before adsorption occured. In addition, it is preferable to have a concentration >1 mol/L HCl to avoid precipitation of Sn when adding CMDTC to the sample solution. A retention of more than 90% was observed for both procedures. The adsorbed complexes were eluted in an opposite direction (0.2 mL/min) with different eluting solutions, namely, 1.5 mol/L NH_4OH, 1 mol/L NaOH, CCl_4, 1.5 mol/L HNO_3, or 90% EtOH/1 mol/L HNO_3. The best Pt recoveries were obtained with NaOH (yield ± 90%) and EtOH (yield ± 90%) followed by NH_4OH (yield ± 85%). An elution diagram was recorded to check in which fraction of the eluate most of the Pt was present. As can be seen in Fig. 1, 70–80% recovery was concentrated in 300 μL of eluent (EtOH/1 mol/L HNO_3). Also the influence of the NH_4OH concentration (1.5, 3, and 6 mol/L) was checked, because this could enhance the elution of Pt. No significant influence could be observed. In general, EtOH elution resulted in better recoveries than NH_4OH. Since measurement of a 90% EtOH solution with ICP-MS requires the addition of O_2 to the carrier gas and 10% EtOH does not, the influence of the EtOH concentration of the eluate on the Pt yield was also investigated. The concentrations were varied from 10 up to 90% EtOH. Good recoveries in 2 mL eluate were found for 40–90% EtOH solution, though the fraction where most of the Pt was present shifted slightly to higher elution volumes.

Since biological matrices must be digested with oxidizing acids, the influence of $HClO_4$ on the adsorption and elution of Pt was investigated. Addition of 5, 10, and 20% (v/v) $HClO_4$ causes only a slight decrease in recovery. When using Kl as a reducing agent just after the destruction, one must, however, be aware of the formation of insoluble perchlorates that may plug the column.

This procedure was investigated in the presence of a liver matrix. Although the recovery decreased, still 56–74% Pt have been eluted in a 300-μL fraction. Because the Pt yield is variable, this enrichment method is only suitable when ID-ICP-MS is performed.

The procedure was not only checked for its efficiency to enrich Pt, but also for its effectiveness to eliminate concomitant major and trace elements. ICP-MS measurement of the liver digest and the nonradioactive Pt eluent showed that the amounts of Ca, Fe, Mg, Mn, Na, and P were reduced to <1%. About 50% of the Cu and Mo was still present, as well as important amounts of the reducing agents Sn or I.

CONCLUSION

Both ^{191}Pt production methods, ^{190}Pt (n,γ) ^{191}Pt and ^{191}Ir (p,n) ^{191}Pt, have advantages and drawbacks. Neutron irradiation provides better chemical purity, whereas proton irradiation offers best radionuclidic purity. In the first case, the experimental conditions (enriched ^{190}Pt tar-

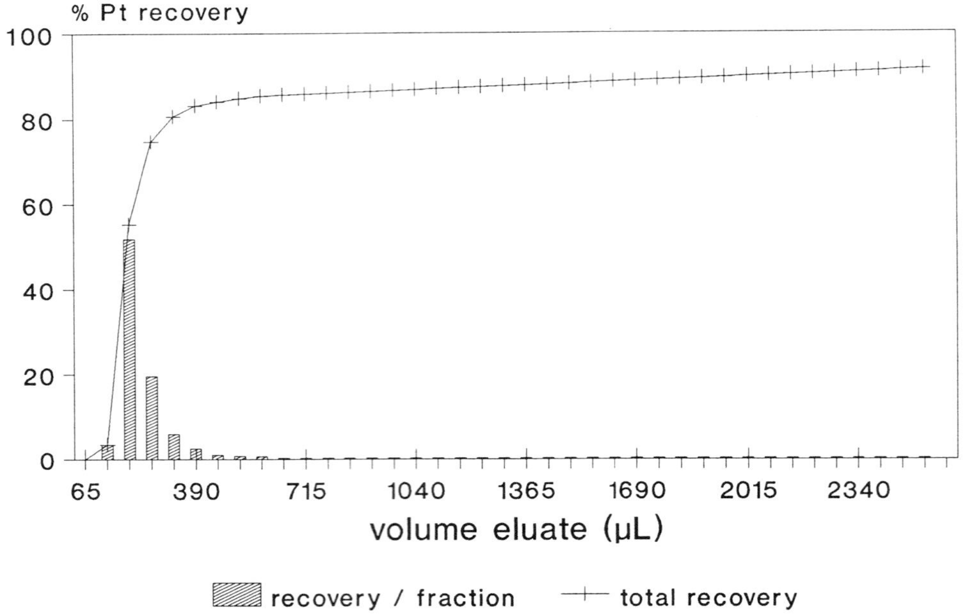

Fig. 1. Elution of ^{191}Pt–CMDTC complex from a 126-µL XAD-4 column with 2 mL 90% EtOH/1 mol/L HNO_3 at a rate of 0.2 mL/min.

get material, 4-d irradiation time, and 2.10^{14} $cm^{-2}{\cdot}s^{-1}$ neutron fluence rate) may be more demanding. The availability of high fluence rate reactors is limited, whereas most of the cyclotrons can accelerate protons in the energy range of interest.

For the development of enrichment procedures, two ways were explored: liquid–liquid extraction and adsorption chromatography. In the case of Pt(IV) extraction, the overall recovery for Pt increases when extraction and back-extraction are performed twice. The extraction with TBP followed by back-extraction with 2 mol/L NH_4OH gives recoveries of almost 90% for solutions with and without matrix. The molarity (1–4 mol/L) of NH_4OH does not seem to influence the back-extraction from TBP. The procedure with TBP + rubeanic acid extraction is hardly suitable for the determination of Pt at trace or ultratrace levels because of the very poor back-extraction with NH_4OH. Residues of HF or $HClO_4$ used for a digestion hinder the extraction of Pt with TBP.

The best enrichment is achieved by adsorption of Pt as Pt–CMDTC complex on a microcolumn packed with XAD-4, followed by an elution with EtOH. It yields reproducible and acceptable recoveries (80–100% in 2 mL, 60–80% in 300 µL), and the separation from matrix elements that might interfere with the ICP-MS measurement is generally good.

Accurate measurement of Pt in samples after isolation through the extraction procedure or enrichment via adsorption chromatography, as described in this study, is only possible by isotope dilution mass spectrometry, since this method allows for correction of the variable yields.

ACKNOWLEDGMENTS

Grateful acknowledgments are made to the Institute for Scientific Research in Industry and Agriculture (IWONL), the Inter-University Institute for Nuclear Sciences (IIKW), and the German Ministry of Research and Technology for financial support. R. Cornelis is a research director of the National Fund for Scientific Research (Belgium) (NFWO).

REFERENCES

1. V. F. Hodge and M. O. Stallard, *Env. Sci. Technol.* **20,** 1058–1060 (1986).
2. F. Alt, U. Jerono, J. Messerschmidt, and G. Tölg, *Microchim. Acta* **III,** 299–304 (1988).
3. S. A. Wood and D. Vlassopoulos, *Anal. Chim. Acta* **229,** 227–238 (1990).
4. O. Nygren and G. T. Vaughan, *Anal. Chem.* **62,** 1637–1640 (1990).
5. M. Parent, K. Strijckmans, R. Cornelis, J. Dewaele, and R. Dams, *Nucl. Instr. Meth.* **B86,** 355–366 (1994).
6. M. Parent, R. Cornelis, R. Dams, and F. Alt, *Anal. Chim. Acta* **281,** 153–160 (1993).
7. K. Brajter and U. Kozicka, *Talanta* **26,** 417–419 (1979).
8. T. Ishimori, K. Watanabe, and E. Nakamura, *Bull. Chem. Soc. Japan.* **33,** 636–644 (1960).
9. A. K. Dé and M. S. Rahaman, *Analyst* **89,** 795–799 (1964).
10. M. R. Plantz, J. S. Fritz, F. G. Smith, and R. S. Houk, *Anal. Chem.* **61,** 149–153 (1989).
11. M. L. Lee, G. Tölg, E. Beinrohr, and P. Tschöpel, *Anal. Chim. Acta,* **272,** 193–203 (1993).

Determination of Cobalt in Biological Samples by Radiochemical Neutron Activation Analysis Employing Reversed-Phase Chromatography

S. SUBRAMANIAN* AND J. R. W. WOITTIEZ

Interfaculty Reactor Institute, Delft University of Technology, Mekelweg 15, 2629 JB Delft, The Netherlands

ABSTRACT

We report the final part of our "on-line yield" pilot program in this article. This last part deals with aspects of accuracy and precision of the determination of cobalt in biological materials by RNAA with on-line yield determination. To obtain an insight into the accuracy of the technique, certified reference materials were analyzed, and for selected materials, results obtained by INAA and RNAA of identical samples have been compared. It shows that the mineralization in our RNAA scheme may be a sample-related source of systematic errors that is not related to the presence or absence of HF. Finally, we present some critical notes on pros and cons of the practice of on-line yield concept and some ideas for further research.

Index Entries: cobalt; high-performance liquid chromatography; radiochemical separation; neutron activation analysis; UV absorbance.

INTRODUCTION

In our previous communications *(1,2)*, we introduced the concept of radiochemical neutron activation analysis with "on-line" chemical yield quantification. The project was initiated in 1990 with a pilot study to investigate the feasibility of the idea. It was chosen to perform investi-

*Author to whom all correspondence and reprint requests should be addressed.

Biological Trace Element Research Editor: G. N. Schrauzer © 1994 by Humana Press Inc.

gations of the applicability of this concept on two single-element assays, namely, Cr using ion chromatography and Co using reversed-phase HPLC. This article concludes our communications on the pilot program of cobalt.

The pilot program so far reported consisted of three phases:

1. Choice of the model system—we chose cobalt because of its good nuclear and chemical properties, like long half-life of the target analyte ^{60}Co and highly stable complexes;
2. Methodology of chromatographic separation—it was decided to perform reversed-phase HPLC of the cobalt Diethyldithiocarbamate complex; and
3. Validation of the concept by radiotracer experiments.

Here, the vital parameter to be checked was the constancy of the ratio between the UV and radioactive signals, independent of the chemical performance *(1)*.

Having studied the feasibility of the idea and having concluded that the concept should be, in principle, applicable to real samples, it remains to be proven that on-line yield RNAA is capable of obtaining the most accurate results possible. This article reports on our efforts to validate the concept for RNAA of Co in real samples. Furthermore, some main conclusions that could be drawn from the pilot program and directions of further research stemming from this will be discussed.

EXPERIMENTAL DESIGN

In order to demonstrate that the method generates accurate results when applied not only to tracer solutions (both academic and digests of biological samples), but also to real samples, two ways of accuracy control were chosen: (1) analysis of certified reference materials and (2) comparison of RNAA and INAA results for certified and candidate reference materials. Since the aim of the study was to check whether or not on-line yield RNAA generates accurate results, rather than investigating trace or ultratrace analytical chemistry of Co, it was decided to avoid analysis of these reference materials where counting statistics plays a major role in the final precision. Thus, materials with cobalt contents lower than 20 μg/kg were not analyzed.

The chromatographic mode of sample processing demands batchwise irradiation, with each batch comprising two samples, a quality-control sample (QCS) and a standard. The chromatographic sequence is QCS, sample 1, sample 2, and standard. Each concentration is calculated from the counting result and UV absorbance of the standard in the appropriate batch. For those batches where INAA and RNAA are performed, the instrumental analysis of each sample is performed prior to processing them, i.e., each sample is counted for 24 h in its quartz capsule before decomposing it for radiochemical processing. The QCS was used to estimate the precision of the technique for a longer period of time, which includes precision between batches.

EXPERIMENTAL

Preirradiation Sampling

The samples (around 200 mg) were packed in suprapure silica ampules and weighed accurately before sealing for irradiation. Cobalt standard solution was pipeted on a clean filter paper equal to an amount of 1 µg, dried, and packed in the ampule.

Irradiation

Test samples were irradiated at the institute's nuclear reactor. Each batch was irradiated for 12 h at a neutron flux of 4.3×10^{12} $cm^{-2} \cdot s^{-1}$ and allowed to decay for 3 wk before processing.

Postirradiation Chemistry

Mineralization

Irradiated ampules were crushed with an automated pneumatic press and transferred to an Erlenmeyer flask containing 100 µg of a Co(II) carrier solution. Destruction of organic matters in the sample and standards was performed in open vessels with H_2SO_4, H_2O_2, HNO_3, and $HClO_4$ until a clear solution was observed *(3)*. For botanical samples, HF was also used.

Complexation

Sodium diethyldithiocarbamate (NaDDTC) was added to the solution (pH 6.0) to form CoDDTC and other dithiocarbamates of minor elements occurring in the matrix. The complexes were extracted three times with 70/30 (v/v) chloroform/acetone. The organic phase collected together was evaporated and redissolved in a mobile phase *(3)* for reversed-phase chromatography.

Reversed-Phase Chromatography

Sample solution (1.0 mL) was injected into the HPLC system. The separation was performed on a reversed-phase semipreparative column with continuous UV monitoring at 254-nm wavelength using 65/35, methanol/acetate buffer *(2)*. The fraction containing 45 mL of ^{60}CoDDTC/CoDDTC was collected in a plastic container (diameter 3.5 cm). The quality-control sample and standard were processed identically as mentioned above.

γ-Ray Spectrometry

The processed samples were measured with a GeLi (Philips) detector whose relative detection efficiency is 12% with a resolution (FWHM) 2.2 keV for the 1332.5 keV γ-ray of ^{60}Co. This was coupled to an Ortec amplifier and a PC-based Nuclear Data "Accuspec"/"ASAP" 8096 channel analyzer. The samples and standards were counted directly on the top of the detector for the maximum counting efficiency, for 24 and 4 h, respectively, in identical geometries. The absolute yield of the whole procedure in terms of ^{60}Co counts was found to be 95 ± 2.9% (*N* = 2). This was determined by comparing the counts of chromatographically processed standards against an unprocessed one.

RESULTS AND DISCUSSION

Results are given for the precision, using the data of the QCS for a time interval of 16 mo, in which 14 batches were analyzed. Also discussed are results for CRMs and the data obtained by comparing INAA and RNAA results for CRMs.

NIST 1577A Bovine Liver, our quality-control sample, yielded an average result of 212 ± 17 μg/kg for a certified value of 210 ± 50 μg/kg (see Fig. 1). Since the purpose of having a quality-control sample is to check for any analytical variability or experimental errors, it can be inferred that considering the time span of more than a year and logistic variations that have occurred during this period, the results obtained are quite satisfactory serving their purpose. Selecting this material was beneficial because the large zinc and copper content in the matrix necessitates good HPLC separation of these DDTC extractable transition elements. The threat of UV interference from these components compels a perpetual need to maintain stringent control over the whole separation scheme.

The average values of cobalt concentrations for other CRMs selected and candidate RMs are given in Table 1. For Bowen's kale, we found 62.5 ± 3.8 μg/kg matching with the certified value of 63.2 ± 11 μg/kg. We suppose that despite a good separation of major components, the sample matrix does play a role in the quantification, since in this case, the zinc and copper content are relatively lower (factor 4× and 40×, respectively) compared to bovine liver, which is reflected in a significantly better precision.

Individual data of comparison between RNAA and INAA results for candidate reference materials are displayed in Table 2. The values are self-explanatory. The unity in ratio between RNAA and INAA results indicates the parallelism between the two modes used. We obtained 0.99 ± 0.11 for peach leaves, demonstrating excellent agreement between the two techniques, whereas we found lower results (0.87 ± 0.09) for apple leaves.

Table 3 displays additional information of NIST in both candidate leaf reference materials. It can be seen from the table that our values for INAA and those from NIST *(4)* are in excellent agreement (79 ± 4 vs 76 ± 2 and 102 ± 3 vs 101 ± 3 μg/kg, respectively). Our INAA results are deviating from our RNAA results for apple leaves, but not for peach leaves, implying that our RNAA scheme suffers from sample-related problems. We believe that insufficient mineralization of the leaf sample may have contributed to the large deviation from unity in the RNAA/INAA ratio.

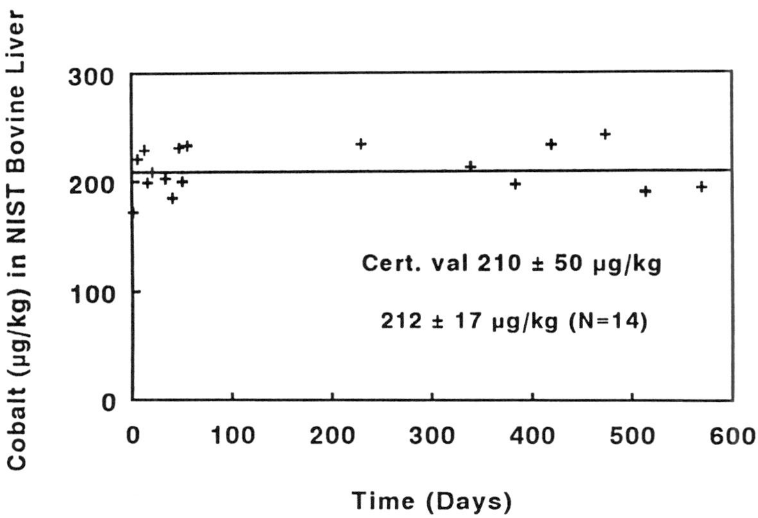

Fig. 1. Control sample over a time interval of 16 mo.

Table 1
Results of Cobalt Determination by RNAA in Reference Materials

Reference Materials	N	Yield Corr. results (µg/kg)	Certified values (µg/kg)
NIST SRMs			
1577A Bovine Liver	14	212±17	210±50
1515 Apple Leaves	8	89 ±9*	(90)
	8	102 ±3.4+	
1574 Peach Leaves	6	78.3±8.8*	(70)
	8	78.6±3.7+	
BCR CRM 274 Single Cell Protein	4	45±2.2	39±3
Bowen's Kale	3	62.5±3.8	63.2±11

Figures in parenthesis are NIST information values.
N = number of determinations.
+ = INAA.

In order to investigate difficulties in the mineralization step, we tried to analyze samples without HF in the acid destruction mixture. The absence of HF appeared to affect peach leaves results (61 ± 6 µg/kg), but not the apple leaves (91 ± 6 µg/kg). Thus, we concluded that the discrepancy between our RNAA and INAA results for apple leaves is not a silica-related effect. We also deduce that for peach leaves, the addition of

Table 2
A Comparison of Individual Results Between INAA and RNAA of NIST Candidate Reference Materials

INAA (μg/kg)	RNAA (μg/kg)	RNAA/INAA	Mean of ratio
Peach Leaves	Peach Leaves		Peach leaves
75.5	80.1	1.06	0.99±0.12
86.2	79.0	0.92	
79.8	63.5	0.80	
78.2	85.6	1.09	
73.7	87.9	1.19	
80.0	74.1	0.93	
78.1	81.6	1.04	
80.7	73.7	0.91	
Apple Leaves	Apple Leaves		Apple Leaves
102.0	102	1.00	0.87±0.09
104.2	78	0.75	
100.0	87	0.87	
108.0	93	0.86	
104.7	84	0.80	
101.5	102	1.00	
98.3	84	0.85	
98.0	80	0.82	

Table 3
Summary of Results for NIST 1515 Apple Leaves and 1574 Peach Leaves

	Peach Leaves (μg/kg)	Apple Leaves (μg/kg)
INAA IRI	79±4 (N=8)	102 ±3 (N=8)
RNAA IRI	78±9 (N=8)	89±9 (N=8)
Info. NIST	70	90
INAA NIST	76±2	101±3
RNAA IRI (No HF)	61±6 (N=6)	91±6 (N=4)

HF is vital for complete destruction. We conclude from the results for QCS, CRMs, and candidate RMs that the idea of on-line yield determination, as applied to cobalt determination by RNAA, is being demonstrated to be sufficiently strong to be further developed and applied for more challenging analytical problems.

CONCLUSIONS AND FURTHER RESEARCH

The HPLC approach to RNAA is relatively new with many areas still needing development. The method proves that the additional analytical step of yield determination is made obsolete with the handy use of a UV detector. The pilot program on application of on-line yield determination to RNAA of both Co and Cr in biological materials has, as intended, resulted in theoretical and practical suggestions for future research. Some advantages and disadvantages are discussed here.

Advantages

The yield of the analyte need not be high, reproducible, or presumptive, for it is actually determined in each sample and for each element. On-line UV spectrometry is sufficiently accurate to quantify the metal–ligand complex of the carrier. No second off-line yield determination step is needed, since UV spectrometry is an integral part of HPLC. This avoids the analysis of carrier in an analytically spoken, nonidentical sample compared to the one used for γ spectrometry (e.g., gravimetry, reactivation, and so forth). Finally, a direct visual impression of the separation performance is obtained by the on-line signal.

Disadvantages

The main problem of the on-line yield application to RNAA is the interfacing of acid digestion and reversed-phase HPLC. In the pilot program, we chose liquid–liquid extraction followed by injection of the volume-reduced extract to the semipreparative column. This step appears to be a major limitation of the technique. At present, we are investigating the possibilities of selective solid-phase extraction and subsequent elution of the element as a complex on an analytical column. Here, several types of chromatography, such as hydrophobic interactive chromatography, ion-pair reversed-phase HPLC, ion chromatography, and size-exclusion chromatography, are considered.

ACKNOWLEDGMENT

The authors thankfully acknowledge the valuable comments and suggestions of Prof. Dr. Ir. J. J. M. De Goeij during the development and execution of this project.

REFERENCES

1. J. R. W. Woittiez, S. Subramanian, K. J. Volkers, and J. J. M. De Goeij, *J. Radioanal. Nucl. Chem., Articles* **168** 5–14 (1993).
2. S. Subramanian and J. R. W. Woittiez, *J. Liquid Chromatogr.* **16,** 553–572 (1993).
3. J. R. W. Woittiez, A. Lagerwaard, M. D. C. Tangonan and K. J. Volkers, *J. Radioanal. Nucl. Chem., Articles* **169,** 229–237 (1993).
4. D. A. Becker, *J. Radioanal. Nucl. Chem., Articles* **168,** 169–183 (1993).

Advanced Short-Lived Nuclide NAA with Application in the Life Sciences

N. N. PAPADOPOULOS*,[1] AND N. F. TSAGAS[2]

[1]*NCSR Demokritos, 153 10 Ag. Paraskevi, Athens, Greece; and*
[2]*Democritus University of Thrace, 671 00 Xanthi, Greece*

ABSTRACT

A new technique for short-lived nuclide activation analysis has been developed that compensates the rapid radioactive decay during the counting period by simultaneous approach of the sample holder to the detector with a mechanical device, permitting prolongation of the counting time and reduction of the required complementary cyclic activation to avoid sample container damage. The operation of the analytical system is automated by a programmable logic controller (PLC). This improvement of short-lived nuclide activation analysis, providing a high throughput, is important in biological and environmental research, where often a large number of samples has to be analyzed for sufficient sampling statistics.

Index Entries: Short-lived nuclide NAA; radioactive decay compensation; sample movement; automation; high throughput.

INTRODUCTION

Neutron activation analysis has already been established as a method especially suitable for the determination of trace and minor elements in various materials, such as biological material where the available sample quantities are often limited and, thus, the sensitivity of the method should be high, whereas the number of samples is often large, necessitating a high analytical throughput *(1)*. Proper sampling and sample preparation are important in avoiding contamination because of the small quantities involved. Long-time activation, waiting for decay, and counting allow the analytical laboratory to be far from the neutron acti-

*Author to whom all correspondence and reprint requests should be addressed.

vation facility, which is usually a reactor or an accelerator, and reduce matrix interferences from short- and medium-lived nuclides, but are time-consuming and allow only long-lived nuclide analysis *(2)*. Short-time measurements increase the analytical throughput and broaden the analytical range, since then also short- and medium-lived nuclides can be determined. The long-lived nuclide interference can be reduced by shortening the activation time to avoid their building up or by enhancing the epithermal neutron flux with a cadmium screen to reduce their neutron cross-section. The analytical system for short-lived nuclide activation has already been described elsewhere *(3)*. However, short-time measurements have to be performed close to the site of the neutron activation source, such as the reactor, and provide low counting statistics for the elements determined, since the counting time has to be short and the count rate has to be between certain limits to avoid dead-time losses, pulse pile-up effects, and background interferences at extremely high and low count rates, respectively. In order to overcome the problems, in addition to the use of the instrumental Loss-Free-Counting System developed at the Atominstitut in Vienna *(4)*, a new technique has been developed to prolong the counting period and, thus, to improve the statistics by compensation of the radioactive decay *(5,6)*.

THE NEW TECHNIQUE

The rapid radioactive decay and count rate decrease during the counting period has been compensated for by approaching the radioactive sample holder to the detector during this period, increasing the counting efficiency by changing the geometry factor. Since the radioactive decay is exponential, whereas the geometry factor is inversely proportional to the square of the distance between the radioactive sample and the detector, the compensation by linear sample movement will not be complete. This does not matter, because the count rate does not have to be constant, but only between certain limits to avoid dead-time losses, pulse pile-up effects, and background interferences at extremely high and low count rates, respectively, but the measurements should be repeatable for comparison with the calibration standards. By reducing the count rate decrease, as is obvious and has been tested, the counting time can be prolonged considerably, practically at least four times, depending on the half-life of the nuclides. Thus, the complementary cyclic activation can also be reduced to a few cycles to obtain sufficient counting statistics for good results. By reducing the number of cycles, the damage of plastic sample containers can be avoided, since they cannot withstand much irradiation and many shocks from the pneumatic tube terminals at the reactor core. Several techniques have been tested to achieve the proper slow movement of the sample holder for a radioactive decay compensation. Tests have shown that the use of a piston, driven by controlled air or even oil pressure, is not adequate for such low speeds. Therefore, a mechanical system, consisting of a slow motor with variable frequency, a

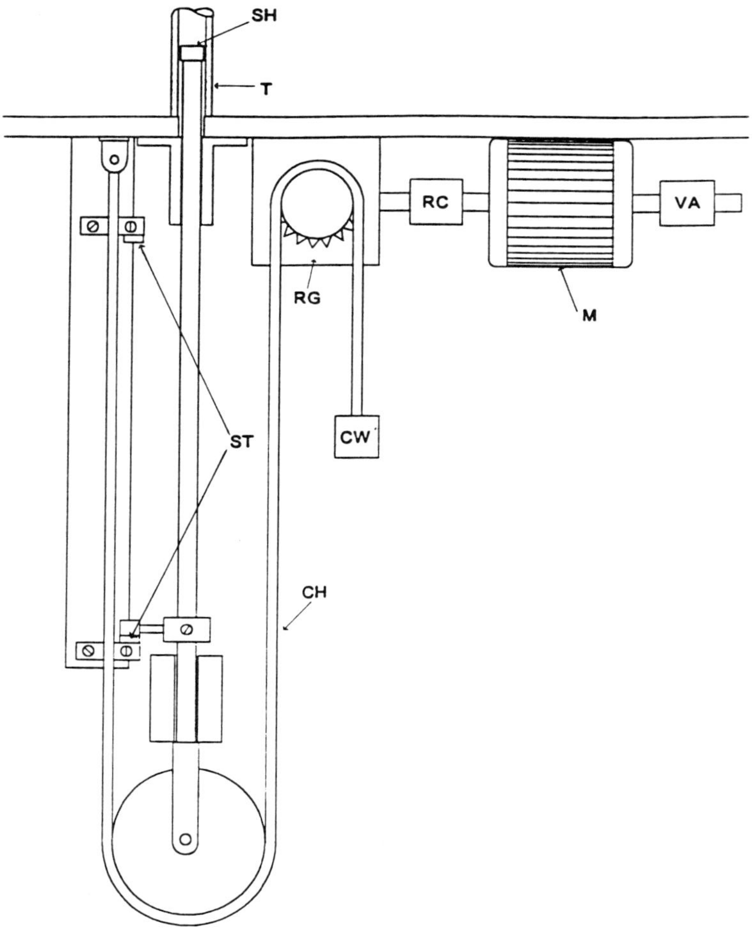

Fig. 1. System of sample holder movement at counting position. C = chain, M = motor, RC = revolution counter, RG = reduction gear, SH = sample holder, ST = stroke terminal, T = tube, and VA = variac.

reduction gear, and a rotary to linear movement converter (Fig. 1), has been designed, constructed, and tested to move the sample holder during the counting period more reliably. A slight observed timing discrepancy has to be checked.

In short-lived nuclide analysis with high throughputs, automation of the system is also important and adequate. For automatic operation of the nuclear analyzer (Fig. 2), a programmable logic controller (PLC) SIMATIC S5 has been installed with homemade main, counter and photocell interfaces (Fig. 3). The PLC, according to a preset time schedule, activates valves, pumps, diverters, timers, counters, and photocells of the analytical system for several operational commands, such as setting of

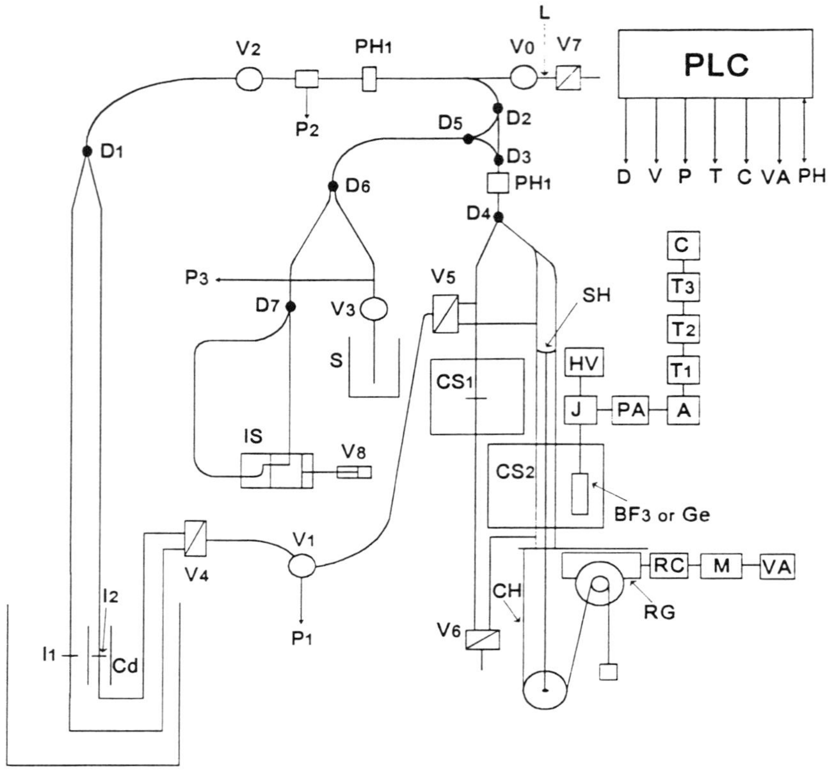

Fig. 2. Modified nuclear analyzer with movable sample holder at counting position and PLC. A = amplifier, BF_3 = BF_3 neutron detector, C = counter-timer, Cd = cadmium, CH = chain, CS = counting station, D = diverter, Ge = Ge γ detector, HV = high-voltage supply, I = irradiation position, IS = intermediate storage, J = junction box, L = sample loader, M = motor, P = pump, PA = preamplifier, PH = photocell, RC = revolution counter, RG = reduction gear, S = storage, SH = sample holder, T = timer, V = valve, and VA = variac.

initial conditions, sample mission, irradiation, receive, counting, and temporary and final storage for normal and cyclic operation. The proper operation of the PLC has been checked successfully.

RESULTS AND DISCUSSION

As mentioned before, some preliminary measurements have been performed to check the reliability of the new system. The automation of the analytical system facilitates the analysis, so that it can be performed even by nonspecialists with some supervision. Thus, laboratories far away from the irradiation facility can use the short-lived nuclide analytical laboratory, located close to the reactor or any other neutron activa-

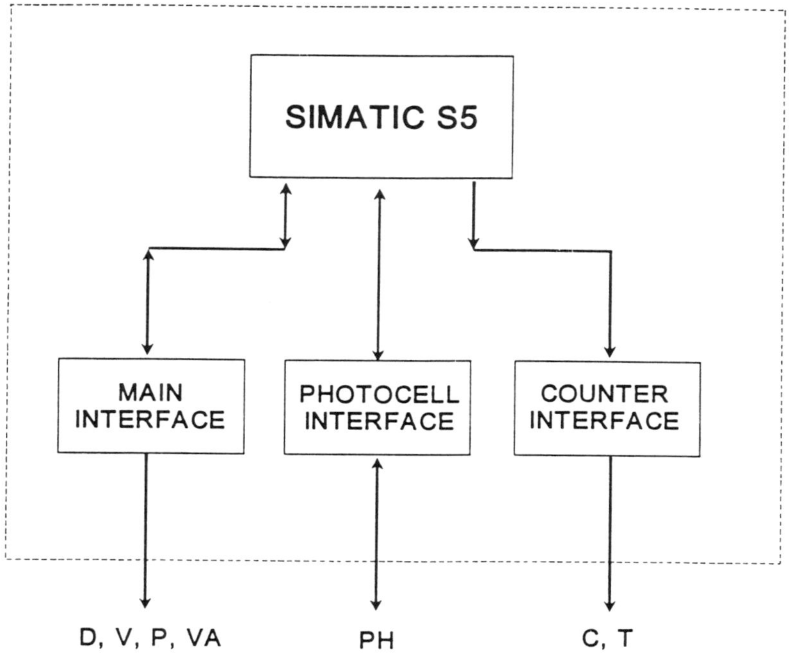

Fig. 3. PLC block diagram *(see* Fig. 2).

tion source, by just sending their samples to the facility to have the measurements made for them.

There are a number of essential trace and minor elements with short-lived radioisotopes, in addition to the medium- and long-lived ones, that are of interest in biological materials, such as F, Sc, Ge, Se, Rh, Ag, Sb, Er, Hf, W, and Pb, and Al, Ti, V, Cu, Rb, Rh, Pd, Gd, Dy, and Ir, with radioisotope half-lives in the second and a few minute range, respectively. An important element in biological material is selenium, whose short-lived $^{77}Se^{m}$ isotope has a half-life of 17.5 s, whereas its long-lived isotope ^{75}Se has a half-life of 120 d. It is obvious that the use of the short-lived nuclide will provide a much higher throughput, which is important in biological and environmental research, where often a large number of samples have to be analyzed for sufficient sampling statistics. Some biological (intercomparison) samples are being analyzed at our laboratory by means of the new proposed system in order to test the new irradiation/counting procedure of neutron activation analysis.

ACKNOWLEDGMENTS

The authors would like to thank G. Lefkopoulos, G. Hatzakis, C. Gatsos, J. Theodorakopoulos, the machine shop staff, A. Salevris, N. Malliaros, and the reactor staff for their contribution to the work.

REFERENCES

1. V. P. Guinn and J. Hoste, Element analysis of biological materials, IAEA Technical Report Series No. 197, Vienna (1980) 105–140.
2. N. N. Papadopoulos and S. Synetos, *Isotopenpraxis* **24,** 288–290 (1988).
3. N. N. Papadopoulos, *J. Radioanal. Chem.* **72,** 463–477 (1982).
4. G. P. Westphal, *Nucl. Instrum. Methods,* **B10/11,** 1047–1050 (1985).
5. N. N. Papadopoulos and N. F. Tsagas, *Analyst,* **119,** 613–616 (1994).
6. N. N. Papadopoulos, E. P. Efstathopoulos, and N. F. Tsagas, Rapid isotopic and elemental uranium analysis in safeguards materials by neutron activation, Proc. 15th ESARDA Symposium on Safeguards and Nuclear Material Management, Rome, Italy, May 11–13 1993, pp. 527–530.

Application of Nuclear Analytical Techniques Using Long-Life Sealed-Tube Neutron Generators

P. BACH, S. CLUZEAU, AND C. LAMBERMONT*

SODERN, 20 avenue Descartes, 94451 Limeil Brevannes, France

ABSTRACT

The new range of sealed-tube neutron generators developed by SODERN appears to be appropriate for the industrial environment. The main characteristics are the high emission stability during the very long lifetime of the tube, flexible pulsed mode capability, safety in operation with no radiation in "off" state, and the easy transportation of equipment. Some applications of the neutron generators, called GENIE, are considered: high-sensitivity measurement of transuranic elements in nuclear waste drums, bulk material analysis for process control, and determination of the airborne pollutants for environmental monitoring.

Index Entries: Neutron generators; sealed neutron tube; pulsed 14-MeV neutron activation; inelastic scattering; γ-ray spectra.

INTRODUCTION

Methods based on neutron activation analysis were originated by G. Hevesy and H. Levi 57 years ago. Up to now, the use of neutron beams for studying elemental composition has been limited mostly to research and a few industrial applications, such as oil logging. The main reason is the lack of transportable neutron generators adapted to the industrial environment: neutron sources that operate a long time without a cumbersome maintenance. If we consider all possible sources miniature reactors are still heavy and nontransportable. They have to comply with very severe regulations. Accelerators, in spite of recent improvements, have the reputation of needing highly qualified people and being difficult to

*Author to whom all correspondence and reprint requests should be addressed.

use. Isotopic neutron sources are small and reliable, but they cannot be turned off. They produce low neutron fluxes, and their transport and storage are strongly regulated. The ideal source would be a "turn-key" sealed generator, with a guaranteed long life, that produces its wanted neutron flux on demand and without delay and trouble.

Sealed-tube neutron generators are characterized by a very narrow neutron energy spectrum at 14 or 2.5 MeV, by a high yield (10^7–5×10^{11} n/s in 4 π sr), by modulation possibilities at frequencies up to 20 kHz. Neutron emission is delivered without any excessive radiological shielding owing to tritium hazard or to a residual activation of the installation.

Using the pulsed mode, more selective activations of tested samples are possible, and by means of synchronized detection techniques, prompt γ-ray spectra resulting from either inelastic, capture interactions or nuclear reactions can be used because of better signal-to-background ratios and more powerful analysis. Now, such neutron generators are available, and tests show a working time of thousands of hours, according to the tube type and the selected neutron output. These neutron generators were designed and realized in transportable versions by SODERN, under the name GENIE (*see* Fig. 1).

There are two main fields of industrial applications of these generators: industrial analysis mostly for nuclear waste drums measurements, on-line process control, ground logging, and airbore pollutants, and nondestructive testing by neutron radiography and by spectrometric analysis.

METHODS

Sealed-tube neutron generators use the following nuclear reactions:

$$D + T \rightarrow {}^4He\ (3.6\ MeV) + n\ (14.1\ MeV) \quad (1)$$

$$D + D \rightarrow {}^3He\ (0.8\ MeV) + n\ (2.45\ MeV) \quad (2)$$

$$D + D \rightarrow T\ (1\ MeV) + H\ (3\ MeV) \quad (3)$$

The first two fusion reactions (1) and (2) are very neutron prolific even at low ion energies. The third one (3), of equal probability as (2), does not produce neutrons. On this basis, two kinds of industrial generators were designed by SODERN *(1,2)*.

The first kind is based on reaction (1), which is the most efficient. In this configuration, a mixture of deuterium and tritium ions is accelerated at more than 100 kV into a tritiated coating target. Monoenergetic neutrons (14 MeV) are isotropically emitted. Neutron emission is controlled either in a continuous mode or in a pulsed mode at frequencies up to 20 kHz. These generators are advantageous for their high output and their high neutron energy. The working time of the sealed tubes is long. Thus, the TN 26 B SODERN tube has a lifetime on the order of thousands of

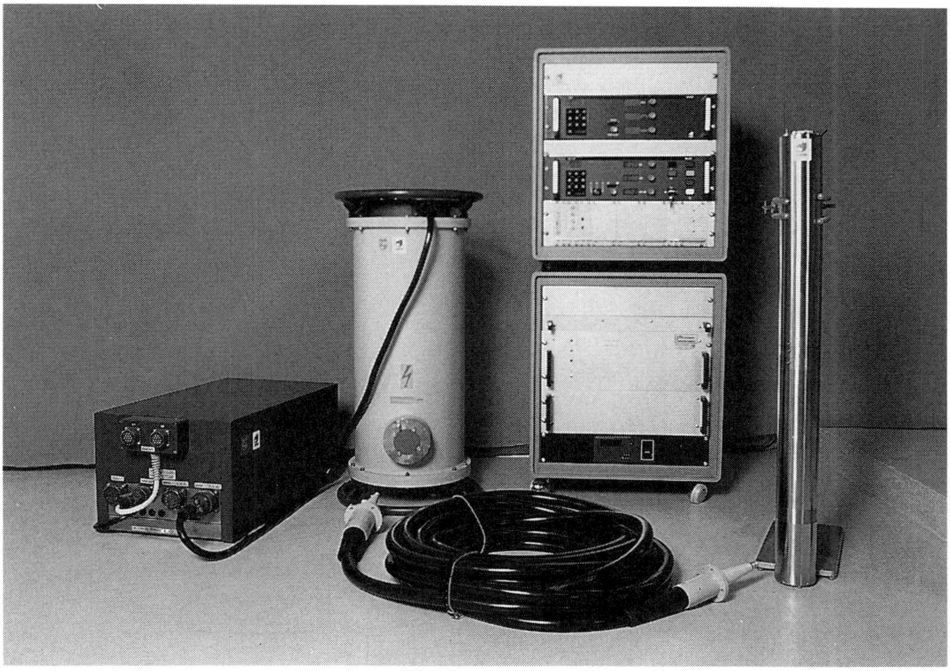

Fig. 1. The neutron generator GENIE 26 (>10^8 n/s).

Table 1
Neutron Emission of Some SODERN Generators

Generator type	Tube type	High voltage power supply	Neutron output (n/s)	
			D - T tube (14 MeV)	D-D tube (2,5 MeV)
GENIE 46	TN 46	225 kV	10^{10} to 4 x 10^{11}	4 x 10^7 to 10^9
GENIE 26	TN 26 B	125 kV	10^7 to 10^8	

hours for a continuous emission at 10^8 n/s, and the TN 46 a lifetime of more than 1500 h.

The second kind of generators is based on reaction (2). In this configuration, no radioactive material is used. The deuterium beam is accelerated at more than 200 kV into a deuterated coating target. Neutrons between 2.45 and 3 MeV are emitted, especially in the direction of the beam; neutron emission is 2–2.5 times higher in the ion-beam direction than in the perpendicular plane. These generators may be used when very high energy neutrons are not desirable. The maximum neutron output of these generators is limited to 10^9 *n*/s, because the cross-section for reaction (2) is lower than for reaction (1). Table 1 summarizes some characteristics of the SODERN tubes.

Nondestructive analytical methods, such as Instrumental Neutron Activation Analysis (INAA) are especially useful when using high-energy neutrons from pulsed sealed tubes. During one short pulse, high-energy neutrons are emitted and interact directly with samples, leading to elastic scattering, to inelastic scattering with prompt emission of characteristics γ-rays, and also to nuclear reactions (if any) with delayed emission of γ-rays or particles. After the pulse, neutrons are slowed down to thermal energy, and interact with materials: characteristic γ-rays from neutron capture are promptly emitted in a number of cases, and a delayed emission of γ-rays or particles can appear from low-energy activation. On fissile materials, low-energy neutrons induce fission neutrons of higher energy. By adjusting the pulse width, the pulse repetition rate, the detection gating, and the geometry of the sample irradiation, it is possible to separate the contributions of these mechanisms.

An example of this application is a generator used in a hidden explosives detection device *(3)*. A parcel containing an explosive is irradiated with 14-MeV neutrons in pulsed mode. During the emission (10–20 μs), a detector is gated synchronously in order to detect only the γ-ray spectra resulting from inelastic scattering events, for detection of carbon (4.43 MeV), nitrogen (5.106 and 2.31 MeV), and oxygen (6.13 MeV). During the blank between pulses, the detector is also gated: signals are directed to a second multichannel electronic, which gives mainly a γ-ray spectra from thermal neutron capture, for the detection of nitrogen (10.83 MeV), hydrogen (2.23 MeV), and so forth.

Neutron tubes are also very useful when used as a more conventional on/off neutron source. One example is the transportable neutron radiography system *(4)*. Another example is the activation of oxygen for detection of corrosion in steel industry products, which is very attractive when using 14-MeV neutrons.

A large number of applications may be considered (*see* Table 2). We will describe here three typical examples.

RESULTS

Nuclear Waste Management

Countries that produce radioactive waste have never, over recent years, ceased to upgrade the enforcement rules applied to their storage. The method founded on the fission process appears to be the only

Table 2
Some Examples of the Applications of Neutron Generators

Applications	**Neutron interactions used for detection**	**Neutron generator**	**Reference**
Nuclear waste management	Induced fission	GENIE 26 (14 MeV-pulsed)	CEA *(5,6,7)*
Bulk material analysis	Inelastic scattering Neutron capture Neutron activation	GNT 00 or GENIE 26 (14 MeV-pulsed)	PECHINEY LCPC *(8)*
Mineral logging	Neutron capture Neutron activation	GNT 00 or GENIE 26 (14 MeV-pulsed)	LCPC *(9)* COGEMA
Physical measurements	Calorimetry	GNT 00 (14 MeV-continuous)	LAPP *(10)*
Neutron radiography	Neutron transmission	GENIE 46 (14 MeV-continuous)	SODERN/ IABG/LTV *(4)*
Explosives detection	Inelastic scattering Neutron capture	GENIE 26 or 46 (14 MeV-pulsed) GENIE 46 (2.5 MeV-continuous)	CEA/ SODERN patent *(3)*
Disarmament verification	Neutron capture	GENIE 26 (14 MeV-pulsed)	ETCA/ SODERN

Table 3
Detection Limit of 239 Pu as Measured in Two Kinds
of 200-L Waste Drums *(6)*

Matrix	Mass	Maximum dispersion* of results	Detection limit for ^{239}Pu
Dismantling material	182 kg	± 49 %	4,5 mg
Graphite	137 kg	± 35 %	1,7 mg

* induced by the dispersion of fissile material location in the drum

method enabling the examination of waste containers with the required accuracy (in particular when high-density wastes are concerned, either in a heterogeneous matrix or in presence of γ-radiating products). This method is based on a measurement of transuranium elements contained in radioactive waste by irradiating the container with a series of neutron bursts and then analyzing the emitted radiation. A drum containing fissile material of unknown quantity and location is put in a thermalization chamber and exposed to neutron irradiation followed by detection of prompt fission neutrons.

Neutrons are supplied by a pulsed neutron generator providing short pulses of high amplitude. These neutrons are thermalized in the graphite and polyethylene chamber, and then trigger fission events are produced in the fissile material. Detection of high-energy prompt fission neutrons, delayed fission neutrons, spontaneous fission neutrons, and background events (*see* Fig. 2) allows measurement a fissile material quantity in the drum. An often-used generator (*5–7*) is the GENIE 26 or the GNT 02 unit. The TN 26 emits pulses of 10^{11} n/s during 80 µs, for example, at a frequency of 40 Hz. For a 10-min analysis and with 200-L drums, results in Table 3 show a strong dependence of the sensitivity on a matrix.

Raw Material in a Process Analysis

In the mineral industry, it is often necessary to have a good knowledge of the chemical composition of matter. Nuclear techniques, owing to the deep penetration of radiation, appear very attractive because no additional sampling is needed. For a rapid and high-accuracy determination of matrix elements in raw materials, pulsed neutron techniques appear very interesting.

In the case of cement production, an on-line analyzer using a neutron tube was studied *(8)* by Ciments Lafarge and LCPC (Laboratoire Central des Ponts et Chaussées) for the determination of the composition of raw materials: silicon, calcium, aluminum, and iron are detected in a dusty and often aggressive environment.

The neutron generator tube is controlled by an electronic system enabling irradiation of the material for 100 µs at 300-µs intervals of time. Two analog-to-digital converters are connected to the γ-ray detector and are controlled in the anticoincidence mode relative to the pulsed neutron emission.

The associated multichannel analyzer used is operated in two acquisition channels. The spectrum obtained on channel 1 corresponds to the activation and capture phenomena, whereas the spectrum obtained on channel 2 is representative of the activation phenomena exclusively. The difference between the spectra obtained on channels 1 and 2 thus enables only the spectrum corresponding to neutron capture to be obtained (*see* Fig. 3).

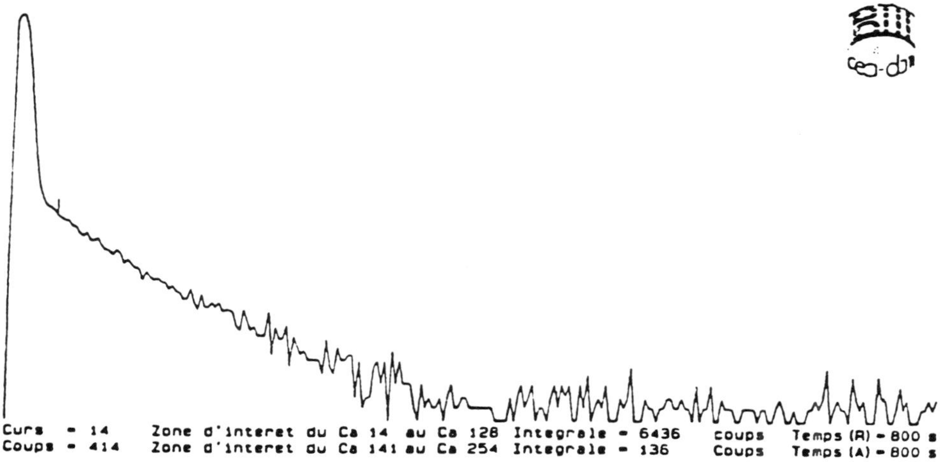

Fig. 2. Prompt and delayed neutron response vs time, as detected by helium 3 counters, in a transuranic waste measurement station (by courtesy of CEA/DAM/B III/SPR).

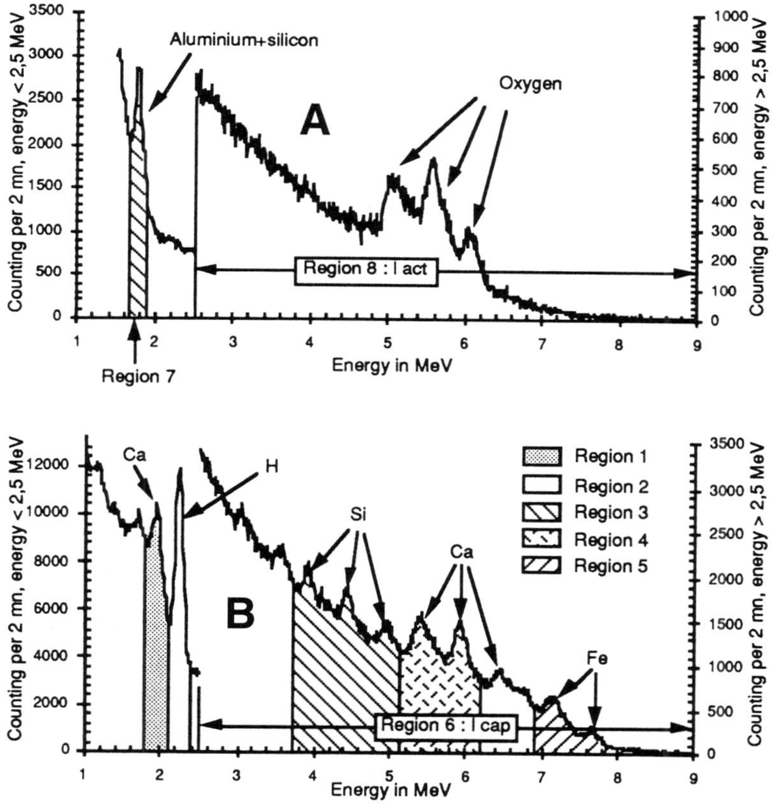

Fig. 3. Typical spectra obtained in the analyzing hopper with raw material (A: exclusive representation of neutron activation, B: exclusive representative of neutron capture [*8*]).

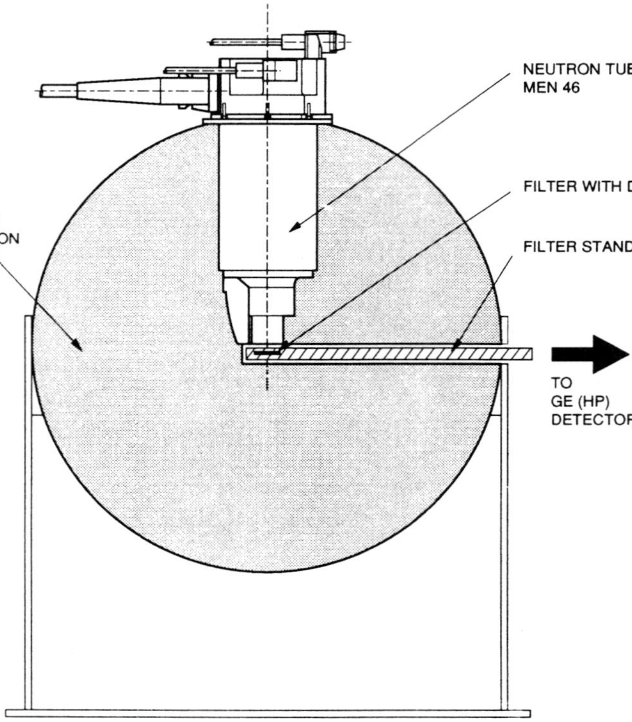

Fig. 4. Possible sketch for a neutron activation device.

Airborne Pollutants

Determination of elemental composition of atmospheric pollutants is rather important, especially with respect to toxic elements.

For example, stack gases usually contain 6–9% oxygen, 7–10% carbon dioxide, 85% nitrogen, and also some dust containing toxic elements.

A flexible method based on the use of a neutron generator for the activation of dust sampled on filters can easily be applied to the determination of a dust chemical composition. This method needs a thermal or fast-neutron yield of about 10^9 n·cm^{-2}·s^{-1} and an exposure time of 2 h (Fig. 4).

The neutron generator GENIE 46 is able to produce such neutron emission; it is possible to use simultaneously the fast-neutron activation to determine cadmium and mercury and the thermal neutrons for chlorine and arsenic.

The cycle "irradiation measurement" of 2 h is given as an illustration and could be improved if necessary. Moreover the equipment could be completely automated.

CONCLUSION

Thanks to the new neutron generators with sealed tube, a large number of applications may be considered, including those in industry, in nondestructive testing, in neutron irradiation for such treatments as dose deposition for hardening studies, in environmental pollution monitoring, and for neutron therapy. Neutron analytical methods could be often very advantageous. However, to get an optimized operating system, it is necessary to develop an appropriate method and to prove its feasibility.

REFERENCES

1. S. Cluzeau and J. R. Huriet, *Nucl. Instr. Methodes,* **B 56/57,** p. 911 (1991).
2. P. Bach, Sealed tube neutron generators for detection and analysis, Proc. Conf. on the Applications of Nuclear Techniques, Heraklio, June 1990, World Scientific edition.
3. CEA Patent No 2.588 969 (Feb. 26, 1988); CEA/SODERN Patent No 2 641 867, Jan. 13, 1989.
4. W. E. Dance, S. Cluzeau, and H. U. Mast, Integration of an advanced sealed tube neutron generator into a mobile neutron radiology system, 11th Int. Conf. on Applications of accelerators, Denton, TX, Nov. 8, 1990.
5. M. Bardy, P. Ruffet, and M. Cance, 9th Annual Symp. on Safeguards an Nuclear Material Management, London, May 12–14, 1987.
6. P. Ruffet and P. Choplin, Non Destructive Assay of Radioactive Waste Meeting, Cadarache, France, Nov. 20–22, 1989.
7. Y. Beroud, J. Romeyer-Dherbey, and R. Berne, CEA-SGN paper for Non Destructive Assay of Radioactive Waste Meeting, Cadarache, France, Nov. 1989.
8. J. P. Baron and L. Debray, Potential of Nuclear Techniques for On Line Bulk Analysis in the Mineral Industry, Int. Conf. Application of Nuclear Techniques, Heraklion, June 1990, World Scientific edition.
9. J. P. Baron, J. Cariou, and S. Cluzeau, Use of new neutron generator for *in situ* determination of elements, Conf. on Applications of Accelerators, Denton, TX, Nov. 1986.
10. B. Aubert, J. Colas, Ph. Ghez, J. C. Lacotte, B. Mours, I. Wingerter, and M. Yvert, Report NIM-A 279, 1989, p. 126.

Optimization of Irradiation Conditions for Photon Activation Analysis of Biological and Environmental Samples Using a High Power Electron Accelerator

H.-W. THÜMMEL,*[,1] C. SEGEBADE,[2] AND K. HIRSCH[3]

[1]*Klinik und Poliklinik für Nuklarmedizin, Universität Leipzig, D-04303 Leipzig, Permoserstr. 15, Germany;* [2]*Bundesanstalt für Materialforschung und-prüfung, Berlin, Germany; and* [3]*Umweltforschungszentrum, Leipzig, Germany*

ABSTRACT

Using the bremsstrahlung flux density data of Tanaka et al. (1983), the PHOTAC code has been developed to calculate the activity induced in a sample during photon activation as a function of the geometrical arrangement of the electron energy and the beam profile. Some results are given for sample volumes from 1 to 1000 cm^3 and sample distances from 1 to 10 cm at 25 MeV. Examples for the application of the PHOTAC code to estimate the maximum activity output and to minimize the heating up of the sample during irradiation are discussed.

Index Entries: Photon activation techniques; bremsstrahlung; electron beams; energy range 10–100 MeV; irradiation conditions; heat transfer; optimization; photonuclear reactions.

INTRODUCTION

Photon Activation Analysis (PAA) is well established for trace analysis *(1)* and radionuclide production. For high sensitivity of nondestructive photon activation analysis, especially at environmental or biological samples and high activity output, a very high power electron accelerator

*Author to whom all correspondence and reprint requests should be addressed.

is likely to be applied. Although when using PAA, the thermal load of the samples is lower compared to other activation techniques (e.g., neutron activation with a reactor), some serious heating problems may arise because about 30% of the beam power is transported by bremsstrahlung and partly absorbed in the sample, i.e., at the BAM* 35-MeV LINAC using an electron beam of 3 kW (30 MeV, 100 μA) about 1 kW bremsstrahlung power. Therefore, during such experiments, an intensive sample cooling is necessary, but this is—because of the low thermal conductivity of water and organic samples—more effective with smaller samples. On the contrary, a highly sensitive analysis without preconcentration of traces *(2)* should be done with rather large samples.

METHOD: THE PHOTAC CODE FOR CALCULATING THE ACTIVITY OUTPUT

For quantitative estimation of the activity output and sensitivity as a function of the irradiation conditions, the code PHOTAC *(3)* has been applied. Fig. 1. shows the irradiation geometry (and the terms used). PHOTAC takes into account the primary electron energy, *E*, beam profile (e.g., Gaussian with fwhm b), the sample distance a from the converter target, sample volume, *V*, the form of the (cylindrical) irradiation vessel (radius *r*, length *h*), and the sample composition. PHOTAC 01 is based on the Monte-Carlo data of Tanaka et al. *(4)* for the differential photon flux density produced by high-energy electrons (1–30 MeV), which yet has been fitted in a rather rough way ($1/k^2$ approximation) in the region of the Giant Resonance peak of the activation cross-section function.

PHOTAC has been confirmed recently by experiments at the BAM LINAC for 17, 25, and 30 MeV using potassium iodate solutions irradiated in glass vials of different volumes located at different distances and irradiated together with Ni and Cu monitors. Typical results obtained are given in Fig. 2. The function K(a,V,E)—an effective volume (dimension [cm^3])—is proportional to the activity induced and depends only on the geometrical parameters *(a, V)* and on the beam parameters *(E*, fwhm b). The saturation activity A of the (elemental) sample (mass *m* [g], density ρ [g/cm^3], mass number *A*) owing to a photonuclear reaction containing an isotope (isotopic abundance or enrichment w) having a second moment of the cross-section σ_{-2} [mb/MeV] at the energy, *E*, after a long-term irradiation (applying a electron beam current *i* [μA]) is given by:

$$A = m\, i\, B\, \mathrm{K(a,V,E)},\ [\mathrm{GBq}] \tag{1}$$

with:

$$B = 11.64\ (w\,\rho/A)\ \sigma_{-2[\mathrm{mb/MeV}]}\ (E_{[\mathrm{MeV}]}/25)^{2.37},\ [\mathrm{GBq\ cm^{-3}/g^{-1}/\mu A^{-1}}] \tag{2}$$

B only depends on the sample properties.

*Bundesanstalt für Material forschung.

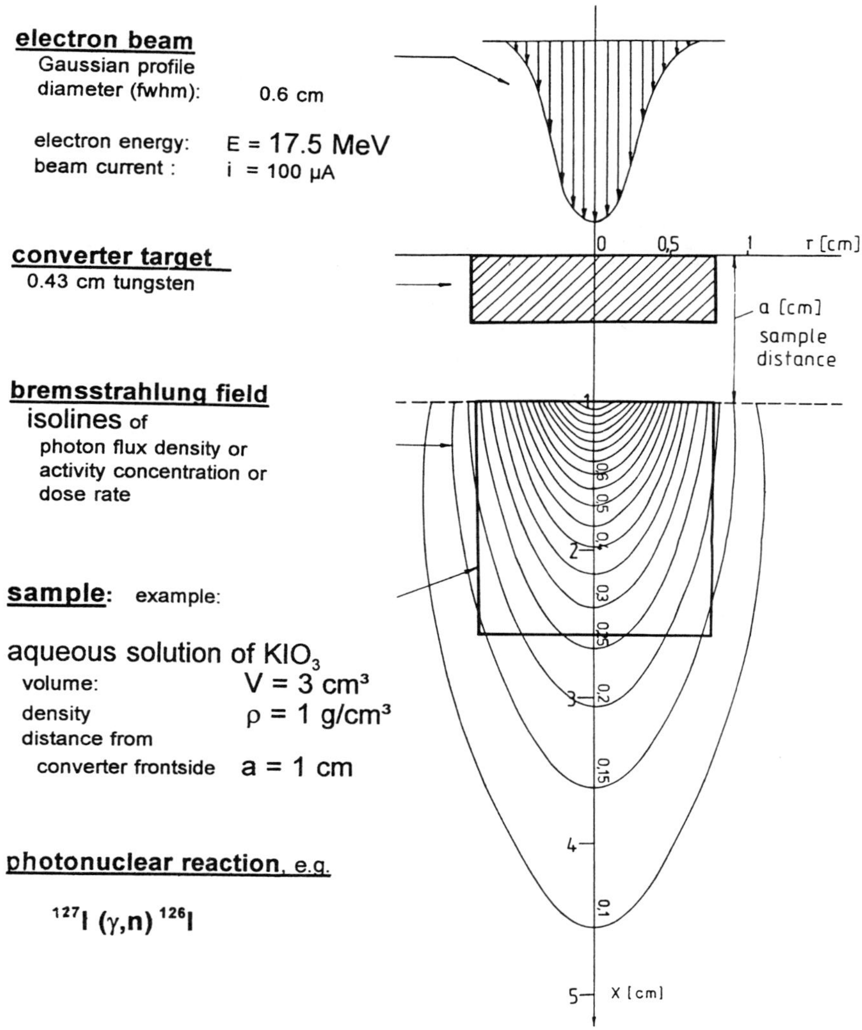

Fig. 1. Schematics of the irradiation geometry used in photon activation techniques.

RESULTS AND DISCUSSION

Using the PHOTAC code, different geometrical arrangements have been compared quantitatively to obtain the highest activity output or to find out the best solution for overcoming heating problems. The following problem has been studied.

Activation of High-Volume Water Samples: Optimum Sample Container Shape

Ultratrace analysis of heavy elements of environmental water samples using preconcentration could be disturbed because of organic com-

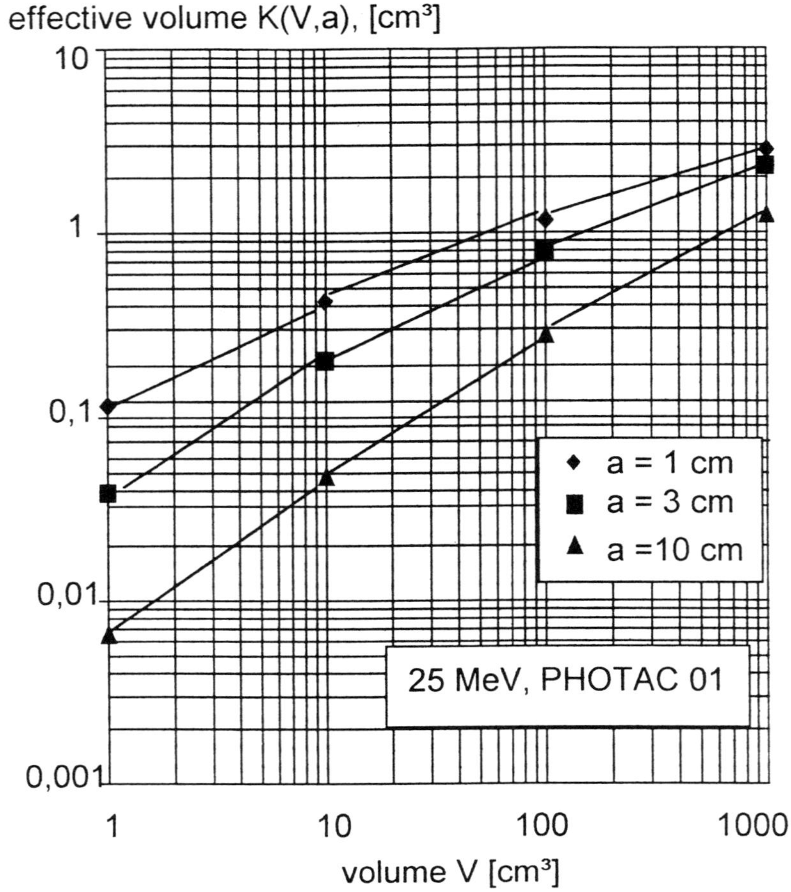

Fig. 2. Typical results of PHOTAC 01 calculations: relative activity output vs sample volume V and distance a of the sample from the converter at E = 25 MeV.

plexants *(2,5,6)*; therefore, water samples of sufficient volume should be irradiated directly. To obtain maximum activity output and sensitivity, the shape of the cylindrical sample vial (up to 5 L) has been optimized using PHOTAC. The optimum diameter-to-length ratio for a given volume as a function of beam properties and sample position is shown in Fig. 3 as a function of sample volume V, distance a, and electron energy E. The activity output and, therefore, the sensitivity in photon activation analysis might be reduced by a factor of up to 3 if the vial shape would be chosen far from optimum. At, e.g., ~17 MeV this optimum is approached if diameter d and length h of the sample container are chosen nearly equal ($d = h$), and at, e.g., 30 MeV if $d = 0.5\ h$.

Heating Up of Water Samples

In Fig. 4, the distribution of the beam power to converter target and sample at a 25-MeV/2.5 kW electron beam is illustrated for a 1000 cm^3 water sample. About 50% (1200 W) of the beam power is converted into bremsstrahlung. From this, 750 W are emitted in the forward direction

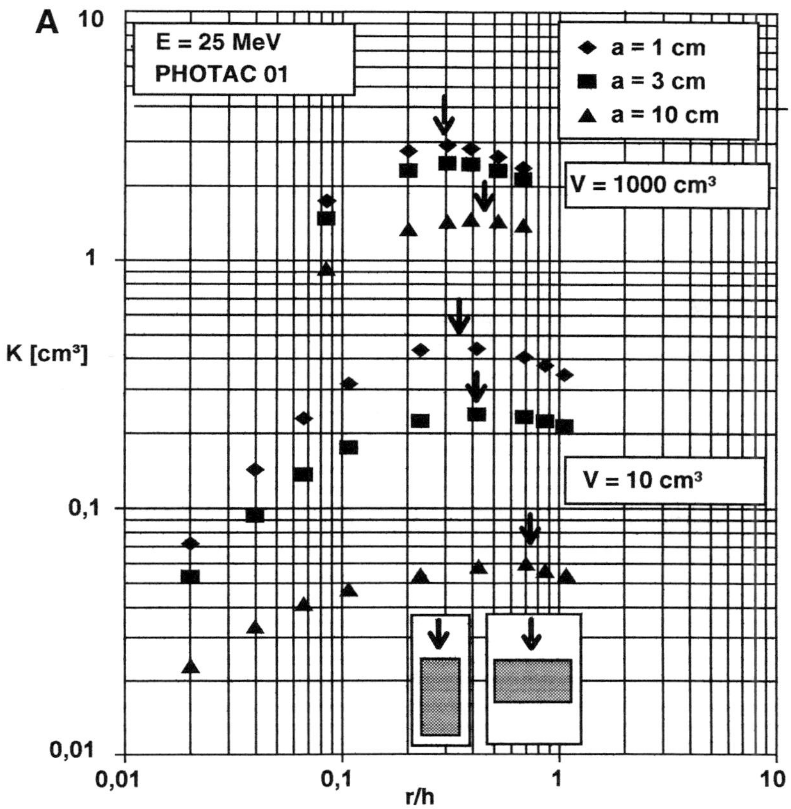

Fig. 3A. Relative activity output vs radius-to-length ratio *(r/h)* of the sample container at different sample distances *a* and sample volumes *V* for a 25-MeV electron beam with Gaussian profile (fwhm *b* = 0.6 cm).

and partly (300 W, i.e., 12% of the beam power) absorbed in the water sample. The power transfer and the corresponding power densities (up to 30 W/cm^3 at the beam axis in near-converter partition of the water sample) are also given. Because of this, the 1-L water sample is heated up (owing to the 300-W power transfer) to boiling within about 1200 s. The effect of cooling the (glass) vial surface by free-air convection is small (<20 W), and by a forced-air convection (blower) below about 200 W; thus, water cooling (using e.g., a double-walled vial) must be applied to avoid boiling in the case of long-term irradiation.

Because of the high-power transfer to the near converter part of the sample, the boiling sets in immediately with the formation of bubbles, which displace the water from this critical volume of highest bremsstrahlung flux density. Calculations by PHOTAC show an activity reduction up to 30% in the case of a 100 cm^3 sample $(d = h = 5$ cm) with an almost empty cylinder of 10 cm^3 $(d = h = 2.3$ cm) near the converter. The sample activity in the case of the disturbed irradiation can be calculated as the difference of the activities of the original volume $V = 100$ cm^3 and the displaced volume $V = 10$ cm^3. In the case of a 1000-cm^3 volume with

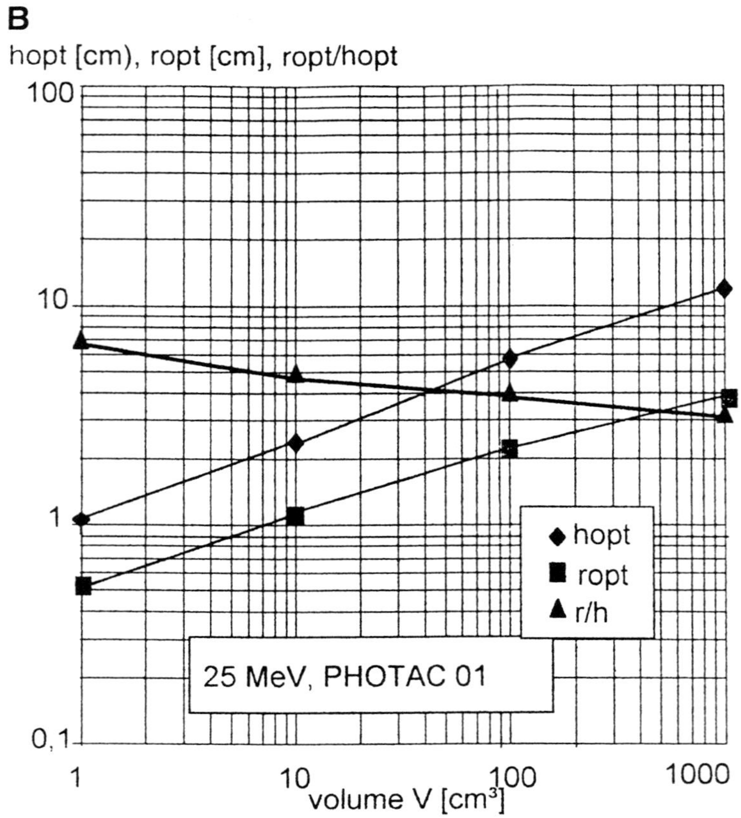

Fig. 3B. Optimum length h_{opt}, optimum radius r_{opt}, and optimum radius-to-length ratio *(r/h)* of the sample container as a function of the volume, *V*, at a sample distance of $a = 1$ cm for a 25-MeV electron beam with Gaussian profile (fwhm $b = 0.6$ cm).

a volume displacement of 10 cm^3, the activity loss is about 10%. Thus, boiling near the entrance window might be tolerated in the case of trace analysis of water samples when using internal standards, but the sensitivity is reduced.

If boiling in the near-target part of the vial should be avoided, a convection of the water within the vial must be supported, because the thermal conductivity of water is low. One way to do this is the use of a flow chamber (e.g., 50 cm^3) combined with a cyclic sample pumping *(2)*. This results in a more effective cooling, but every time only a part of the sample is exposed to the bremsstrahlung, which leads to a reduction of the activity output and of the sensitivity. Using PHOTAC, the activity loss can be estimated. Direct irradiation with 25 MeV bremsstrahlung (at i = 100 μA) of a 1000-cm^3 sample gives an effective volume of $K = 2.5$ cm^3 (at $a = 3$ cm); the irradiation of, e.g., only a 50 cm^3 sample (combined with cyclic pumping) gives $K = 1$ cm^3. The induced activity is lower by a factor of 2.5 using the small vial.

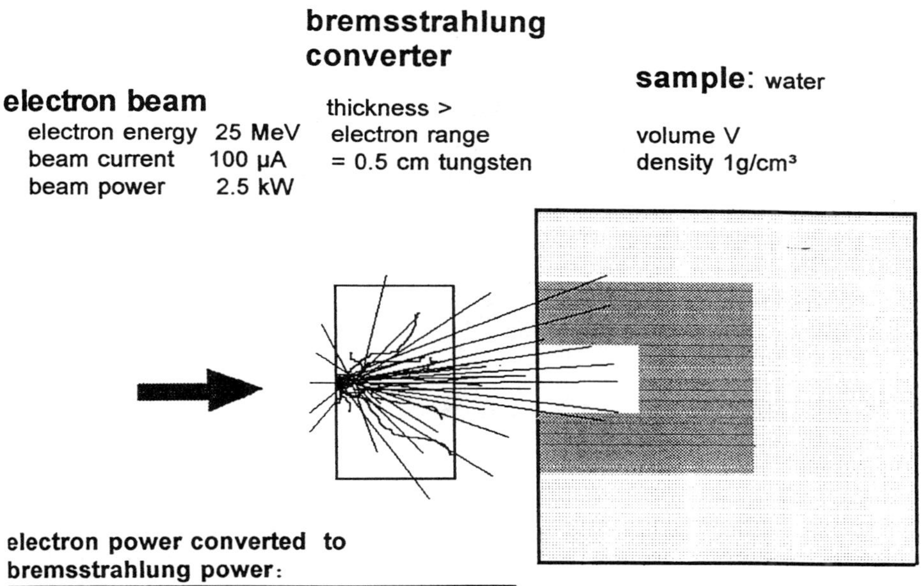

electron power converted to bremsstrahlung power:

	back-scattered	absorbed	transmitted	sum
electrons	5 %	46 %	0	51 %
bremsstrahlung	5 %	15 %	29%	49 %
sum	10 %	61 %	29 %	100 %
power [W]	150	1500	**750**	2500

bremsstrahlung power absorbed in the sample:

V [cm³]	0.1	1	30	1000
P [W]	3	13	80	**300**
P/ V [W/cm³]	**30**	13	3	0.3

Fig. 4. Distribution of the beam power to converter target and sample at a 25-MeV/2.5 kW electron beam.

The flow velocity, v, in the critical zone (beam diameter d_f, $d_f = 0.6$ cm) necessary to avoid boiling and the time during which the sample is heating up by ΔT can be calculated from the heat capacity of water, c, and the absorbed beam power, P, is given in *(7)*. From this it follows that the water flow in the critical zone near the converter must be higher than about $v = d_f/t$ = 0.6 cm/15 s = 0.04 cm/s; i.e., in this case, a slight water flow is sufficient to avoid boiling effects. To reduce the activity loss, a construction for the flow chamber arrangement should be approached were the irradiated volume is as high as possible, i.e., the loop should be small.

A second way to avoid boiling effects is to provide an internal convection in the critical zone near the converter by stirring with the advantage that the whole sample is irradiated permanently. Thus, the sample as a whole is heating up to boiling as discussed above.

Heating Up of Organic Samples

During instrumental trace analysis of organic samples having a very low thermal conductivity and low thermal resistance, the sample must be cooled during an in-axis irradiation by an intensive stream of air or—if this is not sufficient—it must be irradiated in an off-axis position and rotated during the irradiation *(8,9)*. In this case, a strongly reduced activity and sensitivity cannot be avoided. Therefore, in some cases, it is more efficient to keep the photon flux density just at an uncritical level, depending on the sample properties, by applying a limited beam power or an irradiation position far downstream from the converter.

Heating Up of Cryogenic and Gas Samples

A typical example for this problem is the production of iodine-123 by the photonuclear reaction $^{124}Xe(\gamma,n)$ $^{123}Xe \rightarrow {}^{123}I$. To yield an activity output of some 100 GBq, a sample mass of about 20 g highly enriched Xe-124 must be positioned at a very small distance from the bremsstrahlung converter target in a very condensed form (i.e., as an cryogenic sample or as a high-pressure gas sample) and irradiated at a beam power of about 10 kW (i = 400 µA at 25 MeV). Two problems concerning cooling of the bremsstrahlung converter and of the sample itself can arise in this case.

In the case of the croygenic sample, the mass thickness of the converter target must be chosen at least equal to the range of the electrons of the used primary energy (i.e., 12 g/cm^2 at 30 MeV) to prevent electrons from striking the cryo target. Thus, the thickness of the water-cooled lamella converter giving a "cold" bremsstrahlung flux free of electrons is at least about 3 cm for 25 MeV electrons.

In the case of the gas sample, which can be cooled more easily because of the gas convection inside the gas cell (or cyclic pumping), a part of the electrons can be allowed to strike the gas target (and to heat up it). The mass thickness of the converter target can be reduced to about 30–50% of the electron range and the thickness of the lamella converter to about 1.2–1.5 cm for 25-MeV electrons.

Thus, the question to be answered by PHOTAC calculations is which case gives the higher activity output, a cryogenic sample having a relative small volume (3.7 cm^3, because of the high density of 2.7 g/cm^3) at a relative large distance of about a = 3.0 cm or a high-pressure gas sample having a larger volume (10 cm^3; density 1 g/cm^3 at some 100 bar) at a small distance (about 1.5 cm). The results of PHOTAC calculations show that the geometrical yield $K(V)/V$ is almost the same in both cases, i.e., the approach that is more reliable from the technical point of view can be chosen. (In this case, the cryo target system cannot be used at a very high electron beam power, because the bremsstrahlung power absorbed by the Xenon ice sample, i.e., up to 30% of the electron beam power, cannot be drawn off from the sample container surface owing to the very low thermal conductivity of the xenon ice.)

SUMMARY

In estimating the influence of geometrical conditions and beam parameters on the activity output, the simulation of the activation process by the PHOTAC code was a helpful tool in making the best choice for PAA of biological and environmental samples.

ACKNOWLEDGMENTS

This work was supported by KAI E. V. Berlin and by Kernforschungszentrum Karlsruhe, Institut für Kernphysik (Zyklotron-Labor).

REFERENCES

1. C. Segebade, H.-P. Weise, and G. Lutz, *Photon Activation Analysis,* Walter de Gruyter, Berlin, 1988.
2. C. Segebade, H.-W. Thümmel, and W. Heller, *J. Radioanal. Nucl. Chem.* **167,** 83–89 (1993).
3. H.-W. Thümmel and K. Hirsch (1988, 1991, unpublished).
4. S. Tanaka, R. Tanaka, T. Tabata, R. Ito, Y. Nakai, and K. Ozawa, Data on thick target bremsstrahlung produced by electrons, JAERI-M 83-019 (1983), 106 pp.
5. W. Heller, W. Hirsch, C. Segebade, and H.-W. Thümmel, *Zfl-Mitteilungen (Leipzig)* **No. 161,** 63–69 (1990).
6. H.-W. Thümmel, W. Heller, R. Süβ, and C. Segebade, *Zfl-Mitteilungen (Leipzig)* **No. 166,** 101–111 (1992).
7. G. Jahn, Probleme der Wärmeübertragung bei der Bestrahlung von Festkörpern, HMI-B-349 (Hahn-Meitner-Institut Berlin, März 1981).
8. D. Schulze, H. Kupoch, B. Eckert, and G. Böttger, Patent DE 4106362 A1 (H 01 J 35/10).
9. D. Schulze and H. Kupoch, Patent DE 4105871 A1 (G 21 k 5/00).

Environmental Applications

Use of Nuclear Analytical Techniques in Bioenvironmental Studies

AN. A. KIST

Institute of Nuclear Physics, Tashkent, Ulughbek, Uzbekistan, 702132, CIS

ABSTRACT

Bioenvironmental studies remain to be one of the most important fields of applied analytical chemistry. At present time, more than 50% of nuclear analytical studies deal with bioenvironmental investigations. The first period of utilizing nuclear analytical methods in bioenvironmental sciences could be characterized as "purely analytical," in which these methods were used for determination of sample composition in competition with other non-nuclear methods. Later, the outstanding advantages of the former methods were used for more detailed description of systems to be studied, including element speciation, spatial distribution, and so forth. The present period not only develops approaches of previous periods but also considers the bioenvironmental processes more widely and is focused on their dynamics. In this field, a large extent of utilizing various nuclear analytical techniques can be expected as well.

Index Entries: Nuclear analytical methods; activation analysis; trace elements; medicine; environment; biogeochemistry.

INTRODUCTION

Neutron activation analysis (NAA) has been widely used for years as a routine analytical method in bioenvironmental sciences owing to low determination limits for many elements, high potential for accuracy and precision (as virtually blank-free method), multielement pattern, and considerable matrix independence.

The application of NAA for biological samples is almost as old as the method itself. One of the very first studies was devoted to analysis of material of biological origin, i.e., paper. To determine sulfur, the reaction $^{32}S(n,p)^{32}P$ was used *(1)*. For 56 years, NAA became a competitive routine analytical method, which gave much new knowledge in bioenvironmental sciences.

It is necessary to emphasize that against the background of growing differentiation of sciences, analytical chemistry, and especially radioanalytical chemistry, the necessity of integration of biological sciences introducing the term *life sciences* was stressed. According to this tendency it seems to be acceptable to consider the term *bioenvironmental sciences* as part of life sciences. This term stresses the very close connection between biogeochemical, biological, medical, environmental, and related sciences, especially from the point of view of the analyst. Obviously, there is a need in the life sciences to determine numerous analytes in extensive lists of samples of various origins (water, atmosphere, soil, and plant, animal and human tissues, and so on) independently on the final application of data obtained. In this author's opinion, bioenvironmental sciences try to describe the modern "noochemical" processes (processes in noosphere in terms given by Vernadskiy and Teilhard de Chardin). The noosphere is probably a more adequate term than biosphere because of the growing role of human activity on the chemistry of earth surface. The essential and often irreversible changes of processes in biosphere–noosphere caused by human activity need new methodological and analytical approaches.

THE HISTORY OF NUCLEAR ANALYTICAL TECHNIQUES IN BIOENVIRONMENTAL STUDIES

A brief survey of the method in general can be given according to the milestones formulated by "patriarchs" of activation analysis *(2–4)*, with some specific author's additions (Table 1). The main results of NAA are given in ref. *2* as brief formulations, with some additions (in brackets):

1. Opened up the field of trace elements and their importance (and demonstrated high acceptability of NAA for bioenvironmental sciences);
2. Led to the recognition of numerous biologically essential trace elements (dealing with the elements not studied earlier and not determined using other analytical methods);
3. Greatly assisted the early development of the semiconductor industry (also assisted the technology of high-purity materials, space technology, and so on);
4. Provided multielemental analytical methods of great use in environmental studies (and in the case of instrumental NAA, relatively simple and productive methods with acceptable accuracy and precision);
5. In forensic work, helped to establish the importance of trace element composition to determine common or noncommon origin of evidence specimen (including soils, plants, drugs and narcotics, saliva, hair, and other biological specimens).

Table 1
Stages in the Growth of NAA and Selected Nuclear Analytical Techniques Including Their Bioenvironmental Applications

V.P.Guinn *(2)*	R.E.Jervis *(3)*	Additions
1936-1944: The induction period	-	The very first works determining the future directions
1944-1950: The nuclear reactor	PRIMITIVE PERIOD: (late 40's, early 50's): NAA at atomic energy labs (Oak Ridge, Harwell, Chalk River, Saclay, USSR) on "suitable matrices"	The real analysis of biological samples
1950-1960: The NaI(Tl) detector	MIDDLE AGES OF RADIOANAL.CHEMISTRY (late 50's, early 60's) Gamma-ray spectrometry, group separations, accelerator methods, INAA with Ge(Li) of high purity materials, biomedical, forensic samples, etc.	Routine methods in medicine, biogeochemistry, biology, agriculture, forensic sciences, in-vivo analysis
1960-1970: The Ge detector	20th CENTURY (70's and 80's) Multielemental analysis, prompt and ion-beam methods, PIXE, RMs, selected elements: Hg, Cd, As, Se, and Pb, "in-vivo" analysis	Element speciation, combination with related methods, spatial distribution, harmonization, quality assurance, extensive usage of computers
1970-1991: Extensive applications	INTO 21st CENTURY (Future scope, predicted roles)	Detailed studies in dynamics

Summarizing this period, one can conclude that NAA in combination with other nuclear (and non-nuclear) methods is highly important and irreplaceable for development of bioenvironmental sciences. If the first bibliography *(5)* includes about 20% of references devoted to analysis of bioenvironmental samples, the last conference "Modern Trends in Activation Analysis—MTAA-8" in 1991 includes more than 50% of reports connected with bioenvironmental sciences.

The main reason for this statistics is not only the method's widely known advantages (and disadvantages), but the accumulation of very important experience by the foremost centers of NAA. Reviewing the NAA bibliography, one can conclude that experience of these laboratories is at least as important as the theoretical or technical development and facilities.

PRESENT AND FUTURE

It is very difficult and speculative to try to predict the development of nuclear analytical techniques. However, a few possible directions can be presumed. In the nearest future combination of various techniques (including those capable of determining species and spatial distribution), more and more sophisticated methods of raw data treatment as a tools of system approach will be used to describe dynamics of various processes. This presumption could be illustrated by many recent examples and references. However, taking into consideration the limits of spare here, I can only try to give a few examples accompanied by a few references.

Speciation in tissues and biofluids, and spatial distribution: Data on total element concentrations in biosamples give only incomplete information to describe their role. This is why in recent years a growing number of studies was performed in which NAA was combined with methods of preparative biochemistry. Analysis of subcellular fractions, nucleic acids, proteins of various molecular mass, and so forth allowed for better understanding of the biochemistry of Se, Cr, Zn, I, Au, and Hg. It also gained very important information on the possible biological role of other elements, such as Sc, Cd, Sb, REE, Th, and U.

PIXE and related techniques using narrow beams, activation autoradiography, track radiography, and other methods of spatial distribution determination gave new information about differences in behavior of native and incorporated iodine in thyroid gland, element distribution across a single hair, and so forth.

As an example of combined studies of dynamics, consider the problem of coal combustion. As it is widely known, coal combustion is one of most important sources of pollutants, including toxic trace elements. In determinations of composition of coal of various origin, NAA and other nuclear methods play an important role. These methods determined some elements not studied earlier or determined scarcely. These methods were also used to study the influence of sample handling and treatment. For instance, it was shown that the methods that require sample ashing are unacceptable for coal analysis because of hardly predictable losses of elements. These losses resulted in underestimates of some elements, such as Hg. The real concentration were several decades higher and this was the reason to reconsider the technology of combustion of coal from various deposits.

Nuclear analytical methods played a very important role in quality assurance, especially on homogeneity testing and certification of various reference materials prepared at the BCR, IRANT, NBS/NIST, and SABS.

In addition to work on determination of coal composition, speciation of elements in coal and combustion products was tried. Elements like Au, Br, Cr, Hg, Sb, Se, and Zn were found predominantly in the organic fractions of the coal. These elements were also found in higher amounts in gaseous coal combustion products. Neutron activation autoradiography showed very irregular spatial distribution of these elements in coals. The new technique for trapping extra fine particles and sampling the vapor-gaseous phase allowed to detect high amounts of As, Au, Hg, Zn, and so on in these samples in the vicinity of coal burning power stations.

Dybczynski et al. *(6)* found the elements As, F, Mo, and Se very leachable in a study on leaching of elements from fine coal fly ash. This conclusion conforms with results of another study *(7)*, which lists the elements As, Se, Sr, and Ba in groundwater as elements specific for identification of coal-fired power plants as a source of pollutants. Winchester *(8)*, when studying metal solubility in aerosols, concluded that sulfur-containing aerosols (connected with combustion of fossil fuels) and solubilized metals can correlate with medical statistics and, especially, with chronic obstructive pulmonary disease.

Additional sources of atmosphere and biosphere contamination are solid wastes and products of its treatment. Buchholz et al. studied size fractionated municipal incinerator ash *(9)*. Substantial releases to the environment were found for Ag, Cd, Cr, Cu, Hg, Sb, Se, Sr, and Zn. Similar results were obtained by Krishnan et al. *(10)* in their study on binding and leaching of trace elements from solid wastes. Household wastes were also identified as a source of heavy metal pollution, specifically for the elements As, Cd, Cr, Cu, Hg, Pb, and Zn *(11)*.

This block of bioenvironmental studies concludes that very significant sources of atmospheric contamination are combustion products of fossil fuels and solid wastes. These products are mainly contained in a finely dispersed, very leachable, and/or mobile form. Environmental data for these contaminants correlate with medical statistics. Body burden with these metals can be indicated by composition of human hair, and more sophisticated studies in this respect can be expected in the near future. Air passes through the human body in amounts higher by a factor of at least 30,000 (volume related) and not less than 20 (mass related) compared to water and food. Taking into account that the respiratory system is less protected than the gastro-intestinal tract, and that the detected chemical forms of toxic elements are potentially very dangerous, one may conclude that bioenvironmental studies are indeed very important for health risk assessment, and that NAA can be used to contribute to solve this problem.

Productivity of instrumental nuclear analytical methods allows to analyze a large number of samples. The data thus obtained allow to make conclusions with high statistical significance. Studies in this direction will be very important for determination of health status of populations, effects of agriculture intensification, mapping of polluted areas, and so forth.

A very important step in nuclear analytical techniques was the wide application of computers, especially for gamma- and X-ray spectra evaluation. Study of the composition of bioenvironmental samples by nuclear analytical techniques allows determination of up to 83 parameters, *viz.* elements, radionuclides, their species, and compounds. Despite using computers for obtaining multiparameter data, routine multiparameter evaluation of rough data is still not very widely used, and predominant fraction of studies deals with very simple statistical methods and considers very simple (mostly single element) irregularities. Scarce results of studies accomplished by evaluation of data obtained with multiparameter statistical methods (multiple correlation, discriminant analysis, cluster analysis, factor analysis, main components analysis, and so on) show that very important progress can be made in this direction.

Some new methods of radioactivity detection, new sources of neutrons, gamma-rays, and charged particles, new techniques of element speciation, and other expected and unexpected results of nuclear physics, analytical chemistry, and biochemistry will be also very important for development of nuclear analytical techniques in noochemistry.

SUMMARY

This brief review deals with just a few examples and can be considered as an addition to similar reviews *(12–14)*. Nuclear analytical techniques will play a very important role in the future of noochemistry—the science most important for human existence.

REFERENCES

1. G. T. Seaborg and I. J. Livingood, *J. Am. Chem. Soc.* **60,** 1784 (1938).
2. V. P. Guinn, *J. Radioanal. Nucl. Chem.* **160,** 9 (1992).
3. R. J. Jervis, *J. Radioanal. Nucl. Chem.* **160,** 21 (1992).
4. M. De Bruin, J. Radioanal. *Nucl. Chem.* **160,** 31 (1992).
5. G. J. Lutz, R. J. Boreni, R. S. Maddock, and W. W. Meinke, *Activation Analysis: A Bibliography,* NBS Technical Note 467, Gaithersburg, MD, 1968.
6. R. Dybczynski, K. Kulisa, M. Malusecka, M. Mandecka, H. Polkowska-Motrenko, S. Sterlinski, and Z. Szopa, *Biol. Trace Element Res.* **26/27,** 335 (1990).
7. I. Olmez, and M. Hayes, *Biol. Trace Element Res.* **26/27,** 355 (1990).
8. J. W. Winchester, *Biol. Trace Element Res.* **26/27,** 195 (1990).
9. B. A. Buchholz, S. Landsberger, and M. Plewa, *MTAA—8* Abstracts, Vienna (1991), p. 43.
10. S. S. Krishnan, R. E. Jervis, and D. A. Vela, *MTAA—8* Abstracts, Vienna (1991), p. 72.
11. P. Bode, M. De Bruin, Th. G. Aalbers, and P. J. Meyer, *Biol. Trace Element Res.* **26/27,** 377 (1990).
12. R. M. Parr and E. Cortes-Toro, *Biol. Trace Element Res.* **26/27,** 671 (1990).
13. A. A. Kist, *Biol. Trace Element Res.* **26/27,** 661 (1990).
14. A. A. Kist, *J. Radioanal. Nucl. Chem.* **167,** 227 (1993).

Nuclear Analytical Techniques in Environmental Studies

R. E. JERVIS

Department of Chemical Engineering, University of Toronto, Canada

ABSTRACT

Nuclear analytical techniques are particularly suitable for measuring trace components in a wide variety of environmental samples, and for that reason, the techniques have made a significant contribution to environmental research. Presently, at a time when biosphere contamination and threats of global change in the atmosphere are of widespread concern, there exist an impressive array of specialized instrumental methods available to life scientists engaged in environmental studies; however, the nuclear techniques will probably continue to play a useful role in the future, notwithstanding the decreasing availability of necessary facilities, such as research reactors and accelerators. Reasons for the particular suitability of radioanalytical techniques are reviewed and illustrated by examples of recent applications to solid wastes, biomonitoring, and urban aerosol source identification in this laboratory.

Index Entries: Nuclear analytical chemistry; environmental & life science applications; trace heavy metals; solid wastes; leaves as biomonitors.

INTRODUCTION

Nuclear analytical techniques have made a significant contribution to environmental and life sciences studies, particularly in the detection of heavy metal contamination and other toxic inorganics, atmospheric emissions, solid wastes, biomonitoring, and surveys of exposed population groups. Reviews on activation analysis include prominent sections on the environment and applications to life sciences, as do several recent monographs on nuclear analytical topics *(1–4)*. Some reasons for this use of nuclear methods in environmental research and aspects of their suitability are suggested below (in the first half of this article).

Biological Trace Element Research Editor: G. N. Schrauzer © 1994 by Humana Press Inc.

Public interest in preserving the environment has never been greater, although some announced government programs intended to intensify environmental surveillance, to promote abatement, and further to implement remedial measures on announced schedules for emission cutbacks have waned somewhat under recession pressures and other political priorities. However, one such program, the international global change program, IGP, has been progressing with a relatively high level of international participation. At the recent Rio de Janeiro environmental "summit," there was much concern expressed about environmental degradation in general and atmospheric global change in particular, and action accords were agreed to. Clearly, national and international strategies for increased surveillance, monitoring, and control abatement and remedial measures will produce an increased requirement for analytical work. More complex measurements will have to be made in the future in a wide variety of sample media. Probably almost every known toxic substance will have to be measured in air, water, soil, and sediments, biota, wildlife, and human subjects in order to be able to assess and control human health hazards resulting from environmental transport and food chain accumulation of contaminants.

WHAT WILL BE AN APPROPRIATE ROLE FOR NUCLEAR TECHNIQUES IN FUTURE ENVIRONMENTAL STUDIES? *(5)*

(Other articles in this symposium address the future of nuclear analytical methods in general and the problem of the reducing availability of nuclear facilities: reactors and accelerators, needed to accomplish applied radioanalytical measurements.) Environmental analytical chemists have a broad range of instrumental methods available for accomplishing the sophisticated determinations of individual chemical species of organic and inorganic substances in the environment. Several of these techniques, e.g., ICP/MS and laser ablation spectrometry (LAMS), yield large quantities of output data that require comprehensive data handling, compilation, and interpretation—facilities comparable to those required for applied γ-spectrometry. Compared to the tedious lab procedures required for careful sample preparation (cleaning, dissolution, standard additions, and so forth) for some methods, other instrumental methods are less demanding and error prone because they are capable of analyzing samples directly whether they be gases, liquid, or solid—also an attractive feature of INAA.

Notwithstanding the availability of such (conventional) methods and for a variety of reasons—listed below—activation analysis still continues to be much used in environmental studies, not only by NAA specialists in reactor institutes, but also as basic analytical tools for environmentalists. Radioanalytical techniques can yield good information on sample constituents of environmental importance with a minimum of laboratory effort, with adequate precision and accuracy, and without the need for an expensive, dedicated, alternative analytical instrument. Some of the particular reasons why radioanalytical methods compete favorably with other instrumental trace analysis methods, such as AAS and ICP, include the following:

1. Adequate sensitivity for many elements of environmental significance (including several that are less common to analysts) combined with reliability, accuracy, and freedom from the need to establish empirical corrections;
2. Selectivity with relatively little matrix or interelemental interference;
3. Nondestructive analysis possible in many environmental samples, e.g., soils, sediments, biota, and biological specimens;
4. Simultaneous multielement capability for 25–40 elements, some of environmental significance; and
5. Nuclear techniques are advantageously used for certification of analytical reference materials and often serve as reference analytical method for ad-hoc round-robin comparisons with other instrumental methods.

In addition, nuclear-based analytical techniques can be useful for meeting certain particular needs of current environmental studies. They can readily handle the variety of sample types to investigate environmental pathways of trace components from their emission sources through transport media of air, water, and food chains to different receptor and target species and to humans. Nuclear techniques can be applied to speciation studies essential in environmental toxicology. Synergism among contaminants will probably need to be researched more in the future, and multielemental methods will be useful in such studies. Another requirement of some studies is to be able to handle large sample suites, e.g., from air pollution sampling networks. Because some nuclear techniques can often be performed instrumentally, if need arises, they can be automated with robotics and software. Normally, radioanalytical methods are not considered feasible to use for environmental monitoring; however, there are examples of NAA studies in environmental health in which in excess of 5000–10,000 data points were involved *(6)*.

In other situations, nuclear analytical methods can play a role complementary to those instrumental techniques that are more often employed for monitoring, for day-to-day routine testing, and for *in situ* field measurements. Some of the "hyphenated" instrumental techniques, such as GC-MS and ICP-MS, are ideal for trace organics in gases and liquids and for speciation, such as for the carcinogenic organic halides, PAHs, PCBs, and so on. Solid samples are often more ideal for INAA or PIXE without the need for elaborate sample dissolution and preparation. This can obviate the need for tedious chemical procedures with their attendant risks of sample contamination, loss of volatiles, or reagent blanks. On the other hand, liquid or gaseous samples are better handled by other methods.

Reference was made to the capability of INAA to measure 25–35 elements in atmospheric aerosols, or when complemented by IPAA or PIXE, up to 30–40 elements *(6)*. Although not all of these may be directly of environmental importance as known toxins, some recent work indicated the usefulness of trace multielemental data for source identification and source apportionment at receptor sites *(6,7)*.

In environmental studies in our laboratory, techniques used were INAA, PNAA, PIXE, XRF, AAS, and ICP(AES). Although INAA has been a basic method yielding about 70–80% of the results, the other techniques have been used in a complementary way for extension to other heavy metals, such as Pb, Cd, and Ni, to light elements, and for confirmatory analyses. Experience gained has led to some conclusions about an appropriate role for nuclear-based methods for environmental analyses:

1. For analysis of a single contaminant, atomic absorption is usually very suitable whether samples are water or solids; calibrations appropriate to the matrix can be done readily. AAS is often used for routine monitoring.
2. Methods like ICP, AAS, and chromatography techniques work well with liquid samples and standards, and are more suitable than nuclear methods, which can accept only very limited volumes of water or gases and often suffer sensitivity limitations for that reason; also, analysis of volatile components may be done better by these (nonnuclear) methods.
3. XRF seems more suitable for measuring minor constituents and major matrix components, and is less suited to trace analysis.
4. Electron microscopy (SEM) and other microprobe methods serve well to characterize microcomponents, but cannot provide precise and representative compositional analysis of solids.
5. In environmental studies in which material balances are of interest, e.g., in measuring the volatility of trace elements during combustion, incineration, or processing, or in assessing the leachability of contaminants in solids, such as fly ash, incinerator ash, sewage solids, or other solid-waste forms, a combination of INAA of solids before and after, and radiometric measurements with activated solids can facilitate such studies *(see below)*.

CURRENT APPLICATIONS OF INAA

Biomonitors

Measurements have been made of tree leaves from several cities, including a highly industrialized urban site, compared to leaves growing in a forested lakeland recreational area. Previous work by Capannesi et al. *(8)* reported tests on oak leaves as potential biomonitors, whereas

Steinnes *(9)* has reported results on epiphytic lichens used to monitor heavy metal deposition in Norway. Plant material can be suitable for monitoring air particulate matter in contaminated local areas, whereas surface soil testing can present problems of interpretation and, testing exposed population groups may often not be feasible or permitted.

Oak, maple, and birch leaves from the same location were analyzed by INAA after washing, and it was found that oak leaves consistently showed higher concentrations of anthropogenic trace elements. Leaf tissue was separated from leaf veins and stems, and some appreciable differences were found: Al, Cl, Ca, Ti, V, Mn, Cu, Br, and Sr were higher in leaf tissue, whereas veins and stems contained higher Na, K, Mg, Ba, and Dy. Since the objective was to assess surface collection of atmospheric contaminants by leaves during their growing season, as distinct from observing uptake of trace elements from soils through roots, only leaf tissue material was included in further analyses.

Replicate analyses of leaves from a single (Ottawa) tree showed some variability (2–10%) in trace elements' concentrations among 16 tested (Table 1), but this was mainly attributable to counting. Of the different locations listed in Table 2, the Haliburton oak samples were from a forested lakeland area more than 150 km from major industrial sources. Ottawa is a mainly residential government city. Cambridge has light industry. Toronto-A was a central location near incinerators, heavy traffic, and industry, whereas Toronto-B was also a downtown area with traffic, but further from major air pollution sources. The oak leaves reflect differences in local air environment: the halides, V, Cu, Ti, and Al tended to be higher in the more industrialized areas and near motor-vehicle routes (some elements by more than a factor of 10). On the other hand, Mn, which occurs in petrol and vehicle tires, was found to be higher in the remote forested region and was lower in trees near high traffic—as also reported by others *(8)*. Na and Cl were increased also near roadways where deicing salt was applied during winter. Clearly, INAA proved very useful in analyzing tree leaves for possible use for urban biomonitoring.

Encapsulation of Solid Waste

Another current research in which INAA using SLOWPOKE neutrons has been very suitable is a study of environmental hazards from ground water leaching of solid wastes placed in land disposal sites, such as the municipal solid waste (MSW) ash from garbage incinerators. (The IAEA recently conducted a CRP on nuclear-based techniques for study of solid wastes that included research such as this [*10*].) There is much concern in the industrialized world about proliferating solid wastes and the need to identify alternate disposal technologies for MSW. One option is to encapsulate solid waste incinerator ash to produce a solid material that resists leaching, and that may find uses for road building or structural landfill. Waste solidification with portland cement has been reported, but a more promising material, bitumen (asphalt), has been used in this study.

Table 1
Trace Elements in (Oak) Tree Leaf Tissue (Ont., Canada)

Element	Range (µg/g)	Mean Conc. (±s.d.)
Na	15-29	24 ± 5
Mg	1500-3800	3300 ± 800
Al	38-150	97 ± 34
Cl	26-87	69 ± 20
Br	0.07-0.6	0.4 ± 0.2
I	0-0.2	0.06 ± 0.07
K	0.003-0.005	0.003 ± 0.001
Ca	0.009-0.01	0.01 ± 0.002
Ti	1.4-11	4 ± 3
V	0.05-0.1	0.08 ± 0.02
Mn	22-71	61 ± 18
Cu	0.9-4.7	3 ± 1
Sr	5.1-15	10 ± 4
Ba	9.8-33	19 ± 9
Dy	0.002-0.02	0.01 ± 0.01
U	0-0.07	0.05 ± 0.02

Table 2
Comparison of Tree Leaves from Different Locations

	Location†				
	Haliburton	Ottawa	Cambridge	Toronto-A	Toronto-B
Na	42	24	150	200	3800
Mg	1700	3300	3100	3200	1500
Al	99	97	140	460	140
Cl	82	69	240	2000	1500
Br	10	1.7	1.2	25	1.5
I	0.25	0.06	0.05	0.44	0.22
K	0.001	0.003	0.004	0.017	0.007
Ca	0.01	0.01	0.02	0.03	0.02
Ti	30	4.2	10	31	12
V	0.15	0.08	0.13	0.48	0.16
Mn	710	61	120	110	78
Cu	6.5	3.1	3.5	11	3.5
Sr	38	9.5	9.5	27	10
Ba	84	19	20	54	34
Dy	0.06	0.01	0.02	0.04	0.03
U	0.06	0.05	0.03	0.10	0.07

*(concentrations in µg/g)
†Locations: Haliburton (forest); Ottawa (residential); Cambridge (light industry); Toronto-A (central/industrial); Toronto-B (central)

In Table 3 are listed the contents of 25 elements in MSW incinerator bottom ash from Vancouver. (In a separate paper from this laboratory [*11*], details are reported of the physical and chemical behavior of solid wastes, such as MSW and sewage waste ash.) Of concern are the relatively high concentrations of As, Cd, Cr, Cu, and Zn and their leachability (Figs. 1 and 2) over periods from 0.5–10 d under EPA standard leaching test conditions (solid/liquid ratio: 1:20; pH 5). By contrast, some

Table 3
Vancouver Bottom MSW Ash: Elemental Concentrations (μg/g)

Element	Isotope	Concentration
Ag	Ag-110m	0.3±0.02
Al	Al-28	58000±280
As	As-76	7.9±0.1
Au	Au-198	4.4±0.1
Br	Br-82	5.5±0.2
Ba	Ba-131	800±58
Ca	Ca-49	95380±460
Cd	In-115m	33.9±0.03
Ce	Ce-141	32.1±0.13
Cl	Cl-38	2739.2±27
Co	Co-60	26.4±0.21
Cr	Cr-51	694±1.73
Cu	Cu-66	2270±83
Fe	Fe-59	61600.5±200
Dy	Dy-165	2.62±0.03
K	K-42	40800±3000
La	La-140	24.2±0.01
Mg	Mg-27	22600±800
Mn	Mn-56	1000±130
Na	Na-24	31900±520
Sb	Sb-122	76.5±0.03
Sc	Sc-46	7.3±0.01
Se	Se-75	7.5±0.5
Sm	Sm-153	6.0±0.01
Ti	Ti-51	5450±230
V	V-52	71.3±0.4
Zn	Zn-65	890±8.3

preliminary experiments on encapsulation/solidification with bitumen have shown that much of this leachability can be eliminated using concentration ratios ranging from 15:85 to 50:50 bitumen/MSW. Over most of this range, the solidified MSW is a strong, stable, nonfriable material resistant to weathering or winter deterioration. In the figures, in which the leachability of the MSW ash is compared to the bitumenized solid for As, Br, Cd, and Cr, (done under prescribed 4:1 liquid/solid; pH 5 conditions), it can be seen that leachability is reduced by up to a factor of 100. Accessory experiments in analyzing thin sections of the bitumenized

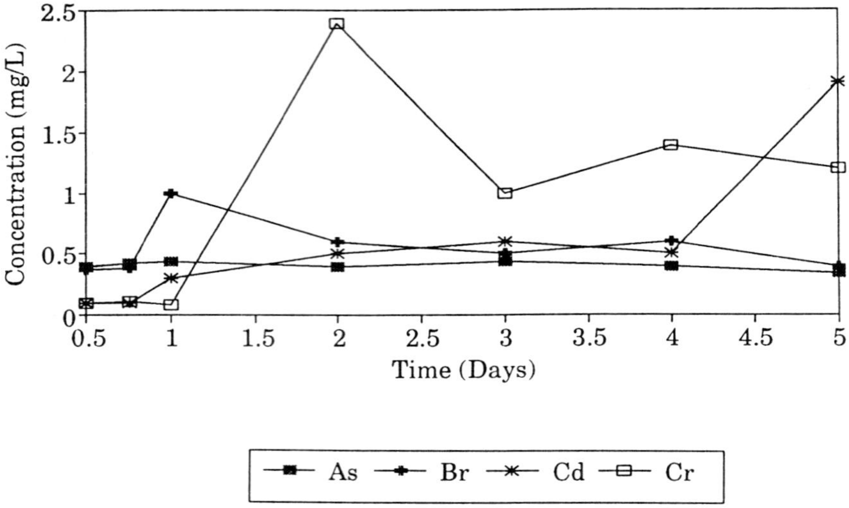

Fig. 1. Leaching of MSW solid waste (Vancouver; solid/liquid = 20/1, pH ~ 5).

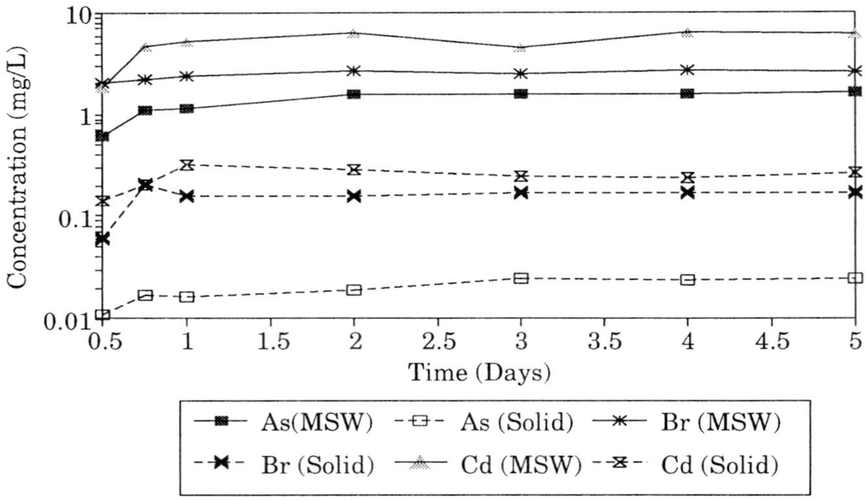

Fig. 2. Effect of bitumen solidification on MSW leaching (Vancouver MSW; bitumen/MSW: 25/75; solid/liquid = 4/1, pH ~ 5).

MSW indicate that most of this residual leaching occurs in a 0.1-cm surface layer and, since it mostly occurs within 1–2 d, a predisposal soak of such solid material could obviate any appreciable leaching by ground water where such solids were used for road building or for structural landfill, such as in retaining walls.

ACKNOWLEDGMENTS

The biomonitor results were obtained by R. Qureshi, B.A.Sc., and the measurements with bitumenized MSW are preliminary, unpublished M.A.Sc. thesis results of Lu Xing. The cooperation and assistance of S. Aufreiter and Dr. R.G.V. Hancock, U. of Toronto SLOWPOKE Reactor Facility are gratefully acknowledged. Financial support for this research and for SLOWPOKE operation was provided by NSERC, the Natural Sciences and Engineering Research Council of Canada.

REFERENCES

1. W. D. Ehmann et al., Nuclear and radiochemical analysis, *Anal. Chem. Fund. Rev.* (1992).
2. W. D. Ehmann and D. E. Vance, *Radiochemistry and Nuclear Methods of Analysis,* Wiley, 1991.
3. T. Weizhi, *J. Radioanal. Nucl. Chem., Articles,* **151,** 7–40 (1991).
4. S. Parry, *Activation Spectrometry in Chemical Analysis,* Wiley, 1991.
5. R. E. Jervis, The appropriate role for different nuclear techniques in the study of environmental problems, in *Proceedings of the 2nd International Workshop: Neutron Activation in the Environment,* Dubna, Russia, JINR publication, pp. 116–127 (1993).
6. M. M. Ko and R. E. Jervis, *J. Radioanal. Nucl. Chem., Articles,* **161,** 159 (1992).
7. T. G. Pringle and R. E. Jervis, *J. Radioanal. Nucl. Chem., Articles,* **123,** 321 (1988).
8. G. Capannesi, A. Cecchi, C. Ciavola, and A. F. Sedda, *J. Radioanal. Nucl. Chem., Articles,* **123,** 713 (1988); and, *J. Radioanal. Nucl. Chem., Articles,* accepted, in press (1993).
9. E. Steinnes, Trondheim University, Norway, personal communication (1988).
10. IAEA Coordinated Research Project, Use of Nuclear and Nuclear-Related Techniques in the Study of Environmental Pollution Associated with Solid Wastes (1987–92); Report: IAEA TEC-DOC, NAHRES-9, Vienna (1992).
11. S. S. Krishnan, R. E. Jervis, and L. D. Vela, Studies of incinerator ashes and environmental effects using radioanalytical techniques, these proceedings.

Studies of Incinerator Ashes and Environmental Effects Using Radioanalytical Techniques

S. S. KRISHNAN,*[,1,2] R. E. JERVIS,[1] AND L. D. VELA[1]

[1]*Department of Chemical Engineering, University of Toronto; and* [2]*The Toronto Hospital, General Division, Medical Physics Laboratory (Room CCRW-G803), 200 Elizabeth Street, Toronto, Ontario, Canada M5G 2C4*

ABSTRACT

We have studied solid waste incinerator ashes to understand the leaching mechanism and speciation of toxic and other elements in them. Leaching media, such as water and acetate buffer, at various pHs were used. Incinerator ashes generally contain concentrations of many toxic elements, such as Cd, As, Hg, and Se. These elements are leached out rather easily. Many of the elements are leached within the first few minutes to an hour, and the majority of the elements reach peak equilibrium concentrations within 200 h. The pH and nature of the leaching medium are important factors in the leaching of the elements.

Index Entries: Municipal solid waste; municipal refuse; incinerator ash(es); leaching; solid waste; toxic elements.

INTRODUCTION

Solid wastes are of diverse matrix structure, which is a result of waste source and incinerator burn technology among other things. Accurate characterization of leachable contaminants in solid wastes prior to landfilling is very important for effective waste management. To date, a number of laboratory leaching tests *(1–6)* have been used to predict the environmental impact of waste leachate. The environmental authorities in several countries have designed and standardized their own leaching tests, and have made them statutory law for regulating the disposal of

*Author to whom all correspondence and reprint requests should be addressed.

hazardous waste generated under their jurisdictions and to serve as protocols for routine waste analysis. The United States Environmental Protection Agency's (US-EPA) current leaching test is known as the Toxicity Characteristic Leaching Procedure (TCLP) *(1)* and that of the province of Ontario, Canada, is known as Regulation 309 *(2)*. The pollutant concentrations in the TCLP and Regulation 309 leachates are compared with listed maximum limits, and if any one of the pollutant concentrations is equal to or greater than these limits, the waste has failed the test. Materials that fail the tests are regulatorily interpreted to present a hazard to the environment and must be disposed appropriately as hazardous wastes.

Most of the leaching test work, including the regulatory protocols, has employed the batch-shaking procedure. It involves contacting loose granular material with a selected leaching medium, such as water or dilute acetic acid, on a mechanical shaker or some other shaking device. Choice of parameters, such as liquid/solid ratio, pH, type of shaking apparatus, and mode and nature of leaching medium, have been largely arbitrary. It is important to consider the various physical and chemical phenomena affecting the leachability of elements in solid wastes. The question of whether equilibrium concentrations of the various elements have been achieved at the end of the leaching period has to be addressed.

EXPERIMENTAL

Sample Collection and Preparation

Fly ash samples from municipal incinerators with different burn technologies and different flue gas clean-up technologies were used. Where possible, an effort was made to collect the fly ash, bottom ash, and the composite mixed ash. The ash samples were air-dried and homogenized to a 200 mesh size.

Laboratory Leaching Test

To 5.0-g samples in 250-ml polyethylene bottle were added 100 mL of the leaching medium. The leaching media employed were water initially at pH 7, and ammonium acetate buffer at pHs 4, 5, 7, 8, and 10. The samples were agitated on a wrist-action shaker for various periods of time ranging from 12 h to several days. At sampling times, a 1-mL aliquot of the leachate was removed, centrifuged for 10 min, and filtrated through 0.45-μm filter.

Instrumental Neutron Activation Analysis (INAA)

The liquid samples from the leaching test were pipeted into polyethylene vials and heat-sealed. Neutron irradiations were performed at the University of Toronto Slowpoke-2 Reactor facility, which provides neutron flux in the range 1×10^{11} to 1×10^{12} neutrons·cm^{-2}·s^{-1}. γ-Ray spectra of samples and standards were acquired by an APTEC/NRD hyperpure Ge detector (active vol 93 cm^3; 21.6% efficiency relative to Na[I] detector; resolution 1.90 Ke FWHM at 1.33 MeV) interfaced to an APTEC/MCARD2 PC-based Multi-Channel Analysis/Windows 286 (PCMCA/WIN) system with 8192 channels.

RESULTS AND DISCUSSION

Total Elemental Concentrations

The municipal incinerator ash studied contained 563 mg/kg cadmium. Arsenic varied from 6 mg/kg in hospital incinerator ash to 106 mg/kg in the case of municipal incinerator ash. Mercury varied from 1 mg/kg in the case of coal fly ash to 7 mg/kg in municipal incinerator ash. Selenium varied from 13 mg/kg in hospital incinerator ash to 50 mg/kg in municipal incinerator ash. The pattern is generally similar with many other elements. The municipal incinerator ash, in particular, has high concentration of many elements, thereby posing a significant environmental health hazard.

Effect of pH on Leaching

The acetate buffer used had such a strong buffering capacity that all samples were brought to and maintained at their respective pH throughout. Only the sample leached with water showed pH fluctuations.

Figures 1–4 show the effect of pH and leaching medium on leachability. For elements like Br, Cl, and Na, as the pH changed there was no significant change in amount leached. For Al and V and other "high-boiling-point" elements, the amount leached is dependent on pH with the lower pH leaching more of the element. For Cd, the least leachable pH is 8. It is more leachable at pH, 10, 4, 5, and 7 than at pH 8. Of the leaching media pH 4, which was the lowest pH studied, was the most potent, and water was the weakest leaching medium.

Leaching Trends

Figures 5 and 6 are representative of the trends observed. For aluminum, as shown in Fig. 1, in some samples, there is a steady increase of element leached with time, and the concentration does not level off. For some samples, the concentration rises, reaches a peak, and then goes down and levels off again. In the case of Cd (Fig. 4), a high concentration is achieved in the early stages, and then it decreases with time. For some elements, like Br, Cl, Na, and Ca, the peak leachable amount is attained within the first few hours.

These above trends indicate that most elements do not achieve equilibrium concentrations at the end of the regulatory protocols' shake extraction times, such as the 24 h for Ontario Regulation 309 and 18 h for the USA-EPA. The leaching trends observed correlate well with the boiling/condensation theory *(7–12)*. A "volatilization-condensation" mechanism occurs when low-boiling-point elements, such as Cd and Hg,

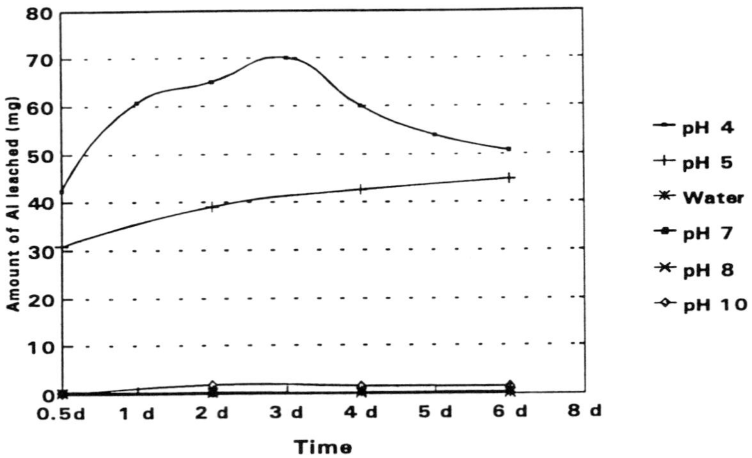

Fig. 1. Al leached with time and pH.

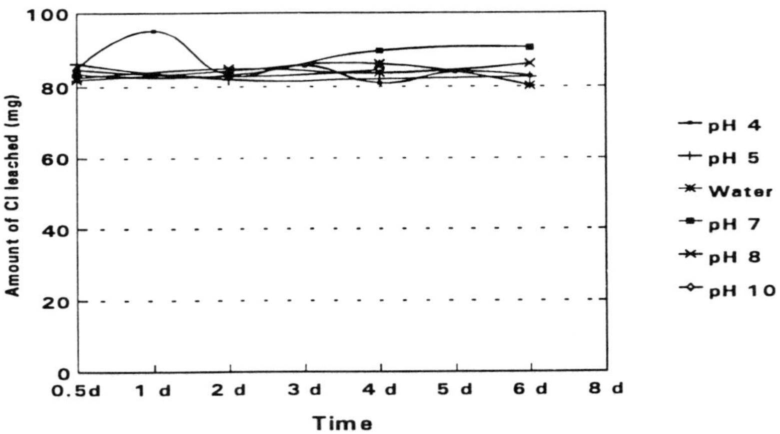

Fig. 2. Cl leached with time and pH.

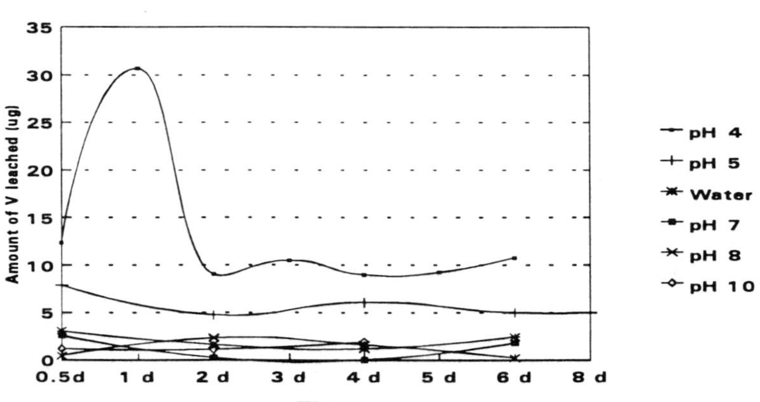

Fig. 3. V leached with time and pH.

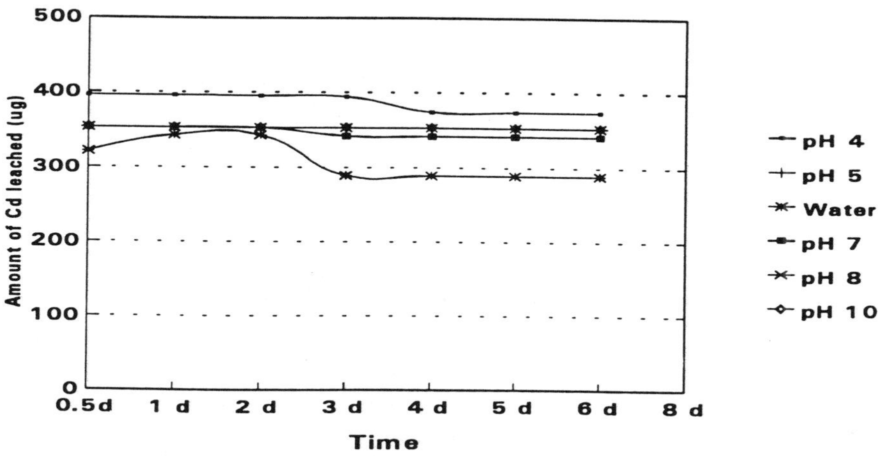

Fig. 4. Cd leached with time and pH.

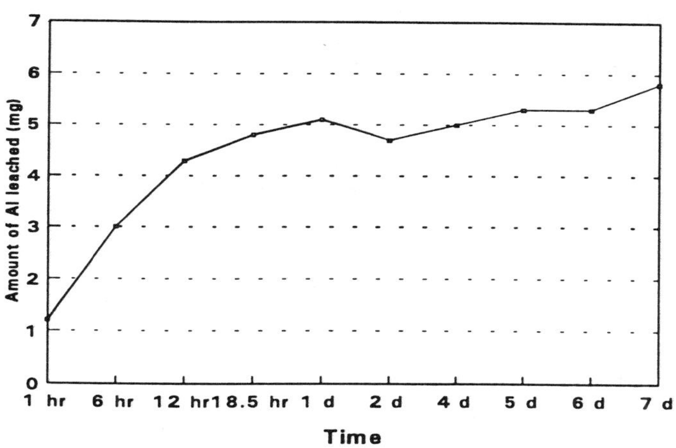

Fig. 5. Al leaching in MSW.

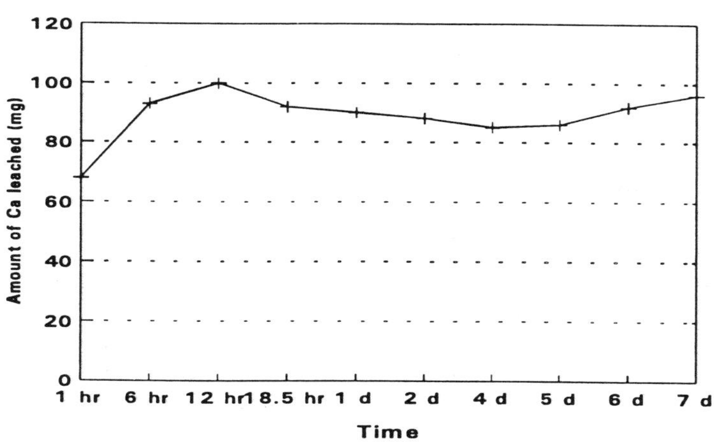

Fig. 6. Ca leaching in MSW.

volatilize in the high-temperature combustion zone and then condense at lower temperatures onto the surfaces of less volatile elements, such as Mn, Si, and Al. The partitioning of the elements between the bottom ash and fly ash determines the matrix structure of the respective ash. The extractive behavior of the respective ash, which determines the leaching of toxic contaminants into the environment, is related to the thermal chemistry. The elements most likely to be leached are those that occur principally as surface-deposited. The amounts of low-boiling-points elements, e.g., As, Br, Cd, Cl, and Hg, actually leached in this study are in the region of 75% and greater, indicating a very high extractability, whereas elements with high boiling points, e.g., Al, Fe and the rare earths, show low extractability. This implies that toxic elements, such as As, Cd, and Hg, are an environmental threat when disposed of in waste disposal sites. When the wastes are dumped in landfill sites and in ash ponds, the concomitant leaching will endanger soils, surface, and ground waters.

Solidifying Solid Waste Ashes

An important aspect of incinerator ash management is the solidification of these wastes so that the toxic elements are fixed to the matrix, thereby rendering them nonleachable. One method we are currently studying is the fixing of these elements to bitumen. Preliminary results show that leachability of all the elements is insignificant when bound to bitumen.

Management of Incinerator Input

In many situations, e.g., hospital wastes containing biological wastes, the incineration process cannot be avoided for a variety of reasons, e.g., the lack of effective alternative processes and the lack of funding for changeover of the system within a short time. In these instances, the legitimate concerns of the hospital waste managers and the community must be addressed. Here a careful inventory of the waste is very useful. Our studies show that among the municipal or hospital wastes, the components that contain the most cadmium are common plastic items, such as colored containers. If ordinary items, such as bottles, plastics, and colored containers, are not added or removed and the waste strictly restricted to essential wastes that have to be incinerated because of biological contamination, the level of cadmium in the hospital incinerator ash and the air can be minimized. Another possibility is to acid or water wash the ashes for an hour by which up to 75% of the cadmium content is leached out. The rest of the solid waste containing the nonleachable residuals can then be sent to landfill sites.

The careful understanding of the distribution of toxins among the waste contents and the control of these are very essential. Similarly, careful control of the conditions to minimize leaching is also important. Procedures of fixing the toxins will also help minimize environmental contamination.

REFERENCES

1. US EPA, 1986, *Fed. Reg.* **51(9),** 1750–1758, (Jan. 14, 1986).
2. Ontario Ministry of the Environment, *Regulation 309, under the Environmental Protection Act amended by O. Regulation 175/83; O. Reg. 574/84; O. Reg. 322/85; O. Reg. 464/85; and O. Reg. 460/88* (1988).
3. A. Tessier, P. G. C. Campbell, and M. Bisson, *Anal. Chem.* **51,** 844–851, (1979).
4. C. L. Perket and W. C. Webster, *Hazardous and Industrial Solid Waste Testing, ASTM STP 805,* American Society for Testing and Materials, Philadelphia, PA, (1981).
5. A. Wadge and M. Hutton, *Environ. Pollut.* **48,** 85–99, (1987).
6. R. M. Harrison, D. P. H. Laxen, and S. J. Wilson, *Environ. Sci. Technol.* **15,** 1378–83, (1981).
7. R. L. Davison, D. F. S. Natusch, J. R. Wallace, and C. A. Evans, Jr., *Environ. Sci. Technol.* **8,** 1107–1113, (1974).
8. D. F. S. Natusch, J. R. Wallace, and C. A. Evans, Jr., *Science* **183,** 202(1974).
9. D. H. Klein, A. W. Andrew, J. A. Carter, J. B. Emery, C. Feldman, W. Fulderson, W. S. Lyon, J. C. Ogle, Y. Talmi, R. I. van Hock, and N. Bolton, *Environ. Sci. Technol.* **9,** 973 (1975).
10. T. R. Keyser, D. F. S. Natusch, C. A. Evans, Jr., and R. L. Linton, *Environ. Sci. Technol.* **12,** 768 (1978).
11. L. D. Vela, R. E. Jervis, and S. S. Krishnan, *J. Radioanal. Nucl. Chem., Articles* **161,** 181–187, (1992)
12. L. D. Vela, M.A.Sc. Thesis, University of Toronto.

Study of a Low-Selenium Environment in China by INAA and Mössbauer Spectrometry

CHIFANG CHAI,*,[1] JIBIN TIAN,[1] QINFANG QIAN,[1] PEIQUN ZHANG,[1] QIN XU,[1] AND DAJUN MAO[2]

[1]*Institute of High Energy Physics and Laboratory of Nuclear Analysis Techniques, Academia Sinica, P.O. Box 2732, Beijing 100080, China; and* [2]*Epidemic Prevention Station of Exi Autonomous Prefecture, Hubei 445000, China*

ABSTRACT

The neutron activation analysis, γ coincidence spectroscopy, nondispersive hydrogen flame atomic fluorescence spectroscopy, and Mössbauer spectrometry were used to study the low-selenium environment of the Exi Autonomous Prefecture, a well-known Keshan disease region. The Se contents in the soil samples there range from 0.075–0.18 mg/kg with the average of 0.13 mg/kg, whereas in the maize from 0.001–0.018 mg/kg with the average of 0.0099 mg/kg. The ^{57}Fe Mössbauer spectrum of the soil indicates an anoxic environment. In addition to the Fe^{3+} species, the compounds containing low-valence iron, e.g., goethite, and so forth, also exist. The rare earth element (REE) pattern obtained by NAA further confirms the reductive soil environment, which causes the selenium deficiency.

Index Entries: Se; low-Se environment; NAA; Mössbauer spectrometry.

INTRODUCTION

Selenium is one of the trace elements essential to human beings. In addition to being an active component of glutathione peroxidase *(1)*, a number of proteins containing Se have been recently reported. Although the correlation of Se deficiency with the Keshan disease has been stud-

*Author to whom all correspondence and reprint requests should be addressed.

ied for quite a few years and some valuable results have been introduced *(2)*, so far it is still a controversial subject.

The Se-deficient regions are widespread in China, exhibiting a broad Se-deficient belt running from northeast toward southwest. The Exi Autonomous Prefecture, Hubei Province, is located in the middle of this belt. It belongs to the northern subtropical zone, and the soil there is of a weakly acidic yellow and brown type with evergreen broadleaf, and deciduous forest and broadleaf-mixed forest. The soil-forming source rock is the brown sandstone.

Parts of Lichuan County, Xianfeng County and Hefeng County of the Prefecture are the well-known Keshan disease regions. In this work, neutron activation analysis (NAA), γ coincidence spectroscopy, and nondispersive hydrogen flame atomic fluorescence spectroscopy (NDAFS) were used to measure the Se contents of the soil and botanic samples in this region, whereas the Mössbauer spectroscopy (MS) was used for study of the valent state of the Fe-containing compounds, in order to elucidate the environmental impact on the Se deficiency.

EXPERIMENTAL METHODS

Sampling

The representative soil and maize samples were taken from the Keshan disease region in Exi. In the meantime, the soil and maize from the normal region were also sampled. The soil samples were subject to air-drying, pulverizing, and homogenizing, and then ground in an agate mortar and 100-mesh sieved. After drying at 105°C for 3 h and cooling, the samples were stored for use. A volatile loss of Se during the sample preparation was examined by analysis of the certified reference material of soil-3 provided by the IAEA. The experimental results indicated that the volatile loss can be neglected.

The maize samples were dried at 60°C, ground, and homogenized for use. Also, the volatile loss of Se in preparation process was negligible.

Analytical techniques

NAA

About 40-mg samples wrapped in a high-purity aluminum foil were irradiated for 8 h in a heavy-water-cooled reactor of the Institute of Atomic Energy, Beijing, at a neutron flux of $7 \times 10^{13}\ n \cdot cm^{-2} \cdot s^{-1}$. The samples were counted by a PC-based multichannel γ spectrometer after being cooled for different times. The energy resolution and relative efficiency of an HPGe detector were 1.8 keV at 1332.5 keV line and 30%, respectively. The quantitative results were automatically printed out by a software package developed in our laboratory. The SRM 1632a coal of NIST, USA, and Soil-3 and Soil-5 of IAEA were used for analytical quality control.

NDAFS

A model WFX-3 atomic fluorescence spectrometer made by Wenzhou Instrument Factory, China, and a Hitachi recorder were used. The chemicals used were 70% KBH_4, HCl3 mol/L, conc. HNO_3 and H_2SO_4, and a Se standard solution.

About 1 g soil or maize sample was weighed and heated rigorously in a 150-mL beaker with 2 mL of dilute H_2SO_4 (1:1) and 10 mL of conc. HNO_3. After cooling for some time, it was further heated to near dryness. After cooling, 5 mL of conc. HCl were added, and the sample solution was transferred to a calorimetric tube. Its fluorescence intensity was recorded, and the Se concentration was read out from a standard working curve.

γ Coincidence Spectroscopy

We used the two peaks (135.9 and 264.5 keV) of ^{75}Se in a γ coincidence way to avoid the interference from some nuclides emitting rays of similar energy, e.g., ^{181}Hf, ^{182}Ta, and so on. The 136-keV γ rays of ^{75}Se were used as a gate signal outputted by an NaI(Tl) crystal, and 264.5 keV γ ray of ^{75}Se were recorded by an HPGe detector, surrounded by an annular plastic scintillator to suppress the Compton scattering and interfering γ rays in an anticoincidence way. The γ coincidence counting system has the Se sensitivity of 0.01 μg/g.

Mössbauer Spectrometry

The methodology and Mössbauer spectrometry equipment were as in ref. 3.

RESULTS AND DISCUSSION

Comparison of Analytical Methods of Se

In this work, three methods were used to determine the Se contents in soil and botanic samples from μg/g to ng/g range. The γ coincidence spectroscopy only serves as a reference method for analysis of Se, mainly because of the long counting time. However, its accuracy and sensitivity are satisfactory. It can be used to determine Se content reliably in a wide concentration range.

The conventional NAA (CNAA), generally, is suitable for the biological and environmental samples containing relatively high Se concentration. In case of low-Se samples, its accuracy is doubtful, especially for the low-Se soil samples, in which some nuclides (e.g., ^{181}Hf, ^{152}Eu, ^{182}Ta, and so forth) constitute severe interference.

As to the NDAFS technique, it has high sensitivity for Se determination and is able to determine Se content as low as 1 ng/g easily with good accuracy. In our practice, the CNAA is mostly used in the Se range of 0.1 to 10 μg/g.

Se Contents in Soil and Maize Samples of Low-Se Region of Exi Autonomous Prefecture

Listed in Table 1 are the Se contents in the soil and maize samples from the low-Se Exi region, determined by conventional, γ coincidence counting NAA and NDAFS, where also enclosed are the Se contents in the normal districts and the literature values *(4)* for comparison.

It can be seen from Table 1 that the Se contents in the Exi low-Se soil samples range from 0.075–0.18 µg/g with the average of 0.13 µg/g. Compared with the normal average value of 4.14 µg/g, it is lower by a factor of 31.8, but close to the Se values of other low-Se belts (0.19, 0.09, and 0.23 µg/g for the northeast, middle, and southwest low-Se soil, respectively).

The Se contents in the Exi low-Se maize samples vary from 0.0010–0.018 µg/g with the average of 0.0099 µg/g. Compared with the normal average value of 0.19 µg/g, it is lower by a factor of 19.2. A significant fact is that, unlike the low-Se soil, the Se contents in the low-Se maize samples are not comparable with those of other low-Se belts, but even lower. This implies that a part of Se in the Exi soil does not exist as a bioavailable species, thus not absorbable by plants.

^{57}Fe Mössbauer Spectra of Low-Se Soil at Exi

The MS technique can be applied to study the valency of iron in environmental samples. In order to decipher the formation of low-Se environment, we used MS to determine the ^{57}Fe Mössbauer spectra of the low-Se and normal-Se soil samples from Exi (Fig. 1A and B). The former consists of a paramagnetic central dual peaks and ferromagnetic six-line spectrum as shown in Fig. 1A, considerably different from the ^{57}Fe MS spectrum of the normal soil as shown in Fig. 1B. The results of evaluation of these MS spectra are listed in Table 2.

Compared with the normal soil, the low-Se soil shows a complicated ^{57}Fe MS spectrum, in which hematite, goethite (α-FeOOH), magnetite, and so on, are responsible for the six-line components, exhibiting a reductive characteristic of the low-Se soil.

It is well-known that the major factor affecting the Se species in soil is whether there exist the crypto-crystalline and amorphous compounds of silicon, aluminum, and ferrous oxide in soil. Under acidic and anoxic circumstance, the goethite, amorphous ferrous oxide, and empholite strongly adsorb selenite. In addition, the adsorption of the selenite on goethite is irreversible. As a result, the Se content in soil environment available to plants severely decreases, aggravating the Se deficiency in the ecosystem *(5)*. This is the reason for the lower Se content in the maize samples from low-Se Exi region, Hubei province.

Rare Earth Element (REE) Pattern of Low-Se Soil of Exi

The REE patterns of the low-Se soil samples from Exi, Hubei Province, determined by NAA are shown in Fig. 2A. For comparison, the

Table 1
Se Contents of Soil and Maize Samples of Low-Se Regions at Exi Autonomous State, Hubei Province (μg/g)

Sample No.	Soil	Sample No.	Maize
S-1	0.0975±0.01	M-1	0.012±0.001
S-2	0.0975±0.01	M-2	0.0066±0.0007
S-3	0.140±0.007	M-3	0.0066±0.007
S-4	0.076±0.005	M-4	0.011±0.002
S-5	0.18±0.01	M-5	0.0051±0.0007
S-6	0.075±0.002	M-6	0.0026±0.0007
S-7	0.13±0.04	M-7	0.015±0.002
S-8	0.140±0.007	M-8	0.0010±0.0005
S-9	0.14±0.01	M-9	0.013±0.002
S-10	0.14±0.02	M-10	0.018±0.002
S-11	0.15±0.01	M-11	0.018±0.002
S-12	0.17±0.02		
S-13	0.15±0.02		
S-14	0.15±0.02		
S-15	0.13±0.02		
S-16	0.17±0.02		
Range	0.075-0.18		0.0010-0.018
Average	0.13±0.03		0.0099±0.006
Range (Normal Region)	2.12-6.67		0.055-0.28
Average (Normal Region)	4.14±1.90		0.19±0.12
Northeast low-Se belt (China)	0.19		0.049
Middle low-Se belt (China)	0.09-0.17		0.013-0.014
Southwest low-Se belt (China)	0.23		0.053

REE patterns of the normal-Se soil samples are diagrammed in Fig. 2B. Their main differences lie in the following two aspects: (1) the total REE contents of the low-Se soil are lower than those of the normal soil by a factor of 2, i.e., the total amount of REE in the former is about 120 ppm, but 240 ppm for the latter, and (2) the low-Se soil exhibits weakly negative Eu-anomaly and positive Ce-anomaly.

The first difference indicates the loss of REE in the low-Se soil, which may be caused by strong leaching and also related to the soil-forming source rock. The rich rainfall and severe soil erosion in this region not only result in the loss of REE, but also in Se. The second difference means a reduction in the type of soil environment, confirming the MS result

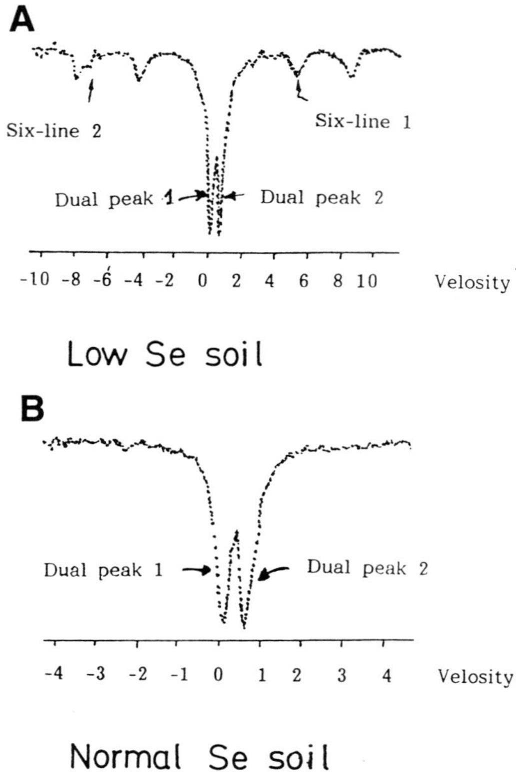

Fig. 1. Fe-57 Mössbauer spectra of the soil samples at Exi Autonomous State, Hubei Province. (**A**) Low-Se region; (**B**) normal region.

Table 2
Fe-57 Mössbauer Spectrum Parameters of Low-Se Soil at Exi, Hubei

Group No.	IS	QS	H_{hf}	$A_i/\Sigma A_i$
Dual Peak 1	0.36mm/s	0.55mm/s		47 %
Dual Peak 2	0.35mm/s	1.0mm/s		18 %
Six-line 1	0.35mm/s		509T	17 %
Six-line 2	0.37mm/s		476T	21 %

Notes : IS-isomer shift ; QS- quadrupole splitting ; H_{hf}-hyperfine magnetic field.

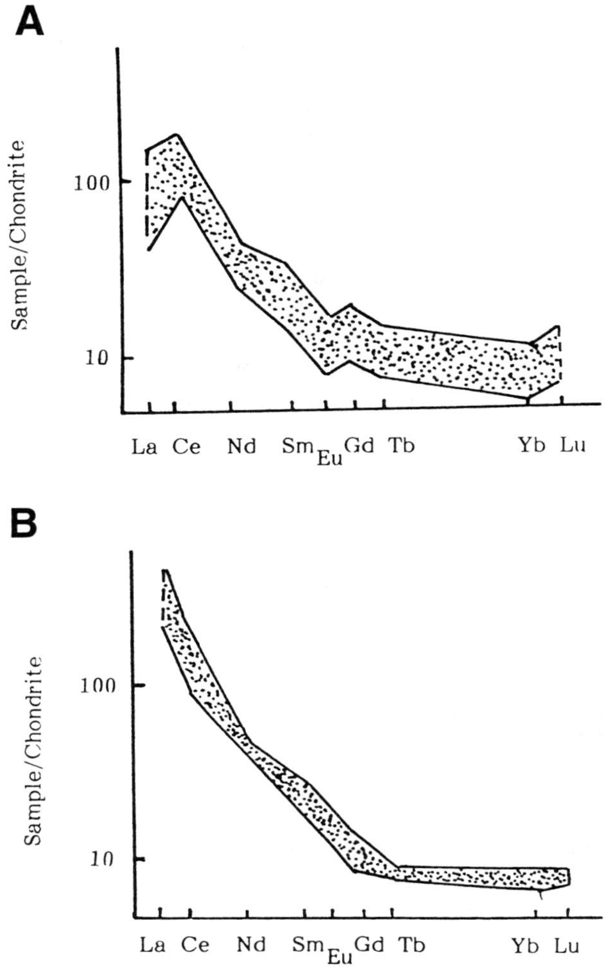

Fig. 2. REE patterns of the soil samples at Exi Autonomous State, Hubei Province. (**A**) Low-Se region; (**B**) normal region.

mentioned in the previous paragraph. The anoxic environment with the weakly acidic soil makes the available Se content for the plant growth in the low-Se soil even lower, causing serious Se deficiency.

SUMMARY

1. The Se contents in the low-Se soil samples from the Exi Autonomous Prefecture range from 0.075–0.18 μg/g with

the average of 0.13 μg/g similar to those of other low-Se soil, whereas in the maize, it ranges from 0.001–0.018 μg/g with the average of 0.0099 μg/g, being lower than the Se contents in the maize samples from other low-Se regions. This indicates that part of Se in the soil exists as a species not absorbed by plants.

2. The ^{57}Fe Mössbauer spectrum of the soil there indicates an anoxic environment. In addition to the Fe^{3+} species, compounds containing lower valence iron, e.g., goethite, and so forth, also exist.
3. The REE patterns of the low-Se soil indicate a strong leaching, resulting in the Se loss. Further, the weak Eu-negative anomaly also confirms the reductive soil environment.

ACKNOWLEDGMENTS

We thank Q. Su, J. Rong, G. Z. Yang, S. L. Tang, Y. Z. Zhu, K. Wand, H. F. Hu and H. R. Shao for their help and support in this work. We also thank the National Natural Science Foundation of China (NSFC) for the grant for this project.

REFERENCES

1. M. Behne, S. Scheid, H. Hilmert, H. Gessner, D. Gawlik, and A. Kyriakopoulos, *Nuclear Analytical Methods in the Life Sciences,* R. Zeisler and V. P. Guinn, eds., Humana Press, Clifton, NJ, 1990, p. 439–448.
2. Keshan Disease Research Group, *Chinese Med. J.* **92,** 471 (1979).
3. J. B. Tian, Study of analytical methods and biological effect of essential trace element Se, Master's dissertation, IHEP, Beijing, 1990.
4. J. A. Tan, *J. Environ. Sci.* **1,** 107 (1989).
5. D. Rai and J. Zachara, Report EA-3356, EPRI by Pacific Northwest Lab. (Battle Institute), Richland, WA, 1984.

Analysis of Membrane Filters and Thick Fly Ash Samples by PIXE

V. HAVRÁNEK,*[,1] V. HNATOWICZ,[1] J. KVÍTEK,[1] AND I. OBRUSNÍK[2]

[1]*Institute of Nuclear Physics, Řež near Prague; and* [2]*Czech Hydrometeorological Institute, Czech Republic*

ABSTRACT

The possibility of the routine PIXE measurement of environmental samples has been tested at the INP in Řež. Both thin samples of aerosols collected on membrane filters Synpor and Nuclepore and thick aerosols and fly ashes samples pressed into pellets were analyzed. The Nuclepore filter was found more suitable for PIXE analysis of aerosols than the Synpor, and the thin aerosol sample analysis was found to be easier than the thick sample analysis. The grain size effect in the case of thick fly ash samples is discussed.

Index Entries: X-ray analysis; PIXE; thick and thin target; aerosol analysis; fly ash analysis.

INTRODUCTION

The particle induced X-ray emission (PIXE) method is now used in a broad field of applications, and among them the aerosol studies play an important role *(1)*. The multielement character and high absolute sensitivity make PIXE very suitable for analysis of the aerosols samples collected on thin-membrane filters. Furthermore, there is a possibility of combining PIXE method with other ion-beam analysis (IBA) methods, such as proton-induced γ-ray emission, (PIGE) and particle elastic scattering (PESA), which provide us with additional information about the sample composition *(2)*, including the hydrogen and some other light-element content. More information about PIXE can be found in the review papers *(3–6)* or in the book *(7)*.

*Author to whom all correspondence and reprint requests should be addressed.

Biological Trace Element Research Editor: G. N. Schrauzer © 1994 by Humana Press Inc.

At the INP Řež, the PIXE method was introduced around the 3-MeV Van de Graaff accelerator approx 10 years ago *(8)*. Since that time, it has been applied for analyses of both thin and thick target samples of biological, geological, technical, and other origins, often together with the Rutherford back scattering (RBS) technique. Recently, the PIXE setup was adapted and tested for aerosol and fly ash sample analysis. We examined two kinds of samples, the thin aerosol samples collected on membrane filters Synpor and Nuclepore, and the thick aerosol and fly ash samples pressed into pellets with specpure graphite as a binder in the weight ratio 1:1.

METHODS

The experimental arrangement of the PIXE vacuum target chamber and data acquisition system is shown in Fig. 1. The proton beam (1–3 MeV) comes from the accelerator, passes through the 1.5-μm thick diffusion foil, then it is collimated by means of two carbon collimators, and passes through a 0.5-μm Ni monitor foil before it hits the sample. The beam is 5 mm in diameter, and the beam current is adjustable in the 0.1–20 nA range at the sample position. Another carbon collimator is placed between the monitor foil and the sample in order to stop protons, which are scattered by the monitor foil to forward angles. Some of the characteristic X-rays produced in the sample by the proton beam escape through a thin exit window (10 μm of mylar) out of the chamber, where they are detected by an Si (Li) detector. Outside the chamber, between the exit window and the detector, a suitable X-ray attenuation filter can be inserted.

The electronics consists of two parts, the monitor line and the main spectroscopic chain, which includes Si (Li) detector fy Enertec (80 mm^2 active area, 200 eV/5.9 keV energy resolution), Canberra Ci 2024 fast amplifier with pile-up rejector, ND 581 analog-digital converter, and a PC computer with ACCUSPEC multichannel buffer board. The monitor counts are corrected on line for the system dead time. It was found that this setup gives good deadtime corrections up to 40% spectrometer dead time.

The PIXE setup has been calibrated by the set of thin standards. Both the commercially available Micromatter standards *(9)* and our own standards calibrated by the RBS method have been used. The X-ray spectra evaluation was performed with the new version of the computer program package PIXE-INP for spectra fitting and quantitative analysis *(10)*. The PIXE-INP code package is based on the digital filter background subtraction, nonlinear least-squares fitting with Gaussian line shape and fixed intensities for shell's multiplets, standard matrix correction for thick samples *(11)*, and filter attenuation calculation. The theoretical sensitivity curves for different filters calculated by the program together with the experimental points obtained from the thin standards are shown in Fig 2.

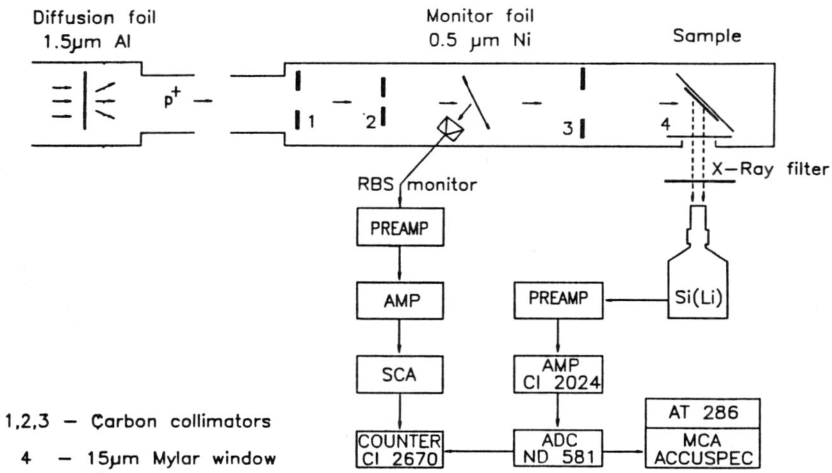

Fig. 1. Schematic view of the PIXE setup at INP.

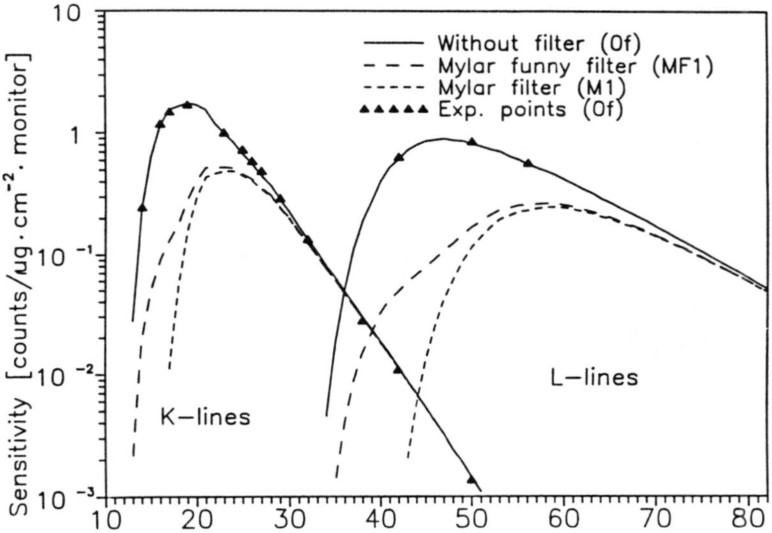

Fig. 2. Theoretical sensitivity curves for the thin target calculated by the PIXE-INP code for three different X-ray attenuation condition: Of—measurement without X-ray filter, MF1—with 0.3-mm mylar pinhole filter with 7% active area central hole, M1—with 0.3-mm mylar filter. The experimental points from the thin standard calibration are shown by filled triangles. The sensitivity is related to the counts of the PIXE monitor based on back scattering of protons. The monitor counts and the total collected proton charge are linked by the scale factor: 10,000 monitor counts ≈ 1 μC collected charge.

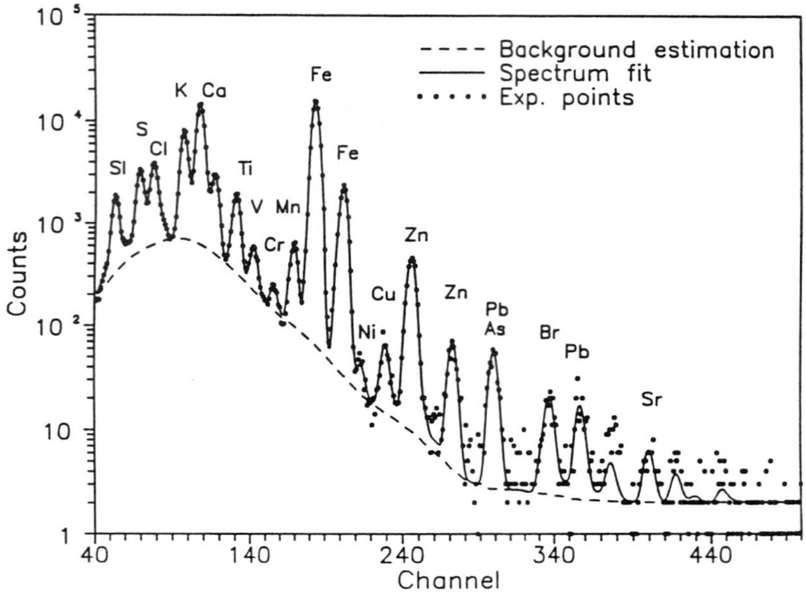

Fig. 3. Typical PIXE spectrum of the aerosol sample collected on the Nuclepore filter measured with mylar funny filter (MF1) and fitted by PIXE-INP code. The proton beam energy was 2.35 MeV and the collected charge 2 μC.

RESULTS AND DISCUSSIONS

All the samples of aerosols, fly ashes, reference materials, and standards were measured at 2.35 MeV proton beam (5 mm in diameter), in the same geometry and with proton current between 1 and 20 nA depending on a sample composition. Different filters were used to reduce intense low-energy X-rays. The advantage of the aerosol samples collected on thin membrane filters (Nuclepore, Synpor) is that they do not need a special sample preparation. They were fixed on the Plexiglas™ frames and placed into the sample holder inside the vacuum target chamber. Figure 3 shows a typical X-ray spectrum of the thin aerosol sample on Nuclepore filter measured with mylar "funny" X-ray filter (0.3-mm thick, 7% central hole). This is one of the samples collected on Synpor and Nuclepore filters near the municipal incinerator at Brno in 1992. The results from Synpor and Nuclepore filters are compared in Table 1. Also, the blank determination of unexposed filters was done (*see* Table 2).

It is seen from Table 1 that the elemental concentrations in the air found in the Synpor samples are generally lower, but the element contents found on the Synpor and Nucleopore filters are correlated, except the sulfur and chlorine for which the relative concentrations differ considerably (higher S on Synpor and Cl on Nuclepore). The same trend can be observed also for the other couple of filters (Synpor and Nuclepore exposed on the same site). Also, the ratio of the mass of elements observable in the PIXE spectra and gravimetric mass of collected aerosol are

Table 1
Elemental Concentrations Found by PIXE Expressed Both in ng/cm^2 and ng/m^3 for Aerosol Samples Collected on Synpore and Nuclepore Filters at the same Place Near the Municipal Incinerator at Brno in 1992[a]

Element	Synpor filter concentration on sample ng/cm^2	Synpor filter concentration in air ng/m^3	Nuclepore filter concentration on sample ng/cm^2	Nuclepore filter concentration in air ng/m^3
Si	3500(500)	2100(300)	10100(1000)	4900(500)
S	4600(600)	2800(400)	3600(400)	1750(190)
Cl	<300	<200	2900(300)	1400(150)
K	1350(300)	820(180)	3000(330)	1440(160)
Ca	1300(300)	800(180)	3600(380)	1750(180)
Ti	82(18)	50(11)	340(40)	170(20)
V			22(4)	11(2)
Cr	18(7)	11(5)	13(4)	6(2)
Mn	40(6)	24(4)	120(15)	60(6)
Fe	1200(130)	730(80)	4300(440)	2100(210)
Ni			12(3)	6(2)
Cu	16(4)	10(4)	32(9)	16(4)
Zn	130(20)	80(14)	280(30)	135(15)
As			21(4)	10(2)
Br			30(6)	15(3)
Rb			19(5)	9(3)
Sr			25(11)	12(5)
Pb	143(34)	87(21)	134(17)	65(8)

[a]The Synpor filter was exposed 2d later than the Nuclepore one. The total air flow/cm^2 and the total collected aerosol mass/cm^2 were 1.64 m^3 and 0.155 mg for the Synpor and 2.06 m^3 and 0.143 mg for Nuclepore filters, respectively.

higher for Nuclepore filters, ca. 22% compared to about 10% for Synpor. This is probably because of different filter efficiencies and the accuracy of the filter weighing, which seems to need additional tests. Nevertheless, the Nuclepore filter appears to be more convenient for PIXE analysis of aerosols. The major advantages of the Nuclepore filter are:

1. The possibility of obtaining lower detection limits (thinner filter has lower spectral background);
2. Lower contamination compared to the Synpor filter (*see* Table 2), which means easier blank corrections;
3. More realistic visible mass (better accuracy of the aerosol elemental concentration in the air);

Table 2
Trace Element Concentrations (Blank) in the Synpore and Nuclepore Membrane Filters used[a]

Synpor concentration				Nuclepore concentration			
Si	350(40)	Ca	500(55)	S	93(7)	Fe	4(1)
S	600(60)	Ti	30(7)	Cl	43(6)	Zn	6(2)
Cl	420(45)	Fe	30(5)	Cr	8(2)		
K	120(16)	Zn	15(5)				

[a]All concentrations and errors (in brackets) are in $ng \cdot cm^{-2}$. The real trace element concentrations in the filter (especially in the Synpor one) are larger because of the filter thickness, but for us only that part that is directly added to the thin aerosol spectrum is important.

4. Higher beam resistance leading in principle to shorter analysis time and
5. Determination of more elements in one measurement with a better precision, for the same air loading and the same analysis time.

Thick samples of fly ash and urban particulate matter require more preparation. The material is mixed in 1:1 ratio with specpure graphite and then pressed into small pellets (12-mm diameter, 1–2 mm thickness). Some of the thick samples were doped by yttrium serving as an internal standard. The results for fine fly ash (RM CTA-FFA-1) and urban particulate matter (SRM 1648) analysis are shown in the Table 3. The measured values were obtained using standard thick-sample corrections *(10)*, assuming that the reference material and the graphite are homogeneously mixed together. These two materials have fine particle structure, so the previous assumption is fulfilled and the measured and certified values agree well.

Different results were obtained for coal fly ash reference materials (RM-EOP, RM-ENO, RM-ECH, and SRM 1633A). The mean measured concentrations (Fig. 4.) for the elements with lower X-ray energy are significantly lower than the certified ones if the standard matrix corrections are applied. This can be explained as a particle size effect. This effect emerges when the characteristic X-ray production thickness (given by the proton stopping power and X-ray attenuation) is similar to the grain size of fly ash. In this case, substantial parts of proton stopping or X-ray attenuation processes might take place inside the fly ash grain, so the thick target correction should be calculated for the pure fly ash matrix (P_i) rather than for homogeneously mixed fly ash and binder matrix (H_i). Generally, for arbitrary but fixed grain size distribution, we can ex-press the thick target correction (K_i) for the characteristic X-ray line i, as a linear combination of these two limit cases using a mixing parameter α:

$$K_i = \alpha \cdot P_i + (1 - \alpha) \cdot H_i \qquad (1)$$

Table 3
Comparison of Elemental Concentrations Determined by PIXE with the Certified/Consensus Values in Urban Particulate Matter SRM 1648 (NIST) and Fine Fly Ash CTA-FFA-1 Reference Materials[a]

Element	Concentration in sample			
	SRM 1648		CTA-FFA-1	
	Certified	Measured	Certified	Measured
Si %	13(1)	12.9(7)	22.5(9)	19.6(10)
S %	5.2 I	7.9(4)		0.58(5)
K %	1.05(5)	1.07(6)	2.20 I	2.17(12)
Ca %	5.83(33)	6.56(33)	2.29 I	1.94(10)
Ti	4070(200)	4020(210)	5800 I	6700(400)
V			260(10)	320(70)
Cr	400(14)	370(35)		
Mn	822(45)	720(40)	1070(40)	1160(100)
Fe %	3.92(24)	3.93(20)	4.89(14)	4.5(3)
Ni	82(12)	44(15)	99(6)	90(25)
Cu	600(23)	570(20)	158(9)	130(25)
Zn	4740(70)	5100(260)	570(60)	720(41)
Ga	40 I	21(10)	49 I	64(15)
Se	24(2)	32(8)		
Br	506(25)	580(40)		
Rb	55(3)	40(10)	185(5)	180(30)
Sr	210(15)	240(20)	250(13)	240(40)
Zr	170..I	165(20)		190(40)
Nb	22 I	19(10)		
Cd	72(2)	100(50)		
Sn	147..I	220(80)		
As			54(3)	60(15)
PB	6520(250)	7480(380)	370(5)	460(60)
Y (internal standard)			845(13)	826(70)

[a]Concentrations of elements that are not marked by % are in μg/g.
[b]Value determined without a correction for Pb M-lines.
I = Non certified (information) value.
Error in brackets includes statistical error from spectrum fitting and 5% error of the calibration (54 [3] means 54 ± 3).

The value of the parameter α depends on an aerosol mean grain size. Using Eq. (1), we fitted the experimental ratios of the measured and certified concentrations for thick reference coal fly ash samples, and we obtained $\alpha = 0.56$. The fit is shown in Fig. 4. The value of the parameter

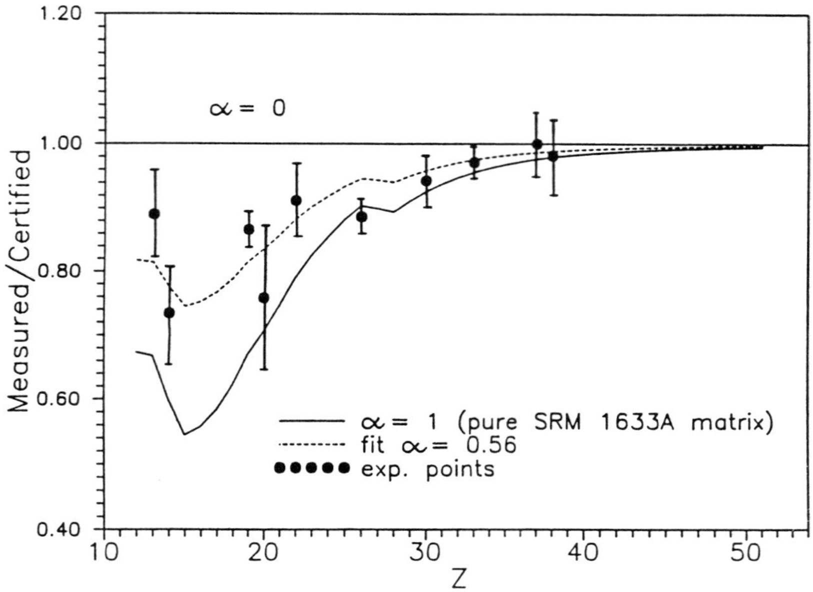

Fig. 4. The particle size effect observed for the coal fly ash pellets (1:1 mixture with graphite). The experimental points (•) are the mean values of the measured to certified concentrations ratios determined from K-lines for four reference material *(see text)*. The lines with various values of parameter α represent three different calculations of the matrix corrections. The best-fit line to the experimental points corresponds to α = 0.56 *(see text* for details).

α can be used to calculate thick target corrections for other fly ash samples with similar grain size and pelletized with binder in the same weight ratio.

Another way to eliminate particle size effect might be to use a binder with X-ray attenuation properties similar to analyzed fly ash. It is a more straightforward way, but it may be difficult to find suitable binder.

SUMMARY

The thin aerosol samples are very suitable for the PIXE method. It was shown that the Nuclepore filter is better than the Synpor one because it is thinner and less contaminated by trace elements. Among different tested attenuation filters used in experimental setup, the 0.3-mm mylar filter with 10% central hole seems to be the most useful for such samples.

On the other hand, thick fly ashes samples are a rather difficult application of the PIXE method, namely because of the complex X-ray spectra and matrix effect corrections. Nevertheless, accuracy (about 10% for most of the elements) can also be achieved for such samples as good as that achieved by analysis of pellets reference materials, if appropriate matrix correction are used. The aluminum "funny" filter (0.2-mm thick, 1.5% pin hole) appears to be the most convenient for thick fly ash and urban aerosols samples.

REFERENCES

1. T. A. Cahill, J. Miranda, and R. Morales, *Int. J. PIXE* **1,** 297 (1991).
2. D. D. Cohen, *Clean Air* **26,** 113 (1992).
3. S. A. E. Johansson and T. B. Johansson, *Nucl. Instr. Methods* **137,** 473 (1976).
4. T. A. Cahill, *Ann. Revs. Nucl. Sci.* **30,** 211 (1980).
5. S. A. E. Johansson, *Chemia Analityczna* **35,** 167 (1990).
6. W. Maenhaut and K. G. Malmquist, *R. E. Van Grieken Handbook of X-Ray Spectrometry,* Marcel Dekker, New York 1992, Part 11, pp. 517–580.
7. S. A. E. Johansson and J. L. Campbell, *PIXE: A Novel Technique for Elemental Analysis,* J. Wiley and Sons, Chichester, 1988.
8. V. Hnatowicz et al., *Czech. J. Phys.* **B34,** 1315 (1984).
9. Micromatter X-Ray Fluorescence Calibration Standards. Micromatter Co., 123 Madrona Lane, P.O. Box 123, Deer Harbor, WA 98243.
10. V. Havránek, V. Hnatowicz, J. Kvítek, and I. Obrusník, *Nucl. Instr. and Methods* **B85**, 637 (1994)
11. J. L. Campbell, J. A. Cookson, and H. Paul, *Nucl. Instr. Methods* **212,** 427 (1983).

Rubidium and Cesium in Spruce Needles

Concentrations and Biodynamics

L. TOBLER,[1] J. BUCHER,[2] V. FURRER,[1] P. SCHLEPPI,[2] AND A. WYTTENBACH*,[1]

[1]Paul Scherrer Institute, CH-5232 Villigen PSI, Switzerland; and [2]Swiss Federal Institute for Forest, Snow and Landscape, CH-8903 Birmensdorf, Switzerland

ABSTRACT

The endogenous concentrations of Rb and Cs have been determined in needles of 56 trees (Norway spruce, *Picea abies*) from eight different sites. Analysis was done by instrumental neutron activation. Concentrations were found to have a very large range (3–28,000 ng Cs/g and 1–190 µg Rb/g). The values on a given site have a tendency toward a log-normal distribution. There is a significant correlation between Rb and Cs, but the correlation is not linear. The concentrations of both elements are a function of the needle age. They decrease smoothly, approaching a constant value, when going from needle age class 1 to 5. It is shown that one algebraic function describes this biodynamic behavior on all sites and at all concentration levels. The function and its parameters are discussed.

Index Entries: Rb; Cs; *Picea abies;* Norway spruce; needles; needle age class; retranslocation; biodynamic function; instrumental neutron activation analysis.

INTRODUCTION

Although there is a large amount of information on the activity of ^{137}Cs in various plant species, very little is known about the stable element Cs. The present work deals with Cs and its chemical homolog Rb in needles of Norway spruce. Its aims are:

*Author to whom all correspondence and reprint requests should be addressed.

1. To establish the concentrations occurring in the foliage of trees from various sites;
2. To check for possible correlations between the concentrations of Rb and Cs; and
3. To establish the concentration changes occurring as the needles get older. Norway spruce keeps its needles for several years, and it is therefore possible to collect, at one date and from one tree, needles differing in age by exact multiples of 1 yr.

The results presented are strictly for endogenous concentrations, meaning that they do not include nonphysiological contributions owing to aerosols present on the needle surface. Surface contaminations were removed by an appropriate washing procedure. Not removing these contaminations would lead to results that are too high, especially with low endogenous Cs concentrations (about a factor of 1.3 at 10 ng/g). At high Cs concentrations and at all Rb concentrations the effect is much smaller. Instrumental neutron activation analysis (INAA) was chosen for this work, because it has the necessary sensitivity to deal with the sometimes very low Cs concentrations (<10 ng/g) and because this multielement method concurrently gives information on other elements.

MATERIAL AND METHODS

Norway spruce trees *(Picea abies)* with an age over 50 yr were sampled at the end of the vegetation period in November by taking branches from the upper third of the crown. Eight different sites in Switzerland and southern Germany, situated over different bedrocks, were sampled. The sites had an area of about 100 · 100 m. They are referred to by their acronyms. The fresh twigs were separated according to their needle age classes (usually from 1 to 5). All subsequent work was done with samples coming from one tree and comprising one needle age class. The twigs were washed with toluene/tetrahydrofuran (1:1), thereby dissolving the epicuticular wax layer and removing any aerosols adhering to the needle surfaces. Leaching of endogenous elements by this washing procedure is <0.1% *(1)*. The washed twigs were dried, and the needles separated from the twig axes and ground. In order to reduce the sample volume for irradiation, the needle powder was ashed at 500°C for 10 h. No losses of Rb were found from ashing, and losses of Cs were <3% *(2)*. All concentrations are reported with respect to needles dried at 85°C.

INAA of the needle ash (120 mg, corresponding to 3 g of dry needles) was used to determine Rb and Cs using the nuclear reactions $^{85}Rb(n, \gamma)^{86}Rb$ (E_γ = 1077 keV, $T_{1/2}$ = 18.6 d) and $^{133}Cs(n, \gamma)^{134}Cs$ (E_γ = 796 keV, $T_{1/2}$ = 2.06 a). The samples were irradiated for 2.5 h at a thermal neutron flux of $2.2 \cdot 10^{13}$ n/s cm^2 in the "SAPHIR" research reactor. Counting was done for 7 h on an HPGe-detector (FWHM 2.1 keV at 1332 keV) after a decay time of 10–30 d. The detection limit was 100 ng/g for Rb and 3 ng/g for Cs. Of the about 250 samples analyzed, none was below the detection limit for Rb, and five were below that for Cs. Precision of analysis (including ashing) was determined by analyzing aliquots of various samples. It was found to be <3% for Rb and 3–10% (according to the concentration) for Cs. Accuracy was checked by the frequent insertion of NIST-1571 (orchard leaves) standard reference material as a control sample. Our data were in agreement, within experimental error, with the recommended or indicated values issued. Statistical calculations were done using the SAS software. The iterative approximation of the nonlinear Eq. (1) below was done by the method of Gauss-Newton.

RESULTS AND DISCUSSION

Individual Trees on a Given Site

Rb and Cs concentrations were found to vary considerably among the individual trees on a given site, even when considering only needles of the same age class (*see* the coefficient of variation CV in Table 1). Actually, of all 25 elements determined in parallel work, the intrasite variation of Cs was always the largest, larger even than the notoriously varying Mn. The coefficient of variation was not the same at every site, but it is remarkable that CV (Rb) and CV (Cs) are positively correlated (Fig. 1). On the average, at a given site, CV (Cs) is about twice as large as CV (Rb). It is therefore probable that the variation between trees at a given site is the result of the variation of a soil parameter and that this parameter acts on both elements in the same way, but much more strongly on Cs than on Rb. CV for Cs is seen to be often larger than 40%, which is a strong indication that the values are not normally distributed. In previous work *(1)*, where a greater number of trees (n = 35) were sampled at a much larger site (12 · 12 km), it could be shown that both Rb and Cs are actually log-normally distributed.

On a given site and for a given needle age class, Rb and Cs in individual trees are not always strongly correlated. An exception is the site GRO, where the Spearman rank correlation gives $r_S = 1.00$ ($n = 5$, $P < 0.001$).

Trees on Different Sites

Table 1 gives the site means for the concentrations in needle age class 1. Sites are seen to differ enormously. The ratio between the highest and the lowest site mean is about 1300 for Cs and 30 for Rb. The ranking of the sites is almost identical for Rb and for Cs. The Spearman rank correlation gives $r_S = 0.8571$ ($P < 0.01$). However, the correlation between Rb and Cs is not linear. The ratio Rb/Cs is seen to decrease when the concentrations increase (Table 1). These observations are in line with the assumption that a common soil parameter is responsible for the variation between sites and that this parameter acts more strongly on the uptake of Cs than of Rb.

It is possible to correlate the values of needle age class 1 of all individual trees in logarithmic form. This correlation is almost linear and highly significant (Fig. 2).

Table 1
Cs and Rb Concentrations in Spruce Needles of Age Class 1
(Site Means and Coefficient of Variation)

Site[a]	n[b]	Cs ng/g[c]	CV[d]	Rb µg/g[c]	CV[d]	Rb/Cs
AUE	8	14.3	37	4.73	24	331
CHA	8	24.0*	58	5.85	29	244
REG	8	24.3*	60	6.28	30	258
ALP	7	25.6*	48	5.60	26	219
VIL	6	324*	59	43.3	32	134
GRO	5	541	36	31.6	16	58
SAR	5	661	19	32.8	12	50
SCH	9	18600	23	134	17	7

[a]Sites are ordered according to increasing Cs values.
[b]Number of trees analyzed individually.
[c]Site mean. Values with an asterisk are geometric means, which are about 20% lower than the corresponding arithmetic means.
[d]Intrasite of variation, CV = $s(x)/x \cdot 100\%$.

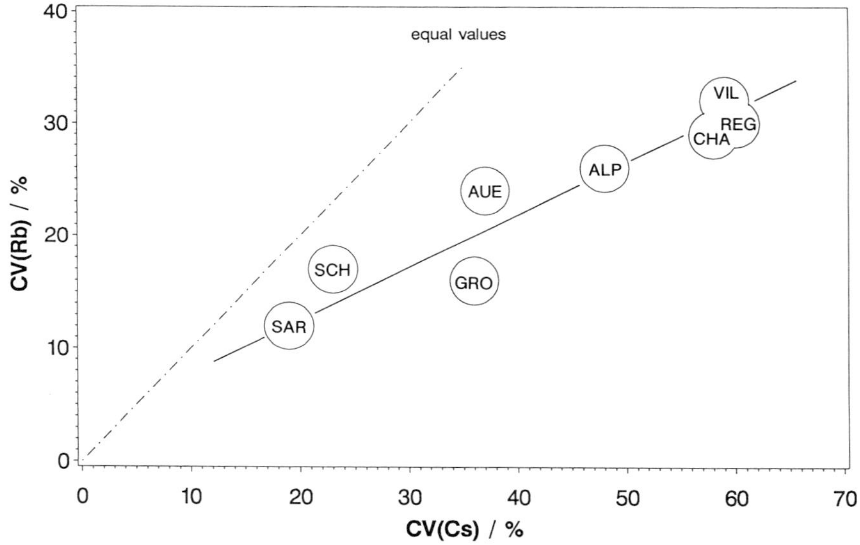

Fig. 1. Intrasite variation of the Rb and Cs consentrations of individual trees (needle age class 1). Values are from Table 1.

Concentrations in Different Age Classes

It is well known that the concentrations of all elements (except C and possibly H and O) change with the age of the needles. Supposing a constant needle weight (as was actually found at all sites except CHA), the measured concentration in a given needle age class is the difference

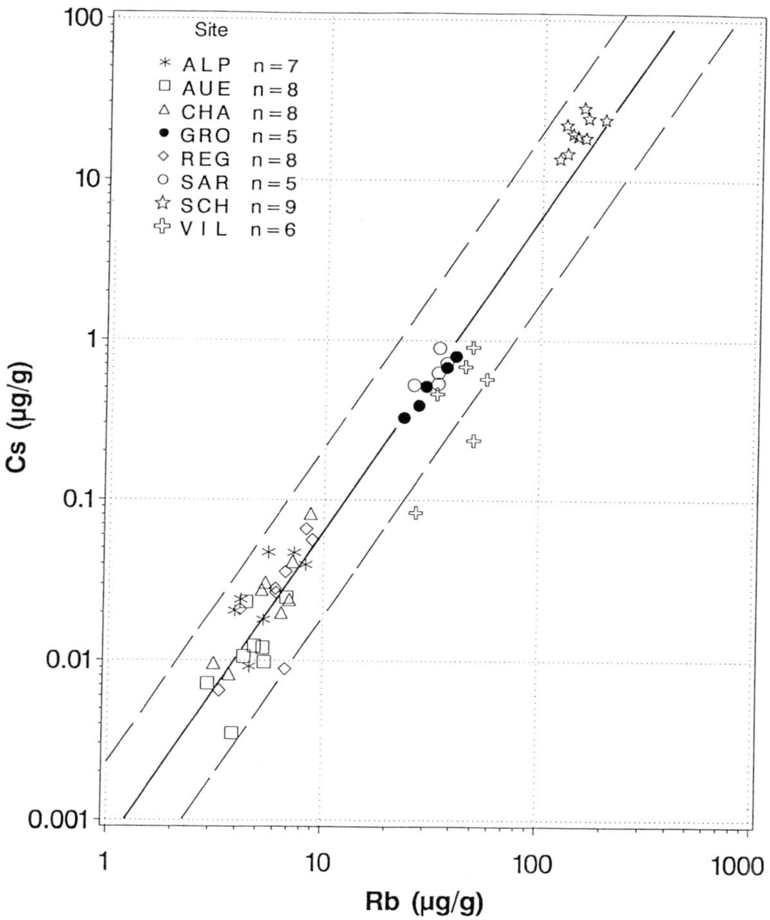

Fig. 2. Correlation of Cs and Rb in needle age class 1 of the individual trees from all sites: log(Cs) = (–3.171 ± 0.080) + (1.970 ± 0.060) log(Rb) (n = 56, r = 0.9758, P < 0.0001). The SD of the residuals (root mean square error) is 0.254, corresponding to a factor of 1.80. The range indicated is the 95% confidence limit for an individual predicted value.

between the total influx (or uptake) into the needle and the total efflux (or retranslocation) out of the needle. Needle concentrations generally will therefore either increase or decrease with time, according to the element considered *(3)*. Table 2 gives the values of Rb and Cs in five successive age classes from one site. Both elements are seen to decrease smoothly with time. The decrease is largest for young needles, and gets smaller and smaller in older needles, until it becomes practically zero, meaning that the concentration reaches a constant value in old needles.

Since the individual trees have greatly different values on one site, site means calculated in the usual way have large errors, and the significance of the decrease is apparently not great. However, since all trees at one site change similarly with time, although on a different concentration

Table 2
Concentrations and Concentration Ratios in Successive Needle Age Classes at Site REG (n = 8 Trees)

Needle	Rb (μg/g)					Cs(ng/g)					Rb/Cs	K/Rb
age class	x(a)	SE(a)		SE*(b)		x(a)	SE(a)		SE*(b)			
1	6.28	0.67 A		0.14 A		24.3	5.8 A		1.2 A		258	1050
2	2.75	0.30	B	0.07	B	12.2	2.6	B	0.7	B	225	1793
3	2.01	0.21	B	0.06	C	9.3	1.7	B	0.4	C	216	2093
4	1.85	0.19	B	0.06	C	8.3	1.6	B	0.4	C	222	2175
5	1.80	0.19	B	0.06	C	8.5	1.6	B	0.5	C	212	2167

[a]Mean and standard error calculated directly from the measured values. Age classes with the same letter are not significantly different (Duncan's multiple range test, α = 0.01).

[b]Reduced error resulting from the normalization process *(see text)*. Significance of difference as above.

level, a suitable normalization will remove the variations between trees *(3)*. The resulting errors for site means are thus reduced by approximately a factor of 4 to about 3% for Rb and 5% for Cs.

These reduced errors are small enough to show that the decrease with time is indeed significant at least up to age class 3 (Table 2). This effect is responsible for the observation of Robarge et al. *(4)*, who investigated age classes 1, 2, and 3 of *P. rubens* and found no significant decrease for Cs, and for Rb between age classes 1 and 2 only.

The decrease of the concentration with time can be described by the function:

$$c_t = A + B \cdot \exp - \lambda(t - 1) \quad (1)$$

where t = needle age class (current year's needle have t = 1) and c_t = experimental concentration of age class t. A, B, and λ are constants that are characteristic for a given tree or a given site mean. c_t, A, and B are in concentration units; t and λ are dimensionless; t is integer.

The fit of Eq. (1) to the experimental site means is very good. Deviations between experimental and fitted values are 3% on average. The deviations are randomly distributed and entirely explained by the (reduced) error of the experimental data (Fig. 3). The resulting errors of the parameters of Eq. (1) are, on average over all stations, 4% for A and B, and 11% for λ.

The biodynamic function (Eq. [1]) is seen to separate the total concentration into two parts: the first part *(A)* does not change with time, whereas the second part ($B \cdot \exp - \lambda[t - 1]$) decreases with time. The change of the second part is regular; it decreases each year by the same factor (exp λ). For the current year's needles (t = 1), Eq. (1) reduces to $c_1 = A + B$ and for old needles (in practice $t > 3$ or 4) to $c_{old} = A$.

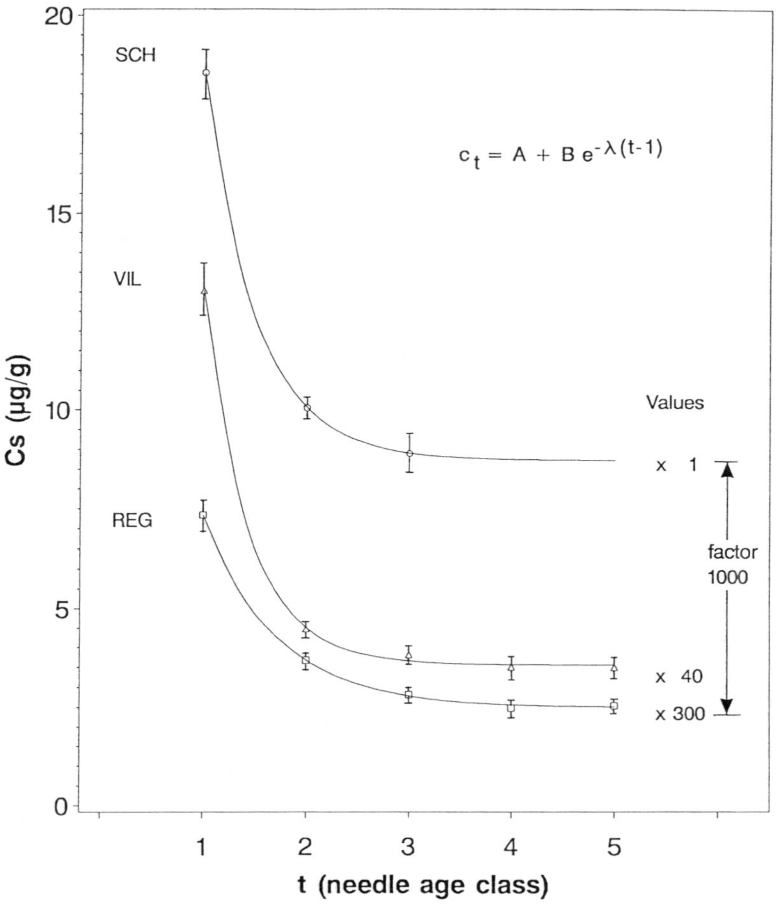

Fig. 3. The comparison of the biodynamic behavior of Cs on sites with high, intermediate, and low concentrations. Points are experimental site means and (reduced) SEs. The line is the fit of Eq. (1) (parameters in Table 3). This equation is only defined for integer values of *t*, but is drawn here as a curve in order to make clear the shape of the function. All values for VIL have been multiplied by 40, and those for REG by 300 for an easier presentation. The shape of the curves for Rb is very similar to that for Cs.

The Biodynamic Function at the Various Sites

Table 1 shows that there are very large differences among the concentrations of the various sites. Nevertheless, it is possible to fit the same function to all sites. Figure 3 shows this for three sites with high, intermediate, and low Cs concentrations.

The resulting parameters of Eq. (1) for all sites are given in Table 3. The parameters *A* and *B* reflect the concentration levels of the various sites and show the corresponding large variations. Their ratio B/A, how-

Table 3
The parameters of Eq. (1), fitted to the station means of the Rb and the Cs concentrations

Site[a]	age classes[b]	A		B		B/A		λ	
		Rb µg/g	Cs ng/g	Rb µg/g	Cs ng/g	Rb	Cs	Rb	Cs
AUE	5	1.9	7.4	2.8	6.5	1.45	.88	1.15	1.95
CHA	5	3.1	10	2.7	14	.87	1.42	.87	1.27
REG	5	1.8	8.3	4.5	16	2.49	1.94	1.54	1.40
ALP	4	2.5	13	3.4	13	1.35	.99	1.45	1.13
VIL	5	14	89	30	237	2.17	2.65	1.69	2.31
GRO	4	13	221	18	318	1.35	1.44	1.11	1.59
SAR	4	12	196	21	462	1.80	2.35	1.28	1.36
SCH	3	66	8730	69	9800	1.05	1.12	1.38	1.99
x						1.56	1.60	1.31	1.62
CV						35 %	41 %	20 %	25 %

[a]Sites are in the same order as in Table 1. The number of trees analyzed per site is as in Table 1.
[b]The number of successive needle age classes analyzed per site.

ever, is much more constant. B/A is the ratio, in age class 1, between the amount formally available for retranslocation and the amount formally not available for retranslocation. This ratio is seen not to depend on the concentration level and to be roughly equal for Rb and for Cs. B/A is about 1.6, implying that of the total concentration present in age class 1, about 40% are in a form not available for retranslocation.

The parameter λ also does not depend on the concentration level. Its value is slightly greater for Cs than for Rb. The value of $\lambda = 1.31$ for Rb indicates that the amount formally available for retranslocation gets smaller each year by a factor of 3.7, whereas $\lambda = 1.62$ (Cs) implies a factor of 5.0. The variation of λ and B/A between sites is larger than their uncertainty, and differences between sites therefore are real.

The Ratio Between Rb and Cs in Successive Age Classes

It can be shown that the ratio Rb/Cs is constant in successive needle age classes only if the biodynamic functions of the two elements fulfill the conditions $B/A(\text{Rb}) = B/A(\text{Cs})$ and $\lambda(\text{Rb}) = \lambda(\text{Cs})$. Table 3 shows that this holds for none of the sites exactly. The experimentally found shift in Rb/Cs with t, however, is not very great; it usually is smaller than 20% over five age classes. The direction of the shift (increasing or decreasing) is not the same on all sites. The specific numbers for one site are given in Table 2.

The small shift in the ratio Rb/Cs shows that both elements are very similar in their accumulation and retranslocation processes. This is contrasted to the different behavior of K, the chemical homolog of Rb and Cs. The ratio K/Rb (or K/Cs) increases on all sites considerably with time. The increase at the various sites is by a factor between 1.3 and 2.0 over five successive age classes; an example is given in Table 2. The observed increase is the result of substantial and systematic differences in the values of (B/A) and of λ between K and Rb or Cs.

CONCLUSIONS

The variation in the concentrations of Rb and Cs is very large, as much between individual trees on one site as between different site means. In both cases, the variations are larger for Cs than for Rb. The range for Cs in needle age 1 is 14–18,600 ng/g (i.e., a factor of 1300) for site means and 3–28,000 ng/g (i.e., a factor of $1 \cdot 10^4$) for all individual trees investigated. These factors are much larger than those found for any of the concurrently measured 23 elements *(5)*. A similar result was obtained with mushrooms, where Cs shows the largest range among 15 investigated trace elements *(6)*. It may therefore be stated that Cs is probably the element showing the largest natural variation in a given land plant species, and that its uptake is regulated little or not at all by the plant.

There is a strong correlation between Cs and Rb, both among site means and among all individual trees. The correlation is approximately logarithmic with a slope of 2, meaning that the Cs concentration changes by 2 orders of magnitude if the Rb concentration changes by one order. In a cluster analysis of all elements analyzed concurrently in spruce needles, we have shown *(5)* that Cs and Rb have the highest correlation and therefore the shortest distance among all possible pairs of elements. This is because of their parallel biodynamic curves as well as their similarities when comparing different sites or individual spruce trees. The greater variability of Cs is also reflected by the fact that at individual sites, CV(Cs) is always larger than CV(Rb). The few cases reported in the literature, in which both elements have been determined in conifers, fall well within the 95% confidence limits indicated in Fig. 2, which therefore seems to be generally valid, irrespective of the actual soils or bedrocks involved. It is assumed that both the observed variation of Cs and Rb and the correlation between their concentrations is the result of a common soil factor, which however acts much more strongly on Cs than on Rb. A prime candidate for this soil factor are certain clay minerals (montmorillonite and illite), which show a much larger selectivity coefficient for Cs than for Rb *(7)*. The clay content at the investigated sites has not yet been determined.

The situation found here with stable Cs (i.e., large differences between individuals and between sites) is very similar to that found for ^{137}Cs from the Chernobyl fallout (e.g., *8)*. Even in cases where ^{137}Cs is known to be taken up through the roots and not directly through the leaves, the observed large variation of ^{137}Cs among plants is usually attributed to presumed, large local variations in the fallout activity, leading to different local soil activities. The present work, however, shows that differences in the uptake from the soil alone are sufficient to induce very large variations in the plant Cs concentration.

The concentrations of Rb and Cs decrease smoothly with increasing needle age, approaching a constant value in old needles. On a given site, the ratio Rb/Cs varies only slightly with the needle age, showing the great similarity of both elements as far as transportation and deposition processes in the plant are concerned. The biodynamic behavior of Rb and Cs does not depend on the concentrations involved, and it is roughly the same at low as at high concentrations (Fig. 3). The same insensitivity of the biodynamic behavior to the concentration has also been noted for other elements (e.g., Si, Sc, and REE [*9*]), but not for Mn *(3)*. Furthermore, the decrease of Rb and Cs with needle age demonstrates that statements, such as "all non-essential elements accumulate with ageing of the leaves" *(10)* are not universally true.

The biodynamic behavior of Rb and Cs can be well described by an algebraic function (Eq. [1]). This functions splits the total inventory of the needle into two parts, one that is constant with t and one decreasing with t. The decrease is the result of the retranslocation of the elements from older into younger needles. The part that is formally available for retranslocation gets smaller in each successive needle age class by the same factor exp λ (or about 3.7 with Rb and 5.0 with Cs). The part that is constant with t formally does not participate in the retranslocation. Since Rb and Cs in the plant do not form insoluble compounds, the distinction between these two parts must be the result of a distribution of the elements into different cell compartments. The cell compartments containing Rb and Cs probably are the same as those for K *(11)*. It should be noted that the part staying constant with t can arise either from the condition that there is no influx into and no efflux from this compartment (which then would be effectively isolated from the other compartments) or from the condition that the influx is equal to the efflux.

A dynamic behavior similar to Rb and Cs is also found with the essential elements K, Mg, Cu, P, and Cl *(3)*. The retranslocation of the essential elements is usually attributed to the need of higher concentrations in the young needles, where the physiological activity is highest. Rb and Cs are not essential, and their transportation roughly parallel to K is probably owing to their sharing of the same carriers in some cases.

The similarity of the essential elements mentioned above is expressed by the fact that they all can be fitted by Eq. (1). The parameters involved, however, are different from those of Rb and Cs. This is reflected by the ratio K/Rb, which increases by about a factor of 2 between young and old needles, whereas the ratio Rb/Cs stays practically constant (Table 2). The parameter λ is smaller for K (about 0.7) than for Rb (1.31) and Cs (1.62), indicating that with K the amount available for retranslocation gets smaller each year only by a factor of 2 (as opposed to a factor of 3.7 for Rb and 5.0 for Cs). Similarly, the ratio B/A is about 0.6 for K as opposed to 1.6 for Rb and Cs, indicating that in age class 1, 60% of the total K (as opposed to 40% for Rb and Cs) is not available for retranslocation. If the ratio B/A is taken as a measure of the relative mobility of an element, then Rb and Cs are more mobile than K. The great shift of the ratio K/Rb as a function of the needle age further shows that the substitution of ^{86}Rb or of stable Rb as a tracer for K in plant physiological studies is a poor approximation, contrary to what is usually assumed. This assumption has been questioned before *(12)*.

ACKNOWLEDGMENTS

We thank S. Brütsch for her help with the preparation of the samples and R. Lorenzen for typing the manuscript. The help of the staff of the SAPHIR reactor at the Paul Scherrer Institute for providing the irradiation facilities is greatly appreciated. The work was partially supported by the Swiss National Foundation (Grant 3-30755.91).

REFERENCES

1. A. Wyttenbach, S. Bajo, L. Tobler, M. Adam, and H. W. Zöttl, *Applications of Isotopes and Radiation in Conservation of the Environment,* IAEA-SM-325/185, Vienna, 1992, pp. 535–546.
2. A. Wyttenbach, S. Bajo, J. Bucher, P. Schleppi, and L. Tobler, *Commun. Soil Sci. Plant Anal.* **24,** 1557–1566 (1993).
3. A. Wyttenbach, P. Schleppi, L. Tobler, S. Bajo, and J. Bucher, *IUFRO—Symposium on Nutrient Uptake and Cycling in Forest Ecosystems,* Kluwer Academic Publishers, Amsterdam, (1993) in press.
4. W. P. Robarge, J. M. Pye, and R. I. Bruck, *Plant Soil* **114,** 19–34 (1989).
5. P. Schleppi, A. Wyttenbach, J. Bucher, L. Tobler, and S. Bajo, *IUFRO—Symposium on Nutrient Uptake and Cycling in Forest Ecosystems,* Kluwer Academic Publishers, Amsterdam, (1993) in press.
6. G. Ingrao, P. Belloni, and G. P. Santaroni, *J. Radioanal. Nucl. Chem. Articles* **161/1,** 113–120, (1992).
7. E. Brouwer, B. Baeyens, A. Maes, and A. Cremers, *J. Phys. Chem.* **87,** 1213 (1983).
8. D. Mascanzoni, *J. Radioanal. Nucl. Chem. Articles* **161/2,** 483–488 (1992).
9. A. Wyttenbach, P. Schleppi, J. Bucher, V. Furrer, and L. Tobler, *Biol. Trace Element Res.* **41,** 13–29 (1994).
10. W. H. O. Ernst, *Element Concentration Cadasters in Ecosystems,* H. Lieth and B. Markert, eds., VCH Verlagsgesellschaft, Weinheim, 1991, pp. 17–40.
11. R. Schmidt, U. Lüttge, and D. Kramer, *Trees* **3,** 154–160 (1989).
12. R. Van den Driessche, *Mineral Nutrition of Conifer Seedlings,* CRC, Boca Raton, FL, 1991, p. 36.

Heavy Metals in *Parmelia sulcata* Collected in the Neighborhood of a Coal-Fired Power Station

M. CARMO FREITAS

Instituto Nacional de Engenharia e Tecnologia Industrial (INETI), Instituto de Ciências e Engenharia Nucleares, Departamento de Química, P-2685 Sacavém, Portugal

ABSTRACT

The epiphytic lichen *Parmelia sulcata* was collected in the neighborhood of a Portuguese coal-fired power station (Sines coal power station) as monitor for heavy metal air pollution. A study of the metal contents variability along 1991 and 1992 was performed. The heavy metals Ag, As, Br, Co, Cr, Fe, Hg, Sb, Se, and Zn were determined by k_0-based instrumental neutron activation analysis. The concentrations found in 1991 and 1992 show an accumulating process of Co and Fe (~5%/mo) and of Cr and Sb (~7%/mo). Low accumulation is observed for Ag, Se, and Zn (~2%/mo), and no concentration variation is observed for As, Br, and Hg. It is concluded that the metal accumulation observed is the result of the nearby ash and coal deposits.

Index Entries: Heavy metals; air pollution; biomonitors; coal station; neutron activation analysis.

INTRODUCTION

The toxic emission from coal-fired power stations have become the subject of growing concern since they may constitute a health risk. The pollutants can be inserted into the air, the surface water, the soil, and the ground water. Lichens are frequently used for monitoring purposes *(1,2)*. Lichens receive their mineral nutrients as particulates and in solution by the interception of atmospheric and aquatic deposition, without the complication of absorptive function by the root systems of higher plants. These organisms also have the ability to accumulate metals to levels far greater than their physiological needs, understandable by their ion-exchange properties and by the fact that surface, structure, and roughness facilitate interception and retention of particles. In addition, the absence of a well-developed cuticule or similar wax minimizes any physical resistance to the diffusion of soluble minerals into the plants *(2)*. In this work, *Parmelia sulcata* was collected from *Quercus suber* in the neighborhood of the Sines coal-fired power station for study of heavy metal pollution. The lichen was picked up from the same tree on April, October (1991), and April, May, June, July, August (1992). The sampling location was south of the coal station near the aerosol collector (about 100 m far from the station) as shown in Fig. 1. The samples were analyzed by k_0-based INAA *(3)*.

EXPERIMENT

In the laboratory, the pieces of *Quercus suber* bark were carefully removed from the *Parmelia sulcata* samples, which were then washed in deionized water until dust was no longer visible. In all cases, care was taken in following the same procedure for the materials of the various samplings.

Next the washed samples were lyophilized for 48 h and then frozen in liquid nitrogen in a Teflon™ capsule and ground in a Mikrodismembrator II mill. Finally, the samples were put into clean polyethylene bottles and homogenized for 48 h.

For irradiation, three samples of ~500 mg each were pelletized and then packed in clean polyethylene cans. Therefore, they were sandwiched between 0.1% Au-Al wires of 1-mm diameter (CBNM). The lichen samples were irradiated for 5 h in channel 54 (neutron flux: 2.2×10^{13} $cm^{-2} \cdot s^{-1}$) of the Portuguese Nuclear Reactor. The samples were counted using a HPGe (active vol: 138 cm^3; FWHM: 1.85 keV at 1.33 keV). Table 1 shows the nuclides, γ-ray energies, and decay time used for INAA measurements.

The areas of the γ-ray peaks in the spectra were determined with computer code SEQAL, and the elemental concentrations were calculated with SINGCOMP. The effective solid angles necessary for the efficiency conversion were computed with SOLANG; the lichen composition was taken from the "average" composition given by Bowen *(4)*.

RESULTS

Table 2 shows the results obtained for *Parmelia sulcata* collected in different periods of 1991 and 1992 south of the Sines coal power station. For the sets shown in Table 2, the concentrations were normalized to 1 for the first sampling period (April 1991). Next, a linear fit was made of these normalized values (ρ) vs the number of months M (M = 0 for April 1991), and finally, the function was normalized to $\rho = 1$ for M = 0 (*see* Figs. 2 and 3).

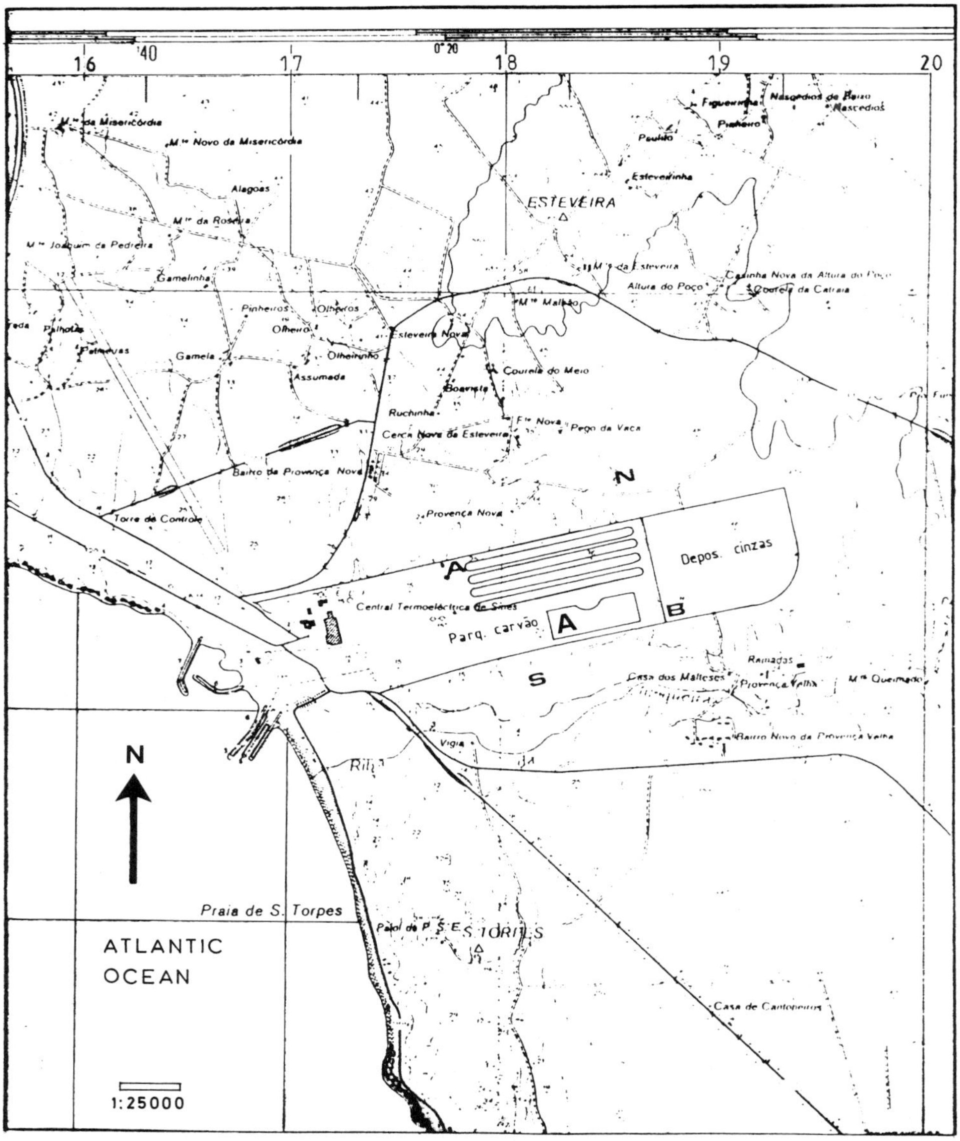

Fig. 1. Localization of the Sines coal power station. A—coal piles and coal deposit; B—ash deposit; S—sampling location with aerosol collector.

There is a significant correlation for Fe, Co, Cr, and Sb, thus meaning that from April 1991 to August 1992, the contents on these metals south of the Sines coal power station were increasing per month 5.9, 5.4, 7.5, and 7.1%, respectively. For the other determined metals, the increase is null (Hg, As, and Br) or very slight (Ag, Zn, and Se). The observed increases of Fe, Co, Cr, and Sb may originate from particulates of the coal and ash deposits being transported through the air *(see* Fig. 1) and falling

Table 1
Nuclides, γ-Ray Energies and Decay Time Used for INAA Measurements

Element	Radionuclide	Energy of gamma-ray used (keV)	Decay time
Ag	^{110m}Ag	658	30 d
As	^{76}As	559	4-7 d
Br	^{82}Br	776	4-7 d
Co	^{60}Co	1173, 1332	30 d
Cr	^{51}Cr	320	30 d
Fe	^{59}Fe	1099	30 d
Hg	^{203}Hg	279	30 d
Sb	^{124}Sb	602, 1691	30 d
Se	^{75}Se	264	30 d
Zn	^{65}Zn	1115	30 d

Table 2
Elemental Concentrations (in μg/g) in *Parmelia sulcata*[a]

	April 91	Oct. 91	April 92	May 92	June 92	July 92	August 92
Ag	0.21±0.01	0.13±0.02	0.15±0.02	0.29±0.03	0.23±0.02	0.18±0.02	0.24±0.01
As	1.97±0.08	2.6±0.1	1.66±0.06	2.8±0.1	1.93±0.08	1.69±0.07	2.6±0.1
Br	42±2	44±2	29±1	46±2	37±1	35±1	37±1
Co	1.34±0.05	2.01±0.08	1.8±0.1	3.0±0.1	2.41±0.09	1.88±0.08	3.0±0.1
Cr	6.8±0.3	8.8±0.4	8.4±0.3	15.2±0.6	12.8±0.5	10.3±0.4	16.6±0.7
Fe%	0.23±0.01	0.32±0.01	0.30±0.01	0.51±0.02	0.43±0.02	0.34±0.01	0.50±0.02
Hg	0.26±0.01	0.34±0.01	0.19±0.02	0.29±0.01	0.33±0.02	0.26±0.01	0.33±0.01
Sb	0.82±0.03	1.67±0.07	1.13±0.04	2.18±0.09	1.67±0.07	1.65±0.07	2.36±0.09
Se	0.54±0.02	0.75±0.03	0.64±0.03	0.75±0.03	0.81±0.04	0.63±0.02	0.91±0.04
Zn	450±18	474±19	386±11	726±29	489±20	454±18	654±26

[a]The uncertainties are 1 s.

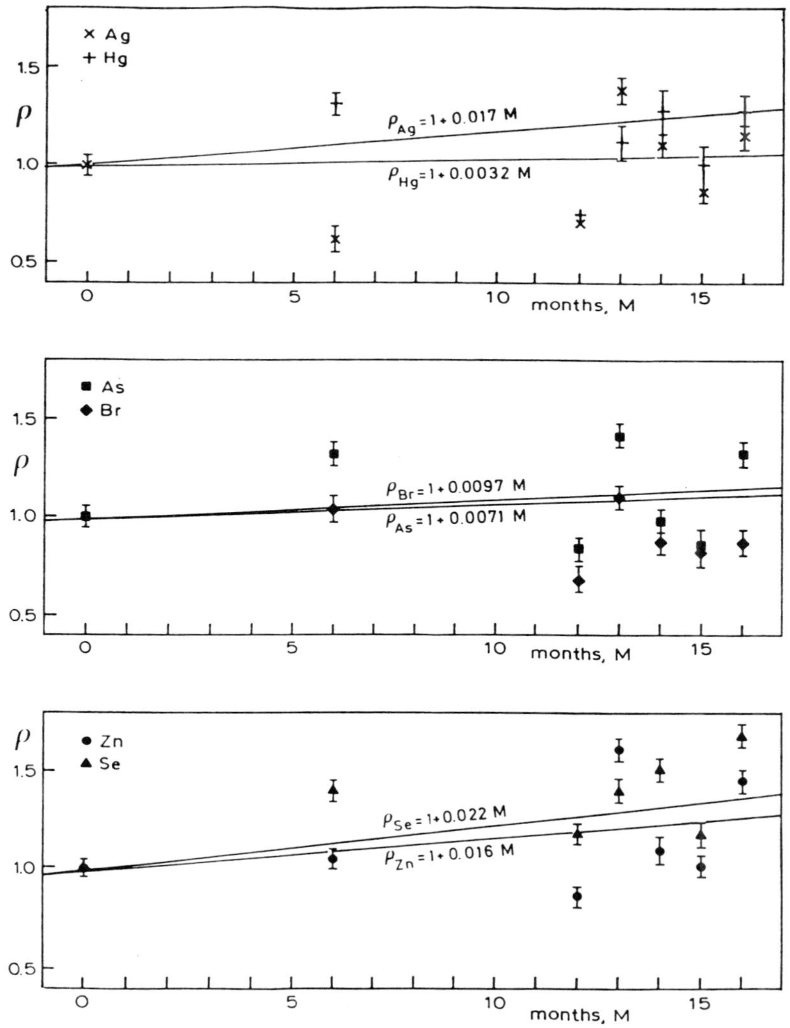

Fig. 2. Increase with time of the contents in Co, Fe, Cr, and Sb of *P. sulcata* south of the Sines coal power station, from April 1991 to August 1992. ρ-normalized values.

out on *P. sulcata* where they are absorbed by it. Similar increases were observed for the lanthanide contents (*see* 5).

The elements Hg, As, Br, Ag, Zn, and Se did not accumulate in *P. sulcata*, probably because of the depletion of these elements in the ash deposits *(6)*. They are usually enriched in the fly ashes emitted by the station chimneys *(6)*; probably because of the height of the station chimneys (225 m) and the proximity of the sampling point (100 m), the fly ashes do not fall out on *P. sulcata*. It can be concluded that, at the chosen sampling location, the elemental enrichment on *P. sulcata* does not seem to be the result of the operation of the power station, but more probably the accumulation of the deposits in the station territory.

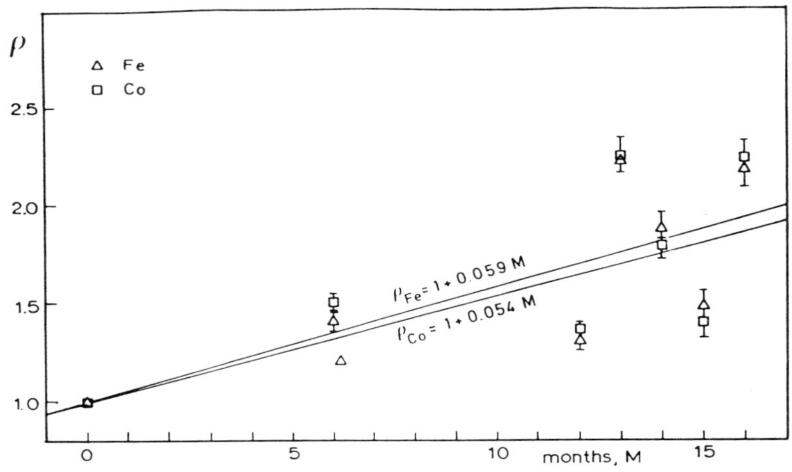

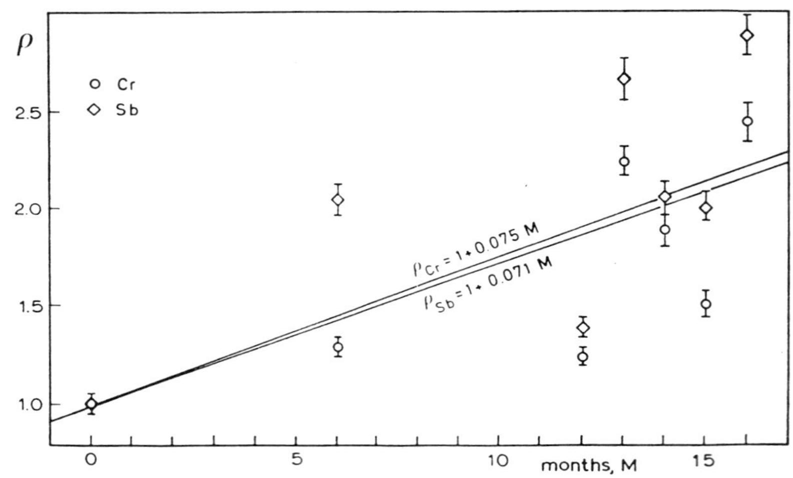

Fig. 3. Increase with time of the contents in Ag, Hg, As, Br, Se, and Zn of *P. sulcata* south of the Sines coal power station, from April 1991 to August 1992. ρ-normalized values.

REFERENCES

1. M. De Bruin, *IAEA Bull.* **4,** 22 (1990).
2. F. Catarino, C. Máguas, C. Branquinho, C. Sérgio, and W. Kratz, *Environ. Pollu. 1-ICEP,* **1**.
3. M. C. Freitas and E. Martinho, *Anal. Chim. Acta* **219,** 317 (1989).
4. H. J. M. Bowen, *Environmental Chemistry of the Elements,* Academic, London, 1979.
5. F. De Corte, M. C. Freitas, Lin Xilei, and A. Simonits, paper presented at this conference (NAMLS 1993).
6. C. Block and R. Dams, *Environ. Sci. Technol.* **10,** 1011 (1976).

Metal Determination by EDXRF in Lichens

A Contribution to Pollutants Monitoring

G. CANIGLIA,*[,1] I. CALLIARI,[2] L. CELIN,[1] AND A. M. TOLLARDO[2]

[1]*Department of Biology and* [2]*C.U.G.A.S., University of Padua, Italy Via Trieste, 75, 35121 Padova*

ABSTRACT

Samples of *Evernia prunastri*, collected in a mountainous zone, were exposed in urban and highly industrialized areas in order to monitor the atmospheric pollution. Amounts of Cr, Mn, Fe, Ni, Cu, Zn, and Pb were determined by EDXRF spectrometry in secondary target excitation mode and in the thin film approach. An increase in metal concentrations was noted on all sites with time exposition increases, but the highest final amounts were observed on those thalli mounted near a steel works. The washing procedure influenced the concentration of all the elements as previously verified on *Pseudevernia furfuracea*.

Index Entries: Transplant lichens; *Evernia prunastri*; bioaccumulation; EDXRF; heavy metal pollution.

INTRODUCTION

A correct determination of atmospheric pollution generally requires a wide network of sampling sites capable of giving both instant and time-averaged data. This kind of approach may take considerable time and money, so for the detection and recognition of air pollution, organisms that respond sensitively and specifically to the pollutants under consideration can be used *(1)*.

Lichens are especially sensitive to air quality *(2,3)*, forming a sensitive symbiosis between algae and fungi, and exhibiting an unshielded

*Author to whom all correspondence and reprint requests should be addressed.

 © 1994 by Humana Press Inc.

surface against air from which they gain the main part of toxic and essential elements. These organisms are also long living plants and exhibit a high tolerance to heavy metals. In several studies *(4–6)*, the transplantation of healthy lichens collected from background environments was employed, since by this method, the development of injuries and accumulation of pollutants are easy to follow in comparable basic material. Moreover, the method can also be applied in areas with lichen desert.

The aim of the present study was to determine the metal contents (Cr, Mn, Fe, Ni, Cu, Zn, and Pb) of the *Evernia prunastri* samples exposed on 14 sites located in the Industrial Zone (I.Z.) of Padua from May 1992 to April 1993. The thalli were placed in small plastic traps, allowing the lichens to be exposed without substrate. Thus, the metal accumulation depends on the atmospheric deposition only.

In order to estimate the role of a pollutant source, most of the lichen thalli were transplanted in the vicinity of a steel works. The effect of exposure time was investigated by collecting samples every 3 mos.

MATERIALS AND METHODS

Willow tree branches covered with *Evernia prunastri* (L.) Ach. were picked in the northeast mountain zone of Alpago (ca. 1000 m altitude). The experiment was carried out from May 1992 to April 1993, and the lichens were mounted in small traps hung on wood or cement poles. Each trap is basically a container made up of a Plexiglas™ frame (15 × 15 × 1.5 cm), covered with a nylon fiber net, with mesh of 1 cm^2. Twelve of the 14 transplantation sites were selected near the steel works (distance range: 0.3–3.5 km), another was near one of the most busiest crossroads of the town, and the last one (site 14) was at the biology department (Fig. 1). None of the sites had growing lichens. Site 1 was in front of the steel works (0.3 km SW), the others were at varying different distances as follows: site 2: 0.8 km NW; site 3: 1.3 km SW; site 4: 1.4 km NW; site 5: 1.4 km SE; site 6: 1.5 km NW; site 7: 1.8 km NE; sites 8 and 9: 1.9 km NW; site 10: 2.0 km NW; site 11: 2.2 km SW; site 12: 3.5 km NW; site 13: near a crossroad 5.0 km NW; and site 14 near department 5.8 km NW. On this last site, the trap was hung at the 4th floor of the building.

The pre-exposure values of the metal amounts were determined in order to use them as reference values. After 3, 5, 8, and 11 mo, samples from each site were collected. The lichens from all sites were washed πin distilled water, dried at 50°C for 48 h *(7)*, and stored in closed bags in the dark. To verify the washing effect, thalli of sites 1 and 13, collected after 11 mo of exposure, were divided in two parts; one of them was washed as the previous ones; the other was dried and stored without washing.

The amounts of Cr, Mn, Fe, Ni, Cu, Zn, and Pb were measured by energy-dispersive X-ray fluorescence (EDXRF). Suitable samples for the

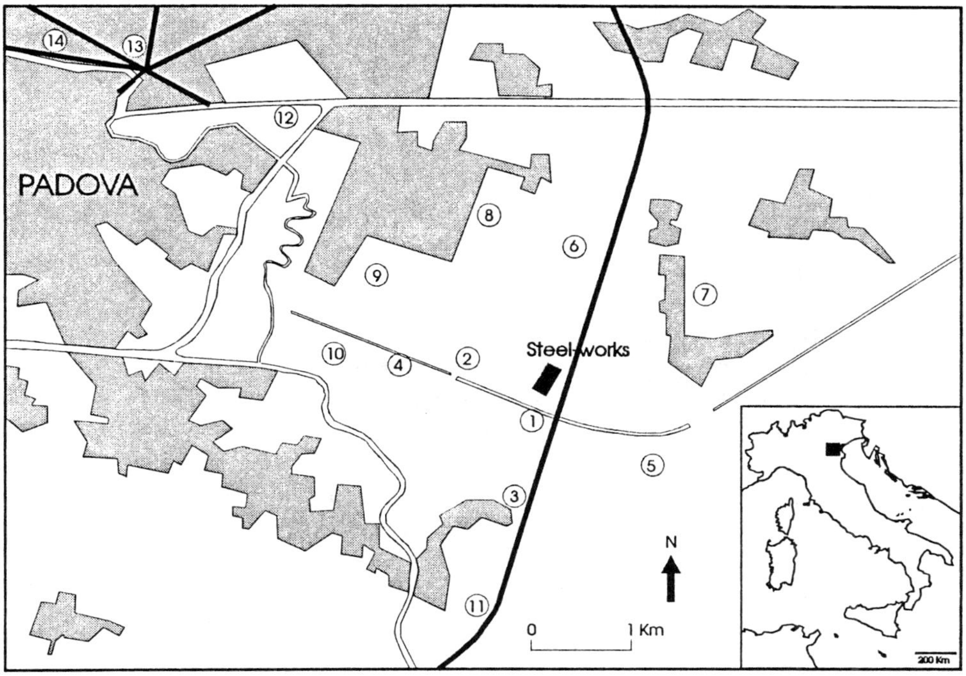

Fig. 1. The IZ of Padova and the monitoring sites.

EDXRF analysis were obtained by an acid digestion procedure using a microwave oven *(8)*. Each sample was divided into at least two subsamples, digested, and analyzed separately. All digestions were done in closed Teflon™ vessels of 120-mL capacity, with 200 mg of lichen samples. As digestion solvents, 3 mL of HNO_3, containing a Zr solution of 200 µg/g as internal reference, 0.2 mL of H_2O_2, and 0.05 mL of HF were employed. The targets were prepared by pipeting 300 µL of each solution on a Mylar film and dried in a vacuum dryer.

Calibration was performed by means of the NBS standard reference material HSRM 1571 (Orchard Leaves), digested by the same procedure used for the samples, with an addition of multielemental standard solutions. Samples and standards were analyzed under vacuum conditions on a Kevex 770 Delta spectrometer equipped with an Rh tube, seven secondary targets, and an Si(Li) detector with the resolution of 160 eV at 5.9 keV.

The working parameters were:

Ge secondary target, 17 keV, 3.3 mA, for Cr, Mn, Fe, Ni, Cu, and Zn;
Zr secondary target, 27 keV, 3.3 mA, for Pb; and
Ag secondary target, 35 keV, 2.0 mA for Zr.

The irradiation time was 500 s.

RESULTS AND DISCUSSION

The minimum detectable concentrations (expressed as μg/mL and as μg/g on dry wt) were:

0.12 μg/mL and 2 μg/g for Cr, Ni and Cu;
0.18 μg/mL and 3 μg/g for Mn and Zn;
0.24 μg/mL and 4 μg/g for Fe; and
0.48 μg/mL and 8 μg/g for Pb.

Tables 1 and 2 present mean concentrations, ranges, and ratios of metals found in *Evernia prunastri*. In the first line, the concentration values in lichen thalli before the exposition are reported; these data are used to determine the ratios for each element between exposed and nonexposed lichens.

Although on all sites, the mean values increase with the length of exposure, the rate of increase is not the same for all elements and for all sites. All metals, except Cu and Ni, increased considerably on sites 1 and 2, showing concentrations from 5 to 20 times the initial ones. Moreover, on these sites, the 3-mo values are generally higher than on the others.

As regards the other sites, the amounts of metals show a continuous increased upon exposure time. The concentrations of Zn, Mn, Pb, and Fe at varying exposure times as a function of a distance from the works are reported in Fig. 2. For all elements, the curves have different slopes from 0 to about 1.5 km distance, after which the curves are almost parallel lines. This means that in the first part (0.3–1.5 km), the rate of accumulation depends on a distance from the source, whereas in the second (1.5–5.8 km), the rate is quite constant. These data seem to confirm, almost qualitatively, the results obtained with other species exposed for different time periods. In these studies *(9,10)* an inverse relationship between the concentrations of several elements in lichens and the distance from the pollutants source has been observed.

Table 3 shows the metal concentrations in washed and unwashed samples of sites 1 and 13. The washing procedure generally induces a loss of element content, but on site 1, close to the works, the loss is lower than on site 13, close to the crossroads. Previous studies have shown that washing procedure may induce element losses on *Xantoria parietina (11)*, whereas no losses were verified on *Cladonia (12)*. The quantitative difference in the effects of washing of *Evernia prunastri* thalli evidenced herein could suggest that each species has different mechanisms of metal retention *(13,14)*. This study confirms that *Evernia prunastri* exhibits a useful capability in monitoring atmosphere pollution, since it gives a sensitive response on both short and long periods of exposure, maintaining a good living status and without reaching a saturated condition.

Table 1
Element Concentrations in *Evernia prunastri*

Sample		n	Cr			Mn			Fe			Ni			Cu			Zn			Pb		
			mean (μg/g)	range (μg/g)	ratio**	mean (μg/g)	range (μg/g)	ratio	mean (μg/g)	range (μg/g)	ratio	mean (μg/g)	range (μg/g)	ratio	mean (μg/g)	range (μg/g)	ratio	mean (μg/g)	range (μg/g)	ratio	mean (μg/g)	range (μg/g)	ratio
Initial values		5	<2	-	-	24	8	-	247	99	-	4	3	-	11	13	-	58	55	-	26	11	-
Site	M*																						
n° 1	3	2	10	1	>5.0	164	15	7.0	1201	158	5.0	3	1	1.0	23	6	2.0	283	50	5.0	47	14	2.0
	5	3	21	10	>10.5	215	93	9.0	2050	600	8.0	7	5	1.5	28	15	2.5	484	232	8.5	134	53	5.0
	8	2	15	1	>7.5	160	26	6.5	2158	231	8.5	9	3	2.0	31	1	3.0	560	162	9.5	228	96	8.5
	11	5	41	18	>20.5	337	70	15.0	4454	794	18.0	11	4	2.5	40	7	3.5	1228	198	21.0	479	203	18.5
n° 2	3	2	3	1	>1.5	113	21	5.0	917	112	3.5	5	0	1.0	18	1	1.5	195	26	3.5	60	7	2.5
	5	2	8	3	>4.0	130	56	5.5	1285	508	5.0	5	4	1.0	23	12	2.0	292	81	5.0	84	38	3.0
	8	2	20	8	>10.0	147	4	6.0	2198	132	9.0	7	0	1.5	38	1	3.5	516	47	9.0	161	2	6.0
	11	2	22	1	>11.0	178	27	7.5	2652	368	10.5	11	4	2.5	34	6	3.0	758	54	13.0	218	16	8.5
n° 3	3	2	2	0	>1.0	82	36	3.5	355	47	1.5	5	0	1.0	24	0	2.5	48	13	1.0	42	5	1.5
	5	2	3	2	>1.5	79	13	3.5	555	11	2.0	4	1	1.0	28	2	2.5	86	12	1.5	46	22	2.0
	8	2	3	0	>1.5	47	33	2.0	631	150	2.5	4	0	1.0	31	15	3.0	109	83	2.0	40	10	1.5
	11	2	15	6	>7.5	139	53	6.0	1316	514	5.5	8	6	2.0	27	12	2.5	296	154	5.0	87	79	3.5
n° 4	3	3	3	1	>1.5	67	62	3.0	483	103	2.0	4	4	1.0	19	13	1.5	125	127	2.0	44	26	1.5
	5	2	3	2	>1.5	64	22	3.0	531	70	2.0	4	0	1.0	17	6	1.5	144	39	2.5	30	7	1.0
	8	2	5	4	>2.5	51	10	2.0	693	104	3.0	7	5	1.5	24	7	2.5	249	67	4.5	50	21	2.0
	11	2	7	2	>3.5	56	0	2.5	899	71	3.5	9	2	2.0	18	4	1.5	269	35	4.5	55	25	2.0
n° 5	3	2	<2	-	-	44	16	2.0	423	79	1.5	2	0	0.5	20	2	2.0	70	55	1.0	24	20	1.0
	5	2	3	3	>1.5	64	3	2.5	551	93	2.0	7	2	1.5	22	5	2.0	97	22	1.5	38	14	1.5
	8	2	2	0	>1.0	56	9	2.0	791	37	3.0	3	0	0.5	19	3	2.0	133	14	2.5	46	12	2.0
	11	2	10	1	>5.0	80	3	3.0	1111	7	4.5	7	2	1.5	19	1	2.0	195	9	3.5	50	4	2.0
n° 6	3	2	<2	-	-	58	18	2.5	240	15	1.0	4	0	1.0	14	1	1.5	28	5	0.5	18	4	0.5
	5	2	<2	-	-	51	1	2.0	343	15	1.5	4	0	1.0	17	7	1.5	31	26	0.5	22	4	1.0
	8	2	<2	-	-	37	1	1.5	504	4	2.0	5	1	1.0	24	3	2.0	118	12	2.0	29	21	1.0
	11	2	9	5	>5.0	70	13	3.0	756	129	3.0	6	5	1.5	20	7	2.0	329	81	5.5	35	16	1.5
n° 7	3	2	2	0	>1.0	56	19	2.5	298	77	1.0	<2	-	-	16	0	1.5	71	7	1.0	31	9	1.0
	5	2	4	1	>2.0	66	9	2.5	392	79	1.5	8	1	2.0	25	1	2.5	89	23	1.5	43	16	1.5
	8	2	<2	-	-	60	23	2.5	547	158	2.0	6	2	1.5	17	0	1.5	76	17	1.5	36	3	1.5
	11	2	8	1	>4.0	82	18	3.5	722	10	3.0	4	0	1.0	17	4	1.5	123	26	2.0	40	2	1.5

*M = number of months exposed; range = difference between maximum and minimum value.
**Ratio = mean concentration divided by a pre-exposure value.

Table 2
Element Concentrations in *Evernia prunastri*

Sample		n	Cr			Mn			Fe			Ni			Cu			Zn			Pb		
Site	M		mean (μg/g)	range (μg/g)	ratio	mean (μg/g)	range (μg/g)	ratio	mean (μg/g)	range (μg/g)	ratio	mean (μg/g)	range (μg/g)	ratio	mean (μg/g)	range (μg/g)	ratio	mean (μg/g)	range (μg/g)	ratio	mean (μg/g)	range (μg/g)	ratio
n° 8	3	2	3	2	> 1.5	64	9	2.5	330	28	1.5	4	0	1.0	17	5	1.5	65	2	1.0	19	1	0.5
	5	2	3	1	> 1.5	72	23	3.0	402	116	1.5	6	1	1.5	45	10	4.0	49	21	1.0	32	8	1.0
	8	2	4	0	> 2.0	43	9	2.0	552	134	2.0	6	3	1.5	48	7	4.5	77	32	1.5	55	25	2.0
	11	2	10	1	> 5.0	63	2	2.5	985	28	4.0	6	0	1.5	44	1	4.0	108	8	2.0	58	13	2.0
n° 9	3	3	< 2	-	-	48	25	2.0	383	195	1.5	10	13	2.5	17	2	1.5	78	115	1.5	37	38	1.5
	5	3	< 2	-	-	53	23	2.0	518	192	2.0	6	6	1.5	23	16	2.0	51	27	1.0	37	28	1.5
	8	2	< 2	-	-	38	3	1.5	633	64	2.5	3	3	0.5	21	3	2.0	101	14	1.5	41	10	1.5
	11	2	11	4	> 5.5	71	14	3.0	1044	92	4.0	10	5	2.0	24	3	2.0	167	29	3.0	53	21	2.0
n°10	3	2	< 2	-	-	47	1	2.0	562	119	2.5	3	4	1.0	76	1	7.0	36	20	0.5	29	24	1.0
	5	2	< 2	-	-	58	1	2.5	586	11	2.5	5	5	1.0	49	7	4.5	55	4	1.0	37	9	1.5
	8	2	< 2	-	-	49	31	2.0	652	182	2.5	5	0	1.0	26	19	2.5	88	41	1.5	48	35	2.0
	11	2	7	0	> 3.5	70	18	3.0	842	194	3.5	10	2	2.0	26	6	2.5	216	85	3.5	48	12	2.0
n°11	3	2	< 2	-	-	61	12	2.5	381	110	1.5	2	1	0.5	20	2	2.0	95	2	1.5	22	19	1.0
	5	2	2	1	> 1.0	83	9	3.5	684	165	3.0	4	3	1.0	22	2	2.0	142	4	2.5	46	11	2.0
	8	2	2	1	> 1.0	53	4	2.0	646	120	2.5	4	2	1.0	21	0	2.0	136	57	2.5	41	16	1.5
	11	2	10	5	> 5.0	72	6	3.0	950	26	4.0	6	2	1.5	21	1	2.0	223	18	4.0	58	13	2.0
n°12	3	2	< 2	-	-	67	10	3.0	355	11	1.5	4	1	1.0	20	1	2.0	37	21	0.5	28	5	1.0
	5	2	< 2	-	-	71	0	3.0	448	34	2.0	5	5	1.0	17	5	1.5	17	7	0.5	46	3	2.0
	8	2	2	0	> 1.0	49	6	2.0	590	16	2.5	7	0	1.5	29	5	3.0	86	2	1.5	60	2	2.5
	11	2	8	4	>4.0	68	15	3.0	906	168	3.5	8	4	2.0	27	4	2.5	101	5	1.5	59	4	2.5
n°13	3	3	3	1	>1.5	35	13	1.5	586	487	2.5	3	2	1.0	28	25	2.5	119	148	2.0	53	54	2.0
	5	2	2	1	>1.0	61	2	2.5	709	158	3.0	7	8	1.5	42	2	4.0	153	67	2.5	84	9	3.0
	8	2	< 2	-	-	34	2	1.5	798	27	3.0	3	3	1.0	36	7	3.5	96	23	1.5	68	6	2.5
	11	5	7	2	>3.5	55	18	2.5	966	391	4.0	5	4	1.0	50	27	4.5	170	104	3.0	100	69	4.0
n°14	3	2	< 2	-	-	28	3	1.0	306	37	1.0	< 2	-	-	12	3	1.0	61	43	1.0	78	2	3.0
	5	2	3	2	> 1.5	74	28	3.0	477	73	2.0	8	1	2.0	20	5	2.0	80	39	1.5	151	96	6.0
	8	2	2	0	> 1.0	28	4	1.0	442	111	2.0	4	5	1.0	19	4	2.0	37	9	0.5	50	3	2.0
	11	2	9	3	> 5.0	61	3	2.5	781	111	3.0	6	4	1.5	19	4	2.0	118	20	2.0	156	3	6.0

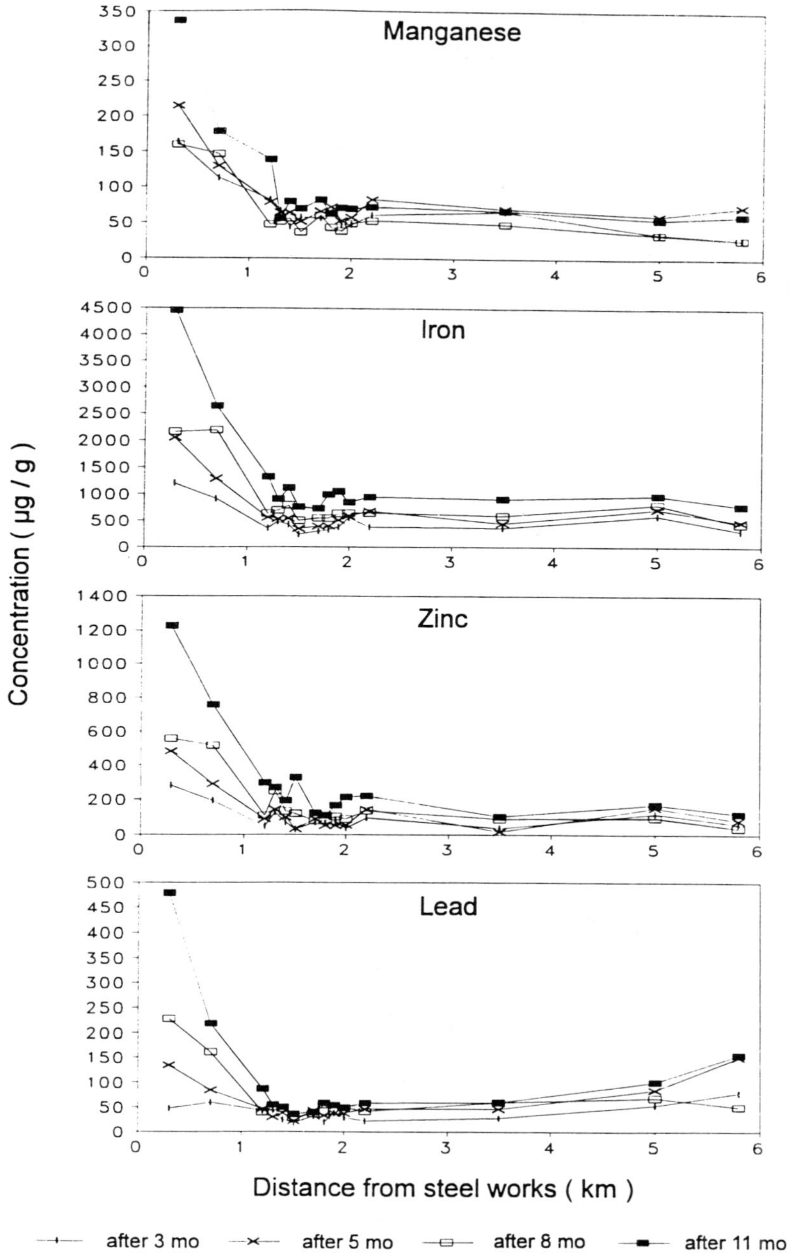

Fig. 2. The metal contents as a function of distance from the steel works.

Table 3
Comparison Between Element Concentrations of Washed and Unwashed Samples (μg/g)

	n	Cr		Mn		Fe		Ni		Cu		Zn		Pb	
		mean	SD	mean	SD	mean	SD	mean	SD	mean	SD	mean	SD	mean	SD
Site n° 1 (steel works)															
Washed	5	41	6.9	337	28	4454	349	11	1.3	40	3.2	1228	74	479	80
Unwashed	5	53	9.0	419	26	5535	352	15	2.4	46	3.8	1409	184	387	30
Site n°13 (cross-road)															
Washed	5	7	0.8	55	7.9	966	187	5	1.5	50	11	170	46	100	27
Unwashed	5	15	4.0	100	7.5	2335	330	11	2.9	78	9.8	267	44	147	31

ACKNOWLEDGMENTS

The authors wish to thank L. De Carvalho, University of Lisboa, for her kind cooperation in supplying NIST standards.

REFERENCES

1. J. Hertz, *Bioindicators for Monitoring Heavy Metals in the Environment,* E. Merian, ed., VCH, Weinheim, 1991, pp. 221–231.
2. J. Garty, *Can. J. Bot.* **66,** 668 (1988).
3. L. Folkeson, *Water, Air, Soil Pollut.* **11,** 253 (1979).
4. I. M. Brodo, *Ecology* **42,** 838 (1961).
5. D. L. Hawksworth, *Air Pollution and Lichens,* B. W. Ferry, M. S. Baddeley, and D. L. Hawksworth, eds., Athlone Press, London, 1973, pp. 38–76.
6. V. Palomaki, S. Tynnyrinen, and T. Holopainen, *Ann. Bot. Fennici* **29,** 25 (1992).
7. H. Olkonnen and K. Takala, *Ann. Bot. Fennici* **12,** 131 (1975).
8. H. M. Kingston and L. B. Jassie, *Anal. Chem.* **58,** 2534 (1986).
9. F. D. Tomassini, K. J. Puckett, E. Nieboer, D. H. Richardson, and B. Grace, *Can. J. Botany* **54,** 1591 (1976).
10. G. Brusasca, D. Anfossi, and G. Tinarelli, *Atmospheric Environ.* **26,** 707 (1992).
11. R. Bargagli, F. P. Iosco, and C. Leonzio, *Inquinamento* **2,** 33 (1985).
12. E. Nieboer, H. M. Ahmed, K. J. Puckett, and D. H. Richardson, *Lichenologist* **5,** 291 (1972).
13. M. G. Glenn, E. V. Orsi, and M. E. Hemsley, *Grana* **30,** 44 (1991).
14. G. Tyler, *Water, Air, Soil Pollut.* **47,** 321 (1989).

Energy-Dispersive X-Ray Fluorescence Analysis Applied to Biomonitoring on Alps

ROBERTA VECCHI,* GIANLUIGI VALLI, AND ANGELA MARIANI

Istituto di Fisica Generale Applicata, University of Milan—Italy, Via Celoria, 16 - 20133 Milano, Italia

ABSTRACT

The results of a research in progress at the Istituto di Fisica Generale Applicata—University of Milan—on natural and anthropogenic elements' determination in mosses as biological indicators are shown. The analytical technique was the energy dispersive X-ray fluorescence (EDXRF) spectrometry. The chosen areas for a preliminary pollution mapping were Italian Central Alps (Valmalenco and Valtellina) and the hills around Como (pre-Alps); in addition, a group of mosses was from Monza Park, near Milan.

Index Entries: Biomonitoring; X-ray fluorescence; mosses; pollution.

INTRODUCTION

The increasing interest about the pollutant loads in the environment promoted the research and the setup of alternative techniques with respect to the traditional ones. Our institute carries out a research on the possibility of applying EDXRF analysis to evaluation of natural and pollutant elements in mosses as environmental monitors.

Mosses were chosen for our monitoring because of their capacity for absorbing metals directly from the atmosphere; many authors (1–3) point out that this is the preferred exchange process with the environment with respect to the possible interchange with their substrate.

In fact, these organisms receive nutritive mineral elements, as particulates or in solution, by atmospheric deposition, since their substrate

*Author to whom all correspondence and reprint requests should be addressed.

anchoring system does not act as a root apparatus. Furthermore, the moss tissue has specific cationic exchange properties, and therefore, many trace metals are efficiently retained.

It is interesting to observe that water and elements absorption happens through the whole bryophytes surface. In fact, they do not have an ion-absorbing vascular system, nor is their surface provided with scarf skin or waxes. These organisms have also shown a good tolerance to high concentrations of various metals (4–6).

This kind of bio-monitoring is characterized by other advantages: (1) the "bio-accumulation" process makes easier the identification of those pollutants that can be found in the atmosphere in very low concentration; and (2) the bioindicator gives a time-integrated measure, and furthermore, it is possible to monitor wide areas. Those characteristics, in addition to the fact that mosses are greatly diffused in the investigated areas, led us to use them for a preliminary pollution mapping on the Alps and to verify EDXRF performance and suitability.

METHODS

In the years 1992–1993 (from May to September), more than 100 moss samples were picked up. Four areas with different typologies have been investigated:

1. Two typical alpine valleys, Valmalenco (1500–2000 m sl [above sea level]) and Valtellina (600–1200 m sl);
2. Como area (200–800 m sl): it is situated on hills of pre-alpine origin. It is an urbanized area and it is characterized by a small industrial plant; and
3. Monza Park (150 m sl); it is 10–20 km N-E from Milan. Its a wide green area in the suburbs of a metropolis and next to important roads and industrial areas.

At each sampling site, six samples of the moss were collected from different parts of the carpet; afterward, they were put together into a plastic bag and then transported to our laboratories for the analysis. Among the different species recognized, we found more frequently *Hylocomium splendens*, *Polytrichum formosum*, *Brachytecium populeum*, *Pleurozium schreberi*, *Hypnum cupressiforme*, and *Atrichum angustatum*. Many of them, such as *Hylocomium s.*, *Hypnum c.*, and *Pleurozium s.*, are mentioned by other authors *(2,7)* as very efficient bioindicators.

For each sampling site, we specified the position, the substrate type, the exposure, and the distance from roads and from possible anthropogenic pollution sources with their typology. All samples were dried at 90°C for 24 h in an airy oven and then pulverized. The measurement geometry used was a Teflon support in which the samples were put and slightly pressed.

The samples were then analyzed by EDXRF. Our spectrometer can detect elements with $Z > 11$. The X-ray tube generator has a rhodium anode and the maximum voltage and current applicable are 30 kV and 300 μA respectively.

The X-rays emitted by atoms of elements in the sample irradiated for 2000 s by the primary beam, on purpose collimated and filtered, were then counted by an Si(Li) detector. Figure 1 shows a typical spectrum of the moss analysis. The spectrometer was calibrated for these elements: S, Al, K, Ca, Ti, Fe, Ni, Cu, Zn, Br, and Pb using Community Bureau of Standards (BCR) and National Institute of Standards and Technology (NIST) standards. Detection limits found are of about 1–2 μg/g of dry wt for elements present in low concentrations.

RESULTS

In total, 80 samples were analyzed: 22 from the Como area, 25 from Valmalenco, 15 from Vatellina, and 18 from Monza Park. In Figs. 2, 3, and 4, the mean values (in μg/g of dry wt) for each sampling site are shown, and elements with similar concentrations ranges are put in the same diagram.

The elements detected in all samples are either of a natural origin, they are moss constituents or terrigenous metals, or they originate from an anthropogenic source. In Table 1 the concentration ranges for each element are summarized.

DISCUSSION

We can conclude from the results that: (1) Different concentration ranges are characteristic for different elements. For example, we found values of a few μg/g for Cr and Ni, and of several thousand of μg/g of dry wt for K, Ca, and Fe, and (2) particular element mean values vary (up to a factor 2) for different monitored areas. Concentrations decrease from zones that are more influenced by pollution sources to those less subjected to pollution. The typical case is lead. This element is often correlated to bromine; the simultaneous presence of both these elements identifies the motor-vehicle petrol combustion as their possible source.

Less evident is the difference in the sulfur concentrations among the four areas monitored; the mean values vary from 1084 μg/g in Valmalenco up to 1740 μg/g in Monza Park. This trend appears to characterize a wide diffusion of this pollutant mostly owing to a heavy oil combustion process.

ACKNOWLEDGMENTS

The authors are grateful to Ugo Facchini for his helpful suggestions, and to Giuseppina Gerosa and Chiara Serenthà for their collaboration to the 1992 measurements.

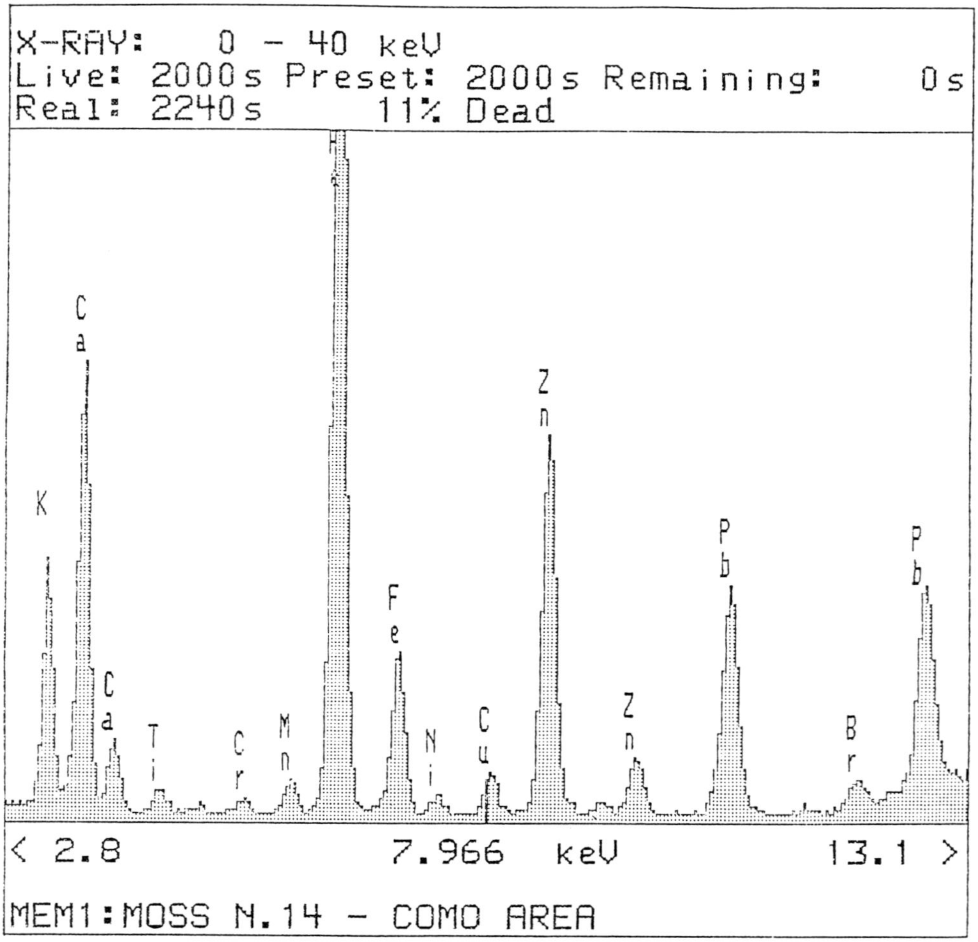

Fig. 1. Typical spectrum of a moss analysis using energy-dispersive X-ray fluorescence.

REFERENCES

1. R. Bargagli, F. P. Iosco, and C. Leonzio, *Inquinamento* **2,** (1985).
2. L. Folkeson, *Water, Air and Soil Pollution* **11,** (1979).
3. J. W. Huckabee, *Atmospheric Environ.* **7,** (1973).
4. M. Spampani, *Le Scienze* **167,** (1982).
5. P. L. Nimis, *EUR 13060 EN,* A. G. Colombo and G. Premazzi, eds., 1990.
6. E. Steinnes, *Control and Fate of Atmospheric Trace Metals.* J. M. Pacyna and B. Atar, eds., Kluwer, Dordrecht, 1989.
7. E. Steinnes, *Kjeller Report No.KR-154,* Institutt for Atomenergi Kjeller Research Establishment, Kjeller, Norway, 1977.

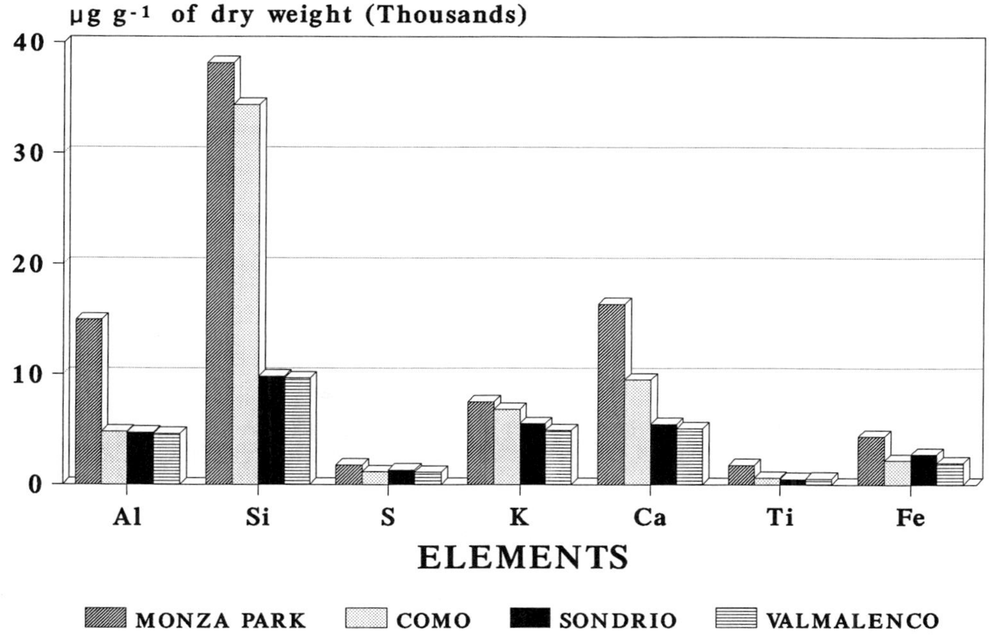

Fig. 2. Mean concentration (in µg/g of dry wt) for Al, Si, S, K, Ca, Ti, and Fe detected in the monitored areas.

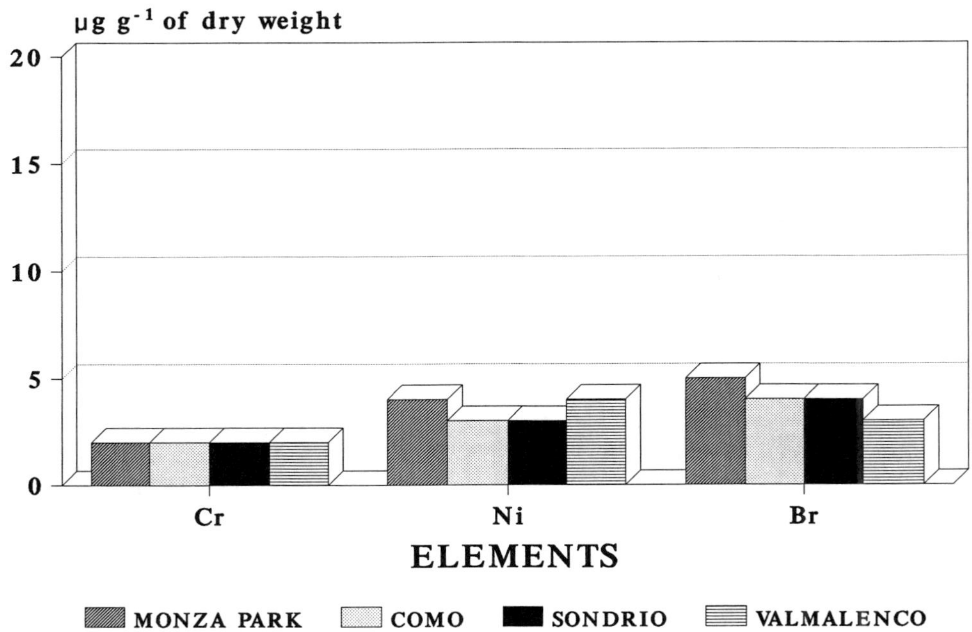

Fig. 3. Mean concentration (in µg/g of dry wt) for Cr, Ni, and Br detected in the monitored areas.

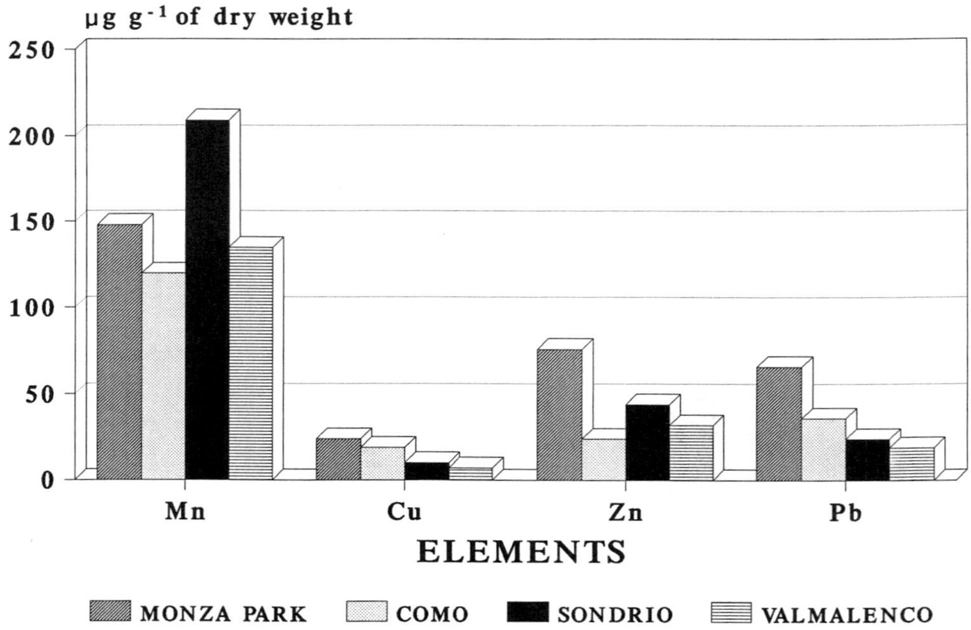

Fig. 4. Mean concentration (in µg/g of dry wt) for Mn, Cu, Zn, and Pb detected in the monitored areas.

Table 1
Elements Concentration Ranges (in µg/g of dry wt)

Element	**VALMALENCO** (25 samples)	**SONDRIO** (14 samples)	**COMO** (22 samples)	**MONZA PARK** (18 samples)
Al	1329-8924	2391-9222	1473-17991	3568-27748
S	580-1735	761-1821	748-1646	1158-2282
K	2876-8266	3910-9701	4266-8651	3838-9730
Ca	1930-8220	3731-8402	2681-17938	3887-25611
Ti	101-935	375-721	225-948	276-1794
Mn	65-277	115-380	59-270	77-311
Fe	620-4592	1182-3946	979-5115	1005-8551
Cu	2-12	3-21	5-62	17-39
Zn	15-48	29-70	24-97	39-115
Pb	4-48	10-48	16-77	19-133
Cr	1-7	1-3	1-3	1-4
Br	1-7	2-7	2-7	3-16
Ni	2-8	1-5	2-6	3-8

A Survey of Trace Elements in Fresh-Water Fish and Rice Along the Han River by Neutron Activation Analysis

K. S. PARK*, N. B. KIM, Y. S. KIM, K. Y. LEE, S. K. CHUN, AND Y. Y. YOON

Korea Institute of Geology, Mining and Materials, 30, Gajeong-dong, Yooseong-gu, Taejeon, 305-350, Korea

ABSTRACT

For a case study of environmental pollution, radiochemical neutron activation analysis (RNAA) was applied to the crucian and rice collected along the Han River. The crucian was analyzed for three times in 1973, 1987, and 1990. Sixteen trace elements (Hg, Cd, As, Br, Cu, Na, K, Se, Cr, Hf, Rb, Fe, Zn, Co, La, and Cs) were determined by RNAA using distillation and diethyldithiocarbamate extraction methods. Contents of Na, K, Se, Hf, Fe, Zn, and Co were almost constant regardless of the sampling place and year. The contents of the other elements showed increasing trends down river, especially in the first investigation. At the lower part of the river, the contents showed decreasing trends with the time of sampling, especially during the first two investigations. These trends were typical for Hg and Cd. Rice was analyzed by the same method for 12 elements, and the results showed no regional trends, but have decreased after 1973.

Index Entries: Trace elements; radiochemical neutron activation analysis; fresh-water fish; rice; river pollution.

INTRODUCTION

Studying environmental pollution owing to industrialization and urbanization activities can be facilitated employing a suitable biological

*Author to whom all correspondence and reprint requests should be addressed.

monitor readily available in both polluted and unpolluted background regions. Concerning elemental pollution, determination of many trace elements in biological monitors is frequently required. Neutron activation analysis is a suitable method for this purpose. Unfortunately, some toxic elements of interest, such as Hg, Cd, and As, can hardly be determined by instrumental neutron activation analysis because of overwhelming activities of several radionuclides, such as ^{24}Na, ^{42}K, ^{32}P, ^{82}Br, and so forth, in the γ-ray spectrum. Therefore, a radiochemical separation is required for determining these elements. The most commonly used separation methods are the ion-exchange method (1–3), the solvent-extraction method *(4,5)*, or a combination of these two methods.

For the environmental case study, the Han River is one of the most suitable areas in Korea, because it flows from East to West through Seoul, the most urbanized and industrialized city. The river contamination was investigated by determining Hg, Cd, As, and four other elements in crucian as a monitor 17 yr earlier (1973) *(2,3)*. In that experiment, distillation and ion exchange were applied for a group separation of the elements. The river contamination was investigated again by later determining 16 elements, including the above elements in 1987 *(6)*. In this experiment, the elements were separated into three groups by distillation and solvent extraction methods, in which the later was based on the extraction of cadmium and copper with diethyldithiocarbamic acid (HDDC) *(7)*.

At the first investigation in 1973, the same analytical method was also applied to 44 rice samples collected throughout the country to estimate the national contamination level. The content of seven trace elements, including mercury, cadmium, and arsenic, did not show any regional trends except for the high content of mercury at Haengju (site No. 7 in Fig. 1).

In the present work, the contamination of Han River was investigated again (in 1990) to evaluate the changes of the contamination levels of toxic elements during the past 17 yr. Crucians were collected at 10 sampling sites, the same as those used previously *(6)*, and the same analytical procedures were applied. In addition, rice samples were collected from the sites near the crucian sampling sites along the Han River to investigate possible correlation with crucians.

EXPERIMENT

Sample Collection and Pretreatment

Fresh-water fish (crucian) were collected from 10 sites along the Han River (as shown in Fig. 1) in 1990. Only crucians of 18–20-cm length were used for this investigation. Samples were cut into pieces with a plastic knife. Skin and bones were removed as completely as possible, and axial muscles were taken for analysis. They were transferred into polyethylene bags and dehydrated by drying for 3 d in a thermovac freeze unit.

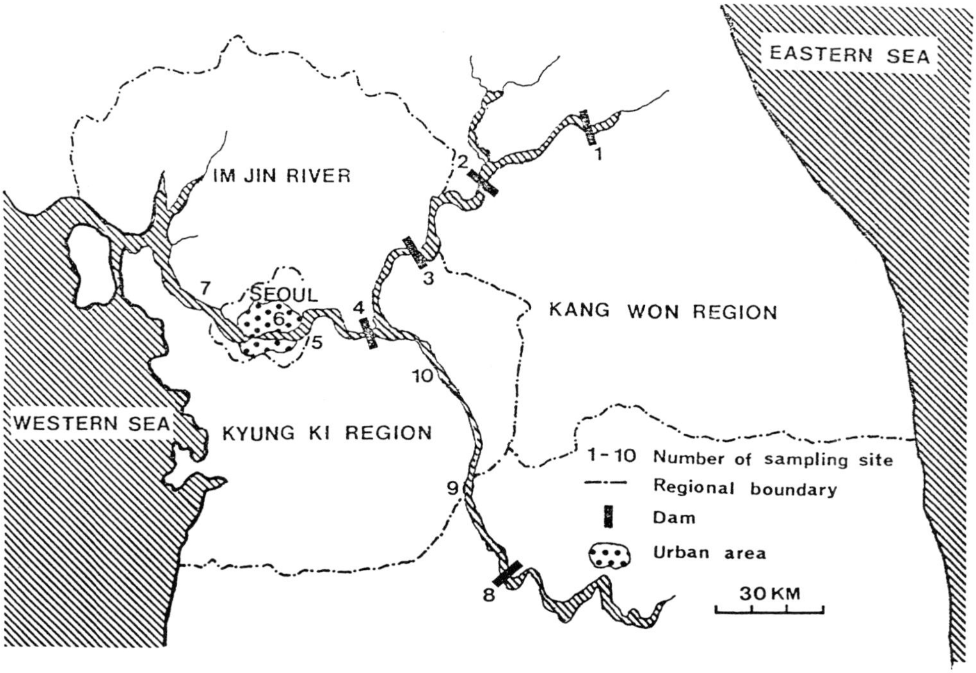

Fig. 1. Sampling sites along the Han River.

Rice samples were collected at 8 of the 10 crucian sampling sites. The two sites of Kwangnaru and Yeoido were excluded because rice is not produced there anymore. The rice was washed three times with deionized water and dehydrated by drying for 3 d.

Neutron Irradiation

Three grams each of the dehydrated crucian and rice samples were weighed into a precleaned silica ampule. The ampule was irradiated in the rotary specimen rack of the TRIGA Mark III reactor with two comparators (Au and Co) for 3 d. the neutron flux was about 5×10^{12} $n/cm^2/s$ at the irradiation position.

Radiochemical Separation

After the comparators were taken off, the irradiated silica ampules were thoroughly cleaned to remove surface contamination and cooled in liquid nitrogen to prevent possible losses of Hg and Br by volatilization *(1)*. Each ampule was inserted into a tygon tubing that was previously sealed at one end. The tubing was then filled with 3 mL of a decomposition mixture of conc. nitric and conc. sulfuric acids (9:1, v/v) in order to cover the ampule completely and also with 1 mL of an $HgCl_2$ solution containing 10 mg of Hg as a carrier. The tygon tubing was closed at the other end with a Hoffman clamp. Pressure was applied to crush the ampule, and the sample was mixed with the decomposition mixture. The released gases were absorbed by shaking the tubing. The tygon tubing was then opened.

The irradiated sample and splinters of the ampule were quantitatively transferred into the distillation flask of a modified Bethge distillation apparatus *(8)*, using a few milliliters of conc. nitric acid for washing. Ten milliliters of decomposition mixture were added to the distillation flask with 1 mL of Cd carrier solution containing 0.2 mg of Cd. The flask was heated at 250°C until the nitric acid distilled off into reservoir. After cooling, the distillate was allowed to run back into the flask. The heating cycle was repeated until the solution was completely clear and colorless. A mixture of 5 mL of 70% perchloric acid and 0.5 g of glycine dissolved in 5 mL of water was then added to the digested sample solution. The flask was heated again. The mercury fraction was collected with distillate in the reservoir. The distillation was interrupted when the temperature rose up to 320°C. The distillate in the reservoir was collected in a 50-mL volumetric flask for counting the activities of Hg and Br.

The digested solution in the flask was transferred into a 250-mL separatory funnel. Forty milliliters of a tartaric acid solution (50 mg/mL) and 60 mL of a 10*M* sodium hydroxide solution were added in the funnel. Thirty milliliters of chloroform, containing $3.4 \times 10^{-3}M$ HDDC, were added in the funnel as an organic phase, and the funnel was shaken for 1 min. Cadmium and copper were extracted into the organic phase and the water phase was washed with 10 mL of chloroform. The HDDC and the chloroform phases were combined in a 50-mL volumetric flask for counting. The aqueous phase was evaporated by heating to reduce the volume and transferred into a 50-mL volumetric flask to count the activities of other elements. The HDDC solution was prepared as follows *(7,9)*: The precipitate ($Zn[DDC]_2$), which was produced by adding NaDDC to a zinc chloride solution, was dissolved in chloroform to give a $4 \times 10^{-2}M$ solution. This solution was stable for at least 5 m if kept in a dark bottle. HDDC was then prepared for every analysis by shaking equal volumes of $Zn(DDC)_2$ in chloroform and 1*M* hydrochloric acid. The shaking time should not be longer than 20 s.

Counting and Evaluation of Element Contents

Three volumetric flasks were counted by a HPGe detector (CANBERRA, USA) coupled to a 4096-channel analyzer (CANBERRA S-90). The activity of nuclides in the γ-ray spectrum was calculated with PIAD (Peak Identification and Activity Determination) *(10)* computer code. The element content in the samples was calculated by the single comparator method, using Au and Co comparators *(11,12)*. The analytical results of crucian and rice are given in Tables 1 and 2.

Table 1
Analytical Results of Trace Elements in Crucian (ppm in Dry Base)

ele-men	Year	Soyang Site 1	Euiam Site 2	Chung-pyung Site 3	Paldang Site 4	Kwang-naru Site 5	Yeoido Site 6	Haengju Site 7	Chungju Site 8	Yeoju Site 9	Yang-pyung Site 10
Hg	73	0.17	0.23	0.18	0.77	0.83	2.7	2.9			
	87	0.12	0.16	0.17	0.25	0.27	0.28	0.39	0.28	0.48	0.65
	90	0.13	0.015	0.13	0.23	0.22	0.25	0.55	0.46	0.23	0.35
Cd	73	0.21	0.23	0.25	0.62	0.71	1.1	1.2			
	87	0.008	0.005	0.006	0.017	0.018	0.024	0.035	0.004	0.002	0.044
	90	0.003	0.010	0.003	0.007	0.004	0.003	0.010	0.002	0.003	0.005
As	73	0.30	0.59	0.62	0.32	0.20	16	0.13			
	87	0.25	0.20	0.17	0.35	0.30	0.44	0.59	0.45	0.23	0.93
	90	0.23	0.10	0.12	0.22	0.18	0.41	0.40	0.50	0.30	0.18
Br	73	5.5	1.4	4.1	5.9	5.9	3.5	6.8			
	87	1.5	1.7	2.1	2.4	3.0	1.9	3.4	3.3	2.5	1.6
	90	2.0	1.9	2.9	1.5	2.7	4.6	3.0	3.9	3.1	7.0
Cu	73	0.071	2.2	4.2	3.5	5.5	7.2	3.5			
	87	0.69	0.69	0.69	0.59	0.54	0.96	1.3	0.72	1.3	0.51
	90	0.54	1.7	0.90	1.4	0.60	2.6	3.0	0.9	1.0	0.61
Na,%	73	0.15	0.14	0.20	0.16	0.21	0.13	0.24			
	87	0.093	0.10	0.098	0.12	0.10	0.095	0.12	0.12	0.14	0.070
	90	0.11	0.20	0.13	0.16	0.23	0.10	0.14	0.16	0.20	0.13
K,%	73	0.67	0.87	1.0	0.75	0.47	1.9	1.1			
	87	1.1	1.1	1.2	1.1	1.2	1.2	1.3	1.3	1.1	1.2
	90	1.0	0.8	1.0	1.0	1.1	0.89	0.90	1.0	1.18	1.0
Se	87	0.70	0.68	0.79	0.48	0.53	0.63	0.40	0.67	0.75	0.52
	90	0.54	0.64	0.65	0.43	0.90	0.33	0.67	0.57	0.65	0.51
Cr	87	0.27	0.31	0.53	0.43	0.39	0.64	0.76	0.58	0.61	0.40
	90	0.12	0.27	0.31	0.40	0.37	1.0	1.0	0.27	0.35	0.35
Hf	87	0.009	0.008	0.007	0.009	0.011	0.015	0.018	.0.004	0.005	0.006
	90	0.009	0.020	0.013	0.016	0.012	0.016	0.015	0.002	0.005	0.009
Rb	87	5.6	7.3	6.3	8.7	8.0	11	10	11	6.5	11
	90	4.9	11	6.3	4.2	11	15	6.5	7.5	6.7	7.2
Fe	87	22	17	17	16	14	18	27	23	22	10
	90	23	15	25	23	26	16	14	16	9.9	24
Zn	87	57	61	61	75	70	39	73	51	51	41
	90	68	96	58	71	75	52	50	29	66	48
Co	87	0.013	0.015	0.015	0.014	0.016	0.019	0.015	0.010	0.013	0.006
	90	0.015	0.008	0.019	0.019	0.014	0.015	0.033	0.015	0.010	0.013
La	87	0.013	0.016	0.022	0.030	0.031	0.036	0.038	0.029	0.020	0.032
	90	0.013	0.019	0.018	0.021	0.017	0.024	0.030	0.020	0.020	0.023
Cs	87	0.010	0.017	0.021	0.023	0.032	0.035	0.030	0.037	0.008	0.049
	90	0.009	0.008	0.020	0.021	0.021	0.015	0.027	0.037	0.024	0.025

RESULTS AND DISCUSSION

The method for pretreatment and digestion used in this work was same as in the previous investigations *(2,3,6)* and the solvent extraction method was also the same as in the second investigation *(6)*. These studies revealed that the Hg loss from biological samples was negligible,

Table 2
Analytical Results of Trace Elements in Rice (ppm in Dry Base)

element	Soyang Site 1	Euiam Site 2	Chung-pyung Site 3	Paldang Site 4	Haeng-ju Site 7	Chung-ju Site 8	Yeoju Site 9	Yang-pyung Site 10	Average Present	Average 1973
Hg	0.0068	0.0087	0.011	0.0060	0.0098	0.0032	0.013	0.015	0.0092	0.040
Cd	0.024	0.021	0.017	0.026	0.017	0.013	0.018	0.022	0.020	0.11
As	0.020	0.056	0.11	0.069	0.055	0.049	0.051	0.079	0.061	0.22
Br	0.43	0.10	0.33	0.26	0.24	0.12	0.35	0.17	0.25	0.43
Na	3.7	4.0	3.7	4.9	3.0	8.0	4.1	3.5	4.3	20
K	770	520	650	590	470	720	680	580	620	690
Se	0.022	0.035	0.012	0.031	0.029	0.032	0.027	0.019	0.026	-
Cr	0.11	0.076	0.21	0.083	0.21	0.10	0.15	0.20	0.14	-
Rb	1.4	2.3	2.0	4.4	1.8	2.0	2.1	4.2	2,5	-
Fe	1.5	1.9	3.0	2.6	1.0	3.9	2.3	3.7	2.8	-
Zn	14	14	9.1	13	12	13	11	12	12	-
Co	0.0016	0.0025	0.0024	0.0021	0.0039	0.0021	0.0023	0.0021	0.0024	-

because the element recoveries were over 98%. The reliability of this method was verified by analyzing NIST biological reference materials.

Contamination of Crucian

Figure 1 shows the middle part of the Korean peninsular, where the Han River flows from East to West through Seoul, the most industrialized and urbanized city in Korea, and 10 sampling sites: Soyang (site No. 1), Euiam (site No. 2), Chungpyung (site No. 3), Paldang (site No. 4), Kwangnaru (site No. 5), Yeoido (site No. 6), Haengju (site No. 7), Chungju (site No. 8), Yeoju (site No. 9), and Yangpyung (site No. 10). Sites 8–10 were not included in the first investigation in 1973, but were included in the second (1987) and in the third (the present investigation). The analytical results for crucians collected in 1990 are shown in Table 1 for 16 elements (Hg, Cd, As, Br, Cu, Na, K, Se, Cr, Hf, Rb, Fe, Zn, Co, La, and Cs). The contents of seven elements (Hg, Cd, As, Br, Cu, Na, and K) determined in the first study and the contents of 16 elements in the second study are also included in Table 1.

Site No. 1 shows the lowest contents of 16 elements of all 10 sampling sites and similar values for all three investigations. This can be explained by the fact that the site is the uppermost place, and therefore, the least contaminated by industry and urbanization. Among the 16 elements, 7 elements (Na, K, Se, Hf, Fe, Zn, and Co) are characterized by similar contents regardless of the sampling place and year, with average contents of 0.13%, 1.1%, 0.61ppm, 0.010ppm, 19ppm, 59ppm, and 0.015ppm, respectively. These elements do not show any relationship to environmental contamination. The contents of the other nine elements increase as the river flows down (regional trend). The upper part of the river divides into two tributaries, northern and southern, and sites Nos. 8, 9, and 10 at the southern tributary are comparable with sites Nos. 2 and 3 at the northern tributary from the environmental point of view. The average contamination level of the southern tributary seems to be about 40% higher than that of the northern one, considering the elements that have the regional trend.

For the nine elements having the regional trend, the contents of seven elements (Hg, Cd, As, Cr, Rb, La, and Cs) in the second investigation are slightly higher than those in the third study for the whole regions, whereas the other two (Br and Cu) are similar. The contents of the elements in the first investigation are much higher than those in the second and/or in the third study at the lower part of the river, whereas they are very similar at the upper part. The ratios of the average content at the lower part (sites Nos. 6 and 7) to that at the upper part (sites Nos. 1, 2, and 3) are 14 for Hg, 6 for Cd, 20 for As, 2 for Br and 5 for Cu in the first investigation, and 2 for Hg, 4 for Cd, 2 for As, 2 for Br, 2 for Cu, 3 for Cr, 2 for Rb, 3 for La, and 3 for Cs in the second one. At the lower part, the ratios of the contents in the first to that in the second study are 8 for Hg, 30 for Cd, 14 for As, 3 for Br, and 5 for Cu.

The contents of Hg and Cd are graphically shown in Fig. 2. It is seen that the Hg content increases as the river flows down. At the first three sites (the upper part), the Hg contents in the three investigations remained constant, in the range of 0.12–0.17 ppm, and can be considered as the natural background level. The contents at the middle part around Seoul (sites Nos. 4 and 5) in the second investigation decreased drastically compared with those in the first one by a factor of 3. After the river passes Seoul (sites Nos. 6 and 7), the factor reached 8. Cd also shows a similar regional trend with Hg. However, at the upper part of the river, the Cd content in the first investigation was 20 times higher than that in the second one, and this abnormality still remains unexplained. In the third investigation, the regional trend almost disappeared. The La and Cr contents exhibit a similar regional increase (by a factor of 3). The content of La in the third investigation is about 20% lower compared to the second study at the lower part of the river. Cr also has the trend that its content in the third study is lower than in the second one, but this trend is reversed at the lower part, which is difficult to explain.

Contamination of Rice

Among 10 sampling sites, rice could not be collected from two sites, Nos. 5 and 6 (Kwangnaru and Yeoido), because of expanding urbanization. The analytical results for 12 elements (Hg, Cd, As, Br, Na, K, Se, Cr, Rb, Fe, Zn, and Co) are shown in Table 2. No regional trend is found for these elements. The rice samples, which were collected from 44 sites through the country, were analyzed in the previous investigation (1973). The contents of eight elements (Hg, Cd, As, Br, Na, K, Cu, and Mn) analyzed in the previous studies did not show any regional trend except for

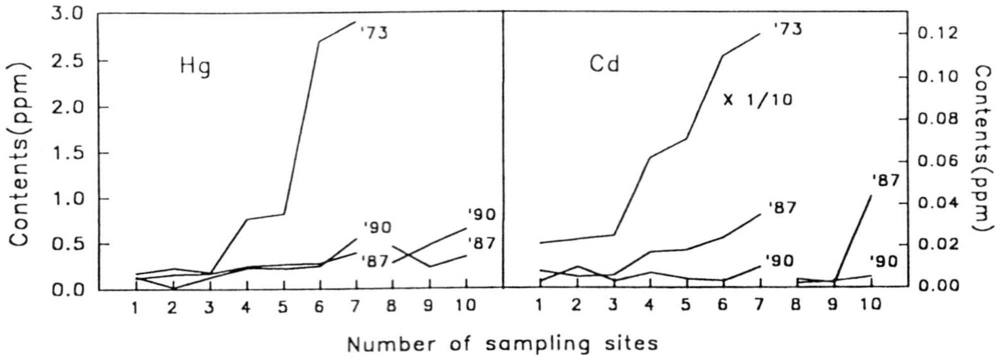

Fig. 2. The trends of mercury and cadmium contents in crucian along the Han River.

Hg at Haengju (site No. 7), which was exceptionally high (0.36 ppm) compared to the other sites and the national average of 0.11 ppm *(3)*. Table 2 contains the national average contents of six elements in the previous investigation. Among the six elements analyzed in both investigations, all elements show two to five times lower contents than those in the first investigation, except for K, which has similar value. Consider that the river water was used for irrigation at the time of previous investigation (the water contamination could be directly transferred to rice), whereas underground water is used nowadays, which is not as seriously affected by regional surface contamination as the river water. At site No. 7 of Haengju, the Hg content, which was exceptionally high, became comparable to the other sites. Therefore, it seems to be certain that the change of the irrigation method caused the contents of five elements in rice to decrease. In case of K, the content has not been changed, because K is one of the compositional elements in fertilizer.

It can be concluded from the above results that the upper part of the Han River has been little contaminated, and that in the middle and lower parts, which had been seriously contaminated, drastical decontamination occurred and is still in progress. This seems to be because of the installation of several sewage treatment facilities and because the contaminated sediment has been cleaned through the project of the redevelopment of the Han River (1983–1985). The rice contamination has also dropped owing to the change of the irrigation method to using underground water instead of river water.

REFERENCES

1. N. B. Kim, C. Lee, and H. I. Bak, *J. Korean Nucl. Soc.* **8,** 41 (1976).
2. C. Lee, N. B. Kim, and E. B. Park, *J. Korean Nucl. Soc.* **5,** 279 (1973).
3. C. Lee, N. B. Kim, and I. C. Lee, Report of IAEA Research Contract No. 1136/RB, 1974.
4. G. Zhuang, Y. Wang, M. Zhi, W. Zhou, J. Yin, M. Tan, and Y. Cheng, *J. Radioanal. Nucl. Chem.* **129,** 459, (1989).

5. E. Taskaev, I. Penev, and L. Kinova, *J. Radioanal. Nucl. Chem.* **120,** 83 (1988).
6. K. S. Park, N. B. Kim, H. J. Woo, Y. Y. Yoon, and K. Y. Lee, *Biol. Trace Element Res.* **26,** 349 (1990).
7. S. Bajo and A. Wyttenbach, *Anal. Chem.* **49,** 158 (1977).
8. B. Sjoestrand, *Anal. Chem.* **36,** 814 (1964).
9. S. Bajo and A. Wyttenbach, *Anal. Chem.* **47,** 1813 (1975).
10. N. B. Kim, Y. S. Kim, J. K. Kim, and K. S. Park, *J. Radioanal. Nucl. Chem.* **125,** 403 (1988).
11. J. I. Kim, *J. Radioanal. Chem.* **63,** 121 (1981).
12. N. B. Kim, K. S. Park, and H. I. Bak, *J. Korean Nucl. Soc.* **18,** 85 (1986).

The Accumulation and Distribution of Trace Metals in Some Localized Marine Species

A. M. YUSOF,*[,1] N. A. RAHMAN,[1] AND A. K. H. WOOD[2]

[1]Department of Chemistry, Universiti Teknologi Malaysia, 80990 Johor Bahru, Malaysia; and [2]Nuclear Energy Unit, PUSPATI Complex, 43000 Bangi, Malaysia

ABSTRACT

Trace elements, such as As, Co, Cr, Hg, Sb, and Zn, were determined by neutron activation analysis (NAA), whereas Cd, Cu, and Pb were determined by graphite furnace atomic absorption spectroscopy (GFAAS) in clam, crab, prawn, swamp cerith, and mussel samples after digestion by microwave heating under controlled conditions before eluting the solutions through a column of a chelating resin, Chelex-100. The standard used in the determination of percentage volatile elements retained by microwave digestion and also in the activation process was *Lobster Hepatopancreas TORT-1*, whereas known mixed standards were prepared from nitrate salts to determine the efficiency of the separation procedure at a controlled pH. Mercury and lead detected in crabs exceeded the maximum permissible level. Some species also showed a high affinity toward certain elements, and their levels of accumulation in the tissues of these species corresponded with the concentration of these elements in sediments, especially at sites in the vicinity of an industrial zone.

Index Entries: Trace elements; clam; crab; swamp cerith; prawn; mussel; microwave digestion; chelating resin; mercury; neutron activation analysis.

INTRODUCTION

In recent years, some metals, such as lead, cadmium, arsenic, cobalt, chromium, and zinc, or their compounds have been used extensively by various industries, thus leading to sharp increases in the contamination

*Author to whom all correspondence and reprint requests should be addressed.

of air, water, and soil. The marine environment is not spared since the open sea is the most exposed area of contamination as a result of industrial wastes discharge. The constant discharge of toxic industrial wastes into the sea through rivers and streams resulted in the accumulation of pollutants in the sediment, where some localized species spawn and feed. Some of these species happen to be served as popular delicacies, and any form of contamination of these species would easily enter the human food chain. The coastal areas of Johore provide the ideal location for this study, since the sprawling fishing villages situated along the coast are in the vicinity of an industrial zone, especially in the southern region of the state. An environmental assessment of coastal sediments from the same area *(1)* showed that some parts of the area were heavily polluted by some heavy and trace metals owing to accumulation over a long period.

This article describes the methods used in a risk assessment study on some popular localized marine species resulting from metal pollution. Neutron activation analysis (NAA) technique was used for the determination of As, Co, Cr, Hg, Sb, and Zn, whereas Cd, Cu, and Pb were determined by graphite furnace atomic absorption spectrometry (GFAAS). The impact of settlement of industrial zones in the study area is also discussed with particular reference to marine species selectivity toward some pollutants. Comparisons were made in terms of concentrations of metal pollutants in the marine species with those recommended to be safe by the Ministry of Health. The distribution and concentration of the elements found were represented on maps for simplicity and identification of abnormalities in metal concentration in the species.

METHODS

Marine species investigated were kuruma prawn *(Panaeus japonicus)*, swimming crab *(Portunus pelagicus)*, blood clam *(Anadara granosa)*, sulcata swamp cerith *(Terebralia sulcata)*, and green mussel *(Perna viridis)*, which were collected randomly by local fishermen between July 1991 and October 1991 at different locations along the southern coastal areas of the state. A total of 30 animals from each species for each location were carefully opened to avoid external contamination, the cavity fluid drained, and the remaining shell contents freeze-dried using an Edward Mudalyo freeze-dryer and ground to <200 mesh with a Herzog HSM-100 grinder.

All reagents were of analytical grade (Analar) and prepared by diluting Analar acids (15.8*M* HNO_3 and 11.8*M* HCl), base (7.1*M* NH_4OH), or salt (NH_4COOCH_3) with quartz-distilled water (Q-H_2O)to the appropriate concentration. The certified reference material (CRM) *Lobster Hepatopancreas TORT-1* was supplied by the National Research Council of Canada (NRCC), whereas mixed secondary metal standards were prepared by dissolving the nitrate salts supplied by Sigma Chemicals. The chelating resin Chelex-100, in the sodium form used in the separation step, was obtained from Sigma Chemicals.

The general procedure for the preparation of the ion-exchange column used to preconcentrate the elements followed that described elsewhere *(2)* with some modifications. The slurry was prepared by soaking 1 g of 50–100 mesh Chelex-100 resin in 30 mL of 2.5 mol/L HNO_3 for about 4 h followed by rinsing with 20 mL of Q-H_2O to get rid of the excess acid, and finally converted to the ammonium form by eluting with 20 mL of 2.0 mol/L NH_4OH. The NH_4^+ addition remove alkaline and alkaline earth metal ions from the sample solution during the elution step. Excess NH_4OH was removed by rinsing with 20 mL of Q-H_2O. Prior to the preconcentration of metal ions, the samples were first buffered by adding a solution of 8 mol/L NH_4COOCH_3 (NH_4Ac) and the pH adjusted to 5.0 ± 0.1 *(2)*.

Sample solutions were prepared by dissolving 1 g of dried material with 5 mL of HNO_3 (70%) in a Teflon™ pressure decomposition vessel of a CEM Model MDS-81D microwave digester. The vessel was closed and heated for 2 min at 48°C at 30% power (190 W). After cooling to ambient temperature, the solution was adjusted to pH 1.5 with 2.5 mol/L HNO_3 to avoid bacterial growth and finally made up to 250 mL in a volumetric flask. A similar treatment was performed on the *Lobster Hepatopancreas TORT-1* CRM to ascertain the quantity of volatile elements retained in the samples after the digestion step.

Prior to elution, the 100-mL samples were adjusted to pH 5.0–5.5 with the dropwise addition of NH_4OH. Any spike (natural or radiochemical) to be used was added prior to the pH adjustment and the solution left to a minimum of 1 hr to equilibrate before any pH adjustment is made. Then about 0.5 mL of NH_4Ac was added to buffer the sample. Approximately 3 mL of the pH-adjusted sample were added to the column to initiate shrinkage of the resin owing to the change in pH and ionic form. The flow rate was controlled not to exceed 0.8 mL/min. The alkali, alkaline earth, and residual halogen elements were selectively eluted by adding 60 mL of 1.0M NH_4Ac to the column in 10-mL aliquots followed by 15 mL of Q-H_2O to remove residual NH_4Ac. The resin was later freeze-dried prior to preparation of samples for the irradiation procedure. The efficiency of the Chelex-100 resin to retain the metal ions was checked by spiking the blank solution with the mixed secondary metal standards containing between 2 and 300 μg/g *(3)* of the metal ions and running it down the column. The quantity of recovered ion was examined before and after the elution, and subsequently, the retention efficiency of the Chelex-100 resin determined.

Between 0.10 and 0.15 g of the dried resin was packed in a polyethylene vial (i.d. 10 mm × 20 mm) supplied by Fischer, heat-sealed, and irradiated for 6 h followed by a cooling time of 3 d for As, Sb, and Zn determination. Co, Cr, and Hg were detected after a cooling time of 4 wk. All irradiations were performed in a 1-MW TRIGA Mk.II reactor of the Nuclear Energy Unit using a rotating rack facility. The thermal neutron flux was 4×10^{12} n/cm^2/s. The samples were counted with an HPGe detector EG&G ORTEC with a resolution of 1.9 kev at 1332 keV ^{60}Co. A

Nuclear Data ND6680 computer-based analyzer system was used for data collection and reduction.

Cd, Cu, and Pb were determined spectrophotometrically with a Perkin-Elmer HG500 atomic absorption analyzer. The same procedure used in the preparation of sample solution was employed to prepare samples for spectrophotometric analysis. Spectrometric-grade standards supplied by Perkin-Elmer were used for calibration. The distribution patterns of elements in the samples investigated were presented on maps, and comparisons were made to evaluate the risk assessment aspect of the species and area under study.

RESULTS AND DISCUSSION

Only 12 suitable locations were identified in this study. However, not all five species investigated could be obtained from these 12 locations. Nevertheless, the analytical results obtained could be used as an indication of the current state of contamination for that particular area. The analyses were carried out on two batches of samples taken at different times, and the average concentrations calculated.

The chelating resin Chelex-100 is an effective resin with strong chelating properties especially for transition metal ions and also with low blank values *(4)*. The fact that this resin is able to concentrate the transition metals and some other elements of interest from the matrix elements makes it a suitable tool for acid-digested biological samples. A minimum pH value of 5.0 was required to retain the metal ions investigated on the Chelex-100 resin while removing the alkali, alkaline earth, and halogen elements by elution with 1 mol/L NH_4Ac. It was found that the chelating efficiency of Chelex-100 increases for the transition metals from pH 4 to 5 and reaches an optimum at approximately pH 5.0 *(2)*. When dried in the column, the 50–100 mesh Chelex-100 resin shrinks to about one-quarter of its hydrated size and forms relatively hard beads that can easily be transferred to the polyethylene vials. The retention efficiencies of the Chelex-100 for the elements investigated are shown in Table 1, and retention corrections were made where applicable. The slight decrease in the percentage retained for As and Hg could be attributed to the fact that both do not belong to the transition series and a pH range of 4 and 5 in the system might not be favorable. Moreover, the digestion method by microwave heating resulted in some loss of As and Hg volatilization.

The *Lobster Hepatopancreas TORT-1* CRM was used with each batch of samples analyzed to provide quality-assurance controls and to test the applicability of this procedure to marine samples. The results obtained are listed in Table 2 and seem to agree well with the certified values. The concentration of Sb is not reported, since this element was not identified in the CRM.

The main aim of this study was to establish information necessary to assess the risk associated with the current "state of the coastal areas" in terms of environmental pollution. Some forms of risk could be made

Table 1
Percentage Retention of Elements by Chelex-100

Elements	Concentration of elements in μg g^{-1}		
	Spiked	Eluted	Retained (%)
^{77}As	200.0	22.24±0.126	88.88
^{113}Cd	16.0	0.0	100.00
^{60}Co	2.0	0.11±0.041	94.55
^{140}Cr	130.0	2.41±0.635	98.14
^{64}Cu	100.0	4.10±0.758	95.91
^{197}Hg	150.0	18.85±2.724	87.43
^{203}Pb	300.0	1.46±0.504	99.51
^{65}Zn	200.0	3.94±0.788	98.03

Table 2
Elemental Determinations *in Lobster Hepatopancreas TORT-1* SRM

Elements	Element certified (μg g^{-1})[a]	Element found (μg g^{-1})[b]
As	24.6 ± 2.2	23.7 ± 1.9
Cd	26.3 ± 2.1	25.8 ± 2.1
Cr	2.4 ± 0.6	2.1 ± 0.5
Co	0.42± 0.05	0.40± 0.04
Cu	439 ±22	428 ±19
Pb	10.4 ± 2.0	9.8 ± 1.7
Hg	0.33± 0.06	0.30± 0.04
Zn	177 ±10	168 ±11

[a]Mean and 95% tolerance limits.
[b]Mean and standard deviation for five determinations.

by comparing the levels of some toxic elements observed throughout the coastal areas to those recommended by the Ministry of Health. The concentrations of elements studied are proportional to the area of the circles.

Samples analyzed for Hg content from all 12 locations showed that only three (locations 1, 6, and 8) exceeded the maximum permissible level (MPL) of 0.5 μg/g *(5)* stipulated by the Ministry of Health, and this was observed only in swimming crab *(Portunus pelagicus)* samples. The concentrations recorded for locations 1, 6, and 8 are 0.64, 0.71, and 1.33

μg/g respectively, as shown in Fig. 1. The maximum concentration reported is 170% higher than the MPL, and this location happens to be in the vicinity of an industrial zone in a bay.

The distribution of Pb in some marine species as depicted in Fig. 2 complemented well the amount accumulated in the sediments collected at the same location as reported in an earlier study *(1)*. Though the MPL is 2.0 μg/g^{-1} *(5)*, some species, such as kuruma prawn *(Panaeus japonicus)*, swimming crab *(Portunus pelagicus)*, and sulcata swamp cerith *(Terebralia sulcata)*, exhibited Pb concentrations of 3.00, 2.00, and 4.75 μg/g, respectively, at location 10, whereas location 9, the Pb content was 2.00 μg/g for kuruma prawn *(Panaeus japonicus)*. These locations are situated well within the bay area, and the species mentioned can be classified as high-risk species in view of the high Pb accumulation exceeding the value recommended by the Ministry of Health.

Only two species, sulcata swamp cerith *(Terebralia sulcata)* and swimming crab *(Portunus pelagicus)* exhibited Cu concentrations exceeding 4.00 μg/g at locations 8 and 9 as shown in Fig. 3. However, this value is far below the MPL of 30.0 μg/g *(5)*. Copper was thought to originate from electroplating and welding activities in the industrial zone, and this is true where ship building is concerned. High levels of copper were also reported in sediment samples in an earlier study *(1)*, ranging from 70.3 μg/g to a maximum of 185 μg/g. None of the other marine species showed any affinity toward the uptake of copper.

The selectivity toward chromium uptake is quite obvious in swimming crab *(Portunus pelagicus)*, sulcata swamp cerith *(Terebralia sulcata)*, and green mussel *(Perna viridis)*, with a maximum concentration of 1.40, 1.24, and 1.45 μg/g, respectively, at three different locations as shown in Fig. 4. The concentrations of Cr recorded for the other species have an average value of 0.5 μg/g. Generally, the presence of As in the species studied did not give any significant representation even though the average As content in the sediments was about 41.4 μg/g *(1)*. The majority of arsenic found in the sea water and some localized marine species indicated the presence of organic arsenic species with the inorganic components existing in ng/kg levels *(6)*. As indicated in Fig. 5, the selectivity for total arsenic by the species studied was quite low, with the exception of blood clam *(Anadara granosa)* exhibiting a value of 1.2 μg/g, which slightly exceeded the MPL of 1.0 μg/g *(5)*. Green mussel *(Perna viridis)* has the lowest selectivity for As with the concentration of <0.1 μg/g. Antimony did not exceed 0.5 μg/g in most species, and these figures are well below the MPL of 1.0 μg/g *(5)*. Thus, there is no significant risk of contamination by this element on the species studied.

Although the concentrations of Cd in most species range between 0.05 and 1.0 μg/g, only locations 7, 8, 9, and 10 (all in the bay area) as shown in Fig. 6 indicated Cd exceeding 0.2 μg/g in green mussel *(Perna viridis)* and sulcata swamp cerith *(Terebralia sulcata)*. Though this is below the MPL of 1.0 μg/g for Cd *(5)*, it gives an indication of the preferential characteristics of these species for Cd in general. On the other hand, cobalt showed no abnormalities in terms of accumulation in the species

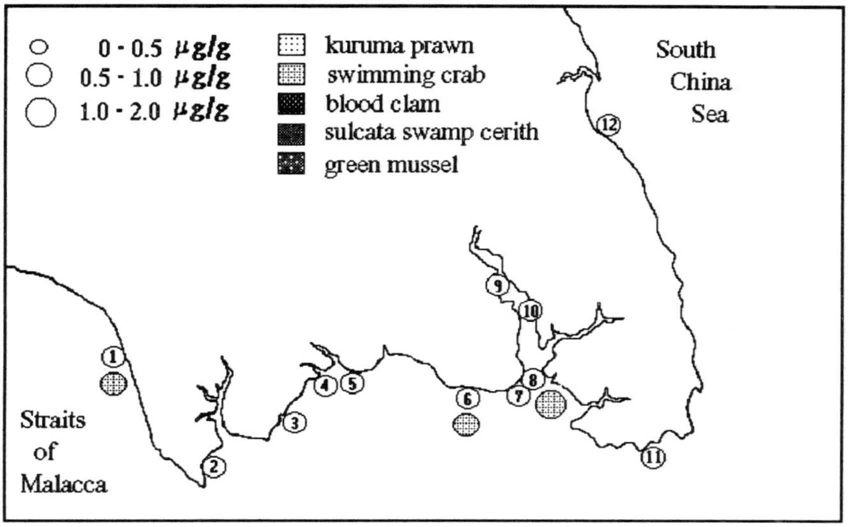

Fig. 1. Concentrations of Hg in localized marine species of South Johore.

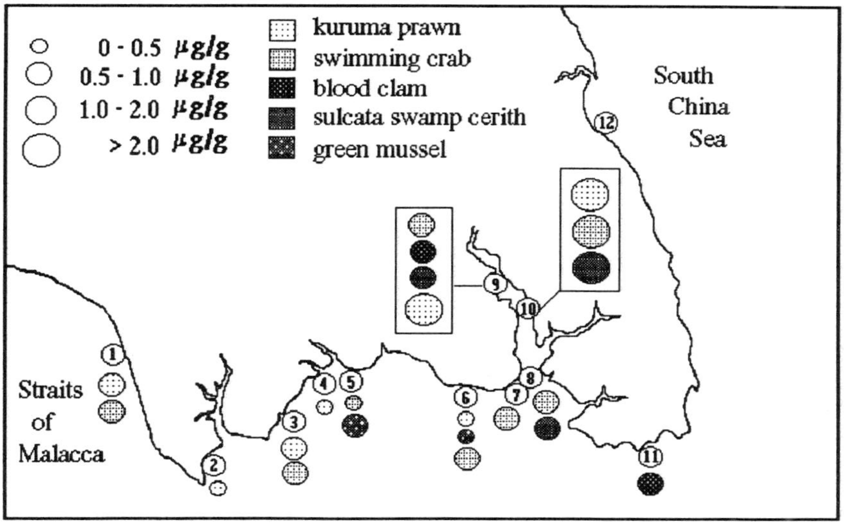

Fig. 2. Concentrations of Pb in localized marine species of South Johore.

studied, with the maximum concentration of 0.17 µg/g in green mussel *(Perna viridis)* and blood clam *(Anadara granosa)*. The general distribution of Co in marine species is shown in Fig. 7.

The concentration of Zn (Fig. 8) was much higher than the other metals studied and was found in all species ranging from as low as 18.4 µg/g to as high as 79.0 µg/g. However, the concentration of Zn in all the species did not exceed the MPL of 100 µg/g *(5)*. Most of the marine species have a higher level of tolerance towards Zn.

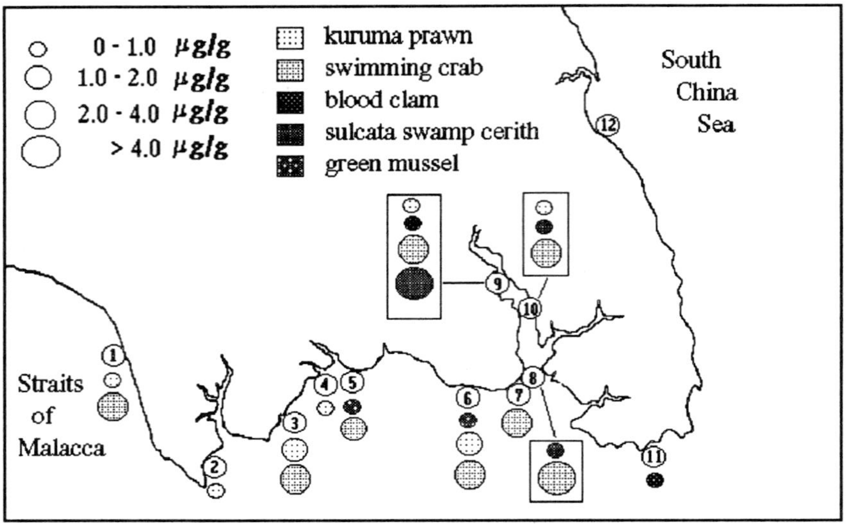

Fig. 3. Concentrations of Cu in localized marine species of South Johore.

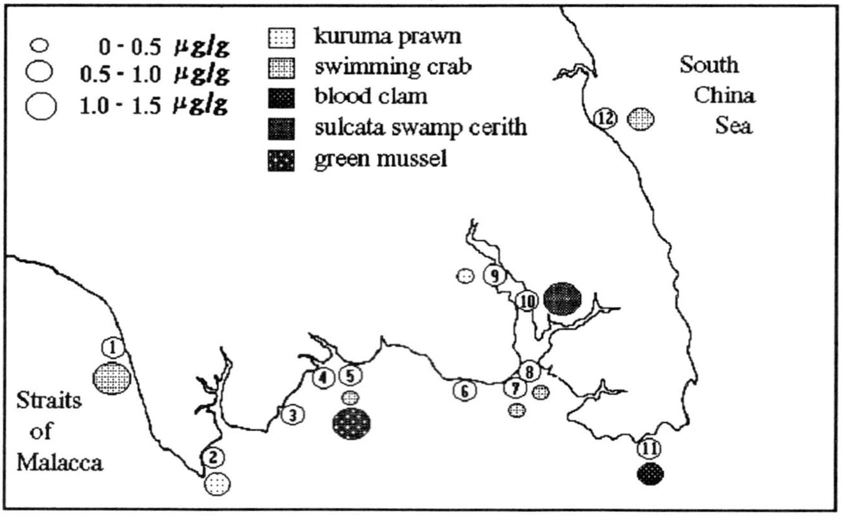

Fig. 4. Concentrations of Cr in localized marine species of South Johore.

Most of the pollutants discharged into the streams or rivers would eventually find their way into the estuaries and accumulate in the sediments *(7,8)*. Consequently, some of the metal species would find their way out from the surface of the sediments via complex biochemistry cycles and pathways *(9)*. These cycles depend on the physicochemical characteristics of the sediments, which influence the metabolism of the microbes. The amount of metal complexes taken up by marine species found on the ocean floor is determined by the chemical form of metals,

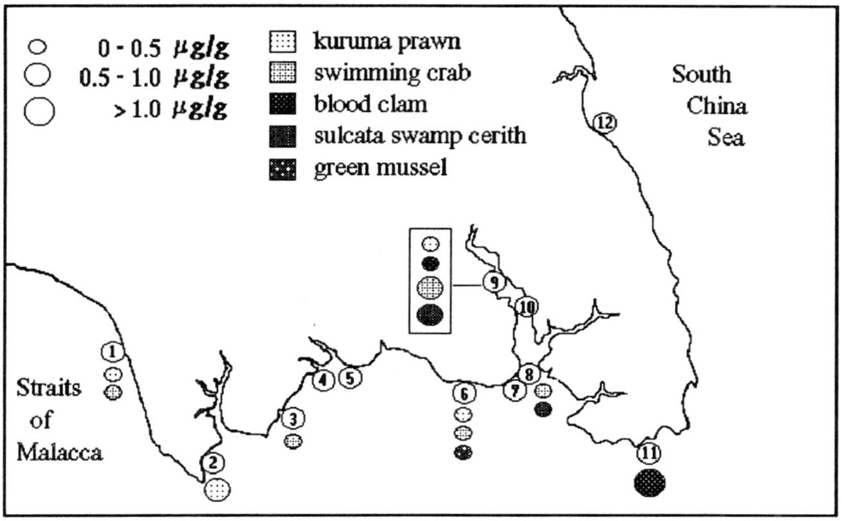

Fig. 5. Concentrations of As in localized marine species of South Johore.

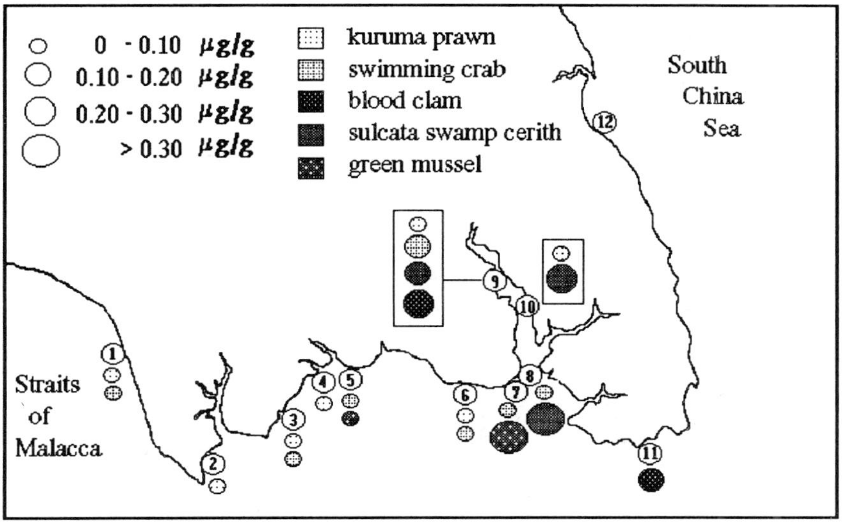

Fig. 6. Concentrations of Cd in localized marine species of South Johore.

water chemistry, and the relative distribution of metals in their soluble form and in particulates. The rate of uptake for these metals depends on the feeding behavior of these species.

Localized marine species have varying degrees of uptake and accumulation of metals in their tissues. This study revealed that some species showed some preferences in accumulating selective metals in their tissues, thus making them potential high-risk species in terms of metal pollution uptake. Swimming crab *(Portunus pelagicus)*, for example, showed

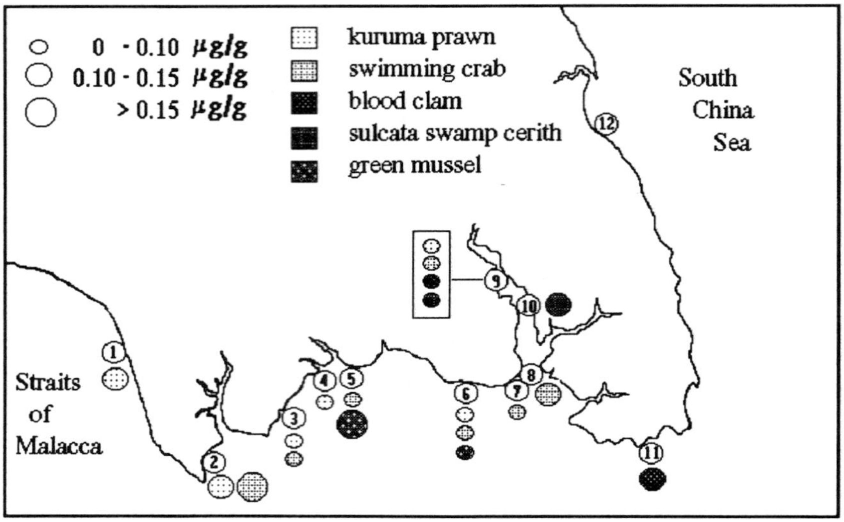

Fig. 7. Concentrations of Co in localized marine species of South Johore.

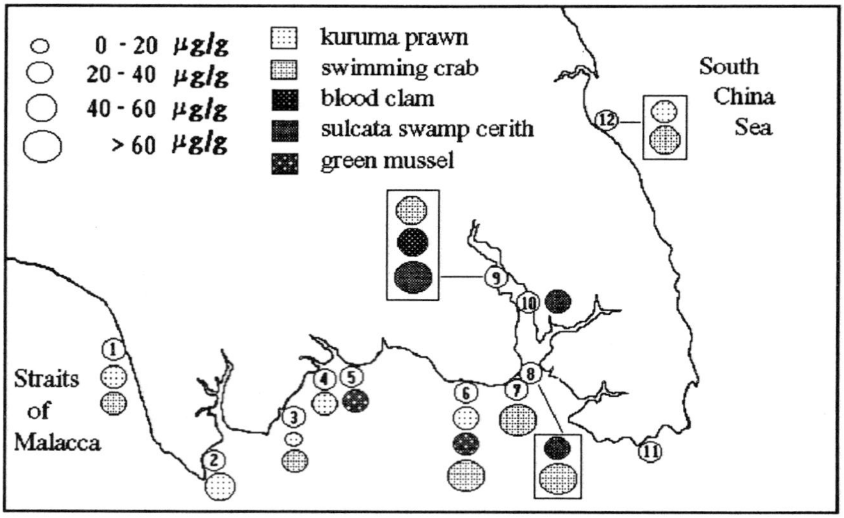

Fig. 8. Concentrations of Zn in localized marine species of South Johore.

concentrations of Hg and Pb higher than the MPL and with a relatively high degree of contamination by Cu and Cr. Concentrations of Pb exceeding the MPL were also evident in kuruma prawn *(Panaeus japonicus)* and sulcata swamp cerith *(Terebralia sulcata)*. In fact, high levels of Cu, Cr, and Cd were detected in sulcata swamp cerith *(Terebralia sulcata)*, thus categorizing this species in the high-risk group. Green mussel *(Perna viridis)*, on the other hand, showed a tendency to accumulate Cr and Cd, whereas blood clam *(Anadara granosa)*, in general, seemed to have a particular affinity toward As and its compounds as reported by several workers *(10–12)*. In this study, blood clam *(Anadara granosa)* accumulated arsenic in levels exceeding the MPL and confirmed the virtual ubiquity of this element in this species. The concentrations of zinc do not indicate any virtual tendency for any of the species to be a potential accumulator, and this trend is also valid for antimony. The apparent absence of Hg in all the species except swimming crab *(Portunus pelagicus)* suggested that this species has a selectivity toward this toxic element.

ACKNOWLEDGMENTS

One of us (A. M. Y.) would like to thank the National Science Development and Research Council (MPKSN) and the Technological University of Malaysia for supporting the grant given under contract No. IRPA 4-07-07-018, the Nuclear Energy Unit, and all those involved directly or indirectly during the tenure of this project.

REFERENCES

1. A. M. Yusof and A. K. H. Wood, *J. Radioanal. Nucl. Chem.* **167,** 341–351 (1993).
2. H. M. Kingston, I. L. Barnes, T. J. Brady, T. C. Rains and M. A. Champ, *Anal. Chem.* **50,** 2064–2070 (1978).
3. C. Lee, N. K. Bae, I. C. Lee, and K. S. Chung, *Talanta* **24,** 241–245 (1977).
4. E. W. Davey and A. E. Soper, *Analytical Methods of Oceanography,* T. R. P. Gibbs, Jr, ed., *Adv. Chem. Ser.,* pp. 147 (1975).
5. Ministry of Health, *Food Act 1983 and Food Regulations 1985 (Revised),* Malaysia, (1985).
6. A. M. Yusof, Z. B. Ikhsan, and A. K. H. Wood, *J. Radioanal Nucl. Chem.* (1994) (in press).
7. J. Gavis and J. F. Ferguson, *Water Res.,* **6,** 989 (1972).
8. J. W. M. Rudd, M. A. Turner, A. Furutani, A. L. Swick, and B. E. Townsend, *Can. J. Fish. Aquat. Sci.* **40,** 2206 (1983).
9. J. Hamilton-Taylor, M. Willis, and C. S. Reynolds, *Limnol. Oceanogr.* **29,** 695 (1984).
10. W. R. Cullen and K. J. Riemer, *Chem. Rev.* **89,** 715–764 (1989).
11. W. R. Cullen and M. Dodd, *Appl. Organomet. Chem.,* **3,** 79–88 (1989).
12. D. J. H. Phillips and M. H. Depledge, *Wat. Sci. Tech.* **18,** 213–222 (1986).

Determination of Antimony in Natural Waters by Preconcentration on a Chelating Sorbent Followed by Instrumental Neutron Activation Analysis

WANDA ŻMIJEWSKA

Institute of Nuclear Chemistry and Technology, ul.Dorodna 16,03-195 Warsaw, Poland

ABSTRACT

Antimony was preconcentrated from natural waters on thionalide-loaded acrylic polymer (Bio-Beads SM-7) from 0.5 m HCl solution. Prior to the preconcentration, Sb(V) was reduced to Sb(III) with potassium iodide. The antimony retained on the resin was determined by neutron activation and γ-spectrometric measurement of ^{122}Sb (564 keV). The lower limit of detection was 0.023 µg/L for a 100-mL sample.

Index Entries: Antimony; preconcentration; chelating sorbent; thionalide; neutron activation analysis; natural waters.

INTRODUCTION

Antimony is a nonessential element to plants, animals, and humans *(1,2)*. However, its concentration in the environment is of interest because of the toxic nature of this element and its compounds *(3)*. The toxic effects of antimony compounds are similar to those of arsenic, trivalent compounds being more toxic than pentavalent ones *(1–3)*. Some antimony compounds are carcinogenic to animals *(2,3)* and are suspected to be carcinogenic to humans *(2)*. Antimony and its compounds have been listed as priority pollutants by the US Environmental Protection Agency *(4)*.

The concentration of antimony in natural waters varies over a wide range, generally from 0.01 to 1 µg/L *(5–10)*. Even the most sensitive modern analytical techniques, such as neutron activation analysis and hydride-generation atomic absorption spectrometry, are incapable of determining antimony at concentration levels below 0.1 µg/L, unless some preconcentraion steps are applied.

The determination of antimony in natural waters by the neutron activation metod without preconcentration is possible only if large samples (several tens of milliliters) are irradiated and radiochemical separation is carried out *(10,11)*.

Various methods have been used for preconcentration of antimony from natural waters in combination with the neutron activation method, e.g., evaporation *(12)*, freeze-drying *(8)*, coprecipitation with manganese dioxide *(13)*, cocrystallization with thionalide *(6)*, and extraction with ammonium pyrolidinecarbodithioate *(4)*. Chelating sorbents have also been used for preconcentration of antimony from waters, but in combination with atomic absorption spectrometry *(14,15)*. In the present work, a chelating sorbent-thionalide-loaded acrylic resin *(16)* has been used for preconcentration of antimony from natural waters prior to its determination by instrumental neutron activation analysis.

EXPERIMENT

Reagents and Solutions

Thionalide, p.a. (Fluka-Buchs);
Bio-Beads SM-7, 20–50 mesh (Bio-Rad);
Sb_2O_3, s.p. (Koch-Light);
$SbCl_5$, 99.99% (Koch-Light);
Hydrochloric acid, Suprapure and A.R. grade; and
Humic acid sodium salt, technical product (Aldrich).

All other reagents used were of A.R. grade. Doubly distilled water (quartz still) was used throughout this work.

A standard solution of Sb(III) of a concentration of 1 mg/mL was prepared by dissolving 0.1197 g of Sb_2O_3 in 2.5 mol/L HCl solution on heating under a reflux condenser and diluting the obtained solution to 100 ml with 2.5 mol/L HCl solution. A stock solution of Sb(V) of a concentration of 40.7 mg Sb/mL was prepared by dissolving 10 g of $SbCl_5$ in concentrated hydrochloric acid and diluting to 100 mL with the acid. Working solutions were prepared by diluting the stock solution with concentrated hydrochloric acid.

A synthetic water solution was prepared of the following composition (mg/L): Na^+(15), K^+(4), Mg^{2+} (20), Ca^{2+}(55), Cl^{-1}(124), and SO_4^{-2}(80).

Radioactive Tracers

Radioactive tracers ^{122}Sb(III) and ^{122}Sb(V) were obtained by irradiation of appropriate amounts of the stock solutions in a nuclear reactor. In order to make sure that the irradiated antimony was in the desired valency state, Sb(V) was separated from Sb(III) by extraction with diisopropyl ether from 3*M* HCl solution *(17)* followed by back-extraction with 0.1M HCl solution.

Apparatus

This consisted of a γ-ray spectrometer with an ORTEC HPGe detector (resolution 1.8 keV at 1332.4 keV, efficiency 40%) and a multichannel analyzer SWAN-31 (Institute for Nuclear Sciences, Świerk, Poland).

Waters Analyzed

Vistula River water samples were collected from the main river stream at the water intake of the waterworks in Warsaw. The samples were brought to the laboratory and filtered through 0.45 µm membrane filters (Sartorius, Type SM) within 4 h, then acidified to pH 1 with hydrochloric acid (Suprapure), and stored in a refrigerator. Well waters were from Legionowo and Warsaw. Tap water was from from Warsaw. The samples of well and tap waters were not filtered, because they were clear. They were acidified and stored as described above.

Conditions of Irradiation

Samples and standards were irradiated in the nuclear reactor EWA at Świerk at a neutron flux of $3 \times 10^{13} n\ cm^{-2}/s$ for 16 h.

Preparation of the Sorbent (16)

Several gramms of Bio-Beads SM-7 resin (20–50 mesh) were washed with several portions of ethanol and allowed to dry in the air. Three grams of the resin were placed in a separatory funnel and shaken with 50 mL of 0.05 mol/L thionalide solution in ethanol for 3 h. The thionalide-loaded resin was filtered off under reduced pressure and allowed to dry in the air. The sorbent (TBB) was stored in a refrigerator.

Conditions of Sorption of Antimony on TBB

Under static conditions, Sb(III) is quantitatively retained on TBB over a concentration range of 0.1–2 mol/L *(16)*. The optimum concentration range for sorption of Sb(V) is the same, but sorption is not quantitative (the maximum retention is 88%).

In this work, the retention of antimony was studied under dynamic conditions. Glass columns (id 0.4 cm) were filled with 0.15 g TBB. Before sorption, the column bed was washed with 5 mL of hydrochloric acid solution of suitable concentration (the same as for sorption). The solution containing the radioactive tracer (0.2–2 µg Sb) was then passed through the column at a definite rate. The effluent was collected. After the sorption, the column was washed with hydrochloric acid solution of the same concentration as for the sorption. The retained antimony together with thionalide was washed from the column with 25 mL of acetone. The percentage retention of antimony was calculated on the basis of measurements of ^{122}Sb activity in the effluent and in the acetone solution. Prior to the sorption, Sb(V) was reduced to Sb(III) using 10% potassium iodide solution, and the liberated iodine was reduced with a few drops of 10% ascorbic acid solution.

Experiments were carried out in order to establish the most suitable conditions of reduction of Sb(V) to Sb(III) (concentration of KI, acidity, time of reduction) and sorption of antimony on TBB (flow rate, sample volume). In all the experiments, the synthetic water was used as sample.

On the basis of the above experiments the following conditions of quantitative sorption of antimony (Sb[V] and Sb[III]) were chosen:

Acidity	0.5 mol/L HCl
KI concetration	0.24%
Reduction time	0.5 h
Sample volume	Up to 250 mL
Flow rate	<2 mL/min

The effect of humic acid on the retention of antimony on TBB under the above conditions was examined using the synthetic water spiked with ^{122}Sb with additions of humic acid (10–20 mg/L).

Determination of Antimony in Water Samples

To 100 ml of water sample, 4.5 mL of conc. HCl (Suprapure) was added followed by 2.5 mL of 10% KI solution. After 0.5 h, a few drops of 10% ascorbic acid solution were added. The sample solution was then passed through a column (id 0.4 cm) filled with 0.15 g of TBB (previously washed with 5 mL of 0.5 mol/L hydrochloric acid solution) at a flow rate of approx 1.7 mL/min. The column bed was then washed with 1 mL of 0.5 mol/L HCl solution followed by 20 mL of water. The bed was then removed from the column, allowed to dry in the air, and transferred to a quartz ampule.

Antimony standards for irradiation were prepared as follows: To 20 mL of 0.5 mol/L HCl solution, containing 100 ng of Sb(III), 0.5 mL of 10% KI solution was added. After 0.5 h, few drops of 10% ascorbic acid solution were added. The solution was transferred to a separatory funnel containing 0.15 g TBB and shaken (mechanical shaker) for 10 min. The solution was filtered off, and the sorbent was washed with approx 20 mL of water, allowed to dry in the air, and transferred to a quartz ampule. The samples and standards were irradiated under the above-given conditions and cooled for 6 d in order to allow ^{24}Na and ^{82}Br to decay. The samples and the standards were transferred to plastic containers (with 1-cm wells in the center), and the activity of ^{122}Sb ($E\gamma = 564$ keV) was measured with the γ-ray spectrometer. The counting time varied from 200 to 3000 s depending on the antimony content.

RESULTS AND DISCUSSION

Since no reference water samples were available, the elaborated method was verified by analyzing samples of the synthetic water and well water to which known amounts of antimony had been added. The obtained results are presented in Table 1. The results of analysis of some samples of natural waters are shown in Table 2. It follows from Table 1 that the recovery of the added antimony is satisfactory. The slightly less recovery of antimony from the samples of the synthetic water (average value 96%) may be the result of inaccurate determination of the blank value.

Surface waters may contain humic and fulvic acids, which form complexes with antimony. The examination of the effect of humic acid on the sorption of antimony under the applied conditions has shown that humic acid over the concentration range 10–20 mg/L has little effect on the sorption of antimony (the sorption decreases by approx 3%). This could be expected on the basis of the studies carried out by Bowen and coworkers *(18)*, who found that Sb(V) and Sb(III) humates are stable over the pH range 7–11 and are highly dissociated below pH 4. Antimony fulvates are not formed below pH 4 *(18)*.

In this work, the determination of antimony was carried out after reduction of Sb(V) to Sb(III) with potassium iodide; hence, the reducing agent was always present in in the sample solution. It should however, be, mentioned that in the absence of potassium iodide, the sorption of antimony at low concentrations is not quantitative. This phenomenon may be the result of oxidation of Sb(III) to the pentavalent state.

Oxidation of tervalent antimony to the pentavalent state in water solutions was also observed by other authors *(9,19)*. Andreae and coworkers *(9)* state that Sb(III) can be oxidized to Sb(V) at unpredictable rate (depending on the quality of water) and attribute this phenomenon to the presence of uncontrolled content of oxidizing agents.

The concentration of antimony in the Vistula river water found in this work (Table 2) is of the same order as that found for other polluted rivers *(7–9)*. The concentration of antimony in the examined tap water is higher than that found by Lieser and Neizert *(8)*. Schneider and Geisler *(20)* determined antimony in drinking water from 11 waterworks in southern Germany and found concentrations of antimony 0.03–5.3 µg/L. The concentration of antimony in tap water determined in this work is about two orders of magnitude lower than the maximum admissible concentration given by WHO (10 µg/L) *(21)*. The concentration of antimony in well waters is an order of magnitude lower than that found by Bunus and coworkers *(12)*.

There is little interference in the γ-spectrometric measurement of the ^{122}Sb activity ($E\gamma$ = 564 keV). The radioisotopes ^{24}Na and ^{82}Br, present in all samples examined, decay to an acceptable level during the applied cooling period. Under the aplied conditions of sorption on TBB, only few elements (Pd, Hg, Au, As) are retained on TBB *(16)*, and their radioiso-

Table 1
Results of the Determination of Antimony in Water Samples with Additions of Known Amounts of Sb (μg/L)

Water analysed	Sb added	Sb found
Synthetic water *	1.0	0.96 0.97 0.95
Well water	0.50	0.52** 0.48**

*Synthetic water of the above-given composition with additions (μg/L): Cu (100), Zn (100), Mn (II) (200), Fe (III) (1000).

**Results obtained after subtraction of the original Sb content in the samples.

Table 2
Results of the Determination of Antimony in Natural Waters

Water analysed	No of determinations	Sb concentration found μg/L (mean value)
Vistula river water (Warsaw)	5	0.154 ± 0.012*
Tap water (Warsaw)	2	0.210
Well water (Legionowo)	2	0.057
Well water (Warsaw)	2	0.046

*Standard deviation of a single determination.

topes except ^{76}As do not interfere with the measurement of ^{122}Sb activity. ^{76}As emits, in addition to 559-keV γ-rays (main peak), 563-keV γ-rays (small peak), which may interfere with the measurement of the 564-keV γ-rays of ^{122}Sb. In the samples examined, arsenic was always present. However, after the applied cooling period, the main peak of ^{76}As was small compared to the 564-keV peak of ^{122}Sb, and the contribution of the 563-keV peak of ^{76}As to the latter was negligible. In the case of higher concentrations of arsenic in water, the determination of antimony could be based on the measurement of ^{124}Sb (604 keV).

The precision of the present method (relative standard deviation) at a concetration of 0.154 μg/L was 7.8%. The lower limit of detection calculated as $3s_B$ (where s_B is the standard deviation of the blank value) is 0.023 μg/L for a 100-mL sample. This is sufficient for the determination of total inorganic antimony in the majority of natural waters. The lower limit of detection could be improved by using purified reagents and larger samples. In the neutron activation method with preconcentration of antimony by extraction, proposed by Mock and Wai *(4)*, a lower limit of detection of 0.01 µg/L was obtained for a 100-mL sample. The authors claim that a limit of 0.001 µg/L can be obtained for a 500-mL sample.

ACKNOWLEDGMENT

Thanks are due to E. Sterlińska for preparation of the sorbent and the synthetic water solution.

REFERENCES

1. R. Iffland, *Handbook on Toxicity of Inorganic Compounds,* G. H. Seiler, H. Sigel, and A. Sigel, eds., Marcel Dekker, New York, 1988, p. 67.
2. C. A. McAuliffe, *Comprehensive Coordination Chemistry,* vol. 3, G. Wilkinson, ed., Oxford Press, Oxford, 1987, p. 278.
3. B. A. Fowler and P. L. Goering, *Metals and Their Compounds in the Environment,* E. Merian, ed., VCH, Weinheim, 1991, p. 743.
4. W. M. Mock and C. M. Wai, *Anal. Chem.* **59,** 233 (1987).
5. G. Baudin, J. Radioanal. Chem. **37,** 119 (1977).
6. S. Gohda, *Bull. Chem. Soc. Jpn.* **45,** 1704 (1972).
7. A. H. Abu-Hilal and J. P. Riley, *Anal. Chim. Acta* **131,** 175 (1981).
8. K. H. Lieser and V. Neizert, *J. Radioanal. Chem.* **31,** 397 (1976).
9. M. O. Andreae, J. F. Asmode, P. Forster, and L. Van'tdack *Anal. Chem.* **53,** 1766 (1981).
10. S. Gladney and J. W. Owens, *Anal. Chem.* **48,** 2220 (1976).
11. L. L. Thatcher and J. O. Johnson, *Nuclear Techniques in Environmental Pollution,* IAEA, Vienna, 1971, p. 323.
12. F. Bunus, P. Dumitrescu, and R. Bulaceanu, *J. Radioanal. Chem.* **27,** 77 (1975).
13. P. Gueguenit, R. Gandon, G. Hemon, and J. Cl. Philippot, *Measurement, Detection and Control of Environmental Pollutants,* IAEA, Vienna, 1976, p. 369.
14. O. G. Kasimova, N. I. Shcherbinina, E. M. Sedykh, L. I. Bolshakova, and G. K. Myasoedova, *Zh. Analit. Khim.* **39,** 1823 (1984).
15. H. Fukunda, J. Tunoda, K. Matsumoto, and K. Terada, *Bunseki Kagaku* **36,** 683 (1987); C. A. **108,** 26739w (1988).
16. J. Chwastowska, W. Żmijewska, and E. Sterlińska, *Anal. Chim. Acta* **276,** 265 (1993).
17. Z. Marczenko, *Spektrofotometryczne Oznaczanie Pierwiastków,* PWN, Warszawa, 1974, p. 136.
18. H. J. M. Bowen, L. Page, I. Valente, and R. J. Wade, *J. Radioanal Chem.* **48,** 9 (1979).
19. M. Metzger and H. Braun, *Anal. Chim. Acta* **189,** 263 (1986).
20. J. Schneider and R. Geisler, *Fresenius Z. Anal. Chem.* **167,** 270 (1973).
21. A. Montiel, B. Welte, and J. Carre, *Analusis* **9,** 1 (1981).

Investigation of Element Speciation in Atmosphere

An. A. Kist

Institute of Nuclear Physics, Academy of Sciences of Uzbekistan, 702132, Ulughbek, Tashkent, Uzbekistan, CIS

ABSTRACT

Elemental composition of rain water was determined with the use of instrumental neutron activation analysis. Solid water-insoluble fractions were separated by filtration through membrane filters. Filtrates were dried, and dry residues as well as solid-phase on filters were analyzed. Concentrations of 20 elements in the samples were determined. Enrichment factors and ratios between element concentrations in liquid and solid fractions were calculated. Data obtained allowed us to suppose that Na, Cl, K, Sc, Cr, Mn, Fe, Se, Br, I, Cs, La, Sm, and Au prevalently exist in the air in the form of relatively coarse aerosol particles; Zn, As, Sb, and Hg prevalently occur in the vapor–gas phase; and Co and W, in the form of finely dispersed aerosol.

Index Entries: Neutron activation analysis; atmosphere; trace elements, aerosol.

INTRODUCTION

In spite of the large variability of element speciation in the atmosphere, the number of methods for sampling of atmospheric trace element contaminants is limited. The most common sampling method for monitoring trace elements in the atmosphere is collection of particulate matter on filters. This method, however, does not provide information on element speciation in air and does not the monitoring of contaminants in finely dispersed aerosol and vapor–gas phases.

An alternative approach to this problem is a study of atmospheric wet precipitations (mist, rain, snow, and so on). It is known that atmospheric wet precipitations are most significantly involved in cleaning the atmosphere and that elemental composition of atmospheric precipitation reflects elemental composition of the atmosphere *(1)*. Efficiency of washing out atmospheric contaminants by wet precipitations obviously should depend on speciation of the contaminants. We aimed at studying elemental composition of rain water for estimation of trace element contents and their speciation in the atmosphere.

EXPERIMENT

Rain water was collected in fractions during a single rain event (5-h duration) into glass vessels with a diameter of 25 cm. Each fraction was sampled for 1 h. A volume of each fraction amounted to 200–500 mL. A solid fraction was separated by filtration through membrane filters Synpor (Czech Republic) with a pore diameter of 2.3 μm. Filtrates were evaporated under an infrared lamp in polyethylene (PE) vials. The vials and filters were sealed in bags made of high-purity PE film.

Elemental composition of filters and dry residues was determined by instrumental neutron activation analysis (INAA). Irradiation of the samples was performed in a nuclear reactor at a neutron flux of 5×10^{13} $n/cm^2/s$ in two stages. At the first stage, the samples were irradiated in PE bags for 15 s. γ-Spectra were measured 15 min after the irradiation. After 1 wk, the samples were packed in aluminum foil and irradiated for 10 h. After this irradiation, the samples were taken out of the foil and packed in PE bags for counting. γ-Spectra were measured two times, 6 and 25 d after the second irradiation. Measurements were carried out using a Ge(Li)-detector coupled to a multichannel analyzer. Spectrum analysis was performed on a microcomputer connected on-line to the analyzer. For quantification, the samples were irradiated together with standards. The standards were prepared by pipeting solutions containing known amounts of elements onto ashless filter paper.

For estimation of blank values, unused clean filters and PE vials were analyzed together with the samples. The highest relative blank value (compared to an element amount determined), not exceeding 10% was found for Au. For other elements, the relative blank values did not exceed 4%. For quality assurance, soil and plant reference samples were analyzed together with the samples studied. Results found were in a good agreement with certified values.

RESULTS AND DISCUSSION

Elemental composition of the samples and volumes of fractions are shown in Tables 1 and 2. Table 1 shows that concentrations of insoluble forms of most elements in rain water were decreasing from the beginning to the end of rain. Concentrations of soluble forms of some elements (Sc, Fe, La, Sm) were decreasing during the rain as well (cf Table 2). However, concentrations of most elements did not seem to depend much on the rain duration. Ratios of element concentrations in liquid phase to concentration in solid phase (cf Table 3) for most elements increased, for some of them 10-fold or even more (Cl, Mn, Zn, W). Enrichment factors

Table 1
Elemental Composition of Solid Phase of Rain Water (ng/mL)

Fraction	1	2	3	4	5	Total	Total
Volume, mL	250	400	200	250	200	1300 ng/mL	EF
Na	74	40	49	20	14	40	0.32
Cl	59	52	46	19	11	40	28
K	280	210	78	79	46	150	1.3
Sc	0.16	0.11	0.081	0.041	0.035	0.090	1
Cr	7.5	1.9	1.2	0.61	0.49	2.4	2.7
Mn	5.5	3.1	2.6	1.5	1.3	2.9	0.64
Fe	550	390	320	140	120	320	1.4
Co	0.28	0.21	0.22	0.11	0.12	0.19	1.8
Zn	40	34	32	18	14	29	58
As	0.28	0.37	0.32	0.20	0.058	0.26	12
Se	0.21	0.090	0.10	0.13	0.14	0.13	320
Br	0.14	0.039	0.18	0.039	0.037	0.080	11
Sb	0.49	0.20	0.17	0.060	0.080	0.21	46
I	0.42	0.16	0.094	0.080	0.13	0.18	130
Cs	0.10	0.050	0.059	0.034	0.027	0.054	2
La	0.18	0.13	0.10	0.05	0.04	0.10	1.3
Sm	0.089	0.058	0.053	0.018	0.017	0.049	1.7
W	0.13	0.087	0.27	0.14	0.14	0.14	21
Au	0.00058	0.00050	0.00086	0.00018	0.00016	0.00046	20
Hg	0.078	0.042	0.065	0.042	0.045	0.053	150

(EF) were calculated for elements in the liquid and solid phases using Sc as the reference element and lithosphere as the reference system.

Since separation of soluble and insoluble fractions was made by filtration through the membrane filter, the soluble fraction represents all the forms of atmospheric trace elements that were able to penetrate through the filter. Thus, the soluble fraction included a soluble part of aerosol particles as well as aerosol particles of a small size, liquid aerosol particles, and the vapor–gas phase of atmospheric contaminants dissolved in rain water. In other words, insoluble fractions represented mainly composi-

Table 2
Elemental Composition of Soluble Phase of Rain Water (ng/mL)

Fraction	1	2	3	4	5	Total	Total
Volume, mL	250	400	200	250	200	1300 ng/mL	EF
Na	370	310	510	390	280	360	18
Cl	290	390	1200	990	860	680	3100
K	300	260	510	260	340	320	17
Sc	0.014	0.018	0.014	0.012	0.009	0.014	1
Cr	1.4	2.4	1.3	1.5	0.88	1.6	12
Mn	0.23	0.21	0.070	0.37	0.78	0.31	0.44
Fe	44	41	19	42	18	35	0.99
Co	0.13	0.11	0.10	0.08	0.10	0.11	6.5
Zn	73	60	240	340	340	190	2400
As	0.31	1.3	0.32	0.34	0.24	0.61	170
Se	0.80	0.24	0.26	0.26	0.27	0.36	5600
Br	5.2	3.9	5.6	6.7	3.9	5	4400
Sb	1.0	0.85	0.56	0.74	0.81	0.81	1100
I	1.8	1.7	3.8	3.4	3.9	2.7	13000
Cs	0.030	0.029	0.15	0.20	0.054	0.085	20
La	0.035	0.031	0.025	0.011	0.010	0.024	1.9
Sm	0.020	0.004	0.003	0.006	0.005	0.008	1.7
W	0.22	0.18	1.1	1.7	1.1	0.76	720
Au	0.0016	0.00073	0.0023	0.0017	0.0020	0.0010	420
Hg	0.82	0.75	1.10	0.57	1.30	0.87	15000

tion of relatively coarse aerosol particles, whereas soluble fractions represented composition of part of coarse aerosol particles leachable by rain water as well as the finely dispersed aerosol and vapor–gas phases of atmospheric contaminants.

Data obtained illustrate the idea that a rain shower washes contaminations out of air. This process includes precipitation of solid particles, leaching elements out of them and dissolving elements of the vapor–gas phase.

Table 3
Ratios of Elemental Concentration in Liquid Phase to Elemental Concentration in Solid Phase

Fraction	1	2	3	4	5	Average
Na	5.0	7.7	10	20	20	9
Cl	4.9	7.5	27	53	76	17
K	1.0	1.2	6.5	3.3	7.4	2.1
Sc	0.09	0.16	0.17	0.29	0.27	0.16
Cr	0.18	1.30	1.0	2.4	1.8	0.67
Mn	0.040	0.066	0.027	0.25	0.62	0.11
Fe	0.081	0.10	0.060	0.29	0.14	0.11
Co	0.46	0.53	0.46	0.75	0.83	0.58
Zn	1.8	1.8	7.4	20	24	6.5
As	1.1	3.5	1.0	1.7	4.1	2.4
Se	3.8	2.7	2.6	2.0	1.9	2.8
Br	38	99	31	170	110	62
Sb	2.1	4.2	3.3	12	10	3.9
I	4.30	10	41	42	29	15
Cs	0.31	0.57	2.6	5.8	2	1.6
La	0.20	0.24	0.26	0.22	0.29	0.24
Sm	0.22	0.07	0.05	0.35	0.29	0.16
W	1.7	2.1	4.0	12	8.4	5.4
Au	2.7	1.5	2.7	9.6	11	2.2
Hg	11	18	16	14	29	16

Substantial changes of elemental composition of solid and liquid phases of rain water in time suggest that occasional sampling of rain (and probably other kinds of wet precipitations) and their analysis can give incomparable results if sampling conditions, such as duration and intensity of rain and time between the beginning of rain and sampling, are not taken into consideration.

Supposing that the main part of atmospheric particulate matter is washed out by rain, it is possible to suggest that by knowing the height of rain clouds, elemental composition of rain water, and intensity of rain, information on average elemental composition of the atmosphere can be obtained.

Table 4
Element Amounts Washed Out from Air and Trace Element Concentrations in Aerosols (ng/m^3)

Element	Liquid phase	Solid phase	Total	Aerosol
Na	2400	270	2670	2700-8700
Cl	4500	260	4760	.
K	2100	1000	3100	.
Sc	0.09	0.60	0.69	0.66-6.6
Cr	11	16	27	30-180
Mn	2.1	19	21	.
Fe	230	2100	2330	3700-13000
Co	0.70	1.3	2	4.4-13
Zn	1200	190	1390	4.2-15000
As	3.4	1.8	5.2	.
Se	2.1	0.86	2.9	.
Br	33	0.53	33	40-600
Sb	5.4	1.4	6.8	4.5-100
I	18	1.2	19	.
Cs	0.56	0.36	0.92	0.2-11
La	0.16	0.69	0.85	0.4-15
Sm	0.05	0.33	0.38	0.18-2.5
W	5.1	0.94	6.0	.
Au	0.010	0.003	0.013	0.08-0.85
Hg	5.8	0.35	6.2	7-100

In our case, the height of rain clouds, including a thickness of the cloud layer, was estimated to amount to 1000 m. We calculated an amount of elements washed out from 1 m^3 of atmosphere during the rain (cf Table 4). For comparison, data on aerosol composition in the same region obtained within the framework of other studies in our laboratory are also given in Table 4. One can see that the liquid fraction constituted more than a half of the total contamination of atmosphere by Na, Cl, K, Zn, As, Se, Br, Sb, I, Cs, W, Au, and Hg, i.e., by elements for which one can suggest prevalence of soluble or volatile forms. The elements Sc, Mn, Fe, La, and Sm were mainly contained in the solid phase, which can also be explained by their chemical properties.

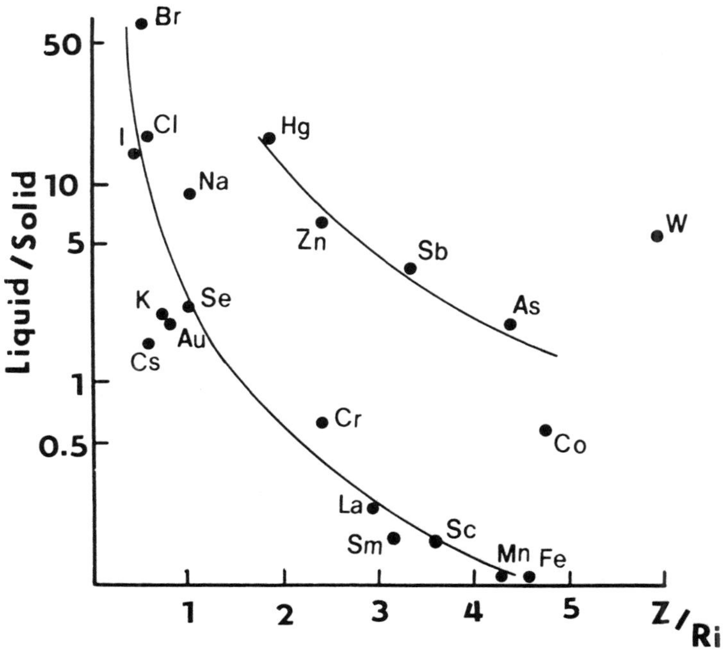

Fig. 1. Ratios of elemental concentration in liquid-to-solid phase vs ion potential, Z/R_i

In general, by comparing data obtained in this study with composition of airborne particulate matter, one can assume that analysis of total wet precipitations can be a tool for evaluation of average atmosphere composition and a certain kind of determination of element speciation. Coarse aerosol particles are washed out from atmosphere relatively rapidly. Their concentrations in air during the rain decreased significantly, because one of their most important source—wind carrying soil dust—is inactive during rain. Elements whose concentrations in soluble fractions decreased during rain at the same rate were probably generated from soil and were in the coarse aerosol phase. A ratio of these elements (Sc, Fe, La, Sm) in soluble to insoluble fractions remained about constant during rain with a value lower than 1. The ratio of soluble to insoluble fraction for Na, K, Cl, and Br had rather high values and increased during rain. An increase of this ratio suggests that submicron and vapor–gas phases of atmosphere contamination are not washed out from the atmosphere as quickly as coarse aerosol. Another conclusion may be that sources generating submicron and vapor–gas phases to the atmosphere are continuously active during rain, unlike soil weathering.

One of the fundamental parameters that can characterize solubility of an element is ion potential, a ratio of ion charge, Z, to crystallographic ion radius, R_i. Ratios of element concentrations in liquid phase to concentrations in solid phase vs ion potential are shown in Fig. 1. One can see that for the majority of elements, the liquid/solid factor correlates with ion potential. On this basis, the elements studied can be divided into several groups. The elements Na, Cl, K, Sc, Cr, Mn, Fe, Se, Br, I, Cs, La, Sm, and Au have similar character of relationship between the liquid/solid factor and ion potential, and form one group. Speciation of these elements is probably similar–they are contained in the coarse aerosol fraction, and their liquid/solid factor is mainly determined by solubility of their chemical forms. The elements Zn, As, Sb, and Hg form an another group for which the liquid/solid factors are much higher than those for the first group. The probable reason is that because of higher volatility, these elements are prevalently contained in the vapor–gas phase. This corresponds to common presumptions about chemical properties and speciation of trace elements in atmosphere. However, the behavior of Co and W is rather unexpected. The liquid/solid factors of these elements are much higher than for the elements in the first group that they should belong to, and the elements do not seem to belong to the second group either. A plausible explanation may be that the sampling point was situated 8 km from a hard-alloy factory, which can release Co and W into the atmosphere in the form of a finely dispersed aerosol. We have shown in our previous work *(2)* that Co and W have high values of enrichment factors in the atmosphere in this region and probably exist in fine aerosol particles or in the vapor–gas phase.

This study can be considered as an initiation of wider and more detailed investigation of the process of washing out contaminants from the atmosphere and relationships between rain water composition and air composition before and after rain. In addition, studies of composition of rain of various durations and intensities and with various heights of rain clouds can be of interest.

REFERENCES

1. E. Schrimpf, O. Klemm, R. Eiden. T. Freiert, and R. Herrmann, *Staub.-Reinhalt. Luft.* **44,** 72 (1984).
2. An. A. Kist, A. A. Kist, and L. I. Zhuk, *J. Radioanal. Nucl. Chem., Articles* **167,** 227 (1992).

Determination of Heavy Metals in Humic Substances by Instrumental Photon Activation Analysis

D. SCHULZE,*,[1] H. KUPSCH,[1] AND C. SEGEBADE[2]

[1]Universität Leipzig, Bereich Geowissenschaften, Abteilung Geochemie, Permoserstr. 15, 04318 Leipzig, Germany; and [2]Bundesanstalt für Materialforschung und -prüfung Berlin, Unter den Eichen 87, 12203 Berlin, Germany

ABSTRACT

Photoactivation analysis is used for the determination of heavy metals in water and humic substances. Isolated humic acids serve as a carrier of environmental pollutants. The investigation of the drinking water catchment area in the middle German conurbation Leipzig/Halle shows it to be endangered from old neglected deposits of toxic wastes (chemical factory destroyed during World War II). Elevated concentrations of the elements zinc, cadmium, lead, and mercury in humic acids from surface waters were found.

Index Entries: Activation analysis; heavy metals; trace elements; humic acids; water.

INTRODUCTION

A systematic investigation of the basic interactions of microcomponents in natural systems is necessary for a better understanding of geochemical processes in a region under the influence of manmade pollutants. Humic substances (humic and fulvic acids) comprise a class of biogenic, organic substances in water and soil, and play different roles in the environment. They are complexing agents or sorbants for pollutants and naturally occurring substances, and also a precursor of toxic substances in natural processes *(1)*.

Because of their polyfunctionality (high part of phenolic OH^- and carboxyl groups) and their colloidal dissolving behavior humic sub-

*Author to whom all correspondence and reprint requests should be addressed.

 © 1994 by Humana Press Inc.

stances influence the transport, reactivity, and migration of pollutants in the natural system. The ability to form water-soluble and insoluble complexes, associates, and hydrated oxides, as well as the interaction with minerals and organic substances (pesticides, fatty acids, and so forth) is a key feature of the humic substances. Therefore, they are mobilized carriers of environmental pollutants.

We have investigated a series of groundwater and surface-water samples from a drinking water region near the Elbe River in Saxony (Germany). One specific feature of this region is that residual pollution was found from a chemical factory (explosives) destroyed during the World War II. The aims of our studies were to determine the heavy metal ions bonded to aquatic humic acids and to characterize the path of pollutants.

EXPERIMENT

Resin Isolation Procedure

The water sample is pressure-filtered through a 0.45-µm membrane filter to remove particulate material. The filtered water is acidified with HCl to pH 2 in order to protonate the humic substances to an uncharged state. The acidified water is pumped onto a preparative glass column with XAD-8 resin. The uncharged humic substances and the neutral organic solutes are sorbed onto the resin column. The capacity of the column to concentrate humic substances from water has a *k* value (column distribution coefficient) of 50; that means approx 95% of the humic substances in water is retained on the XAD-8 resin column *(2)*. After saturation the resin column is rinsed with 0.1*M* NaOH, and the humic substances are ionized and then eluted. The pH is lowered to 1 with HCl. Humic acids will precipitate, and the fulvic acids will remain in solution. Humic acids are separated from fulvic acids by a high-speed centrifugation, washed with 0.1*M* HCl and subsequently dried. The content of humic acids was approx 0.3–2.0 mg/L of water.

Photon Activation Analysis

Groundwater samples have been irradiated in large cylindrical glass containers (1–5 L) *(3)*. Aliquots of calibrated scandium and samarium solutions were added to the water samples as internal radiation dose monitors before bremsstrahlung exposure to correct for deviations of the analytical results owing to several sources of error, e.g., inhomogeneous activation influenced by the beam geometry, uncontrolled flux and energy shifts of the accelerator, and others *(4)*.

The dried humic acids were also prepared with an internal standard (Sc, Sm) and irradiated directly. Irradiations were carried out with the electron linear accelerator (30 MeV, 150 µA) of the BAM in Berlin and the microtron (22 MeV, 20 µA) in Leipzig. Exposure periods varied from 2–4 h (BAM Linac) and 5–8 h at the microtron. Large-volume coaxial germanium γ-ray detectors, as well as planar low-energy germanium diodes were used for a measurement of the photon emission of the product activities. Spectra were counted in 2048 resp. 4096 channels, and processed with appropriate programs.

RESULTS AND DISCUSSION

Figure 1 shows the localities of the examined area in Germany, namely the drinking water region around Torgau in Saxony. This drinking water reservoir is important for the support of the middle-German conurbation Leipzig/Halle. First we investigated a series of groundwater samples near the Elbe River. Table 1 presents a comparison between analyzed data for a representative groundwater sample and an Elbe River sample. Although the groundwater sample was taken only 400 m from the Elbe River, it is obvious that the trace element pattern is different. This means that there is probably no transfer between the Elbe and the groundwater. The trace contents of the groundwater is predominantly determined by geochemical interactions of the surrounding soil. The direct photon irradiation of untreated large water samples is a significant advantage in comparison with neutron activation analysis (NAA). On the other hand, investigations by NAA show another nuclide spectrum because of different nuclear reactions and are favorable for the determination of rare earth elements *(5)*.

Our special interest was in which way heavy metal ions are bound to humic acids and if there is an influence of residual pollution from the chemical factory destroyed during the World War II. Therefore, we investigated the isolated humic substances of surface-water samples taken in the region around the factory (Torgau-North) and in a neighboring drinking water region (Torgau-South). The summary of the examinations is given in Fig. 2. In this figure, only the mean values are given. In the single samples, the concentrations of several heavy metals are greatly different. However, there are no significant characteristics for interpretation at this point. This is caused, first, by the different interaction with the surrounding soil and, second, by the different extension of the residual pollutants. Nevertheless, obviously, the concentrations of the elements zinc, cadmium, lead and mercury are sometimes larger (logarithmic scale in Fig. 2.) in the polluted area around Torgau-North in comparison with the other. This is of importance, because toxic heavy metals bonded to humic acids are mobilized in water and present in transport processes in the region. On the other hand, the differences in heavy metal concentrations are significant between Torgau-North and Torgau-South. This allows the conclusion that there is no transfer between the polluted area around the former chemical factory and the drinking water area. Furthermore (not explained in detail here), from the single values follows:

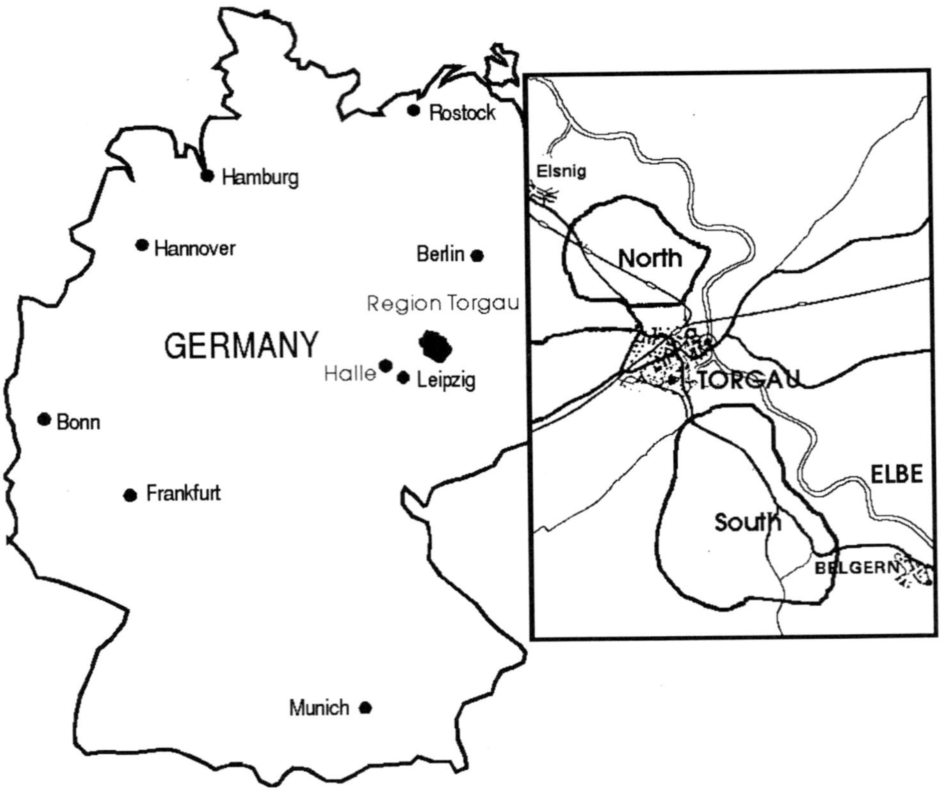

Fig. 1. Location of sampling in the Torgau region in Saxony (Germany).

the migration of the heavy metal ions does not take place in the direction to the Elbe River, but further inland.

Other investigations of organic pollutants in this context have shown that the concentrations of nitroaromates are some orders of magnitude larger in Torgau-North (up to 100 mg/L water) than in Torgau-South (approx 0.1 mg/L) *(6)*. This emphasizes the risk potential of the residual pollution from the destroyed chemical factory and calls for practicable solutions to avoid contamination of the drinking water reservoir. Further examinations are in progress.

SUMMARY

Photon activation analysis is a suitable technique for the determination of heavy metals in water and humic substances. Water-isolated humic acids are carriers of environmental pollutants. Through an examination of a region with old neglected deposits of toxic waste near a drinking water catchment area in Saxony (Germany), it could be shown that enrichments of the heavy metals zinc, cadmium, lead, and mercury

Table 1
Data for Environmentally Relevant Trace Elements Detected by PAA in a Groundwater and River Water Sample (All Values in μg/L)

Element	Nuclide	E keV	groundwater	river water
Ti	Sc-48 (a)	175	7	75
Cr	Cr-51	320	7.4	14
Mn	Mn-54	835	418	91
Co	Co-58	811	2.6	1.0
Ni	Ni-57	1379	10	3.5
Cu	Cu-64	1346	<1	17.4
Zn	Zn-65	1116	110	130
As	As-74	596	30	11
Se	Se-75	265	<1	1.0
Y	Y-88	1836	54	1.1
Zr	Zr-89	909	4.4	4.3
Mo	Mo-99 (b)	140	<0.1	0.2
Ag	Ag-106m	451	<1	1.0
Cd	In-115m	336	<0.5	3.0
Sn	Sn-113	392 (c)	<1	19
Sb	Sb-122	564	4.7	0.7
Te	Te-121	508	0.64	<0.1
Ba	Ba-135m	268	75	47
Ce	Ce-139 (b)	166	4.1	0.6
Hg	Hg-197 (b)	67	0.004	0.004
Tl	Tl-202	439	0.1	0.1
Pb	Pb-203	279	1.3	3.5
U	U-237	208	0.14	0.07

[a]From (γ,p) reaction.
[b]Measured by LEPS.
[c]From In-113m daughter nuclide.

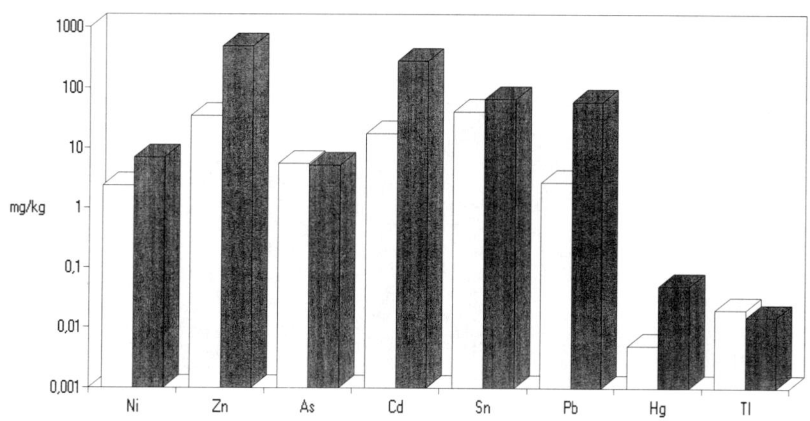

Fig. 2. Distribution of heavy metal species in humic acids of surface water, Torgau-South, white; Torgau-North, black.

that enrichments of the heavy metals zinc, cadmium, lead, and mercury take place in surface waters.

REFERENCES

1. Proc. Int. Symp. Humic Substances in the Aquatic and Terrestrial Environment, Linköping, Sweden, August 21–23, 1989, in *Lecture Notes in Earth Sciences,* vol. 33, S. Bhattacharji, G. M. Friedman, H. J. Neugebauer, A. Seilacher, eds., Springer Berlin, Heidelberg, 1991.
2. L. Malcolm, *Lecture Notes in Earth Sciences,* pp. 9–36
3. C. Segebade, H. W. Thümmel, and W. Heller, *J. Radioanal. Chem.* **167,** 383 (1992).
4. C. Segebade, H.-P. Weise, and G. J. Lutz, *Photon Activation Analysis,* de Gruyter Berlin, New York, 1988.
5. I. Kim, B. Delakowitz, P. Zeh, T. Probst, X. Lin, U. Ehrlicher, and C. Schauer, *CEC-Report RCM 00692,* TU München, 1992.
6. H. Kupsch, A. Rietzschel, T. Dittmar, D. Rößler, M. Richter, and C. Franke, *Report 02/93,* Universität Leipzig, 1993.

Accumulation of Cesium and Radiocesium in Forest Litter in Selected Regions of Poland and Its Influence on Litter-to-Mushroom Transfer Factor

PAWEŁ ZAGRODZKI,*,[1] JERZY W. MIETELSKI,[1] MIROSŁAW KROŚNIAK,[2] AND BARBARA PETELENZ[1]

[1]*Henryk Niewodniczański Institute of Nuclear Physics, 31-342 Kraków, Radzikowskiego 152, Poland; and* [2]*Department of Food Chemistry and Nutrition, Jagiellonian University, Kraków, Poland*

ABSTRACT

The aim of this work was to check whether the stable cesium content in forest litter affects the value of radiocesium from litter-to-mushroom transfer factor *Tf* or not. Total cesium in litter, measured by AAS, varied from 0.1–2.7 µg/g. These data, combined with earlier results for mushrooms, showed no simple correlation for *Tf*. More complex relationships provided very high correlation coefficients, but their validity needs further investigation.

Index Entries: Stable cesium; radiocesium; forest contamination; transfer factors; mushrooms.

INTRODUCTION

The present work is a continuation of studies on environmental transfer processes of trace elements. Artificial radionuclides originating from nuclear weapons tests or from failures of nuclear power plants are well known as useful indicators of such processes. The tendency toward taking up trace elements by plants or mushrooms from the soil is char-

*Author to whom all correspondence and reprint requests should be addressed.

 © 1994 by Humana Press Inc.

acterized by the quantity called transfer factor *Tf*. The uptake of radiocesium into mushrooms depends on many parameters of the soil or litter, such as the organic matter *(1)*, clay component *(2)*, exchangeable potassium content of soil (litter), and pH *(3)*, but no satisfactory model of the dependence has been established as yet.

The aim of this work was to answer the question whether the content of stable cesium in litter also affects the value of radiocesium from litter-to-mushrooms *Tf* or not. To quantify our considerations, we introduced a concept of a "specific" transfer factor *sTf*. Like *Tf*, it was defined as the ratio between the specific activity of the mushroom (expressed per unit of the dry mass) and the specific activity of the layer of litter (also per unit of the dry mass). The additional assumption was that the activity ratio ^{134}Cs-to-^{137}Cs in the litter must be practically identical as that in the mushroom. This assumption is not a trivial one, and it was deduced from the hypothesis that the ^{134}Cs-to-^{137}Cs ratio for a litter layer (decreasing with the depth of the layer) is reflected by the ratio in the fruit body of the mushroom *(4,5)*. It is true only for a "fresh" radiocesium contamination because the amounts of "old" ^{137}Cs (deposited around 1960) are fixed and not available for the mushroom. Taking "old" ^{137}Cs radiocesium into account would result in more complicated relationships.

Our previous work on environmental radioactivity in Poland after the Chernobyl accident resulted in selecting 20 sites (out of 280) where the mushrooms revealed the highest radioactivity (^{137}Cs, ^{134}Cs). The interesting observation was that there were two regions where the mushrooms grew under similar (highest) fallout conditions, but at a quite different industrial contamination. Thus, we examined mushrooms and forest litter from two representative regions of Poland, looking for the interpretation of the *Tf*s that we had determined. Whenever possible, the use was made of the measurements performed earlier in our laboratory *(6–8)*.

MATERIALS AND METHODS

The samples originated either from the heavily polluted Silesia region or from the region of Mazurian Lakes (northeastern Poland), which is practically not threatened by the industry. In each case, the samples were those of two upper layers of forest litter (A0—holorganic, A1—hemiorganic), taken from 10 different sites of each region. Therefore, the experimental material consisted of 20 pairs of samples. The measurements of ^{137}Cs and ^{134}Cs were performed earlier as described elsewhere *(6,9)*. The total cesium was assayed by AAS. To do this, the samples were mineralized in a CEM MDS-2000 microwave digester. The analytical procedure was a modification of that used to determine cesium in *Xerocomus badius* mushrooms *(7)*. The solutions obtained from the mineralization were freed from suspensions by filtering them through a Filtrak No. 390 filter paper and then concentrated to a volume of 10–20 mL by evaporation in PTFE vessels. The losses of cesium caused by filtration were evaluated by a γ-spectrometric measurement of the residual material trapped in the filter. The losses did not exceed 10%. The atomic absorption analyses of cesium were performed using the AAS-30 Carl-Zeiss Jena spectrometer with the ETA-30 graphite furnace module, MPE autosampler, and Massman graphite tube equipped with L'vov platform.

RESULTS AND DISCUSSION

The results of total cesium determined in this work for the A0 and A1 layers of forest litter are summarized in Table 1 and compared with those obtained previously for the mushrooms *(7)*, as well as with the results of radioactivity measurements for both mushrooms and litter *(6,9)*. The total cesium content varied from <0.1 to 2.7 µg/g. Although the mean values are equal in both layers (0.75 µg/g), the differences in the content level are smaller for A1 samples than for A0 samples (SD: 0.49 and 0.75, respectively). The Silesian samples showed a higher level of total cesium (mean = 1.18; SD = 0.84 in A0, mean = 0.94; SD = 0.56 in A1) than the samples collected at the Mazurian Lakes (mean = 0.33; SD = 0.30 in A0, mean = 0.56; SD = 0.33 in A1). It should be mentioned that the observed levels of total cesium were far above the content of ^{137}Cs expressed in µg/g.

The data from Table 1 were used to calculate *Tf* and *sTf*. Neither *Tf* nor *sTf* showed statistically significant correlations with such simple parameters like total cesium, specific activity of cesium-to-stable-cesium ratio, or potassium-to-cesium ratio. However, very high correlation coefficients were found for more complicated relations. To describe the dependence of *Tf* on environmental parameters, the following model was suggested as a reasonable compromise between simplicity and accuracy:

$$Tf_{i,M} = 10^{\gamma} A^a B^b C^c D^d \qquad (1)$$

where $Tf_{i,M}$ = litter-to-mushroom transfer factor for *i*th layer and *M*th isotope (i = 0,1; M = 134, 137), A = total cesium in the litter layer (Bq/g), B = content of the other radiocesium isotope in litter layer (Bq/g) (^{134}Cs for $Tf_{i,137}$ calculation and vice versa), C = density of the litter layer (g/cm^3), D = ^{40}K content in the litter layer (Bq/g), and γ, a, b, c, and d are the adjustable parameters of the model.

The *sTf*s defined in the Introduction are indexed only by the cesium isotopes. The parameters of all obtained correlation functions are presented in Table 2. It can be seen that *sTf* had much lower correlation coefficients, and moreover, the fitted parameters bore much bigger errors.

Our model was built on the intuitive expectation that the *Tf*s depend mainly on soil parameters. We neglected the dependence of *Tf* or *sTf* on pH because of the small range of pH variations (3.22–4.38) in the litter layers. We presumed significance of the variable *B*, which ought to reflect the competition between ^{137}Cs and ^{134}Cs in their migration from soil to mushroom. Density of soil (variable *C*) reflects the proportion between

Table 1
Total Cesium, ^{134}Cs, and ^{137}Cs Content of Mushrooms; Total Cesium, ^{134}Cs, ^{137}Cs, ^{40}K, and Density of Two Upper Litter Layers

Mushroom species			Litter									
Xerocomus Badius			Leaf Layer (A0)					Humus Layer (A1)				
Total Cs μg/g	Cs-137 Bq/g dm	Cs-134 Bq/g dm	Total Cs μg/g	Cs-137 Bq/g dm	Cs-134 Bq/g dm	K-40 Bq/g dm	Density g/cm³	Total Cs μg/g	Cs-137 Bq/g dm	Cs-134 Bq/g dm	K-40 Bq/g dm	Density g/cm³
3.3±0.1	23.7±0.5	2.18±0.10	2.39±0.41	2.69±0.53	0.27±0.04	0.12±0.03	0.27	0.86±0.08	1.42±0.18	0.13±0.01	0.14±0.03	0.35
4.3±0.2	121.1±3.6	12.5±1.0	1.48±0.15	3.31±0.15	0.35±0.01	0.07±0.01	0.34	1.26±0.08	1.11± 0.09	0.11±0.01	0.14±0.02	0.41
2.5±0.1	29.1±0.9	2.96±0.24	1.43±0.17	2.03±0.12	0.22±0.01	0.12±0.02	0.25	1.96±0.18	2.28± 0.27	0.23±0.02	0.09±0.02	0.31
8.8±0.2	22.4±0.7	1.79±0.14	1.21±0.10	0.36±0.04	0.04±0.01	0.08±0.02	0.29	1.67±0.05	0.23± 0.03	0.01±0.01	0.14±0.03	0.48
1.2±0.1	13.6±0.4	1.34±0.11	2.72±0.08	2.14±0.38	0.20±0.03	0.24±0.05	0.39	1.20±0.05	1.76± 0.36	0.16±0.02	0.42±0.10	0.71
10.5±0.7	156.7±6.3	16.26±0.20	0.44±0.09	3.22±0.24	0.34±0.02	0.07±0.01	0.30	0.26±0.12	0.28± 0.07	0.02±0.01	0.24±0.06	0.54
6.8±0.1	18.2±0.4	1.63±0.10	0.60±0.18	0.89±0.04	0.09±0.01	0.07±0.02	0.29	0.53±0.22	0.86± 0.07	0.08±0.01	0.26±0.03	0.80
3.9±0.4	27.5±1.1	2.80±0.25	0.49±0.05	1.33±0.04	0.14±0.01	0.11±0.02	0.28	0.58±0.08	1.69± 0.11	0.17±0.01	0.13±0.2	0.57
2.2±0.1	12.2±0.4	1.13±0.19	0.47±0.08	1.22±0.17	0.12±0.01	0.53±0.09	0.65	0.49±0.05	0.53± 0.08	0.04±0.01	0.72±0.13	1.01
1.4±0.2	20.9±0.4	2.14±0.13	0.53±0.09	2.66±0.17	0.27±0.02	0.23±0.02	0.82	0.54±0.11	1.52± 0.28	0.14±0.02	0.46±0.10	1.16
1.7±0.1	20.5±1.0	1.98±0.24	0.89±0.16	1.64±0.07	0.18±0.01	0.07±0.02	0.29	0.82±0.16	0.97± 0.05	0.09±0.01	0.15±0.02	0.41
1.7±0.1	21.2±1.0	1.87±0.22	0.01±0.14	1.04±0.03	0.11±0.01	0.17±0.02	0.33	0.43±0.09	0.17± 0.01	0.01±0.01	0.27±0.02	0.78
1.4±0.2	20.7±0.8	2.03±0.18	0.11±0.06	0.49±0.01	0.05±0.01	0.13±0.03	0.20	0.74±0.44	0.44± 0.01	0.05±0.01	0.01±0.02	0.33
2.2±0.1	27.0±0.5	2.50±0.10	0.66±0.12	1.92±0.13	0.21±0.01	0.10±0.02	0.28	0.72±0.17	0.42± 0.02	0.02±0.01	0.14±0.02	0.39
1.6±0.3	20.7±0.4	2.01±0.12	0.55±0.24	2.35±0.07	0.24±0.01	0.05±0.02	0.28	0.05±0.06	0.37± 0.01	0.03±0.01	0.25±0.02	0.75
1.8±0.1	11.5±0.5	1.07±0.10	0.01±0.08	0.48±0.02	0.05±0.01	0.05±0.01	0.33	0.53±0.11	2.35± 0.18	0.24±0.02	0.15±0.02	0.38
2.0±0.1	22.6±0.5	2.27±0.14	0.47±0.12	1.72±0.13	0.18±0.01	0.11±0.02	0.27	1.00±0.25	0.78± 0.06	0.07±0.01	0.25±0.03	0.58
2.2±0.3	22.5±0.5	2.19±0.13	0.27±0.07	1.54± .09	0.15±0.01	0.14±0.02	0.29	0.31±0.17	1.02± 0.24	0.08±0.01	0.16±0.04	0.36
0.8±0.2	15.8±0.8	1.50±0.17	0.09±0.05	0.11±0.01	0.01±0.01	0.01±0.01	0.29	0.92±0.20	1.29± 0.08	0.14±0.01	0.09±0.02	0.29
1.3±0.1	3.8±0.8	0.25±0.02	0.24±0.08	3.29±0.17	0.31±0.08	0.30±0.10	0.31	0.12±0.01	0.29± 0.04	0.02±0.01	0.52±0.09	1.13

Table 2
The parameters of model (1) applied to $Tf_{0,137}$, $Tf_{1,137}$, $Tf_{0,134}$, $Tf_{1,134}$, sTf_{134} and sTf_{137}

Transfer factor	Parameters of model					r^2	d.f.
	γ	a	b	c	d		
$Tf_{0,137}$	0.30	0.08	-0.59	-0.48	-0.29	0.87	5,13
$Tf_{1,137}$	-5.59	-1.81	-5.19	-12.7	7.28	0.91	5,13
$Tf_{0,134}$	0.87	0.07	-0.45	-0.43	-0.29	0.80	5,13
$Tf_{1,134}$	-0.41	-1.11	-4.38	-9.53	3.82	0.95	5,13
sTf_{134}	5.03	0.40	-0.68	0.07	-0.75	0.58	5,13
sTf_{137}	4.59	-0.44	-0.69	-0.06	0.86	0.62	5,13

the organic and clay content of the litter, so it can be a measure of the fixation of cesium ions. The role of total cesium (variable *A)* is not clear. Stable cesium can be regarded either as a carrier of radiocesiums or as their rival. The same would be true for potassium (variable *D)*. At the present moment, it is hard to explain why the parameters related to total cesium and potassium always had opposite signs or why these signs alternate from one to another litter layer. Some inconsistencies among the absolute values in each type of parameters seem too big and suggest that their adjustment requires further investigation. Maybe some discrepancies are the result of experimental errors so that another analytical method for measuring stable cesium would be necessary.

REFERENCES

1. J. Andolina and O. Guillitte, in *Transfer of Radionuclides in Natural and Semi-Natural Environments,* Elsevier Applied Science, London, 1990, pp. 135–142.
2. G. S. Linsley and S. M. Haywood, and J. Dionian, in *International Atomic Energy Agency Report* IAEA-SM-257/8, 1980, Vienna.
3. P. Schuller, J. Handl, and R. E. Trumper, *Health Physics.* **3,** 575 (1988).
4. Y. Muramatsu, S. Yoshida, and M. Sumiya, *Sci. Total Environ.,* **105,** 29 (1991).
5. O. Guillitte, A. Fraiture, and J. Lambinon, in *Transfer of Radionuclides in Natural and Semi-Natural Environments,* Elsevier Applied Science, London, 1990, pp. 468–476.
6. J. W. Mietelski, M. Jasinska, B. Kubica, K. Kozak, and P. Macharski, in Proceedings of the Seminar on the Dynamic Behaviour of Radionuclides in Forests, Stockholm, May 18–24, 1992 (to be published in *Sci. Total Environ.*).
7. P. Zagrodzki, J. W. Mietelski, and M. Krośniak, in Proceedings International Seminar, Ecological Approaches on Environment and Health in Urban Areas in Central and Eastern Europe, Katowice/Kraków, 1992.
8. J. W. Mietelski, P. Macharski, M. Jasińska, and R. Broda, this vol.
9. J. W. Mietelski, M. Jasinska, B. Kubica, K. Kozak, and P. Macharski, in Proceedings of the Third International Conference on Nuclear and Radiochemistry, 7–11 Sept., 1992, Vienna (to be published in *J. Radioanal. Nucl. Chem., Articles*).

The Radioanalytical Laboratory at Pelindaba

Some Applications for Environmental and Health Monitoring

F. J. FOUCHÉ* AND A. FAANHOF

Atomic Energy Corporation of South Africa Ltd., P.O. Box 582, Pretoria, 0001, Republic of South Africa

ABSTRACT

This article deals with the radioanalytical work performed at Pelindaba, Atomic Energy Corporation of South Africa. It reports on the activities of a commercial laboratory, employing analytical techniques, such as total α/β counting, α spectrometry, liquid scintillation counting, γ spectrometry, and neutron activation analysis. Techniques discussed in more detail include semicyclic activation analysis for the determination of fluorine in vitamin mixes, as well as the determination of uranium and radium in urine samples for personnel monitoring.

Index Entries: α Spectrometry; delayed neutron counting; fluorine; radium; semicyclic activation analysis; uranium; urine.

INTRODUCTION

The radioanalytical laboratory at Pelindaba, Radioanalysis applies a variety of analytical techniques in measuring radioactivity, i.e.:

Gross α/β counting;
α Spectrometry;
Liquid scintillation counting;
γ Spectrometry; and
Neutron activation analysis.

*Author to whom all correspondence and reprint requests should be addressed.

Biological Trace Element Research Editor: G. N. Schrauzer © 1994 by Humana Press Inc.

These techniques are applied either after preconcentration of the elements/radionuclides of interest through (radio)chemical separation or as purely instrumental techniques.

The various fields of application include:

1. Assisting in environmental programs for mining companies and industries where the impact of these activities on the environment is studied in air, water, solids, and soil. All the available techniques can be exerted to offer either an in-depth environmental study, or just a scanning program for preliminary studies, for example, uranium, thorium, and radium determination in environmental water with NAA and α spectrometry.
2. The certification of materials, ranging from different foodstuffs, building materials, minerals, and drugs for export and safety purposes. All the available techniques are applied here, depending on the requirements needed, for example, $^{134}Cs + ^{137}Cs$ in food by γ spectrometry.
3. Personnel monitoring for actinides, tritium, radium, and uranium contamination in urine. The techniques applied here are α spectrometry, liquid scintillation counting, and NAA, for example ^{241}Am, ^{239}Pu, and ^{237}Np in urine by preconcentration (electroplating) and α spectrometry.
4. Multielement analysis of trace elements, i.e., uranium and thorium in environmental samples, fluorine in vitamin mixes, and so forth, by making use of neutron activation analysis, for example, determination of uranium and thorium in "export materials," such as zircon, titanium, gypsum, coal ash, and so forth.

The following analytical procedures are discussed in more detail.

SEMICYCLIC ACTIVATION FOR THE DETERMINATION OF FLUORINE IN VITAMIN MIXES

Vitamin supplements, administered to women during pregnancy, are subjected to strict regulations. Among others, the limit for fluorine per capsule is < 30 mg.

Neutron activation analysis offers a quick and cost-effective method for analyzing fluorine, but interfering radionuclides resulting from the matrix elements, such as magnesium, manganese, and sodium, cause a serious loss of sensitivity. Since radiochemical separation techniques are labor-intensive and therefore expensive, a semicyclic activation method was evaluated.

The laboratory operates a fully automatic, pneumatic rabbit system called Routine Instrumental Neutron and Gamma Analysis System (RINGAS), for the analyses of short-lived radionuclides by neutron activation. As the name suggests, RINGAS is employed for neutron activation and measurement by either delayed neutron counting, primarily for the determination or uranium and to a lesser extent also thorium, as well as γ spectrometry for the determination of a suite of short-lived radionuclides. Another feature of RINGAS is that two irradiation positions are available for either unshielded or cadmium-shielded irradiation. This system offers a minimum irradiation time of 1 s and minimum delay time of 11 s, allowing the analysis of such short-lived radionuclides as fluorine-20 with a half-life of 11 s.

The RINGAS system was not designed for cyclic activation analysis, since each sample has to complete the full cycle of irradiation, cooling, and counting before the next sample can be analyzed. Although cyclic activation analysis offers the advantage that only one sample is needed, then irradiated, and analyzed a few times in succession, it suffers from a greater disadvantage in that there is a build-up of longer-lived radionuclides, becoming more pronounced with each additional irradiation cycle. Semicyclic activation was therefore more preferable for this specific situation, even though sample preparation was more labor-intensive in that 10 or more subsamples had to be prepared for every analysis.

For the specific example of the vitamin supplements, the semicyclic analysis was performed on 20 separate capsules. Irradiation was carried out in the cadmium-shielded position in order to enhance the activity of fluorine relative to interfering radionuclides, such as sodium-24 and manganese-56. Furthermore, short irradiation, delay, and counting times of 5, 11, and 22 s, respectively, were used to determine fluorine-20 optimally with respect to the interfering radionuclides. Since the main γ-energy peak of fluorine-20 has a relatively high energy of 1634 keV, samples and element standards were counted behind a 1-mm lead shield to eliminate γ energies at the lower end of the energy spectrum, thereby keeping instrument dead times to a minimum.

Individual spectra were accumulated and added, thereby enhancing the fluorine-20 peak with respect to the interfering radionuclides. Figure 1 illustrates the advantage of semicyclic activation over one single analysis. It can clearly be seen that there is a great improvement in the signal-to-noise ratio for the accumulated 20-sample spectrum. Figure 2 illustrates the accumulated spectrum, identifying the fluorine-20 γ-energy peak, as well as the interfering γ-energy peaks, such as aluminum-28 and manganese-56.

RADIUM AND URANIUM ANALYSIS IN URINE SAMPLES FROM DECONTAMINATION WORKERS

Decommissioning of acid and uranium plants at South African gold mines has led to the need for monitoring contract workers employed at these plants for possible contamination from uranium and its decay

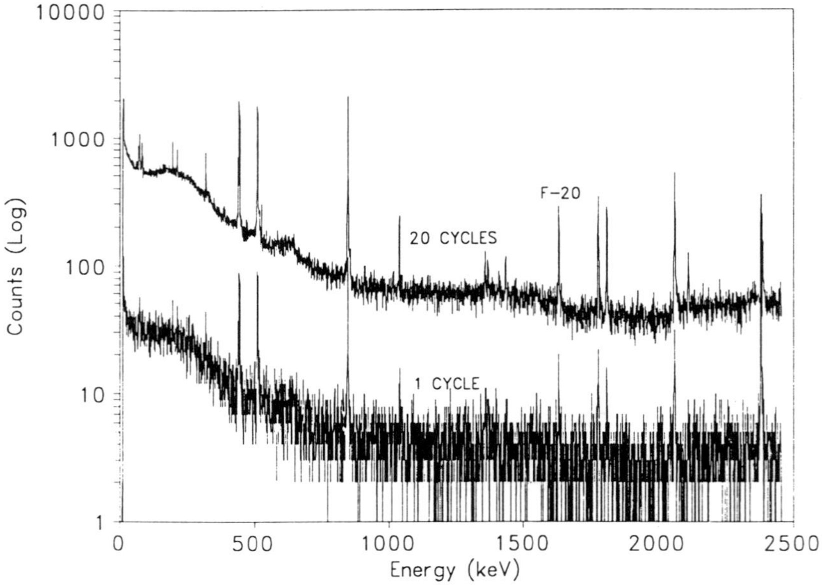

Fig. 1. Improvement of signal-to-noise ratio with semicyclic activation analysis.

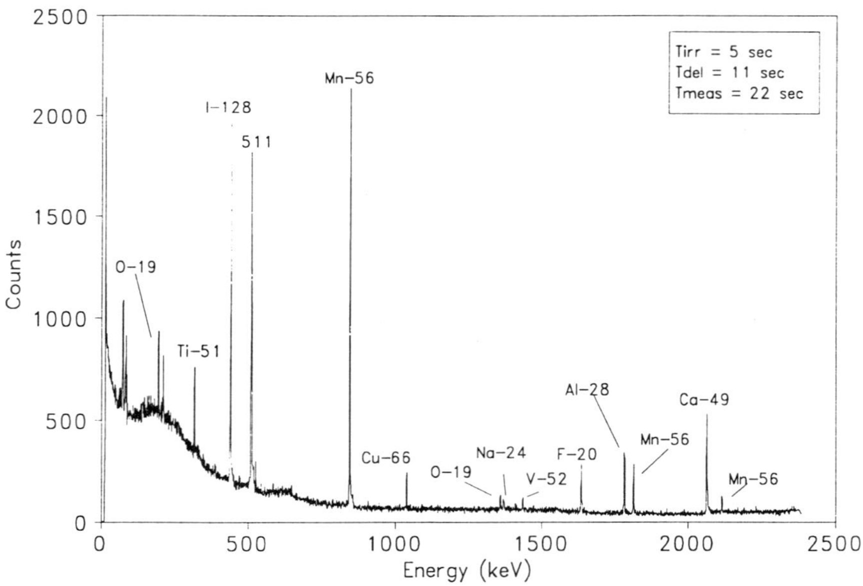

Fig. 2. Accumulated spectrum for fluorine-20 determination.

products. These radioelements are present in great quantities, and the dusty conditions cause great concern for internal contamination by inhalation.

The regulatory body in South Africa requires strict control of workers by monitoring urine samples on a regular basis and has set the observable lower limits as low as 0.72 mBq/L and 8.2 mBq/L for uranium-238 and radium-226, respectively. Special analytical procedures had to be developed to meet these demands.

A special laboratory had to be established for the analysis of low levels of uranium, radium, and thorium in environmental and urine samples, in order to avoid contamination from the high activity level samples, also analyzed in our laboratories. Presently, this laboratory handles 120 water and urine samples/mo for uranium and radium analyses and eventually should be able to cater to at least 200 samples/mo for uranium, radium, and thorium analysis.

In order to reach the low limit set for uranium-238 (0.72 mBq/L or ~60 ng/L), an instrumental neutron activation method cannot be employed. A preconcentration step is therefore introduced, whereby 250 mL of urine is chemically treated to form the uranium thiocyanate, which is absorbed on a column containing an anion-exchange resin. The resin is dried and ultrasonically sealed in special irradiation capsules for analysis by neutron activation and delayed neutron counting. Irradiation, delay and measuring times are on the order of 60, 11, and 60 s, respectively. The lower limit of determination attainable with this routine method for uranium-238 is 0.15 mBq/L.

Urine samples that are preserved by freezing and then thawed later tend to develop a large amount of deposits. These deposits cause problems in that they block the resin columns, making the analysis impractical and tedious.

As for uranium, a preconcentration step needs to be performed for the determination of radium-226 in order to reach the level of 8.2 mBq/L in liquids. Radium is coprecipitated with barium sulfate and then determined quantitatively by α spectrometry. Figure 3 illustrates an α spectrum of radium-226 and its daughters. After preparation, the sample was left for 3 wk in order for equilibrium to be reinstated after loss of radon-222 during the preconcentration step. The sample was then counted for 15 h. From Fig. 3, it can be seen that radium-226 is in equilibrium with its daughters and that there is no loss of radon-222 from the sample because of the vacuum conditions during the counting period.

Radium-226 is determined directly by measurement at 4.78 MeV. In certain cases, though, it is better to measure it indirectly through the daughter-product polonium-214 at 7.69 MeV, with corrections for the state of radioactive equilibrium. Polonium-214 produces an α-energy peak at the higher end of the spectrum, where interference from other α emitters, such as ^{228}Ra, ^{224}Ra, and ^{223}Ra from the natural uranium and thorium decay series, is less.

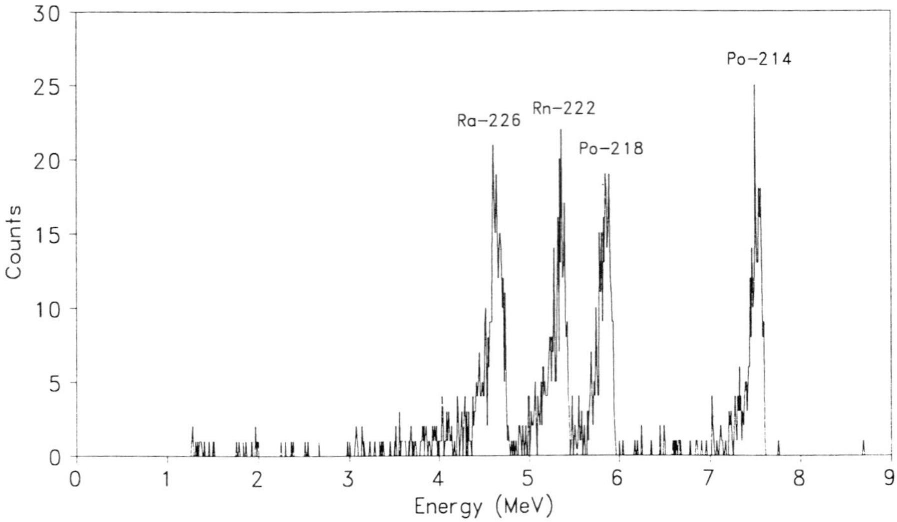

Fig. 3. α Spectrum of radium-226 after preconcentration from urine.

New developments that are currently considered are fast ashing of urine samples prior to analyses, which enables direct uranium analysis on the ash by DNC, as well as a fast column separation method for the determination of radium by liquid scintillation counting and α spectrometry.

SUMMARY

Radioanalysis is a commercial laboratory, striving to provide clients with a cost-effective but also a quality service. Although very low determination levels are obtainable by involved (radio)chemical procedures to suit every client's needs, these tend to be extremely labor-intensive and time-consuming, and therefore quite expensive, a situation that few clients are prepared to accept. The challenge that we, as an analytical laboratory, therefore face is to introduce fast, simple, and cost-effective methods of analysis, still maintaining international acceptable standards of quality.

Biomedical Applications

Application of Nuclear Analytical Methods in the Investigation and Identification of New Selenoproteins

D. BEHNE,* C. WEISS-NOWAK, M. KALCKLÖSCH, C. WESTPHAL, H. GESSNER, AND A. KYRIAKOPOULOS

Department of Trace Elements in Health and Nutrition, Hahn-Meitner-Institut Berlin, 14109 Berlin, Germany

ABSTRACT

Nuclear methods have been applied in the investigation of selenium-containing proteins in rat tissues. Selenium was determined in tissues, cells, and cellular compartments by instrumental neutron activation analysis via ^{77m}Se or ^{75}Se. For tracer studies, the selenium compounds were labeled in vivo by administering ^{75}Se with a high specific activity to rats. Quantitative determination of very small amounts of the element in protein fractions was achieved by measurement of the tracer after replenishment of selenium-depleted animals with the labeled element. The application of the nuclear methods in the detection, characterization, and identification of new selenium-containing proteins is shown with the help of some examples.

Index Entries: Neutron activation analysis; ^{75}Se-tracer, selenium, selenoprotein; rat.

INTRODUCTION

Selenium is an essential element, and selenium deficiency is related to several diseases in animals and humans *(1)*. The pathological changes found in animals include growth retardation, skin lesions and hair loss, visual defects, reproductive disorders, pancreas atrophy, liver necrosis, and dystrophies of the skeletal muscle and of the heart muscle. In humans, a low-selenium state may lead to cardiomyopathy and muscular disorders. A role of selenium in cardiovascular disease, osteoarthropathy,

*Author to whom all correspondence and reprint requests should be addressed.

Biological Trace Element Research Editor: G. N. Schrauzer © 1994 by Humana Press Inc.

and cancer has also been discussed, but here direct relationships between selenium deficiency and these diseases have not yet been established.

With these effects, the question of how selenium functions at a molecular level has been of great interest. One explanation for the essentiality of the element was found when glutathione peroxidase was identified as a selenoprotein *(2)*. This enzyme catalyzes the reduction of peroxides and thus protects the cells from oxidative damage. However, evidence has now accumulated that shows that this is not the only function of the element and that there are several other biologically active selenoproteins. By combining tracer procedures and protein separation methods, it was shown in several studies that selenium is present in mammalian tissues in a larger number of compounds *(3–7)*. A main part of our research is concerned with the investigation of these compounds, and here nuclear analytical techniques have proven to be very useful tools. In this article, with the help of some examples, several applications of these methods in the detection, characterization, and identification of selenium-containing proteins are shown and discussed.

METHODS

The nuclear methods used in these studies include instrumental neutron activation analysis, both via the short-lived ^{77m}Se (half-life 17.5 s) and the long-lived ^{75}Se (half-life 120.4 d). With the determination via ^{77m}Se (irradiation 20 s at a thermal neutron flux of 2×10^{13} cm^{-2}/s, decay time 0.4 s), the limit of detection was 5 ng. The analysis via ^{75}Se is more time-consuming since it involves the irradiation of the samples for 5 d, followed by a decay period of 2–3 mo to reduce sufficiently the phosphorus bremsstrahlung *(8)*. However, with this method, using a thermal neutron flux of 7×10^{13} cm^{-2}/s, a lower limit of detection could be reached that for most biological materials was around 0.1 ng. In the measurement of ^{75}Se the γ-rays with energies at 121.1, 136.0, 264.7, 279.5, or 400.7 keV can be used for the evaluation, providing that no interfering radionuclides with similar γ lines are present in the sample. The lowest limit of detection can be achieved by measuring the 400.7-keV line in a well-type detector, since then the intensity of this line is considerably enhanced by the coincident summing of the other γ-rays.

The second nuclear method comprises the different tracer techniques that were applied in the detection and characterization of selenium-containing proteins and in the investigation of their distribution, as well as in quantitative selenium analysis in protein fractions when very low limits of detection were required. As a selenium tracer, ^{75}Se was used. ^{75}Se with the high SA of 1000 Ci/g Se was produced by long-term irradiation of enriched ^{74}Se (Research reactor, University of Missouri, Columbia, MO).

In the investigation of the selenium-containing proteins, the nuclear methods were combined with several biochemical procedures as is outlined schematically in Fig. 1. For the experiments, rats were used that were fed either a selenium-deficient diet with a content of 2–5 μg Se/kg *(9)* or the same diet with 300 μg Se/kg added as sodium selenite. In the tracer experiments, the animals were labeled in vivo by administration of ^{75}Se-selenite. The distribution of the element among tissues and subcellular fractions was determined either by neutron activation analysis or by the measurement of the tracer. In histological sections of the labeled tissues, selenium was also localized by autoradiography. In the next stage, the tissue proteins were separated using several electrophoretic and chromatographic methods. In this way, information was obtained on new selenium-containing proteins, on their distribution in tissues and cellular compartments, and on some of their characteristics, such as relative molecular mass and isoelectric point. From the results of these experiments, suitable procedures for their purification are being worked out. Selenium analysis of the purified compounds and of their fragments, peptide fingerprints, amino acid analysis, immunoassay, and determination of the amino acid sequence are being used for further characterization and identification.

INVESTIGATION OF SELENIUM-CONTAINING PROTEINS

First information on the existence of several selenium-containing proteins was obtained by administrating ^{75}Se-selenite to selenium-depleted rats and separating the tissue proteins by SDS-polyacrylamide gel electrophoresis (SDS-PAGE) *(6)*. After autoradiography, 13 selenium-containing proteins or protein subunits could be distinguished that were numbered in the order of their increasing molecular mass. The subcellular fractionation showed that these compounds were unevenly distributed among the cellular compartments and might therefore be involved in different processes of intracellular metabolism *(10)*.

Selenium Analysis in Protein Fractions

In the investigation of these compounds, we have started to measure the selenium content in the different protein fractions separated by SDS-PAGE. In this way, if the molar ratio of selenium in these compounds was known, their tissue level could be determined directly from the selenium content. Conversely, with an unknown selenium-containing protein, the number of selenium atoms per molecule could be calculated, if information on the protein concentration could be obtained in another way. Since most of the selenium compounds are present in mammalian tissues in very small amounts, methods with a low detection limit are needed. In Table 1, this parameter is listed for the selenium determination by instrumental neutron activation analysis via the long-lived ^{75}Se and by tracer measurements.

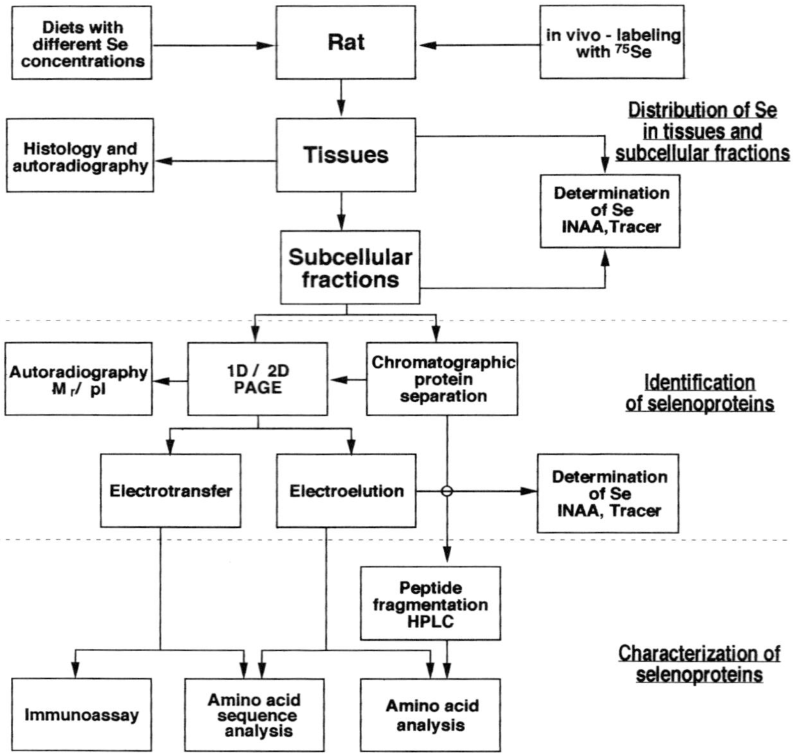

Fig. 1. Methods used in the investigation of selenium-containing proteins.

With neutron activation analysis, the limit of detection for selenium in an enriched protein fraction was around 20 pg and, thus, lower than in the case of most other biological samples, since with the enrichment procedures the amounts of interfering elements were reduced. However, quantitative analysis by tracer measurements was even more suitable. It could be carried out by giving the labeled element with a known specific activity to extremely selenium-depleted rats. Because then the amounts of selenium present in the tissues stemmed almost exclusively from the tracer, the selenium content of a sample could be determined directly from the activity. Using a low-level counting device, tracer activities down to 1×10^{-3} Bq could be determined. Therefore, when a small amount of the labeled element with the high SA of 3.7×10^{13} Bq/g Se (1000 Ci/g Se) was administered to the selenium-deficient animals, the limit of detection was below the femtogram range.

Quantitative selenium determination in animals with a normal selenium status by means of tracer analysis could be carried out after repletion of the deficient rats with the labeled element. Here the specific activity had to be fairly low, because otherwise, with the relatively large amounts of selenium to be administered, there would have been a danger of radiation damage. In this study, the selenium pools of rats were emptied by feeding the deficient diet for four generations and then

Table 1
Selenium Determination in Protein Fractions Obtained from Rat Tissues

Method	Limit of detection	
	Se (g)	Se (Mol)
INAA (via ^{75}Se) [a]	2×10^{-11}	3×10^{-13}
Tracer [b]		
1000 Ci ^{75}Se/g Se	3×10^{-17}	3×10^{-19}
Se-deficient rats		
Tracer [c]		
1.25 Ci ^{75}Se/g Se	2×10^{-14}	3×10^{-16}
Replenished rats		

[a]Instrumental neutron activation analysis.

[b]Determination of ^{75}Se after administration of a small amount of ^{75}Se-labeled selenium to extremely selenium-depleted rats.

[c]Determination of ^{75}Se after replenishment of extremely selenium-depleted rats with oral doses of ^{75}Se-labeled selenium twice a week over a period of 3 mo. The amount administered was equivalent to the selenium intake from a diet with 300 µg Se/kg.

repleted by giving an oral dose of ^{75}Se-labeled selenite (20 µCi, 15.8 µg Se) by a stomach tube twice a week over a period of 3 mo. With a food consumption of 15 g/d, the amount of selenium administered was equivalent to the intake of the element from a diet with the adequate content of 300 µg Se/kg. As is shown in Table 1, with these experimental conditions, the detection limit was below the picogram range and thus three orders of magnitude lower than with neutron activation analysis.

The validity of this approach was tested in several experiments. Neutron activation analysis of selenium in skeletal muscle, liver, plasma, and erythrocytes, which together contain nearly 90% of the element present in the rat tissues *(11)*, showed that the selenium contents in the main pools of the deficient animals were <1% of those in the selenium-adequate controls (Table 2). After repletion, the selenium levels calculated from the tracer distribution were compared with the values obtained by neutron activation analysis in tissue samples from the control animals fed a diet with 300 µg Se/kg (Fig. 2). The highly significant correlation (R = 0.986) found between the two sets of values indicated that under these conditions, the ^{75}Se activity reflected the distribution of the element in the normal state.

With this method, we were in a position to measure quantitatively the levels of the different selenium-containing proteins in tissues, cells, or subcellular fractions. After SDS-PAGE of the samples, the gel sections that contained the labeled proteins were cut off, and the tracer activities measured by means of a Ge(Li) well-type detector. Figure 3 shows, as an

Table 2
Effects of a Low-Selenium Diet on the Selenium Content in the Main Selenium Pools of the Rat

Tissue	Se content (µg/kg dry sample) [a)]		
	Se+ [b)]	Se- [b)]	Se-/Se+ (%)
Muscle	620 ± 80	< 9 [c)]	< 1.5
Liver	2660 ± 270	17 ± 3	0.6
Erythrocytes	1540 ± 110	< 10 [c)]	< 0.6
Plasma	4760 ± 150	34 ± 4	0.7

[a]Mean ± SD for three male animals fed the diet for six generations.
[b]Se+, 300 µg Se/kg diet; Se–, 2–5 µg Se/kg diet.
[c]Limit of detection.

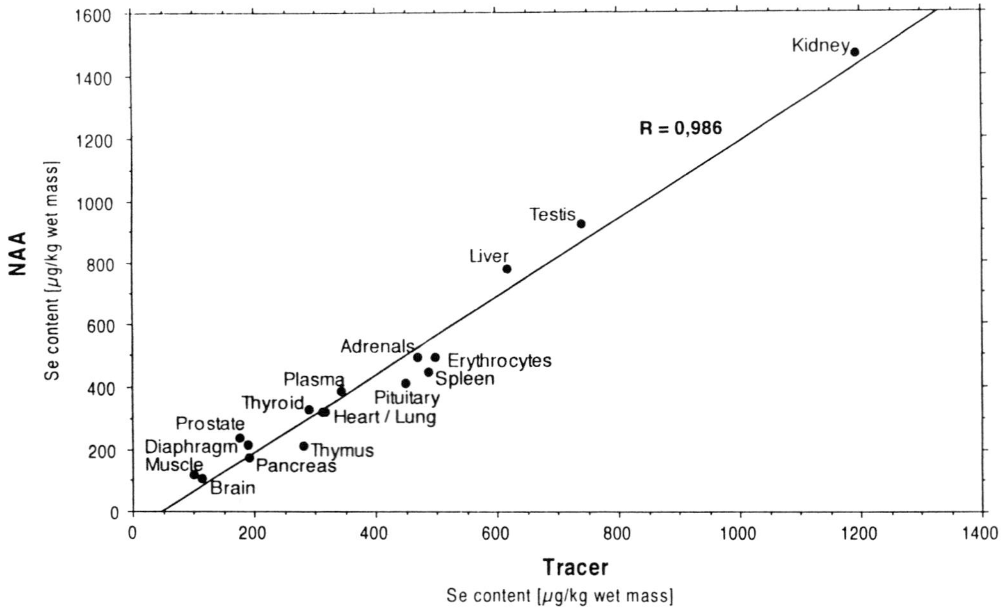

Fig. 2. Selenium content in tissues of the rat. Comparison of the results obtained by neutron activation analysis in rats fed a diet with 300 µg Se/kg and by determination of the ^{75}Se-tracer after replenishment of extremely selenium-depleted rats with oral doses of the ^{75}Se-labeled element twice a week over a period of 3 mo. The amount of selenium administered was equivalent to the selenium intake from a diet with 300 µg Se/kg.

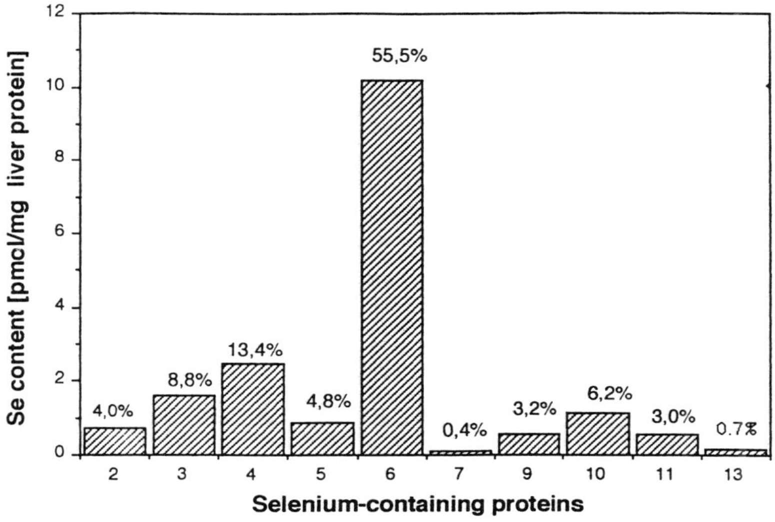

Fig. 3. Selenium distribution among selenium-containing proteins in rat liver separated by SDS-PAGE. The selenium analysis was carried out by determination of the ^{75}Se-tracer after replenishment of extremely selenium-depleted rats with oral doses of the ^{75}Se-labeled element. The amount of selenium found in each fraction was expressed as pmol Se/mg liver protein and also as a percentage of the total amount present in these selenium compounds.

example, the selenium concentration in several selenium-containing proteins in the liver, expressed as pmol Se/mg total liver protein. In the cases where only one selenium atom is present per molecule, this value also represents the concentration of the protein in question. The predominant selenium compound in the liver homogenate was the glutathione peroxidase subunit (protein No. 6), which contained 55% of the total protein-bound amount of the element. This agrees well with the value of 58% calculated from the liver selenium content and the enzymatic glutathione peroxidase activity in the cytosol *(10)*.

In the liver homogenate, the protein No. 7 had a relatively low concentration of 0.1 pmol Se/mg liver protein. It had, however, several interesting characteristics that prompted us to choose it as the first selenium compound to be investigated in more detail. Here nuclear analytical methods played a major part, and these studies may therefore serve as further examples of the use of nuclear techniques in selenium research.

Identification of a Seleno-Enzyme

Protein No. 7, which in SDS-PAGE had a relative molecular mass of 27.8 kDa, was found to be one of the main selenium-containing proteins in the thyroid, and was otherwise only detected in the liver and in the kidney *(6)*. In the liver, it was shown to be membrane-bound *(10)*. These

Table 3
Tracer Distribution in Protein Fragments After Trypsin Digestion of Liver Microsome Proteins in the 27-kDa Range Labeled with ^{75}Se and ^{125}I

Gel section [a]	^{125}I (cps x 10^{2})	^{75}Se (cps x 10^{-3})	^{125}I / ^{75}Se (x 10^{5})
1	4.2	2.0	2.1
2	4.4	2.2	2.0
3	2.1	1.2	1.8
4	3.4	1.8	1.9
5	4.8	2.1	2.3
6	7.6	3.8	2.0

[a]After digestion and separation of the fragments by SDS-PAGE the gel was cut into several sections *(12)*.

three tissues have the highest activities of the type I iodothyronine deiodinase. This enzyme catalyzes the deiodination of thyroxine to the biologically more active thyroid hormone triiodothyronine and, thus, has a key role in thyroid hormone metabolism. It is likewise membrane-bound and has a subunit of 27 kDa. Because of these similarities, we suggested that the selenium-containing protein No. 7 and the deiodinase subunit are identical.

Nuclear methods using two tracers were applied to prove this hypothesis *(12)*. Protein No. 7 was labeled in vivo with ^{75}Se. Since the deiodinase subunit binds covalently to the bromoacetyl-derivative of thyroxine, it was labeled in vitro via ^{125}I-thyroxine.

With the labeled compounds, it was shown by SDS-PAGE that both have the same velocity in the electric field. The final proof that they not only have the same relative molecular mass, but are in fact identical was obtained by peptide fingerprints. After double-labeling with ^{75}Se and ^{125}I, the 27-kDa proteins from rat liver microsomes were isolated, and either partly digested with trypsin or cleaved chemically by cyanogen bromide. With these procedures, specific individual sets of peptide fragments were produced, representative of the original molecule, which were separated by SDS-PAGE. The SDS-gels were then cut into sections for the measurement of the activities of ^{75}Se and ^{125}I. The data in Table 3 show the results obtained in this way in the trypsin digestion. It was found in both experiments that the tracers had identical quantitative distribution patterns and that in each set the ratio of their activities in the fragments was constant. This indicated that the deiodinase subunit and protein No. 7 are identical, and that selenium and the thyroxine compound were present at the same site in the molecule.

By using the double-labeling technique, it was also possible to obtain information on the molar content of selenium in the deiodinase subunit. This determination was carried out with liver microsomes during the replenishment of selenium-deficient rats with ^{75}Se-selenite. The selenium content of the protein could therefore be calculated directly from the ^{75}Se activity in the 27-kDa range. The content of the deiodinase subunit was determined by titration with bromoacetyl-thyroxine. From these values obtained from animals at different stages of repletion and in the replenished state, it was estimated that the deiodinase subunit contained one selenium atom per molecule.

The studies suggest that the element is present at the active site of the enzyme, most probably in the form of selenocysteine. Our finding that the deiodinase is a selenoprotein was confirmed by the work of Arthur et al. *(13)* and Berry et al. *(14)*. In selenium deficiency, the level of the deiodinase is diminished, which leads to a decrease in the production of triiodothyronine. This means that not only iodine, but also selenium is necessary for normal thyroid hormone metabolism, and that in thyroid-hormone-related diseases, the selenium status of the patients has to be taken into consideration.

Detection of Further Selenium-Containing Proteins

By means of tracer techniques using in vivo labeling of selenium-deficient rats with ^{75}Se-selenite with a high specific activity, followed by SDS-PAGE and autoradiography of the separated proteins, detailed investigations of the distribution of selenium-containing proteins were carried out on a large number of tissues. The results of these experiments are shown in Table 4. The first column contains a list of the 13 compounds described earlier *(6)*. In the second, there are several additional selenium-containing proteins or protein subunits that we have found since. Most of those were only detected in one or a few tissues, or were only visible after enrichment, for instance, after subcellular fractionation. In the last column, those selenoproteins are listed that have already been sequenced and identified. They include the monomeric phospholipid hydroperoxide glutathione peroxidase *(15)*, which in the SDS-gel is located at 20 kDa. In the sperm, a mitochondrial capsule selenoprotein was detected *(16)* that in SDS-PAGE has a similar relative molecular mass in the 20-kDa range. The subunits of two glutathione peroxidases are found in the 23–24 kDa range. These are the classical tetrameric glutathione peroxidase present in the erythrocytes and tissues *(2)* and the plasma glutathione peroxidase *(17)*, which has a similar tetrameric form, but shows differences in its amino acid sequence. A further tetrameric glutathione peroxidase was recently described *(18)*, which does not cross-react with antisera against the other two compounds and was found to be mainly present in the gastrointestinal tract. The subunit of the type I iodothyronine deiodinase *(12)* is situated in the 27–28 kDa range, and the

Table 4
Selenium-Containing Proteins or Protein Subunits in Rat Tissues (After In Vivo Labeling with ^{75}Se and Protein Separation by SDS-PAGE)

Study 1988 (6)		M_r [a]	Additional	M_r [a]	Identified
			+	10	
1	+++[b]	12.1			
			+	14	
2	+++	15.6			
			++	16.5	
3	+++	18.0			
4	+++	19.7			PHGPx[c], MCS[d]
5	+	22.2			
			++	23	
6	+++	23.7			GPx [e]
			+	27	
7	+	27.8			I-D [f]
			+	30	
			+	32	
8	+	33.3			
			+	38	
			+	44	
			+	47	
9	+++	55.5			Se-P [g]
10	+++	59.9			
11	+++	64.9			
12	+	70.1			
13	++	75.4			
			+	116	

[a]Relative molecular mass (kDa).
[b]Found in a few (+), nearly all (++), or all (+++) of the tissues investigated.
[c]Phospholipid hydroperoxide glutathione peroxidase *(15)*.
[d]Mitochondrial capsule selenoprotein *(16)*.
[e]Subunits of the cellular glutathione peroxidase *(2)* and the plasma glutathione peroxidase *(17)*.
[f]Subunit of type I iodothyronine deiodinase *(12)*.
[g]Plasma selenoprotein P *(19)*.

plasma selenoprotein P *(19)* in the 57–58 kDa range. A further selenoprotein with a molecular mass of 10 kDa has recently been partly sequenced *(20)*.

Besides the few selenoproteins identified so far, about 20 selenium-containing proteins or protein subunits have now been found in the tracer experiments after separation by SDS-PAGE. Some of them are possibly linked to each other and may be part of the same metabolic chain. However, there is also a possibility that there are several selenium-containing compounds with the same relative molecular mass in SDS-PAGE, so that the number might prove to be even greater once other separation methods have been applied. Considering these findings and the information already available on the differences in their distribution among tissues, cells, and cellular compartments, it may be assumed that in addition to the compounds already identified, several more biologically important selenoproteins do exist. From the examples described in this article, it can be seen that nuclear techniques may be valuable tools in the characterization and identification of these newly found selenium compounds.

REFERENCES

1. O. A. Levander, in *Trace Elements in Human and Animal Nutrition,* vol. 2, W. Mertz, ed. Academic, Orlando, FL., 1986, pp. 209–279.
2. J. T. Rotruck, A. L. Pope, H. E. Ganther, A. B. Swanson, D. G. Hafeman, and W. G. Hoekstra, *Science* **179,** 588–590 (1973).
3. W. C. Hawkes, E. C. Wilhelmsen, and A. L. Tappel, *J. Inorg. Biochem.* **23,** 77–92 (1985).
4. K. G. Danielson and D. Medina, *Cancer Res.* **46,** 4582–4589 (1986).
5. H. I. Calvin, K. Grosshans, S. R. Musicant-Shikora, and S. I. Turner, *J. Reprod. Fert.* **81,** 1–11 (1987).
6. D. Behne, H. Hilmert, S. Scheid, H. Gessner, and W. Elger, *Biochim. Biophys. Acta* **966,** 12–21 (1988).
7. J. K. Evenson and R. A. Sunde, *Proc. Soc. Exp. Biol. Med.* **187,** 169–180 (1988).
8. D. Behne and H. Jürgensen, *J. Radioanal. Chem.* **42,** 447–453 (1978).
9. D. Behne, A. Kyriakopoulos, S. Scheid, and H. Gessner, *J. Nutr.* **121,** 806–814 (1991).
10. D. Behne, S. Scheid, A. Kyriakopoulos, and H. Hilmert, *Biochim. Biophys. Acta* **1033,** 219–225 (1990).
11. D. Behne and W. Wolters, *J. Nutr.* **113,** 456–461 (1983).
12. D. Behne, A. Kyriakopoulos, H. Meinhold, and J. Köhrle, *Biochem. Biophys. Res. Commun.* **173,** 1143–1149 (1990).
13. J. R. Arthur, F. Nicol, and G. J. Beckett, *Biochem. J.* **272,** 537–540 (1990).
14. M. J. Berry, L. Banu, and P. R. Larsen, *Nature* **349,** 438–440 (1991).
15. F. Ursini, M. Maiorino, and C. Gregolin, *Biochim. Biophys. Acta* **839,** 62–70 (1985).
16. H. I. Calvin, G. W. Cooper, and E. Wallace, *Gamete Res.* **4,** 139–149 (1981).
17. K. Takahashi, N. Avissar, J. Within, and H. Cohen, *Arch. Biochem. Biophys.* **256,** 677–686 (1987).
18. F.-F. Chu, J. H. Doroshow, and R. S. Esworthy, *J. Biol. Chem.* **268,** 2571–2576 (1993).
19. K. E. Hill, R. S. Lloyd, J.-G. Yang, R. Read, and R. F. Burk, *J. Biol. Chem.* **266,** 10,050–10,053 (1991).
20. S. C. Vendeland, M. A. Beilstein, C. L. Chen, O. N. Jensen, E. Barofsky, and P. D. Whanger, *J. Biol. Chem.* **268,** 17,103–17,107 (1993).

Detection and Determination of Selenoproteins by Nuclear Techniques

S. F. STONE,*[,1] G. BERNASCONI,[1] N. HASELBERGER,[1] M. MAKAREWICZ,[1] R. OGRIS,[1] P. WOBRAUSCHEK,[2] AND R. ZEISLER[1]

[1]*Division of Human Health and Agency's Laboratories Seibersdorf, International Atomic Energy Agency, POB 100, A-1400 Vienna, Austria;* [2]*Atominstitut der Österreichischen Universitäten, Schüttelstrasse 115, A-1020 Vienna, Austria*

ABSTRACT

Microbeam X-ray spectrometry, energy-dispersive X-ray fluorescence analysis, and neutron activation analysis were evaluated for the detection of selenium contained in the selenoprotein glutathione peroxidase. The glutathione peroxidase had been previously separated using polyacrylamide gel electrophoresis. The use of Bragg-reflected polarized X-ray beams was employed in the X-ray fluorescence measurements to minimize the problem of scatter owing to the gel matrix. Current detection limits of selenium in a gel matrix are 2.1 ng in the bench-top microbeam X-ray system and 30–60 ng using XRF with polarized beams. Neutron activation analysis was used for quality-control measurements, with a detection limit here of < 0.08 ng. The work has in principle established the feasibility of such an approach.

Index Entries: Polyacrylamide gel electrophoresis (PAGE); protein; metalloprotein; selenoproteins; energy-dispersive X-ray fluorescence (EDXRF); microbeam X-ray spectrometry; instrumental neutron activation analysis (INAA)

*Author to whom all correspondence and reprint requests should be addressed.

INTRODUCTION

Research on selenium has been extensive since the element was established about 35 years ago as essential *(1)*, but the knowledge about it and its role in biological functions is still limited. Only in 1973 was selenium shown to be an essential structural component of glutathione peroxidase, a selenoenzyme, which became the first known biochemical moiety with bound selenium *(2)*. The sophistication of the studies on selenium has increased since this time from the emphasis of determining elemental concentrations in various whole organs, tissues, and body fluids to making these determinations in smaller and smaller subunits, first in various cellular components and at present in elements associated with biological macromolecules.

Aided by new analytical procedures, selenium metabolism has been discovered in recent years to be increasingly complex: when ingested at high levels, it can be toxic and produce immunosuppression effects, but when administered as a supplement at nontoxic levels, it has also been shown to counteract heavy metal toxicity, especially in cases of high levels of cadmium, mercury, methylmercury, and silver *(3)*. There have also been numerous studies investigating whether selenium can counteract certain chemical carcinogens or have some protecting effect against coronary heart disease *(3)*. These investigations are still continuing. Selenium deficiency has been indicated as one of the causes of two human diseases, both found in China: the cardiomyopathy, Keshan disease, and the Kaschin-Beck disease, involving osteoarthropathy. Selenium, as part of glutathione peroxidase, has an important role in the immune system, catalyzing the reduction of organic peroxides, especially hydrogen peroxide, which protects the cells against oxidative damage. Recently, a new selenoprotein was identified, Type I Iodothyronine 5´deiododinase, which is involved in thyroid hormone metabolism *(4)*. To date, five mammalian selenoproteins have been recognized, and the existence of more are definitely indicated *(5)*.

In much of this research, the use of radioactive labeling has played an invaluable role. Radiolabels have been employed to detect and identify possible new selenoproteins. The powerful separation tool polyacrylamide gel electrophoresis (PAGE) is one of the biochemical techniques normally used in separation of samples containing radiolabeled proteins to be studied. In PAGE, hundreds to thousands of proteins can potentially be separated in a two-stage separation *(6)*. An additional dimension for the identification of the proteins can be added, since the trace element is specific for certain proteins, and the radioactive label may be in the form of a radioactive isotope of this trace element. In our case, the selenium contained in the radiolabeled selenoproteins can be used for such an identification. However, it is not always convenient, or even possible, to introduce radiolabeled compounds to humans, animals, or tissues being studied. For a direct detection of elements associated with proteins, new analytical techniques need to be developed. They must be of at least comparable specificity and sensitivity, utilizing the most sensitive techniques for trace element detection.

The combination of PAGE and neutron activation analysis (PAGE-NAA) for the determination of elements associated with proteins has previously been demonstrated *(7,8)* in the analysis of separated phosphoproteins. However, there are limitations to whole-gel irradiations at adequate neutron fluences, although quantitative analysis of the autoradiographed phosphoproteins was competitive with other methods *(8)*. A more readily available method would be one that uses laboratory sources for excitation. Literature data on the sensitivity of X-ray fluorescence (XRF) systems ranging from total reflection XRF to bench-top microbeam XRF suggest that a direct determination with such a laboratory tool should be possible *(9,10)*. Also, proton-induced X-ray fluorescence (PIXE) might be considered as a complementary tool because of its sensitivity *(11)*.

The work on developing a capability for direct analysis of trace elements in electrophoretically separated proteins included microbeam XRF, energy-dispersive XRF, and instrumental NAA. The microbeam XRF appears to be particularly suitable because it would combine sensitivities that have been reported in the 10-fg range *(10)* with a scanning capability. This capability would be ideal to apply to PAGE separations.

METHODS

Polyacrylamide Gel Electrophoresis (PAGE)

Samples of glutathione peroxidase from bovine erythrocytes (Sigma, St. Louis, MO) were separated by PAGE with sodium dodecyl sulfate (SDS) according to the procedure described by Laemmli *(12)*. Slab gels of 1.5-mm thickness were prepared with a stacking gel of 5% and a separating gel of 15% polyacrylamide. Low-molecular-weight markers (Sigma) were used for calibrating the separation. Care was taken to use clean reagents and materials to minimize the trace element background. The reagents (from Sigma) have been previously investigated and found to have acceptably low trace element content. All water used in the gels and buffers was of HPLC grade, and preparation and sample handling was done as much as possible in HEPA-filtered air to minimize contamination. A clean nitrocellulose backing (BioTrace, Gelman Sciences, Ann Arbor, MI) was used as gel support for dried gels. The protein separations were duplicated in order to have similar samples for all the tests.

Individual sample loadings ranged from 10–50 μL, of solutions containing approx 1 mg/mL protein. Electrophoresis was run at constant amperage, between 15 and 20 mA/gel, until the leading buffer line was close to the bottom of the gels. Following electrophoresis, the proteins were separated in bands in approx 10-mm wide strips on the gel according to their molecular weight. The bands were visualized by a nonspecific protein stain, Coomassie brilliant blue. The gels were vacuum-dried at 80°C for at least 4 h. For the microbeam XRF, a strip containing the bands, was excised for scanning across the bands. For EDXRF and NAA, sections of approx 2 × 10 mm were excised, including individual bands.

X-Ray Microbeam Spectrometry

The microbeam spectrometry system of the PCI Laboratory Seibersdorf *(13)* consisted of an X-ray tube with a Mo anode (fine focus, FK 60-04, 2 kW, Siemens, Vienna, Austria), a conical glass capillary with a length of 10 cm and an end diameter of 30 µm, and an Si(Li) detector (80 mm^2 with a 7-µm Be window). The location of the detector was 90° in reference to the incident beam. The capillary to sample distance was 1 mm, and the distance between the sample and detector was 20 mm. Measurement times ranged from 1000 s/scanned point to 5000 s for spot analysis.

Energy-Dispersive X-Ray Fluorescence

A plane polarized beam was utilized to irradiate the samples. The primary excitation source was a molybdenum anode (Siemens fine focus, FK 60-01, 0.8 kW), at 50 kV, 16 mA, and the secondary target was single crystal molybdenum cut to the (611) plane, providing a Bragg reflection through an angle of 90°. Gel sections were placed on a sample holder and positioned 4 cm from the detector. A Si(Li) detector (12 mm^2, with a 7-µm Be window) with an FWHM of 165 eV was used for X-ray detection. A PC-based multichannel analyzer was used (Nucleus PCA-2, Oxford Instruments, Oakridge, TN). Measurement times ranged from 300 to 1000 s.

Neutron Activation Analysis

Samples of excised portions of the strips containing protein bands and other gel areas and samples of multielement comparator standards were sealed in ultrapure quartz prior to irradiation at the ASTRA reactor in Seibersdorf. Irradiation times ranged from 4 to 24 h at a neutron flux of $7 \cdot 10^{13}/cm^2 \cdot s$. The samples were counted in the well of an HPGe detector, rated at 20% relative efficiency and of 1.9 keV FWHM at 1332 keV. Counting times ranged from 3600 to 90,000 s, after a decay time of 5–7 d. Additional counts were done after a decay of 3–4 mos. The INAA procedure employed is described in ref *(14)*.

RESULTS AND DISCUSSION

In the PAGE separations, the glutathione peroxidase sample was separated into two different bands, one at approx 22 kDa and the other at 24 kDa. These are known to be monomeric plasma glutathione peroxidase and the subunit of the cellular glutathione peroxidase.

The microbeam XRF studies involved the analysis of blank gels and backings, as well as gels containing separated selenoproteins. The gen-

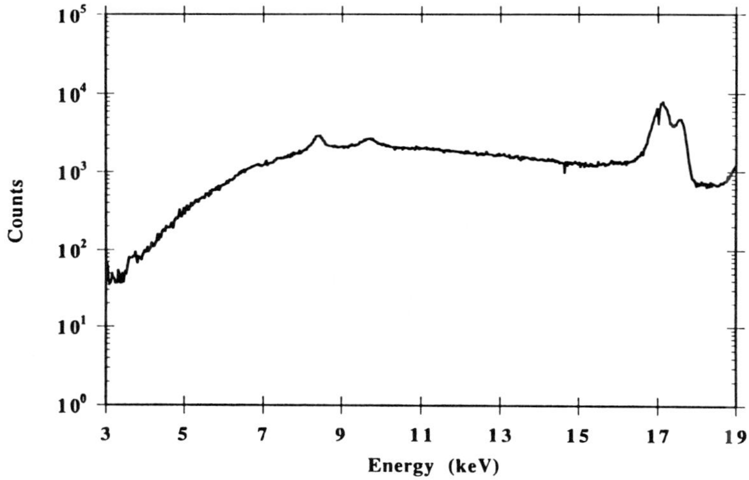

Fig. 1. Spectrum of one point of a 5 × 11 point scan across the middle of a gel section containing a band of separated glutathione peroxidase, using microbeam X-ray spectrometry. The problem of scatter by the gel matrix is shown in the extremely high background (2 counts/s/channel) at the location of the Se X-rays (11.2 keV).

eral background of trace elements in the selected backing and in the blank gels has been found to be fairly low. The nitrocellulose backing, BioTrace, was found to have a very low trace element background and appears to be more stable than the previously used Whatman filter paper. For these first trace element studies, a Cr X-ray tube was used.

A scan of 55 points (5 rows of 11 points), separated by about 500 µm, was performed on a gel section containing a glutathione peroxidase band, using the microbeam with a Mo X-ray tube. Some of the points showed indications of Se, however, not above the detection limit of $3 \cdot 10^{-4}$ g/cm^2. Considering the area that is excited, that is $7 \cdot 10^{-6}$ cm^2, this detection limit corresponds to an absolute mass of $2.1 \cdot 10^{-9}$ g. This is on the order of the amount that can be estimated from the known sample loadings. The problem in this and other point measurements has been an unexpected increase in the spectral background caused by scatter from the gel matrix. The extent of the problem of scattering from the gel matrix is illustrated in Fig. 1. There are only limited means available to reduce the scatter background.

To explore whether a polarized beam could be used to improve the detection of selenium, sections of a gel, referred to as Gel 1, were excised and analyzed by XRF using a polarized Mo beam. This technique reduces the scatter of the source radiation into the detector and, thus, minimizes the background. The excised protein bands of GSH-Px in Gel 1 at approx 22 and 24 kDa are identified as sections 2 and 4, respectively. Gel sections of similar size were taken at about 23 kDa (section 3), at

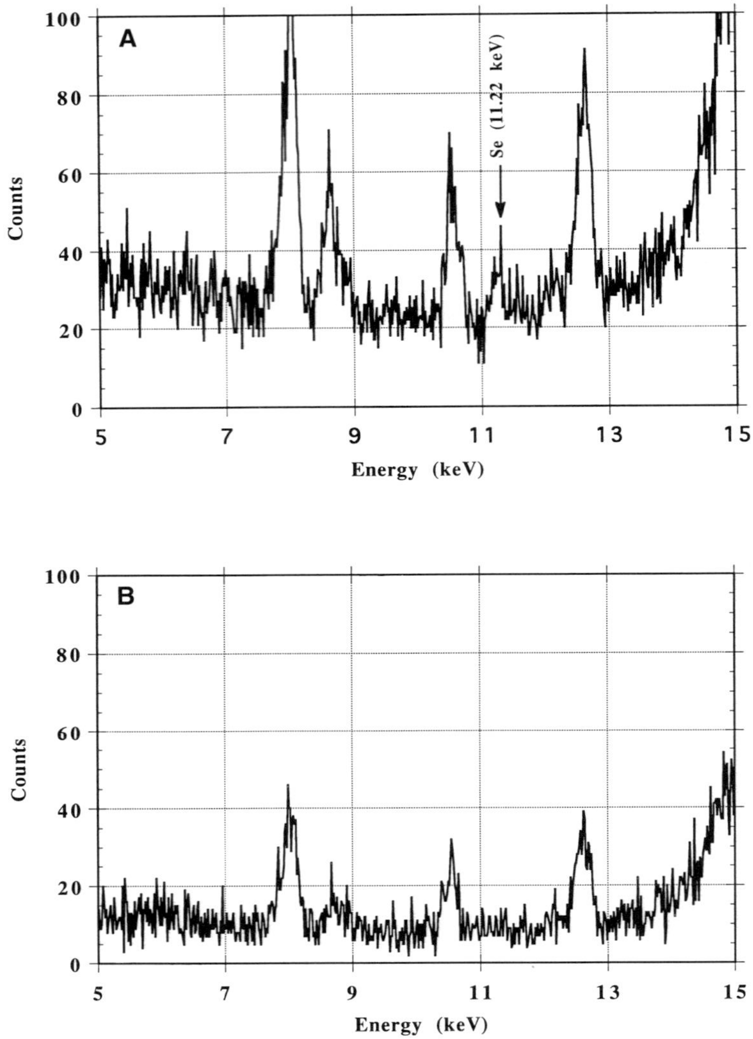

Fig. 2. EDXRF spectra of excised gel sections with the band of separated gluathione peroxidase **(A)**, and at slightly higher molecular weight outside the band **(B)**. (A) Gel 1, section #2, showing a statistically significant selenium peak relating to the protein band at 22 kDa, (B) Gel 1, section #3, showing no selenium peak at approx 23 kDa.

about 21 kDa (section 1), at the high-molecular-weight separation limit, approximately at 200 kDa (section 5), and a gel blank below the buffer front, which is before the lower-molecular-weight separation limit (section 6). In Fig. 2A and 2B, portions of two XRF spectra are shown from the analysis of two of these sections. Figure 2A represents the analysis of gel section 2, and Fig. 2B represents gel section 3. In gel section 2, the

Table 1
NAA Results of Se Determinations in 2 × 10 mm Sectioned Gel Strips Containing Separated Glutathione Peroxidase (GSH-Px), Gel 1, Strip with a Loading of 25 μL Protein Solution

	Sample		Se amount (ng)
Direction of molecular weight separation ↓	#5	approx. 200 kDa (beginning of separation)	0.291±0.014
	#4	24 kDa (band of subunit of tetrameric GSH-Px)	28.54±0.23
	#3	approx. 23 kDa (between bands)	15.55±0.21
	#2	22 kDa (band of monomeric GSH-Px)	32.98±0.33
	#1	approx. 21 kDa (below protein band)	15.70±0.23
	#6	below buffer front (blank gel)	≤0.08

area for the Kα peak of Se at 11.22 keV was found to be statistically significant (30% uncertainty). For gel section 3, there was no statistical indication of an Se peak. These samples were later analyzed by NAA, with the results described below.

The theoretical detection limit of 11 ng Se by EDXRF using a polarized beam *(9)* indicated that this would be a useful technique for determining selenium in the separated protein bands. However, the scattering is still too high, obscuring the selenium X-ray signal. The use of the polarized beam reduced the background, compared to that seen in the microbeam XRF spectrum, but the detection limits were not improved as much as expected. The detection limit under our conditions was 30–60 ng Se, which is much higher than in a nonscattering sample.

Instrumental NAA has sufficient sensitivity for the analysis of gel sections containing separated proteins. The Se results from the NAA determinations of the sections from Gel 1 are shown in Table 1. The major amounts of Se were found in sections #2 and #4, at the approximate locations of the separated GSH-Px bands at 22 and 24 kDa. However, an appreciable amount of Se was also found in the section at about 23 kDa (#3) and in the section at about 21 kDa (#1). The total Se content of these four sections was 92.8 ± 0.3 ng, compared with an expected Se amount of 86 ± 3 ng, calculated from the protein loading. Even though there is a good agreement between the expected and determined amounts of Se, the individual results indicate that the sectioning of the gel according to the protein staining was not precise enough. Exact sectioning of a gel containing a more complex separation of dozens of proteins would be extremely difficult, if not impossible. These results re-emphasize the need for an analytical technique for Se determination in separated proteins that is not dependent on the sectioning of the gel.

CONCLUSIONS

From the work presented here, it is feasible to utilize X-ray beams for the detection of Se in PAGE-separated selenoproteins. Currently, the level of Se detection is 30–60 ng absolute ($1.4–2.5 \cdot 10^{-7}$ g/cm^{-2} for EDXRF with polarized beams and 2.1 ng absolute ($3 \cdot 10^{-4}$ g/cm^2) for microbeam XRF, which is too high for the routine analysis of separated selenoproteins. However, it is clear that improvements can be made to lower detection limits in all the detection techniques that were employed in this work. Beam shape, sample-detector geometry, and measurement times have not yet been fully exploited for optimization. Other means can also be used to complement the techniques used in the current work. It has been shown previously with radioactive labeling experiments that it is feasible to detect and quantify ^{75}Se using autoradiography *(15)*. Autoradiography could be applied to neutron-irradiated gels containing selenoproteins not labeled prior to separation. After locating the appropriate gel area containing a selenoprotein, this area could be quantified through densitometry, or excised and analyzed by γ spectrometry to determine the selenium quantity. The possibility of the utilization of proton-induced X-ray emission (PIXE), including microbeam scanning, for the analysis of the gels is also being investigated. The determination of Fe and Ni in separated metalloproteins has been previously demonstrated *(16)*. Preliminary results on our gels with separated glutathione peroxidase have shown a lowering of detection limits from the XRF analyses by at least a factor of 50 and determination of Se appeared feasible.

ACKNOWLEDGMENTS

The authors wish to thank the staff at the ASTRA reactor, Austrian Research Centre in Seibersdorf for their help in the irradiations of the NAA samples. We also wish to thank Dr. Christine Streli of the Atominstitute of the Austrian Universities for help on the XRF data.

REFERENCES

1. K. Schwarz and C. M. Foltz, *J. Am. Chem. Soc.* **39,** 3292, 3293 (1957).
2. J. T. Rotruck, A. L. Pope, H. E. Ganther, A. B, Swanson, D. G. Hafeman, and W. G. Hoekstra, *Science* **179,** 588–590 (1973).
3. P. D. Whanger, *J. Trace Elem. Electrolytes Health Dis.* **6,** 209–221 (1992).
4. D. Behne, A. Kyriakopoulos, H. Meinhold, and J. Köhrle, *Biochem. Biophys. Res. Commun.* **173,** 1143–1149 (1990).
5. B. A. Zachara, *J. Trace Elem. Electrolytes Health Dis.* **6,** 137–151 (1992).
6. P. H. O'Farrell, *J. Biol. Chem.* **250,** 4007–4021 (1975).
7. S. F. Stone, R. Zeisler, and G. E. Gordon, *Biol. Trace Element Res.* **26–27,** 85–91 (1990).
8. S. F. Stone, R. Zeisler, G. E. Gordon, R. P. Viscidi, and E. H. Cerny, in ACS Symposium Series #445, Biological Trace Element Research: Multidiscipli-

nary Perspectives, K. S. Subramanian, G. V. Iyengar, and K. Okamoto eds., American Chemical Society, 1991, pp. 266–277.
9. H. Aiginger and P. Wobrauschek, *J. Radioanal. Chem.* **61,** 281–293 (1981).
10. P. Engström, S. Larsson, A. Rindby, and B. Stocklassa, *Nucl. Instr. Methods Phys. Res.* **B36,** 222–226 (1989).
11. W. Maenhaut and K. G. Malmqvist, in *Handbook of X-ray Spectrometry; Methods and Techniques,* R. E. Van Grieken and A. A. Markowicz, eds., Marcel Dekker, New York, Basel, Hong Kong, 1993, pp. 517–581.
12. U. K. Laemmli, *Nature* **227,** 680, 681 (1970).
13. G. Bernasconi, N. Haselberger, A. Markowicz, and V. Valkovic, *Nucl. Instr. Methods B,* **B86,** 333–338 (1994).
14. M. Makarewicz and R. Zeisler, *Biol. Trace Element Res.* this vol.
15. S. F. Stone, R. Zeisler, and G. E. Gordon, in *Trace Element Analytical Chemistry in Medicine and Biology,* vol. 5, P. Brätter and P. Schramel eds., Walter de Gruyter, Berlin, 1988, pp 157–166.
16. Z. Szökefalvi-Nagy, Cs. Bagyinka, I. Demeter, K. L. Kovács, and Le Huen Quynh, *Biol. Trace Element Res.* **26,** 93–101 (1990).

A Study on Children's Condition Thalassemia Using Neutron Activation Analysis and Other Techniques

S. KRISHNAN,*[,1] W. C. STURTRIDGE,[1] N. F. OLIVIERI,[2] A. F. COLLINS,[2] R. Y. QURESHI,[1] M. KRISHNAN,[1] N. BELLUZZO,[2] E. BOGDANOVICH,[1] AND S. M. LUI[1]

[1]The Medical Physics Laboratory, The Toronto Hospital, General Division, Room CCRW-G-803, 200 Elizabeth Street, Toronto, Ontario, M5G 2C4, Canada; and [2]Division of Haematology/Oncology, The Hospital for Sick Children, Toronto, Ontario, Canada

ABSTRACT

In this study, 50 thalassemia patients were tested using dual-energy X-ray absorptiometry (DEXA) and in vivo neutron activation analysis (IVNAA) to determine their bone mineral status. Both techniques were suitable for this purpose. Lower age was found to correspond to lower liver iron content and higher bone mineral content in the normal range. Patients undergoing treatment with transfusion had higher bone mineral content. Osteopenic patients had higher hepatic iron content than those with normal bone status. In the case of DEXA, bone mineral content (BMC) divided by height cubed was found to be a better indicator of bone mineral status than the BMD usually given. Liver density as determined by DEXA correlates well with hepatic iron

Index Entries: Bone density; dual-energy X-ray absorptiometry; neutron activation; thalassemia; transfusion.

*Author to whom all correspondence and reprint requests should be addressed.

INTRODUCTION

Two of the most devastating blood diseases in terms of cost of treatment as well as human suffering, are homozygous β thalassemia (HBT) and sickle cell disease (SCD). These are the single most common gene disorders in the world. According to WHO estimates, by the year 2000, about 7% of the world population will be carriers for one of these diseases. At present, over 200,000 children are born annually who will be affected by these diseases. Most of the affected children in the Third World die within the first decade of life. In North America, The Hospital for Sick Children in Toronto manages the largest program for patients with thalassemia.

These two disorders of human hemoglobin cause such anemia that, after the first year of life, monthly red blood cell transfusions are necessary for survival. This, in turn, leads to body iron loading, which itself is fatal and so necessitates removal by daily chelation therapy. Thus, monthly blood transfusions and lifelong, cumbersome, irritating, daily overnight infusions of desferroxamine are essential. The cost of this therapy has restricted its use to only a small portion of these patients in developing countries. The search for an easier oral therapeutic agent is currently under way.

The careful management of these patients under transfusion and chelation therapy has increased their life expectancy to the second decade. This has brought on concerns of other secondary effects when these patients survive many years. One of the concerns is their bone status, since bone formation is intimately connected with blood and bone marrow. Since many of these patients are susceptible to osteopenia and fractures associated with it, we began a study of their bone mineral status. The results are reported in this article.

METHODS

In this study, we followed 50 thalassemia patients with an average age of 24 yr (range 15–39 yr). All but seven of them were under transfusion. A superconducting quantum-interference device (SQUID) susceptometer *(1)* at another site was used to measure hepatic iron stores. We determined the bone mineral status of these patients using both IVNAA *(2)* and DEXA techniques.

IVNAA using thermal neutron activation of ^{48}Ca to ^{49}Ca in the bone is an established direct method of estimating bone mineral status *(2)*. In this work, we have considered this technique as the gold standard. It involves partial body irradiation using plutonium-beryllium sources. Decay of ^{49}Ca to ^{48}Ca is measured using γ counters. The patient's calcium count is a measure of total body calcium and is normalized for height by dividing by the cube of height. The test takes about 1 h to complete, gives total body status, and the radiation dose to the patient is on the order of 3.5 mSv.

DEXA, on the other hand, is a relatively new technique based on X-ray attenuation. A self-contained X-ray source mounted beneath the patient provides alternating pulses at 70 and 140 kVP. An X-ray detector is mounted above the patient. Controlled by a computer, both components move across the patient in a serpentine pattern. The soft tissues that are contained within the area of interest are automatically subtracted, and only the bones are imaged and measured. Soft tissues do not have a substantial effect on the bone mineral density (BMD) calculations, so the results are accurate for most sizes of patients in the anatomical area desired. The BMD is the BMC (in g) divided by the projected area of the bone and is measured in units of g/cm^2. Projected area is used because the scanner measures in only two dimensions. Using the Hologic QDR 1000™ X-ray Bone Densitometer, a test comprising scans of the lumbar spine, hip, and whole body, takes less than three-quarters of an hour to complete, gives site-specific data, and the radiation dose is under 0.1 mSv.

RESULTS AND DISCUSSION

Correlation of CaBI and DEXA Data

Calcium Bone Index (CaBI), a measure of bone mineral status determined by IVNAA, correlates poorly with lumbar spine BMD data.

Calcium Bone Index (CaBI), a measure of bone mineral status determined by IVNAA, correlates poorly with lumbar spine BMD data ($R = 0.31$) and somewhat better with femoral neck BMD data ($R = 0.50$) obtained using the DEXA method. It should be noted that in IVNAA the bone mineral content measured is normalized to body size by dividing it by height raised to the third power, whereas DEXA gives bone mineral density (BMD, g/cm^2).

The correlation with CaBI data is improved if the lumbar spine BMC is divided by height cubed ($R = 0.54$), similar to data from IVNAA (Fig. 1). This indicates that BMD divided by height cubed is a better indicator of bone mineral status than the BMD usually given.

Relationship Between Liver Iron Content and CaBI

For osteopenic patients (CaBI <0.8), the liver iron content is higher than in patients with CaBI >0.8 (Fig. 2).

Effect of Transfusion on Bone Mineral Status

Patients undergoing transfusion have higher CaBI than those not under treatment (Fig. 3).

Effect of Age on Hepatic Iron Store

Patients under 21 yr of age have lower liver iron content than older patients (Fig. 4).

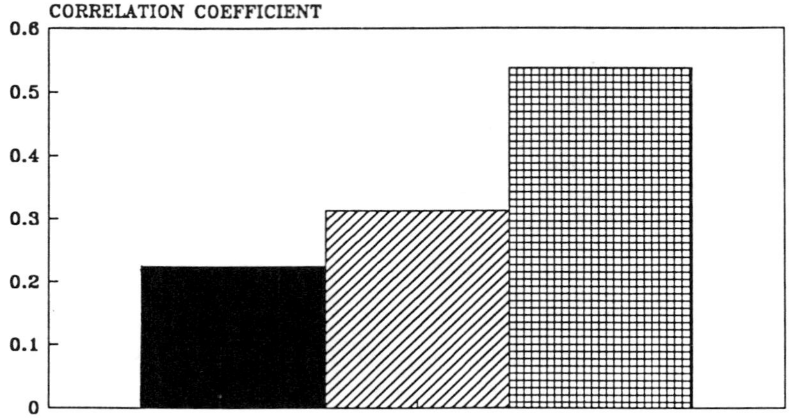

Fig. 1. Correlation of BMC and BMD with CaBI. (■) Lumbar BMC; (▨) Lumbar BMD; (▦) Lumbar BMC/(H-3)

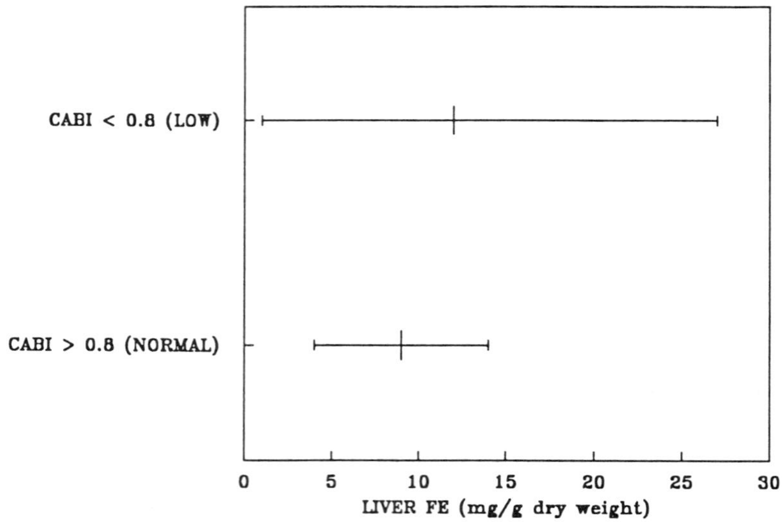

Fig. 2. Relationship between liver Fe load and bone mineral status.

Regional Liver Analysis

A regional analysis was also performed on a selection of the patient whole-body scans obtained (12 low liver iron and 12 high liver iron) as well as on six normal volunteer scans using the DEXA technique in order to determine their utility in estimating hepatic iron stores. The region of interest selected was a rectangle of fixed size, bordering the liver. The measured BMD is an indication of the density of the liver tissue. The liver biopsies for iron were not performed at the same time as the CaBI and DEXA tests and may have been performed up to a few months in

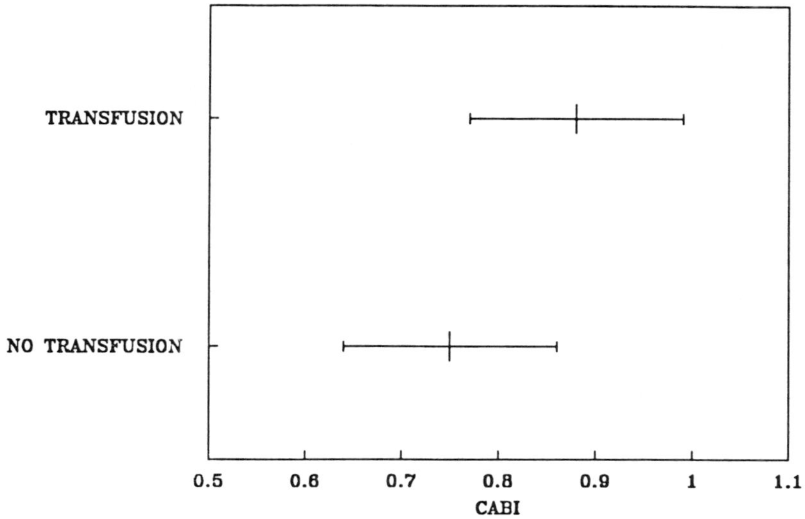

Fig. 3. Effect of transfusion on CaBI.

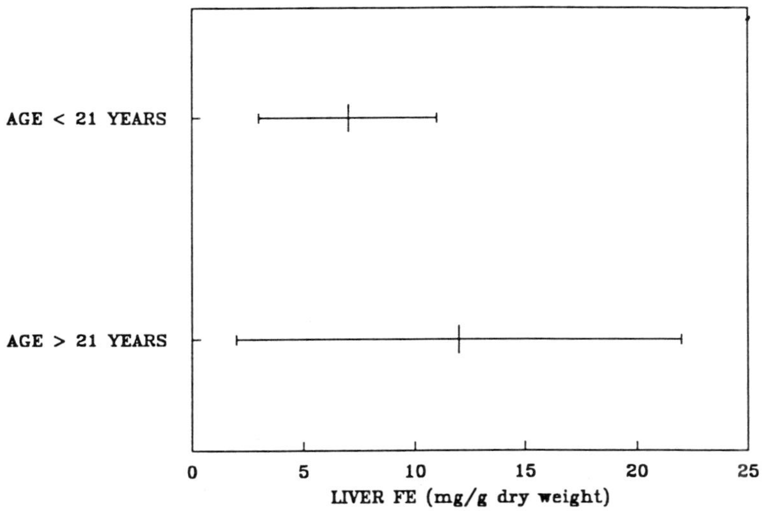

Fig. 4. Effect of age on liver iron load.

advance, depending on the patient. Comparison of results from the biopsies and DEXA scans showed that the latter technique was able to differentiate between the differences in liver density on the basis of BMD and hence liver iron loading (Fig. 5).

Errors in the BMC results for the regional liver analysis using the DEXA technique may be the result of the occurrence of ribs in the region of interest (which vary in BMC from person to person) as well as the variance in liver size. This technique may be useful as a screening technique to detect transfusional iron overload.

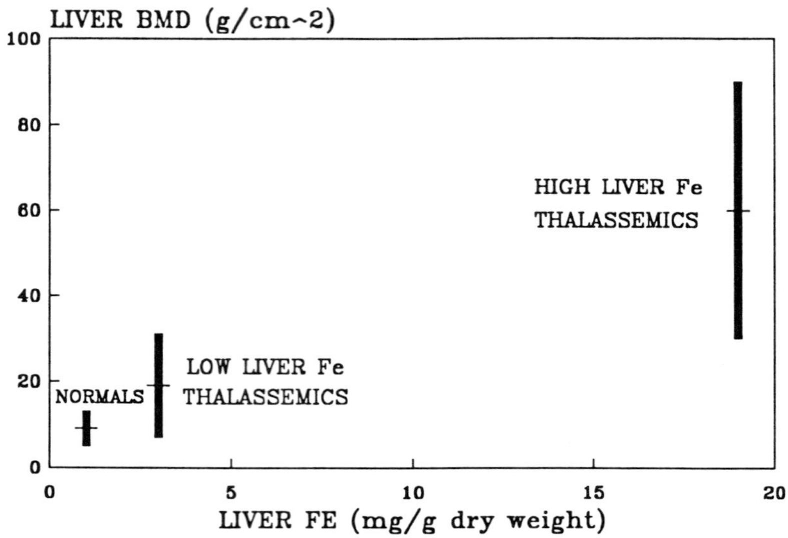

Fig. 5. Comparison of hepatic Fe store evaluation using biopsy and DEXA.

SUMMARY

The results indicate that IVNAA and DEXA are both suitable techniques for determining the bone mineral status of thalassemia patients. DEXA is fast and allows regional analyses. In addition, the radiation dose to the patient is very low (0.1 mSv). Problems may be encountered with patients who have difficulty lying flat for the duration of scanning because of physical deformities of the spine.

IVNAA requires a slightly longer test time and involves exposing the patient to a larger radiation dose (3.5 mSv). CaBI measures the amount of calcium in the body and so is a more direct indicator of the actual bone condition.

Treatment with transfusion corresponds to higher bone mineral content, as does lower age. Lower age also corresponds to lower liver iron content. It therefore appears that patients who are long survivors of thalassemia are at risk of developing osteopenia.

REFERENCES

1. G. M. Brittenham, D. E. Farrell, J. W. Harris, E. S. Feldman, E. H. Danish, W. A. Muir, J. H. Tripp, and E. M. Bellon, *N. Engl. J. Med.* **307(27),** 1671–1675 (1982).
2. S. S. Krishnan, K. G. McNeill, J. R. Mernagh, and J. E. Harrison, *Biol. Trace Element Res.* **26–27,** 415–421 (1990).

Kinetics of Elemental Content Changes of Bone Tissue of Mice During Evolution Under Hypokinetic Stress

T. E. BURKOVSKAYA,[1] M. V. FRONTASYEVA,*[,2] AND S. F. GUNDORINA[2]

[1]*Institute of Medical Biological Problems, Moscow, Russia; and* [2]*Frank Laboratory of Neutron Physics Joint Institute for Nuclear Research, Dubna, Russia*

ABSTRACT

Concentration of 13 elements in bone tissue of mice held in tightly spaced cages for 3 wk, which caused an acute stress reaction, was determined by means of neutron activation analysis. Functionally different bone tissues of mice skeletons—the femur, accomplishing both supporting and dynamic functions; the parietal bone, being practically immovable; and the ectopic bone, newly formed under kidney capsule in the place of syngeneic bone marrow implantation—were analyzed. Similar dynamics of the elemental composition of investigated bones was found: the progressive demineralization owing to the loss of Ca, P, Mg, and Fe is accompanied by the compensatory inclusion of Sr in the bone tissues. In the ectopic bone, it was not as high. During evolution under hypokinetic stress, the microelement concentrations (Zn, Cr, Rb, Ru, Br, Co, Sb) change significantly. Results obtained form the evidence for some system character of osteoporosis at limited mobility.

Index Entries: Bone tissue of mice; elemental content; hypokinetic stress; demineralization; bone evolution.

*Author to whom all correspondence and reprint requests should be addressed.

Biological Trace Element Research Editor: G. N. Schrauzer © 1994 by Humana Press Inc.

INTRODUCTION

One of the experimental models simulating the effects of weightlessness that is widely used by investigators in space biology and medicine is clinostatic hypokinesia caused by the limitation of mobility *(1)*. Experimental hypokinesia causes an acute stress reaction and metabolism violation, which in turn, lead to the suppression of osteogenesis, the increase of resorption, and change of mineral content in the bone tissue *(1,2)*. To study the mechanisms of the mineral metabolism violation, one finds a reasonable comparative study of the peculiarities of mineral elements in functionally different bones of mice skeletons: the femur, accomplishing both supporting and dynamic functions; the parietal bone, being practically immovable, and the ectopic bone, which was newly formed in the process of hypokinesia. Ectopic osteogenesis was reproduced by the method of getherotropic transplantation of syngeneic bone marrow under a kidney capsule *(3,4)*. The investigation of mice characterized by a high level metabolism, including mineral metabolism, allows one to reproduce the model of osteogenesis in the mature organism and in a short time to obtain a definite impact of the hypokinesia *(5)*. The hypokinesia was achieved in mice by placing them tightly in limited-space cages. The use of neutron activation analysis allows simultaneous multielement analysis of samples of small mass with high sensitivity to be carried out. The microelement information of metabolism in these organisms is extremely scarce *(6)*.

METHODS AND EXPERIMENT

One-half of the bone marrow, contained in the femur of the intact donor, was implanted in male mice of the CBA line and hybrids F_1(CBA × C57BL) under the kidney capsule. On the average $(1.3 \pm 0.1)\ 10^7$ cells were introduced. At the place of implantation of the bone marrow from the donor's osteogenic precursor cells, the bone tissue grows in the form of a shell, under which the ectopic hemopoietic center is further formed. On the fifth to sixth day after implantation, i.e., in the primary phase of the bone matrix formation, mice were held in individual tightly spaced cages. Three experiments were carried out with 37 mice recipients after 3 wk of hypokinesia. The 120 animals that were kept in normal cages after implantation served as a control. One month after the surgery, the animals were sacrificed; parietal, femur, and ectopic bones were extracted. In addition to studying the elemental content changes of bone tissue under acute stress, nonsurgically altered animals were exposed to a 3-d (51 mice) and a 7-d (12 mice) hypokinesia. In this case, the parietal and femur bones were investigated. Soft tissues and bone marrow were removed and the bones were dried to a constant weight. To avoid errors connected with the geometry of a sample, bones of 5–10 animals were pressed in tablets, which were measured under the same geometry.

Table 1
Relevant Analytical Data *(8)* for Elements Studied in This Work

Element	Radionuclide	Half-life	E_γ (keV)	Activation cross-sections of stable nuclide (10^{-24} cm^2) σ_o	I_o
Na	^{24}Na	15.02 h	1369	0.13±0.03	0.32±0.005
Mg	^{27}Mg	9.462 min	1014.4	0.035±0.002	0.027±0.002
P	^{28}Al	2.241 min	1778.9	fast neutrons	
Ca	^{49}Ca	8.716 min	3084.4	1.09±0.14	0.9±0.1
Fe	^{59}Fe	44.5 d	1291.6	1.28±0.05	1.4±0.1
Cr	^{51}Cr	27.70 d	310.1	15.9±0.2	8.1±0.5
Co	^{60}Co	5.271 y	1332.4	18.8±1.5	39.7±4.3
Zn	^{65}Zn	243.9 d	1116.0	0.726±0.020	1.42±0.05
Br	^{82}Br	35.30 h	776.5	3.0±0.15	48.0±2.4
Rb	^{86}Rb	18.66 d	1976.6	0.053±0.005	1.74±0.1
Sr	^{85}Sr	64.84 d	514.0	0.35±0.07	6.6±1.1
Ru	^{103}Ru	39.26 d	497.1	1.21±0.07	4.6±0.9
Sb	^{124}Sb	60.20 d	1691.0	4.1±0.1	128±10

INAA was carried out at the IBR-2 reactor using a pneumatic system with a thermal neutron flux density of 1.1×10^{12} n/cm^2/s, and a resonance neutron flux density of 0.23×10^{12} n/cm^2/s and fast neutron flux density of 1.4×10^{12} n/cm^2/s *(7)*. Samples of about 20–30 mg were irradiated for 15 min to determine short-lived isotopes, and for 7–10 d to determine long-lived isotopes. Samples for short irradiation were packaged in polyethylene (PE) bags, which were not removed before measurements. For long irradiations, samples were wrapped in Al foil and transferred to clean PE bags for counting. γ-ray spectra were recorded two times after short irradiation, after 3 and 20 min, and two times after long irradiation, after 4 and 20 d, using counting times depending on the total activity of the sample. Measurements of the induced radioactivity were performed using Ge(Li) detectors with a resolution of 2 keV for the ^{60}Co 1333 keV γ-line, with an efficiency of about 6% compared to a 3 × 3 in. NaI detector for the same line. The radionuclides used and the γ lines selected are listed in Table 1, along with relevant cross-section data *(8)*. γ spectra were processed using the program for activation analysis ACTIV *(9)*.

The quantification was based on the use of the Certified Reference Material *"Animal Bone"* H-5 (International Atomic Enenrgy Agency, Vienna) irradiated along with the bone samples. Data processing was

Table 2
Elemental Content of Bone Tissue of Mice in Control Experimental Animals and During Evolution Under Hypokinetic Stress (g/g)

Element	Control			3-day hypokinesia		7-day hypokinesia		3 week hypokinesia		
	femur	parietal	ectopic	femur	parietal	femur	parietal	femur	parietal	ectopic
Ca 10^{-3}	n=8 234.2±23	n=7 231±12	n=11 223±14	n=10 247.3±10	n=12 233±11	n=4 225±16	n=4 * 202±8.3	n=6 * 171±24.6	n=6 * 143.6±21	n=2 * 175±8.4
P 10^{-3}	n=10 104±4.2	n=7 105.3±9.0	n=9 108±26.6	n=4 99.3±11.2	n=12 101±4	n=6 98.8±16.7	n=4 * 84.7±5	n=6 * 71±6 7	n=6 * 84.4±9.4	n=5 89.8±16.7
Na 10^{-3}	n=8 3.99±0.2	n=7 3.68±0.21	n=10 □ 3.31±0.22	n=10 3.75±0.2	n=12 3.21±0.23	n=4 4.11±0.17	n=4 3.84±0.48	n=6 3.56±0.36	n=6 3.55±0.39	n=3 3.91±0.65
Mg 10^{-3}	n=8 4.82±0.36	n=6 4.48±0.32	n=8 5.18±0.35	n=10 4.9±0.34	n=12 3.85±0.21	n=4 5.03±0.6	n=4 3.95±0.17	n=6 * 3.77±0.31	n=6 * 3.15±0.51	n=5 5.11±0.49
Zn 10^{-5}	n=14 40.8±4	n=18 59.4±12.2	n=7 69.3±18.4	n=10 * 20.9±1.8	n=8 □ 40.6±7.3	n=4 31.9±4.1	n=4 56.1±12	n=7 37.3±6.3	n=4 40.4±16.7	n=4 □ 58.5±3.2
Sr 10^{-5}	n=9 8.4±5.9	n=6 □ 50.7±6.4	n=10 9.6±0.6	n=8 * 20.6±1.7	n=5 □ * 67.6±5.7	n=7 * 27.8±2.0	n=4 * 93.9±2.5	n=5 * 51.2±10	n=4 * 93.1±16.1	n=7 □ * 20.5±0.3

Fe 10^{-5}	n=9 14±2	n=10 □ 54.8±9.8	n=6 □ 48.9±12.7	n=3 12.3±2.6	n=5 □ * 26.1±3.6	n=5 * 5.4±0.9	-	n=5 * 6.9±1.2	n=5 □ * 28.1±8.1	n=3 * 17.3±2.6
Br 10^{-6}	n=8 23.0±2.8	n=7 □ 35.5±6.0	n=5 □ 8.0±2.0	n=4 17.4±3.0	n=5 * 74±10	n=4 23.0±5.1	n=4 37.6±3.7	n=3 28.0±9.5	n=5 * 20.0±4.5	n=3 □ 8.28±3.27
Cr 10^{-6}	n=6 5.0±0.35	n=4 7.0±0.54	-	n=3 □ * 0.62±0.23	n=4 * 3.0±0.5	n=3 5.83±1.7	-	n=3 5.62±0.66	n=5 5.4±0.5	-
Rb 10^{-6}	n=7 7.1±1.13	n=7 6.62±2.0	n=5 □ 27.8±9.0	n=5 * 2.1±0.45	n=4 * 1.45±0.21	n=2 7.5±0.6	-	n=4 6.8±0.83	n=5 7.92±2.41	n=3 □ * 3.59±1.12
Ru 10^{-6}	n=6 5.82±1.76	n=8 4.65±1.08	-	n=4 * 1.63±0.4	n=7 * 1.56±0.28	n=3 2.26±0.08	n=3 4.1±2.13	n=3 2.51±0.41	n=4 2.79±1.47	-
Co 10^{-6}	n=6 0.56±0.2	n=6 0.52±0.13	-	n=7 0.3±0.07	n=5 * 1.07±0.12	n=4 * 0.57±0.07	-	n=4 0.39±0.06	n=5 0.44±0.07	-
Sb 10^{-6}	n=6 0.09±0.03	<0.01	n=5 □ 0.5±0.19	n=3 0.05±0.03	<0.01	n=3 0.07±0.01	<0.01	n=3 0.11±0.02	<0.01	n=3 0.12±0.02

* Significant difference compared to control animals.
□ Significant difference compared to femur.

performed using standard Student's method at a significance level 0.05. Number of measurements for each point varied from 3 to 14.

RESULTS AND DISCUSSION

Experimental results obtained for four groups of experimental animals in control and during evolution under hypokinetic stress are given in Table 2. The high fast neutron flux gives rise to substantial interferences owing to (n,p) and (n,α) reactions in neighboring elements to Na, Mg, P, and Ca. Therefore, results for these elements are to be considered as relative values only. Other reactions that should be checked are ^{56}Fe(n,p)^{56}Mn and ^{54}Fe(n,α)^{51}Cr, although they did not affect the present results appreciably, since the interfering effect for both reactions is < 1% for the given concentration of Fe.

Under normal conditions, content of the main components of hydroxyapatite Ca and P does not differ greatly for parietal, femur, and 30-d ectopic bones of intact mice. Comparison of functionally different bones of the skeleton showed that the content of Sr in immovable parietal bone is six times greater, Fe is four times greater, and Br, Zn, and Cr 1.5 times greater than in supporting femur bones (Fig. 1). In the newly formed bone, a higher content of Mg, Zn, Fe, Rb, and Sb was determined. This is perhaps caused by the high demand for elements used as activators in enzyme systems and catalysts of physical and chemical reactions of the apatite junctions of newly forming bones *(6)*. At the same time, concentration of Ca and P was the same, and Na and Br were even lower in comparison with the skeleton bones.

Limitation of mobility caused acute two-phase stress reaction in mice: a mobilization phase for the first 3–4 d and resistent phase for the following weeks. The mass of ectopic bones of such animals at the place of bone marrow implantation under the kidney capsule was significantly lower. (Fig. 2)

In the evolution under hypokinetic stress, the elemental content of bones was changing. Progressive leaching of elements with large specific weight in the mineral component of bone (Ca, P, Mg, Fe) took place. The content of Ca and P diminished by 16–38% in all bones investigated. However in the ectopic bone, it was not as high. Concentration of Mg diminished in the skeleton bones only. Evidently, the active mineralization of newly formed bone corrected the loss of elements to a certain extent. The loss of Fe was the largest. Its concentration diminished by a factor of 2 in the skeletal bones, and in the ectopic bones, by a factor of 3. The deficit of Fe that plays an important role in the metabolism of osteoblasts and in the red cells production might be the reason for the geterotropic osteogenesis inhibition as well as in the ectopic hemopoietic tissue *(3)* (Fig. 2).

Bone tissue demineralization was accompanied by the significant increase (by a factor of 2–6) in concentration of Sr, close to Ca in its physical and chemical properties, and able to substitute for Ca in the hydrox-

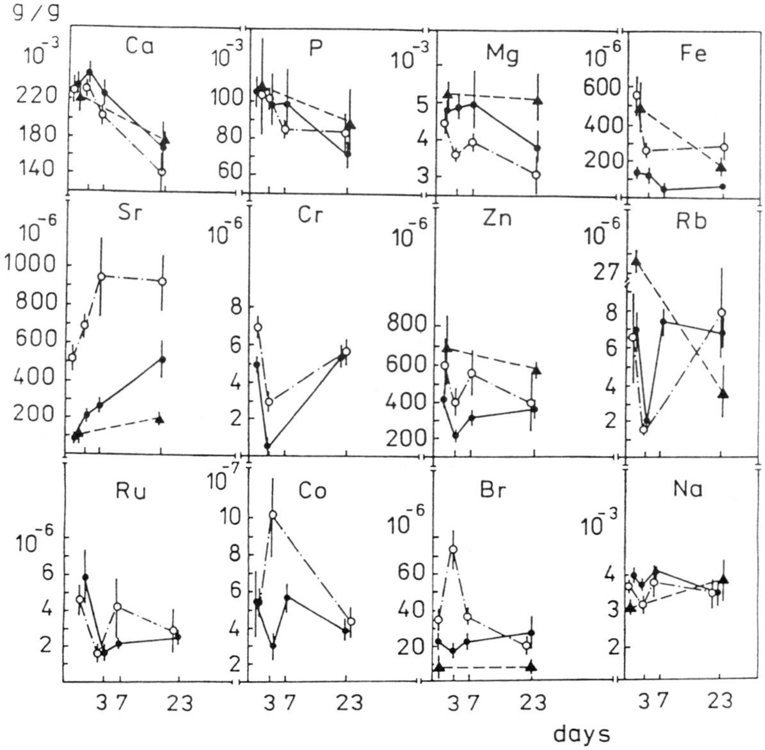

Fig. 1. Elemental content changes of bone tissue of mice during evolution under hypokinetic stress. ——— Femur —— ectopic bone ·—·—·— parietal bone.

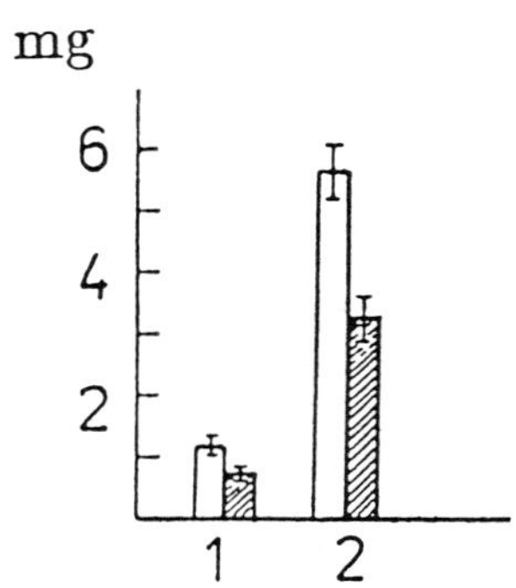

Fig. 2. The effect of hypokinesia on the ectopic bone mass (1) and on the ectopic hemopoietic tissue (2) □ in normal recipients, and ▨ in recipients under hypokinesia.

yapatite crystal lattice. Accumulation of Sr is promoted apparently by the activation of the hypophysial-adrenal system, caused by the hypokinesia.

Kinetics of changes of the majority of the trace elements investigated (Zn, Cr, Rb, Ru, Sb), as well as the macroelement Na, was characterized by their temporary mobilization from the bone tissue in the first stage of the stress followed by restoration of their concentration. Dynamics of changes of Br and Co content in parietal and femur bones went in the opposite direction. These biologically active elements, being the co factors of many ferment systems, participate in the formation of the complex "enzyme substrate," and bind with enzymes by hormones and vitamins, and affect all types of metabolism *(6)*. Apparently, fast mobilization of elements from the bone tissue, a depletion of mineral resources, is caused by the high demand for them in connection with activation of hormone and enzyme systems in the first stage of the hypokinetic stress. Similar dynamics of the elemental composition of functionally different bones of a skeleton is the evidence for some system character of osteoporosis at limited mobility.

ACKNOWLEDGMENTS

The authors express their deep gratitude to Dr. S. V. Vorozhtsova for help with sample preparation and to Dr. V. M. Nazarov for useful discussions.

REFERENCES

1. E. A. Kovalenko, and N. N. Gurovsky, *Hypokinesia*. Moscow, 1982.
2. I. V. Fedorov, *Kosmicheskaya Biologia* **5,** 74–77 1982.
3. T. E. Burkovskaya, and S. V. Vorozhtsova, *Kosmicheskaya Biologia* **5,** 50–54 1980.
4. I. P. Chertkov and O. A. Gurevich, *Stvolovaya krovetvornaya kletka i ee mikrookruzhenie*, Moscow, 1984.
5. Yu. I. Moskalev, *Mineral Metabolism*, Moscow, 1985.
6. A. P. Skoblin and A. M. Belous, *Microelements in Bone Tissue*, Moscow, 1968.
7. V. M. Nazarov, S. S. Pavlov, E. Herrera, and M. V. Frontasyeva, *J. Radioanal. Nucl. Chem., Articles* **167,** 11–21 (1993).
8. E. M. Grintakis and J. I. Kim, J. *Radioanal. Nucl. Chem.* **29,** 175; 1976 *J. Radioanal. Nucl. Chem.* **42** 181; (1978) *Handbook on Nuclear Activation Cross-Section*, IAEA, No. 273, 1986.
9. V. B. Zlokazov, *Comp. Phys. Commun.* **28,** 27 (1982).

Determination of Sodium-to-Calcium Ratio in Mouse Femora by INAA

M. RAKOVIČ,*[,1] J. KUČERA,[2,3] N. PILECKÁ,[1] AND J. POLÍVKOVÁ[1]

[1]*Department of Biophysics, 1st Medical Faculty, Charles University, Salmovská 3, CS-120 00 Prague;* [2]*Czech Ecological Institute, NAA Laboratory, Řež, Czech Republic; and* [3]*Present address: Institute of Nuclear Physics, Academy of Sciences of Czech Republic, CS-250 68 Řež near Prague, Czech Republic*

ABSTRACT

Within the scope of the study of the heteroionic exchange in the bone mineral, the sodium-to-calcium ratio was established in femora of 55 male ICR mice. The purpose of the work was to establish the precision of the method as well as biological variations, since the authors intend to investigate the sodium-to-calcium ratio on pathophysiological models in the future with a special regard to osteoporosis. The resulting mean value was of 0.02515 ± 0.00053. The small standard deviation indicates that the ratio is not subject to considerable individual variations. The accuracy of the method was checked by analyzing the reference bone sample IAEA H-5. The results obtained exerted a good agreement with the certified sodium and calcium concentrations.

Index Entries: Bone; bone mineral; calcified tissues; calcium; femur; mice; neutron activation analysis; sodium.

INTRODUCTION

Nuclear analytical methods can serve as a very useful tool in the study of the calcified tissues *(1)*. Our recent observations demonstrating the presence of two compartments of sodium ions in the bone mineral *(2)* suggested that a role of the heterogenic exchange of these ions in the

*Author to whom all correspondence and reprint requests should be addressed.

 © 1994 by Humana Press Inc.

bone hydroxyapatite should be studied under pathological conditions. Before starting with the work on pathophysiological models and on human (bioptic and or necroptic) material, it was necessary to elucidate the problem of physiological values and their variations.

For this purpose, in the present work, the sodium-to-calcium ratio was determined in a typical compact bone (the femur) of mice. Instrumental neutron activation analysis (INAA) was considered as a suitable approach to this problem.

EXPERIMENTAL

Fifty-five ICR mice of body weights of 20–25 g delivered from Velaz, Prague received the Velaz standard diet and water *ad libitum* for 14 d. After that, they were sacrificed using the ether narcosis, and femora were removed. After mechanical removal of soft tissues from these bones, one femur of each animal was dried at 105°C until the constant weight was achieved, packaged into polyethylene vials, and forwarded for the neutron activation. The femurs (28.6–49.6 mg), three samples of certified reference material (CRM) H-5 Animal Bone *(4)* (45.3–52.2 mg) and synthetic Na and Ca standards containing 200 μg and 30 mg of the respective elements were subsequently activated in the nuclear reactor core at a neutron flux density 6.10^{13} $n/cm^2/s$ for 1 min. A neutron fluence monitor (20 μg of Au) was irradiated simultaneously with each sample and standard to check possible variations of the neutron fluence. After cooling times of 7–19 min, the γ-ray spectra were measured using a HPGe detector (relative efficiency of 23%, resolution FWHM 1.9 keV for the 1332.5 keV γ-rays of ^{60}Co) coupled to a Nuclear Data 683 γ-ray spectrometric system.

RESULTS

Results are presented in Table 1. Calcium contents, sodium contents, and sodium-to-calcium ratios are given in relevant columns for all the 55 samples. Sample No.55 was analyzed three times to show the precision of the method, which is characterized by the standard deviation SD (55). Under this value, there is a mean value of the 55 samples with a relevant standard deviation (SD 1-55) characterizing the variation, including the random errors of the determination as well as the biological variability. High precision and accuracy of the INAA procedure employed are apparent from a comparison of our results with certified values of sodium and calcium in the CRM IAEA H-5, which are also given in Table 1.

DISCUSSION AND CONCLUSIONS

The narrow range of the values of the Na/Ca ratio observed demonstrates that under physiological conditions, the incorporation of sodium

Table 1
Results of Determination of Ca and Na in Mouse Femora by INAA

Sample No.	Ca, %	Na, %	Na/Ca	Sample No.	Ca, %	Na, %	Na/Ca
1	23.56	0.5888	0.02499	31	21.84	0.5523	0.02529
2	22.70	0.5772	0.02543	32	22.51	0.5735	0.02548
3	23.37	0.5782	0.02474	33	22.96	0.5618	0.02447
4	21.81	0.5462	0.02504	34	22.58	0.5606	0.02483
5	21.60	0.5412	0.02506	35	22.69	0.5797	0.02555
6	22.58	0.5802	0.02569	36	21.99	0.5554	0.02525
7	21.48	0.5500	0.02560	37	23.17	0.5667	0.02446
8	23.40	0.5713	0.02441	38	22.66	0.5620	0.02480
9	22.76	0.5690	0.02500	39	21.75	0.5650	0.02597
10	22.48	0.5745	0.02555	40	21.63	0.5445	0.02517
11	23.17	0.5952	0.02569	41	22.21	0.5552	0.02500
12	23.57	0.5887	0.02498	42	23.12	0.5716	0.02472
13	22.45	0.5737	0.02555	43	21.83	0.5582	0.02557
14	21.81	0.5900	0.02705	44	21.92	0.5440	0.02482
15	22.31	0.5641	0.02528	45	22.88	0.5572	0.02435
16	22.72	0.5713	0.02515	46	22.47	0.5689	0.02532
17	23.35	0.5734	0.02456	47	21.84	0.5428	0.02485
18	23.02	0.5593	0.02429	48	22.51	0.5639	0.02505
19	21.96	0.5453	0.02483	49	23.25	0.5800	0.02495
20	22.51	0.5651	0.02510	50	22.15	0.5625	0.02540
21	22.45	0.5552	0.02473	51	21.77	0.5517	0.02534
22	22.42	0.5806	0.02590	52	23.64	0.5940	0.02513
23	22.26	0.5684	0.02553	53	23.12	0.6059	0.02620
24	22.87	0.5597	0.02447	54	23.35	0.5842	0.02502
25	23.12	0.5681	0.02457	55	22.12	0.5646	0.02552
26	22.02	0.5557	0.02523	55	22.20	0.5632	0.02537
27	23.23	0.5722	0.02463	55	22.37	0.5608	0.02507
28	22.22	0.5846	0.02631	MEAN (55)	22.23	0.5628	0.02532
29	23.42	0.5987	0.02556	SD (55)	0.128	0.00192	0.00023
30	22.98	0.5631	0.02450				
IAEA H-5	20.85	0.5122	0.02457				
	20.97	0.5086	0.02425				
	21.31	0.5072	0.02380				
MEAN	21.04	0.5093	0.02421	MEAN(1-55)	22.58	0.5679	0.02515
SD	0.239	0.00258	0.00038	SD (1-55)	0.594	0.01503	0.00053
	Certified (4)						
MEAN	21.20	0.5000	0.02358				
95 % CL	0.806	0.0280	0.03474				

ions into the bone mineral through the heteroionic exchange with hydroxyapatite crystals occurs based on well-established kinetics of the exchange processes among the crystal interior, crystal surface, bound hydration shell of polarizable ions, and diffuse oriented layer + bulk solution *(3)*, and thus, it is subject only to minimum individual variations. It will be of interest to study this ratio under pathological conditions of the bone mineral apposition and resorption processes.

ACKNOWLEDGMENT

The present work was financially supported by Grant No. 0528-3 from the Internal Grant Agency, Ministry of Health of Czech Republic.

REFERENCES

1. M. A. Chaudhri, *Biol. Trace Element Res.* **26–27,** 149 (1990).
2. M. Rakovič, and N. Pilecká, *Phys. Med. Biol.* **37,** 1399 (1992).
3. W. Neuman and N. Neuman, *The Chemical Dynamics of Bone Mineral,* University of Chicago Press, Chicago, 1958.
4. Information Sheet for Reference Material H-5, Animal Bone, IAEA, Vienna, October 1982.

Vanadium Levels in Urine and Cystine Levels in Fingernails and Hair of Exposed and Normal Persons

J. KUČERA,*[1,2] J. LENER,[3] AND J. MŇUKOVÁ[3]

[1]Czech Ecological Institute, NAA Laboratory, CS-250 68 Řež near Prague; [2]Present address: Institute of Nuclear Physics, Academy of Sciences of Czech Republic, CS-250 68 Řež near Prague, Czech Republic, and [3]National Institute of Public Health, Šrobárova 48, CS-100 42 Prague 10, Czech Republic

ABSTRACT

Vanadium was determined by radiochemical neutron activation analysis (RNAA) with proven accuracy in urine of workers occupationally exposed to vanadium-rich dust in a vanadium pentoxide production plant, and values in the range of 3.02–762 ng/mL (median 33.0 ng/mL) were found. In a control group consisting of administrative workers of the plant, urinary vanadium levels were found in the range of 1.05–53.4 ng/mL (median 2.53 ng/mL), whereas in an another control group of occupationally nonexposed persons, these values amounted to 0.066–0.489 ng/mL (median 0.212 ng/mL).

Accuracy of the results was tested by analysis of reference material IAEA A-13 Animal Blood and NIST SRM-1515 Apple Leaves, and very good agreement was found with literature and the NIST certified values, respectively. Unlike urine, no significant differences were found for cystine levels in fingernails and hair of exposed and control persons.

Index Entries: Occupational exposure to vanadium; urinary vanadium; base-line levels, fingernails and hair cystine, exposure test, accuracy of analysis.

*Author to whom all correspondence and reprint requests should be addressed.

Biological Trace Element Research Editor: G. N. Schrauzer © 1994 by Humana Press Inc.

INTRODUCTION

There is increasing interest in the biological functions and metabolic roles of vanadium, because the element was reported in the early 1970s to be an essential trace element for chick and rat *(1,2)*, and speculations appeared later as to whether vanadium could also be essential for humans. Vanadium is released in large quantities into the environment, mainly by burning oil and also from various industrial processes. Well-known occupational exposure sources involve boiler cleaning and production of vanadium pentoxide. Therefore, the assessment of potential effects of vanadium on human health should be considered. Moreover, recent discussions about the essential nature of vanadium seem to suggest that this issue is still an open question *(3,4)*.

In the assessment of health risk arising from environmental, occupational, or accidental exposure to toxic metals, the use of biological monitoring of metal concentrations in human tissues and/or body fluids is steadily increasing *(5,6)*. In studies on occupational exposure to vanadium, the elevated element levels in blood, serum, urine, hair, and nails were discussed as possible bioindicators *(7,8)*. However, the sensitivity of the particular tests is not very well known, and, moreover the reported reference ("normal") values vary widely from 0.24–31 ng/mL for urine and from 0.03–800 ng/mL for blood and/or serum, presumably because of the analytical difficulties involved *(7,9)*. Contradictory findings were also reported on decreased cystine levels in fingernails of workers exposed to vanadium *(10,11)*. Therefore, in this work, RNAA was employed for vanadium determination in urine and also cystine levels were determined in nails and hair from exposed and control persons to investigate the sensitivity of these tests for the assessment of occupational exposure to vanadium, and to elucidate the baseline levels of the analytes, as well.

SAMPLES

Spot urine, fingernail, and hair samples were obtained from 58 workers, only men, aged 23–58 yr (average 41 yr) exposed to vanadium containing dust for 6–398 mo (average 110 mo) in a vanadium pentoxide production plant and from two groups of control persons. In the factory, occupational exposure limits for vanadium in work place air (0.5 and 1.5 mg/m^3 for the average and peak concentrations, respectively) were exceeded at several work places, as was determined by vanadium determination in aerosols by atomic absorption spectrometry. The first control group (C-1) consisted of 14 administrative workers of the factory, 7 men and 7 women, aged 28–58 yr (average 46 yr) who were not supposed to be occupationally exposed, whereas the second control group (C-2) was formed by 21 workers of a research institute, 12 men and 9 women, aged 22–63 yr (average 47 yr), chemists or radiochemists without occupational exposure to vanadium.

The 2-mL urine samples were pipeted with a clean glass pipet from precleaned collection polyethylene (PE) flasks into precleaned 10-mL PE irradiation vials (by leaching with high-purity dilute HCl 1:1 and washing with deionized water) and heat-sealed. The nail and hair samples were analyzed as collected without any special treatment.

Aliquots of reference materials amounting to 150–200 mg in the "as received state" were heat-sealed into PE capsules prepared from PE tubes of 8 mm internal diameter, which were cleaned as described above. Their dry weight was ascertained by drying at 80°C to constant weight on nonanalyzed portions.

METHODS

Radiochemical Neutron Activation Analysis

The urine samples, reference materials, and vanadium standards (a solution with known vanadium concentration heat-sealed in PE vials) were irradiated separately, with gold fluence monitors, in a thermal neutron fluence rate of 5–6 × 10^{13} $n/cm^2/s$ for 1 min. Urine was washed out from the vials with about 2 mL of water, and the total postirradiation procedure, consisting of rapid wet ashing in a mixture of concentrated H_2SO_4 and HNO_3, and then $HClO_4$, followed by addition of dilute HCl and $KMnO_4$ solution, and extraction of vanadium with a 0.2% solution of *N*-benzoyl-*N*-phenylhydroxylamine (N-BPHA) in toluene *(12,13)* was carried out. A 100-μL aliquot was weighed out from the irradiated standards and adjusted to the same volume as vanadium extracts with N-BPHA (5 mL) for γ-ray spectrometry measurements. The measurements of the 1434.4-keV γ-rays of ^{52}V (half-life 3.75 m) in the 5-mL fractions were carried out on a coaxial HPGe detector (Canberra, relative efficiency 36%, resolution FWHM 1.9 keV for the 1332.5 keV γ-rays of ^{60}Co) coupled to a Nuclear Data 683 γ-spectrometer with a PDP 11/73 computer for 7 min followed by 8–10 min after the end of irradiation. The chemical yield of the vanadium separation was occasionally measured during analysis runs using the ^{48}V-radiotracer and, in agreement with our previous experience, was very high and stable, amounting to 98.3 ± 1.5%.

Determination of a Sum of Cysteine and Cystine in Nails and Hair

About 10 mg of the nail or hair samples were finely cut and treated with 3–5 mL of peroxoformic acid in a closed flask at about 200°C for 1–2 h to oxidize cysteine and cystine to stable cysteic acid to prevent cystine losses on hydrolysis by racemization. The oxidizing agent was evaporated using a vacuum evaporator at 40°C, 10 mL of HCl 6 mol/L were added, and the samples were hydrolyzed by boiling under reflux for 24–26 hr. The hydrolase was filtered, the filter washed twice with hot water, and the solution was evaporated using a vacuum evaporator at 60°C. The residue was taken up with 10 mL of a citrate/HCl buffer solution of pH 2.2, and cysteic acid was determined by an automatic amino acid analyzer (AAA-339, Mikrotechna, Prague) using pure cystine as a standard and taurine as an internal standard. The separation yield amounted to 95.2–99.8%, and repeatability expressed as the relative standard deviation was 4.8%.

Peroxo-formic acid was always prepared fresh by mixing of 30% hydrogen peroxide and formic acid 1:9 (v/v), and the mixture was left to react in a closed vessel for 30 min. The agent should not be used after 4 hr from the start of its preparation when kept at 20°C, because it decomposes.

RESULTS AND DISCUSSION

The vanadium contents determined in urine of exposed and control persons are summarized in Table 1. A wide range of urinary vanadium was found in occupationally exposed workers. This data set was log-normally rather than normally distributed, as can be seen on comparing arithmetic and geometric means and median, and as was ascertained by the Shapiro-Wilk test *(14)*. About 70% of the values were below 50 ng/mL, which is the maximum admissible limit for occupationally exposed workers in the Czech Republic. Wide variations of urinary vanadium obviously reflect various levels of vanadium in work place air, which amounted up to about 5 mg/m^3 at several dusty work places, such as crushing, milling, and smelting vanadium-rich slag, a starting material for vanadium pentoxide production. Association of urinary vanadium with length of exposure of particular workers and with vanadium levels in work place air was not examined in this work. This would need more detailed measurements of vanadium in work place air, which is planned in the near future, as well as close examination of how particular workers stick to occupational hygiene regulations (using respirators, and so forth).

In the control group C-1, one very high value of urinary vanadium (53.4 ng/mL) was found, which was not used for evaluation of mean values in this group (cf Table 1), because it turned out that this person frequently visited the polluted work places as a part of his duties. Using the remaining 12 values, more than one order of magnitude lower median value of urinary vanadium was found in this group compared to the group of exposed workers. There were no significant differences (α = 0.05) between urinary vanadium values in men (4.16 ± 2.81 ng/mL) and women (3.35 ± 3.62 ng/mL).

Unlike the exposed and control C-1 groups, much more consistent values of urinary vanadium were found in the control group C-2 with a median value about one and two orders of magnitude lower compared to the controls C-2 and the exposed workers, respectively. This seems to suggest that employing administrative workers of a plant where an occupational health problem is studied as a control group, which is a not so rarely used approach in such studies, cannot be recommended if it is not sure that the administrative workers are completely isolated from expo-

Table 1
Urinary Vanadium in Exposed and Control Persons in ng/mL

Subjects	Range	N[a]	AM[b]	SD[b]	GM[c]	SD[c]	Median
Exposed	3.02-762	58	88.2	126.9	28.3	3.40	33.0
Controls C-1	1.05-53.4	12	3.75	3.12	3.59	2.93	2.53
Controls C-2	0.066-0.489	21	0.223	0.096	0.203	1.61	0.212

[a]Number of results evaluated.
[b]Arithmetic mean, standard deviation.
[c]Geometric mean, standard deviation.

Table 2
Comparison of Urinary Vanadium Values as Determined by Various Analytical Techniques (According to ref. *(9)*)

Analytical method	Range of mean or median values (ng mL^{-1})	N[a]
RNAA with proven accuracy and minimum sample handling	0.212[b]	1
ETA-AAS with pre-separation	0.24 - 0.8	4
Other analytical and sample preparation techniques	2 - 31	7

[a]Number of reported data sets.
[b]This work.

sure sources. Otherwise, this approach can lead to wrong conclusions about "normal" levels of particular analytes. The values determined in the controls C-2 belong to the lowest urinary vanadium values reported, as can be seen in Table 2. Also in this group, no significant differences ($\alpha = 0.05$) were observed between urinary vanadium levels in men (0.233 ± 0.062 ng/mL) and women (0.233 ± 0.134 ng/mL).

The accuracy of the vanadium determination by RNAA was tested by concurrent analyses of biological reference materials (RM) IAEA A-13 Animal Blood and NIST SRM-1515 Apple Leaves. The values found from three replicate analyses of 1.11 ± 0.03 ng/g and 237 ± 15 ng/g (dry wt) compared very well with our previous results of 1.00 ± 0.15 ng/g for the former RM *(8,13)* and with the NIST certified value of 260 ± 30 ng/g *(15)* for the latter RM. Considering that the 2 mL urine samples and 150–200 mg RMs' aliquots were analyzed, both the lowest absolute vanadium

Table 3
Comparison of Blood or Serum Vanadium Values as Determined by Various Analytical Techniques (According to ref. *(9)*)

Analytical method	Range of mean or median values[a] (ng mL^{-1})	N[b]
RNAA with proven accuracy and/or minimum sample handling	0.031 - 0.070	4
NAA with pre-irradiation separation in a clean laboratory Class 100	0.023	1
ETA-AAS with pre-separation	< 0.080 - 0.240[c]	1
NAA with pre-irradiation separation	0.35 - 0.94	2
Other analytical and sample preparation techniques	0.77 - 800	80

[a]Unless otherwise indicated.
[b]Number of reported data sets.
[c]Range of individual results.

amounts in urine and the elevated ones matched those in the RMs aliquots.

Concerning the baseline vanadium levels in urine and blood or serum of occupationally nonexposed persons, a very similar picture can be seen on comparing Tables 2 and 3. The lowest vanadium values in the above body fluids, which are about one order of magnitude lower in serum or blood than in urine, were determined using RNAA with proven accuracy at the ultratrace level and with minimum sample handling. Thus, it can be inferred that vanadium determination in biological materials below the ng/mL level by other analytical techniques than RNAA is prone to be hampered by either contamination problems or the results can presumably be biased owing to unrecognized interferences. This is because the results obtained by RNAA with proven accuracy are in good mutual agreement, and this method should be considered superior for this purpose. Thus, increasing evidence indicates *(8,9,12,13)* that the lowest end of the results for vanadium determination in blood, serum, and urine is probably correct, although the possible influence of exposure to vanadium from polluted environments, namely by oil burning, on baseline vanadium levels in some studies cannot be excluded *(9)* and needs further investigation.

In our previous study *(8)*, up to four orders of magnitude differences were found between blood vanadium levels in the workers exposed to vanadium (range 3.1–217 ng/mL) and control adults (range 0.032–0.095 ng/mL), whereas in this work, urinary vanadium values differing to about the same extent were determined (cf, Table 1). Thus, it seems that

Table 4
Results of Cystine Determination in Fingernails and Hair in mg/g

Subjects	Range	N[a]	AM[b]	SD[b]	GM[c]	SD[c]	Median
Fingernails							
Exposed	67.7-142.5	44	109.5	13.5	108.6	1.14	110.5
Controls C-1	85.5-121.4	11	107.6	10.8	107.1	1.11	110.3
Controls C-2	104.6-143.1	10	118.5	11.9	118.0	1.10	117.6
Hair							
Exposed	122.5-184.3	50	154.9	14.0	154.2	1.10	156.4
Controls C-1	128.7-169.8	11	154.7	16.2	153.9	1.11	156.1

[a,b,c]See Table 1.

vanadium determination in both blood and urine can be used as a test of occupational exposure to vanadium with approximately the same sensitivity. However, since the baseline vanadium levels in nonexposed persons are about one order of magnitude higher in urine than in blood and are therefore easier to determine, measuring urinary vanadium can be recommended as a more suitable exposure test both from the analytical point of view and easier availability of urine compared to blood as concerns sampling. Moreover, it seems that in addition to RNAA, other analytical techniques, such as ETA-AAS, can be used more successfully for this purpose compared to blood analysis (cf Tables 2 and 3).

Results of fingernail and hair cystine determinations in the samples from the exposed and control groups are summarized in Table 4. Unlike the urinary vanadium levels in the exposed persons, the results of cystine determination in both tissues in all groups examined were normally distributed, and no significant differences ($\alpha = 0.05$) were detected between the analyte levels in the exposed and control groups (only a limited number of fingernail samples were analyzed in the group C-2, and no attempt was made to analyze hair samples of this group for technical reasons). A direct proof of accuracy of the cystine determination in fingernails and hair cannot be given, because no reference material for this purpose exists. However, using a method, based on a different principle for the fingernail cystine determination in this work, similar values were found compared to those determined by a colorimetric method *(11)*, which were in the range of 106–162 mg/g with no significant difference between workers exposed to vanadium pentoxide dust and control persons. Thus, results of this study confirm the finding of Kiviluoto et al. *(11)* that the cystine content in fingernails (and presumably also in hair) cannot be used as an exposure test for vanadium pentoxide dust.

ACKNOWLEDGMENTS

The authors want to acknowledge financial support for this work provided by the Grant Agency of Czech Republic (Grant No. 203/93/2326) and Internal Grant Agency of the Ministry of Health of Czech Republic (Grant No. 1633-3).

REFERENCES

1. L. L. Hopkins and H. E. Mohr, in *Newer Trace Elements*, W. Mertz and W. Cornatzer, eds., Marcel Decker, New York, 1971, pp. 195–213.
2. K. Schwartz and D. B. Milne, *Science* **174,** 426–428 (1971).
3. F. H. Nielsen, in *Essential and Toxic Trace Elements in Human Health and Disease*, A. S. Prasad, ed., A. R. Liss, New York, 1988, pp. 288–289.
4. B. R. Nechay, L. B. Nanninga, P. S. E. Nechay, R. L. Post, J. J. Branthan, I. G. Macara, L. F. Kubena, T. D. Philips, and F. H. Nielsen, *Fed. Proc. Fed. Am. Soc. Exp. Biol.* **45,** 123–132 (1986).
5. T. W. Clarkson, L. Friberg, G. F. Nordberg, and P. R. Sager, *Biological Monitoring of Toxic Metals*, Plenum, New York, 1988.
6. H. K. Dillon, and M. H. Ho, eds., *Biological Monitoring of Exposure to Chemicals, Metals*, Wiley, New York, 1991.
7. WHO, Vanadium: Environmental Health Criteria 81 (1988), Geneva.
8. J. Kučera, A. R. Byrne, A. Mravcová, and J. Lener, *Sci. Total Environ.* **15,** 191–205 (1992).
9. E. Sabbioni, J. Kučera, and R. Pietra, *Sci. Total Environ.*, submitted.
10. J. T. Mountain, F. R. Stockell, and H. E. Stokinger, *Arch. Ind. Health* **12,** 494–502 (1955).
11. M. Kiviluoto, L. Pyy, and A. Pakarinen, *Int. Arch. Occup. Environ. Health* **46,** 179–182 (1980).
12. A. R. Byrne and J. Versieck, *Biol. Trace Element Res.* **27,** 529–540 (1990).
13. A. R. Byrne and J. Kučera, *Fresenius J. Anal. Chem.* **340,** 48–52 (1991).
14. S. S. Shapiro and M. B. Wilk, *Biometrika* **52,** 591–611 (1965).
15. US National Institute of Standards and Technology, Certificate of Analysis, Standard Reference Material 1515, Apple Leaves, Gaithersburg, January 22, 1993.

Hair Trace Elements in Kidney Dialysis Patients by INAA

R. E. JERVIS,*[,1] B. T. KUA,[1] AND G. HERCZ[2]

[1]*Department of Chemical Engineering and Applied Chemistry, University of Toronto, and* [2]*Nephrology, Wellesley Hospital, Toronto, Canada*

ABSTRACT

INAA was used to determine selected trace elements—Ca, Al, P, and S—in 104 cleaned scalp hair samples from kidney dialysis patients ($n = 54$) and healthy controls ($n = 50$) in order to explore any differences in these elements that might be related to prolonged dialysis and/or associated medication in comparison with blood serum levels of Al and P measured in the same clinic at the time of hair sampling. After correction for P (and Si) interference in Al content, it was observed that there were no significant differences (at 95% confidence level) in hair Al and Ca, which had been expected, whereas while there were definite increases in P and S. Multivariant factor analysis applied to the same data set, however, showed some multiple correlations among four variables: serum Al, duration of dialysis, medication, and hair Al.

Index Entries: Kidney patients' hair; INAA; blood serum Al, P levels; effects of prolonged dialysis.

INTRODUCTION

During the past decade, there have been differing reports and speculations concerning the biological effects of trace aluminum in humans *(1)*, and its possible role in Alzheimer's, kidney dialysis dementia, and associated bone disorders *(2)*. There are some indications, supported by blood serum monitoring, that Al may accumulate in dialysis patients after prolonged periods of dialysis (>1 yr), possibly from the dialysis fluids or from medication with oral antacid preparations. On the other

*Author to whom all correspondence and reprint requests should be addressed.

hand, the role of Al and any relationship to Ca in bone disorders are not clear, since there is some clinical evidence that dialysis dementia and bone disorders occur in some kidney patients and that a condition of Ca deficiency might increase Al absorption *(3)*.

The main objective of this study was to assess hair trace element levels by instrumental NAA in kidney dialysis patients (and controls) and to assess the significance of hair analysis for monitoring such patients compared to blood analysis. The measurement of Al in blood serum by atomic absorption is still essentially a research procedure, and the test is not routinely available to nephrologists. On the other hand, considerable experience in hair INAA and its precleaning in this laboratory *(4,5)* gave confidence that a definitive study could be done to establish the clinical significance of hair analysis with a typical patient pool.

A secondary objective of the study was to look for significant correlations among hair levels of Al, Ca, P and the corresponding blood concentrations, the severity of kidney disfunction, duration of dialysis, use of medications, antacids, and so forth. The technique of multivariant statistical factor (or cofactor) analysis, such as is available commercially in a STATGRAPHICS software package *(6)* was considered potentially helpful to apply for this purpose.

EXPERIMENTAL METHODS

Hair samples included in this study were collected from side and rear scalp areas of 54 outpatients of the Kidney Dialysis Clinic, Wellesley Hospital, Toronto, and 50 healthy subjects (somewhat lower in mean age and with proportionately more males). Most of the patients had 4 h/d dialysis, 3×/wk, to cleanse their blood. Among them, the duration of dialysis ranged from < 1 yr to as long as 25 yr, with a modal time of 1.7 yr (mean time: 3.2 yr). Extensive clinical data were provided for the patient group, viz., current serum levels of Al and phosphate, length of dialysis, and history of medications given.

All hair samples, about 0.5–1 g mass, were cleaned prior to trace analysis by a modified IAEA procedure *(7)* employing acetone/superpure water/acetone washes, each for 10-min periods, followed by warm air-drying. For INAA analysis, hair masses of 200–300 mg each were placed in plastic irradiation vials. Neutron activation was performed in the SLOWPOKE reactor at a thermal neutron flux of $5 \times 10^{11}/cm^2/s$ for 3–5 min followed by γ-spectrometric determination for 300 s after 50–100 s delay.

A number of analytical quality-control measures were used in this study:

1. Assessment of precision in replicate analyses of a homogenized hair sample;
2. Examination of the effect of the IAEA hair cleaning procedure on Al, P, and Ca concentrations, and
3. Accuracy comparisons with some hair standards.

Table 1
Results for Hair Standards

	NIES-5	Chinese Std.
Al(corr.)	241 (240)*	13.4 (13.3)
P	n.d.	226 (184)
Ca	748 (728)	767 (790)

(–) indicates published reference values.
*19% correction for high Si: 1.20% (cf. 1.25%).

The results of replicate analyses of a single homogenized hair sample (from a control subject) showed that, under the irradiation and counting conditions used in most of this study, the precision of the element determination (measured repeatedly) was: Al (after correction for P contribution) ±3.8%; Ca: ±2.5%; P: ±6.7%; and S: ±2%. The counting error in Al was typically much lower, and much of this error resulted from an interference correction procedure using Cd-covered capsules, described below.

Two typical, large hair samples were analyzed before and after successive washings by the IAEA hair cleaning procedure to ascertain the effect, if any, of cleaning on hair trace element content. The first washing removed 6–8% of Al from unwashed hair. Other elements, such as Ca and Zn, were not appreciably different, and subsequent washings had a negligible effect.

Comparison of results obtained in this study for a Chinese hair standard (Shanghai Inst. Nuclear Studies, 1988) *(8)* and the Japanese NIES-5 hair standard (Table 1), with published results from the labs providing the standards, and some literature values *(9)*, indicated reasonable agreement ranging from 5% for Al and Ca down to 1% for S and Si; P differed by 18% in the Chinese hair. Corrections for epithermal neutron interference in Al from P amounted to 20% in the Chinese standard and an 18% correction in Al for NIES-5 from Si interference, since NIES-5 hair had a very elevated Si level of 1.2%.

Correction for P Interference on Al Determination in Hair

When a sample matrix contains appreciable phosphorous content, the epithermal and fast neutron reaction $^{31}P(n,\alpha)^{28}Al$ can contribute to ^{28}Al activation some "apparent Al" content that may be attributed incorrectly to natural Al content. However, corrections to the total observed ^{28}Al activity to account for the epithermal activation interference can be made by reference to pure P irradiated bare in the same reactor spectrum and also when covered with 0.8 mm cadmium. After irradiating hair, Al and P "bare," and "Cd-shielded," the correction factors for fast neutron interference were calculated as follows. The total (observed) Al-28 activity is given by:

$A_{tot} = A_{Al} + A_P$ for samples irradiated "bare" and by:

$A'_{tot} = A'_{Al} + A'_P$ for Cd-covered samples, where (′) refers to Al-28 activity induced in Cd-covered Al and P.

Relating the latter Al-28 activities to specific activities of Al and P measured under the same conditions in the same irradiation position of the reactor, say, S_{Al}, S_P, S'_{Al}, and S'_P, it can be shown that the mass of Al in a given hair sample, corrected for any P fast reaction interference, is given by

$$m_{Al}=A'_{tot}S_P - A_{tot}S'_P/(S'_{Al}S_P - S_{Al}S'_P)$$

The precision obtainable in the derived result for m_{Al} is a function mainly of the counting errors involved in evaluating the specific activities, S_i, and A_{tot} and the corrected Al values had errors of from 5 to 20% among the hair samples analyzed. The magnitudes of the corrections for "apparent Al" in the hair samples attributable to P interference were negligible in some hair, (1–3%), but were as high as 29% in others (Table 2); on the average, corrections were 5.2%.

Fast neutron interference from the reaction: $^{28}Si(n,p)^{28}Al$ was also examined and was found to be negligible, except in the case of the Japanese hair standard, NIES-5, which apparently, has a much increased Si content (of about 1.2%) compared to the Canadian patient and control hair samples. For NIES-5, a correction of 19% was required for the Al content, which was 241 ppm (corrected) compared to the certified reference value of 240 ppm *(9)*. These corrections for epithermal and fast neutron interferences in INAA of Al applied to the in-reflector irradiation positions of SLOWPOKE-2, where about 15% fast flux is found, are typical of light water-moderated pool-type reactors.

RESULTS AND DISCUSSION

A comparison of means, medians, and ranges of hair concentrations for Al (corrected), Ca, P, and S for the kidney dialysis patients and healthy controls is listed in Table 3. The sulfur concentrations are listed mainly as a consistency check of the INAA procedures, because sulfur itself is a major component of the keratin material in the hair matrix and is normally found at 4–5% in human scalp hair.

It can be observed from Table 3 that there were statistically significant differences in Ca and P in patients' hair: Ca was lower, and P was higher. However, only a slight increase was observed in hair Al, whereas a much larger effect had been expected since serum Al levels were elevated in the same patients (825 nmol/L average) compared to levels usually observed in controls (approx 200 nmol/L). In the case of a subgroup of 10 patients on prolonged dialysis, their hair Al was consistently > 15 ppm, and their serum levels had a mean of 1000 nmol/L; their hair P was also elevated, above 250 ppm.

Table 2
Extent of P Interference in Hair Al by INAA

Sample	Al(total)	Al(corrected)	% Al correction
K1	2.9 (ppm)	2.3 (ppm)	21
K2	3.4	3.1	9
K3	5.8	5.3	8.5
K4	8.5	8.0	6
K5	6.6	6.3	4.5
K6	26.7	26.1	2
K7	5.0	4.5	10
K8	12.0	11.6	3
K9	4.3	3.7	14
.	.	.	.
.	.	.	.
K54	8.1	6.6	18
		(median: 7.7 ppm) (cf. 4.0 controls)	(mean: 5.2%)

Table 3
Elemental Concentrations in Scalp Hair of Normal Controls and Dialysis Patients

Element		Normal Controls (n = 50)	Dialysis Patients (n = 54)
Aluminium (μg/g)	Mean	9.4	9.7
	Median	7.5	6.6
	Range	2.4 - 27.9	1.6 - 28.7
Phosphorus (μg/g)	Mean	228	385
	Median	199	289
	Range	101 - 486	63 - 2155
Calcium (μg/g)	Mean	1149	988
	Median	881	592
	Range	141 - 4464	104 - 5040
Sulfur (%)	Mean	4.6	5.1
	Median	4.5	5.1
	Range	3.9 - 5.5	3.7 - 6.6

Concentration distribution diagrams (Figs. 1–3) illustrate the slight differences in hair Al and Ca in the patients, compared to controls, but a considerable increase in phosphorus. Attempts to correlate these hair levels with other patient clinical information, such as blood serum Al, serum phosphate, duration of dialysis, history of medication, age, or sex, were made, and mostly weak correlation coefficients were obtained. Significant correlations were evident only for: (1) serum Al and duration of dialysis, as well as use of medication; and (2) serum phosphate and age.

It was suspected that there might be interrelationships among these multiple variables involved that could be reflected also in hair trace element concentrations. To assess this, multivariant statistical factor analy-

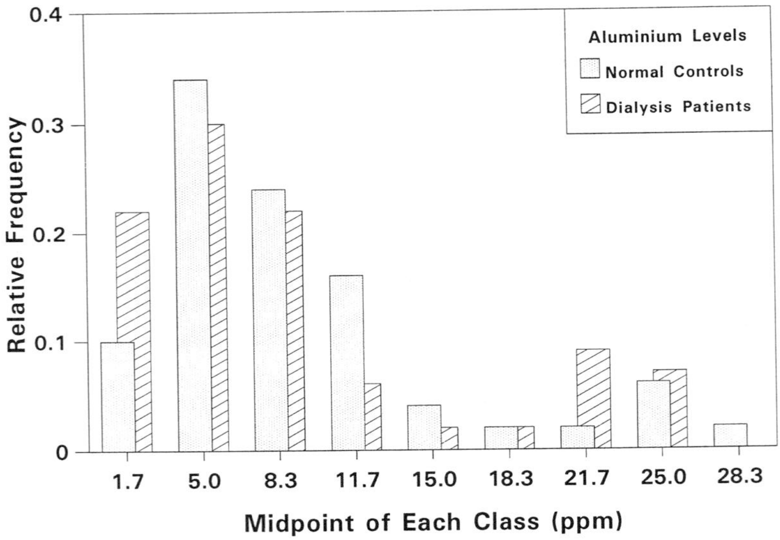

Fig. 1. Aluminum in hair of dialysis patients and controls.

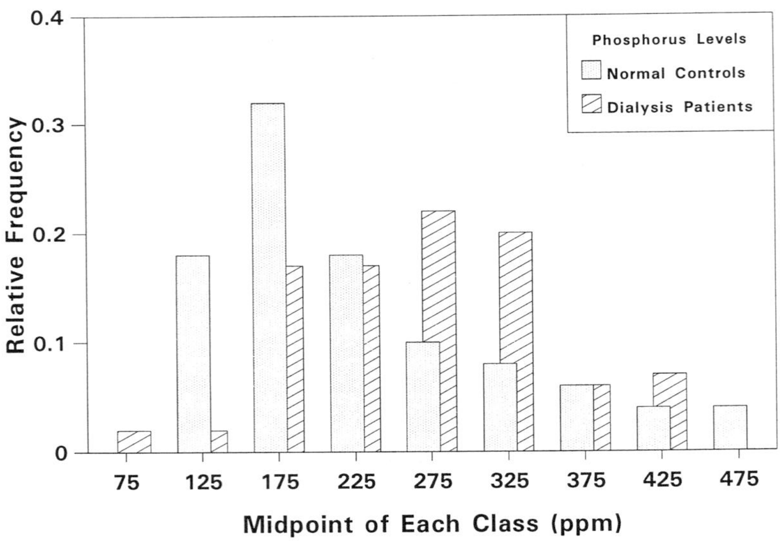

Fig. 2. Phosphorus in hair of dialysis patients and controls.

sis was performed on the patients' data set using the STATGRAPHICS software package *(6)*. Some of the input variables could not strictly be quantified, such as the use of medications, e.g., in the form of aluminum hydroxide gel, and sex, and these were entered as +1 or –1 values only for two categories of information: male/female and use (or not) of medication. The results of the multivariant cofactor correlations involving the whole-patient data set (n = 54) are summarized in Table 4. It can be seen that four groups of multicorrelated variables are revealed having significant factor loadings (which were somewhat magnified through the

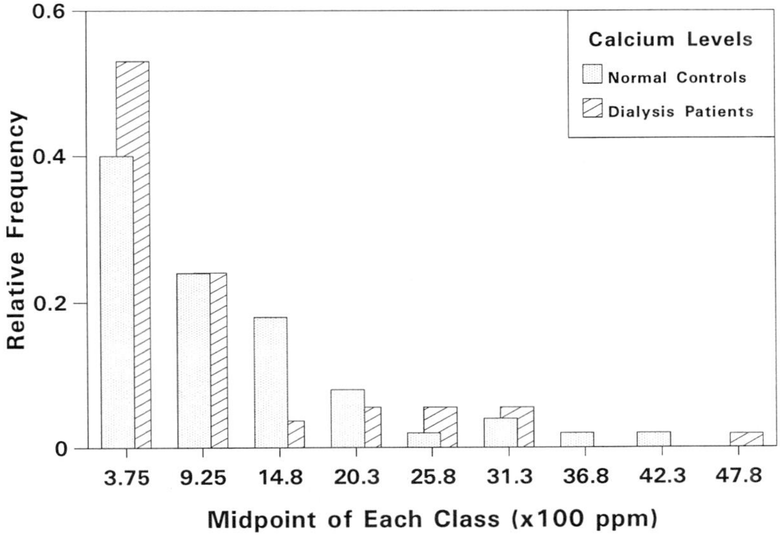

Fig. 3. Calcium in hair of dialysis patients and controls.

Table 4
Varimax Rotated Factors[a]

Variables	Factor I	Factor II	Factor III	Factor IV
Serum Aluminium	0.84			
Duration Dialysis	0.79			
Medication	0.55			
Hair Aluminium	0.44	0.51		
Sex		0.80		
Hair Calcium		0.71		
Hair Phosphorus			0.82	
Hair Sulphur			-0.81	
Serum Phosphate				-0.81
Age Group		0.31		0.69

[a]All factor loadings < 0.3 are not shown.

"Varimax rotation" option in STATGRAPHICS). In such factor analysis, variables have either positive or negative factor loadings, and values of < 0.3 are usually neglected.

The variables cocorrelated in factor I were serum Al (0.8), duration of dialysis (0.8), use of medication (0.55), and hair Al (0.4), the latter being statistically quite significant. This is of interest because only slight increases in hair Al were evident in the patients although this had been an expected outcome of the study. Factor II grouped the variables—sex (–0.8), hair Ca (0.7)—to highlight some otherwise obscure relationships between hair levels and other clinical information.

These results indicate that INAA can be successfully applied in a clinical setting to determine trace elements in scalp hair of patients. However, the actual clinical or diagnostic value of such hair analysis in nephrology or in associated osteoporosis and dialysis dementia is not clear at present.

REFERENCES

1. A. C. Alfrey, A. Hegg, and P. Craswell, *Am. J. Clin. Nutr.* **33,** 1509–1516 (1980).
2. A. C. Alfrey, *Kidney Int.* **29,** S8–S11 (1986).
3. The Sunday Sun "Calcium Counteracts Aluminum," Toronto, July 15, 1990.
4. A. C. Brown and R. G. Crounse, eds, *Hair, Trace Elements, And Human Illness*, Praeger, New York, 1980.
5. A. Chatt and S. A. Katz, *Hair Analysis: Applications in the Biomedical and Environmental Sciences*, VCH Publishers, New York, 1988.
6. **STATGRAPHICS** Software Package, STSC, Inc., Rockville, MD.
7. I.A.E.A. *Activation Analysis of Hair as an Indicator of Contamination of Man by Environmental Trace Element Pollutants*, RL/41 H, 1977.
8. Y. Cheng, Shanghai Inst. Nuclear Studies, personal communication, 1988.
9. K. Okamoto, *Anal. Chem.* **332,** (1988) 524–527.

Determination of Platinum by Neutron Activation Analysis in Nerve Tissue from Rats Treated with Cisplatin

BERND RIETZ,*,[1] KAJ HEYDORN,[1] AND ANDERS KRARUP-HANSEN[2]

[1]Risø National Laboratory, Isotope Division, DK-4000 Roskilde, Denmark; and [2]Department of Oncology, The Finsen Institute, University Hospital of Copenhagen, DK-2100 Copenhagen Ø, Denmark

ABSTRACT

The analytical method used for the determination of traces of platinum and gold in different tissues of Wistar rats is based on neutron activation analysis with radiochemical separation of gold. This separation is performed by electrolytic deposition of gold on a niobium cathode, which ensures the highest radiochemical purity without any spectral interference from calcium or other major elements. Corrections for the nuclear interference from double neutron capture, caused by the gold content of the samples, were found to be insignificant at the levels reported.

The following neural tissues were analyzed for their content of platinum: the dorsal root ganglions as well as the dorsal and ventral part of the spinal cord. The highest level of platinum was found in dorsal root ganglions.

Index Entries: Platinum; gold; neural tissues; neutron activation analysis; radiochemical separation; electrolytic deposition; cisplatin; rats.

*Author to whom all correspondence and reprint requests should be addressed.

INTRODUCTION

Cisplatin—*cis*-diamminedichloroplatinum—is a widely used cytostatic agent. The mechanism of cisplatin antitumor effect is still not completely clear, but it seems most likely that the agent makes a direct binding to the DNA of the tumor cells. Since the introduction of cisplatin in the treatment of testicular cancer in the mid-1970s, the probability of 5 yr of survival has increased by a factor of two to almost 95% *(1)*. The price has been a significant number of chronic side effects, in particular nephro-, oto-, and neurotoxicity. To date, its major toxic and dose-limiting effect is peripheral neuropathy. New cytostatics are on the way, but none of these appear to be able to provide a substitute for cisplatin.

Neurotoxicity is observed mainly in the form of a sensory neuropathy with, e.g., reduced feeling and lack of sensing of vibration stimuli in the hands and feet. A number of studies have attempted to describe this mechanism. Most probably, it consists of a central effect with damage to the nerve cells of the dorsal horns of the spinal cord or to those of the dorsal root ganglions *(2)*, which are located close to the spinal cord. A concentration of platinum is thought to be a probable cause of this effect. Concentration measurements of platinum can possibly support this proposition, even though analyses have not revealed higher platinum concentrations in the central nerve tissue than found in other localities. Since the spinal cord is enclosed by the blood–brain barrier, the concentration of platinum is expected to be lower.

The presence of platinum *per se* even in minimal quantities might alone explain the neurotoxicity owing to a potential vulnerability of the nerve cells. Relative differences in concentration may therefore be of importance.

The establishment of an animal research model combined with platinum measurements, with a view to describing the effect of cisplatin on nerve tissue may simulate the human pathophysiologic condition. Furthermore, potential neuro-repairing agents could be tested before being administered to patients. It is well known that cells of a dorsal root ganglion may die after peripheral nerve injury. Therefore, the spinal cord was investigated for physiologic disturbances, at the level of the distal lumbar segments, which supply the dorsal root ganglion LV, which in turn supplies specific peripheral nerves—in this case the sciatic nerve. Furthermore, the sural nerve, a branch of the sciatic nerve, has been studied *(3)*.

The ^{199}Au daughter of ^{199}Pt formed by the reaction:

$$^{198}\text{Pt}\ (n, \gamma)\ ^{199}\text{Pt} \xrightarrow{\beta^-} {}^{199}\text{Au} \quad (1)$$

is a favorable indicator for the determination of platinum in biological materials, especially when radiochemical separation is applied.

Haverland and Wiebe *(4)* applied the same type of reaction for the determination of platinum at mg/kg levels in tissues of rats treated with cisplatin using INAA. However, information concerning the kind of tissues and the Pt concentration found was not reported.

Zeisler et al. *(5)* studied the time-dependent retention of Pt in blood, liver, kidney, and cells of tumor-bearing rats after administration of platinum coordination complexes used in cancer chemotherapy. They used INAA, and the detection limits for platinum were claimed to be at nanogram levels.

To our knowledge, radiochemical neutron activation analysis has thus far not been used for the identification of the major sites of platinum accumulation in neural tissues of rats following treatment with cisplatin. Analyzing BCR Reference Materials for platinum by using radiochemical neutron activation analysis with radiochemical separation of gold, detection limits of about 1 ng/g could be obtained *(6)*. Therefore, we used the same technique as mentioned in *(6)* for the determination of platinum in neural tissues of rats.

EXPERIMENTAL

Samples

Wistar adult rats weighing about 200 g housed in plastic cages were fed with commercial rat cubes and tap water ad libitum. Intraperitoneal injection of cisplatin (Platinol® 0.5 mg/mL) at a dosage of 2 mg/kg was given once a week in some rats for 7 wk and in other rats for 12 wk. The animals were anesthetized with Halothan^R for the terminal perfusion fixation. All nerve tissue was fixed *in situ* by perfusion through the heart. The vascular bed was rinsed for 2 min with Ringer's solution containing heparin (5000 U/L) and procaine (100 mg/L). For a further 10 min, the perfusion was maintained with 2.5% glutaraldehyde in Ringer's solution to a total amount of 1.0 L, pH 7.4 at 20°C *(3)*. Within 2 h the spinal cord and the dorsal root ganglions were dissected. A slice of the relevant nerve tissue was cut, placed in a preweighed polyethylene vial, and the final wet weight was determined. Altogether, 0.01–0.08 g of neural tissues, 0.3 g kidney tissue, and 0.8 g fresh heart blood were weighed (uncertainty of weighing: ±0.5 mg). The vials containing the materials were then frozen at –80°C, and served as irradiation containers for later neutron activation analysis. The main part of the nerve tissue was processed for histologic and electron microscopic examination. These investigations have yet to be completed.

Irradiation

The neural tissues, kidney, and blood of the Wistar rats were irradiated individually in the same weighed half-dram polyethylene vials previously used for sample storage. Two half-dram polyvials, containing 0.02 μg of Au and 8 μg of Pt in 1 cm^3 of 2 mol/L HNO_3 (the Au and Pt standard, respectively) were placed in two-dram polyethylene vials. All vials were heat-sealed, placed in separate aluminum irradiation containers, and irradiated together for 6 h at 4.5×10^{17} $n/m^2/s^4$ in the rotating activation facility of the DR 3 heavy-water moderated and cooled reactor operating at 10 MW.

After decaying for about 45 h, the half-dram polyvials, containing the Au and Pt standards, were removed from the two-dram polyethylene vials for counting. The samples were removed from the polyvials prior to radiochemical separation.

Radiochemical Separation

Reagents: hydrochloric acid (HCl), 37%, p.a., nitric acid (HNO_3), 65%, p.a., perchloric acid ($HClO_4$), 60%, p.a.

Au-carrier: 0.208 g of gold wire (Advent, GB) 99.99%, dissolved in 10 cm^3 aqua regia.

Platinum comparator: 0.084 g of platinum wire, pure (Danish-Dutch Noble Metal Co. Copenhagen), dissolved in 100 cm^3 of aqua regia and diluted 1:100 with 2 mol/L HNO_3.

Gold comparator: 100 μL of Au carrier diluted to 100 cm^3 with 2 mol/L HNO_3 in a volumetric glass flask. After shaking, 100 μL of this solution was diluted to 100 cm^3 with 2 mol/L HNO_3. This comparator solution was stored in darkness and used within 1 wk.

Digestion and Separation

After irradiation, complete decomposition of the irradiated materials is achieved by digestion with a mixture of hydrochloric, nitric, and perchloric acids under strict temperature control in a modified Bethge digestion apparatus for 3 h *(6)*. Before starting the digestion procedure, an inactive Au-carrier solution with approx 5 mg of Au is added to the sample.

In our previous investigation *(6)*, we used sample sizes of about 2–3 g of material. In the present experiments, amounts of 0.01–0.8 g of biological materials were used. Consequently, all digestions of rat tissues were performed by using a modified digestion apparatus, reduced by a factor of about 3 in comparison with the digestion equipment described previously. However, because the digestion procedure has not been changed, we refer to *(6)*.

After digestion, gold is separated from other radioactive species in the irradiated sample by electrolysis of the sample solution (1 mol/L hydrochloric acid).

Also, the electrolysis procedure was performed as described *(6)*. The only change was the use of a rotating niobium cathode (0.5×3 cm^2), driven by an electric motor.

Counting

All samples and standards were counted at a distance of 1.5 cm from a 60 cm^3 γ-X Ge-detector, using 8192 channels at a gain of 0.2 keV/channel. The detector has a resolution of 1.62 keV at 1333 keV and a relative efficiency of 10%.

The Au standard was counted for 24 h during the radiochemical separation of the sample, and subsequently, the separated sample was counted for 24 h. Finally, the Pt standard was counted for 1 h.

Radiochemical Yield

The yield of the separation was determined by reirradiation of the separated sample together with a reference containing the same amount of gold as was added to the sample. Irradiation for 30 s takes place in the pneumatic tube of the DR 3 reactor with a fluence rate of 2–3 × 10^{17} $n/m^2/s$. Sample and reference were counted for 5 min at a distance of 17 cm from the detector.

RESULTS AND DISCUSSION

An adequate sensitivity for the determination of platinum requires a long irradiation time at a high neutron flux density. However, the radiation decomposition of biological materials limits the duration of the irradiation. The limit of detection for platinum is determined by the formation of ^{198}Au, from which the ^{199}Au indicator of Pt cannot be separated. Therefore, an irradiation time of 6 h at approx 4 × 10^{17} $n/m^2/s$ in the carbon-dioxide cooled, rotating facility of the Danish DR 3 reactor was chosen as a compromise. These conditions give a modest radiation decomposition and an *a priori* detection limit for platinum of LD < 0.01 µg for quantities of gold up to <0.17 µg Au *(7)*.

The electrolytic deposition of gold on a niobium cathode over a period of 5 h ensured the highest radiochemical purity, with a complete elimination of the 159-keV γ-ray interference from ^{47}Sc, at a potential of 800 mV between a graphite anode and the niobium cathode. In a separate experiment, the probable interference from calcium was investigated by addition of a carrier-free ^{47}Sc to nonirradiated BCR Reference Material No. 185 bovine liver. A decontamination factor of 1.58 × 10^7 was found at a chemical yield of 74%, which corresponds to a separation factor of 0.85 × 10^{-7}. Therefore, interference from calcium could be disregarded *(6)*.

When complete separation is achieved, the only significant interference comes from the second-order reaction of the naturally occurring gold in the sample matrix (double neutron capture):

$$^{197}\text{Au}\ (n, \gamma)\ ^{198}\text{Au}\ (n, \gamma)\ ^{199}\text{Au} \quad (2)$$

Correction for this interference is based on the counting of a gold comparator for 24 h, so that the precision of the Pt determination for the tissue sample is not significantly affected. Increased sensitivity can be obtained by longer irradiation in higher neutron flux, but this leads to a greater Au correction.

Four different biological materials (spinal cord, dorsal root ganglions, kidney, and heart/blood) were analyzed for their content of platinum by the method described. The results for the determinations of platinum and gold are presented in Tables 1 and 2. The samples of the kidney and the blood showed the highest level of platinum, followed by the dorsal root ganglions, the dorsal part of the spinal cord, and finally the ventral part of the spinal cord (range: 0.06–8.90 mg/kg). There seems

Table 1
Platinum and Gold Found in Samples of Neural Tissues, Kidney, and Heart Blood from Four Rats

Rat No.	Material	Sample size mg*	Apparent Pt ng ± SD	Observed Au ng ± SD	Pt-correction ng ± SD	Chemical yield %	Actual Pt ng ± SD
1	Spinal cord ventral part	29	3.4±0.2	0.0391±0.0001	0.777±0.006	93±3	2.8±0.3
	Dorsal root ganglions	12	16.8±0.2	0.0557±0.0004	0.227±0.003	95±3	17.4±0.6
2	Spinal cord ventral part Lumbar V	33	8.7±0.3	0.3095±0.0005	4.429±0.041	95±3	4.5±0.3
	Spinal cord dorsal part Lumbar V	33	5.9±0.2	0.1112±0.0003	0.804±0.005	96±3	5.3±0.3
	Kidney	321	1933.3±5.4	0.384 ±0.009	0.614±0.020	68±2	2846.4±99.9
	Dorsal root ganglions Lumbar V	42	20.2±0.3	0.0537±0.0004	0.318±0.006	88±3	22.5±0.9
	Blood	804	534.4±1.6	0.4044±0.0008	2.022±0.019	87±3	612.0±21.5
3	Spinal cord ventral part Lumbar V	79	9.3±0.3	0.507 ±0.001	2.849±0.026	83±3	7.8±0.4
	Spinal cord dorsal part Lumbar V	32	5.1±0.1	0.1030±0.0006	0.499±0.006	81±3	5.7±0.3
	Kidney	386	2487.5±5.7	0.675 ±0.002	3.745±0.042	75±3	3333.9±117.0
	Dorsal root ganglions Lumbar V	25	19.9±0.2	0.0601±0.0002	0.644±0.005	82±3	23.5±0.9
	Blood	832	744.5±1.9	0.0891±0.0005	0.569±0.006	71±2	1044.9±36.7
4	Spinal cord ventral part Lumbar V	45	2.5±0.2	0.0369±0.0004	0.177±0.003	89±3	2.7±0.2
	Spinal cord dorsal part Lumbar V	45	9.7±0.3	1.406 ±0.002	6.355±0.086	95±3	3.6±0.3
	Dorsal root ganglions Lumbar V	33	8.9±0.2	0.322 ±0.001	1.667±0.017	64±2	11.3±0.5

* Fresh sample weight.

Table 2
Platinum Content in mg/kg Fresh Weight in Four Rats Treated with Cisplatin up to an Accumulated Dose of 24 mg/kg rat

Rat No.	Spinal cord		Dorsal root ganglions	Kidney	Blood
	ventral	dorsal			
1	0.098±0.009*		1.448±0.081*		
2	0.136±0.010	0.161±0.009	0.537±0.021	8.867±0.311	0.761±0.027
3	0.098±0.005	0.179±0.009	0.941±0.039	8.637±0.303	1.256±0.044
4	0.059±0.005	0.079±0.007	0.342±0.017		

*Accumulated dose = 14 mg cisplatin/kg rat.

to be no effect of the dosage of cisplatin administered and the platinum concentrations found. It can be clearly seen from Table 1 that the content of gold in the sample is very variable, but the corrections for the Au content are insignificant. To our knowledge no biological reference material has been certified for platinum. Therefore, the reliability of our method could not be documented yet. The uncertainty given in Table 1 represents 1 SD, and comprises counting statistics and flux variations between the sample and the comparators. The uncertainty given in Table 2 includes also the uncertainty of weighing.

In spite of the small number of rats analyzed it is clear that:

1. Since the kidney is the main route of elimination of cisplatin, it is not surprising that the kidney concentration is much higher than the concentrations in the CNS;
2. In all samples, the concentration in the dorsal ganglions is significantly higher than in the spinal cord; and
3. In all cases, the concentration in the dorsal part of the spinal cord is higher than in the ventral part.

The consistency of these observations lend support to the reliability of the experimental results.

The results of neutron activation analysis seem to be within a reasonable range compared to the results, described by Gregg et al., *(8)*. In *(8)*, major sites of platinum accumulation within human neural tissues were identified after treatment with cisplatin using atomic absorption spectrophotometry. The limit of detection of the method was 0.02 μg Pt/g. Whether or not the presence of platinum compounds has any pathophysiologic importance is at the present time uncertain. Until a number of animal experiments are performed, we cannot yet on the basis of the current data draw any firm conclusion.

SUMMARY

The radiochemical neutron activation analysis method described permits the simultaneous determination of Au and Pt in biological materials with high accuracy and reasonable sensitivity. The analytical technique used in this investigation has provided quantitative information about the distribution of platinum among different kinds of rat tissues. Therefore, this study supports the performance of investigations concerning the distribution of platinum in human tissues following the administration of cisplatin.

ACKNOWLEDGMENTS

Support was given by the Danish Cancer Society and the E. Willumpen Foundation.

REFERENCES

1. G. Daugaard, H. H. Hansen, and M. Rørth, *Ann. Oncol.* **1,** 195–202 (1990).
2. A. Krarup-Hansen, K. Fugleholm, S. Helweg-Larsen, E. N. Hauge, H. Schmalbruch, W. Trojaborg, and C. Krarup, *Brain* **116,** 1017–1041 (1993).
3. H. Schmalbruch, *Anat. Rec.* **215,** 71–81 (1986).
4. G. W. Haverland and L. I. Wiebe, *Appl. Radiat. Isot.* **42,** 775–776 (1991).
5. R. Zeisler, F. Lux, H. Seidenberger, H. Schönenberger, and W. Beck, *Proceedings Series, Nuclear Activation Techniques in the Life Sciences, IAEA-SM-227,* 467–477 (1978).
6. B. Rietz and K. Heydorn, *J. Radioanal. Nucl. Chem.* **174,** 49–56 (1993).
7. X. Lin, K. Heydorn, and B. Rietz, *J. Radioanal. Nucl. Chem.* **160,** 85–99 (1992).
8. R. W. Gregg, V. J. M. Molepo, V. J. A. Monpetit, N. Z. Mikael, D. Redmond, M. Gadia, and D. J. Stewart, *J. Clin. Oncol.* **10,** 795–803 (1992).

Study of Trace Elements in Blood of Cancer Patients by Proton-Induced X-Ray Emission (PIXE) Analysis

M. A. DUROSINMI,[1] J. O. OJO,[2] A. F. OLUWOLE,*[,2]
O. A. AKANLE,[3] AND N. M. SPYROU[3]

[1]*Department of Haematology, Obafemi Awolowo University, Ile-Ife, Nigeria;* [2]*Department of Physics, Obafemi Awolowo University, Ile-Ife, Nigeria; and* [3]*Department of Physics, University of Surrey, Guildford, GU2 5XH, UK*

ABSTRACT

Proton-induced X-ray emission (PIXE) analysis was employed to determine the concentrations of six elements in the plasma and erythrocytes of 18 cancer subjects (15 males and 3 females) with neoplastic disorders and in 70 controls (35 males and 35 females). It was found that the concentrations of Br, K, and Zn were significantly elevated in the erythrocytes of the cancer subjects compared to the controls, whereas the concentration of Fe was significantly depressed, but with no difference observed in the concentration of Ca. In the plasma, the concentrations of Br, Cu, Ca, and K were significantly elevated, whereas the concentrations of Fe and Zn were found to be significantly depressed compared to the controls.

Index Entries: Trace elements; plasma; erythrocytes; cancer; PIXE analysis.

INTRODUCTION

Trace elements are present in <0.005% of the body weight. They are required in small concentrations as essential components of biological enzyme systems. Some of the elements are believed to be of importance

*Author to whom all correspondence and reprint requests should be addressed.

 © 1994 by Humana Press Inc.

in the natural history of neoplastic diseases. They are sometimes useful as diagnostic or prognostic tools in cancer patients. For example, serum Cu is a useful indicator of response of patients with Hodgkin's disease to chemotherapy as well as an early indicator of disease relapse *(1)*.

Trace elements may also be carcinogenic *(2)*. Excessive oral ingestion of Zn was thought by some to be related to cancer of the stomach *(3)*, whereas others found that Zn ingestion inhibited development of tumors *(4)*. Similarly, nickel compounds have been implicated in the etiology of the cancer of the lung and nasal sinuses *(5)*. It is therefore not surprising that scientists have developed interest in the blood levels of trace elements of cancer patients and normal controls *(6–11)*. Determination of elemental levels has been performed by various techniques, most often, Atomic Absorption Spectrophotometry (AAS). Zn, Cu, and to some extent Fe and Mg have been the elements most studied.

This article examines the levels of Br, Cu, Fe, and Zn found in the blood of Nigerian subjects with different types of hematological cancers using PIXE. To the best of our knowledge, this is the first of such studies in this country.

METHODS OF ANALYSIS

Our subjects include 18 cancer subjects, aged 5–62 (29 ± 20) yr, 15 males and 3 females. There are 14 cases of lymphoreticular malignancies, eight cases of Burkitt's lymphoma, four of chronic granulocytic leukaemia (CGL), two each of chronic lymphocytic leukemia (CLL) and multiple myeloma (MM), and one each of Hodgkin's disease and non-Hodgkin's lymphoma. Blood samples were collected before commencing any form of definitive cancer therapy. Controls were healthy subjects, 35 males and 35 females with ages ranging between 3 and 60 (26 ± 10) yr.

Sterile scalp vein needle with polyethylene catheter, polyethylene syringes, and prewashed heparinized polyethylene screw-cap tubes were used for blood collection and storage. Ten milliliters of blood were collected from each subject after allowing the first 1–2 mL to run into a sequestrene tube for blood count, in order to reduce contamination from the elements of the steel needle. The blood was separated by high centrifugation into plasma and erythrocytes, and subsequently frozen until ready for freeze-drying.

The freeze-dried samples were crushed with a polyethylene spatula into a fine powder in their individual containers and pellets made from the powder, and these were stuck side by side on a 3-mm thick aluminum plate with the help of double-sided adhesive tape. They were then carbon-coated with a thin layer of carbon and carbon dagged on the sides to make the samples conducting.

Analysis of samples was carried out using a 2-MeV proton beam under vacuum, incident on the samples with a diameter of about 3 mm, using the University of Surrey's Van de Graaf accelerator. PIXE spectra were collected for about 300 s with the low-energy photon Si(Li) detec-

Table 1
Elemental Concentrations in Plasma and Erythrocytes of Cancer Subjects and Controls (μg/mL)

Element	ERYTHROCYTES		PLASMA	
	Subjects Mean ± SD	Controls Mean ± SD	Subjects Mean ± SD	Controls Mean ± SD
Br	2.12 ± 0.60	1.63 ± 0.49	3.83 ± 1.60	2.48 ± 1.03
K	4287.8 ± 417.8	2946.2 ± 434.2	292.6 ± 64.6	230.4 ± 94.6
Fe	868.2 ± 96.1	917.9 ± 137.0	1.37 ± 0.94	2.55 ± 0.66
Zn	11.6 ± 2.2	9.4 ± 2.1	0.51 ± 0.10	0.72 ± 0.19
Ca	34.3 ± 10.1	29.6 ± 14.0	128.7 ± 10.4	111.3 ± 50.2
Cu	2.08 ± 1.68	1.32 ± 0.40	1.56 ± 0.27	1.16 ± 0.52
S	nd	nd	1581.4 ± 697.3	911.4 ± 422.2

tor fitted with a 350-μm mylar filter to suppress signals from low-energy X-rays in order to reduce dead-time problems. Simultaneous RBS spectra were also collected with a surface-barrier detector. Analysis of PIXE spectra for determination of peak areas was carried out by the PIXAN computer code.

RESULTS AND DISCUSSION

The concentration of the elements in μg/mL are summarised in Table 1, whereas the level of significance is given in Table 2. The concentrations of Br, Cu, K, and Zn were significantly elevated in the erythrocytes of the cancer subjects compared to the controls, whereas the concentration of Fe was significantly depressed. The concentration of Ca shows no significant difference. In the plasma, the concentrations of Br, Ca, Cu, and K were significantly elevated, whereas the concentrations of Fe and Zn were significantly depressed in the cancer subjects as compared to the normal controls.

Table 2
t-Test for Significant Differences in Elemental Concentrations in Plasma and Erythrocytes of Subjects and Controls[a]

Element	ERYTHROCYTES			PLASMA		
	D.F.	t statistic	p value	D.F.	t statistic	p value
Br	57	-3.3306	<0.001	66	-4.0635	<0.0005
K	49	-10.6836	<0.0005	55	-2.5251	<0.01
Fe	64	+1.4113	<0.1	24	+3.1930	<0.005
Zn	65	-3.8472	<0.0005	63	+4.4780	<0.0005
Ca	44	-1.2213	>0.1	54	-1.4530	<0.10
Cu	54	-2.6761	<0.01	56	-3.0206	<0.005
S	-	-	-	45	-4.0266	<0.0005

[a]Positive *t* statistic implies depression in subjects.

Our findings with respect to plasma Zn levels in cancer subjects are consistent with those of others *(6,7)*. The reduction in plasma Zn level is thought to be the result of the hyperzincuria *(2)* and also inadequate Zn intake *(7)* commonly seen in cancer subjects. Other workers have not found any statistical difference between the levels of Zn in cancer subjects and normal controls *(8,11)*. The multicentric (Burkitts and non-Hodgkins lymphoma) and diffuse (multiple myeloma, chronic lymphocytic, and granulocytic leukaemias) nature of the hematological malignancies investigated is similar to the metastatic cancers reported by Gorodetsky et al. *(11)*, in which the concentration of Zn was found to be significantly elevated, as was also observed in this study. Br is significantly elevated in the plasma and erythrocytes of cancer subjects compared to the controls, although there is very little information on Br in hematological cancers.

The presence of significantly elevated plasma Cu levels in our cancer subjects as compared to the controls is consistent with previous reports *(1–2,8–11)*. Plasma Cu is usually elevated in most patients with active cancers, including lymphomas *(9)* and leukemias *(8)*. The similarity between the erythrocyte levels of Cu in the patients and the controls is also consistent with previous reports *(11)*. Fe is significantly reduced in the patients compared to the controls (Tables 1 and 2); this may be because of the generally low hemoglobin concentrations in cancer subjects compared to healthy controls, although Gonodetsky et al. *(11)* found no significant difference in the erythrocyte Fe of cancer subjects and normal controls. Plasma Ca levels are elevated in the cancer subjects probably because of some bone involvement.

Although the number of cancer subjects studied is rather small, the findings are revealing. There is need to study a larger population of cancer patients. We also need to study trace element composition of cancer tissues and compare this with the patients' blood samples.

ACKNOWLEDGMENT

This work is part of an EEC-sponsored project on Environmental Monitoring and Impact Assessment in Nigeria.

REFERENCES

1. M. Hygrovic, C. F. Tessmer, T. M. Mincler, B. Mosler, and G. H. Taylor, *Cancer* **21,** 743 (1968).
2. M. K. Schwartz, *Cancer Res.* **35,** 3481 (1975).
3. P. Stocks and R. I. Davies, *Brit. J. Cancer* **18,** 14 (1964).
4. D. E. Poswillo and B. Cohen, *Nature* **231,** 447 (1971).
5. R. Doll, L. G. Morgan, and F. E. Spiezen, *Nickel* **24,** 623 (1970).
6. I. J. Davies, M. Mansa, and T. L. Dormandy, *J. Clin. Pathol.* **21,** 359 (1968).
7. K. L. Okudo and T. Ohatsuki, *Cancer* **56,** 918 (1985).
8. A. Tanyeli, L. Kayrin, M. Satar, S. Altinbasak, B. Yilmaz, and S. Okten, in *Proceedings of the Third International Congress of Trace Elements in Health and Disease,* G. I. Yuregir, O. Dorma, and L. Kayrin, eds., Gukurova University Publishing Co., Turkey, pp. 623–624, 1989.
9. M. Hygrovic, C. F. Tessmer, F. B. Thomas, P. S. Ong, J. F. Gamble, and C. C. Shullenberger, *Cancer* **32,** 1512 (1971).
10. P. Tani and L. Kokkola, *Scand. J. Respirant. Dis. Suppl.* **80,** 121 (1972).
11. R. Gorodetsky, Z. Fuks, A. Sulkes, H. Ginberg, and Z. Wechler, *Cancer* **55,** 779 (1985).

Elemental Concentrations in Nigerians with Affective Disorders Using Proton-Induced X-Ray Emission

M. A. DUROSINMI,[1] J. O. OJO,[2] A. F. OLUWOLE,*[,2] A. F. ONONYE,[3] O. A. AKANLE,[4] AND N. M. SPYROU[4]

Departments of [1]Haematology and Immunology, [2]Physics and [3]Mental Health Obafemi Awolowo University, Ile-Ife, Nigeria; and [4]Department of Physics, University of Surrey, Guildford, England

ABSTRACT

The concentrations of a number of elements are determined in the plasma and erythrocytes of 21 Nigerians (11 females, 10 males) with symptomatic affective disorders (11 depressives, 10 manics) and in 40 normal controls using proton-induced X-ray emission (PIXE) analysis. The study shows that there is significant elevation of plasma K and Zn, as well as the erythrocyte S in the patients relative to the controls. The plasma and erythrocyte Cu, and the erythrocyte P, Ca, Fe, and Zn are significantly lower in the patients compared to the controls. However, the plasma levels of Ca, S, Fe, and Br are similar in both the patients and the controls. Similarly, the concentrations of K, Br, and Rb show no significant difference in the erythrocytes of patients and controls

Index Entries: Trace elements; affective disorders; plasma; erythrocytes; PIXE analysis.

*Author to whom all correspondence and reprint requests should be addressed.

Biological Trace Element Research Editor: G. N. Schrauzer © 1994 by Humana Press Inc.

INTRODUCTION

Measurements of tissue concentrations of trace elements have become important as diagnostic tools in the investigation of some neurological and psychiatric disorders. There is a strong link between the increasing environmental pollution in the world and the tissue levels of trace elements, and therefore, the incidence of diseases associated with abnormal metabolism of these elements. The abnormal accumulation of Cu in the brain of patients with Wilson's disease has been associated with psychiatric disorders including schizophrenia and manic-depression *(1)*. High intake of dietary V is also known to interfere with the normal sodium pump in susceptible individuals and induces manic-depressive symptoms in the affected subjects *(2–4)*. Higher levels of Al are also found in the tissues of patients with senile dementia (Alzheimer's disease) relative to normal controls *(2,6–7)*, whereas Niklowitz and Mangbar *(8)* and McDermottt et al *(9)* found no significant differences between controls and diseased brains, but suggested a positive correlation of brain aluminum with age. The aim of this work is to investigate the concentrations of certain elements, including K, Ca, Fe, Cu, Zn, Br, Rb, P, and S, in the plasma and erythrocytes of Nigerians subjects with affective disorders.

METHODS OF ANALYSIS

There were 21 subjects, aged 18–45 (31 ± 7) yr, 11 females and 10 males. Ten of the subjects were manic, whereas the other 11 were depressives. There were 40 control subjects who were of similar age to the patients. All the patients were investigated during the active phase of their disease.

Sterile scalp vein needle with polyethylene catheters, polyethylene syringes, and prewashed heparinized polyethylene screw-cap tubes were used for blood collection and storage. Ten milliliters of blood were collected from each subject after allowing the first 1–2mL to run into a sequestrene tube for blood count in order to reduce contamination from the elements present in the steel needle. The blood was separated by high centrifugation of the plasma and erythrocytes, and subsequently frozen until ready for freeze-drying. The freeze-dried samples were then made into pellets and stuck side by side on a 3-mm thick aluminum plate with the help of double-sided adhesive tape. They were carbon-coated with a thin layer of carbon and carbon dagged on the sides to make the samples conducting.

Analysis of samples was carried as described elsewhere in this volume *(10)*.

RESULTS AND DISCUSSION

The results obtained are given in Tables 1 and 2. Seven and nine elements were determined in the plasma and erythrocytes, respectively.

Table 1
Elemental Concentrations in Blood of Mental Patients (mg/L)

Element	ERYTHROCYTES				PLASMA			
	Controls		Patients		Controls		Patients	
	Mean ± SD	No	Mean ± SD	No	Mean ± SD	No	Mean ± SD	No
P	902 ± 347	53	475 ± 75	17	-	-	-	-
S	1017 ± 461	17	1940 ± 150	8	914± 429	50	957 ± 550	21
K	3351 ± 1133	62	3500 ± 270	17	240± 93	55	425 ± 228	21
Ca	32 ± 16	40	25.2 ± 4.4	17	101.5± 48.5	56	112.1 ± 5.5	21
Fe	1024 ± 385	60	752 ± 14	17	1.9± 1.5	64	1.71 ± 0.63	21
Cu	1.5 ± 0.7	58	0.5 ± 0.1	17	1.27± 0.60	59	0.85 ± 0.55	20
Zn	12.4 ± 5.2	61	8.9 ± 2.0	17	0.79± 0.42	64	0.99 ± 0.38	21
Br	2.6 ± 2.4	41	2.1 ± 0.6	17	3.45± 1.97	62	3.21 ± 1.45	21
Rb	5.7 ± 3.2	70	5.7 ± 1.6	17	-	-	-	-

The elements were detected in not less than 17 of the patients. Tests of significance, using Student's *t*-test (Table 2) showed that the levels of K and Zn were significantly elevated and that of Cu significantly depressed in the plasma of the manic-depressive patients relative to the controls, whereas the plasma concentrations of Ca, S, Fe, and Br were similar for the patients and the controls. The erythrocyte concentrations of P, Ca, Fe, Cu, and Zn were significantly lower, but S level was significantly higher in the patients as compared to the controls. However, the levels of K, Br, and Rb were not significantly different in the patients and the controls.

Using the neutron activation analysis technique, Akanle et al. found no significant differences in the erythrocyte levels of K, Fe, Zn, Br, and Rb of their Caucasian depressives; the plasma Zn level of their patients was also similar to that of the controls *(2)*. However, the plasma levels of Fe and Br were significantly higher in their patients relative to normal controls. Our results (Tables 1 and 2) were different from those of Akanle et al. probably because of some environmental factors, such as diet. We have previously established satisfactory correlation with PIXE and neutron activation analysis for the blood levels of a number of elements in Nigerian subjects *(5)*. The results of the elemental concentrations in affective disorders call for further investigations into the importance of trace

Table 2
T-test for Significant Differences in Elemental Concentrations in Blood of Patients and Controls[a]

Element	ERYTHROCYTES			PLASMA		
	D.F	t statistic	p value	D.F	t statistic	p value
P	68	+5.0133	<0.0005	-	-	-
S	25	-5.4654	<0.0005	69	-0.3609	>0.1
K	77	-0.5490	>0.1	74	-5.0703	<0.0005
Ca	55	+1.6720	<0.1	75	-0.9917	>0.1
Fe	75	+2.9073	<0.005	63	+0.5769	>0.1
Cu	73	+5.6623	<0.0005	78	+2.7872	<0.005
Zn	76	+2.7757	<0.005	83	-1.9750	<0.05
Br	60	+0.9246	>0.1	81	+0.5064	>0.1
Rb	85	-0.0197	>0.1	-	-	-

[a]Positive *t* statistic implies depression in patients.

and macroelements in the possible etiopathogenesis of other psychiatric disorders.

ACKNOWLEDGMENT

This work is part of an EEC-sponsored project on Environmental Monitoring and Impact Assessment in Nigeria.

REFERENCES

1. A. G. Bearn, in *Oxford Textbook of Medicine,* D. J Weatherall, J. G. G. Ledingham, and D. A. Warrell, eds., Oxford University Press, Oxford, 1984, pp. 9.114, 9.115.
2. O. A. Akanle, N. M. Spyrou, A. A. Damyanova, D. M. Shaw, and L. Alli, *J. Radioanal. Nucl. Chem., Articles* **113(2),** 405 (1987).
3. E. J. Underwood, in *Trace Elements in Human and Animal Nutrition,* Academic, New York, 1977.

4. M. A. Durosinmi, J. O. Ojo, A. F. Oluwole, O. A. Akanle, W. Arshed, and N. M. Spyrou, *J. Radioanal. Nucl. Chem., Articles* **168(1),** 233 (1993).
5. J. O. Ojo, A. F. Oluwole, M. A. Durosinmi, O. J. Ogunsola, O. A. Akanle, and N. M. Spyrou, Baseline levels of elemental concentrations in whole blood, plasma, and erythrocytes of Nigerian subjects, *Biol. Trace Element Res.*, this vol.
6. D. Crapper, S. Krishnan, and A. Dalton, *Science* **180,** 511 (1973).
7. D. Crapper, S. Krishnan, and A. Onittkart, *Brain* **99,** 67 (1976).
8. W. Niklowitz and T. Mangbar, *J. Neuropathol. Neurol.* **34,** 445 (1975).
9. T. McDermott, A. Smith, K. Igbal, and H. Wisniewski, *Lancet* **2,** 710 (1977).
10. M. A. Durosinmi, J. O. Ojo, A. F. Oluwole, O. A. Akanle, and N. M. Spyrou, this vol.

Trace Element Content in Breasts with Fibrocystic Disease

G. D. KANIAS,*[,1] E. KOURI,[2] H. ARVANITI,[2] H. KARAIOSIFIDI,[2] AND S. KOUNELI[2]

[1]*Radioanalytical Laboratory, Institute of Physical Chemistry, N. C. S. R. "DEMOKRITOS," Aghia Paraskevi Attikis, Athens 153 10, Greece; and* [2]*Pathology Department, Gynaecological Center, "Helena Venizelou," H. Venizelou square 2, Athens 115 21, Greece*

ABSTRACT

The trace elements antimony, bromine, cesium, cobalt, iron, rubidium, scandium, strontium and zinc were determined by instrumental neutron activation analysis in breast tissue samples with fibrocystic disease and in samples with fibroadenoma tumors. The histological lesions of each breast sample with fibrocystic disease were recorded, and a statistical analysis of the lesions in combination with the determined trace elements was carried out. The results showed that the element mean values in fibroadenoma tumors are higher than those of fibrocystic disease. Some other remarkable results of statistical examination are also presented.

Index Entries: Breast; fibrocystic disease; fibroadenoma tumors; trace elements; neutron activation analysis; statistical analysis.

INTRODUCTION

The epithelium of the female breast may undergo a wide variety of benign alterations that are often grouped together under such terms as fibrocystic mastopathy, cystic hyperplasia, benign breast disease, and so on. Although there is some evidence that the alterations may have an

*Author to whom all correspondence and reprint requests should be addressed.

endogenous or exogenous hormonal basis, the etiology is poorly understood (*1*).

It is known that trace elements act at the cellular or subcellular level through different mechanisms. One of these mechanisms could be described as the interaction between trace elements and hormones that regulate the metabolism of higher biochemical substrates. Thus, a possible relationship between trace elements and etiology of the diseases could be expected, and an investigation in this field has been developed during last decade.

Benign breast disease or fibrocystic disease encompasses a wide range of histological lesions. Some are well-defined entities, but a large number of biopsies have a mixture of changes, such as adenosis, epithelial hyperplasia, cyst formation, epithelial metaplasia, and papilloma. Contrary to previously held views, only a minority of the lesions are now considered to carry an increased risk of subsequent carcinoma. This risk is further increased in patients with a family history of breast cancer (*2*). Fibroadenoma is a benign breast tumor that appears in young women. It occurs alone or with fibrocystic disease in the same breast tissue.

Based on all the reasons mentioned above, in this work, we determined the trace elements antimony, bromine, cesium, cobalt, iron, rubidium, scandium, strontium, and zinc in breast tissue samples with fibrocystic disease and/or fibroadenoma. We also tried to relate statistically the determined trace element contents with the recorded breast tissue alterations in each sample. Fibroadenoma alterations were included in this work, although our objective was to study fibrocystic disease only. This was done because sometimes there is a coexistence of fibroadenoma tumors with fibrocystic lesions in the same sample. We also included pure fibroadenoma tumors in order to find out if there is any difference in trace element content between breast samples with fibroadenoma only and breast samples with fibroadenoma and fibrocystic disease.

The analytical method employed in this work was instrumental neutron activation analysis. Correlation coefficients, multiple correlations, and *t*-test as applied to the raw analytical data were evaluated in order to study the possible relationship between trace elements and breast tissue alterations.

EXPERIMENTAL

Sampling

Tissue samples were collected from 23 patients who underwent a breast biopsy owing to mamographic or clinical findings. The excised surgical specimens were immediately transferred to the pathology laboratory for histological examination. The pathologist removed an aliquot for trace element analysis when the macroscopical examination showed signs of fibrocystic disease or fibroadenoma tumors that were reconfirmated microscopically. The histological report for each patient was recorded.

Microscopical Histological Examination

The following benign alterations of breast epithelium were found in the examined samples: adenosis, apocrine metaplasia, epitheliosis, cysts, fibrosis, lobular hyperplasia, papillary apocrine metaplasia, and papilloma. Also, fibroadenoma tumors were recognized in some samples with fibrocystic disease and in some samples without fibrocystic disease.

INSTRUMENTAL NEUTRON ACTIVATION ANALYSIS

Apparatus

Irradiation of samples and standards was carried out using a rotation system of a nuclear reactor of the NCSR "Demokritos" with a maximum thermal neutron flux of $3.9 \cdot 10^{13}$ $n \cdot cm^{-2} \cdot s^{-1}$.

γ-Ray spectrometry of the irradiated samples and standards was done with a high-purity Ge detector with a relative efficiency of 27.8% connected to a Canberra 85 series, 8192 channel analyzer.

Standards for Analysis

Spinach leaves SRM 1570 standard reference material of NIST as well as Milk powder A-11 certified reference material of IAEA were used as analysis standards.

Procedure

The collected breast tissue samples were lyophylized, and about 0.3 g of each breast sample was sealed in quartz tubes for irradiation. Also, the same weight of the standards was placed into geometrically identical tubes. Samples and standards were irradiated simultaneously in the rotation system of the reactor for 30 h. After a 20-d cooling time, the quartz tubes were opened after their freezing in liquid nitrogen, and their radioactive content was transferred into 5-mL culture tubes. The emitted radioactivity was measured for 45 min by means of the multichannel analyzer.

After counting the areas of the photopeaks corresponding to the γ-rays of ^{85}Sr at 514 keV, ^{82}Br at 554 keV, ^{134}Cs at 796 keV, ^{46}Sc at 889 keV, ^{86}Rb at 1077 keV, ^{59}Fe at 1099 keV, ^{65}Zn at 1115 keV, ^{60}Co at 1333 keV, and ^{124}Sb at 1691 keV, samples and standards were compared. The processing of γ-ray spectra was carried out using a PC/AT Samsung S 550 computer and the SPAN (V 3.0) software package of the IAEA.

Statistical Analysis

The statistical treatment of the raw analytical data and of the recorded histological breast alterations was done by the same PC/AT computer, and the commercial software packages SYSTAT and SPSS.

RESULTS AND DISCUSSION

The obtained analytical data of antimony, bromine, cesium, cobalt, iron, rubidium, scandium, strontium, and zinc contents in breast tissue samples are summarized in Table 1 as ranges, mean values, and standard deviations together with the age of patients. In this table, a classification of the data was done according to the breast tissue alterations. The first group comprises the samples with fibrocystic disease only, whereas the second one comprises the samples of pure fibroadenoma tumors. In the third and last group are included the samples in which there is coexistence of fibrocystic disease and of fibroadenoma tumors.

Considering bromine, it must be noted that this element is not detected in all samples analyzed above the detection limit, which is 1 μg. Therefore, minimum and maximum values are given for information only. Also it must be pointed out that none of the breast samples contained strontium concentration above the detection limit of 0.1 μg. The occurrence of each fibrocystic lesion in percent is shown in Table 2.

As we can see in Table 1, fibroadenoma mean values are higher than those reported for fibrocystic disease, and especially for the elements cesium, cobalt, rubidium, and zinc, they are about two times higher. This pattern seems to have the same trend with the variation pattern in the literature, which is on the ratio of malignant to normal breast tissue concentrations for the same elements *(3)*.

A direct comparison of the results found with literature data cannot be done, because there has been lack of data concerning trace elements in fibrocystic disease so far. However, we can make an approximation if we suppose that the fibrocystic disease is not a neoplastic situation. Thus, we can correlate our fibrocystic disease results with those reported for normal breast tissue and our fibroadenoma results with those reported for malignant breast tissue. The comparison showed that our results are found to be at the same concentration level with literature data for iron, rubidium, and zinc. The same is shown even for bromine where we do not have results for all the samples.

The raw analytical data for antimony, cesium, cobalt, iron, rubidium, scandium, and zinc, the age of patients, and the recorded breast tissue lesions apocrine metaplasia, adenosis, cysts, and epitheliosis were treated statistically by calculation of correlation coefficients, multiple correlations, and the *t*-test. The data for all the other recorded tissue alterations were not included in the tests because of the absence of a required minimum number of samples to form two groups. A confidence level of 95% was accepted for all the statistical tests.

Correlation Coefficients

The application of this test showed 11 statistically significant correlations. The main characteristic is that all these correlations have a strong linear trend (r values between 0.87 and 0.97) and especially in the following groups.

Table 1
Ranges, Mean Values, and Standard Deviations (SD) of the Trace Element Contents in Breast Tissue Samples and Age of the Patients (μg/g dry wt)

	AGE	Co	Cs	Fe	Rb	Zn	Sc	Sb	Br	Sr
				FIBROCYSTIC DISEASE						
Min.	23	0.0083	0.0028	39	1.3	12	0.00010	0.011	14	-*
Max.	58	0.17	0.028	200	11	110	0.00051	2.5	40	-*
Mean	42	0.051	0.0088	87	4.1	38	0.00028	0.54		
S.D.	10	0.045	0.0072	50	2.8	28	0.00013	0.79		
				PURE FIBROADENOMA						
Min.	16	0.012	0.0030	62	2.6	18	0.00012	0.011	17	-*
Max.	54	0.44	0.032	260	16	130	0.0011	0.14	41	-*
Mean	34	0.10	0.017	114	8.8	62	0.00036	0.064		
S.D.	15	0.17	0.012	79	5.8	39	0.00037	0.049		
				FIBROADENOMA AND FIBROCYSTIC DISEASE IN THE SAME SAMPLE						
Min.	42	0.010	0.0031	33	1.5	9.1	0.00011	0.030	14	-*
Max.	47	0.071	0.031	70	3.3	28	0.00079	0.42	17	-*
Mean	44	0.027	0.010	48	2.2	15	0.00027	0.17		
S.D.	2.5	0.025	0.011	18	0.68	7.7	0.00029	0.17		

*Below detection limit

Fibrocystic disease only: cesium with rubidium (r = 0.93), cesium with zinc (r = 0.91), and rubidium with zinc (r = 0.90);
Pure fibroadenoma: cobalt with scandium (r = 0.96), cesium with rubidium (r = 0.95), iron with zinc (r = 0.93), and rubidium with zinc (r = 0.87);
Fibroadenoma and fibrocystic disease in the same sample: cobalt with scandium (r = 0.96), cobalt with zinc (r = 0.97), rubidium with zinc (r = 0.91), and scandium with zinc (r = 0.90).

Table 2
Occurrence of Lesions in Breast Tissue Samples with Fibrocystic Disease

Histological lesion	Number of samples with lesion	Percent of the total number of samples
Adenosis	11	64.7
Apocrine metaplasia	11	64.7
Cysts	14	82.3
Epitheliosis	6	35.3
Fibrosis	16	94.1
Lobular Hyperplasia	1	5.9
Papilloma	2	11.7
Pappilary apocrine metaplasia	2	11.7

The above mentioned results show that the correlations between cesium and rubidium and between rubidium and zinc are significant independently of the kind of breast alterations. Also, it seems that the correlation between cobalt and scandium could be characteristic of fibroadenoma tumors, independently of the presence or absence of fibroadenoma lesions in the same sample. The presence of the same significant correlations to both studied breast tissue alterations could lead to the hypothesis that these correlations still exist even in healthy breast tissue.

Multiple Correlations

In this statistical test, a significant correlation means that a group of components of a matrix (with at least two members) is responsible for the variation of the concentration of another single component out of this group. This test is more powerful when data from different groups of variables are correlated. In our case, all the variables are chemical elements. Therefore, we separated the elements in groups according to their chemical properties (periodic table position). The first group is formed by the elements cesium and rubidium (alkalines), and the second one by the elements cobalt, iron, scandium, and zinc (first series of transition metals). Finally, a third group is formed by the most essential elements of the human life (iron and zinc). Thus our multiple correlation test was to examinate if there was any statistically significant correlation of the remaining elements with the above mentioned groups.

The results showed that in the fibrocystic disease, there is a statistically significant multiple correlation between zinc and the alkaline metals (R = 0.92). The first series of transition metals shows a significant correlation with rubidium (R = 0.90) and with cesium (R = 0.94). The same elements have a significant correlation with the group of the most essential elements of human life (rubidium R = 0.90, cesium R = 0.92).

Considering samples with pure fibroadenoma, we can see that there is one significant multiple correlation between scandium and the alkalines group (R = 0.95). Finally, the samples with coexistence of fibroadenoma and of fibrocystic disease showed significant correlation of cobalt (R = 0.99) and of scandium (R = 0.98) with the group of the most essential elements for the human life.

As is shown above, all of the found significant multiple correlations have R higher than 0.90. That means these correlations have a strong linear trend in the multidimensional space.

T-test

This test was applied to trace element data after their classification into two groups according to the presence or the absence of particular histological lesion of the breast. This was carried out in order to examine whether the presence of an alteration in the breast tissue was accompanied by an increase or a decrease in the concentration of particular trace elements at a chosen significant level. Also in our data, this test is applied in order to find out if the concentration of the trace elements differ significantly between the group with pure fibroadenoma tumors and the group with the coexistence of fibroadenoma and of fibrocystic disease in the same sample.

The results of this test showed that for the fibrocystic disease samples, there is a statistically significant difference for the rubidium content between the groups with cyst lesion and without cyst lesion. For all the other lesions, no significant differences were found between the groups with lesion and without lesion.

Application of t-test to the groups with pure fibroadenoma and with coexistence of fibroadenoma and fibrocystic disease showed a significant difference for rubidium and zinc contents. On examining all the samples with fibrocystic disease, we can form two groups, one with fibrocystic lesions only and one with the coexistence of fibrocystic lesions and fibroadenoma tumors. T-tests in these two groups showed significant differences for the elements iron, rubidium, and zinc.

CONCLUSIONS

On evaluation of the analytical data and from all statistical tests applied, we can deduce the following:

1. Mean values of the trace elements determined in fibroadenoma tumors are higher than those in fibrocystic disease;
2. There are statistically significant correlations between some trace elements independent of the kind of breast tissue alterations;
3. There are significant multiple correlations between the alkaline metals (cesium and rubidium) and zinc in fibrocystic disease and with scandium in fibroadenoma tumors;
4. All the significant correlations have a very strong linear trend;
5. Fibrocystic disease samples with cyst lesion differ significantly from those without cyst lesion as concerns rubidium content; and
6. Samples with fibrocystic lesions only and samples with the coexistence of fibroadenoma tumors and fibrocystic lesions differ significantly in the contents of iron, rubidium, and zinc.

These conclusions seem to suggest that rubidium and zinc have a principal role in the breast tissue life process. For zinc, this is something that could be expected because of its responsibility for more than 100 enzymatic reactions. However, elucidating the role of rubidium would need a more detailed study. This also follows from literature data where higher rubidium values in breast cancer tissue than in the normal ones appear.

REFERENCES

1. J. P. Sloane, *Biopsy Pathology of the Breast*, Chapman and Hall Medical Ltd., London, 1985, pp 69–92.
2. Consensus meeting, *Arch. Pathol. Lab. Med.* **110,** 171 (1986).
3. S. L. Rizk and H. H. Sky-Peck, *Cancer Res.* **44,** 5390 (1984).

Mapping Using Human Blood Composition Data

L. I. Zhuk,* I. N. Mikholskaya, E. A. Danilova, and An. A. Kist

Institute of Nuclear Physics Tashkent, Ulughbek, 702132, Uzbekistan, CIS

ABSTRACT

The mapping of elemental composition using INAA on the base of whole blood and hair of inhabitants (2500 samples) was made in Uzbekistan (CIS). The average concentrations of 15 elements (24 for hair) were determined. The results obtained were compared to regional medical statistics. The correlations for various diseases were obtained. The maps of human blood composition in comparison to those for human hair composition seem to be less significant in terms of regional contamination (probably because of stronger homeostatic control of blood). On the other hand, specific changes of blood composition were detected for some occupational groups. Relationships of blood and hair elemental composition and their relative significance are discussed.

Index Entries: Elemental human blood composition mapping; neutron activation analysis; medical statistics.

INTRODUCTION

Studies of human body elemental status seem to be important to estimate the environmental situation, as well as to search for correlations between health status and elemental composition of body tissues and fluids. Drawing maps using data obtained allows detection of some geographical peculiarities. Maps of Uzbekistan inhabitants' hair composition

*Author to whom all correspondence and reprint requests should be addressed.

(1) illustrate productivity of this approach. Within the frame of this study, samples of human whole blood were collected and analyzed. Serious differences are expected for hair and blood composition owing to differences in their properties. The extreme relative changes of chemical composition of hair are far higher than those of blood. Hair is a metabolically nonactive tissue, and changes of its composition remain recorded along the hair. Being an excretory organ, hair reflects the change of body status and can be considered as a chemical status averaging tissue. Blood is strongly homoeostatically controlled against composition changes. The changes around average values are sometimes very rapid and are associated with a huge number of various factors. Thus, data on the population level and maps of elemental composition of hair and blood may have different significance.

METHODS

Sampling Procedure and Preparation of Samples

Uzbekistan is located in the south of CIS. Its area is 447,000 km^2, and the population is about 18 million. The country is quite attractive in terms of climate, geography, economy, and ecology. The territory of the republic was divided into 70 zones, from which the samples of whole human blood and hair were collected. The total made 2000 hair samples and 2500 blood samples, enough for the determination of elemental blood and hair composition characterizing the population *(2)*. Particulars of a donor, such as age, sex, resident time in given area, ethnic group, occupation, hemoglobin content, blood group, and hair color were recorded. Blood samples were taken from elbow vein and placed in sterile flared vials that were immediately sent for lyophilization.

Elemental Analysis

Instrumental neutron activation analysis (NAA) and also NAA with radiochemical separation (for copper) were applied to study the samples collected. Lyophilized blood samples were homogenized in an agate mortar. From the homogenate, two aliquots, 20 and 200 mg, were taken. The aliquots were placed in polyethylene capsules and irradiated together with appropriate standards. Each sample was first irradiated in a flux of 1.8×10^{13} $n/cm^2 \cdot s$ for 15 s. After a few minutes of cooling, chlorine (^{37}Cl) and sodium (^{24}Na) were determined using a Ge(Li)-detector-based computerized γ-ray spectrometry system. After 3–5 da, the samples were irradiated again for 3 min, and after 10–12 h β-activity was measured using aluminum filters to absorbe less energetic β-rays to determine potassium (^{42}K).

The 200-mg samples were additionally packed into aluminum foil and irradiated together with standards in the same flux for 15 h. After irradiation, the aluminum foil was removed and samples packed into counting polyethylene capsules. One week after irradiation γ-spectra were measured for bromine and gold determination, and after 1 mo, γ-spectra were measured again for scandium, chromium, iron, cobalt, zinc, selenium, rubidium, silver, antimony, cerium, and mercury determination.

Copper was determined in 20–40 mg samples, which were irradiated in a flux of 1.8×10^{13} $n/cm^2/s$ for 30 min. After 12 hs, samples were dissolved in hot nitric acid in the presence of 1 mg of copper ions as a carrier. Then, the solution was conditioned by a NH_4OH solution to pH 2–3. Copper was extracted by a solution of sodium diethyldithiocarbamate in CCl_4. Extracts were washed twice by water, and γ-spectra of extracts were measured. The chemical yield was determined by reactivation of extract aliqouts.

The quality of standards and that of the laboratory reference material were evaluated by analyzing the IAEA (Animal Muscle and Animal Blood) reference materials. The agreement obtained with the IAEA values was quite satisfactory.

RESULTS AND DISCUSSION

Mean Values

Table 1 lists arithmetical means for all data (data below determination limit are substituted by 0.5 of a determination limit value), and for data excluding the content below the determination limit and extremely high values. This table also includes data for whole human blood taken from compilation *(3)* and for "standard man" *(4)*. All values given refer to the dry weight. With a few exceptions connected, probably, with biochemical differences of studied areas, these data are in a satisfactory agreement.

The distribution of the results obtained (frequency histograms) was normal or close to normal for Na, Fe, Zn, Se, and Rb. For K, Cl, Co, and Au, the distribution was bimodal. For other elements, the distributions were log normal. For Sc, Cr, Co, Ag, Sb, Ce, Au, and Hg, the histograms were "cut" at the left side, because concentrations of these elements were in part of samples below the determination limit.

Correlations

Very significant positive correlations were found for the pairs Na-Cl, Na-K, K-Cl, Sc-Ag, Co-Zn, and Zn-Se. For Cr-Rb, Cr-Se, Sc-Co, Sc-K, Rb-Ce, Rb-Se, Zn-Cl, Zn-K, Cl-Br, and K-Se less significant positive correlations were found, whereas for Sb-Cl, Zn-Ce, Na-Ce, and Ce-Se, less significant negative correlations were detected.

Concentrations of Na, Cl, Cr, Zn, and Se were significantly higher, and those of Sc and Fe lower in female blood compare to male blood. Concentrations of Rb and Ce were found to be increasing with age. Con-

Table 1
Concentrations of Elements in Blood of Uzbekistan Inhabitants in Comparison with Literature Values (μg/g dry wt, unless otherwise indicated)

	This work		Literature values	
Element	a	b	[3]	[4]
Na(100)*	0.720±0.004	0.720±0.004	0.98	0.96
Cl(100)*	1.220±0.006	1.220±0.006	1.44	1.44
K(100)*	0.930±0.005	0.930±0.005	0.81	0.84
Sc(46)**	3.2±0.10	5.8±0.14	38	-
Cr(46)	0.22±0.01	0.36±0.01	0.033-0.54	0.13
Fe(94)*	0.210±0.005	0.200±0.001	0.22	0.24
Co(37)	0.0150±0.007	0.027±0.006	0.0015-0.045	0.003
Cu(100)	8.2±0.27	8.2±0.27	5.0	5.3
Zn(93)	20.8±0.3	20.6±0.16	35	33
Se(90)	0.480±0.005	0.490±0.005	0.86	1.1
Br(88)	18±1	16.0±0.4	23	25
Rb(85)	4.57±0.06	4.93±0.06	12	13
Ag(17)	0.022±0.04	0.056±0.002	0.017-0.60	0.95
Sb(77)	0.10±0.06	0.085±0.002	0.011-0.024	0.02
Ce(46)	0.290±0.006	0.47±0.01	-	-
Au(55)	0.110±0.001	0.049±0.003	0.00021-0.0021	0.002
Hg(67)	0.020±0.001	0.054±0.001	0.0075-0.1	0.01

[a]All the results.
[b]The results excluding the content below the determination limit and extremely high values.
*in %.; ** in ng/g.
The percentage of samples in which the element was detected is given in brackets—*see text*.

sidering the groups of inhabitants of local origin and of non local origin, statistically significant differences were found only for a few elements (elevated Zn and Br, and decreased Rb and Sb in inhabitants of local origin).

Some differences were observed for inhabitants of rural and urban regions, and for various occupational groups. Rural region inhabitants had elevated Na and Zn, and decreased Sc, Rb, and Sb. High concentration of Se was found in blood of housewives from rural regions. Various groups having occupational contact with metals had elevated concentrations of Cl, K, Sc, Zn, Cr, Co, Ag, Sb, and Hg depending on their occupation.

In some cases, blood elemental composition was found to depend on blood group and hemoglobin concentration. Differences connected with Rh-factor were found for only two elements (significantly elevated Rb and Sb in the negative Rhesus factor group). Blood elemental composition changes as a function of the time of sampling, which can be considered to be associated with biological cycles (including circadian ones), are shown in Fig. 1.

In order to summarize the changeability of blood composition, parameters are given in Table 2 that were found to cause statistically significant changes of blood composition. It is commonly known that elemental composition of biological samples is connected with a huge number of various factors. Usually, the studies are focused on detecting the most important factors influencing the background values. Our data illustrate a known thesis that one of the most important of factors connected with blood elemental composition is the time of sample collection.

Maps

Ranges of elemental concentrations were used for mapping. Similar to many other biological samples, the relative changeability in general, the higher the average element concentration, the lower λ will be. Nevertheless, even for an element of such high concentration, like chlorine, the map indicates significant changes (Fig. 2). The highest value was found near to the drying-out Aral Sea. Probably the salinization at the Aral Sea vicinity reached a level that changed the blood salt balance.

Stronger changes were observed for trace elements. The map for scandium is given as an example in the Fig. 3. The high concentrations are connected probably with:

1. Geochemical anomalies;
2. Technologies in which scandium can be considered as a byproduct;
3. Localities polluted by water containing high concentrations of phosphoric fertilizers (also containing Sc, REE, and U). Low concentrations of copper were connected with technologically more developed and ecologically polluted areas (Fig. 4).

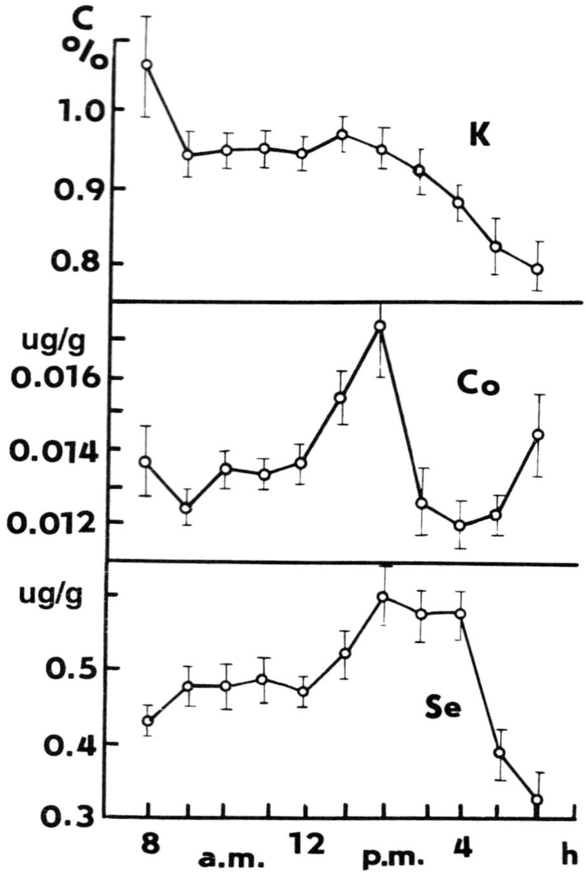

Fig. 1. Daily changes of concentrations of some elements in human blood.

Table 2
Parameters Causing Blood Composition Changeability

Element / Parameter	Na	Cl	K	Sc	Cr	Fe	Co	Zn	Se	Br	Rb	Ag	Sb	Ce
Sex	x	x		x	x	x		x	x					
Age	x										x			x
Nationality				x				x			x		x	
Resident time						x								
Occupation	x			x			x	x	x	x	x	x		x
Blood group	x								x	x	x			
Hemoglobin concentration	x	x		x			x				x		x	
Rh-factor											x		x	
Time	x	x	x	x	x	x	x	x	x	x	x	x	x	x
Meal			x	x						x	x			x

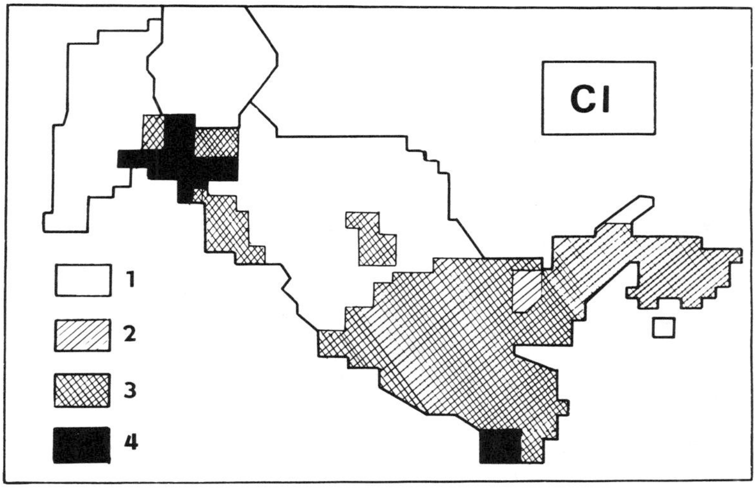

Fig. 2. Chlorine distribution (1—little inhabited area, 2—<0.95%, 3—0.95–1.5%, 4—more than 1.5%).

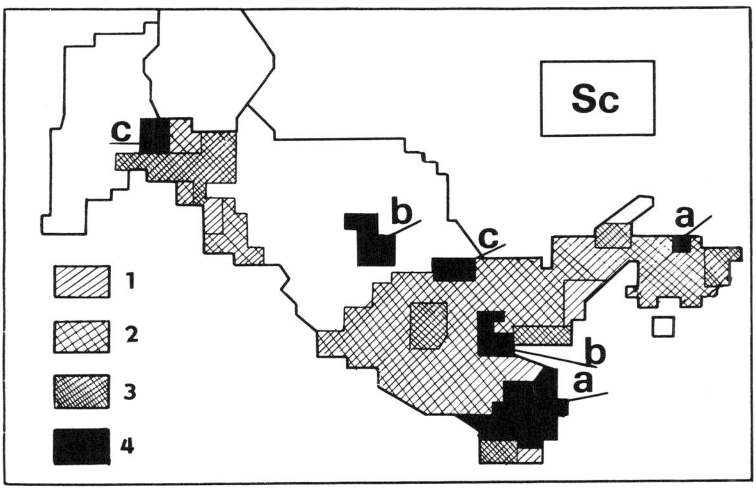

Fig. 3. Scandium distribution (1—<0.8 ng/g, 2—0.8–4.8 ng/g, 3—4.8–7.2 ng/g, 4—more than 7.2 ng/g, a—geochemical anomalies, b—technological anomalies, c—water containing phosphoric fertilizers).

It is well known that the health of populations is connected with environment contamination. Thus, some maps indicate the health status rather than environment situation. Figure 5 gives the correlation of regional statistic of cerebrovascular diseases with zinc concentration. Other correlations can be illustrated by Table 3.

Much stronger correlations were found when data for two elements were used. For example, the correlation coefficient for a Rb-to-K ratio and iron deficiency anemia was equal to −0.58.

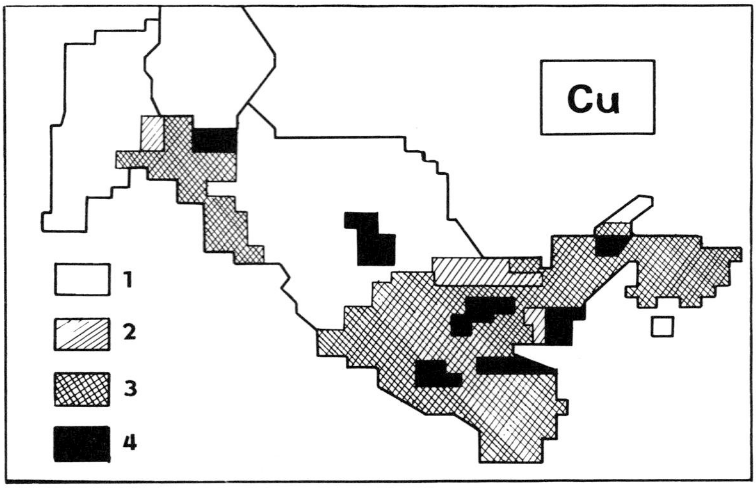

Fig. 4. Copper distribution (1—little inhabited area, 2—<8 μg/g, 3—8–10 μg/g, 4—more than 10 μg/g).

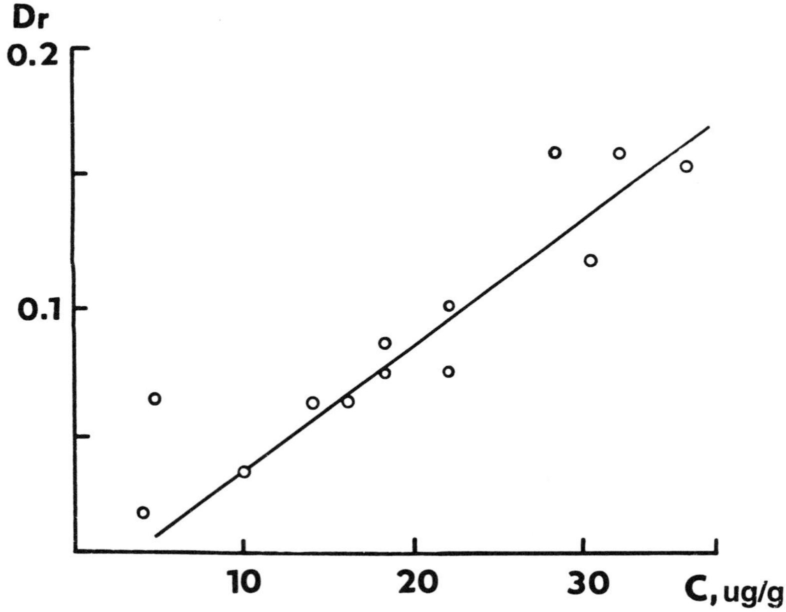

Fig. 5. Cardiovascular sick rate vs zinc concentration.

More sophisticated methods were also used. Among them, the method of canonical multiple correlation was used. This method summarizes the relationship between two sets of variables by finding a small number of linear combinations of each sets that have the highest possible between-set correlations. The results are given in Table 4.

Table 3
Single Element Correlations with Regional Medical Statistics

	Weak correlations		Strong correlations	
Diseases	Positive	Negative	Positive	Negative
Thyreotoxicosis	Zn, Na, Cl, Sc		K	Sb, Ce
Hypertensive	Na	K, Hg	Au	
Ishaemia	Co, Rb, Au, Sc	K, Br, Cu		
Cerebrovascular	Cl	Br, Cr, Ce, Hg	Na, K, Zn	
Anaemias	Zn, Se, Br	Cl, Cr	Sb, Au	Rb, Hg
Diabetes	Na, Sc, Sb, Hg	Cr, Fe, Co, Cu		Br
Oesophagus cancer	Sc	Se, Rb, Sb, Cu	Cl, Zn	Cr, Ce
Stomach cancer	Sc, Zn	Se, Br, Sb, Ce, Cu	Cl	Cr
Rectum cancer	Sc, Rb, Hg	Se, Br, Sb,, Cu	Cl	
Breast cancer	Sb	Se, Na, Cr, Fe, Zn, Cu	Cl, Rb, Hg	
Bladder cancer	Br, Ce		Cl	Na

In order to compare blood and hair data, the correlation coefficients for the regions studied were calculated (Table 5). A few significant values were found, but a tendency (with the exception of bromine) for the decreased element concentrations in hair to be correlated with element accumulation in the human body was noticed.

SUMMARY

These data provide evidence of how informative elemental blood composition may be, especially for human health screening. This work as well as the previous one *(1)* can serve as a background for mapping of wider areas and for global mapping. We can only speculate on additional information that may be obtained by world mapping. In any case, an attempt to draw regional, continental, and worldwide maps of the ele-

Table 4
Canonical Multiple Correlation

No	Disease	Corelation coefficient	Significance Level	Elements
1	Diabetes	0.894	0.082	Na, Fe, Co, Sb, Au, Hg
2	Goiter	not found		
3	Hypertensive	not found		
4	Ischaemia	0.986	0.0062	Na, Cl, Sc, Fe, Co, Cr, Cu, Zn
5	Cerebrovascular	0.859	0.075	Se, Br, Sb, Au, Hg,
6	Anaemias	0.938	0.022	Na, Cl, Sc, Cr, Fe, Zn,
7	Oesophagus cancer	0.981	0.012	Na, Cl, Sc, Cr, Fe, Co, Cu, Zn
8	Stomach cancer	0.908	0.125	Cl, Sc, Cr, Cu, Zn, Au, Hg
9	Rectum cancer	0.829	0.120	Cl, Sc, Cr, Zn, Hg
10.	Breast cancer	0.855	0.080	Na, Cl, Fe, Co, Hg
11.	Bladder cancer	0.874	0.121	Cl, Fe, Co, Cu, Br, Hg

Table 5
Human Blood and Human Hair Composition Correlation

Element	Correlation Coefficient	Significance Level	Element	Correlation Coefficient	Significance Level
Na	-0.451	0.141	Zn	-0.210	0.514
Cl	-0.281	0.376	Se	-0.054	0.870
Sc	-0.582	0.047	Br	0.799	0.0018
Cr	0.052	0.874	Sb	0.286	0.367
Fe	-0.255	0.424	Au	0.026	0.935
Co	-0.0734	0.821	Hg	-0.036	0.912
Cu	-0.195	0.544			

mental blood composition, as well as blood analysis on population level, seems to be quite justified and important.

REFERENCES

1. L. I. Zhuk and A. A. Kist, *Biol. Trace Element Res.* **26–27,** 307 (1990).
2. Yu. S. Ryabukhin, *J. Radioanal. Chem.* **60,** 7 (1980).
3. G. V. Iyengar, W. E. Kollmer, H. J. M. Bowen, *The Elemental Composition of Human Tissues and Body Fluids,* Weinheim, New York; Verlag Chemie, 1978.
4. *Report of the Task Group on Reference Man,* Pergamon, New York, 1975.

Determination of Trace Elements in Normal and Diabetic Whole Blood by Neutron Activation Analysis

F. A. EL-AMRI,*[,1] H. A. EL-SALTANI,[1] N. N. DOGADKIN,[1] AND I. A. ASHOUR[2]

[1]*Department of Chemistry, Al'Fateh University, P.O. Box 13361, Tripoli, Libya; and* [2]*Tajoura Nuclear Research Center, Tripoli, Libya*

ABSTRACT

The concentration of 10 trace elements in samples of normal and diabetic whole blood were investigated by neutron activation analysis (NAA). Instrumental NAA (INAA) was used for determination of Co, Cr, Hg, Rb, Se, and Zn, whereas NAA with radiochemical separation (RNAA) was employed for determination of Al, Cu, Mg, and Mn. The results for normal blood compare well with literature values. Co, Cr, Se, and Zn show lower values in diabetes than in the normal blood samples.

Index Entries: Trace elements; normal whole blood; diabetic whole blood; neutron activation analysis.

INTRODUCTION

The importance of trace elements in body metabolism is receiving increasing attention, because toxicity of some elements as well as the beneficial effects of others to humans have been recognized *(1–3)*. The human body continuously maintains a variety of elements; these elements are closely related to human health and health disorders, since their health deficiency or excess induces physiological changes *(4)*. From this point of view, it is necessary to study changes of element levels in various tissues during various diseases. The element levels in the human organism depend mainly on local environmental factors and dietary

*Author to whom all correspondence and reprint requests should be addressed.

Biological Trace Element Research Editor: G. N. Schrauzer © 1994 by Humana Press Inc.

habits. Therefore, for biomedical studies, it is necessary to establish trace element levels in the tissues and blood of normal population in a region under study. Such data from various regions of the world are now available in the literature *(5)*.

Several modern instrumental methods exist for determination of trace elements in biological material *(6,7)*. The authors of this work have used neutron activation analysis (NAA) for this study. NAA is a very sensitive method that often serves as a reference for other methods, because it does not involve any chemical treatment of samples before neutron activation. It is therefore less prone to blank and contamination problems.

The purpose of this work was to investigate the possibilities of thermal NAA for determination of trace elements in human whole blood. Knowledge of these parameters may be a useful tool either in the investigation of physiological effects or later in the diagnosis of specific diseases.

EXPERIMENTAL

Three hundred and sixty samples (about 2 mL each) of whole blood were collected in polyethylene vials after flushing the stainless-steel needle by the first mL of blood in a diabetic treatment center. Control samples were obtained from donors aged between 10–75 yr in Tripoli city. No anticoagulants were added to the vials. Lyophilized samples were ground to powder in conditions reducing the risk of contamination with elements to be determined. Each sample was subjected to two irradiations. Short irradiation: 50 mg samples were but into clean plastic bags, placed in polyethylene vials, and irradiated with a standard for a period of 2–5 min using a pneumatic tube facility at a thermal neutron flux of 2.5×10^{12} $n/cm^2/s$ at the Tajoura Nuclear Research Center. Radiochemical separation was used for determination of Al, Cu, Mg, and Mn based on solvent extraction as follows: Lyophilized samples were wet-ashed in a conical flask with 2 mL of HNO_3 and $HC10_4$ (2:1) by heating until the evolution of brown gases ceased. Ten milliliters of deionized water were added, and the solution was transferred to a separatory funnel. The flask was washed with two more milliliters of H_2O. The content of the funnel was treated with 10 mL of a saturated solution of tartaric acid, One milliliter of hydroxyquinoline, 10 mL of a mixture of $CHC1_3$, CH_3OH, and $C_4H_{11}N$ (10:2:1), and 20 mL of a buffer solution (pH 10) were added. The organic phase was separated and measured. The time from the end of irradiation until the start of measurement was 7–8 min for each sample.

Long irradiation: 100-mg samples were sealed in clean polytheylene bags, placed in an aluminum container with the standard, and irradiated for a period of 6 h at the same flux as for the short irradiation. Activities of all samples were counted on a Ge/Li detector connected to a multichannel analyzer (SEN Model 8192). The FWHM resolution of the detector was 2.2 keV for the 1332-keV photons of Co while its relative efficiency was 11%. The samples were placed directly on a cap of the Ge/Li detector. The detector was housed in a cubic lead shield to suppress background. Measurement of long-lived radionuclides was done after 2–3 wk of a decay time.

The nuclear parameters of the radioisotopes measured are summarized in Table 1. The activity of the samples was compared with that of the standards, and from the elemental concentration of the standards, trace element contents of the samples were calculated. In this calculation, necessary corrections for decay and counting times were applied.

RESULTS AND DISCUSSION

Table 2 shows that the values of Zn, Se, Cr, and Co are lower in diabetics as expected, because all the above elements are associated with glucose metabolism. It is also clear that the values of Mn, Hg, and Rb are unchanged, probably because these elements have no direct connection with glucose metabolism in the human body, whereas Al and Cu are slightly elevated. Table 3 shows a comparison of our results with some literature values reviewed by Zhuk and Kist *(10)* Mosulishvili and Shinia *(11)* and Iyengar et al. *(12)*

Cobalt is an essential element as part of vitamin B_{12} and must be supplied in the diet in the biologically active form. Nevertheless, it is a metal and is toxic in other forms. It is able to develop or raise hypertension *(13)* and cardiomyopathy. *(14)* The cobalt concentration in our samples was found to amount to 0.19 µg/g, which is in agreement with Iyengar et al. *(12)*, but in disagreement with other authors. *(10,11)*

Chromium has been shown to play an important role in the human organism as the glucose-tolerance factor for the transfer of glucose across the membrane, and it probably also plays a role in atherosclerosis and other physiological processes and biochemical pathways. *(15)* It is a difficult element to be determined reliably. In the literature, there are a wide variety of concentrations published from 0.0014 to 1.85 µg/g *(8,18)*. Our chromium values are in agreement with Mosulishvili *(11)* and Shinia and Iyengar et al., *(12)* but higher than Zhuk and Kist's values. *(10)*

Rubidium is quickly distributed by the blood to other tissues that are being maintained by the red cells. Rubidium in whole blood has been estimated to amount to about 14 µg/g *(17)*, and the element is known to affect the brain in altering behavior and manic-depressive states; it appears to act by altering basic metabolic functions *(18,19)*. Our result of 12.8 µg/g is in agreement with values reported by Mosulishvili and Shinia *(11)* and Iyengar et al. *(12)*

Selenium is a typical essential element with a narrow range of optimum concentrations. A low concentration causes a deficiency in the body, and overdosing leads to poisoning. Selenium is an essential constituent of glutathione peroxidase and other enzymes, but it also binds with some toxic heavy metals like Hg *(13)*. There are data on mean Se concentration in whole blood between 0.27–1.28 µg/g in dry blood, *(12)*

Table 1
Nuclear Parameters[9] of the Nuclides Used

Element	Isotopic abundance %	Cross section barn	Nuclide	Half life	Energy KeV
Aluminium	100.00	0.24	^{78}Al	2.3 min	1780
Copper	30.90	2.30	^{66}Cu	5.2 min	1039
Magnesium	11.30	0.027	^{27}Mg	9.5 min	1014
Manganese	100.00	13.30	^{56}Mn	2.56 h	847
Cobalt	100.00	19.00	^{60}Co	5.26 y	1173
Chromium	4.31	17.00	^{51}Cr	27.8 d	320
Mercury	29.80	4.00	^{203}Hg	46.6 d	280
Rubidium	72.2	0.90	^{86}Rb	18.7 d	1080
Selenium	0.87	30.00	^{75}Se	120 d	265
Zinc	48.9	0.46	^{65}Zn	244 d	1115

Table 2
Trace Element Concentration in Normal and Diabetic Whole Blood
in µg/g dry wt

Element	Normal	Diabetic
Al	1.23 ± 0.18	2.43 ± 0.33
Cu	3.17 ± 0.14	3.43 ± 0.09
Mg	104.00 ± 5.2	112.33 ± 8.3
Mn	0.059 ± 0.005	0.066 ± 0.001
Co	0.19 ± 0.003	0.05 ± 0.001
Cr	0.36 ± 0.006	0.25 ± 0.007
Hg	0.01 ± 0.001	0.01 ± 0.002
Rb	12.80 ± 1.3	12.50 ± 1.1
Se	1.05 ± 0.07	0.89 ± 0.08
Zn	32.70 ± 2.2	18.20 ± 0.91

Table 3
Comparison of Trace Element Concentrations in Normal Whole Blood with Literature Values[a]

Element	This work	Kist *(10)*	Mosulishvili *(11)*	Iyengar *(12)*
Co	0.19±0.003	0.015±0.001	0.024±0.022	(0.012–0.40)
Cr	0.36±0.006	0.22 ±0.01	0.38 ±0.25	(0.028–0.42)
Cu	3.17±0.14	8.20 ±0.27	-	5 (1.4–6)
Hg	0.01±0.001	0.02 ±0.001	0.04 ±0.02	(0.012–0.08)
Rb	12.80±1.3	4.57 ±0.06	15.60 ± 6	(4.7 - 24)
Se	1.05±0.07	0.48 ±0.01	0.55 ±0.2	(0.27- 1.28)
Zn	32.7 ±2.2	20.80 ±0.3	32.8 ±6.1	(17 - 37.20)

[a]The levels are in μg/g dry wt.

and these differences can possibly be explained by geographical factors. Our result of 1.05 μg/g in dry whole blood fit well with literature data. *(12)*

Zinc is one of the most important essential elements for humans. It is the essential component of more than 100 systems of biochemical pathways, mostly of metallo-enzymes, but also of nucleic acids, hormones, immunological, and other physiological systems. *(16)* Deficiency of zinc leads to growth retardation, various diseases, and in the case of total deficiency, to death. *(2,15,16)* Our result of mean zinc concentration in dry whole blood of 32.7 μg/g is in agreement with literature values. *(11,12)*

REFERENCES

1. I. E. Dreosti, *Med. J. Austral.* **2,** 117 (1980).
2. A. S. Prasad and D. Oberias, *Trace Elements in Human Health and Disease*, Academic, New York, 1976.
3. W. Mertz, *Trace Elements in Nutrition of Children*, Raven, New York, 1985.
4. M. J. Pinchin and J. Newham, Anal. Chim. Acta **90,** 77 (1977).
5. J. Versieck and R. Cornelis, *Anal. Chim. Acta* **116,** 217 (1980).
6. W. D. Ehmann and S. W. Yates, *Anal. Chem. Fundam. Rev.* **58,** 47R (1986).
7. R. M. Parr, Intercomparison of minor and trace elements, in IAEA Animal Muscle (H-4) IAEA, Vienna, 1980.
8. N. I. Ward and D. E. Ryan, *Anal. Chim. Acta* **106,** 185 (1979).
9. R. A. Nadkarni and G. H. Morrison, *Anal. Chem.* **45,** 1957 (1973).
10. L. I. Zhuk and A. A. Kist, *J. Radioanal. Chem.* **120,** 369 (1988).

11. L. M. Mosulishvili and N. I. Shinia, *J. Radioanal. Chem.* **88,** 121 (1985).
12. G. V. Iyengar, W. E. Kollmer and H. J. M. Bowen, in *The Elemental Composition of Human Tissues and Body Fluid*, Verlag Chemie, Weinheim, New York, 1978 p. 7.
13. H. K. Asteloot, *Circulation* **37,** 854 (1968).
14. C. S. Alexander, *Ann. Int. Med.* **70,** 411 (1969).
15. M. Simanoff, *Cardiovasc. Res.* **18,** 591 (1984).
16. A. S. Prasad, *Acta Pharmacol. Toxicol.* **5957,** 94 (1986).
17. E. J. Underwood, *Trace Elements in Human and Animal Nutrition*, Academic New York, 1977.
18. B. I. Carson, H. V. Elli, and J. I. McCann, in *Toxicology and Bioligical Monitoring on Metals in Humans*, Lewis Publishers, 1987, p. 200.
19. H. E. Stokinger, in *The Metals in Patty's Industrial Hygiene and Toxicology*, Wiley, New York, 1981, p. 1493.

Changes of Concentrations of the Elements Co, Cr, Sb, and Sc in Tissues of Persons with Joint Implants

C. SCHNABEL,[1] U. HERPERS,*[,1] R. MICHEL,[2] F. LÖER,[3] G. BUCHHORN,[4] AND H.-G. WILLERT[4]

[1]*Abt. Nuklearchemie, Universität zu Köln, Otto-Fischer-Str. 12–14, D 50674 Köln;* [2]*Zentraleinr. f. Strahlenschutz, Universität Hannover, D 30167 Hannover;* [3]*Orthopäd. Klinik, Universität Essen, D 45147 Essen; and* [4]*Orthopäd. Klinik, Universität Göttingen, D 37075 Göttingen*

ABSTRACT

Elevated levels of Co and Cr were found in several organs of deceased implant bearers (CoCr-alloy/polyethylene joint prostheses) by means of instrumental and radiochemical neutron activation analysis as compared to normal persons. For comparative purposes, concentrations of the elements Co, Cr, Sb, and Sc were measured in heart, kidney, liver, and spleen of the patients and normal persons. For Cr determination, a new radiochemical separation technique with sufficiently low determination limit was employed. The importance of such investigations for studying possible carcinogenic effects of corrosion products and wear particles of metallic prostheses is mentioned.

Index Entries: Trace elements; neutron activation analysis; human tissues; systemic effects of joint implants.

INTRODUCTION

The health risk of metallic prostheses owing to their corrosion and possible cancerogenic actions of corrosion products and wear particles was reported in several studies *(1–3)*. Systemic effects of joint prostheses

*Author to whom all correspondence and reprint requests should be addressed.

made of CoCr-alloys were reported by Michel et al. *(1)*, who observed drastic enrichments of Co and Cr in some distant organs of one deceased implant bearer. Black *(2)* stated that "conventional stainless-steel and Co-bases alloy components in use today can sensitize patients as well as elicit immune system responses in previously sensitized individuals." Sunderman concluded in a review *(3)* that local sarcomas may be a rare complication of prostheses made of Ni, Co, or Cr alloys.

The knowledge of normal concentrations of Ni, Co, and Cr in many human tissues is still insufficient. The compilations by Iyengar et al. *(4)* and Snyder et al. *(5)* have shown wide ranges of "normal" concentrations of these elements, and concluded that contamination problems were mainly responsible for existing variations.

Reliable determination of the abovementioned elements in human tissues requires analytical methods with sufficiently low determination limits and demonstrated accuracy. De Bruin *(6)* states that the accuracy reached by neutron activation analysis (NAA) is not reached by any other routine spectroscopic technique. Radiochemical NAA (RNAA) is advantageous in providing better determination limits than instrumental NAA (INAA). Schramel *(7)* recommends RNAA for Cr determination and INAA for Co determination in biomedical and environmental samples. Furthermore, he emphasises that the essentially blank-free nature and low sensitivity to chemical interferences are responsible for high accuracy of NAA.

For quantification of systemic effects of metallic implants, knowledge of both trace element concentrations in organs of deceased implant bearers and normal ranges of elements is required. Contamination-free sampling and sample handling is a prerequisite for the accurate determination of naturally occurring levels of some metals, such as Cr and Co. For determination of normal Cr values, employing a single-element radiochemical separation is usually mandatory to be able to achieve a sufficiently low determination limit.

EXPERIMENTAL

Tissues from deceased implant bearers who had joint prostheses made of CoCr-alloy/polyethylene for 1 mo up to 7 yr were analyzed. One patient (K) had a knee joint prosthesis for 4 yr, all other implants were hip joint prostheses. Samples were taken with standard surgical devices. To remove possible surface contamination, the surfaces of the tissues were cut off using a Suprasil knife and cleaned plastic devices in a clean bench. Portions of 35–75 mg of the lyophilized tissues were sealed in precleaned Suprasil AN vials under dust-free conditions. Irradiations were carried out at FRG-1 in Geesthacht for 18–24 h in a thermal neutron flux of 4.0–5.4 · 10^{13} n/cm^2/s. Owing to a relatively high fast-to-thermal neutron flux ratio, Cr concentrations had to be corrected for the ^{54}Fe(n,α)^{51}Cr interfering reaction. The elements Co, Sb, and Sc were determined by INAA whereas Cr was assayed using a new RNAA procedure with the Cr yield determination, which has been described elsewhere *(8)*.

RESULTS AND DISCUSSION

For the reliable determination of Cr and Co determination in mammalian tissues, strict contamination control and adherence to quality-assurance procedures are necessary. The accuracy of our RNAA procedure for Cr determination is demonstrated by control analyses of NBS SRM-1549 Milk Powder and NBS SRM-1577 Bovine Liver. Our Cr values of 3.1 μg/kg (n = 1) and 72 ± 11 μg/kg (n = 7) for the NBS Milk Powder and Bovine Liver *(8)*, respectively, compare well with the respective certified values of 2.6 ± 0.7 μg/kg *(9)* and 88 ± 12 μg/kg *(10)*. For Co, our value of 152 ± 12 μg/kg (n = 8) for NBS SRM-1571 Orchard Leaves is in very good agreement with a value of 160 ± 47 μg/kg compiled by Gladney et al *(11)*. Favorable conditions for analysis on using Suprasil AN vials can be demonstrated by low blank values found for Co (58 ± 14 ng/kg) and Sb (upper limit of 21 ng/kg). These values can be considered as proof that sample preparation was performed in dust-free conditions.

Results of the element determination in heart tissue of five implant bearers and four normal persons are shown in Table 1 together with some other normal literature values. The Cr concentrations determined in normal persons are in agreement with those reported by Michel et al. *(1)* and Wester *(12)*. On the contrary, the Co levels determined in our normal persons are higher than the range given by Wester *(12)* and the values reported by Michel et al. *(1)*. The Cr values of patients A, D, and E compare well with the normal range, whereas those of patients B (duration of implantation of a hip joint prosthesis 6 mos) and C (hip prosthesis for 1 mo) are significantly elevated and indicate burdening with the element. For Co in heart, no considerable differences were found between the patients and control persons.

In the case of kidney (cf Table 2), our normal values for Cr are not in contradiction with the upper limit determined by Michel et al. *(1)*. For Co, our data for persons without implants are in the range observed by Ehmann et al. *(13)*, but exceed the values of Michel et al. *(1)*. Data for Cr in patient C and Co in patient E can be interpreted as elevated levels.

The element concentrations in liver, which should be considered as the best analyzed human organ, are listed in Table 3. Our data for Cr in liver of normal persons (mean value 25.7 μg/kg, range 8.1–47.7 μg/kg) are in excellent agreement with those found by Lievens et al. *(14)*. For Co concentrations, our values for normal persons are in the range given by Ehmann et al. *(13)* and not in contradiction with the results of Zeisler et al. *(15)*, but higher than the range determined by Lievens et al. *(14)*. Liver of patient K (knee joint prosthesis) exhibits by nearly two orders of magnitude higher Cr and Co concentrations than those of normal persons, and slightly elevated concentrations of both elements were also found in liver of patient C.

Table 1
Concentrations of Cr, Co, Sc, and Sb in Heart of Deceased Implant Bearers in Comparison with Normal Values (μg/kg, dry wt)

Subject	Cr	Co	Sc	Sb	Ref.
Patient A	17.4±6.0	119±4	0.68±0.23	5.54±1.42	
Patient B	658±90	270±49	1.57±0.25	2.77±0.54	
Patient C	211±81	153±8	1.13±0.33	12.6±2.1	
Patient D	36.1±7.0	211±6	0.44±0.14	4.30±0.58	
Patient E	76±36	71.7±3.3	<0.57	6.44±1.96	
Normal (4)[a]	<39.2±8.2	139±40	<0.47	6.35±2.32	
Literature values					
Normal (5)[a]	28±12	31±36	0.51±0.27	2.4±1.8	1
Normal	6.2-73.3	3.3-66	0.01-0.37	2.2-14.7	12

[a]Number of subjects.

Table 2
Concentrations of Cr, Co, Sc, and Sb in Kidney of Deceased Implant Bearers in Comparison with Normal Values (μg/kg, dry wt)

Subject	Cr	Co	Sc	Sb	Ref.
Patient A	61±29	60.9±2.2	<0.47	4.69±0.68	
Patient B	47.5±7.0	60.5±3.1	0.69±0.24	8.57±1.28	
Patient C	130±43	49.4±3.2	1.24±0.37	19.8±1.8	
Patient D	<88.8	81.8±3.2	0.98±0.26	4.21±0.71	
Patient E	66±26	251±8	<1.03	37.9±2.3	
Normal (4)[a]	26±12	65.9±8.3	<0.67	4.17±0.86	
Literature values					
Normal (5)[a]	<39	27±25	0.92±1.0	5.2±2.8	1
Normal (9)[a]		8-125	3.3-5.0	12.5	13

[a]Number of subjects.

Results of analysis of spleen samples are shown in Table 4. Our values for Cr in spleen of normal persons (mean value 41 μg/kg) are lower than those of Michel et al. *(1)* (mean value 134 μg/kg, range <34–258 μg/kg), whereas there is agreement between both data sets for Co concentrations. Distinct elevation of Cr concentration was found for patient C compared to normal persons. No drastically changed Co concentrations were observed in spleen of deceased implant bearers.

Table 3
Concentrations of Cr, Co, Sc, and Sb in Liver of Deceased Implant Bearers in Comparison with Normal Values (μg/kg, dry wt)

Subject	Cr	Co	Sc	Sb	Ref.
Patient A	<12.4	222±8	<2.36	6.25±0.90	
Patient B	24.3±4.8	85.7±2.9	2.24±0.48	4.69±0.58	
Patient C	106±45	274±11	<1.60	25.7±2.1	
Patient D	19.9±6.7	64.9±2.4	<0.59	1.82±0.56	
Patient E	<121	75.3±4.4	6.10±1.28	6.74±2.02	
Patient K	1405±93	7160±240	4.62±1.01	2.46±0.25	
Normal (4)[a]	25.7±8.8	192±35	2.15±1.10	25.9±12.8	
Literature values					
Normal (4)[a]	<14	117±100	3.7±3.0	5.6±8.1	1
Normal(10)[a]		66-310	1.0-11.0	22-107	13
Normal (5)[a]	6.9-34.5	79-134	0.34-1.38	10-69	14
Normal(36)[a]	379±379	145±35	0.20±0.21		15

[a]Number of subjects.

Table 4
Concentrations of Cr, Co, Sc, and Sb in Spleen of Deceased Implant Bearers in Comparison with Normal Values (μg/kg, dry wt)

Subject	Cr	Co	Sc	Sb	Ref.
Patient A	34.2±8.5	75.5±3.0	1.24±0.29	3.46±0.67	
Patient B	29.8±10.0	56.0±4.0	8.64±1.76	15.7±1.8	
Patient C	351±94	63±12	5.57±0.82	3.28±0.64	
Patient D	35.8±7.5	33.8±1.9	5.29±1.07	5.38±1.28	
Patient E	50.7±10.0	61.3±3.5	1.39±0.34	8.81±1.53	
Normal (4)[a]	41±15	22.6±3.7	<0.8±0.2	4.65±2.17	
Literature values					
Normal (4)[a]	134±117	28±21	10±12	2.4±1.3	1

[a]Number of subjects.

Thyroid gland and lymphatic tissues were also analyzed for patient K. Elevated Co concentration of 973 μg/kg was found in thyroid gland as compared to normal values of 87 ± 15 μg/kg determined by Kvíčala et al. *(16)*. Very high Co enrichment was also found in two samples of lymphatic tissues of this implant bearer (2952 and 3166 μg/kg), because Michel et al. *(1)* reported 10 ± 9 as normal values. For Cr, normal concentrations in thyroid gland and lymphatic tissues have not been established yet, so no comparison can be made.

It can be summarized that Co and Cr concentrations in organs of deceased implant bearers were elevated by up to two orders of magnitude compared to normal values. This especially applies to results for patient K with a knee prosthesis. It can be considered that a higher mechanical stress in knee is responsible for the higher elevation of the element concentrations in the organs analyzed compared to those of patients with hip prostheses.

CONCLUSIONS

Systemic effects of joint prostheses made of CoCr-alloy/polyethylene were demonstrated by elevated concentrations of Co and Cr in various organs rather distant from the implants. Because of possible toxic, allergene, and cancerogenic effects of compounds of Cr, Co, and Ni at increased levels, corrosion of metallic prostheses should be considered as a severe problem if frequency and duration of implantations are to be increased in the near future. Contribution was also made to the knowledge of reliable normal Cr concentrations in several human organs, which should facilitate detection of elevated Cr levels as a consequence of systemic effects of metallic implants. This was achieved by employing severe precautions to prevent (or remove) sample contamination prior to analysis and applying a new RNAA procedure with a sufficiently low limit of determination for Cr.

ACKNOWLEDGMENTS

The authors are grateful to the staff of the GKSS, Geesthacht. This work has been partly supported by the Deutsche Forschungsgemeinschaft, Bonn.

REFERENCES

1. R. Michel, M. Nolte, M. Reich, and F. Löer, *Arch. Orthop. Trauma Surg.* **110,** 61 (1991).
2. J. Black, *J. Bone Joint Surg. (Br.)* **70,** 521 (1988).
3. F. W. Sunderman, Jr., in *Biocompatibility of Co-Cr-Ni-Alloys*, H. F. Hildebrand and M. Champy, eds., NATO-ASI Series, Plenum, New York, 1987, pp. 11–19.

4. G. V. Iyengar, W. E. Kollmer, and H. J. M. Bowen, eds., *The Elemental Composition of Human Tissues and Body Fluids*, Verlag Chemie, Weinheim, 1978.
5. W. S. Snyder, M. J. Cook, E. S. Nasset, L. R. Karhausen, G. P. Howells, and I. H. Tipton, eds., *Report on the Task Group on Reference Man ICRP Publication 23*, Pergamon, Oxford, 1975.
6. M. de Bruin, *J. Radioanal. Nucl. Chem.* **160,** 31 (1992).
7. P. Schramel, *J. Radioanal. Nucl. Chem.* **168,** 215 (1993).
8. C. Schnabel, U. Herpers, and R. Michel, *J. Radioanal. Nucl. Chem.,* **178,** 19 (1994).
9. National Bureau of Standards, Certificate of Analysis for Standard Reference Material 1549, Milk Powder, Gaithersburg, 1985.
10. National Bureau of Standards, Certificate of Analysis for Standard Reference Material 1577, Bovine Liver, Washington, DC, 1977.
11. E. S. Gladney, B. T. O'Malley, I. Roelandts, and T. E. Gills, NBS Spec. Pub. 260–11, Washington DC, 1987.
12. P. O. Wester, *Acta Med. Scand.* **178,** 765 (1965).
13. W. D. Ehmann, X.-X. Ding, S. S. Khare, M. A. Lovell, B.-F. Ni, L. Tandon, D. E. Vance, and D. E. Wenstrup, *J. Radioanal. Nucl. Chem.* **168,** 223 (1993).
14. P. Lievens, J. Versieck, R. Cornelis, and J. Hoste, *J. Radioanal. Chem.* **37,** 483 (1977).
15. R. Zeisler, S. H. Harrison, and S. A. Wise, NBS Spec. Pub. 656, Washington DC, 1983.
16. J. Kvíčala, J. Havelka, J. Zeman, and J. Nëmec, *J. Radioanal. Nucl. Chem.* **149,** 267 (1991).

Analysis of Elemental Absorption and Excretion in Mice Bearing Malignant Ascites

YUAN-YAW WEI AND CHIEN CHUNG*

Institute of Nuclear Science, National Tsing Hua University, Hsinchu, 30043 Taiwan, R.O.C.

ABSTRACT

Feces, kidney, and small intestine were sampled from the mice bearing malignant ascites at different stages of tumor growth to investigate the kinetics of elemental distribution in the body. The contents of 14 elements in samples were determined by instrumental neutron activation analysis (INAA). The elements Br, Cl, Cu, Na, Se, and Zn increased, whereas Mn and Rb decreased in kidney with the growth of the tumor. The elements Cl, Na, and Fe, however, appeared to have significantly different behavior in the small intestine.

Index Entries: Malignant ascites; INAA; trace element determination; metabolism; apparent absorption.

INTRODUCTION

The biological role of minor and trace elements in cancer has been extensively investigated in recent years. Statistically significant differences of some elements, including Ca, Cl, Cu, Fe, K, Mg, Mn, Na, Rb, Se, V, and Zn, between malignant and normal tissues have been reported in patients with various forms of cancer *(1–3)*. These results indicate that minor and trace elements play an important role in tumor pathology.

Although the function of trace elements in health and disease has received relatively much attention in recent years, the investigations mainly aimed at assessing the nutritional bioavailability and essentiality of trace elements in organism. From the imbalance of elements between

*Author to whom all correspondence and reprint requests should be addressed.

Biological Trace Element Research Editor: G. N. Schrauzer © 1994 by Humana Press Inc.

normal and tumor cells, however, it seems that a different biodistribution of trace elements has correlation with metabolism. The kinetic analysis of elemental metabolism should therefore be further studied.

The elemental analysis of malignant ascites tumor, liver, and blood of tumor-bearing mice has been reported in our lab *(4)*. It has been found that some elements appeared to have significantly different concentrations in ascites tumor and liver during various tumor-growth stages. These results suggest that the biodistribution of elements in the body of tumor-bearing mice is influenced. In this study, the feces and the kidney and small intestine, which have relationships with excretion and absorption of elements, were analyzed by INAA to investigate the correlation between elemental metabolism and tumor growth.

MATERIALS AND METHODS

Male ICR mice about 6–8 wk old were fed a normal diet for 1 wk after purchase. Malignant ascites of experimental animals were induced by ip injection of 0.1 mL S180 tumor cells from 10-d old tumor-bearing mice. The life-span of tumor-bearing mice is about 15–21 d.

The growing time of the tumor was divided into four stages, with 4 d in each of the first three stages and the final stage lasting until tumor cells inoculated. In every stage, the daily amounts of diet and feces were recorded and analyzed by INAA to assess the metabolic relationship of apparent absorption of minor and trace elements in mice bearing malignant ascites. The changes of body weight, urine excretion, and water intake of animals were also recorded to understand excretion and ingestion differences between normal and tumor-bearing mice during the tumor-growing period. Samples of kidney and small intestine were collected after the animals were killed and dissected. All samples of diet, feces, kidney, and small intestine were crushed in polyethylene vials with glass Pyrex^{tm} pestles after freeze-drying for further homogenization. All glassware, polyethylene vials, and bags used in this work were precleaned by soaking in Merck 4 mol/L nitric acid overnight followed by rinsing several times with deionized water to prevent contaminations during pretreatment.

About 200–300 mg of feces and diet as well as 50–150 mg of kidney and small intestine were prepared. On the other hand, aliquots of NBS SRM 1577a, bovine liver 100–200 mg were used to evaluate element contents in the samples. Each sample was double encapsulated in polyethylene bags to prevent contamination prior to irradiation. The samples were irradiated in a Tsing Hua Open-Pool Reactor at a neutron flux around 3.5×10^{12} $n/\text{cm}^2/\text{s}$ for 5 min followed by short counting, and again at 1.0×10^{12} $n/\text{cm}^2/\text{s}$ for 30 h, followed by long counting. To monitor the possible fluctuation of the neutron flux, each sample package was paired with a 5–15 mg nickel powder monitor in the short irradiation and with a 4-µg cobalt monitor (pipeted onto filter paper) in the long irradiation for the correction of local flux variations.

Activated samples were analyzed by a gamma spectrometer consisting of a 15% relative efficiency HPGe detector, with a resolution of 2.31 keV at the 1332.5-keV photopeak of ^{60}Co, a CANBERRA Series 35 plus multichannel analyzer, and a personal computer with SAMPO 90 software. To evaluate significance of differences between studied data in the control and experimental mice, the two-tailed Student's *t*-test was applied, in pair, 95% ($p < 0.05$) or 99% ($p < 0.01$) confidence levels.

RESULTS

The results concerning the body weight, the amount water and food ingested, and excrements during the tumor-growing stages of mice are shown in Table 1. The body weight of tumor-bearing mice increases significantly after stage II in comparison with controls. The amounts of urine and feces significantly decrease during the entire period of tumor growth. The intake of diet and water, however, also significantly decreases in stages III and IV. The effects of apparent absorption of elements, defined as a difference of element amounts in intake to those in fecal excretion, were not proportional to the decrease of ingestion and excretion. Figure 1 shows that the apparent absorption of Br, Cl, K, Mg, Na, Rb, Se, and Zn significantly decreases at stage III with Cl, K, and Na leveling off in the final stage. For other elements, there are no significant differences.

For the excretion and absorption organs, the element concentrations in the kidney and small intestine show different patterns for different elements as depicted in Figs. 2 and 3. Figure 2 shows that the elements Br, Cl, Cu, Na, Se, and Zn increase significantly during the whole tumour growth, whereas Mn and Rb decrease significantly. The concentrations of Cl, Mn, Na, and Rb show significant differences after stage II, but Cu, Zn, and Se only at the final stage. In the small intestine, the elements Na, Cl, and Fe show significant differences, as illustrated in Fig. 3. Both Na and Cl increase significantly to different extents. The reverse trend is observed for Fe, however. We also analyzed the total amount of elements in diet and feces, and calculated the ratio of excretion to intake for each element. There is no significant difference in the ratio of tumor to normal mice.

DISCUSSION

The present study aims at understanding the metabolic correlation of elements between ingestion and excretion for tumor-bearing mice. Among elements analyzed, the concentration of Na and Cl is higher in both the kidney and small intestine of tumor-bearing mice than in normal mice, as illustrated in Figs. 2 and 3. It has been reported that the high Na and water reabsorption are involved in the pathogenesis of ascites and hepatorenal syndrome in liver cirrhosis *(5)*. On the other hand,

Table 1
Average Body Weight, Ingestion, and Excretion During the Tumor-Growing Stages of 20 Mice

Item	Group[a]	Stage of tumor growth I	II	III	IV
Body weight, g	N	30.8 (26)[b]	32.6(37)	34.3 (44)	34.5 (29)
	T	30.6 (29)	39.9 (53)**	45.6 (38)**	47.6 (55)**
Urine, mL/day	N	2.4 (2)	2.0 (5)	2.6 (7)	2.0 (3)
	T	1.1 (8)	0.6 (2)**	0.3 (2)**	0.3 (2)**
Feces, g/day	N	1.43 (42)	1.46 (40)	1.42 (18)	1.23 (13)
	T	1.59 (38)	1.23 (28)	1.10 (38)	0.73 (38)*
Diet, g/day	N	5.6 (4)	6.1 (7)	6.1 (10)	4.9 (5)
	T	6.2 (13)	4.9 (10)	3.6 (11)**	3.0 (13)*
Water, mL/day	N	7.5 (21)	6.6 (22)	7.9 (27)	6.8 (23)
	T	7.5 (17)	7.2 (16)	4.3 (10)*	3.9 (16)*

[a]N = normal mice, T = tumor-bearing mice.
[b]The uncertainty of the last digits of the mentioned values.
*$p < 0.05$; **$p < 0.01$ (*t*-test).

Epstein et al. *(6)* have reported that the cirrhotic patients with ascites maintain similar sodium concentration in serum to that of the patients without ascites. However, in our previous report, we found that a high concentration of Na and Cl does appear in blood of the mice bearing malignant ascites *(4)*. Other results also indicate the biochemical levels of creatinine and uric acid in serum of these tumor-bearing mice do not change significantly in comparison with normal mice. Hence, the renal function in tumored mice should be normal. These results suggested that the increasing Na concentration in organs of the mice bearing malignant ascites is closely correlated with malignant ascites, unlike in nonmalignant ascites.

On the other hand, significant changes of Br, Cl, Na, and Rb in kidney appear in Fig. 2. These results imply that an electrolytic imbalance exists in tumor-bearing mice. Other significant changes of concentrations of the elements Cu, Mn, Se, and Zn can be viewed as cofactors for enzymes that play essential roles in biochemical and physiological function.

Several reports have indicated that Na is associated with tumor growth *(2,7)*. In this work, an increasing Na concentration was found in malignant ascites, blood, and liver as well as in kidney and small intestine. Although apparent absorption of Na was lower in tumored mice, the Na concentration in analyzed organs was still higher than in normal mice. These results form strong evidence of the correlation between Na retention and tumor growth. The main excretory pathway of Br, Cl, K, Na, Rb, and Se is not by feces *(8,9)* the excreted amounts of these elements in feces are deeply influenced by the ingested amounts. The tumor-bearing mice have less significant apparent absorption of these

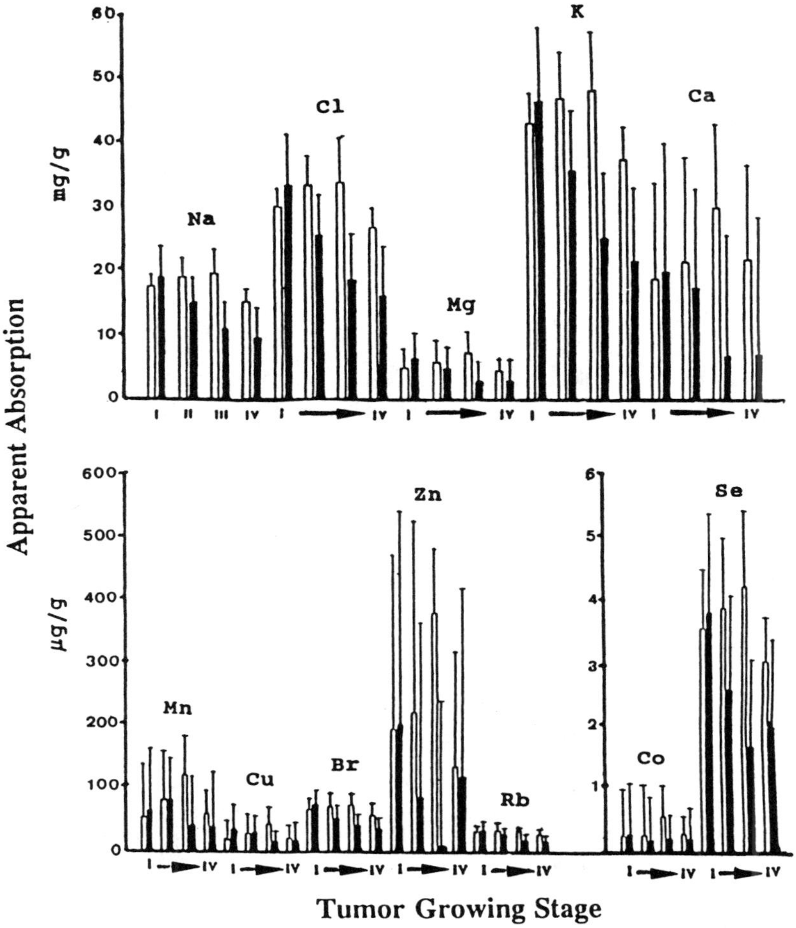

Fig. 1 Average apparent absorption of elements for five normal □ and five tumor-bearing ■ mice during the tumor-growing stages.

elements when the ingested diet decreases. It is suggested that the increase or decrease of elements in tumor-bearing mice has probably no direct association with the intake from intestine.

The Na concentration significantly changes in small intestine, but the Cl content has no parallel change until the final stage. Because of the secretion of gastric juice, consequently chloride mixing with foods and moving along the intestinal tract, the chloride is readily absorbed into the circulation *(8)*. Hence, Cl in small intestine does not significantly change at the earlier stages.

In conclusion, the present study shows that the decrease of elemental apparent absorption in experimental mice has absolutely no correlation with organ biodistribution of elements. Some trace elements, such as

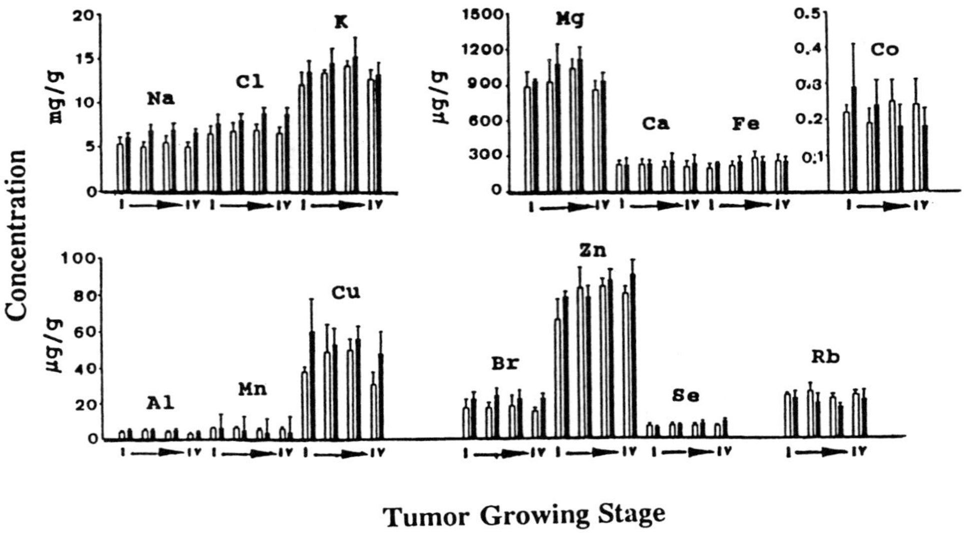

Fig. 2 Average elemental concentration in kidney for six normal □ and six tumor-bearing ▩ mice during the tumor-growing stages.

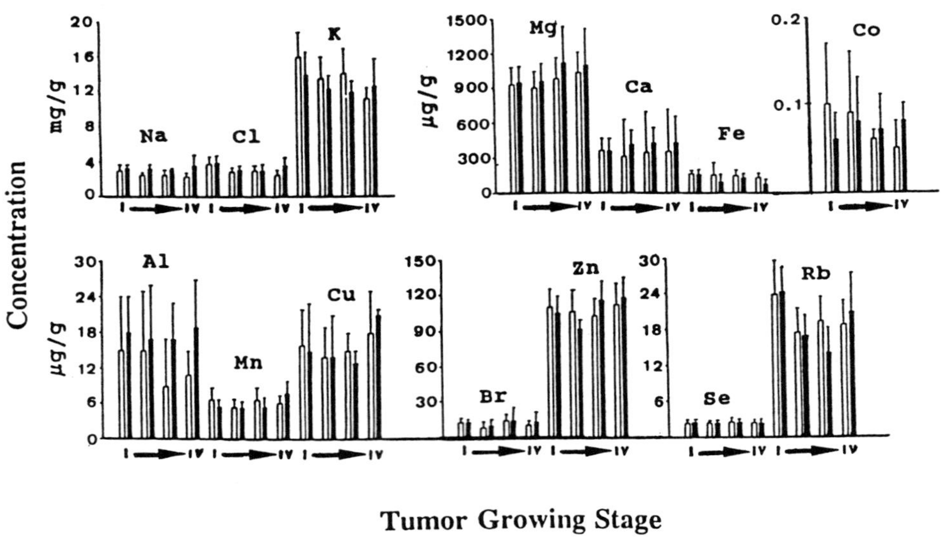

Fig. 3 Average elemental concentration in small intestine for six normal □ and tumor-bearing ▩ mice during the tumor-growing stages.

Cu and Se, appeared to increase in kidney, whereas they decreased in liver. The difference may be associated with the rebiodistribution of trace elements among different organs. These observations merit further exploration of the mechanism of biochemical metabolism of trace elements in tumor pathology.

ACKNOWLEDGMENT

The authors wish to thank the staff members of the Nuclear Science and Technology Development Center, National Tsing Hua University, for their support of this work.

REFERENCES

1. M. K. Schwartz, *Cancer Res.* **35,** 3481 (1975).
2. I. Zs. Nagy, G. Lustyik, G. Lukacs, V. Zs. Nagy, and G. Balazs, *Cancer Res.* **43,** 5395 (1983).
3. E. N. Drake II and H. H. Sky-peck, *Cancer Res.* **49,** 4210 (1989).
4. Y. Y. Wei and C. Chung, *J. Radioanal. Nucl. Chem.* **171,** 383 (1993).
5. P. Gentilini and G. Laffi, *Digestion* **43,** 1 (1989).
6. M. Epstein, R. Loutzenhiser, P. Norsk, and S. Atlas, *Am. J. Nephrol.* **9,** 133 (1989).
7. R. L. Sparks, T. B. Pool, N. K. R. Smith, and I. L. Cameron, *Cancer Res.* **43,** 73 (1983).
8. D. L. Tsalev and Z. K. Zaprianov, in *Atomic Absorption Spectrometry in Occupational and Environmental Health Practice,* vol. I, CRC, Florida, 1983, pp. 117–215.
9. C. H. Robinson, M. R. Lawler, W. L. Chenoweth, and A. E. Garwick, in *Normal and Therapeutic Nutrition,* 17th ed., Macmillan, New York, 1986, pp. 111–155.

Analysis of Biological Samples

Neutron Activation Analysis for the Determination of Trace Elements in Biological Materials

JACQUES VERSIECK

Department of Internal Medicine, Division of Gastroenterology, University Hospital, De Pintelaan 185, B-9000 Ghent and Institute of Nuclear Sciences, Laboratory of Analytical Chemistry, Proeftuinstraat 86, B-9000 Ghent, Belgium

ABSTRACT

Neutron activation analysis, in both its radiochemical and instrumental forms, is a precious technique for the determination of trace elements in biological materials. Probably its most important advantage is its relative freedom from errors resulting from contamination of the samples. Invaluable characteristics are also its excellent sensitivity, outstanding selectivity, and remarkable multielement capability. It is, however, necessary to warn against uncritical expectations. This is best illustrated by the seriously inconsistent results obtained in several laboratories.

Because of the necessity to have access to a nuclear research reactor, the stringent safety rules to be observed, the rather high costs of the analyses, the relatively low sample throughput, and the sometimes long delay between the taking of a sample and the obtaining the final result, the use of neutron activation analysis remained restricted to a few—essentially research—laboratories. It found its main application in solving arduous problems and in paving the way for other analytical techniques better suited to routine applications.

Index Entries: Neutron activation analysis; trace elements; biological materials; biological reference materials; blood plasma; blood serum.

INTRODUCTION

Over the last 30 years, neutron activation analysis grew to be an extremely powerful analytical technique for trace element determinations in biological materials. It started as a single-element technique at a moment when only Na(Tl) detectors were available, but with the advent of high-resolution Ge(Li) detectors or, more recently, high-purity (HP) Ge detectors and associated multichannel analyzers, it really became a multielement technique.

A review of the literature demonstrates that neutron activation analysis was used with success by numerous researchers to measure trace element levels in biological materials. As an example, the technique played a decisive role in the establishment of the "normal levels" in human blood plasma or serum of several "difficult" elements, such as manganese *(1–3)*, chromium *(4)*, molybdenum *(5)*, and vanadium *(6,7)*. It is also worth mentioning that somewhat more than 50% of the selected reference values advanced by Versieck in his 1985 critical review *(8)* were obtained by neutron activation analysis. That neutron activation analysis proved to be capable of providing highly accurate and precise results is also illustrated by the role it played in the certification of biological reference materials *(9)*, especially in those with extremely low levels of many elements *(10,11)*.

On the other hand, a comparison of the published results makes it clear that flagrant inconsistencies were obtained by many investigators using neutron activation analysis. This is shown in Table 1, which catalogs some reportedly normal plasma or serum concentrations published by Astrug et al. *(12)*, Nakahara et al. *(13,14)*, Ward and Pim *(15)*, and Versieck and colleagues *(3–7,16)*. The figures tell their own story: it is evident that many values were largely overestimated, and a few underestimated when compared to the results obtained in Ghent. Obviously, the elements occurring at the lowest levels caused the greatest problems, although it should not pass unnoticed that, apparently, Nakahara et al. *(13)* had difficulties even in determining zinc. That the discrepancies should not be ascribed to so-called physiological variations (age, sex, geographical influences, and others) is illustrated by the fact that comparable inconsistences were found in biological reference materials: for example, in NBS bovine liver (SRM 1577) values for chromium published before the level was certified to be 0.088 ± 0.012 μg/g vary from <0.005 μg/g *(17)* up to 0.5–3.5 μg/g *(18)*.

ADVANTAGES

Sensitivity

It is generally agreed that neutron activation analysis has shown great sensitivity for many elements. Fixed values determining the sensitivities of detection are the atomic weight of the element, the isotopic abundance of the target nuclide, and its cross-section for thermal neutrons; variable parameters are the neutron flux, the irradiation time, the decay period, and the counting efficiency of the detector. Experimental sensitivities may be idealized because of matrix problems, difficulties in radiochemical separations, and other problems associated with the analysis of complex, real samples.

Table 1
Comparison of Trace Element Concentrations (Mean ± SD) Determined in Normal Human Blood Plasma or Serum by Different Investigators Using Neutron Activation Analysis

Element` (unit)	Astrug et al. (1984)	Nakahara et al. (1979)	Nakahara et al. (1980)	Ward and Pim (1984)	Ghent University (1974 - 1991)
Vanadium (ng/mL)				9.5 ± 1.3	0.047 ± 0.012 (♂)
					0.024 ± 0.005 (♀)
Chromium (ng/mL)	13 ± 11	782 ± 495	2.14 ± 11.4	6.1 ± 1.0	0.160 ± 0.083
Manganese (ng/mL)		14.2 ± 26.2	3.49 ± 2.10	2.2 ± 0.4	0.57 ± 0.13
Cobalt (ng/mL)	1.9 ± 0.6	89.3 ± 71.3	5.01 ± 3.09	0.37 ± 0.06	0.108 ± 0.060
Copper (μg/mL)		2.60 ± 3.57	1.01 ± 0.38	1.01 ± 0.07	1.07 ± 0.24
Zinc (μg/mL)	1.38 ± 0.43	3.40 ± 3.88	1.13 ± 0.28	1.13 ± 0.09	0.94 ± 0.13
Selenium (μg/mL)	0.027 ± 0.007	0.340 ± 0.135	0.196 ± 0.071	0.115 ± 0.012	0.13 ± 0.02
Molybdenum (ng/mL)				2.4 ± 0.4	0.58 ± 0.21
Tin (ng/mL)				34 ± 6	0.502 ± 0.096

Guinn and Hoste *(19)* published a series of "best detection sensitivities" under well-defined irradiation and counting conditions and in the absence of interfering activities. Additional information has been published by Erdtmann and Petri *(20)*. In the author's opinion, Erdtmann's *(21)* neutron activation tables give particularly useful information. They catalog the activation products in decays/s and decays/μg of an element after varying irradiation times (up to saturation, when S = 1).

Selectivity

In radiochemical neutron activation analysis (RNAA), the selectivity can be excellent with decontamination factors of more than a millionfold from other elements. This implies considerable skill of the investigator in the choice and execution of the separation method. Instrumental neutron activation analysis (INAA) using Ge(Li) detectors or high-purity (HP) Ge detectors is also highly selective *(22)*. As soon as their energies differ by some 5 or 6 keV (e.g., the 563.2- and 569.3-keV and the 795.8- and 801.9-keV peaks of ^{134}Cs), photopeaks are sufficiently resolved to allow peak integration (calculation of the net counts of the peak) with reasonable reliability; only when their energies differ by 2 or 3 keV (e.g., the 602.7-keV peak of ^{124}Sb and the 604.7-keV peak of ^{134}Cs), they largely overlap. In such cases, however, it is often possible to resolve such peaks by counting repeatedly to see how they decay. It should also be noted that numerous radioisotopes emit several γ-rays, which can also help to unravel the overlapping photopeaks.

Relative Freedom of Contamination

Probably the most praised advantage of neutron activation analysis is its relative freedom from errors resulting from contamination of the samples with exogenous material. Indeed, with a few exceptions, e.g., if an element can only be determined via a very short-lived radioisotope as in the case of vanadium, mineralization and chemical separations can be postponed until after the irradiation when impurities in the reagents, additions from the laboratory equipment, or airborne particulate matter no longer affect the final result. However, during the collection, transport, and manipulation of the samples, the exact same amount of care is needed as in the case of all other techniques (judiciously selected instruments, high-purity collection vessels cleaned with meticulous care, clean room conditions, and so forth). Neglecting these precautions may have disastrous consequences. At the moment of the analysis, *stricto sensu*, no amount of care will produce an accurate result when an error has been introduced earlier *(23)*. As mentioned already, the freedom from contamination is only relative! Some problems are even specific for neutron activation analysis, namely irradiation and wet-ashing blanks *(24,25)*.

Multielement Capability

A much publicized advantage is also that neutron activation analysis is ideally suited for multielement determinations. This is true. However, apparently, this characteristic has been seriously overestimated by some investigators. In certain papers *(15,26)*, the impression is created that numerous elements can easily be measured even in difficult matrices (e.g., human blood plasma or serum) using relatively simple experimental conditions. This is utopia. Indeed, a close look reveals that the vast majority of the figures reported to support this statement are erroneous: most are seriously overestimated so that no valid conclusions can be drawn!

Quality Assessment

Another interesting feature is that all sources of random variation can be identified, in other words, that it is possible to predict the standard deviation of a result: this has been defined as the "*a priori* precision." In the "analysis of precision" *(27)*, it is compared with the actually observed variation. The results may serve as a method of continuous quality assessment.

SOURCES OF ERROR

Flux Gradients and Self-Shielding

Common sources of error are horizontal or vertical flux gradients. For accurate determinations, they must be compensated for or eliminated, e.g., by rotating the samples or by introducing a mathematic correction.

Self-shielding may also introduce significant errors, namely, when there is a difference in absorption by the matrices of sample and standard. Biological matrices being poor neutron absorbers, generally they do not pose a great problem; however, the phenomenon must be kept in mind when designing the standard.

Interfering Reactions

One of the best documented sources of error is interference or threshold reactions by fast or fission neutrons. The potentially resulting errors may easily be calculated according to a formula that may be found in all textbooks on neutron activation analysis. Although the cross-sections (average fast neutron cross-sections or σ_f) for these reactions are low when compared to those for thermal activation (σ_{th}), serious errors may ensue. Some of the most important reactions to be remembered by biologically interested researchers are $^{31}P(n,\alpha)^{28}Al$, $^{54}Fe(n,\alpha)^{51}Cr$, $^{56}Fe(n,p)^{56}Mn$, and $^{65}Cu(n,p)^{65}Ni$. When human blood serum (with a phosphorus content of approx 135 µg/mL) is irradiated in a neutron flux with a thermal-to-fast ratio (Φ_{th}/Φ_f) of 10, the $^{31}P(n,\alpha)^{28}Al$ reaction gives rise to an "apparent" aluminum concentration of approx 97 ng/mL (estimated true normal value: approx 3 ng/mL)!

Spectral Interferences

Spectral interferences are mutual interferences of radionuclides emitting γ-rays with similar energies so that they cannot be resolved, e.g., the 279.5-keV photopeak—one of the six main energies of ^{75}Se—interferes with the 279.2-keV photopeak of ^{203}Hg, the only energy emitted by that isotope. In seeking out possible spectral interferences, it is necessary to have a catalog of γ lines of the radionuclides arranged according to their energies and a table of radionuclides showing all γ-rays associated with their radioactive decay. Unfortunately, sum peaks (peaks appearing from the coincident counting of two γ-rays or of a X-ray and a γ-ray) and single or double escape peaks (of high-energy γ-rays) are practically never listed. When overlooked, some interferences may introduce very serious errors: in the presence of high ^{82}Br activity sum peaks of, 1173.4 keV (554.3 + 619.1 keV) and 1330.8 keV (554.3 + 776.5 keV) emerge in the spectrum and may give highly erroneous results for cobalt measured via its ^{60}Co radioisotope (γ-ray of 1173.2 and 1332.5-keV photopeaks).

Other Problems

Still other problems can arise when a radionuclide has not been equilibrated with its carrier, when a radiochemical separation is carried out imperfectly, or when the counting efficiency differs for sample and standard (different shapes or sizes, different γ-ray attenuation). In September 1982, the International Atomic Energy Agency convened an advisory group meeting on quality assurance in biomedical neutron activation analysis. The published report *(28)* considers potential errors at all stages of the analytical process.

CONCLUSIONS

Neutron activation analysis is a powerful technique for the determination of trace elements in biological samples. Indeed, it is capable of generating highly accurate and precise results even for elements occurring at the ng/mL or ng/g level and below—at least in competent hands. Some of the figures quoted in this article will have made it also clear, however, that it is not a kind of panacea that guarantees reliable results without effort. Other techniques have become competitive in sensitivity and selectivity. However, neutron activation analysis still retains a special place because of its relative freedom from contamination and its largely matrix-independent multielement capability.

ACKNOWLEDGMENT

The author thanks Lidia Vanballenberghe for her more than 20-year-long invaluable assistance in the laboratory, as well as for her dedicated secretarial help in preparing this and several other manuscripts.

REFERENCES

1. G. C. Cotzias, S. T. Miller, and J. Eduards, *J. Lab. Clin. Med.* **67,** 836 (1966).
2. E. Damsgaard, K. Heydorn, N. A. Larsen, and B. Nielsen, *Simultaneous Determination of Arsenic, Manganese and Selenium in Human Serum by Neutron Activation Analysis*. Risø Report No. 271. Research Establishment Risø, Roskilde, 1973.
3. J. Versieck, F. Barbier, A. Speecke, and J. Hoste, *Clin. Chem.* **20,** 1141 (1974).
4. J. Versieck, J. Hoste, F. Barbier, H. Steyaert, J. De Rudder, and H. Michels, *Clin. Chem.* **24,** 303 (1978).
5. J. Versieck, J. Hoste, L. Vanballenberghe, F. Barbier, R. Cornelis, and I. Waelput, *J. Lab. Clin. Med.* **97,** 535 (1981).
6. R. Cornelis, J. Versieck, L. Mees, J. Hoste, and F. Barbier, *J. Radioanal. Chem.* **55,** 35 (1980).
7. A. Byrne and J. Versieck, *Biol. Trace Element Res.* **26–27,** 529 (1990).
8. J. Versieck, *CRC Crit. Rev. Clin. Lab. Sci.* **22,** 97 (1985).
9. H. J. M. Bowen, *Atomic Energy Rev.* **13,** 451 (1975).
10. R. R. Greenberg, *Anal. Chem.* **58,** 2511 (1986).
11. J. Versieck, L. Vanballenberghe, A. De Kesel, J. Hoste, B. Wallaeys, J. Vandenhaute, N. Baeck, H. Steyaert, A. R. Byrne, and F. W. Sunderman, Jr., *Anal. Chim. Acta* **204,** 63 (1988).
12. A. Astrug, V. Kuleva, I. Kuleff, Z. Kiriakov, A. Tomov, and I. Djingova, *Trace Elements Med.* **1,** 65 (1984).
13. H. Nakahara, Y. Nagame, Y. Yoshizawa, H. Oda, S. Gotoh, and Y. Murakami, *J. Radioanal. Chem.* **54,** 183 (1979).
14. H. Nakahara, H. Kudo, H. Muramatsu, K. Masago, and Y. Murakami, in *Trace Element Analytical Chemistry in Medicine and Biology*, P. Brätter and P. Schramel, eds., de Gruyter, Berlin, 1980, pp. 365–380.
15. N. I. Ward and B. Pim, *Biol. Trace Element Res.* **6,** 469 (1984).
16. J. Versieck and L. Vanballenberghe, *Anal. Chem.* **63,** 1143 (1991).
17. K. Kasperek, H. Schicha, and H. Weese, in *Trace Elements in Relation to Cardiovascular Diseases*. Technical Report 157. International Atomic Energy Agency, Vienna, 1973, pp. 71–76.
18. R. A. Nadkarni and G. H. Morrison, *Anal. Chem.* **45,** 1957 (1973).
19. V. P. Guinn and J. Hoste, in *Elemental Analysis of Biological Materials*, International Atomic Energy Agency, Vienna, 1980, pp. 105–140.
20. G. Erdtmann and H. Petri, in *Treatise on Analytical Chemistry*, vol. 14 (part 1), P. J. Elving, ed., Wiley, New York, 1986, pp. 419–643.
21. G. Erdtmann, *Neutron Activation Tables*, Verlag Chemie, Weinheim, 1976.
22. J. Versieck, in *Methods in Enzymology*, vol. 158, part A, J. F. Riordan and B. L. Vallee, eds., Academic, 1988, pp. 267–286.
23. J. Versieck, F. Barbier, R. Cornelis, and J. Hoste, *Talanta* **29,** 973 (1982).
24. R. Cornelis, J. Hoste, and J. Versieck, *Talanta* **29,** 1029 (1982).
25. B. Maziere, A. Gaudry, J. Gros, and D. Comar, in *Trace Analysis: Sampling, Sample Handling, Analysis*, vol. 1, P. D. LaFleur, ed., National Bureau of Standards, Washington, DC, 1976, pp. 593–604.
26. N. I. Ward and D. E. Ryan, *Anal. Chim. Acta* **105,** 185 (1979).
27. K. Heydorn and K. Nørgård, *Talanta* **20,** 835 (1973).
28. Advisory Group of the International Atomic Energy Agency, *Anal. Chim. Acta* **165,** 1 (1984).

Toxic Heavy Metals and Other Trace Elements in Foodstuffs from 12 Different Countries

An IAEA Coordinated Research Program

E. CORTES TORO,[1] H. A. DAS,[2] J. J. FARDY,[3] Z. BIN HAMZAH,[4] R. K. IYER,[5] SUN LAIYAN,[6] N. LEELHAPHUNT,[7] Y. MURAMATSU,[8] R. M. PARR,*,[1] I. H. QURESHI,[9] S. M. RESNIZKY,[10] S. SURTIPANTI,[11] S. A. TARAFDAR,[12] AND M. B. A. VASCONCELLOS[13]

[1]International Atomic Energy Agency, Vienna, Austria; [2]Netherland Energy Research Foundation (ECN), Petten, The Netherlands; [3]ANSTO Isotope Technology Program, Menai, Australia; [4]Nuclear Energy Unit, PUSPATI Complex, Bangi, Selangor, Malaysia; [5]Bhabha Atomic Research Centre, Trombay, India; [6]Institute for Application of Atomic Energy, Beijing, China; [7]Office of Atomic Energy for Peace, Bangkok, Thailand; [8]National Institute of Radiological Sciences, Nakaminato, Japan; [9]Pakistan Institute of Nuclear Science & Technology, Islamabad, Pakistan; [10]Comision Nacional de Energia Atomica, Buenos Aires, Argentina; [11]National Atomic Energy Agency, Jakarta Selatan, Indonesia; [12]Atomic Energy Centre, Dacca, Bangladesh; [13]Instituto de Pesquisas Energeticas e Nucleares, Sao Paulo, Brazil

ABSTRACT

A research program related to the assessment of toxic heavy metals and essential trace elements in foodstuffs has been carried out in 12 countries under the auspices of the IAEA. The main purpose of this program was to obtain data on the elemental concentrations of potentially toxic elements in foodstuffs in various countries, and to compare them with the maximum permissible levels specified in

*Author to whom all correspondence and reprint requests should be addressed.

national legislation and international guidelines. High-priority elements for this study were As, Cd, Cr, Pb, Hg, and Se. Also of interest, but of lower priority, were Sb, Cu, and Zn. Emphasis was placed on the use of nuclear and nuclear-related analytical techniques, complemented by conventional methods, and on quality assurance.

Index Entries: Foodstuff analysis; nuclear analytical techniques; essential elements; toxic elements; analytical quality control; reference materials.

INTRODUCTION

Human growth and metabolism depend mainly on a well-balanced diet in terms of macronutrients (proteins, lipids, carbohydrates and so on). However, traces of some inorganic elements are also an important component of human nutrition because of their essential role in various metabolic processes *(1)*, or because their presence in excessive or deficient amounts may disturb normal biochemical functions of the body. Human exposure to these elements is mainly through the diet.

The IAEA recently organized a 5-yr research program to determine and assess the concentrations of toxic heavy metals and some other—mainly essential—trace elements in foodstuffs in 12 countries. The main purpose of this Coordinated Research Program (CRP) was to obtain analytical data on the elemental concentrations of potentially toxic elements in foodstuffs and to compare them with the maximum permissible levels specified in national legislation and international guidelines *(2)*. The matrices of interest were foodstuffs that comprise together more than 50% of the average daily intake. Drinking water was also considered to be highly relevant and was studied by some participants. An important supplementary purpose of the program was to help establish analytical expertise for work of this kind in individual countries, allowing such laboratories to offer analytical quality-control services and to provide validation support for their own national food monitoring programs. A detailed research protocol was prepared by the IAEA *(3)*. It contains guidelines on the types and amounts of food to be collected to obtain a representative analytical sample, how to prepare it for analysis, and procedures for analytical quality control and data evaluation. Emphasis was placed on the use of nuclear and nuclear-related analytical techniques, complemented by conventional methods in cases where these were considered more appropriate. Quality assurance was emphasized, and the IAEA organized external analytical quality-control exercises to ensure that the analytical data produced by the laboratories was reliable.

EXPERIMENTAL

Sampling and Sample Preparation

Food items that were important components of the diet of the selected populations groups were sampled and analyzed. Some participants collected the samples at local markets, whereas others received samples from local nutritional institutes that identified representative diet samples from selected population groups according to statistical information available to them. The amount of food collected was normally a few kilograms, from which subsamples for analysis were taken. Only edible parts of the foods were used for analysis. The samples were cleaned and prepared as they normally would be in the home. Foods that are eaten raw were just cleaned and washed. All food items were of the same degree of freshness as for normal consumption. Homogenization or mixing was carried out using a suitable food blender and, in some cases, also by the brittle fracture technique *(4)*. Moisture was determined from a small aliquot taken from the mixed homogenous sample. The samples were in most cases dried by freeze-drying. Prior to analysis, the samples were stored in clean polyethylene bottles with screw caps.

Elements of Interest

According to the protocol issued for this CRP, each participant had to determine as many of the following elements as possible, depending on the availability of analytical methods of proven reliability: As, Cd, Hg, Pb, and Se. A second group of elements to be determined, depending on national perceptions of their importance, were: Br, Cr, Cu, Fe, I, Mn, Sb, and Zn.

Analytical Techniques

Purely instrumental analytical methods could not be used for most of the elements and samples of interest because the concentration levels were too low. Most of the analytical methods used by the different laboratories were therefore based on the destruction of the sample followed by a preconcentration, a preseparation, or a postirradiation radiochemical separation procedure. Using these approaches, most of the participants were able to determine the actual concentrations of the elements of interest in all the selected food items and diets. The analytical techniques used were neutron activation analysis (NAA), either instrumental (INAA) or radiochemical (RNAA), anodic stripping voltammetry (ASV), inductively coupled plasma emission spectrometry (ICP-ES), atomic absorption spectrometry (AAS), direct current plasma emission spectrometry (DCP-ES), inductively coupled plasma mass spectrometry (ICP-MS), and proton-induced X-ray emission (PIXE). One of the participants studied arsenic speciation in environmental waters and described suitable techniques to be used for the characterization of dissolved arsenic species in such samples.

Analytical Quality Assurance

In a CRP such as this, where several analytical techniques were to be used in different laboratories with diverse infrastructures and levels of experience, efforts had to be made to assure similar quality of the analytical data. Participants were requested to validate their analytical methodology **before** the analysis of actual samples and to comply with prescribed analytical quality-control procedures. Several mechanisms were implemented to assure accurate and precise results. The determination of the limit of quantitation for each element and sample type, the routine use of certified reference materials during the analytical phase of each individual project, and the analysis of "blind" quality-control samples (actually certified reference materials provided by the IAEA) were ways of assessing the quality of the analytical data.

RESULTS

Only an overview of the large amount of data supplied by the participants is presented here. A technical document containing full reports of all the participants is in preparation and will be issued shortly by the IAEA. Table 1, showing data for Japanese foodstuffs, illustrates the wide range of elements and samples studied.

Rice was one of the major staple foodstuffs within this CRP, since it is an important component of the diet in many of the participating countries. Participants analyzed various types of rice (e.g., raw, boiled, or after different industrial processes). Table 2 shows the results obtained for the analysis of rice by one participating country.

Among the target elements for the CRP, Se was of particular interest to several of the participants on account of its essentiality to humans as well as its potential toxicity. Tables 1 and 3 give values for the Se contents of foodstuffs in two of the countries involved in this program. Particular care is needed with the analysis of foodstuffs that have been homogenized by the brittle fracture technique in Teflon™ containers, since this can lead to fluorine contamination. When analyzed by INAA via the $^{77}Se^{m}$ activation product, a careful correction has to be applied for interference from ^{20}F, which has a similar half-life.

The quality of the data reported was monitored throughout the CRP. Table 4 illustrates the kinds of data collected. It shows the results reported by a participant for the analysis of one set of "blind" quality-control samples supplied by the IAEA.

Results from one participant permitted an assessment of the relative merits of NAA and ICP-MS for the determination of toxic and other trace elements in foodstuffs and diets. NAA gave precisions of 10% or better for the determination of As, Br, Ca, Cl, Cr, Co, Cu, Fe, Hg, K, Mn, Mo, Na, Rb, Se, and Zn in IAEA H-9 (a mixed human diet reference material). In most cases, the analysis could be done by INAA. However, some elements present at low concentrations (e.g., As, Hg) and those subject to interferences (e.g., Cu, Mo) required the use of RNAA, which can be very time-consuming. In contrast, the analysis of solutions by ICP-MS is extremely rapid. A typical semiquantitative, multielemental analysis (70 elements, 20% precision) can be undertaken in 5 min and quantitative analysis (5% precision, 10 elements) within 15 min (these times do not

Table 1
Representative Values of the Elemental Concentrations in Selected Foods for Each Food Category†

Food category**	Food consumption (g)	As	Ca	Cd	Cu	Fe	Hg	K	Mg	Mn	P	Pb	Se	Zn
Cereals (dry)														
Rice	208.8	0.08	100	0.1	2.2	3.9	0.005	810	370	10	1000	0.05	0.02	18
Wheat and others	93.3	0.01	130	0.01	1.5	10	0.005	1300	290	4.2	950	0.05	0.1	4.3
Potatoes	61.3	0.01	150	0.01	1.3	5.1	0.003	4000	200	2.7	400	0.05	0.03	3
Pulses and their products	64.4	0.02	100	0.03	4	40	0.02	5000	800	9	3000	0.01	0.03	30
Vegetables														
Green and yellow vegs.	71.1	0.01	1300	0.05	1	15	0.01	4000	500	5	590	0.1	0.02	8
Other vegetables	183.5	0.001	500	0.01	0.4	4	0.003	3000	130	1.5	500	0.06	0.03	2.5
Fruits	137.9	0.005	100	0.005	0.6	3.5	<0.01	1900	130	1.3	180	0.05	0.001	0.6
Seafood														
Seaweed (dry)	5.5	35	5000	0.25	15	250	<0.01	10000	2500	20	750	1	0.1	40
Fish and shellfish	92.7	2	200	0.01	0.7	10	0.15	3900	360	0.2	2500	0.2	0.6	5
Meat and poultry	69.1	0.05	60	0.01	1	20	0.005	2500	150	0.4	1400	0.1	0.3	20
Eggs	40.1	0.05	600	0.01	0.9	30	<0.01	1300	100	0.3	2000	0.1	0.5	20
Milk and milk products	121.5	0.003	1100	0.001	0.1	1.0	0.001	1500	100	0.05	1000	0.05	0.02	3

†Data from Japan; food consumption in g/day; concentration values in mg/kg. Values are expressed on a wet wt basis excluding rice, wheat, and seaweeds (wet-to-dry ratio of seaweed is about 5:1).

**Major (representative) foods for each food category are: potatoes, potato and sweet potato; pulses and their products, soybeans and tofu; green and yellow vegetables, spinach and carrot; other vegetables, cabbage and chinese cabbage; fruits, orange and apple; seaweed, wakame-algae and hijiki-algae; fish, tuna, bonito, and salmon; meat and poultry; pork, beef, and chicken; milk and milk products, market milk.

Table 2
Elemental Concentrations[a] in Rice Grains by INAA[b]

ELEMENT	POLISHED GRAINS LONG FINE 1 CERGAL	LONGFINE 1 CAMPEAO	LONGFINE 2 CAMIL	LONGFINE 2 CEREJEIRA	YELLOW LONG LUMA	JAPANESE/CATETE YAMAGUI	PARBOILED GRAINS LONG FINE MINGOTE
As	0.272 ± 0.022(11)	0.279 ± 0.027(5)	0.271 ± 0.035(6)	0.093 ± 0.014(2)	---	0.362 ± 0.025(4)	0.310 ± 0.021(4)
Br	0.46 ± 0.04(8)	2.11 ± 0.18(4)	0.36 ± 0.02(6)	0.43 ± 0.04(6)	0.36 ± 0.06(3)	0.93 ± 0.06(4)	0.33 ± 0.02(4)
Na	15.0 ± 1.4(12)	8.9 ± 0.7(4)	13.5 ± 0.6(5)	12.3 ± 1.6(5)	6.6 ± 1.0(4)	17.2 ± 1.0(4)	16.2 ± 0.7(4)
K	328 ± 16(8)	419 ± 6(4)	418 ± 10(4)	654 ± 41(6)	651 ± 29(3)	385 ± 10(2)	1079 ± 58(6)
Rb	2.32 ± 0.27(11)	3.28 ± 0.21(6)	3.67 ± 0.39(7)	3.81 ± 0.22(7)	5.17 ± 0.62(6)	3.67 ± 0.19(6)	11.0 ± 0.7(8)
Zn	14.0 ± 1.6(10)	15.01 ± 0.63(6)	15.2 ± 0.5(6)	16.4 ± 0.4(6)	17.9 ± 2.0(5)	15.9 ± 0.6(6)	7.61 ± 0.66(4)
Fe	---	---	5.76 ± 0.13(2)	7.58 ± 0.83(2)	20.5 ± 0.8(2)	88.0 ± 5.5(4)	12.6 ± 1.2(3)
Co ng/g	---	---	---	25.4 ± 5.7(2)	25 ± 7(2)	19.6 ± 0.3(2)	27.9 ± 4.3(2)
Sc ng/g	---	---	---	2.16 ± 0.08(4)	---	---	1.11 ± 0.08(2)

[a]Means and standard deviation of *(n)* individual determinations.
[b]Data from Brazil. Concentrations in mg/kg dry wt unless otherwise indicated.
---Not detected.

Table 3
Selenium Concentrations in Australian Foodstuffs (mg/kg wet wt basis)

Food Item	Range	Mean	Food Item	Range	Mean
White bread	0.04-0.15	0.08	Full cream milk	<0.01-0.03	0.02
Wholemeal bread	0.08-0.13	0.10	Skim milk	<0.01-0.03	0.02
Infant cereal	0.06-0.12	0.09	Lactogen	<0.01-0.01	0.01
Corn flakes	0.07-0.09	0.08	Butter	<0.02	
Wheat biscuits	0.09-0.35	0.15			
White rice	0.01-0.05	0.03	Potatoes	<0.02	
Bran	0.09-0.24	0.16	Onions	<0.02	
			Carrots	<0.01	
Minced steak	0.06-0.19	0.12	Green beans	<0.02	
Lamb	<0.02-0.13	0.13	Celery	<0.02	
Pork	0.09-0.23	0.15	Pumpkin	<0.02-0.02	0.01
Chicken	0.14-0.23	0.18	Baked Beans	<0.02-0.04	0.02
Canned meat	0.09-0.14	0.12			
			Fruit	<0.02	
Fish	0.22-0.34	0.29			
Salmon	0.22-0.38	0.27	Infant dinner	<0.02	
Sardines	0.25-0.78	0.57	Peanuts	0.06-0.21	0.14
			Peanut butter	0.05-0.11	0.08
Eggs	0.15-0.23	0.19			
Liver	0.20-0.54	0.30			

Table 4
IAEA External Analytical Quality-Control Program†

Element	Sample STD-1 Rice flour NIST-SRM-1568		Sample STD-2 Copepoda IAEA-MA-A-1/TM		Sample STD-3 Fish flesh IAEA-MA-A-2/TM	
	Reported	Expected	Reported	Expected	Reported	Expected
As	0.43	0.41	7.0	6.7	2.8	2.6
Br	0.98	---	950	---	21.9	---
Cu	1.96	2.2	7.5	7.6	3.6	4.0
Co*	22	20	126	120	82	80
Cr	---	---	1.0	1.1	1.5	1.3
Fe	9.0	8.7	61.3	60.0	52.2	54
Hg*	8.0	6.0	310	280	550	470
Mn	18.9	20.1	3.0	2.9	0.92	0.81
Rb	7.8	7.0	3.5	---	8.2	---
Sb*	---	---	80	70	---	5
Se	0.48	0.4	2.8	3.0	1.9	1.7
Zn	22	19.4	166	158	34	33

*values in μg/kg
†Concentrations expressed in mg/kg unless otherwise specified.

include the time needed for dissolving solid samples). Spectral interferences occurring in the argon plasma owing to the presence of chlorides and sulfates in the samples can reduce the effectiveness of the technique for As, V, Cr, Se, Ti, Zn, and Ca. Analysis of H-9 by ICP-MS produced significantly more data than NAA in much less time, but sample dissolution made this technique more labor-intensive than INAA. Although NAA cannot match the superior sensitivity obtained by ICP-MS for a wider range of elements, its usefulness has been verified for a wide range of food samples. Although NAA is a time-consuming technique for many applications, it does not suffer from blank problems after irradiation of the sample, and it becomes the preferred technique where low limits of detection are required for trace concentrations in solid samples.

CONCLUSIONS

A large variety of foodstuffs were selected, collected, and analyzed in different countries using analytical techniques, such as INAA, RNAA, ASV, ICP-ES, ICP-MS, AAS, DCP-ES, and PIXE. All the concentration levels of toxic elements in foodstuffs and diets were within the limits specified in national and international guidelines. Nuclear analytical techniques proved to be sensitive, accurate, and precise for the analysis of this type of matrix. Improvement in detection limits (compared with INAA) was achieved by using destructive methods to eliminate potential interferences. New analytical methodologies, in particular, radiochemical separation procedures, were developed, and the feasibility of using other new analytical techniques (e.g., ICP-MS) was studied. Through this CRP, the participating laboratories were able to validate and improve the quality of their analytical data through extensive analytical quality-control procedures and exercises.

REFERENCES

1. World Health Organization. Report of a WHO/FAO/IAEA Expert Consultation on Trace Elements in Human Nutrition and Health, WHO, Geneva (in press, due 1994).
2. Codex Alimentarius Commission. Contaminants, CAC/VOL. XVII-Ed. 1, FAO, Rome (1984).
3. International Atomic Energy Agency. Nuclear Techniques for Toxic Elements in Foodstuffs: Final Report on a Co-ordinated Research Programme, IAEA, Vienna (in press, due 1994).
4. G. V. Iyengar and K. Kasperek. *J. Radioanal. Chem.* **39,** 301–316 (1977).

Total and Methyl Mercury Levels in Human Scalp Hairs of Typical Populations in China by NAA, GC(EC), and Other Techniques

CHIFANG CHAI,*,[1] WEIYUE FENG,[1] QINFANG QIAN,[1] MING GUAN,[2] XINJI LI,[3] YILUN LU,[4] AND XIOUMEI ZHANG[4]

[1]*Institute of High Energy Physics and Laboratory of Nuclear Analysis Techniques, Academia Sinica, P.O. Box 2732, Beijing 100080, China;* [2]*Norman Bethume University of Medical Sciences, Changchun, China;* [3]*Beijing Environmental Monitoring Station, Beijing, China; and* [4]*Institute of Geography, Academia Sinica, Beijing, China*

ABSTRACT

The contents of total and methyl mercury in scalp hair samples of 1179 fishermen living in a typical Hg-polluted region in northeast China and 27 lying-in women and their newborns in Beijing have been determined by INAA, GC(EC), and other techniques. Only 18 of all fishermen had Hg contents above 5 μg/g, which indicates that the Hg pollution there has been substantially alleviated. The longitudinal Hg patterns of the lying-in women showed a gradually lowering tendency during pregnant period. Further, the Hg contents of the newborn babies generally were above or close to those of their mothers, confirming the mechanism that the methyl Hg, an organic species of Hg with high toxicity, is readily able to penetrate the placental barrier and to accumulate in the fetus. Thus, the mercury poison has occurred at the early stage of pregnancy.

Index Entries: Chemical species of mercury; total Hg; methyl Hg; neutron activation analysis; gas chromatography; human hair; Hg pollution.

*Author to whom all correspondence and reprint requests should be addressed.

Biological Trace Element Research Editor: G. N. Schrauzer © 1994 by Humana Press Inc.

INTRODUCTION

Mercury is a toxic element to the human being. Its specific toxicity lies in the methylated species, i.e., methylmercury (Me-Hg), which is readily able to enter the human body via the food chain *(1,2)*. Recent clinical observation has indicated that the Me-Hg has effects on the early stage of newborn baby development and mental ability of children whose mothers were exposed during pregnancy to three to four times the tolerable weekly intake set by WHO and FAO.

Hg pollution is still an environmental problem to be solved in some regions of China *(3)*. In order to study the level of Hg pollution in the Second Songhuajiang River System, a typical Hg-polluted area, we systematically determined the Hg contents in 1179 hair samples taken from fishermen living in the immediate vicinity of the river by NAA and GC(EC) in the framework of the IAEA Coordinated Research Project on assessment of environmental exposure to mercury in selected human populations as studied by nuclear and other techniques.

It is known that during pregnancy, the methylated species of Hg is easily transferred from mothers to the fetus through placental tissue. In order to study the congenital toxicity of Hg, the correlation between Hg contents in scalp hairs of lying-in women and their newborn babies has been investigated in this work. Also, the variation of Hg contents in pregnant women's hairs during their pregnancy was studied by NAA and synchronous radiation-based XRF as well.

EXPERIMENTAL METHOD

Sampling

We followed the sampling procedure outlined in the United Nations Environment Programme (UNEP) on the determination of methylmercury, total mercury, and total selenium in human hair, Reference Methods for Marine Pollution Studies No. 46 (draft), October 1987, prepared in cooperation with WHO and IAEA *(4)*.

One thousand one hundred and seventy-nine head hair samples of persons living at the Second Songhuajiang River System in northeast China and 27 pairs of mothers' and their newborn babies' hairs immediately at or a couple of days after delivery in the Beijing Zhong-Guan-Chen Hospital were collected, together with the necessary information, e.g., name, age, sex, occupation, and nutritional habits. All hair samples were taken from the occipital area and as close as possible to the scalp. At least 10 cm long and about 10 g hairs were cut from the lying-in women without waving or dying in the last 10 mo. Because of the little hair available, all hairs of the babies were sampled, and their weights ranged from 0.5 to 1.0 g. The hair-washing procedure recommended by IAEA was followed to remove dirty materials from the hair samples, i.e. washing by acetone-water (bidistilled)-water-water-acetone. Then, the hair was left to dry in a clean desiccator for use.

Determination of Total Mercury

Neutron activation analysis (NAA), mercurymetry (MM), and atomic fluorescence spectrometry (AFS) were used to determine the total mercury in the hairs in this work. Here only the methodology of NAA will be briefly described.

Hg has two stable nuclides, Hg-196 and Hg-203. Both of them are able to be readily determined by NAA. The two peaks of 68.8 keV and 77.3 keV of Hg-197 produced by the Hg-196 (n,γ) reaction in a heavy-water reactor with a neutron flux of $4 - 5 \times 10^{13}$ $n/cm^2/s$ were counted after a 30-min irradiation. The counting was done after 3 d of cooling by using a planar HPGe detector.

Because of the high volatility of Hg, we first studied losses of Hg during sample preparation and irradiation. Our experimental results showed no significant loss of Hg at 150°C drying for 3 h, but considerable loss of Hg occurred above 150°C, particularly for hair samples. The lower loss of Hg for the chemical standard of Hg is attributed to the addition of thioacetamide as a stabilizer of Hg. In our nuclear reactor, the volatile loss of Hg can be neglected within 1 h irradiation time. However, the extension to over 2 h causes severe loss of Hg. Thus, the irradiation time of hairs is kept within 30 min. The detailed results have already been given elsewhere *(5)*.

Determination of Methylmercury

Our procedure for Me-Hg analysis of hairs is shown in Fig. 1. The first step is to prepare the sulfhydryl cotton as follows: 15 g defatted cotton were rinsed in a mixing solution of 50 mL thioglycollic acid, 35 mL acetic oxide, 16 mL 36% acetic acid, 0.15 mL conc. H_2SO_4, and 5 mL bidistilled water for 4 d at 37–38°C. Then, the cotton was washed to pH 7 by bidistilled water, squeezed, and dried at 38°C in an oven. The final cotton was stored in a desiccator for use.

Two sets of gas chromatography devices are available. One is Model Varian Vista 6000. The other is Model Shimadzu GC-9A with a C-92A data processor. The chromatographic column is 10% PEG-20M (ϕ3.2 mm × 1.1 m). The 60–80 mesh Chromosorb W is used as carrier. The detector is Ni-63 of 10 mCi. The detailed procedure of Me-Hg analysis has been reported elsewhere *(6)*.

Synchronous Radiation-Based X-ray Fluorescence Spectrometry (SRXRF)

In this work, we also attempted to use SRXRF to study longitudinal variation of Hg contents along hairs in comparison with sectional NAA. A Beijing Electron-Positron Collider (BEPC) is available with 2.8 GeV energy and 150 mA maximum currency intensity. A two-dimensional scanning unit with a size-adjustable slit from 10 × 10 μm to 1 × 1 mm was used. The preliminary experiments indicate that the 9.987-keV LαX line of Hg can be applied.

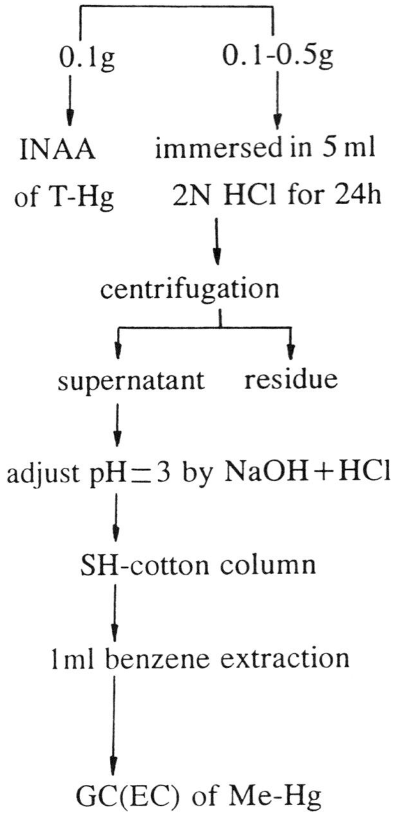

Fig. 1. Experimental procedure for Me-Hg analysis of hair sample.

RESULTS AND DISCUSSION

In-House Analytical Quality Control

We prepared five working hair samples as in-house quality-control materials. Further, two Chinese human hair reference materials (GBW 09101 and GBW 07601) were used to examine the reliability of NAA, MM, and AFS for total Hg analysis. The comparison of results of three analytical methods as given in Table 1 was found satisfactory.

Table 2 lists the results of INAA for two Chinese hair CRMs and two other NIST biological CRMs, citrus leaves SRM-1572 and pine needle SRM-1575. It can be seen that the INAA technique used in this work yielded reliable results for Hg determination at the mg/kg to sub-mg/kg levels.

External Analytical Quality Control

Because the biological CRM for Me-Hg was not available in China, we participated in the external analytical quality-control project orga-

Table 1
In-house Analytical Quality Control of Total Hg (μg/g)

Working Sample No	INAA	MM	AFS
1	0.53 ± 0.05	0.58 ± 0.02	0.68 ± 0.03
2	1.14 ± 0.03	0.99 ± 0.03	1.08 ± 0.03
3	0.43 ± 0.01	0.44 ± 0.01	0.43 ± 0.02
4	0.48 ± 0.02	0.52 ± 0.03	0.51 ± 0.03
5	0.48 ± 0.03	0.51 ± 0.01	0.52 ± 0.01

Table 2
Examination of Analytical Accuracy of Total Hg by INAA (μg/g)

Sample	This work	Certified
Human hair(GBW09101)	2.18 ± 0.03	2.16
Human hair(GBW07601)	0.38 ± 0.05	0.36 ± 0.05
Citrus leaves(SRM1572)	0.070	0.080
Pine needle(SRM1575)	0.20	0.15 ± 0.05

nized by IAEA. Table 3 lists the results obtained in our laboratory and in the "J. Stefan" Institute at Ljubljana, Slovenia, which was appointed as the reference laboratory of the CRP. The data in Table 3 illustrate that the agreement for total Hg is excellent, but for Me-Hg is good for two samples (A, B), fair for D and E, and poor for C and F.

Hg Distribution of Hair Samples of 1179 Fishermen at the Second Songhuajiang River System

The T-Hg and Me-Hg contents in 1179 hair samples taken from the human population living at the Second Songhuajiang River System were determined by INAA, MM, AFS, and GC(EC). Our results indicate that most of the hair samples contain low Hg, <5 μg/g. Only 18 samples have

Table 3
External Analytical Quality Control of Total and Methyl Hg (μg/g)

Sample No.	Total Hg Beijing	Total Hg Ljubljana	Me - Hg Beijing	Me - Hg Ljubljana
A	0.53 ± 0.05	0.57 ± 0.02	0.43 ± 0.04	0.44 ± 0.03
B	1.14 ± 0.03	1.09 ± 0.03	0.89 ± 0.06	0.85 ± 0.07
C	0.43 ± 0.01	0.40 ± 0.01	0.42 ± 0.04	0.27 ± 0.02
D	0.48 ± 0.02	0.49 ± 0.04	0.42 ± 0.03	0.34 ± 0.03
E	0.48 ± 0.03	0.46 ± 0.03	0.42 ± 0.03	0.33 ± 0.02
F	2.18 ± 0.03	1.92 ± 0.25	1.54 ± 0.04	0.90 ± 0.04
G	0.38 ± 0.05	0.38 ± 0.01	---	0.16 ± 0.01

high Hg contents (*see* Table 4). The No. 5 sample contains the highest Hg content, amounting to 113 μg/g. Thus, the Hg pollution is still present, even though it has been substantially alleviated in comparison with the 1970s.

Correlation of Hair Hg Contents of Lying-in Women with Those of Their Newborns

The hair Hg contents of 27 lying-in women and their newborns determined by INAA are listed in Table 5. Figure 2 shows the correlation between them. It can be seen that the hair Hg contents of 22 babies are close to or above those of their respective mothers. Only five babies have a slightly lower Hg contents. The average Hg value ± SD amounted to

Table 4
Total and Methyl Hg Levels in Hair of 18 Fishers Living at the Second Song-Hua-Jiang River System (μg/g)

Sample No.	NAA	T-Hg MM	AFS	ME-Hg(as Hg) GC(EC)
1	9.90 ± 0.30	11.9		7.92
2	9.82 ± 0.30	10.7		8.68
3	11.3 ± 0.3	12.0		6.96
4	9.70 ± 0.36	12.0		7.22
5	113 ± 10	121	101	60.9
6	10.2 ± 0.2	11.3	10.9	5.35
7	7.86 ± 0.34	10.0		6.84
8	9.11 ± 0.27	10.4		6.10
9	10.5 ± 0.3	11.6		1.50
10	16.7 ± 0.2	15.6	14.9	13.7
11	14.5 ± 0.2	16.8		1.90
12	8.45 ± 0.34	13.0		4.64
13	16.2 ± 0.1	15.0		7.64
14	8.64 ± 0.34	10.1	9.9	3.96
15	13.2 ± 0.1	13.3		13.2
16	9.87 ± 0.20	12.7		4.57
17	34.6 ± 0.1	54.1	35.5	--
18	13.0 ± 0.1	20.4	15.5	--

0.66 ± 0.31 μg/g and 0.59 ± 0.25 μg/g for babies and mothers, respectively and no significant difference between these values could be found ($p > 0.05$). Lauwerys et al. *(7)* reported higher Hg contents of newborn blood (umbilical cord) than those of mother blood (venous) by 10–15% in a non-Hg-polluted area in Belgium. Similar results were also given by

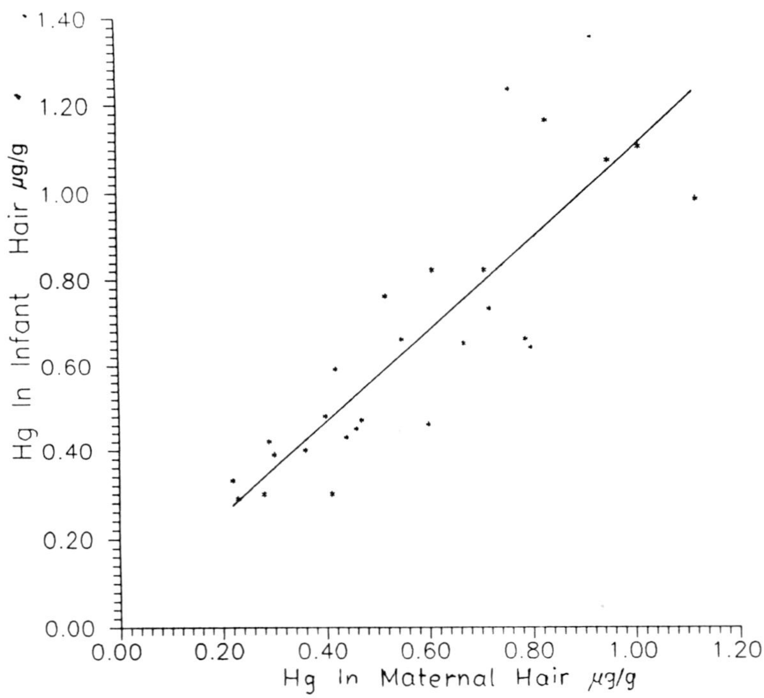

Fig. 2. Correlation between hair Hg contents of lying-in women and their newborns. $N = 27$; $T = +0.862$ ($p < 0.001$); $Y = 0.04 + 1.06X$.

Pitkin et al. *(8)*. Thus, our results agree with those of blood samples. At Hg-polluted districts, the hair Hg contents of babies are likely even higher (*9*).

Longitudinal Variations of Hg Contents in Hairs of Pregnant Women

It is known that the growing rate of human head hair is about 1 cm/mo. Thus, a strand of 10-cm-long hairs can cover the whole time of pregnancy. INAA was used to analyze the Hg contents of lying-in women's hairs of each 1 cm, whereas SRXRF was used for scanning analysis of Hg along the hairs with a step length of 5 mm. Table 6 lists the longitudinal variations of hair Hg contents of three representative lying-in women obtained by INAA. Our results show that the hair Hg contents of the lying-in women decreased with increasing time of pregnancy. The regression analysis of three specimens indicates a significant correlation between them ($\tau_1 = -0.966$, $\tau_2 = 0.900$, and $\tau_3 = -0.871$, $P < 0.001$) (*see* Fig. 3). The SRRF gave the similar tendency, even though its quantitative evaluation was difficult.

Table 5
Hair Hg Contents of 27 Lying-in Women and Their Newborns
Determined by INAA (μg/g)

Sample No.	Mother	Baby
1	0.83	1.16
2	0.52	0.76
3	0.55	0.66
4	0.42	0.59
5	0.36	0.40
6	0.30	0.39
7	0.92	1.35
8	0.67	0.65
9	0.46	0.45
10	1.01	1.10
11	0.79	0.66
12	0.61	0.82
13	0.76	1.23
14	1.12	0.98
15	0.60	0.46
16	0.40	0.48
17	0.47	0.47
18	0.29	0.42
19	0.95	1.07
20	0.22	0.33
21	0.44	0.43
22	0.41	0.30
23	0.71	0.82
24	0.23	0.29
25	0.80	0.64
26	0.72	0.73
27	0.28	0.30
X ± SD	0.59 ± 0.25	0.66 ± 0.31

Our results confirm the mechanism that Hg is gradually transferred from mothers to their fetuses during pregnancy. It follows from animal experiments that the Hg transfer strongly depends on Hg speciation. The inorganic Hg is basically unlikely to go through the placental tissue,

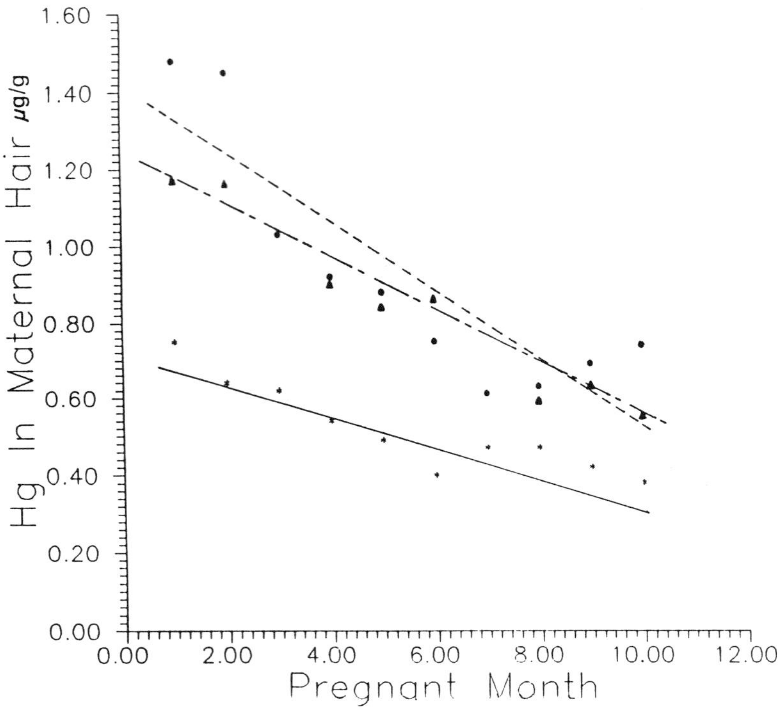

Fig. 3. Variations of hair Hg content of pregnant women with pregnant time. — - —$Y_1 = 1.25 - 0.7X$, $T_1 = -0.966$, $p < 0.001$ — - — $Y_2 = 0.71 - 0.04X$, $T_2 = -0.900$, $p < 0.001$; $Y_3 = 1.42 - 0.09X$, $T_3 = -0.871$, $p < 0.001$.

whereas the methyl Hg, an ester-loving material, readily penetrates it to enter the fetus. It is likely that the fetus is easily able to absorb ester-like materials from placenta *(10)*. Since the early stage of pregnancy is an important period in which infants develop their cerebrum and nervous systems, Hg pollution is more dangerous to the fetus than to the mother.

ACKNOWLEDGMENTS

This work was funded by the International Atomic Energy Agency (Contract No. 6332/R2/RB) and National Natural Science Foundation of China (Contract No. 1939501).

REFERENCES

1. T. Suzuki, T. Miyama, and H. Katsunuma, *Bull. Environ. Contam. Toxicol.* **5,** 502 (1971).
2. H. Tsuchiya, K. Mitani, K. Kodama, and T. Nakata, *Arch. Environ. Health* **37,** 11 (1984).

Table 6
Variation of Hair-Hg Contents of Lying-in Women During Pregnancy (μg/g)

pregnant time (Month)	Sample No. 1	2	3
1	1.17	0.75	1.48
2	1.16	0.64	1.45
3	1.03	0.62	1.03
4	0.90	0.54	0.92
5	0.84	0.49	0.88
6	0.86	0.40	0.75
7	0.61	0.47	0.61
8	0.59	0.47	0.63
9	0.63	0.42	0.69
10 (at delivery)	0.55	0.38	0.74

3. Y. H. Lin, *Environ. Chem.* (in Chinese) **2,** 10 (1983).
4. United Nation Environmental Programme, The determination of methylmercury, total mercury, and total selenium in human hair, Reference Methods for Marine Pollution Studies No. 46 (draft), October, 1987.
5. C. F. Chai, Q. F. Qian, W. Y. Feng. J. G. Sun, J. B. Tian, S. Z. Li, Y. L. Wang, X. J. Li, S. Zhang, X. M. Zhang, and M. Guan, NAHRES-7, IAEA, Vienna, 1991, pp. 41–48.
6. C. F. Chai, Q. F. Qian, W. Y. Feng, J. G. Sun, M. Guan, X. J. Li, Y. L. Lu, X. M. Zhang, and S. Zhang, NAHRES-13, IAEA, Vienna, 1992, pp. 51–58.
7. R. Lauwerys, J. P. Buchet, H. Roels, and G. Hubermont, *Environ. Res.* **15,** 278 (1978).
8. R. M. Pitkin, J. A. Bahns, L. J. Filer, and W. A. Reynolds, *Soc. Exp. Biol. Med.* **151,** 565 (1976).
9. S. Skerfving, *Bull. Environ. Cont. Toxicol.* **41,** 475 (1984).
10. Y. H. Lin, in *Trace Elements in Life Sciences* (in Chinese), vol. 1, K. Wang, ed., Chinese Meteorological Press, Beijing. 1991, pp. 300–337.

Mercury Determination in Hair of Malaysian Fishermen by Neutron Activation Analysis

S. B. SARMANI,* A. Z. KIPRAWI, AND R. B. ISMAIL

Department of Nuclear Science, National University of Malaysia, 43600 Bangi, Malaysia

ABSTRACT

Fish has been known as a source of nonoccupational mercury exposure to fish-consuming population groups. In this study, hair samples collected from fishermen and their families residing in an industrialized area in Penang and a nonindustrialized area in Terengganu were analyzed for mercury by neutron activation. The range, arithmetic mean, geometric mean, and median of the mercury concentrations for the groups in Penang and in Terengganu were 0.45–16.68, 3.61, 3.49, and 2.96 and 6.79–18.31, 12.08, 11.69, and 12.05 mg/kg, respectively. Somewhat lower values than from the Penang group were found in a group from Selangor consisting mainly of office workers. The group in Penang took about 40–100 g of fish/d, whereas the group in Terengganu consumed twice as much. This shows that hair mercury levels depend on a fish consumption pattern, and not on the location of the population. The levels of mercury found in this study were similar to those reported by other workers for fish-consuming population groups worldwide.

Index Entries: Neutron activation analysis; mercury; hair.

INTRODUCTION

Hair has been known as one of the deposition sites for the excretion of metals from the body. Since mercury is excreted from the body partly by deposition in hair, the analysis of hair provides a convenient tool for estimating body burdens of mercury, and thereby also provides an indirect measure of dietary intake. A weekly mercury intake of 0.2 mg, the

*Author to whom all correspondence and reprint requests should be addressed.

Provisional Tolerance Weekly Intake (PTWI) set by WHO and FAO, is expected to result in the hair mercury contents of 5–6 mg/kg *(1)*. An excellent review of the applications of hair analysis in biomedical and environmental studies has been presented by Chatt and Katz *(2)*. Airey *(3)*, who analyzed hair samples from 32 locations in 13 countries for mercury, found that mean hair concentrations were significantly different for the group that ate fish once a month or less (1.4 mg/kg), once a fortnight (1.9 mg/kg), once a week (2.5 mg/kg), and once or more daily (11.6 mg/kg). Several other reports have also shown the positive relationships between fish consumption and the concentration of mercury in hair *(4–6)*.

Because Malaysia is peninsular state, fishing is an important industry, providing people with employment as well as supplying them with protein from the fish. The consumption of fish comprises about 70% of the protein intake of the population, and the daily per capita fish consumption is 100 g. The average mercury levels in common commercial sea fish markets in Malaysia range from 0.06 to 0.42 mg/kg fresh weight *(7)*. With such a high intake of fish, it is apparent that fish is a possible source of nonoccupational mercury exposure to the general population. Earlier work by Silvalingam and Sani *(8)* showed that the mean mercury level of Malaysian fishermen was 8 mg/kg. This suggested that their weekly intake of mercury exceeded the PTWI level.

In this study, hair samples were collected from fishermen and their families residing in two fishing villages in Penang and Terengganu. For comparison purposes, hair samples were also obtained from a group of nonfishermen in Selangor. The aim of this study was to identify certain population groups that may have ingested mercury higher than the PTWI as set by the WHO. This study is part of a coordinated research program on assessment of environmental exposure to mercury in selected human populations as studied by nuclear and other techniques organized by the IAEA. The aim of the project was to evaluate the exposure of selected population groups to mercury and methyl-mercury, and to estimate potential health risks in these groups.

METHODS

Sample Preparation

Hair samples from fishermen's families were collected from residents of the Kuala Juru Village in the state of Penang on the West Coast, and the Chendering Village in the state of Terengganu on the East Coast of Peninsular Malaysia. Penang is one of the most industrialized states in Malaysia, and the Kuala Juru Village is surrounded by a large industrial estate. There are about 100 households in the village with a total population of about 500. The Chendering Village, on the other hand, is located on a prestigious, idyllic coastline free from any industrial activity. Most of the adult male residents in the two villages are fishermen. Hair samples were also collected from residents in Kajang, a small urbanized area with light industrial activities in the state of Selangor. This population group represents the nonfishermen's community. Particulars of each donor, such as name, sex, age, occupation, and dietary habits, were gathered and compiled in a questionnaire form.

Hair samples were washed according to the IAEA washing procedures: Wash hair successively in acetone, thrice in water, and once more in acetone. Add sufficient amounts of the solvents to cover the sample entirely. At each wash, allow the sample to stand at room temperature for 10 min with constant stirring. After each wash, decant the liquid, and add fresh solvent. Carry out the washing in the dust-free environment.

Sample Irradiation

About 150 mg of each sample were packed in a precleaned polyethylene vial. Triplicate analysis was carried out for each sample. Standard was prepared from standard chemical solution pipeted onto ashless filter paper and packed in a polyethylene bag after careful drying. About 50 μL of 10% solution of thioacetamide was pipeted onto the filter paper to prevent mercury losses. Samples, standards, and blanks were irradiated in a PUSPATI TRIGA Reactor for 6 hs in a neutron flux of 2.3×10^{12} $n/cm^{-2}/s^{-1}$.

Measurement of Activities

Mercury was determined using both ^{197}Hg and ^{203}Hg radionuclides. The determination via 77 keV of ^{197}Hg was carried out using an LEPD. The irradiated samples were cooled for 3 d and counted for 30 min. The 279.6 keV of ^{75}Se interfered with the determination of mercury via the 279. 1-keV photopeak of ^{203}Hg, thus a correction for the selenium contribution was necessary. The samples were cooled for at least 2 wk prior to counting.

The γ-ray activities of the samples and standards were measured by a high-resolution γ-spectroscopy system comprising a high-purity coaxial N-type germanium detector with a thin Be window (Tenelec), microcomputer-based MCA system with 80386 SX data processor, 8192 channel ADC and MCA Card (Tenelec), and Tc-244 spectroscopy amplifier (Tenelec).

Accuracy and precision of the INAA technique were evaluated by analyzing IAEA standard reference materials MAA-1 (Copepoda), MAA-2 (fish homogenate), and MAB-3 (fish tissue).

RESULTS

Accuracy of the mercury determination in the standard reference materials is shown in Table 1. Results of the mercury determination in the three population groups are shown in Table 2. Frequency distribution of the mercury concentrations in hair samples of the Kuala Juru fishermen is shown in Fig. 1, whereas the distribution according to age groups is shown in Table 3.

Table 1
Results of Mercury Determination in IAEA Reference Materials (mg/kg)

Reference Material	This work	Certified value
MAA-1	0.33 ± 0.05	0.28 ± 0.01
MAA-2	0.49 ± 0.14	0.47 ± 0.02
MAB-3	0.54 ± 0.19	0.51

Table 2
Total Mercury in Hair (mg/kg)

	Pulau Pinang	Selangor (non-fishermen)	Terengganu
No. of sample	106	45	33
Range	0.45-16.87	0.66-6.90	6.79-18.31
Arith. mean	3.61	3.01	12.08
Median	2.96	2.69	12.05
Geom. mean	3.49	2.95	11.69

DISCUSSION

Accuracy and precision of the analytical method used for the determination of mercury in the IAEA Standard Reference Materials were good, as shown in Table 1. Mercury in most marine organisms and hair samples can be analyzed quite easily and conveniently by instrumental neutron activation, so that tedious radiochemical separation steps can be avoided. By employing a low-energy photon detector, mercury can be determined much faster through the 69- and 77-keV photopeaks of ^{197}Hg. However, the spectrum in this region is quite complex, and the analysis of spectrum becomes complicated. Very convenient conditions can be achieved by analyzing the 279-keV photopeak of ^{203}Hg. The γ spectra of the irradiated hair samples are less complex than other biological tissues, so that favorable detection limits of mercury in hair can

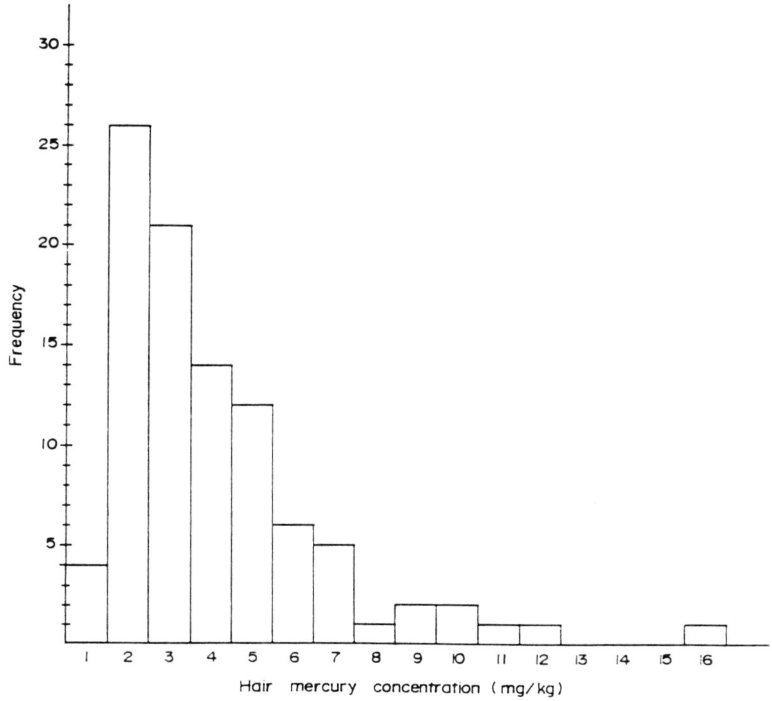

Fig. 1. The frequency distribution of hair mercury concentrations for Penang.

be achieved. The detection limit of the technique used amounted to 0.01 mg/kg.

The mercury levels in the three population groups studied were significantly different (Table 2). However, the narrow range of mercury concentrations for the population in Selangor lies within the range for the group in Penang. The group in Selangor consists mainly of office workers, but their fish-consumption pattern is quite similar to the Penang group. Generally, they consumed fish once a day, but quite different types of fish. Different protein sources were also available to the group in Selangor, since they were economically more sound. Other protein sources, such as meat and poultry products, were generally low in mercury concentrations.

The frequency distribution of mercury concentrations for the Penang group is shown in Fig. 1. About 30% of the samples contained mercury in higher levels than 5 mg/kg. This suggests that a part of the population studied might have exceeded the PTWI of mercury. From the information gathered, some of the donors, especially adult males, consumed as much as 2 kg of fish weekly. This is also reflected in the distribution of mercury concentrations in relation to age of the donors as shown in Table 3, where generally, mercury hair concentrations in adults were higher as compared to children. Based on the concentrations of mercury

Table 3
Distribution of Mercury According to Age of Donor in mg/kg

Age	Range	Arithmetic Mean
1-9	0.45-8.47	3.27
10-19	0.58-10.43	3.16
20-29	0.87-16.87	3.83
30-39	1.04-6.75	3.09
40-49	1.91-12.55	5.22
50-59	1.94-8.33	4.05
>60	1.29-16.65	6.25

in fish marketed in Malaysia, the PTWI limit may be achieved by consuming about 500 g of fish weekly *(7)*.

The hair mercury levels of fishermen and their families in Terengganu were the highest among the groups studied. This is a selected population group that consumed between 200 and 300 g of fish daily. This group preferred tuna to other types of fish. The Malaysian tunas, such as *Euthynnus affinis, Auxis thazarg,* and *Thunnus tonggol,* generally contain higher levels of mercury than other fish. Our study showed that the mercury levels ranged from 0.81 to 3.35 mg/kg fresh weight. The lowest concentrations were found in the white flesh, whereas the highest were found in the liver. Even though this group lives in a nonindustrialized area where mercury pollution is expected to be absent, the mercury body burden, as reflected by the hair mercury, was high. This showed that the amount and type of fish consumed would affect the hair mercury concentrations.

The concentrations of mercury in the hair samples analyzed showed positive correlation with the amount of fish consumed, and is within that established by Airey *(6)*. Based on the national per capita consumption of fish at 100 g/person/d, the expected hair mercury concentration of the general population is 6 mg/kg. This study showed that about 30% of the hair samples from Penang, all from the group in Terengganu, and 15% of the samples from Selangor contained mercury at more than 6 mg/kg. This also indicated the annual per capita consumption of fish of these subject was 35 kg. Our previous work shoved that the mean concentration of hair mercury concentration of the general population in the capital city was 6.6 mg/kg *(9)*, suggesting that the more affluent city population consumed a higher amount of fish than expected. The hair mercury levels found in this study were within the range of mercury concentrations in the head hair of the fish-eating population worldwide as reported by Airey *(3)*.

SUMMARY

The hair mercury levels found in the Malaysian fishermen depend on the amount and type of fish consumed. Industrial activities do not affect the mercury levels in the hair of the population studied. This suggests that industrial activities in Malaysia do not discharge mercury into the environment. The mercury concentration range is within those reported by other researchers for fish-consuming population groups worldwide.

ACKNOWLEDGMENT

The research grant provided by the IAEA through Contract Number MAL 6335/R1/RB is acknowledged. Thanks are due to the Nuclear Energy Unit for providing irradiation facilities, and to Idris Fakih Mokhar, Aziz Ramzan, Hamzah Pardi, Roslan Din, and Ahmad Takim Saring for their competent technical assistance.

REFERENCES

1. International Atomic Energy Agency, *NAHRES-13*, Vienna, 1992, pp. 1–3.
2. A. Chatt and S. A. Katz, *Hair Analysis*, VCH Publishers, New York, 1988, p. 134.
3. D. Airey, *Sci. Total Environ.* **31,** 157 (1983).
4. J. H. Kyle and N. Ghani, *Arch. Environ. Health* **37,** 266 (1988).
5. J. C. Hansen, H. C. Wulf, N. Kromann, and K. Aloge, *Sci. Total Environ.* **26,** 233 (1983).
6. D. Airey, *Environ. Health Prospec.* **52,** 303 (1983).
7. S. Sarmani, A. K. Wood, Z. Hamzah, and A. A. Majid, *J. Radioanal. Nucl. Chem.* **169,** 255 (1993).
8. P. M. Silvalingam and A. B. Sani, *Marine Pollut. Bull.* **11,** 188 (1980).
9. S. Sarmani, *J. Radioanal. Nucl. Chem.* **110,** 627 (1987).

Elemental Composition of Head Hair and Fingernails of Some Nigerian Subjects

A. F. OLUWOLE,*[,1] J. O. OJO,[1] M. A. DUROSINMI,[2] O. I. ASUBIOJO,[3] O. A. AKANLE,[4] N. M. SPYROU,[4] AND R. H. FILBY[5]

[1]Department of Physics, Obafemi Awolowo University, Ile-Ife, Nigeria; [2]Department of Haematology and Immunology, Obafemi Awolowo University, Ile-Ife, Nigeria; [3]Department of Chemistry, Obafemi Awolowo University, Ile-Ife, Nigeria; [4]Department of Physics, University of Surrey, Guildford, England; and [5]Department of Chemistry, Washington State University, Pullman, WA

ABSTRACT

In this article, we present the elemental concentrations determined by INAA for 30 elements measured in some or all head hair samples of 100 Nigerian subjects and 20 elements in the fingernails of some of the same subjects. Measurements of the skewness of the distribution of each element in both tissues confirm previous reports that many tend toward a log-normal distribution. Thus, their concentrations in the tissues may not be under any homeostatic control. The ranges of elemental concentrations together with the medians, and the arithmetic and geometric means, with their respective standard deviations are presented and compared with literature values for other populations. Correlations between elements detected in hair are also sought.

Index Entries: Elemental baseline levels; head hair and nails neutron activation analysis; Nigerian subjects.

*Author to whom all correspondence and reprint requests should be addressed.

INTRODUCTION

The use of the metabolically inactive tissues, hair, and nails in biomonitoring has become well established, especially in investigating the levels and changes of many trace elements accumulated in the body over long periods of time. The ease of sample collection, transport, storage, and preparation, coupled with the usually elevated levels of several trace elements in these tissues have compelled further research even in the face of the well-known shortcomings of these tissues as biomonitors. Such shortcomings include problems with exogeneous contaminants and interpretation of the results. At the present time, hair is being used to study human trace element metabolism *(1,2)* assessing environmental pollution and occupational exposures *(3–5)*, in practical interpretation of the elemental distributions of immediate "human" environment (water supply, cosmetics, smoking habits, and so forth) *(6,7)*, and generally in epidemiological studies *(8,9)*.

The use of nails in biomonitoring is not as widespread as hair, but it is increasing in recent times. The applicability of nails to some of the uses mentioned above has been reviewed by Jenkins *(10)* and also Berlin *(5)*, whereas the possibility of substituting or complementing hair with nails has been investigated by Oluwole and coworkers *(11,12)*.

Before these tissues could be reliably used in any of the applications highlighted above, the baseline levels of the trace elements of interest in "normal" subjects within the same geographical and socioeconomic environment had to be established first. In this article, we have determined the baseline levels of the 30 elements, Ag, Al, Au,* Ba, Br,* Ca, Ce,* Cl, Co,* Cr,* Cs, Cu,* Eu,* Fe,* Hg,* I, La,* Mg, Mn, Na,* S, Sb,* Sc,* Se,* Sm,* Tb, Ti, V, Zn,* and Zr, from a population comprising 100 normal Nigerian subjects, using instrumental neutron activation analysis (INAA). The 16 asterisked elements together with K, Th, Hf, and Rb were also determined in the toenails of some of the same subjects, predominantly the children (age < 15; $n = 21$). The distribution pattern of each element, as well as correlations between the different elements in hair were also investigated. Corresponding values in nails have been reported elsewhere *(12,13)*. The baseline levels recorded in this work for both head hair and fingernails are compared with available data from other regions of the world *(14–17)*.

MATERIALS AND METHODS

Population and Sampling

Scalp hair samples were collected from 100 volunteers (45 males, 55 females), whereas fingernail samples were also collected from 21 of the pedriatric subjects. The ages of the subjects ranged from 3–60 yr. Hair samples were obtained consistently from the left side of the nape, 3–5 mm from the scalp. Nails were obtained from the fingers using a pair of nail clippers. Our procedure for washing of hair samples is similar to the one described previously by Othman and Spyrou *(18)*. It comprised repeated washing and decantation for about 2 h in an ultrasonic bath using only deionized water. A few of the hair samples were viewed under a Scanning Electron Microscope to assess the extent of external contamination. A similar protocol was observed for the nails, except that the nail surfaces were first scraped clean of any exogeneous material using a plastic toothbrush, before the washing.

Analysis

Irradiations were carried out both at the Washington State University (WSU) TRIGA III-fueled research reactor (thermal neutron flux ≈6 × 10^{12} neutrons/cm^2/s) and at the Imperial College Reactor Centre (ICRC) (thermal neutron flux ≈2 × 10^{12} $n/cm^2/s$ for short irradiations and ≈1 × 10^{12} $n/cm^2/s$ for long irradiations). The samples were analyzed as three sets. Set A, comprising 21 pediatric subjects, and Set B, comprising 52 adults, were analyzed for both "intermediate" and "long-lived" elements at WSU, whereas Set C, comprising 27 individuals, was analyzed at the ICRC using both short and long irradiation.

At the WSU, the intermediate-lived isotopes ^{198}Au, ^{47}Ca, ^{152m}Eu, and ^{24}Na were determined by irradiating the samples for 1 h and counting for 1800 s each after allowing initial decay of about 3 d. The long-lived isotopes ^{131}Ba, ^{141}Ce, ^{60}Co, ^{51}Cr, ^{134}Cs, ^{152}Eu, ^{59}Fe, ^{75}Se, ^{160}Tb, ^{65}Zn, and ^{95}Zr were determined by counting for 5000 s after about 2 wk of decay following an 8-h irradiation. NBS fly-ash was used both as neutron flux monitor and a comparator standard.

At the ICRC, the relatively short-lived isotopes ^{24}Na, ^{27}Mg, ^{37}S, ^{38}Cl, ^{49}Ca, ^{51}Ti, ^{52}V, ^{56}Mn, ^{66}Cu, ^{80}Br, ^{85m}Sr, and ^{128}I were determined by irradiating the samples for 360 s, followed by a waiting time of 60 s and a counting time of 360 s. For the intermediate and long-lived isotopes ^{46}Sc, ^{60}Co, ^{59}Fe, ^{82}Br, ^{110m}Ag, ^{153}Sm, ^{131}Ba, ^{203}Hg, ^{140}La, and ^{198}Au, about 30 h of irradiation were carried out, followed first by 5 d of initial decay and 3600 s of counting, as well as a further 1 wk of decay and 5000 s of counting. Bowen's Kale and single-element standards deposited on a filter paper (for Ti, Cu, I, Hg, Sc, Cs, Co, Ag, Cr, Sb, and Au) were used as comparator standards, whereas pure thin zirconium wires were used as flux monitors, thus allowing us the possibility of carrying out "absolute" analysis, at a later stage.

RESULTS AND DISCUSSION

The arithmetic mean, geometric mean, and their respective standard deviations together with the median, range, and skewness of elemental distribution, γ_1, of the concentrations of each element in hair is shown are Table 1, whereas the levels of the elements in fingernails in pediatric subjects are shown in Table 2, which also compares the levels with those

Table 1
Concentrations (μg/g) of Elements in Head Hair of Nigerian Subjects

Element	No.*	Arith Mean (±SD)	Geom Mean (x+GSD)	Median	Range	γ_1
Ag	6	1.79 (0.94)	1.533 (1.810)	1.735	0.45-2.80	- 0.18
Al	27	93.7 (58.7)	71.6 (2.2)	69.17	9.48-210.0	0.51
Au	75	0.047 (0.105)	0.012 (25.6)	0.017	0.006-0.88	2.00
Ba	6	24.5 (6.8)	23.3 (1.3)	21.2	15.6-36.3	0.22
Br	63	16.8 (36.7)	2.73 (6.14)	33.73	0.16-282.2	2.71
Ca	27	2659.7 (1252.8)	2295.1 (1.8)	2534.0	487.4-5543.0	0.41
Ce	27	1.44 (1.51)	0.81 (3.00)	1.17	0.25-6.87	1.77
Cl	27	654.3 (573.9	430.8 (2.5)	446.6	68.39-2724.2	1.65
Co	77	0.399 (0.273)	0.309 (2.039)	0.437	0.06-2.009	1.77
Cr	66	2.594 (2.373)	1.909 (2.435)	3.54	0.31-14.42	1.25
Cs	36	0.221 (0.133)	0.178 (1.922)	0.184	0.034-0.604	1.03
Cu	23	29.95 (18.94)	22.97 (1.99)	22.8	2.12-64.77	0.62
Eu	50	0.051 (0.037)	0.040 (1.98)	0.042	0.01-0.145	1.40
Fe	77	202.6 (133.1)	176.8 (42.8)	213.0	26.7-559.9	0.63
Hg	9	0.17 (0.09)	0.30 (0.01)	0.13	0.09-0.33	1.04
I	27	5.565 (6.482)	3.109 (2.777)	2.30	0.45-37.62	2.97
La	55	0.900 (1.280)	0.180 (4.460)	0.370	0.015-6.109	2.02
Mg	22	269.1 (150.7)	239.3 (1.635)	276.6	79.16-599.0	0.75
Mn	26	8.62 (4.74)	7.44 (1.71)	8.08	2.72-21.93	1.09
Na	60	165.9 (165.9)	89.34 (3.66)	96.3	4.38-216.0	0.10
S	27	26700 (6570)	25800 (1)	26500	3180-40000	- 0.62
Sb	74	0.397 (0.586)	0.202 (3.225)	0.690	0.003-3.94	1.22
Sc	75	0.058 (0.047)	0.040 (32.561)	0.059	0.004-0.25	0.34
Se	23	0.559 (0.438)	0.419 (2.113)	1.156	0.10-1.863	- 0.28
Sm	53	0.013 (0.012)	0.010 (2.223)	0.013	0.001-0.016	1.73
Tb	39	0.084 (0.110)	0.050 (2.624)	0.047	0.006-0.608	3.12
Ti	19	25.89 (22.32)	18.564 (25.891)	20.92	2.64-84.35	1.27
V	27	0.323 (0.244)	0.257 (1.966)	0.21	0.08-0.99	1.19
Zn	77	124.8 (79.0)	129.4 (1.7)	178.5	8.89-550.5	1.86
Zr	15	92.79 (50.76)	72.2 (1.9)	95.6	21.2-220.3	**0.53**

*No. is the no. of samples in which element is detected. Extraneous values are not included in means, but included in median.

obtained elsewhere. Our results were compared with elemental levels in hair obtained from Japanese *(14)*, Bulgarian *(15)*, and Chinese *(16)* subjects, as well as the averages of worldwide mean values compiled by Iyengar et al. (*17*). It has been suggested by Liebscher and Smith *(19)* and also Othman and Spyrou *(18)* that the distribution of elemental concentrations in tissues could be used to determine whether or not those elements are essential in body metabolism. Also Evans and Jervis *(20)* have shown that the distribution should determine the way that the results obtained should be interpreted. Usually, the distribution is determined by plotting for each element, a frequency distribution diagram, and then guessing the function that fits the shape *(18,21)*. Since several elements have been shown to have log-normal distribution in hair and nails, another approach has been to transform the concentrations measured logarithmically and see how good the fit is to a straight line on a log-linear sheet *(14)*. All these could be tedious, in addition to their not being truly quantitative. In this work, we have determined the skewness of the distribution of each element γ_1, using a formula proposed by Pearson *(22)*. γ_1, was obtained first by transforming the individual concentrations X into a z-score ($\approx[X - \bar{X}]/\sigma$, where $\bar{X}$ is the mean and s the standard deviation of the distribution). Cubing the z-scores and then finding their means yields γ_1, as shown below:

$$\gamma_1 = \sum_{i=1}^{N} z_i^3/N \qquad (1)$$

The various values for γ_1 shown in Table 1 can be better appreciated by considering the transformations given by Glass and Hopkins *(22)* for a given normal distribution X ($\gamma_1 = 0.0$), which when transformed into the distributions of X^2, X^3, $1/X$, $\sqrt{X}$, and log X, respectively, have the skewness parameters $\gamma_1 = 1.2$, $\gamma_1 = 2.1$, $\gamma_1 = -0.3$, and $\gamma_1 = -0.7$.

Values obtained in this work agree well with the distribution diagrams of Qureshi et al. *(21)* for the elements Au, Co, Cr, Cu, Fe, Hg, Mn, and Sb, and also with results plotted by Othman and Spyrou *(18)* for Ca, Mg, and V. However, whereas the shapes for the distribution of Zn by Qureshi and Cu, Hg, Mn, and Zn by Othman and Spyrou agree with our values, the degrees of skewness (magnitudes) do not agree well. There is outright disagreement in Qureshi's distribution of Ag and ours, but ours has been obtained from only six individuals, which is not large enough to determine a distribution reliably.

The Nigerian subjects in this work have comparable levels of trace elements in their hair relative to subjects from other countries. In comparing these values, allowance should be made for differences in sample preparation, especially washing, which could introduce significant differences. For the Bulgarian subjects *(15)* for which results for different washing techniques were available, we have chosen the results obtained under conditions closest to ours.

In general, the elements Ag, Br, Ca, Co, Cr, Cu, I, Mn, Na, Sb, and V have higher mean values relative to the literature data (Mn is lower than Kenyan values, *[18]*), but the ranges are in good agreement. In sum, the differences in mean values are not statistically significant, since they

Table 2
Trace Elements in Fingernails of Nigerian Pediatric Subjects (in μg/g)[a]

Elt	THIS WORK (NIGERIA)				BANGLADESH (PIXE)			Other Available References
	No.	Range	Arith. Mean (±std dev)	Geom. Mean (x+std dev)	Range	Arith Mean	Geom Mean	
Sm	19	0.021-0.413	0.148 (0.098)	0.118 (1.732)	-	-	-	-
Au[b]	19	8.05 - 38.89	20.57 (8.58)	18.65 (1.47)	-	-	-	-
La	19	0.169-3.297	1.114 (0.750)	0.896 (1.733)	-	-	-	-
K	19	732.4-3657	2043.9 (858.7)	1842.2 (1.5)	25 - 241	91.1 ± 41.0	82.4 (1.56)	357-2800[*,23]
Ce	15	0.446-8.729	3.223 (1.971)	2.650 (1.704)	-	-	-	-
Th	15	0.064-1.58	0.402 (0.371)	0.291 (1.934)	-	-	-	-
Hf	8	0.139-1.731	0.626 (0.491)	0.458 (2.208)	-	-	-	-
Eu	19	0.072-0.701	0.274 (0.174)	0.227 (1.734)	-	-	-	-
Cr	19	2.233-20.38	9.651 (5.393)	8.031 (1.848)	-	-	-	-
Sc[b]	18	56.74-614	270.58 (140.88)	233.4 (1.6)	-	-	-	-
Fe	17	411.4-2434	1276.7 (570.1)	1150.3 (1.5)	15.6-360.0	77.8 (67.2)	64.4 (1.89)	27-347[*,23]
Zn	19	104.0-261.6	162.6 (40.6)	157.7 (1.2)	72.0-171.0	113 (20)	112 (1.19)	73-304[*,23]; 83-200[*,13]; 129[+,13]
Co	19	0.189-3.388	1.016 (0.684)	0.841 (1.685)	-	-	-	-
Sb	18	0.3-6.0	1.497 (1.441)	0.992 (2.337)	-	-	-	-
Cu	18	26.97-104	54.9 (22.3)	50.565 (1.439)	1.67 - 27.2	6.85 (4.40)	5.26 (1.65)	31[+,14]; 7.5[+,15], 3-18.6[*,15];
Br	19	5.21-36.7	15.15 (8.42)	13.06 (1.62)	0.68 - 4.03	1.65 (0.71)	1.52 (1.50)	-
Na	18	835.2-3261	1640.9 (610.0)	1537.7 (1.35)	-	-	-	152[+,14]
Se	10	0.634-3.536	1.482 (0.763)	1.333 (1.425)	0.41-2.00	1.28 (0.34)	1.24 (1.36)	-
Hg[b]	4	99.56-972.4	375.2 (348.0)	258.3 (2.4)	-	-	-	-
Rb	5	4.381-9.357	7.31 (1.66)	7.09 (1.28)	0.54 - 2.68	1.41 (0.54)	1.29 (1.54)	-

*Range
†Arithmetic Mean.
[a]If not otherwise stated.
[b]μg/g.

can be accounted for by the standard deviations. On the other hand, the elements Zn, Se, Cs, and S, although also agreeing in the ranges, have mean values lower than those found in the literature. Mean values for La in Nigerian subjects are comparable to those in the literature, but a few high values extend the high limit far beyond literature values. In all cases, however, the mean values are within the range of average mean values compiled by Iyengar et al. *(17)*. Good agreement between both means and ranges of levels in Nigerian subjects and literature was observed for Au, Cl, and Mg. However, Al, Ce, Ti, Eu, and Fe are much elevated in Nigerian subjects. Other elements, including Eu, Ba, and Zr, could not be compared because of unavailability of literature data.

Intercomparison of the nail data is difficult also because of unavailability of literature data. The few available references usually involve just a few elements. The only available baseline measurements on nails was on a Bangladeshi population *(23)* with the analysis carried out by PIXE. In general, the values obtained in the nails of Nigerian subjects for K, Fe, Br, and Rb are higher than the Bangladeshi one by about an order of magnitude. Na was compared with values for Kenyan subjects *(18)*, whereas Cu was compared with both Kenyan *(18)* and German *(24)* data. For both elements, values in Nigerian subjects are higher. K, Zn, Se, and As levels are, however, similar in Nigerian results and the literature data. It should be noted that whereas the Bangladeshi population considered here comprised adults only, the Nigerian subjects were pediatric subjects from low-income parents whose fingernails are prone to considerable contamination through occupational exposure, as has been observed in a previous IAEA report *(25)*. Moreover, for most elements where other references are available, levels in the Bangladeshi population were found to be much lower.

Correlations between elements in nails have been previously reported by us *(12,13)*. In hair, correlations were sought between elements detected in the three sets of subjects used in this work. However, correlations between elements that are log-normally distributed might be better reflected by considering rather the logarithmic concentrations of the elements *(14,24)*. We have also done this (Table 3).

Strong correlations were observed within the rare earth elements Sm, Eu, and La, and the transition elements Fe, Sc, Co, and Sb; among the halogens Br, Cl, and I together with Cu and Na; and also among other elements, such as Mg, S, Ca, V, Mn, and Al. Many of these interelemental correlations coincide well with correlations of elements found in Japanese hair *(14)*. In the Japanese, the elements K, Mn, V, Al, Cl, Br, and I also clustered together and were found to have medians that decrease gradually with increasing age, whereas another group comprising Cu, Mg, Ca, Zn, and S has medians that increase gradually with increasing age. These correlation patterns could also suggest the form of the compound in which most of the elements are stored. For instance, considering the table, Na would be complexed mostly with Cl (NaCl) and to a smaller extent with Br, whereas most of I would be complexed in hair with Cu.

Table 3
Statistically Significant ($p < 0.05$) Elemental Correlations in Hair Using (a) Concentrations and (b) Logarithm of Concentrations in Hair

Elt	Correlating Elements (using concentrations)	Correlating Elements (Log concentrations)
Al	V*, Mn*	Mn***, V^{+}
As	Sc, Fe, Zn, Au*	Co
Au	Sc, Zn, As**	La**, Sc, Fe, Zn, Sm, Co**
Br	Co, Eu, La, Na, Sc**, Cl***	Ti, Co*, Eu*, Fe***
Ca	S*, Mg***	Mg***
Co	Br, Cr*, Sc**, Sm**, La***, Sb***, Eu^{+}, Fe^{+}, Zn^{+}	Sb, As*, Br*, Sm***, Au***, La***, Sc***, Fe^{+}, Zn^{+}, Eu^{+}, Sb*
Cl	Br***, Na^{+}	S*, Cu**, Na***
Cu	Zn*, Na^{+}, I^{+}	I*, Mg*, Cl**
Cr	Zn*	Sb*
Eu	Br, Sm**, La***, Sc***, Fe^{+}, Co^{+}	Br*, La*, Sm*, Sb*, Sc*, Fe***, Co^{+}
Fe	Sm^{+}, Eu^{+}, La^{+}, Sc^{+}, Zn^{+}, Co^{+}, Sb^{+}	Au, Sb*, Br***, Eu***, Co^{+}, Sc^{+}, Sm^{+}, La^{+}, Sc^{+}
I	Cu^{+}	Cu*
La	Br, Co***, Eu***, Fe^{+}, Sc^{+}	Sb, Eu*, Au**, Co***, Sc^{+}, Fe^{+}, Sm^{+}
Mg	S*, Ca***	S, Cu*, Ca***
Mn	Al*, V***	V**, Al***
Na	Br, Eu**, Cu^{+}, Cl^{+}	V, Cl***
S	Mg*, Ca*	Mg, Cl*
Sb	Fe^{+}, Zn^{+}	La, Co*, Cr*, Fe*, Eu*
Sc	Ag, Au, Br**, Co**, Eu***, Sm^{+}, La^{+}	Au, Eu*, Co**, Sb***, Fe^{+}, Sm^{+}, La^{+}
Sm	Fe, Co**, Eu**, La^{+},Sc^{+}, Fe^{+}	Au, Eu*, Co***, La^{+}, Sc^{+}, Fe^{+}
Ti	-	Br
V	Al*, Mn**	Na, Mn**, Al^{+}
Zn	Au, Cr, Cu*, Fe^{+}, Co^{+}, Sb^{+}	Au, Sc, Co*

*Implies $p < 0.025$, **implies $p < 0.01$; ***implies $p < 0.005$; †implies $p < 0.0005$.

It is interesting to note that although most patterns are retained, a few changes are seen in the correlation patterns when the logarithms of the concentrations are correlated together. This agrees with the observed fact that the correlation between elements could be destroyed when the distribution is distorted, as for example in permanent hair wave treatment. Perhaps extensive sample washing, especially with chemicals, will produce the same effect.

SUMMARY

Baseline for 30 elements in the head hair and 20 elements in the fingernails of some Nigerian subjects has been presented and compared with similar results from other countries. There is no evidence of any major deficiency or excess accumulation of elements in our Nigerian subjects relative to these other results. The skewness of the elemental distributions has also been quantitatively measured and compared with available data. Elemental correlations found between elements in hair were noted. For some elements, it might be more appropriate in seeking correlations to transform first the concentration values appropriately to reflect their distribution in hair.

ACKNOWLEDGMENT

This work is part of the EEC-sponsored project on Environmental Pollution Monitoring and Impact Assessment studies in Nigeria.

REFERENCES

1. R. B. Bradfield, T. Yee, and J. M. Baertl, *Am. J. Clin. Nutr.* **22,** 1349–1353 (1969).
2. S. B. Deeming and C. W. Weber, *Am. J. Clin. Nutr.* **31,** 1175–1180 (1978).
3. A. Chatt, C. A. Secord, B. Tiefenbach, and R. E. Jervis, in *Hair, Trace Elements and Human Illness*, A. C. Brown and R. R. Crouse, eds., Praeger, New York, 1980, Chapter 3, p. 46ff.
4. V. Valkovic, D. Rendic, and G. C. Phillips, *Environ. Sci. Techno.* **9,** 1150 (1975).
5. A. Berlin, A. H. Wolff, and Y. Hasegawa *Proceedings of the International Workshop at Luxembourg*, April 18–22, 1977, Martinus Nijhoff Publishers, 1979.
6. M. D. Tshiashala, K. M. Kabengele, and B. M. Lumu, *Biol. Trace Element Res.* **26–27,** 287–294 (1990).
7. S., Aufreiter and R. G. V. Hancock *Biol. Trace Element Res.* **26–27,** 721–728 (1990).
8. T. Takeuchi, Y. Nakano, A. Aoki, S. Ohmori, and M. Kasuya, *Biol. Trace Element Res.* **26–27,** 263–268 (1990).
9. Y. Wei-Han, *Jpn. Circ. J.* **46,** 1201–1207 (1982).
10. D. Jenkins, Biological monitoring of toxic metals in human populations using hair and Nails, private communication.
11. A. F. Oluwole, O. A. Asubiojo, A. D. Adekile, R. H. Filby, A. Bragg, and C. I. Grimm, *Biol. Trace Element Res.* **26–27,** 479–484 (1990).
12. J. O, Ojo, A. F. Oluwole, O. I. Asubiojo, and A. D. Adekile, To be published in *Proceedings of the First National Conference on Nuclear Methods*, September 2–4, 1992, Ahmadu Bello University, Zaria, Nigeria, 1993.
13. J. O. Ojo, A. F. Oluwole, O. I. Asubiojo, M. A. Durosinmi, and O. J. Ogunsola, To be published in the *Proceedings of the International Workshop on Environmental Monitoring and Impact Assessment/Energy Masterplan for Rural Development*, April 6–10, 1992, Obafemi Awolowo University, Ile-Ife, Nigeria, 1993.

14. T. Takeuchi, T. Hayashi, J. Takada, Y. Hayashi, M. Kozama, H. Kozuka, H. Tsuji, Y. Kusaka, S. Ohmori, M. Shinogi, A. Aoki, K. Katayama, and T. Tomiyama *J. Radioanal. Chem.* **70,** 29–55 (1982).
15. N. I. Ward, N. M. Spyrou, and A. A. Damyanova, *Proceedings of the 7th MTAA, Holland*. Vol. 2, pp. 1087–1095, 1986.
16. L. S. Chuang and J. F. Emery, *J. Radioanal. Chem.* **45,** 169–180 (1978).
17. G. V. Iyengar, W. E. Kollmer, and H. J. M. Bowen, *The Elemental Composition of Human Tissues and Body Fluids*. Verlag Chemie Publishers, Weinheim, 1978.
18. I. Othman and N. M. Spyrou, *Sci. Total Environ.* **16,** 267–278 (1980).
19. K. Liebscher and H. Smith, *Arch. Environ. Health* **17,** 881–890 (1968).
20. G. J. Evans and R. E. Jervis, *J. Radioanal. Nucl. Chem., Articles* **110,** 613–625 (1987).
21. I. H. Qureshi, M. S. Chaudhary, and S. Ahmad, *J. Radioanal. Chem.* **68,** 209–218 (1982).
22. G. V. Glass and K. D. Hopkins, *Statistical Methods in Education and Psychology*, 2nd ed. Prentice-Hall, New Jersey 1984.
23. S. K. Biswas, M. Abdullah, S. Akhter, S. A. Tarafdar, M. Khaliquzzaman, and A. H. Khan, *J. Radioanal. Nucl. Chem., Articles* **82/1,** 111–124 (1984).
24. M. Wilhelm, D. Hafner, I. Lombeck, and F. K. Ohnesorge *Sci. Total Environ.* **103,** 199–207 (1991).
25. IAEA Report CVD\73.4, February 19–23, 1973 and Geneva (WHO), April 2–6 1973.

Correlations Between Trace Element Levels in Head Hair and Blood Components of Nigerian Subjects

J. O. Ojo,[1] A. F. Oluwole,*,[1] M. A. Durosinmi,[2] O. I. Asubiojo,[3] O. A. Akanle,[4] and N. M. Spyrou[4]

[1]Department of Physics, Obafemi Awolowo University, Ile-Ife, Nigeria; [2]Department of Haematology and Immunology, Obafemi Awolowo University, Ile-Ife, Nigeria; [3]Department of Chemistry, Obafemi Awolowo University, Ile-Ife, Nigeria; and [4]Department of Physics, University of Surrey, Guildford, England

ABSTRACT

Elemental concentrations in head hair and blood (whole-blood, erythrocytes and plasma) of 28 Nigerian subjects were determined using instrumental neutron activation analysis (INAA) and proton-induced X-ray emission (PIXE) techniques. Correlations found between different elements in hair and blood are reported. Hair levels of both zinc and chlorine were found to be significantly correlated with their levels in erythrocytes.

Index Entries: Elemental levels; correlations between head hair and blood levels; Nigerian subjects.

INTRODUCTION

The importance of trace elements in various fields pertaining to human well-being has been well established. For instance levels of essential trace elements are important in nutrition, health, and disease, whereas those of elements of anthropogenic origin could be useful as indices of environmental pollution *(1–3)*. In all these and similar fields, the ultimate goal is often to assess the levels of these elements in the different body tissues of interest. The direct way of doing this is to sample the tissue in question and analyze it for trace elements *(4)*. However, this

*Author to whom all correspondence and reprint requests should be addressed.

approach is invasive and hardly offers any promise for use in routine or diagnostic applications.

It has become acceptable to assess body burden of several trace elements indirectly by considering their levels in blood *(3,5,6)*. The functions and distribution of blood in the body make this practice biologically reasonable, and the constant flow of blood ensures a better homogeneity of the trace elements in blood than in organs. This is especially true when blood is separated into its components. In general, the erythrocytes give better information on elements with long resident times, whereas the plasma is suitable for those resident only for short terms.

However, in recent times, the determination of elemental levels in hair is becoming more and more reliable. Considering that hair is even more easily and atraumatically obtained than blood, with little or no medical restrictions to its collection, efforts are being directed in determining what correlations exist between elemental levels in blood and in hair. Where this is proven, hair could then conceivably be used in place of blood. In some circumstances, the body burden might even be better represented in hair than in blood, for example, as in cases where persistent and cumulative metals (e.g., Cd, Hg, and Pb) are ingested either in single massive doses or irregular doses. It will be very difficult to interpret the results of hair analyses reliably, from the human health point of view, without first establishing the correlation pattern for elemental levels in hair and body burden.

In general, such correlations have been difficult to find. Zhuk and Kist *(7)* reported that they found no reliable correlations between hair and blood in a study of Uzbekistan residents, though they found some "weak relationship" between elemental hair composition and hemoglobin content in the blood, and a stronger relationship between hair composition and blood group. Also Yoshinaga et al. *(8)* found no correlation between elemental levels in hair and internal organs, and Halsted was reported *(9)* as not to have found any correlation between hair zinc and plasma zinc levels. Similarly, Vaughan and Florence *(10)* reported a lack of relationships between Pt levels in blood and hair. Also Kijewski *(11)* did not find any relationship between Tl in hair and blood. However, a few correlation patterns have been positively identified, especially for the heavy metals. Kollmer and Wessenberg and coworkers in separate works were reported *(5)* to have found significant correlations between Cd in hair and blood. Zhuang *(4)* found significant correlations between Se and As in hair and some internal organs (liver, kidney cortex, and lung), whereas the relationship between Pb and Hg in hair and blood has been reported by several workers *(5)*. In other cases (e.g., Cd), contradictory reports have been published on blood–hair elemental correlations *(5)*.

In this work, we have investigated the correlations between 18 elements in hair and 21 in blood (whole blood as well as the plasma and erythrocyte components) obtained from 28 Nigerian subjects using INAA. Correlations were sought by calculating Pearson's product moment correlation coefficients for each possible pair of elements.

MATERIALS AND METHODS

Hair and blood samples were obtained from 28 Nigerian subjects aged between 3 and 60 yr. The details of sample collection and preparation are as previously described *(12,13)*. Eighteen elements—Ti, I, Br, Mg, Cu, Na, V, Cl, Al, Mn, Ca, S, Hg, Au, Sb, Sc, Zn, and Co—were determined in the hair samples by INAA, whereas the eight elements, Br, Na, K, Fe, Se, Rb, Zn, and Cr, and the 20 elements, P, S, Cl, K, Ca, V, Cr, Mn, Fe, Co, Ni, Cu, Zn, Ga, As, Se, Br, Rb, Sr, and Pb, were determined in the blood samples by INAA and PIXE, respectively. The blood samples were analyzed as erythrocytes and plasma, as well as whole blood. Details of our analytical techniques have also been previously described *(14,15)*. Correlations among 109 variables, comprising the various elements listed above in hair and the blood components and other variables, such as age, height, and body weight, were investigated by a user-written Fortran program executed on an IBM PS-2 computer.

RESULTS AND DISCUSSION

Statistically significant correlations ($p < 0.05$) found between the elements in blood and hair are shown in Table 1. These correlations reflect the well-known interrelationships between trace elements in biological trace elements research and could be useful in the study of trace element metabolism. Positive correlations point at reinforcement of one element by the presence of the other, whereas negative correlations point at competition (possibly substitution or blockage) between the pair of elements.

The correlations between the heavy and toxic metals Au, Hg, and Sb in hair and selenium in blood as seen in Table 1 is interesting. This is consistent with literature findings by other workers. Selenium is known to antagonize the heavy metals, thus reducing their toxicity *(16,17)*. Unfortunately, we have not measured the levels of the three heavy metals in blood to check on the existence of hair–blood correlations in their levels.

Table 1 also reveals that correlations between elements in hair and Cr in blood (plasma) are always negative. The only exception was found for Sc in hair and Cr in the erythrocytes. The same observation is true for Cl in the erythrocytes, except for those involving Mn and Hg in hair. The implications of these correlations are not clear to us at the present time.

The major interest in this work, that of establishing correlations between hair and blood levels of the same elements, has yielded little positive results. This is consistent with findings by others in the literature. In particular, we had expected some hair–blood correlations in copper, since in previous works *(14,18)*, we have established differences in both hair and blood levels of copper in sickle cell anemia patients with respect to normal controls. The hair–blood correlation coefficients obtained in this work for copper are $r = 0.2875$, $n = 12$ (Whole blood); $r = -0.4026$, $n = 11$ (red blood cells); and $r = -0.1006$, $n = 16$ (plasma). All are statistically not significant.

Table 1
Correlations Among Elements in Hair and Blood

Hair	Plasma	Erythrocytes	Whole Blood
Ti	Fe, -Cr, Ga	-Rb	-
Br	Mn, Zn, Rb, Se	Se,-Cl, -K, Mn,	Se, Fe, Rb
Mg	-	-Rb	-
Cu	Fe, -Cr, Rb	-Cl, Zn	-
Na	Fe, -Cr, Zn, Rb	K, -Cl, Zn	-
V	Ca	-	-
Cl	-	-Cl	-
Al	Fe, -Cr, Zn	-	-
Mn	Fe, Ca	Cl	-
Ca	-	Zn, -Se	-
Hg	Se, Fe, Ga	Se, Cl	Se
Au	Fe, Se, -Cr	Se, -Cl	Se, Zn
Sb	-	Se,	Br
Sc	-K, Ga	Cr, Br	Br
Zn	Fe, Se, -Cr, -Ca, Ga	Se, Zn, P, -Cl	Se, Zn
Co	Se	Se	Se,Zn
Na	-	Se	Cu

However, significant hair–blood correlations have been found for zinc and chlorine, but only consistently in the erythrocytes. Considering the results obtained using INAA, significant correlation is also found between Zn hair and Zn whole blood, but this result is not reproduced when PIXE is used for the analysis. It is quite possible that the large standard deviations (poor precision) of the PIXE result are responsible for the masking of this correlation. The total results obtained both with INAA and PIXE are shown in Table 2, whereas the scattergraphs for hair–erythrocytes levels of Zn and Cl are shown in Fig. 1.

Zinc is probably the most reported of trace elements, and several important roles have been established for it in nutrition and in health and disease *(19)*. The present finding is therefore important. We have not found any similar findings of hair–blood correlations involving zinc or chlorine (or any essential element), but several reports have established the absence of significant correlation between hair and plasma, or hair and serum zinc levels *(8)*. Partially pointing to the same conclusions we have drawn here are the results by Houtman, who was reported to have noted *(8)* that zinc is one of the most regulated elements in the hair, and also Prasad *(20)*, who reported lower zinc levels in both hair and erythrocytes of sickle cell anemia patients.

Even though the levels of the correlations found here may not be significant enough for clinical purposes, an interesting question arises as to the mechanisms involved even for these weak correlations. Laker *(8)* explained the lack of correlation between plasma zinc and hair zinc based on the fact that about 75–85% of the blood zinc is associated with carbonic anhydrase, a cofactor he considers too important to be discarded into the hair. Although our own result confirms this lack of significant association between plasma and hair zinc levels, the explanation above is not supported by the correlation observed between erythrocytes

Table 2
Correlation Between Zinc and Chlorine in Hair and Blood

	Plasma	Erythrocytes	Wholeblood
INAA Zn	n=11 r=0.2508 NS	n=11 r=0.6253 p<0.05	n=11 r=0.7338 p<0.02
Cl	-	-	-
PIXE Zn	n=8 r=0.6087 NS	n=7 r=0.6885 p<0.1	n=8 r=-0.1736 NS
Cl	n=15 r=-0.2079 NS	n=11 r=-0.7254 p<0.02	n=12 r=0.2420 NS

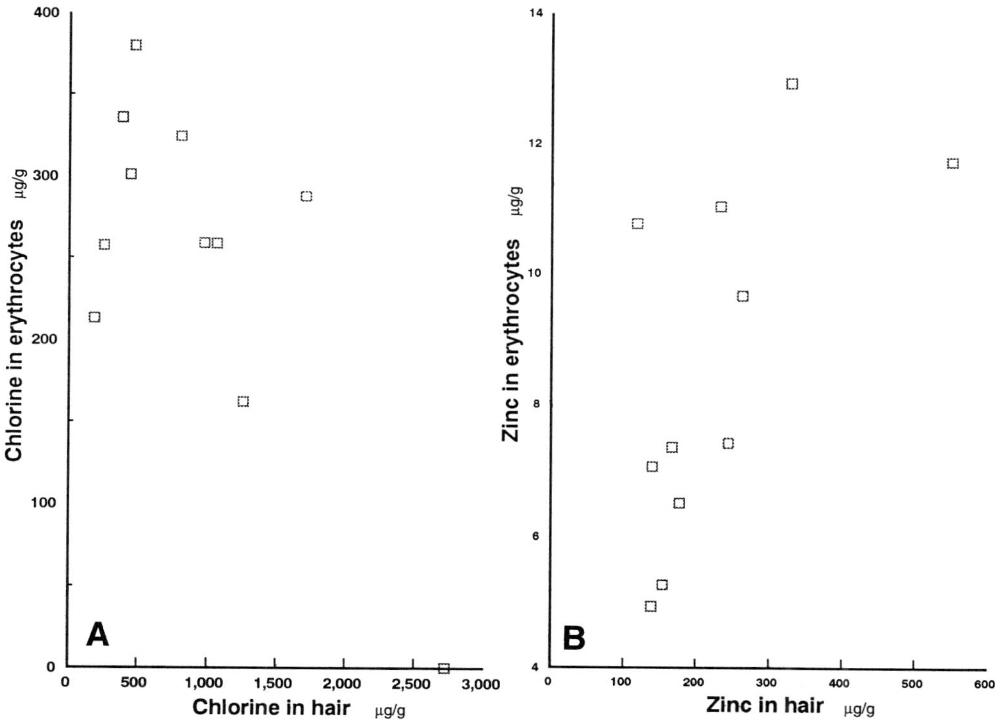

Fig. 1. Scattergraphs for chlorine and zinc in hair and erythrocytes.

and hair zinc levels in this work. Prasad and coworkers *(20)* have previously reported a strong correlation ($p < 0.001$) between the erythrocyte zinc concentration and its carbonic anhydrase content—except it can be shown that zinc is stored in hair in another form, related but not identical to carbonic anhydrase.

More carefully controlled work on a larger population will be done to establish if any correlation exists between hair zinc and whole-blood zinc. Perhaps it might be possible to generalize that correlations between hair and blood are better sought in erythrocytes rather than whole blood or plasma. The short resident times of elements in the plasma pool might make plasma unsuitable, since there could be a very wide gap in the different time scales reflected in both media, although the levels in the whole blood might not be sensitive enough to be correlated with hair levels.

ACKNOWLEDGMENT

This work is part of the EEC-sponsored project on Environmental Pollution Monitoring and Impact Assessment studies in Nigeria.

REFERENCES

1. H. J. M. Bowen, in Trace Elements in Biochemistry, Chapters 7, 8, and 11, Academic Press, 1966.
2. K. T. Smith, ed, *Trace Minerals in Foods*, Marcel Dekker Inc., 1988.
3. A. Berlin, A. H. Wolff, and Y. Hasegawa, *Proceedings of the International Workshop at Luxembourg*, April 18–22, 1977. Martinus Nijhoff Publishers, 1979.
4. G. Zhuang, Y. Cheng, M. Tan, Y. Wang, and M. Zhi, *J. Radioanal. Nucl. Chem., Articles* **149,** 305–315 (1991.)
5. S. A. Katz and A. Chatt, in *Hair Analysis: Applications in the Biomedical and Environmental Sciences*, VCH, 1988, pp. 37–47.
6. D. B. Tuey, in *Introduction to Biochemical Toxins*, E. Hodgson and F. E. Guthrie, eds., Blackwell Scientific Publications, 1980, p. 55.
7. L. I. Zhuk and A. A. Kist, in *Nuclear Analytical Methods in the Life Sciences*, R. Zeisler and V. P. Guinn, eds., Humana Press, New Jersey, 1990, pp. 307–320.
8. J. Yoshinaga, H. Imai, M. Nakazawa, T. Suzuki, and M. Morita, *Sci. Total Environ.* **99,** 125–135 (1990).
9. K. B. Bradfield, T. Yee, and J. M. Baertl, *Amer. J. Clin. Nutr.* **22,** 1349–1353. (1969).
10. G. T. Vaughan and T. M. Florence, *Sci. Total Environ.* **111,** 47–58 (1992).
11. H. Kijewski, *Arch. Kriminol.* **173,** 36–44 (1984).
12. J. O. Ojo, A. F. Oluwole, M. A. Durosinmi, O. J. Ogunsola, O. A. Akanle and N. M. Spyrou, this vol.
13. A. F. Oluwole, J. O. Ojo, M. A. Durosinmi, O. I. Asubiojo, O. A. Akanle, N. M. Spyrou, and R. H Filby, this vol.
14. M. A. Durosinmi, J. O. Ojo, A. F. Oluwole, O. A. Akanle, W. Arsed, and N. M. Spyrou, *J. Radioanal. Nucl. Chem., Articles* **168,** 233–242 (1993).
15. J. O. Ojo, M. A. Durosinmi, A. F. Oluwole, W. Arsed, O. A. Akanle, and N. M. Spyrou, *Nucl. Instr. Methods* **B79,** 408–412 (1993).
16. T. Aoi, T. Higuchi, R. Kidokoro, R. Fukumura, A. Yagi, S. Ohguchi, A. Saba, H. Hayashi, N. Sakamoto, and T. Hanaichi, *Human Toxicol.* **4,** 637–642 (1985).
17. G. Ingrao, P. Belloni, S. Di Pietro, and G. P. Santraroni, *Biol. Trace Element Res.* **26–27,** 699 (1990).
18. A. F. Oluwole, O. I. Asubiojo, A. D. Adekile, R. H. Filby, A. Bragg, and C. I. Grimm, *Biol. Trace Element Res.* **26–27,** 479 (1990).

19. A. S. Prasad and D. Oberleas, eds., *Trace Elements in Human Health and Disease. Vol. I: Zinc and Copper*. Academic, 1976.
20. A. S. Prasad, E. B. Schoomaker, J. Ortega, G. J. Brewer, D. Oberleas and J. Oelshlegel, *Clin. Chem.* **21,** 582 (1975).

Baseline Levels of Elemental Concentrations in Whole Blood, Plasma, and Erythrocytes of Nigerian Subjects

J. O. OJO,[1] A. F. OLUWOLE*,[1] M. A. DUROSINMI,[2] O. J. OGUNSOLA,[1] O. A. AKANLE,[3] AND N. M. SPYROU[3]

[1]Department of Physics, Obafemi Awolowo University, Ile-Ife, Nigeria; [2]Department of Haematology and Immunology, Obafemi Awolowo University, Ile-Ife, Nigeria; and [3]Department of Physics, University of Surrey, Guildford, England

ABSTRACT

In this pioneering effort on Nigerians, the arithmetic means, ranges, and standard deviations for the concentrations of the 13 elements, Br, Ca, Cl, Cu, Fe, K, Na, P, Pb, Rb, S, Se, and Zn, detected in freeze-dried whole blood, erythrocytes, and plasma of 120 subjects drawn from different geographical regions of Nigeria are reported. Elemental analyses have been carried out largely using instrumental neuron activation analysis (INAA) with both proton-induced X-ray emission (PIXE) and graphite furnace atomic absorption spectrometry (AAS) serving as complementary techniques. Our values, compared with similar data from other parts of the world, show a general good agreement. Values in male and female subjects are compared. Correlations between different elements are also noted.

Index Entries: Baseline levels; INAA; PIXE; elemental analysis of blood components; Nigerian subjects; elemental analysis.

INTRODUCTION

Trace elements studies continue to assume important proportions in many fields of scientific endeavors, including human nutrition, environmental pollution, and industrial exposure, and also in health and dis-

*Author to whom all correspondence and reprint requests should be addressed.

 © 1994 by Humana Press Inc.

eases where disorders in trace elements metabolism are reflected as elevation, depression, or imbalances in tissue elemental concentrations *(1)*. For such studies, it is necessary to have first, baseline values for "normal" subjects (which may be roughly defined as healthy, unexposed average native residents). Since the elements are continuously assimilated in the human body through food, drinks, and the ambient air, it is clear that such baseline levels in a particular population will depend on local factors, such as geochemical variability of the environment, dietary habits, culture, and social habits.

Other spin-offs from baseline measurements of elemental levels include the possibility of studying the many known metabolic interactions among trace elements, as well as determination of elemental resident times and other similar parameters of interest in health physics as discussed by Snyder et al. *(2)*.

For most elements, the usual route of entrance is the intestinal tract and, to a lesser extent, the lungs. They are then distributed within the body by the circulating blood. The various body tissues obtain their supply of nutrients from the plasma. However, there is a rather high turnover of the elements in the plasma, and so, elemental levels in erythrocytes often give better indication of body burdens. In this work, elemental levels for some 13 elements are reported in the whole blood, plasma, and erythrocytes of Nigerian subjects.

METHODS

Sample Collection and Preparation

The subjects consisted of 120 individuals (60 males, 60 females) aged between 3–60 yr drawn from three geographical regions of Nigeria: Ibadan-Ife in the southwest; Jos in the middle belt region; and Sokoto in the northwest. Jos is situated on a plateau at an altitude of 1300 m above sea level. The subjects were volunteers, from different backgrounds, including school pupils, college students and professors, civil servants, policemen, and prisoners. Particulars of each donor, such as age, sex, profession, residency, body weight, and height, and records of any disease or any drug taken regularly were obtained together with the blood samples.

Sterile scalp vein needles with polyethylene catheters, polyethylene syringes, and prewashed heparinized (20 μL/mL of blood) polyethylene screw-cap tubes were used for blood collection and storage. Ten milliliters of blood were collected from each subject after allowing the first 1–2 mL to run into a sequestrene tube in order to reduce contamination. This initial blood was used for biochemical tests, including checks for hepatitis and HIV viruses. About 2–3 mL of the blood were retained as whole blood, whereas the rest was separated into the different components within 6 h of collection by centrifugation at 3000 rpm for 15 min. The clear plasma was then drawn out into a clean polyethylene tube, whereas the other blood components (platelets and leukocytes) were pipeted out and thrown away leaving the erythrocytes. Each of the three products (whole blood, erythrocytes, and plasma) were frozen and then freeze-dried for 24–36 h in their individual containers using an Edwards Vacuum Freeze drier. Samples of the heparin used were also pelletized and analyzed for elemental contents. The wet and dry weights of the samples were recorded before and after the freeze-drying process.

The freeze-dried samples were crushed with a polyethylene spatula into fine powder and homogenized thoroughly. Seven-millimeter pellets (about 50–100 mg each) were then made from the fine powder in order to ensure reproducible neutron irradiation and counting geometries. For PIXE work, 5-mm pellets were made, and these were stuck on an aluminum plate with a double-sided adhesive for presentation into the target chamber for analysis. The pellets were coated with ≈5-nm layer of pure graphite in order to facilitate charge conduction. Sample preparation was carried out in the Clean Rooms at the Haematology and Immunology Department, O.A.U., Ile-Ife, Nigeria and at the Physics Department, University of Surrey, Guildford, England.

Analysis

Both INAA and PIXE were used as complementary techniques for elemental analysis. INAA was carried out at the Imperial College Reactor Centre, Ascot, England, whereas the PIXE facilities used were those found at the University of Surrey. In addition, a Hydride Graphite furnace Atomic Absorption Spectrophotometer (Perkin-Elmer) was used in assaying for lead, with IAEA reference material A13 first being used to check the accuracy of the results. Full details of these analytical methods have been described in previous works *(3–5)*.

RESULTS AND DISCUSSION

The baseline levels of the 13 elements, Br, Ca, Cl, Cu, Fe, K, Na, P, Pb, Rb, S, Se, and Zn, detected in whole blood, erythrocytes, and plasma of 120 Nigerian subjects are shown in Tables 1–3. Each table consists of the mean, standard deviation, range, and number of subjects in which the elements has been detected. The base references were those on a Bangladeshi *(6)*, European *(7)*, and worldwide average values *(8)*.

Distribution Patterns

Although a total of 120 subjects were involved in this study, the analyses were carried out in batches. Not all the samples collected have been subjected to the four different analytical modes: short-irradiation protocol by INAA, long-irradiation protocol by INAA, PIXE analysis, and AAS. Hence, not all the elements here have been assayed in all 120 subjects. The number of samples in which each element has been assayed and detected are indicated in Tables 1–3. In general, the elements Br, Cl, Rb, and Zn occur in practically all samples in concentrations higher than

Table 1
Baseline Elemental Concentrations in Erythrocytes (μg/mL)

Elt	THIS WORK (Nigerian Subjects)			LITERATURE	
	Mean ± SD	Range	No	Mean ± SD	Range
Br	1.63 ± 0.46	0.81 - 3.02	48	1.45 [8]	-
Ca	3.2 ± 1.6	5.3 - 84.3	40	3.2	0.63-5.2[8]
Cl	639 ± 290	119 - 1382	28	2038	1760-2477[8]
Cu	1.32 ± 0.4	0.52 - 2.61	38	0.98	0.75-1.31[8]
Fe	907 ± 137	186 - 1830	55	1082	985,1140[8]
K	2950 ± 434	1720 - 3790	33	3546	3030-4230[8]
Na	942 ± 350	411 - 1940	52	284	177-360[8]
P	902 ± 347	126 - 1574	53	706	656-978[8]
Rb	7.27 ± 1.59	4.56 - 12.04	57	-	-
Se	0.14 ± 0.02	0.03 - 0.53	81	-	-
Zn	13.6 ± 4.3	5.8 - 15.4	45	14.2	7.6-16.1[8]

Table 2
Baseline Elemental Concentrations in Plasma (μg/mL)

Elt	THIS WORK (NIGERIAN SUBJECTS)			LITERATURE	
	Mean ± SD	Range	No	Mean ± SD	Range
Br	2.38 ± 1.04	0.92 - 6.42	87	-	-
Ca	102 ± 49	27 - 239	56	96	89-106 [8]
Cl	2350 ± 1120	776 - 4909	51	3610	3260-4280[8]
Cu	1.12 ± 0.52	0.23 - 2.33	40	1.12	0.61-1.41[8]
Fe	1.4 ± 0.9	0.2 - 5.4	30	1.10	0.71-1.27[8]
K	230 ± 95	53 - 483	39	157.0	137-208[8]
Na	2800 ± 456	1840 - 4478	82	3162	2990-3330[8]
P	80 ± 44	17 - 283	26	112[8]	-
Rb	0.39 ± 0.15	0.15 - 2.36	79	-	-
Se	0.050 ± 0.031	0.001 - 0.189	83	-	-
Zn	0.72 ± 0.18	0.42 - 2.69	54	1.14	0.79-1.70 [8]

Table 3
Baseline Elemental Concentrations in Whole Blood (μg/mL)

Elt	THIS WORK (Nigerian Subjects)			LITERATURE	
	Mean ± SD	Range	No	Mean ± SD	Range
Br	2.29±0.95	1.07 - 5.47	88	1.10±0.35 4.7	0.2-2.0 [6] 1.3-8.1 [8]
Ca	63.2±21.4	0.9 - 93.4	35	52.9±7.4 60.5	30-80 [6] 57.5-78.0 [8]
Cl	1290 ± 806	454 - 3180	51	2886	2590-3290 [8]
Cu	1.01±0.36	0.07-1.99	40	0.98±0.4 1.01 1.225±0.064	0.4-2.76 [6] 0.64-1.28 [8] 0.807-1.643 [7]
Fe	435 ± 76	106 - 577	58	339±23 447	280-410 [6] 301-530 [8]
K	1700 ± 335	1170 - 2450	28	1622	1450-1920 [8]
Na	1890 ± 319	1170 - 2930	55	1969	1710-2050 [8]
P	610 ± 218	55 - 913	39	345	311-510 [8]
Pb	0.120 ± 0.060	0.017 - 0.325	37	0.214	0.088-0.40 [8]
Rb	3.75±0.93	1.15 - 5.84	50	4.18±0.71 2.49 2.805± 0.408	2.5-6.0 [6] 1.17-5.98 [8] 0.90-4.145 [7]
Se	0.08±0.04	0.03 - 0.26	85	0.171 0.1075± 0.0064	0.057-0.32 [8] 0.076-0.140 [7]
Zn	4.65 ± 1.01	2.26 - 7.09	58	4.39±0.96 7.0 6.340±0.210	3.0-10.8 [6] 4.8-9.3 [8] 4.076-7.594 [7]

their detection limits in both INAA and PIXE, in the three different matrices considered. Also, Fe and K are well detected in both whole blood and erythrocytes; and Se in whole blood by both techniques. In all these cases, the values obtained by INAA have been chosen for the tables. In addition, Na (in whole blood and plasma) and Se (in erythrocytes and plasma) are well detected only by INAA. On the other hand, the elements Ca and Cu (in all three matrices), P (in erythrocytes only), and Fe and K (in plasma only) are well detected only by PIXE.

The tables further highlight the compartmentalization modes of the various elements in the different blood components. They show that Fe, K, P, Rb, Se, and Zn are predominantly packed in the erythrocytes, whereas Ca, Cl, and Na are more packed in the plasma. Br, Cu, and S are, however, not so compartmentalized—the ratios of the concentrations in both the erythrocytes and plasma to that in the whole blood being similar and close to unity. Pb was assayed only in whole blood.

Another useful index of the essentiality/biological importance of an element is the range ratio, defined as the ratio of the maximum level detected to the lowest concentration found in a tissue. Biologically important elements are well regulated homeostatically and have small range ratios. From Tables 1–3, the following elements are found with range ratios of <5.

In whole blood: Br, K, Na, and Zn;
In erythrocytes: Br, Cu, Fe, Na, Rb, and Zn; and
In plasma: Na, Se, and Zn.

Intercomparison of Baseline Data

The tables under discussion also provide intercomparison of baseline data between the Nigerian subjects used in this work and other populations. Unfortunately, the bulk of work published in the literature was just on whole blood and sometimes blood serum. As such, in addition to the base reference works chosen here (based on different cultures involved and spectrum of elements reported), a few values have been incorporated from other authors, especially to supplement the data on erythrocytes and plasma. Clearly, a more meaningful comparison is achieved when not just single elements, but also a wide spectrum of elements are compared between authors. The base reference works used are those by Khan et al. *(6)*, the Ispra Joint Research Centre *(7)*, and Iyengar et al. V. *(8)*. In general, there is good agreement between values obtained in this study and those reported in the literature, and there is no strong reason to suggest Nigerian subjects suffer from any serious trace element imbalances. Values for Pb in whole blood are agreement with Iyengar et al.'s data as well as those reported by Haeger-Aronsen et al. V. *(9)*, for a Swedish population, and Saito et al. V. *(10)* for a Japanese population, but quite lower than those for India *(11)* and Iran *(12)*.

There is no difference in the levels picked from the different geographical regions in Nigeria, except that somewhat higher mean values (though all within range) are found for the Jos population for the elements Zn, Fe, K, and Rb in the plasma and Fe and Zn in the erythrocytes. Previous reports from other regions suggest a decrease in elemental concentrations with altitude owing to depletion of trace elements out of the soil, for example, by podzolization. Where such effect is significant, the resulting reduction in plant intake of trace element is expected to be reflected in tissues of humans resident in the region. We are not able at present to explain the higher values in the Jos population, but it might as well be the result of other factors, such as diet, that we have not studied in this work.

Table 4
T-test for the Significance of Difference Between Arithmetic Means of Elemental Concentrations in Blood of Female and Male Nigerian Subjects

	WHOLE BLOOD			ERYTHROCYTES			PLASMA		
Element	DF	t-value	p value	DF	t-value	p value	DF	t-value	p value
Cl	49	-3.1791	<0.005	26	-0.3959	NS	49	1.9743	<0.05
Fe	56	-1.6353	<0.1	54	-1.4769	<0.1	19	-0.0376	NS
K	36	-1.3560	<0.1	52	-1.5927	<0.1	37	1.5196	<0.1
Rb	77	-0.2769	NS	77	-1.0658	NS	45	-1.3826	<0.1

Male–Female Characterization

It is known that for some trace elements, body requirements may differ for the two sexes *(13,14)*. The differences in elemental levels obtained in this work for both male and female Nigerian subjects are shown in Table 4. These results indicate that trace element levels are generally similar in male and female Nigerian subjects. However, the levels of the essential elements, including Zn, Rb, Fe, and Ca, are somehow depressed, whereas others, like K and Cu, are elevated in females relative to the males. Most of these depressions and elevations are not found to be statistically significant. In the whole blood, significant differences (depressions) were found in the levels of Cl and Fe in females. In the red blood cells, K concentrations were significantly elevated, whereas Fe was significantly depressed. In the plasma, Cl and K were significantly elevated in female subjects, whereas Rb was significantly depressed. These findings are consistent with the few literature data available for some of the elements involved *(13,14)*.

Elemental Interactions in Blood

Elemental interactions both within each blood component and between the different components are studied first by finding in each individual elemental ratios for all possible elemental pairs. The dispersion of these ratios around their mean values in all the subjects was then used as a measure of the regulation or otherwise of each elemental pair. Table 5 lists the various elemental ratios that show evidence of some regulation in the three blood compartments. In particular, the ratios Ca/Zn, Cu/Zn, P/Se Na/Rb, Na/Zn, and Rb/Zn are fairly well regulated in all three compartments. In plasma, standard deviations of about 20% are found for the ratios Cl/K, Cl/Ca, and Ca/Zn. In the erythrocytes, standard deviations as low as about 27% are found for the ratios K/Zn, Fe/Cu, and Fe/Zn, whereas P/K and Fe/K have standard deviations <20%. In whole blood, the regulation is less strict, but standard devia-

Table 5
Correlations[1] Between Elements in Blood ($p < 0.1$)

ERYTHROCYTES		PLASMA		WHOLE BLOOD	
P:	Ca′, Fe*, Se*, Cl**, Zn**, K+, Cu+	P:	Ca′, Se*, Cl*, Zn*, K**	P:	Cl**, Fe*, K+, Ca+, Cu+, Zn+
Cl:	Zn′, K*, Cu*, P**	Cl:	P*, K+, Ca+, Cu+, Zn+	Cl:	Se*, P**, Cl+, Ca+, Cu+, Zn+
K:	Ca′, Fe′, Cl*, Cu**, Zn**, P+	K:	[2]<u>Rb</u>′, P**, Cl+, Ca+, Cu+, Zn+	K:	-<u>Na</u>′, Fe*, P+, Cl+, Ca+, Cu+, Zn+
Ca:	K′, Fe′, P*, Zn*, Cu**	Ca:	P′, Cl+, K+, Cu+, Zn+	Ca:	Fe′,P+, Cl+, K+, Cu+, Zn+
Cu:	Fe′, Cl*, K**, Ca**, Zn**, P+	Cu:	Cl+, K+, Ca+, Zn+	Cu:	Fe**, P+, Cl+, K+, Ca+, Zn+
Zn:	Cl′, Ca*, K**, Cu**, P**	Zn:	<u>Rb</u>*, P*, Cl+, K+, Ca+, Cu+	Zn:	<u>Fe</u>**, <u>Rb</u>**, P+, Cl+, K+, Ca+, Cu+
<u>Br</u>:	Se′	<u>Br</u>:	-	<u>Br</u>:	-
<u>Na</u>:	Se′	<u>Na</u>:	-	<u>Na</u>:	-K′, -Rb′
<u>Rb</u>:	-	<u>Rb</u>:	K*, Zn*	<u>Rb</u>:	-Na′, Fe*, Zn**
Fe:		Fe:	<u>Se</u>**	Fe:	<u>Rb</u>*, <u>Zn</u>**
Se:	Br′, Na′	Se:	P*, <u>Fe</u>**	Se:	-

[1]Implies $p < 0.1$; *implies $p < 0.05$; **implies $p < 0.01$; †implies $p < 0.001$.
[2]Underlined elements are detected by INAA.

Table 6
Regulation of Elemental Ratios in Blood

	PLASMA	ERYTHROCYTES	WHOLE BLOOD
Standard deviations ≤50%	+ P/Ca, P/K, P/Se, Cl/Se, K/Se, Ca/Fe, Ca/Se, Cu/Zn, Zn/Se, S/Cl, S/K * Na/Rb, Na/Zn,	+ P/Zn, P/Se, P/Pb, K/Ca, K/Se, K/Pb, Ca/Zn, Fe/Se,Zn/Se, Zn/Pb, Rb/Pb * Na/Fe, Na/Zn	+ P/Cl, P/Zn,* P/Se,Ca/Zn, Fe/Cu, Cu/Zn, * Na/Zn, K/Fe, K/Rb, K/Zn, Se/Zn
Standard deviations ≤ 33%	* S/Ca, S/Zn, Rb/Zn	+ P/Cu, K/Cu, Ca/Cu, Cu/Zn * Na/Zn,	+ K/Zn, Ca/Cu * Na/K, Na/Fe,
Standard deviations ≤ 29%	+ Zn/Cl	+ K/Zn, Fe/Cu, Fe/Zn * Na/K, Fe/Rb, Fe/Zn	+ - * Na/Rb
Standard deviations ≤ 25%	+ K/Cl	* Rb/Zn	+ Fe/Zn * Fe/Zn,Fe/Rb,Rb/Zn
Standard deviations ≤ 20%	+ Ca/Zn, Ca/Cl	+ P/K * K/Fe	-

†Determined by PIXE. *Determined by INAA.

tions of <25% were found for the ratios Fe/Zn, Fe/Rb, and Rb/Zn. These clearly indicate the existence of fundamental interactions between the pairs of elements, many of which are chemically similar. All the elements above belong to the class of essential elements and are no doubt under homeostatic regulation.

A more general approach to studying elemental interactions was undertaken by finding the Pearson's product moment correlation coefficients between the elements using the CLUSTAN software. However, since not all elements were picked in every subject, not all the subjects available were used because of some restrictions imposed by the software. Correlations were sought by withering down the number of subjects involved as more elements were incorporated into the calculation. Table 6 shows significant correlations ($p < 0.1$) observed in the three blood compartments and is consistent with the results obtained in Table 5.

ACKNOWLEDGMENT

This work is part of the EEC-sponsored project on Environmental Pollution Monitoring and Impact Assessment studies in Nigeria.

REFERENCES

1. A. S. Prasad and D. Oberleas, *Trace Elements in Human Health and Disease,* vols. 1 and 2, Academic, 1976.
2. W. S. Snyder, M. J. Cook, E. S. Nasset, L. R. Karhausen, G. P. Howells, and J. Tipton, *Report of the Task Group on Reference Man*, Pergamon, London, 1976.
3. J. O. Ojo, M. A. Durosinmi, A. F. Oluwole, W. Arsed, O. A. Akanle, and N. M. Spyrou, *Nucl. Instr. Methods* **B79,** 408–412 (1993).
4. M. A Durosinmi, J. O. Ojo, A. F. Oluwole, O. A. Akanle, W. Arshed, and N. M. Spyrou, *J. Radioanal. Nucl. Chem., Articles* **168,** 233–242 (1993).
5. O. J. Ogunsola, A. F. Oluwole, O. I. Asubiojo, M. A. Durosinmi, A. O. Fatusi, and W. Ruck, *Sci. Total Environ.,* in press (1994).
6. A. H. Khan, M. Khaliquzzaman, M. B. Zama, M. Hussain, M. Abdullah, and S. Akhter, *J. Radioanal. Chem.* **57,** 157–167 (1980).
7. Joint Research Centre, Commission of the European Communities, Ispra, Italy, Trace Element reference values in tissues from inhabitants of the European Community, SP-1.90.06 March 1990.
8. G. V. Iyengar, W. E. Kollmer, and H. J. M. Bowen, *The Elemental Composition of Human Tissues and Body Fluids,* Verlag Chemie, Weinheim, 1978.
9. B. Haeger-Aronsen, M. Abdulla, and I. B. Fristedt, *Arch. Environ. Health* **23,** 440–445 (1971).
10. K. T. Saito, Y. Sato, and K. Jiizuma, *Industrial Health* **11,** 85–87 (1973).
11. A. L. Aggarwal, C. V. Patel, C. V. Rayani, and S. K. Chatterjee, *Indian J. Med. Res.* **70,** 650–656 (1979).
12. H. Farsam, G. H. Salari, and A. Nadim, *Am. Ind. Hyg. Assoc. J.* **43,** 373–378 (1982).
13. S. B. Deeming and C. W. Weber, Amer. *J. Clin. Nutr.* **31,** 1175–1180 (1978).
14. G. Ingrao, P. Belloni, S. Di Pietro, and G. P. Santraroni, *Biol. Trace Element Res.* **26–27,** 699 (1990).

A Comparative Study of Elemental Composition of Human Breast Milk and Infant Milk Substitutes

F. A. BALOGUN,[1] O. A. AKANLE,*,[2] N. M. SPYROU,[2] AND J. A. OWA[3]

[1]CERD, Obafemi Awolowo University, Ile-Ife, Nigeria; [2]Physics Department, University of Surrey, Guildford, GU2 5XH, England; and [3]Paediatrics Department, Obafemi Awolowo University, Ile-Ife, Nigeria

ABSTRACT

Instrumental neutron activation analysis (INAA) and proton-induced X-ray emission (PIXE) analysis have been employed to determine the concentration of 13 elements in human breast milk, various infant formulas, and locally produced cereals from Nigeria, as well as from various infant formulas and natural cow and goat milk available in the UK. The study shows that if the locally produced cereal is to be used on a regular basis for babies in Nigeria, then their diet must be supplemented with essential trace elements. Furthermore, parents should be discouraged from giving their infants cow and goat milk because of the high concentration of major elements compared to human breast milk.

Index Entries: Breast milk; infant milk formula; cereals; trace elements.

INTRODUCTION

There has been a growing interest among international and governmental organizations, such as WHO and UNICEF, over the past two decades to encourage breast feeding of infants all over the world *(1–4)*. It is estimated that more than 1.5 million babies die annually from malnutrition. This nutritional deficiency is not just in the protein content, but also in the essential trace and minor elements of importance to infant

*Author to whom all correspondence and reprint requests should be addressed.

 © 1994 by Humana Press Inc.

growth. It has been demonstrated through extensive animal experiments that there is a need for an adequate supply of essential trace elements, such as Cu, Fe, Mn, and Zn, for the growth and development of infants. Deficiencies of these elements have been known to result in poor growth, abnormal overall development, skin disorders, bone fractures, and increased neonatal morbidity *(5)*.

Human breast milk and its substitutes are usually the only sources of food for infants during their early stages of life. It is therefore essential that the breast milk and substitutes contain all necessary nutrients in adequate amounts. However, although deficiencies of trace elements are generally believed to be rare in breast-fed infants, those infants that are fed on milk substitutes or infant formulas may experience deficiencies in essential trace elements.

The situation is, however, different in Nigeria. As a result of the economic decline in the country, the commercially available infant formulas have become very expensive so that parents, particularly in the rural areas, now rely almost exclusively on locally produced cereals to feed their babies, and these cereals are believed to be adequate in essential nutrients.

The aim of this study is, therefore, to investigate whether these locally produced cereals and the commercial infant milk products contain adequate levels of essential trace elements when compared with trace element levels in human breast milk. Various infant formulas and cow and goat milk used in the UK have also been analyzed for comparison. Initial findings of the ongoing study are presented.

METHODS OF ANALYSIS

Human breast milk was collected after morning feed from 22 lactating mothers, at various stages of lactation from a Nigerian population, into clean prewashed polyethylene containers. They were kept in a low-temperature freezer and then freeze-dried. Such factors as the age of the mother, occupation, and collection date were also noted. The freeze-dried samples were then crushed with a polyethylene spatula into fine powder in their individual containers ready for analysis, while maintaining clean conditions.

Four infant formulas based on cow milk and produced by different food industries were also purchased. They are SMA, SIMILAC, NAN, and COW&GATE. These infant formulas are already in powder form, hence requiring no further treatment. In addition to the infant formulas, three locally produced cereals were also purchased. One of these cereals, called Babeena cereal, is commercially available in powder form like the infant formulas, whereas the other two were purchased from local markets in Nigeria and ground into fine powder. One of the cereals is corn-based, and the other is wheat-based.

Various infant formulas were also supplied by two main infant formula product companies in the UK, SMA and COW&GATE Nutrition. The formulas analyzed are broadly:

1. Those used from birth onwards;
2. Those used from birth onward, but for hungrier babies, and these contain more protein;
3. Those used from 6 mo onward; they are used in conjunction with solids and given to babies instead of cow milk;
4. Those that are based on soya protein and used for babies allergic to cow milk protein lactose; they are also used from birth onward;
5. Those that are based on hydrolyzed whey protein, and used for babies with proven lactose, sucrose, and whole-protein intolerance, steatorrhea associated with cystic fibrosis, and other malabsorption syndromes; and
6. Those used for low-birthweight babies.

All the infant formulas were in powder form and required no further treatment, except the low-birthweight formulas available in liquid form, which were freeze-dried using an Edwards Vacuum freeze-dryer. It should also be noted that the low-birth formulas and the hydrolyzed whey protein formulas are prescription-only formulas.

In addition to the infant formulas, pasteurized cow milk and goat milk were also purchased, and these were also freeze-dried. The freeze-dried samples were then crushed as before.

Instrumental Neutron Activation Analysis (INAA)

Powdered samples from the human breast milk, the infant formulas, the local cereals from Nigeria, and the cow and goat milk were all made into pellets to give reproducible irradiation and counting geometries. They were then individually placed in small polyethylene capsules, and these were double contained in a bigger polyethylene container, heat-sealed ready for irradiation.

The samples were irradiated using the In-Core-Irradiation System (ICIS) facility at the Imperial College Reactor Centre, Ascot, UK. A short irradiation of 180 s duration was used followed by a waiting period of 60 s and counted for a period of 300 s for the detection of the short-lived isotopes induced in the milk samples. Accuracy of the method was checked by analyzing IAEA milk powder standard (A-11) along with the samples and standard. Sample masses varied between 30 and 50 mg.

Proton-Induced X-ray Emission (PIXE) Analysis

Samples and standards were also pelletized, and these were stuck side by side on a 3 mm thick aluminum plate with the help of double-sided adhesive tape. They were then carbon-coated with a thin layer of carbon and carbon-dagged on the sides to make the samples conducting.

Analysis of the samples was carried out using a 2 MeV proton beam under vacuum, incident on the sample with a diameter of about 3 mm, using the University of Surrey's Van de Graaf accelerator. PIXE spectra were collected for 300 s with the low-energy photon Si(Li) detector fitted with a 350 μm mylar filter to suppress the signal from low-energy X-rays in order to reduce dead time and other problems. Simultaneous RBS spectra were also collected with a surface barrier detector for matrix correction.

RESULTS AND DISCUSSION

The elemental concentrations determined in the human breast milk, the Nigeria infant formulas, and locally produced cereals using both techniques are presented in Table 1 for 13 elements. Both techniques allow the determination of the concentrations of Br, Ca, and K, and their values were found to be very close, but in addition, INAA allows the determination of the concentration of Al, Cl, I, K, Mg, Mn, and Na, whereas PIXE, used as a complementary technique and because only short irradiations were carried out for INAA, allows the determination of the concentration of Cd, Cu, Fe, Rb, and Zn.

It can be observed from Table 1 that for most of the elements detected, the infant formulas have concentrations in the same range as the values found for the human breast milk, particularly for the essential trace elements, like Cu, Mg, and Zn, as well as for the major elements, like Ca, Cl, K, and Na. This observation should be expected, because even though the infant formulas are based on cow milk or soya protein, the manufacturers have modified their elemental content to fit closely to the metabolic needs of the infants, as well as the concentration found in human breast milk. However, the bioavailability of these elements in the infant formulas is much less than that in the breast milk. The concentration of Fe in the infant formulas was found to be higher than that of the breast milk, which can only be explained by the fact that because of the low availability of Fe in cow milk, the formulas had been supplemented by a high content of this element. This high concentration of Fe in the infant formula can, however, be disadvantageous, since a high concentration of Fe has been suggested to compromise the naturally occurring bacteriostatic mechanisms that are present in infants *(6)*, since incidence of such diseases as gastrointestinal infections have been found to be lower in breast-fed infants than in bottle-fed infants *(7,8)*.

The situation is different for the locally produced cereals from Nigeria. Corn- and wheat-based cereals have low concentration of all the major and trace elements, with the exception of Al and Mn, where the concentrations were found to be higher in the cereals than in the breast milk and the infant formulas. The high level of Al may be a result of the aluminum grinder used to grind the cereals into powder form. Babeena cereal, however, had values in the ranges of the infant formulas and the breast milk. This is because this cereal formula is available commercially, and supplements have been added to compensate for the low values of these elements in the cereals. The Fe value was also very high compared to the breast milk.

Table 1
Range of Concentrations, Arithmetic Mean, and Standard Deviations in Breast Milk, Mean Concentrations (μg/g) in Various Infant Formulas and Local Cereal Formulas from Nigeria

	HUMAN BREAST MILK		NIGERIA INFANT FORMULA				NIGERIA LOCAL CEREALS		
Element	Range	Mean±sd	SMA	Cow&Gate	NAN	Similac	Babeena	Corn based	Wheat based
Al	7.5-24.5	12.6±4.3	23.6±1.7	15.7±1.1	15.8±1.1	22.9±1.8	65.6±3.9	45.5±3.6	33.9±3.4
Br	3.4-17.6	9.6±4.6	59.7±9.8	49.3±10.5	14.7±4.9	32.8±6.7	ND	ND	ND
Ca	1322-3144	2182±105	3312±112	662.3±13.5	3901±125	524±187	6536±138	189±28	123±22
Cl	2582-11262	5315±2341	3886±205	4899±189	3321±269	3871±158	1930±138	201±79	188±89
Cu	3.0-10.1	5.9±1.9	2.9±0.4	3.1±0.3	3.8±0.4	3.8±0.3	1.9±0.8	0.7±0.3	0.2±0.1
Fe	4.2-19.3	10.2±4.8	79±8	105±14	78±12	47±8	118±16	3.7±0.1	4.9±0.2
I	0.2-1.3	0.6±0.3	0.2±0.1	0.2±0.1	ND	0.6±0.2	ND	ND	ND
K	3480-7526	5708±685	4444±328	7114±847	6169±524	6780±587	4071±367	ND	ND
Mg	233-684	429±135	292±98	560±104	395±126	364±140	589±117	63±26	137±48
Mn	0.58	-------	0.2±0.1	ND	ND	ND	0.2±0.1	0.6±0.2	4.6±1.4
Na	1439-9667	3315±2288	1412±483	1228±277	1423±366	2202±147	904±105	165±62	32.2±3.7
Rb	6.9-17.9	11.8±3.5	9.8±1.7	10.6±2.4	14.9±1.8	8.9±1.3	13.7±2.5	1.4±0.6	1.6±0.2
S	4017-18289	7120±3347	ND	ND	ND	ND	ND	ND	ND
Zn	8.2-59.9	33.8±12.6	34.9±6.7	27.9±8.5	28.1±4.9	34.7±7.8	17.9±6.4	7.7±2.5	5.9±1.7

In Table 2 is presented the mean concentration for 13 elements obtained from the analysis of the various infant formulas supplied by two infant product companies in the UK. Also presented are the values obtained from the analysis of natural cow and goat milk. The values obtained for most of the elements in the infant formulas compared well with those obtained for the Nigerian infant formula and breast milk, with the exception of the concentration of Ca, Cl, Fe, and K in SMA Progress, which is an infant formula used for babies after 6 mo. The concentrations of Fe, Mg, and Na obtained for both the Cow&Gate and SMA soya-based milk were also found to be lower than other values, whereas the Zn concentration was found to be higher. The high Zn values may be a result of high-Zn supplement added to the soya-based formula because of the soya protein's adverse effect on Zn absorption as pointed out by Bostrom *(9)*. Other special infant formulas, like the low-birthweight infant formula, have a lower concentration of such elements as Cu, Fe, and Mg, and even the concentrations of major elements, such as Ca, Cl, and Na, were lower in this infant formula when compared with other formulas, but the values are closer to the concentration found in breast milk, whereas the concentrations of the elements found in the formula used for babies with proven lactose, sucrose, and whole-protein intolerance were in the same range as other infant formulas. Morishima et al. *(10)*, however, reported a Zn deficiency in this type of formula. It should be pointed out that those infants receiving these special formulas have an increased risk of trace element deficiences, since many of the disorders are characterized by fecal and urine loss with severe diarrhea.

Table 2 also gives the concentration of the elements found in natural cow and goat milk, which are usually given to infants after 6 mo. The concentrations of the major elements Ca, Cl, K, and Na were found to be considerably higher than those of human breast milk and the infant formulas, whereas the concentrations of Cu, Fe, and Zn were found to be lower. However, as pointed out by Flynn *(11)*, even though the concentrations of Na, K, and Cl in milk are of physiological importance in the feeding of the infants, clinical problems can arise if there is an excessive intake of these nutrients, since the kidneys of the infants have a limited capacity to concentrate solids. Though infant formulas are based on cow milk, a recommended limit has been set for these elements. Despite the high concentration of Ca in cow and goat milk, it has been suggested by Fomon and Strauss *(12)* that the bioavailability is higher in the breast milk, but that the absolute amount of Ca absorbed is greater from cow milk because of the higher concentration of the element in the infant formulas. It is, however, essential that parents be discouraged from giving their infants these types of natural milk, since they are also high in saturated fats and low in vitamins C and D.

Attempts have also been made to divide the breast-milk samples into two categories based on postpartum dates: colostrum (days 1–10) and mature milk (after day 10), indicating the collection days of milk samples after the birth of the baby. It is known that colostrum milk is richer in minerals and trace elements, whereas the transition milk has an

Table 2
Arithmetic Mean and Standard Deviation of Various Infant Milk Formulas from UK (in μg/g)

Element	SMA Gold	C&G Premium	SMA White	C&G Plus	SMA Wysoy	C&G Soya	SMA Progress	C&G Pepti	SMA Low Birth	C&G Low Birth	Cow Milk	Goat Milk	IAEA A-11
Al	15.6±2.2	15.3±1.8	22.6±2.6	26.4±3.5	31.7±4.8	20.6±2.7	33.8±2.8	17.7±1.9	8.1± 2.1	9.0± 2.7	33.6±3.8	47.0±3.7	27±3.9
Br	17.5±3.6	26.1±2.9	35.9±5.6	41.8±8.7	ND	ND	23.4±2.7	ND	12.9±2.6	ND	80.2±9.8	43.3±5.6	13±1.9
Ca	3500 ±118	4407 ±146	3683 ±124	6400 ±132	4750 ±156	4464 ±142	7556 ±118	4997 ±120	2300 ±107	1812 ±112	10180 ±789	11376 ±798	12993 ±1040
Cl	3208 ±258	3396 ±217	3563 ±204	3728 ±278	3493 ±226	3037 ±289	5293 ±384	2990 ±206	1625 ±179	1808 ±163	7860 ±899	14566 ±1202	9445 ±1145
Cu	4.2±1.0	3.5±0.8	4.4±0.9	3.8±0.5	3.6±0.4	3.1±0.6	4.6±0.4	3.5±0.4	2.0±0.7	1.8±0.5	1.6±0.3	1.8±0.6	ND
Fe	54.2±5.6	46.6±7.9	51.5±5.8	42.5±6.6	60.5±4.8	65.7±3.8	89.8±7.9	50.5±6.8	21.4±1.8	19.7±3.9	15.9±3.3	10.5±3.7	4.0±0.6
I	0.4±0.2	ND	ND	0.3±0.1	ND	ND	ND	ND	0.3±0.1	ND	0.3±0.1	ND	ND
K	4163 ±489	4685 ±508	4900 ±464	7451 ±387	5958 ±428	5731 ±512	7235 ±407	5199 ±528	1745 ±245	1774 ±312	8566 ±1168	14153 ±1538	17589 ±1061
Mg	301±125	442±118	309±137	452±128	415±120	382±118	519±107	579±110	203±89	281±78	950±113	1265 ±195	1119 ±80
Mn	1.8±0.3	0.7±0.1	1.3±0.2	0.8±0.1	1.2±0.1	3.1±0.7	0.8±0.2	0.6±0.2	0.3±0.1	ND	ND	ND	ND
Na	1211 ±256	1614 ±318	1403 ±307	2116 ±158	1562 ±263	1408 ±289	2463 ±145	1784 ±105	715 ±87	790 ±79	4008 ±212	3465 ±214	4275 ±331
Zn	42±6	32±6	74±3	38±6	63±3	66±3	46±6	45±5	36±6	38±5	19±3	18±4	39±3
Cd	1.8±0.2	1.5±0.3	ND	1.6±0.2	1.7±0.1	ND	2.8±0.2	2.2±0.3	1.8±0.2	1.5±0.2	ND	ND	ND

Table 3
Mean Elemental Concentration and Standard Deviation (μg/g) in Breast Milk Divided into Two Categories, Colostrum and Mature Milk

Element	Colostrum	Mature Milk
Al	11.6 ± 4.7	13.7 ± 3.8
Br	11.7 ± 4.1	7.8 ± 4.4
Ca	2262 ± 627	2152 ± 423
Cu	6.6 ± 1.2	5.3 ± 2.5
Fe	11.1 ± 5.1	9.4 ± 4.8
I	0.7 ± 0.3	0.5 ± 0.3
K	5687 ± 1196	5740 ± 1304
Mg	426 ± 156	436 ± 123
Cl	5173 ± 1876	5737 ± 1304
Na	3211 ± 1771	3660 ± 1816
Rb	13.1 ± 3.0	10.6 ± 3.8
S	6949 ± 2817	7432 ± 2062
Zn	38.6 ± 9.3	29.1 ± 5.8

increased concentration of some elements, but a decreased concentration of some other elements. However, mature milk is thought to have a constant and stable composition of trace elements with minimum daily change. The division into the two categories is presented in Table 3, and it was observed that the concentrations of Ca, Cu, Fe, I, Rb, and Zn are higher in the colostrum than in the mature milk, whereas the concentrations of Cl, K, Mg, Na, and S are higher in the mature milk.

SUMMARY

Our findings show that formulas used for infants should be adapted to the nutrient needs of the infants, and the overall aim should be to ensure that the infants absorb and retain trace elements from the formula in amounts equivalent to that obtained from the breast milk.

ACKNOWLEDGMENT

This work would not have been possible without the EEC Linkage Programme under the Lome-III agreement, which fostered exchange and collaboration between our universities.

REFERENCES

1. Canadian Paediatric Society and American Academy of Pediatrics, Commites on Nutrition, *Pediatrics* **62,** 591–601 (1978).
2. Great Britain Department of Health and Social Security, Report on Health and Social Subjects No 9, HMSO, London 1974.
3. International Pediatric Association, *Acta Paediatr. Scand.* **65,** 275–277 (1977).
4. Twenty-Seventh World Health Assembly, WHO Official Record No. 217, 1978.
5. B. Lonnerdal, C. L. Keen, M. Ohtake, and T. Tamura, *Am. J. Dis. Child.* **137,** 433–437 (1983).
6. J. J. Bullen, H. J. Roger, and L. Leigh, *Br. Med. J.* **1,** 69–75 (1972).
7. A. S. Cunningham, *J. Pediatr.* **95,** 685–689 (1979).
8. M. E. Fallot, J. L. Boyd, and F. A. Oski, *Pediatr.* **65,** 1121–1124 (1980).
9. K. Bostrom, in *Textbook of Paediatric Nutrition,* R. M. Suskind, ed. Raven, New York, 1981, pp. 41–64.
10. T. Morishima, S. Yagi, and T. Takamura, in *Trace Element Metabolism in Man and Animal,* J. M. Howell, J. M. Gawthorne and C. L. White, eds. Griffin Press, Netley, South Australia, 1981, pp. 487–490.
11. A. Flynn, in *Advances in Food and Nutrition Research,* vol 36, Academic, San Diego, California, 1992, pp. 209–252.
12. S. J. Fomon and R. G. Strauss, *N. Engl. J. Med.* **299,** 355–357 (1978).

Trace Element Contents in Food Determined by Neutron Activation Analysis and Other Techniques

M. F. REIS,*[,1] M. ABDULLA,[2] R. M. PARR,[3] A. CHATT,[4] H. S. DANG,[5] AND A. A. S. C. MACHADO[6]

[1]INETI-ICEN-DEEN, Sacavém, Portugal; [2]Department of Clinical Pharmacology and Therapeutics, Hamdard University, New Delhi, India; [3]Division of Life Sciences, International Atomic Energy Agency (IAEA), Vienna, Austria; [4]Department of Chemistry, Dalhousie University, Canada; [5]BARC-Bhaba Atomic Reactor Centre, Bombay, India; and [6]Laquipai, Faculty of Science, University of Oporto, Portugal

ABSTRACT

Advances in analytical methodology and sophisticated instrumentation introduced during the last few decades have not only helped to recognize the presence of a large number of essential and toxic trace elements in biological materials and food stuffs, but also added a new dimension in our understanding of their role in health and disease. In deficiency states, most essential trace elements cause health problems. The trace element problem as a public health issue has a very low priority in developing countries. Daily dietary intake data based on well-conducted studies are limited in many parts of the world. The present authors are in the process of generating data for the intake levels of a number of major and minor inorganic elements both in developed and developing countries. The results so far obtained show wide variations. The intake levels of several elements, including potassium, magnesium, zinc, copper, and selenium, are below the current recommended levels. The concentration of toxic metals, such as aluminum, cadmium, lead, and mercury, are within acceptable limits in most of the diets analyzed.

Index Entries: Essential and toxic trace elements; food; analytical techniques; neutron activation analysis.

*Author to whom all correspondence and reprint requests should be addressed.

INTRODUCTION

Advances in analytical methodology and sophisticated instrumentation introduced during the last few decades have certainly helped to bring about a breakthrough in the field of trace element research. Using powerful analytical techniques, such as neutron activation analysis (NAA), inductively coupled plasma emission spectrometry (ICP-ES), and atomic absorption spectrometry (AAS), researchers have demonstrated the presence in living systems and their food chain of a number of trace elements that were considered until now as contaminants or accumulated material.

The nutritional importance of trace elements has grown rapidly during the last century mainly because of a better understanding of their biological functions. In deficiency states, most essential trace elements create health problems. Iron-deficiency anemia and goiter owing to iodine deficiency are good examples illustrating the impact of certain trace element deficiencies. In terms of people afflicted, the above mentioned disorders are close to that of people suffering from protein-calorie malnutrition *(1)*. Several other pathological conditions in animals, including humans, are implied to be associated with deficiencies of trace elements *(2)*. Although severe deficiencies of trace elements that require immediate medical attention are rare, it has become increasingly evident that a marginal deficiency of several trace elements, including that of copper, iron, selenium, and zinc, is fairly common, even in parts of the population of affluent countries *(3,4)*. A lack of characteristic signs and symptoms is the main reason why these deficiencies are not detected at an early stage. It is a common belief that the minimum requirements of some of the trace elements are so low that a pure deficiency of trace elements rarely occurs in humans.

From a public health point of view, it is important to assure the general population that the intake of all trace elements in the average, daily diet is adequate to meet the requirements. At the same time, the ideal diet should not contain more than the permitted levels of toxic heavy metals. One of the basic requirements of nutritional trace element research is the knowledge concerning the true intake levels from prepared meals. Barring occupational exposure, the food chain remains the major pathway through which the essential and toxic trace elements enter the human body. Only limited information is available at present concerning the intake of trace elements from prepared meals, especially in the developing countries. As a public health issue, trace element nutrition has a low priority in most developing countries, where 75% of the world's population lives. Without knowing the true exposure levels, the clinical and public health significance of trace elements cannot be assayed adequately. The present study deals with the analysis of a number of essential and toxic trace elements in duplicate diets collected from populations living in various parts of the world.

MATERIAL AND METHODS

The material in this study consisted of 24-h dietary samples collected from populations living in various parts of Sweden and Portugal. The sampling collections were made by the duplicate-portion dietary sampling technique *(3)*. All individual items collected during the day were homogenized, and suitable aliquots of the homogenate were lyophilized. The lyophilized powder was used for the determination of trace elements. Table 1 shows the summary of the study groups used in the present in-vestigation. For the sake of comparison, we have also included the re-sults of the dietary samples collected by different countries of the world under a research contract organized by the International Atomic Energy Agency *(5)*. Most of the samples in this study were also collected employing the duplicate-portion dietary sampling technique.

The lyophilized dietary samples were wet-ashed prior to analysis by atomic absorption spectrometry and ICP-ES. For NAA, aliquots of about 200 mg of the lyophilized material were weighed into quartz ampules made from ultrapure quartz (Suprasil AN, Heraeus). Elemental comparator standards were also prepared. For analytical quality control, samples of reference materials (IAEA H-9 and NIST RM-8431) were used. These materials were dried according to the procedures prescribed by the issuing agencies. The details of collection and sample preparation are given elsewhere *(3,6–8)*.

Neutron activation analysis (NAA) in its various forms is particularly well suited for the simultaneous determination of trace elements in dietary samples. In the present study, a combination of conventional instrumental NAA (INAA) and several methods based on the principles of cyclic activation (cyclic activation analysis and replicate or cumulative analysis) was used for the determination of ten trace elements (Co, Cr, Cu, Fe, Hg, Mn, Rb, Se, V, and Zn). Flame AAS (FAAS) was used to determine the macrominerals K and Na, whereas Cr, Ni, Cd, and Pb were determined by graphite furnace AAS (GFAAS). The determination of Ca, Cu, Mg, Mn, P, and Al was performed by ICP-ES *(7,9)*. All these methods were also used for analyzing a number of certified reference materials to evaluate accuracy and precision. The samples collected by different countries under the IAEA research contract were analyzed by reference laboratories in various parts of the world using the most advanced techniques, including NAA *(10,11)*.

RESULTS

The concentration of essential trace elements found in the diets is shown in Table 2. As can be observed from this table, the dietary intake of most of the trace elements analyzed is almost the same for adults, pregnant women, and teenagers in Sweden. The intake of trace elements,

Table 1
The Details of the Study Groups Included in the Present Investigation

Study group	Country	Number	Technique used
Adults	Sweden	79	AAS, NAA
Adults	Portugal	15	AAS,NAA,ICP-ES
Teenagers	Sweden	13	AAS,NAA
Pregnant women	Sweden	45	AAS

Table 2
Daily Dietary Intake of Essential Trace Elements

Element	Country	Daily Intake mean ± SD median	IAEA Data median (range)	Unit
Copper	Sweden	1.5 ± 0.3 [a]	1.4	mg
	"	2.0 ± 0.8 [p]	(1.1 - 2.0)	
	"	1.5 ± 0.4 [t]		
	Portugal	1.2		
Chromium	Sweden	52 ± 13 [a]	50	µg
	"	48 ± 14 [t]	(20 - 285)	
	Portugal	70		
Iron	Sweden	12.1 ± 3.0 [a]	13	mg
	"	15.0 ± 4.2 [p]	(5.1 - 47)	
	"	16.8 ± 2.7 [t]		
	Portugal	9.1		
Selenium	Sweden	33.0 ± 4.0 [a]	61	µg
	"	39.0 ± 6.0 [p]	(34 - 133)	
	"	77.0 ± 5.0 [t]		
	Portugal	65.0		
Zinc	Sweden	8.1 ± 2.8 [a]	10	mg
	"	11.8 ± 3.6 [p]	(8 - 14)	
	"	8.3 ± 3.1 [t]		
	Portugal	10.1		

a, adults; p, pregnant women; t, teenagers.

such as copper, chromium, iron, selenium, and zinc, in the average normal diets is almost the same in Sweden and Portugal.

We have also investigated the intake levels of sodium, potassium, calcium, and magnesium, and the results are shown in Table 3. The intake levels of magnesium and potassium are below the current recommendations.

Table 3
The Dietary Intake of Minerals in Sweden and Portugal

Element	Country	Intake (mg/day) mean ± SD median	IAEA Data median (range)
Sodium	Sweden	2875 ± 729	3500
	Portugal	3967	(1570 - 6900)
Potassium	Sweden	2108 ± 850	2735
	Portugal	2609	(500 - 4500)
Calcium	Sweden	800	770
	Portugal	812	(209 - 1650)
Magnesium	Sweden	150	300
	Portugal	262	(120 - 680)

Table 4
Toxic Metal Contents of the Daily Diets

Element	Country	Daily Intake median	IAEA Data median (range)	Unit
Aluminium	Sweden	3.8	4.4	mg
	Portugal	4.6	(2.2 - 17.0)	mg
Cadmium	Sweden	12.0	14	µg
	Portugal	10.4	(8 - 200)	µg
Lead	Sweden	36.0	51	µg
	Portugal	49.0	(7 - 515)	µg
Mercury	Sweden	7.5	4.1	µg
	Portugal	7.1	(0.7 - 76)	µg

The intake of the toxic metals aluminum, cadmium, lead, and mercury is shown in Table 4. For these trace elements, the levels found in the Portuguese diets are also very similar to those found in the Swedish diets.

DISCUSSION

The nutrient composition of individual food items, especially that of trace elements, may vary in different parts of the world because of seasonal changes and differences in geographic distribution of the elements in the soil *(3)*. Information concerning the dietary intake of trace elements can be obtained by either direct or indirect techniques. Only direct analysis of the actual food consumed during 24 h can provide an accurate estimate of the dietary intake of both essential and toxic elements. Indirect methods are based on computation from standard food tables. Many food tables have no data at all concerning the concentration of some trace elements (e.g., chromium and selenium) in individual food items. The food tables are generally based on the analysis of raw food stuffs, and as such, they may not be accurate in estimating the intake of trace elements from prepared meals ready for consumption.

In the present study, diet samples were collected employing the duplicate-portion technique. The duplicate portion is a copy collected while eating normally to represent the food and fluids consumed during 24 h by visual judgment *(3,6)*. Although slightly inferior to the direct method of weighing all the actual food consumed during a 24 h period and the subsequent analysis of the samples, the duplicate-portion technique is probably the most satisfactory and practical technique by which one can study the actual intake of trace elements in small and well-defined population groups. The validity of the technique can be checked by correlating the intake of sodium, potassium, and nitrogen with the 24-h urinary excretion *(3,12)*.

Although the sample size in the present study is small, the results provide some indication of the trace element status in the general population. Unlike the population in developing countries, such as India, the general population in most of the European countries is usually homogenous, and the eating habits do not vary significantly between different regions of the country. From the results of the present study, it appears that the intake of several essential elements in prepared meals in Sweden and Portugal is almost the same and that it is below the current recommendations *(13)*. This is specially the case for elements, such as potassium, magnesium, zinc, and selenium. As regards toxic elements, none of the countries exceed the provisional tolerable intakes (PTI) as specified by FAO/WHO *(14)*.

We are in the process of generating data from developing countries, and it would be interesting to see if there are significant differences in the intake levels between developed and developing countries. When the IAEA study is completed, it will be possible to compare the intake levels in developing countries with those in industrialized countries.

REFERENCES

1. M. Abdulla, R. M. Parr, and G. V. Iyengar, in *Essential and Toxic Trace Elements in Human Health and Disease*, A. S. Prasad, ed., Wiley-Liss, New York, 1983, pp. 311–328.
2. E. J. Underwood, in *Trace Elements in Human and Animal Nutrition*, 4th ed., Academic, New York, 1977.
3. M. Abdulla, Ph. D. Dissertation, Lund University, Sweden, 1985, pp 6–98.
4. W. Mertz, *Science* **213,** 1332–1338 (1981).

5. R. M. Parr, M. Abdulla, N. K. Aras, A. R. Byrne, C. Camara-Rica, S. Finnie, A. G. Gharib, G. Ingrao, G. V. Iyengar, F. A. Khangi, S. S. Krishnan, J. Kumpulainen, S. Liu, R. Schelenz, S. Srianajuta, J. T. Tanner, and W. Wolf, in *Proceedings of the 7th International Symposium on Trace elements in Man and Animals—TEMA-7* Dubrovnik, Yugoslavia, May 1990, published by the Institute for Medical Research and Occupational Health, University of Zagreb, Croatia, 1991.
6. M. Abdulla, M. Jagerstad, K. Kolar, A. Norden, A. Schutz, and S. Svensson, in *Trace Element Analytical Chemistry in Biology and Medicine*, P. Bratter and P. Schrammel, eds., Walter de Gruyter, Berlin, 1983, pp. 75–86.
7. M. F. Reis, J. M. Pereira Miguel, A. A. S. C. Machado, M. Abdulla, E. Martinho, R. Parr, M. Makerowicz, R. Ogris, E. Zeiler, and E. Wehrstein, in *Proceedings of the 1st Int. Conf. on Trace Elements in Health and Disease with Special emphasis on Atherosclerosis*, Lisbon, 1992 (M. F. Reis, J. M. Pereira Miguel, A. A. S. C. Machado, eds., Smith-Gordon and Company Limited, London, 1994, *in press*).
8. M. F. Reis, J. Holzbecher, E. Martinho, and A. Chatt, *Biol. Trace Element Res.* **26–27,** 629–635 (1990).
9. M. F. Reis and E. Martinho, *J. Radioanal. Nucl. Chem., Articles* **167,** 39–44 (1992).
10. L. Kosta, *Fresenius Z. Anal. Chem.* **324,** 649 (1986).
11. R. M. Parr, H. Crawley, M. Abdulla, G. V. Iyengar, and J. Kumpulainen, in *Inventory of Analytical Results in Order of IAEA Sample Code*, IAEA, Vienna, (1992).
12. B. Isaksson, *Am. J. Clin. Nutr.* **33,** 4 (1980).
13. National Research Council, *Recommended Dietary Allowances*, 10th ed., National Academy Press, Washington, DC, 1989.
14. WHO Technical Report Series N. 751 and 776, 1987; 1989.

Determination of Trace Elements in Human Lung Samples

SIZUE O. ROGERO,*[,1] MITIKO SAIKI,[1] PAULO H. N. SALDIVA,[2] AND MIRIAN L. DALIBERTO[2]

[1]*Instituto de Pesquisas Energéticas e Nucleares, IPEN-CNEN/SP Radiochemistry Division, CEP 05422-970 São Paulo, Brazil*
[2]*Faculdade de Medicina, Universidade de São Paulo, CEP 01246-900, São Paulo, SP, Brazil*

ABSTRACT

Lung samples from smokers, nonsmokers, and one stillborn were analyzed by an instrumental neutron activation analysis (INAA) method. Pulmonary tissue and hilum lymph node samples were obtained separately from autopsies, and then submitted to cryogenic homogenization, lyophilization, and sterilization. Short and long irradiations were performed in an IEA-R1 nuclear reactor, and γ-ray activities were measured using a Ge(Li) or hyperpure Ge detector. Precision of results was evaluated by analyzing one lung tissue in replicates and their accuracy by analyzing reference materials. Comparisons were carried out between results obtained in pulmonary tissues and lymph nodes, as well as those obtained in samples from different groups of individuals.

Index Entries: Instrumental neutron activation analysis; lungs; lymph nodes; pulmonary tissues; smokers; nonsmokers; stillborn.

INTRODUCTION

Determination of trace elements in lung samples has been of great interest, since the inhalation of toxic atmospheric contaminants presents risk not only to the respiratory tract, but also to other organs of the body. Since the air and blood are in contact in the lungs, the inhaled toxic substances are transferred into the bloodstream, and dispersed in the body by both lymphatic and circulatory systems or else by mucocilliar depurations and swallowing.

*Author to whom all correspondence and reprint requests should be addressed.

Epidemiological studies *(1–3)* and the reviews of several papers *(4,5)* have also shown that various inhaled particles or gases are capable of producing harmful health effects. Consequently, lungs are considered to be a critical organ.

In this context, the determination of normal levels of elements in pulmonary tissues of a selected population group has been very useful for a comparative study with results obtained for individuals exposed to pollutants or in the study of pulmonary diseases related to trace elements. In addition to direct determination of trace elements in the lung tissue, bronchoalveolar lavage has been suggested to be an indicator of lung disorders of occupational origin *(6)*. Analysis of the bronchoalveolar lavage is advantageous from the diagnostic point of view, since the collection of lung biopsies is traumatic and can present risk of complications to the patient. In the present work, INAA was used to determine the trace elements in pulmonary tissues and lymph nodes from smokers, nonsmokers, and from one stillborn.

EXPERIMENTAL

Lung Sample Collection and Treatment

Lung samples were obtained on autopsies performed at the Institute of Forensic Medicine of the São Paulo University. They were collected from seven smokers with age varying from 34 to 69 yrs and from six nonsmokers aged from 43 to 74 yr. As an internal control, a sample was collected from one unborn child of about 9 mo of age deceased by intrauterine events. The *causa mortis* of these donors were not chronic pulmonary diseases. The lung samples were collected within 6–12 h postmortem. Information concerning the residence, profession, smoking habits, and length of work of each individual was provided by their closest relatives.

About 10–30 g of pulmonary tissue macroscopically normal from the right inferior lobe and samples of lymph nodes from pulmonary hilum were wrapped in clean polyethylene foil, stored in liquid nitrogen, and sent to the laboratory, where these samples were stored in a freezer at a temperature below –10°C until cryogenic homogenization was carried out as follows.

The pulmonary tissue and lymph node samples were separately placed in a Teflon™ vessel immersed in liquid nitrogen. By using a Teflon™ pestle, the frozen solid material was fractured, ground, and then lyophilized. During freeze-drying, a mean weight loss of 82% was found. The resulting fine powder of each sample was sterilized by irradiation in a ^{60}Co source and placed in a polyethylene container that was stored in a dessicator.

To avoid contamination of the samples, a series of precautions were taken. A titanium knife was used for cutting the samples. A polyethylene spatula was used for their transfer. Handling of the samples was performed inside a glove box or a class 100 laminar flow hood.

Preparation of Standards

Multielement and single-element standard solutions of appropriate concentrations were pipeted onto a small sheet of Whatman No. 42 filter paper using an Eppendorf pipet. After drying at room temperature, these sheets were placed into polyethylene bags and irradiated together with the samples.

Instrumental Neutron Activation Analysis

Irradiations were performed at the IEA-R1 research nuclear reactor at the IPEN-CNEN/SP. Aliquots of 50–100 mg of pulmonary tissues or lymph nodes were weighed and heat-sealed in clean polyethylene bags. Two irradiations were carried out for each sample: Short irradiations (30 min) using pneumatic system facility, in a thermal neutron flux of 3.7×10^{11} $n/cm^2/s$ to determine Cl, K, Mn, and Na, and long irradiations (16 h) in a thermal neutron flux of about 10^{13} $n/cm^2/s$ to determine Br, Cr, Cs, Fe, Hf, La, Rb, Sb, Sc, Se, Th, and Zn.

The irradiated samples and standards were fixed in stainless-steel planchets and counted using hyperpure Ge or Ge(Li) detector coupled to an EG & G Ortec 4096 channel pulse height analyzer connected to a Monydata PC 200 Plus microcomputer. The detectors used had a resolution (FWHM) better than 1.35 keV for the 122-keV γ-ray of ^{57}Co and 2.4 keV for the 1332-keV γ-ray of ^{60}Co. The spectral data were processed using VISPECT software. Element concentrations were calculated by the comparative method.

Analysis of Reference Materials

Four certified reference materials (CRMs), bovine liver 1577a from NIST USA, Bowen's kale from IUPAC, vehicle exhaust particulates No. 8 from NIES, Japan, and animal muscle H-4 from IAEA Austria, were analyzed for the evaluation of the accuracy of the method. The moisture content in reference materials was ascertained by drying in accordance with the producer's recommendations. The following values (in percent) of the weight loss were found and used for correcting the final results: 5.48 for bovine liver, 6.95 for H-4 animal muscle, 1.35 for vehicle exhaust particulate, and 12.77 for Bowen's kale.

RESULTS AND DISCUSSION

Table 1 shows that our results obtained for the CRMs agree well with the respective certified or information values. The relative errors for most of the elements determined were found to be <10%. Precision of the results was also satisfactory for most elements. The relative standard deviations varied from 0.4 to 11%. Less precise results were obtained for Br and Fe in vehicle exhaust particulate reference material and for some elements present at the concentrations lower than 1 μg/g.

Table 1
Results of Control Analyses of Certified Reference Materials (µg/g Unless Otherwise Indicated)

Element	Bovine Liver 1577a This Work	Bovine Liver 1577a Literature Values (7)	Bowen's Kale This Work	Bowen's Kale Literature Values (7)	Vehicle Exhaust Partic. This Work	Vehicle Exhaust Partic. Literature Values (7)	H-4 Animal Muscle This Work	H-4 Animal Muscle Literature Values (8)
Br	9.2+-0.6 (4)*	(9)**	26.7+-1.7 (6)	24.9+-2.5	67 +-12 (5)	(56)	4.5+-0.4 (4)	4.1+-0.6
Cl	2769+-141 (6)	2800+-98	4443+-453 (8)	3560+-427			2019+-55 (4)	1890+-83
Cr	0.84+-0.19 (5)		0.89+-0.14 (4)	(0.369)	23.5+-1.3 (4)	24.5+-1.5		(0.010)
Cs, µg/kg	18.56+-0.08 (2)		95+-10 (8)	76.3+-5.9	246+-7 (4)	(240)	135+-13 (5)	120+-10
Fe	194+-12 (5)	194+-19	120+-12 (9)	119.3+-14.3	4510+-687 (4)		51.7+-4.6 (6)	49+-2
Hf, µg/Kg					194+-4 (4)			
K, %	0.992+-0.063 (8)	0.996+-0.007	2.64+-0.16 (7)	2.436+-0.146		0.115+-0.008	1.56+-0.13 (3)	1.58+-0.06
La, µg/kg			87+-20 (8)	86.4+-13.0	1190+-50 (5)	(1200)		
Mn	10.7+-0.8 (9)	9.9+-0.8	14.3+-3.2 (7)	14.82**			0.70+-0.30 (4)	0.52+-0.04
Na	2447+-154 (9)	2430+-129	2583+-258 (8)	2366+-284			2073+-87 (4)	2060+-126
Rb	12.1+-0.4 (5)	12.5+-0.1	54+-6 (9)	53.4+-5.3	4.0+-0.2 (5)	(4.6)	19.5+-1.6 (6)	19.0+-1.5
Sb, µg/kg			70+-18 (4)	68.5**	5990+-360 (4)	6000+-40		
Sc, µg/kg			15.1+-1.0 (9)	9.48**	564 +-21 (4)	(550)		
Se	0.72+-0.07 (5)	0.71+-0.07			1.45+-0.04(5)	(1.3)	0.32+-0.05 (4)	0.28+-0.03
Th, µg/kg			12.2+-2.2 (5)	10.4**	344+-8 (4)	(350)		
Zn	127+-4 (5)	123+-8	33.7+-2.3 (9)	32.26+-2.74	879+-39 (5)	1040+-50	90+-6 (6)	86+-3

*Number of determinations.
**Information values.

Table 2
Precision of Analysis of One Sample of Pulmonary Tissue in
(μg/g, dry wt, Unless Otherwise Indicated)

Element	XM +- S (N)	SR , %
Br	16.8 +- 1.5 (4)	8.9
Cl, %	1.73 +- 0.18 (5)	10.4
Cs	0.96 +- 0.01 (4)	1.0
Fe	2423 +- 73 (4)	3.0
K, %	0.82 +- 0.05 (6)	6.0
Mn	1.46 +- 0.15 (3)	10.2
Na, %	1.42 +- 0.14 (5)	9.8
Rb	32.8 +- 0.7 (4)	2.1
Sb	0.30 +- 0.02 (4)	6.7
Sc μg/kg	1.39 +- 0.07 (4)	5.0
Se	0.66 +- 0.03 (4)	4.5
Zn	69 +- 2 (4)	2.9

$X_M \pm S$—arithmetic mean and standard deviation.
S_R—relative standard deviation.
n–number determinations

The precision of the results was also evaluated by analyzing one sample of pulmonary tissue in replicate as shown in Table 2. Element concentrations found in the pulmonary tissues from smokers, nonsmokers, and stillborn are presented in Table 3 together with the data published by Vanoeteren et al. *(5)* for comparison. It can be seen from this table that the elements determined exhibit considerable intersubject variability. However the results for nonsmokers are within the range reported in the literature, except for the elements Cl and Hf.

Table 4 shows the results obtained for lymph node samples from nonsmokers, smokers, and stillborn. No literature data for this tissue were found for comparison.

The concentration levels of the trace elements in pulmonary tissues from smokers and nonsmokers are, in general, of the same order of magnitude. One exception was Sb, for which a higher value was found for smokers. The results for lymph nodes (Table 4) indicate the same pattern as observed for the lung samples. This finding indicates that hilum

Table 3
Elemental Concentrations of Pulmonary Tissue of Nonsmokers (NS), Smokers (S), and Stillborn (SB) (dry wt)

Element	Smoking Habit	X_M +- s	X_G x: S_G	Median	Range	Vanoeteren et al[5] Range
Br (ug/g)	NS S SB	29 +- 11 29 +- 8 22.4 +- 0.4**	27 x: 1.46 28 x: 1.30	28 27	17—46 19—40	2—120
Cl (%)	NS S SB	1.4 +- 0.4 1.1 +- 0.2 1.46 +- 0.04	1.3 x: 1.33 1.1 x: 1.22	1.2 1.0	1.0—1.9 0.9—1.6	0.5—1.1
Cr (ug/g)	NS S SB	1.6 +- 1.0 1.4 +- 0.7 0.6 +- 0.1	1.4 x: 1.76 1.3 x: 1.72	1.3 1.2	0.6—3.4 0.5—2.5	0.01—2.5
Cs (ug/g)	NS S SB	0.29 +- 0.07 0.18 +- 0.05 0.15 +- 0.01	0.28 x: 1.33 0.18 x: 1.29	0.31 0.18	0.16—0.35 0.12—0.24	0.015—1.5
Fe (%)	NS S SB	0.22 +- 0.09 0.17 +- 0.08 0.106 +- 0.001	0.20 x: 1.64 0.16 x: 1.50	0.23 0.14	0.08—0.32 0.10—0.32	0.02—0.25
Hf (ug/kg)	NS S *	22 +- 8 24 +- 17	21 x: 1.41 20 x: 1.87	20 17	15—38 12—49	(0.01)
K (%)	NS S SB	0.9 +- 0.1 1.0 +- 0.2 0.86 +- 0.12	0.9 x: 1.16 0.9 x: 1.24	0.9 0.9	0.8—1.2 0.7—1.3	0.25—1.0
La (ug/g)	NS S	0.3 +- 0.1 0.28 +- 0.09	0.3 x: 1.48 0.26 x: 1.48	0.2 0.29	0.2—0.5 0.13—0.38	0.005—1
Mn (ug/g)	NS S SB	1.3 +- 0.3 1.3 +- 0.4 1.2 +- 0.5	1.2 x: 1.24 1.2 x: 1.44	1.2 1.2	1.0—1.6 0.7—1.8	0.05—15
Na (%)	NS S SB	6.0 +- 0.4 7.0 +- 0.3 1.19 +- 0.01	1.0 x: 1.39 1.0 x: 1.31	1.0 1.0	0.6—1.6 0.6—1.5	0.5—1.5
Rb (ug/g)	NS S SB	32 +- 6 27 +- 7 33.9 +- 0.5	31 x: 1.19 27 x: 1.30	31 23	26—39 19—37	2.5—50
Sb (ug/g)	NS S SB	0.29 +- 0.07 0.7 +- 0.5 0.04 +- 0.01	0.28 x: 1.29 0.5 x: 2.52	0.31 0.9	0.20—0.36 0.14—1.2	(0.01—0.5)
Sc (ug/g)	NS S SB	0.03 +- 0.03 0.02 +- 0.02 0.002 +-0.0007	0.025 x: 1.97 0.013 x: 2.71	0.02 0.02	0.01—0.08 0.003—0.05	0.0005—0.035
Se (ug/g)	NS S SB	0.6 +- 0.1 0.6 +- 0.1 0.6 +- 0.1	0.6 x: 1.22 0.6 x: 1.22	0.6 0.6	0.5—0.9 0.5—0.8	0.25—25
Th (ug/g)	NS S	0.03 +- 0.01 0.04 +- 0.03	0.03 x: 1.41 0.03 x: 2.29	0.03 0.02	0.02—0.05 0.01—0.10	(0.005—0.1)
Zn (ug/g)	NS S SB	73 +- 21 57 +- 4 67.7 +- 0.8	71 x: 1.28 57 x: 1.08	66 58	58—113 52—65	5—150

Literature values given on wet wt basis were converted to dry wt basis using a factor of 0.2 according to Ref (5)

Values in parenthesis are speculative data.

*Results for four samples.

**Statistical counting errors.

$X_M \pm S$—Arithmetic means and their standard deviations.

X_G X:S_G—Geometric means and their standard deviations.

lymph nodes may be used as an indicator of elemental profile in the lungs because of their physiological function of draining and clearance of exogenous materials. Especially for epidemiological purposes, it is important to choose a determined sampling point because of the possible differences in distribution of elements in lungs.

Elements Hf, La, Sb (in lymph node), and Th were not detected in the sample from stillborn. The levels of Cr, Cs, Sb, and Sc were significantly lower in stillborn than those for adults. This comparison indicates

Table 4
Elemental Concentrations of Lymph Nodes for Nonsmokers (NS), Smokers (S), and Stillborn (SB) (dry wt)

Element	Smoking Habit	x_M +- s	x_G x: s_G	Median	Range
Br (µg/g)	NS	17 +- 6	16 x: 1.37	14	12—25
	S	11 +- 7	9 x: 1.73	8	5—26
	SB	13.8 +- 0.5			
Cl (%)	NS	0.89 +- 0.11	0.88 x: 1.14	0.88	0.7—1.0
	S	0.65 +- 0.21	0.61 x: 1.51	0.67	0.26—0.88
	SB	1.03 +- 0.05			
Cr (µg/g)	NS	5.9 +- 1.5	5.8 x: 1.28	5.6	4.2—8.2
	S	8.8 +- 11.0	5.3 x: 2.82	6.6	1.7—33.2
	SB	0.62 +- 0.08			
Cs (µg/g)	NS	0.49 +- 0.25	0.44 x: 1.72	0.40	0.20—0.82
	S	0.21 +- 0.08	0.19 x: 1.60	0.21	0.07—0.29
	SB	0.10 +- 0.01			
Fe (%)	NS	0.23 +- 0.15	0.19 x: 2.05	0.21	0.06—0.47
	S	0.15 +- 0.08	0.13 x: 1.66	0.13	0.06—0.30
	SB	0.073 +- 0.001			
Hf (µg/g)	NS	0.63 +- 0.86	0.35 x: 3.15	0.33	0.1—2.36
	S	0.13 +- 0.15	0.07 x: 3.51	0.06	0.01—0.45
K (%)	NS	0.70 +- 0.17	0.68 x: 1.33	0.77	0.42—0.89
	S	0.59 +- 0.18	0.56 x: 1.39	0.56	0.33—0.84
	SB	0.65 +- 0.04			
La (µg/g)	NS	1.5 +- 0.5	1.4 x: 1.56	1.5	0.6—2.2
	S	1.1 +- 0.8	0.8 x: 2.29	0.9	0.4—2.2
Mn (µg/g)	NS	7.1 +- 3.4	6.4 x: 1.68	7.1	3.4—10.6
	S	5.0 +- 3.5	3.8 x: 2.39	5.0	0.9—11.7
	SB	0.15 +- 0.05			
Na (%)	NS	0.9 +- 0.1	0.9 x: 1.12	0.9	0.8—1.0
	S	0.7 +- 0.2	0.7 x: 1.41	0.7	0.36—0.93
	SB	0.967 +- 0.006			
Rb (µg/g)	NS	23 +- 7	22 x: 1.37	22	14—34
	S	17 +- 7	16 x: 1.59	19	8—27
	SB	22.4 +- 0.4			
Sb (µg/g)	NS	1.7 +- 1.1	1.3 x: 2.37	1.7	0.3—3.2
	S	5.7 +- 8.5	2.6 x: 3.68	2.2	0.5—24.1
Sc (µg/g)	NS	0.43 +- 0.25	0.36 x: 1.98	0.45	0.14 - 0.66
	S	0.17 +- 0.16	0.09 x: 4.09	0.13	0.01 - 0.42
	SB	0.0042+-0.0006			
Se (µg/g)	NS	0.94 +- 0.23	0.92 x: 1.31	1.0	0.6—1.2
	S	0.78 +- 0.29	0.74 x: 1.44	0.7	0.5—1.3
	SB	0.49 +- 0.08			
Th (µg/g)	NS	0.35 +- 0.14	0.33 x: 1.56	0.37	0.18—0.50
	S	0.20 +- 0.17	0.13 x: 3.02	0.12	0. 3—0.45
Zn (µg/g)	NS	67 +- 7	67 x: 1.10	67	57—77
	S	68 +- 50	58 x: 1.73	59	36—178
	SB	68.3 +- 0.4			

that several elements accumulate in the lungs from air pollution and/or from tobacco smoke.

ACKNOWLEDGMENTS

The authors are grateful to CNPq and FAPESP from Brazil for financial support.

REFERENCES

1. N. Neuberger and M. Kundi, *Arch. Geschwulstforch* **61,** 47–53 (1991).
2. J. T. Hodgson and R. D. Jones, *Br. J. Ind. Med.* **47,** 665–676 (1990).
3. J. C. Mc Donald, B. W. Case, P. E. Enterline, V. Henderson, A. D. Mc Donald, M. Plourde, and P. Sebastien, *Ann. Occup. Hyg.* **34(5),** 427–441 (1990).
4. M. Chiba and R. Masironi, Toxic and trace elements in tobacco, World Health Organization/TOH/CLH/91.1, 1991.
5. C. Vanoeteren, R. Cornelis, and E. Sabbioni, Critical evaluation of normal levels of major and trace elements in human lung tissue, EUR 10440 EN, March 1986.
6. E. Sabbioni, R. Pietra, F. Mousty, F. Colombo, G. Rizzato, and G. Scansetti, *J. Radioanal. Nucl. Chem., Articles* **110(2),** 595–601 (1987).
7. E. Cortes Toro, R. M. Parr, and S. A. Clements, Biological and environmental reference materials for trace elements and organic microcontaminants. IAEA/RL/128 (Rev.1) Vienna, IAEA, 1990.
8. Y. Muramatsu and R. M. Parr, Survey of currently available reference materials for use in connection with the determination of trace elements in biological and environmental materials, IAEA/RL/128 Vienna, IAEA, Dec. 1985.

Levels of Some Trace Elements in Serum of Prague Inhabitants Measured by INAA

J. KVÍČALA,*,[1] V. ZAMRAZIL,[1] AND S. ČERMÁK[2]

[1]*Institute of Endocrinology, Národní 8, 116 94 Prague 1; and* [2]*Adico, Ltd., Radiová 1, 102 27 Prague 10, Czech Republic*

ABSTRACT

Sera of 240 randomly selected Prague inhabitants of both sexes from ages 6 to 65 yr were analyzed by instrumental neutron activation analysis (INAA) for Fe, Zn, Se, Rb, Cs, and Sc. Significant differences were found between serum Fe and Zn levels of men from Prague and from rural regions of Central Bohemia. Concentration of Se was one of the lowest in Europe and exhibited age-dependent, but not sex-dependent changes. Age-dependent changes of serum concentrations were also found for Rb, Cs, and Sc, whereas sex-dependent changes were detected for Zn, Rb, Cs, and Sc.

Index Entries: Serum; iron; zinc; selenium; rubidium; cesium; scandium; age differences; sex differences.

INTRODUCTION

Some trace elements have been recognized as essential for human beings, and some of them are suspected of playing this role *(1,2)*. Changes of trace element levels may be associated with various diseases *(3,4)*, and therefore, it is necessary to study this phenomenon. Since concentrations of trace elements are also dependent on local environmental factors, there is a need for comparative studies on healthy persons from the same region. In this context, sex and age should also be taken into consideration *(1)*.

The aim of this study was to determine average blood serum concentrations of iron, zinc, selenium, rubidium, cesium, and scandium in the Prague population as "normal values" for biomedical research, pos-

*Author to whom all correspondence and reprint requests should be addressed.

sibly as a diagnostic tool, to evaluate age and sex differences, and also to compare the obtained values with those for a population from a rural region of central Bohemia *(5)*.

MATERIALS AND METHODS

Materials

Inorganic standards for INAA were prepared by dissolving high-purity compounds in deionized water (17.5–18 MΩ) prepared in an apparatus NANOpure (Barnstead, Wilhelm Werner GmbH, Berg). $Fe_2(SO_4)_3$ (99.999%), SeO_2 (99.999%), $RbNO_3$ (99.998%), $CsNO_3$ (99.999%), Sc_2O_3 (99.99%), and $ZnSO_4 \cdot 7H_2O$ (Alfa Products, Karlsruhe) were used for this purpose. For wet-ashing of serum and its preparation for measurement, suprapure HNO_3, $HClO_4$, and HCl (Merck, Darmstadt) were used. Human serum (the second-generation reference material) prepared by Versieck et al. *(6)* and International Atomic Energy Agency (IAEA) reference material (RM) H-4 (animal muscle) were used as control materials.

Subjects

The subjects were males and females of the ages 6, 10, 13, 18–35, 36–49, and 50–65 yr from Prague.

Methods

Blood serum was analyzed by neutron activation analysis (NAA) as described elsewhere *(5)*. Quality assurance was obtained by coanalyzing the Versieck's human serum *(6)* and the IAEA RM H-4 *(7)*.

A γ-spectrometric system was employed that consisted of a multichannel analyzer MB 7329/S (Silena, Milano) and a coaxial HPGe detector (relative efficiency 34.6% and FWHM 1.87 KeV for γ-rays of 1.33 MeV). For peak area evaluation, the program Silgamma (Silena, Milano) was used.

RESULTS AND DISCUSSION

Mean element levels (± SD) in serum of all subjects examined amounted to (in µg/g, dry wt) 28.1 ± 7.8 of Fe, 10.9 ± 2.2 of Zn, 0.69 ± 0.18 of Se, and 1.61 ± 0.35 of Rb, whereas these values for Cs and Sc (in ng/g, dry wt) were 7.9 ± 2.8 and 1.01 ± 0.42, respectively. The results for groups of boys, men, girls, and women are given in Table 1. No significant sex or age differences were found for Fe. Highly significant differences were detected for serum Zn between girls and boys and between girls and women, but not between boys and men. Borderline significance of differences ($p < 0.1$) was found between men and women. Sex-related differences were proven for Rb, Cs, and Sc, but not for Se. Significant differences were also found among age groups for Se, Rb, Cs, and Sc as

Table 1
Serum Concentrations of Trace Elements—Age- and Sex-Different Groups

	Prague								Central Bohemia	
	Boys		Men		Girls		Women		Men	
Element*	n	Mean±S.D.	n	Mean±S.D.	n	Mean±S.D.	n	Mean±S.D.	n	Mean±S.D.
Fe	56	26.9±8.3	61	28.6±8.3	58	27.7±8.0	59	29.1±6.6	106	24.4±8.4
Zn	56	10.8±2.5	63	11.2±2.4	59	11.6±1.7	60	10.5±1.9	108	9.6±1.6
Se	55	0.62±0.12	63	0.74±0.20	59	0.61±0.11	58	0.78±0.19	109	0.71±0.14
Rb	56	1.47±0.31	63	1.70±0.39	59	1.69±0.28	59	1.57±0.36	62	1.69±0.38
Cs	46	6.30±2.73	57	7.86±2.85	57	7.09±1.50	55	10.10±2.86	107	11.90±3.40
Sc**	20	0.83±0.22	38	1.27±0.55	58	0.93±0.28	57	0.97±0.41	93	1.20±0.50

*Values in μg/g of dry serum for Fe, Zn, Se, and Rb, and in ng/g of dry serum for Cs and Sc; the values can be recalculated to liquid serum by dividing them by a factor of 11.

**Sc was measured for all groups of females, for 13-yr-old boys, and 36–65-yr-old men.

depicted in Figs. 1–4. These differences may be associated with sexual maturing in the age group of 13 yr, but also with increasing age in the cases of Se, Rb, and Sc.

Results for Sc were obtained for females in all age groups (Fig. 4), and for males only in the age groups of 13, 36–49, and 50–65 yrs (Table 1). To compare data from Prague with those from central Bohemia, concentrations of trace elements in serum of men of the latter region are also shown in Table 1. Significant differences were detected using Student's *t*-test between men of the two regions for Zn, Fe, and Cs serum levels.

It is obvious from the obtained results that the changes in the content of trace elements in human body are not only according to localities, but also according to age and sex. This fact is probably associated with the overall rate of metabolism, which is different in various times of life (childhood, maturity, old age), as well as with various eating habits in these age periods. Metabolism of some trace elements is also controlled by hormones. This is apparent not only from the differences between men and women, but also from the significant changes in boys and girls in the age of 13 yr (Figs. 1–4).

All these factors should be accounted for in the biomedical research as well as for diagnostic purposes, prognosis, and medical care of patients. Normal values for comparative purposes should be chosen according to a locality, sex, and age of individuals or groups under the research or medical care.

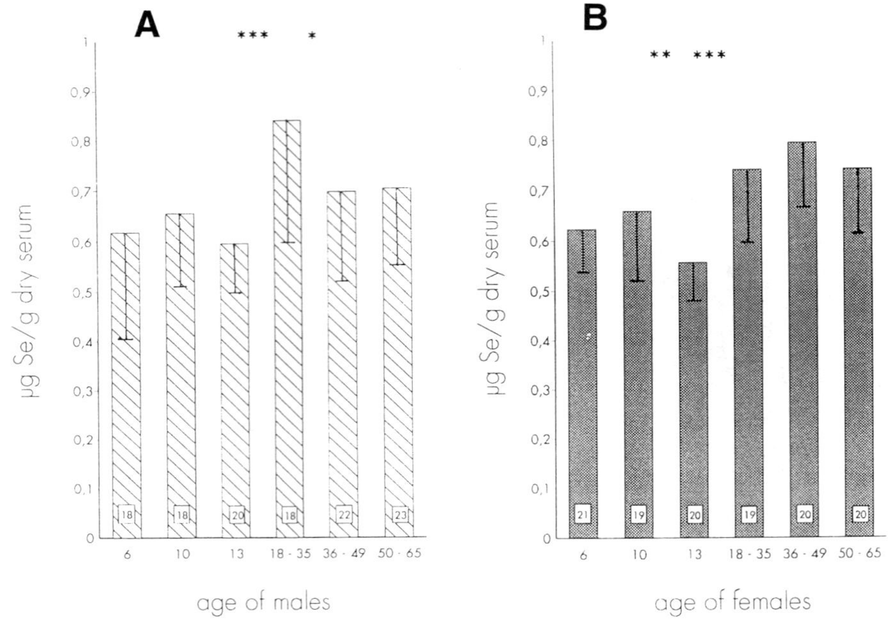

Fig. 1. Sex- and age-related concentrations of Se in serum of Prague inhabitants (males—1A; females—1B) ⊥ = standard deviation, figure in column = number of subjects; significance of differences: * = $p < 0.05$;** = $p < 0.01$;*** = $p < 0.001$.

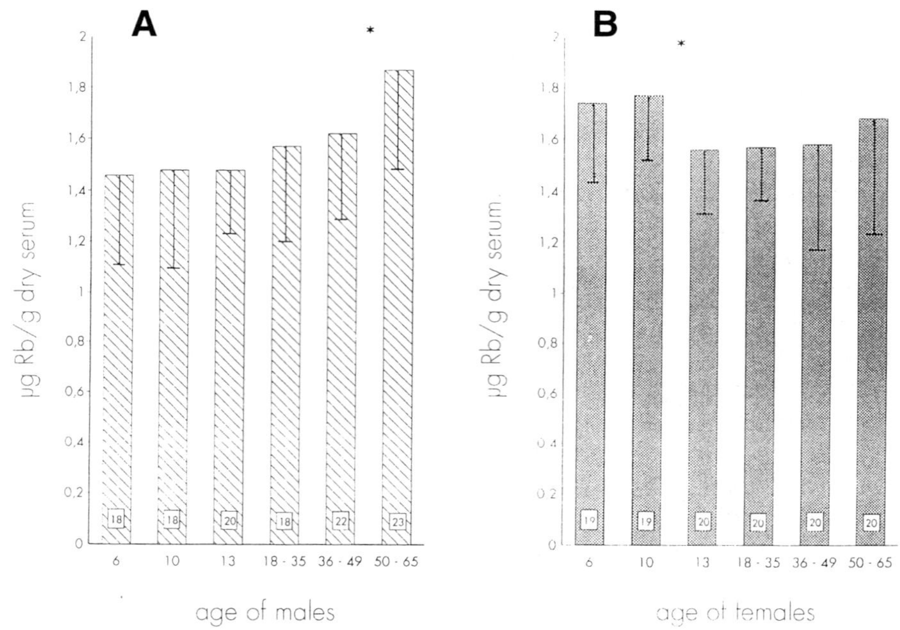

Fig. 2. Sex- and age-related concentrations of Rb in serum of Prague inhabitants (males—2A; females—2B). Symbols as shown in Fig. 1.

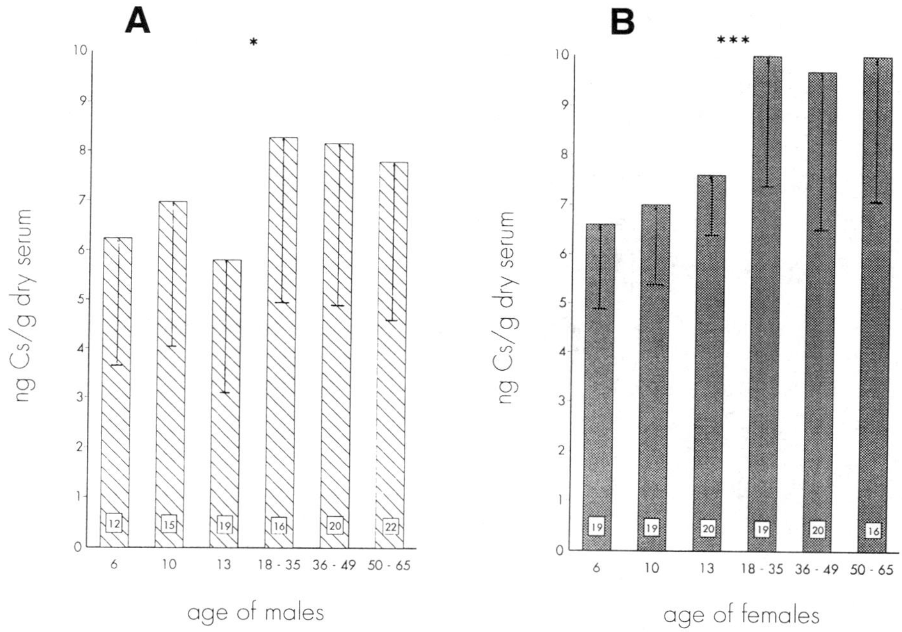

Fig. 3. Sex- and age-related concentrations of Cs in serum of Prague inhabitants (males—3A; females—3B). Symbols as shown in Fig. 1.

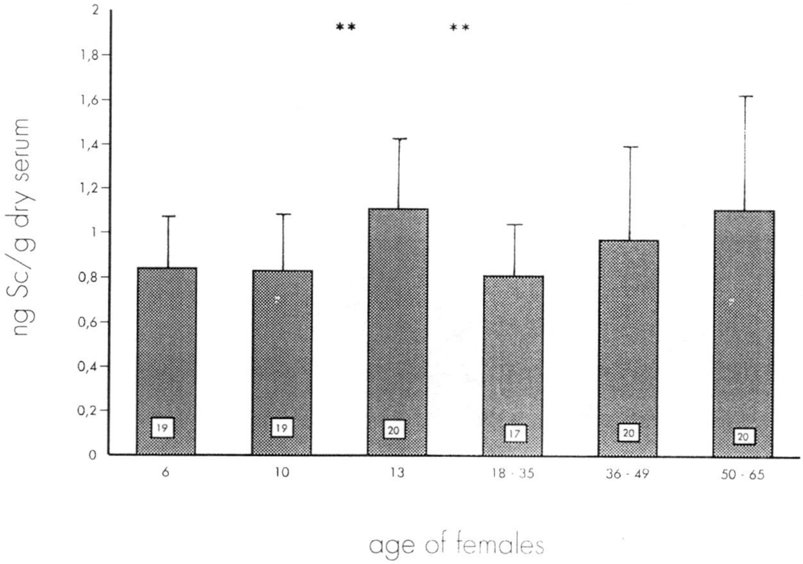

Fig. 4. Sex- and age-related concentrations of Sc in serum of Prague females according to their age. Symbols as shown in Fig. 1.

Noteworthy is the very low content of Se in all age groups of both sexes and both regions. The Se levels between 5.9 and 8.2 μg/g in dry serum, i.e., between 54 and 75 μg/L are very low in both the global and European scales (*8–10*). This deficiency may contribute to higher mortality and to lower life-span of inhabitants of the Czech Republic by higher radical damage of the cells and tissues *(11)*.

ACKNOWLEDGMENTS

The kind gift of biological reference material of human serum from Professor J. Versieck Gent, (Belgium) is acknowledged. This work was supported by the grant No. 1428-3 of the IGA MZ ČR and 0986 of the GA ČR.

REFERENCES

1. B. Momčilovič, ed., *Trace Elements in Man and Animals* **7**, IMI, Zagreb, 1991.
2. M. Anke, D. Meissner, and C. F. Mills, eds., *Trace Elements in Man and Animals*, Verlag Media Touristik, Gersdorf, 1993.
3. A. S. Prasad and D. Oberleas, *Trace Elements in Human Health and Disease*, Academic, New York, 1976.
4. I. E. Dreosti, *Med. J. Aust.* **2,** 117 (1980).
5. J. Kvíčala and J. Havelka, *J. Radioanal. Nucl. Chem.* **121,** 261 (1988).
6. J. Versieck, L. Vanballenberghe, A. DeKesel, J. Hoste, B. Wallaeys, J. Vandenhaute, N. Baeck, H. Steyaert, A. R. Byrne, and F. W. Sunderman, Jr., *Anal. Chim. Acta* **204,** 63 (1988).
7. R. M. Parr, IAEA RL 69, Report No. 2, IAEA, Vienna, 1980.
8. H. Robberecht, H. Deelstra, and R. Van Grieken, *Biol. Trace Element Res.* **25,** 149 (1990).
9. O. Oster and W. Prellwitz, *Biol. Trace Element Res.* **24,** 91 (1990).
10. J. Néve, *Trace Element Electr. Health Dis.* **5,** 1 (1990).
11. Abstracts of the Trace Elements and Free Radicals in Oxidative Diseases, Chamonix (France), April 5–9, 1993.

Determination of Aluminum and Phosphorus in Biological Materials by Reactor Activation Analysis Using Germanium as Integral Flux Monitor and Comparator

J. C. FURNARI AND I. M. COHEN*

Comisión Nacional de Energía Atómica, Avenida del Libertador 8250, 1429 Buenos Aires, República Argentina

ABSTRACT

A method for determination of aluminum and phosphorus in biological materials, based on activation in a nuclear reactor and measurement of ^{28}Al, produced by the $^{27}Al(n,\ \gamma)^{28}Al$ and $^{31}P(n,\ \alpha)^{28}Al$ reactions, is described. Irradiations in the undisturbed and epicadmium spectra provide a two-equation system in order to determine the contributions of aluminum and phosphorus to the total activities. Germanium is used as an integral flux monitor and comparator, through the reactions: $^{74}Ge(n,\gamma)^{75}Ge$, $^{76}Ge\ (n,\gamma)^{77}Ge$, and $^{72}Ge\ (n,p)^{72}Ga$.

Index Entries: Aluminum; phosphorus; activation analysis; biological materials; nuclear reactors; germanium; neutron flux.

INTRODUCTION

The increasing need for fast and reliable determinations in reactor activation analysis led to the use of simple methods of standardization. Thus, for instance, multi-isotopic comparators have been successfully used, since they allow the simultaneous measurement of the thermal and epithermal components of the reactor spectrum by irradiation of a single element. However, the analysis can also be influenced by the fast neutrons in the reactor. One of the examples of this situation occurring in an analysis of biological materials is the determination of aluminum based

*Author to whom all correspondence and reprint requests should be addressed.

on the $^{27}Al(n,\gamma)^{28}Al$ reaction, which is interfered with by phosphorus, through the reaction, $^{31}P(n,\alpha)^{28}Al$. On the other hand, since the capture reaction on phosphorus leads to the pure β^- emitter ^{32}P, the (n,α) reaction offers an interesting possibility for its assay.

The authors have explored the feasibility of the determination of both aluminum and phosphorus in biological materials through measurement of ^{28}Al, after two irradiations at the RA-6 reactor (Centro Atómico Bariloche) in the undisturbed and epicadmium spectra. The individual components to the whole activity can be calculated from the two-equation system obtained for the measurement after the abovementioned two irradiation conditions, but the use of comparators other than those based on the studied elements is practically mandatory, because of the short half-life of ^{28}Al (2.24 min).

The possibilities of using germanium as an integral flux monitor and comparator have been studied by the authors and coworkers. Since the conclusions of the study will be published elsewhere, *(1)* only a few comments are provided in this work. Briefly, the utilization of germanium for such purpose is based on the reactions: $^{74}Ge(n,\gamma)^{75}Ge$, $^{76}Ge(n,\gamma)^{77}Ge$, and $^{72}Ge(n,p)^{72}Ga$. The characteristics of the use of Ge as a comparator in activation analysis, with emphasis on NAA of biological materials, are as follows. a) The difference between resonance integrals to cross-section ratios for ^{75}Ge and ^{77}Ge (13.8–1.68) is comparable to that of the pair: ^{198}Au-^{60}Co, frequently used for monitoring the thermal and epithermal components of the reactor spectrum. Therefore, germanium is an acceptable flux monitor. b) Because of the half-lives of ^{75}Ge and ^{77}Ge (1.3797 and 11.30 h, respectively), germanium is especially useful for flux determinations after short irradiations. c) The ^{75}Ge resonance integral to cross-section ratio is approximately corresponding to other medium and short-lived nuclides, ^{24}Na, ^{27}Mg, ^{28}Al, ^{38}Cl, ^{42}K, ^{49}Ca, and ^{56}Mn, which are formed from biologically important elements. For all of them, ^{75}Ge is an excellent comparator. d) Despite the low value of the fission averaged cross-section of the $^{72}Ge(n,p)^{72}Ga$ reaction, some of the peaks from ^{72}Ga ($t_{\frac{1}{2}}$ = 14.10 h) can be detected even in the presence of the high ^{77}Ge background, thus allowing the determination of the fast component of the reactor spectrum.

STATE OF THE ART OF THE DETERMINATIONS OF THE NUCLEAR CONSTANTS FOR THE PARAMETRIC CALCULATIONS

1) $^{74}Ge(n,\gamma)^{75}Ge$ and $^{76}Ge(n,\gamma)^{77}Ge$ Reactions

Although the isotopic abundances and the intensities of the γ-rays are reasonably well determined, and the atomic weight of germanium and the half-lives of the radionuclides are known with low error, the uncertainties involved in the cross-sections are far from satisfactory. Moreover, since ^{77}Ge is also partially formed by the decay of ^{77m}Ge, the error of the transition factor increases the total uncertainty for the cross-section of the latter reaction. Instead, the k_0 factors and the resonance integral to cross-section ratios have been determined with good precision *(2)*. The determinations performed at the authors' laboratory *(1)* show a reasonable agreement with the published values.

2) $^{72}Ge(n,p)^{72}Ga$ Reaction

All the nuclear constants involved are well known, with the only exception being the fission averaged cross-section; very discrepant values are found in the literature, so that a new determination has been carried out *(3)*, using the RA-3 and the RA-6 reactors, located respectively, at Centro Atómico Ezeiza and Centro Atómico Bariloche, both belonging to Comisión Nacional de Energía Atómica. The value obtained, which was used for the calculations in the present work, is (0.0959 ± 0.0049) mb.

3) $^{27}Al(n,\gamma)^{28}Al$ Reaction

Both the k_0 factor and the resonance integral to cross-section ratio have been determined with good precision *(4)*.

4) $^{31}P(n,\alpha)^{28}Al$ Reaction

The atomic weight of phosphorus is very well known, whereas both the ^{31}P isotopic abundance and the intensity of the 1178.9-keV γ-ray from ^{28}Al are 100%. With respect to the fission averaged cross-section of the reaction, Calamand *(5)* compiled two experimental values, the average of which, after renormalization, is (1.9 ± 0.6) mb. Gryntakis quoted these values in a new compilation *(6)*, and no other data have been found in the literature. Since the uncertainty of the Calamand's value would have resulted in a high error for the phosphorus determination, a new determination was performed at Centro Atómico Bariloche *(7)*. The new value for the fission averaged cross-section of the $^{31}P(n,\alpha)^{28}Al$ reaction, obtained as a result of five independent measurements, is (1.852 ± 0.071) mb. This value was used in the calculation of the phosphorus contents.

EXPERIMENTAL

Samples weighing 100–200 mg, wrapped in polyethylene bags, and hyperpure germanium slices (ca. 200 mg) were sequentially irradiated with and without cadmium covers for 180 s, using the pneumatic facility, at one of the position of the RA-6 reactor (thermal flux: 4×10^{12} $n/cm^2/s$; epithermal flux/U ln E: 8×10^{10} $n/cm^2/s$; fast flux: 7×10^{11} $n/cm^2/s$). Samples and standards were measured by high-resolution γ spectrometry with a hyperpure Ge detector and a 4096 multichannel analyzer. Peak areas were obtained by the Nuclear Data VAX/VMS software for the spectral analysis. After a short decay (40–50 s), the ^{28}Al peak at 1778.9 keV was measured for 300 s in the samples. Then the germanium standards were measured (30 min) for ^{75}Ge and ^{77}Ge, and following the decay of the former, another measurement (3–5 h) was performed for ^{77}Ge and ^{72}Ga. The peaks of the germanium comparators measured were:

^{75}Ge: 198.6 keV;
^{77}Ge: 211.0, 215.5, 367.4, 416.3, and 558.0 keV;
^{72}Ga: 834.1 and 2201.7 keV.

All these measurements were corrected for self-attenuation, the factors being on the order of 6% for the 198.6-keV radiation to 1% for the most energetic peak, at 2201.7 keV.

The eventual contribution of the polyethylene wrapping to the activities formed in the samples was controlled by irradiations of all the bags to be used, some days before the determinations. The ^{28}Al blank activities were consequently subtracted from the whole activities.

As mentioned before, the calculations of the concentrations were performed from a two-system equation, the basic formula being:

$$[Csp^{(Al-28)}/S \cdot D \cdot M] = [Csp^{(Ge-75)}/S \cdot D \cdot M]\,[k_0^{(Al-28)}\,\varepsilon^{(Al-28)}/k_0^{(Ge-75)}\,\varepsilon^{(Ge-75)}]\,\{[f + Q(\alpha)^{(Al-28)}]/[f + Q(\alpha)^{(Ge-75)}]\} \times [Al] +$$

$$+ (Csp^{(Ga-72)}/S \cdot D \cdot M)\,\{[\sigma(n,\alpha) \cdot \theta]^{(P-31)}/[\sigma(n,p) \cdot \theta]^{(Ge-72)}\}\,[i^{(Al-28)}/i^{(Ga-72)}]\,[\varepsilon^{(Al-28)}/\varepsilon^{(Ga-72)}]\,[W^{(P)}/W^{(Ge)}] \times (P) \qquad (1)$$

Where: Csp = specific counting; $S = 1 - \exp - \lambda \cdot t(i)$, saturation factor; t(i), = irradiation time; $D = \exp - \lambda \cdot t(d)$, decay factor; t(d), = decay time; $M = [1 - \exp - \lambda \cdot t(m)]/\lambda \cdot t(m)]$, factor of decay during the measurement; t(m), = real measurement time; k_o, = k_o factors, as defined by Simonits and De Corte *(8)*, measured for a specific radiation (the superscript *Al – 28* is referred to the constants corresponding to the capture reaction on ^{27}Al) · ε = efficiency for the measured radiation. f = $\Phi(s)/\Phi(e)$, ratio of the conventional subcadmium flux and the epithermal flux /ln E; f was calculated from ^{75}Ge and ^{77}Ge measurements by:

$$f = \{[Csp/S \cdot D \cdot M]^{(75)} \cdot [k_0 \cdot \varepsilon \cdot Q(\alpha)]^{(77)} - [Csp/S \cdot D \cdot M]^{(77)} \cdot [k_0 \cdot \varepsilon \cdot Q(\alpha)]^{(75)}/Csp^{(77)}[k_0 \cdot \varepsilon]^{(75)} - Csp^{(75)}[k_0 \cdot \varepsilon]^{(77)}\} \qquad (2)$$

$Q(\alpha) = I(\alpha)/\sigma$, resonance integral to cross-section ratio; $I(\alpha)$ is the resonance integral corrected on the influence of the α factor, according to what suggested by De Corte et al. *(9)*. [Al], [P], = aluminum and phosphorus concentrations. $\sigma(n,p)$ $\sigma(n,\alpha)$ = fission averaged cross-section of the threshold reactions θ, = isotopic abundance of the target nuclide; i, = intensity of the measured radiation; and W, = atomic weight.

RESULTS AND DISCUSSION

The results of the determination of aluminum and phosphorus in two reference materials (Bowen's kale and SRM 1577a bovine liver) are, respectively: Al, mg/kg: 38.8 ± 2.1 and 1.50 ± 0.30 (recommended and indicative values: 39.9 ± 9.9 and 2), and P, g/kg: 4.15 ± 0.25 and 10.6 ± 1.3 (recommended and certified values: 4.48 ± 0.16 and 11.1 ± 0.40). A reasonable agreement is observed between the concentrations found in this work and the certified or informative values.

The precision of the data depends primarily on the statistical dispersion of the measurements: high contents of chlorine and sodium can affect the counting of the short-lived ^{28}Al in the undisturbed spectrum. The detection of ^{72}Ga at the same conditions is sometimes difficult, especially for reactor positions with the high thermal to fast flux ratios. A simple radiochemical method, developed in the course of a previous work *(3)*, can be applied to remove the interference of ^{77}Ge, whenever necessary.

As mentioned before, all the parameters involved in the calculations are known with good precision, so that the final uncertainty is not high. The possible interference of silicon, via the ^{28}Si(n,p)^{28}Al reaction, which can affect the whole analysis process, but mainly the phosphorus assay, should be taken into account. Silicon contents are usually low in biological materials, in comparison with the relatively high concentrations of phosphorus. Nevertheless, the eventual influence of silicon on phosphorus determination should be specifically controlled. Some other threshold reactions induced on silicon, e.g., ^{29}Si(n,p)^{29}Al, could be used for that purpose.

We can also conclude that ^{75}Ge can be used as comparator for the determination of other elements of biological interest, namely sodium, magnesium, chlorine, potasium, calcium, and manganese. The irradiations performed in the undisturbed reactor neutron spectrum have also been employed for assay of these elements, including a second measurement of the samples. It goes beyond the scope of this article to formulate comments about the results and the certified or recommended values for the reference materials, but it should be mentioned that the agreement gives additional evidence of the usefulness of germanium as flux monitor and comparator.

ACKNOWLEDGMENT

The authors wish to thank M. A. Arribere for her valuable help in many aspects of this work.

REFERENCES

1. I. M. Cohen, J. C. Furnari, A. J. Kestelman, and M. A. Arribere, *J. Radioanal. Nucl. Chem., Articles* to be submitted
2. F. De Corte, A. Simonits, F. Bellemans, M. C. Freitas, S. Jovanovic, B. Smodis, G. Erdtmann, H. Petri, and A. De Wislapelaere, *J. Radioanal. Nucl. Chem., Articles* **169,** 125 (1993).
3. J. C. Furnari, I. M. Cohen and A. J. Kestelman, *Appl. Rad. Isotopes* to be published.

4. F. De Corte and A. Simonits, *J. Radioanal. Nucl. Chem., Articles* **133,** 43 (1989).
5. A. Calamand, in *Handbook on Nuclear Activation Cross-Sections*, IAEA Technical Report Series No. 156, 1974, p. 273.
6. E. M. Gryntakis, *J. Radioanal. Chem.* **52,** 219 (1979).
7. I. M. Cohen, M. A. Arribere, A. J. Kestelman, and J. C. Furnari, to be presented at the *21th Meeting of the Asociación Argentina de Tecnoligía Nuclear*. Mar del Plata, Argentina, November 8–12, 1993.
8. A. Simonits and F. De Corte, *J. Radioanal. Chem.* **24,** 31 (1975).
9. F. De Corte, L. Moens, A. Simonits, and A. De Wispelaere, *J. Radioanal. Chem.* **52,** 295 (1979).

Effect of Enhanced Bromide Intake on the Concentration Ratio I/Br in the Rat Thyroid Gland

MILOSLAV VOBECKÝ* AND ARNOŠT BABICKÝ

Academy of Sciences of Czech Republic, Institute of Analytical Chemistry, 142 20 Praha 4, Czech Republic

ABSTRACT

Interaction of bromine with iodine was studied in the rat thyroid gland under the conditions of different bromide intake. Bromine and iodine in the thyroid dry weight were determined by instrumental neutron activation analysis (INAA). It was found that with increased bromide intake the bromine concentration in the thyroid gland increased with simultaneous decrease in the iodine concentration. The change in the I/Br concentration ratio depends on a number of halogen binding positions and on the bromide supply. The I/Br parameter reacts sensitively to the changes of bromide intake already in the region of low bromine concentration levels.

Index Entries: INAA; bromine; iodine; rat thyroid gland; I/Br concentration ratio.

INTRODUCTION

Bromine is one of the pollutants of anthropogenic origin. Its main sources are emissions from coal-fired plants and exhausts from the combustion of leaded gasoline. Bromine also enters the environment in a number of compounds used as pesticides (fumigation of food and tobacco), pigments, medicines, disinfectants, preparations for photographic technologies, as working liquids in hydraulic and fire-extinguishing systems, and so forth. The contamination from the coal combustion shows global impact, whereas the contamination from the exhaust gases has local character. Increased burdening of living organisms with bromine as a consequence of higher levels of the element in the environment can evoke various processes. One of the organs in which an interaction of

*Author to whom all correspondence and reprint requests should be addressed.

Biological Trace Element Research Editor: G. N. Schrauzer © 1994 by Humana Press Inc.

chemically allied halogens could occur is the thyroid gland. It may be presumed that increased bromine intake is capable of influencing the composition of the thyroid gland hormones by producing their bromine analogs. The aim of this study was to find out experimentally how the content of bromine and iodine changed in the rat thyroid gland during increased bromine intake.

EXPERIMENTAL

Instrumental Neutron Activation Analysis

Iodine was determined by short-term irradiation by means of the radionuclide ^{128}I (half-life 25.0 min) and bromine by long-term irradiation by means of the radionuclide ^{82}Br (half-life 35.3 h). The activation of samples and elemental standards was carried out in the nuclear reactor LWR-15 (Nuclear Research Institute, Řež) at the power of 8 MW. The determination of iodine was performed by using the pneumatic irradiation facility in the working regime: irradiation time 1 min, decay time 4 or 15 min, measurement time 10 min. The long-term activation of these samples was carried out 10 d after the short-term activation by 10 h of irradiation in the reactor core. Computerized spectrometric systems equipped with HPGe detectors (Ortec—efficiency 11% and Canberra—efficiency 20%) were used for γ-spectrometric measurements. Accuracy of determination was assured by means of reference materials (IAEA-V-8 Rye Flower and NIST-SRM 1549 Milk Powder).

Exposition of Rats to Bromide

Two series of male rats (Wistar, breed Inst. of Physiol. Acad. Sci. of Czech Rep.) at the age of 21 d were used. Each series consisted of four groups of six animals. Three groups drank water containing KBr (AR, Lachema) with Br concentrations of 10, 50, and 100 µg/g, respectively, for 15 or 57 d. The fourth group in each series served as a control group and drank bidistilled water. Water and standard pelletized diet were available to all groups *ad libitum*. The bromine content in the diet was 4.7 µg/g (diet A) and 1.7 µg/g (diet B); the iodine content in both diets was lower than 0.8 µg/g. At the end of the experiment, the animals were sacrificed by the overdose of diethylether narcosis; the thyroid glands were collected, lyophilized, weighed, and sealed in polyethylene foils.

RESULTS AND DISCUSSION

In the thyroid gland tissue, iodine is above all present in the molecules of thyronine in positions 3, 5, 3′, or 5′. With respect to the chemical relationship of iodine and bromine, it may be assumed that bromine analogs of the abovementioned compounds also arise during increased intake of bromine. From this presumption, it follows that the number of iodine and bromine atoms in the studied sample of the thyroid gland also represents the number of substituted and/or substitution-capable positions with these halogens in the thyronine components. The evaluation of the degree of the iodine substitution in the rat thyroid gland by bromine necessitates employment of such an analytical method capable of sufficiently accurate determination of both elements in the same sample, the mass of which amounts to only a few milligrams. Application of inadequate chemical analyses in studies of the bromine–thyroid gland relationship was already pointed out by Wolff *(1)*. With respect to usual concentrations of bromine and iodine in the rat thyroid gland and to the radiochemical character of these elements, nondestructive method of INAA was therefore used.

In addition to the lines of dominant activities of ^{128}I, ^{24}Na, and ^{38}Cl, the γ-ray spectrum of the short-term activated samples of the thyroid gland also shows the lines of the radionuclides ^{42}K, ^{41}Ar, and (^{80}Br), which are listed in Table 1. In this way, it is possible to determine iodine, chlorine, sodium, and potassium in the samples of rat thyroid gland with mass of a few milligrams (dry wt of the whole organ). However, it is not possible to determine bromine in concentrations lower than 20 µg/g because of high background below the photopeak of ^{80}Br formed mainly by the ^{24}Na and ^{38}Cl activity. Since activation by epithermal neutrons, which would improve the Br signal-to-background ratio, was not feasible, long-term activation (10 h) of the same sample was chosen for the bromine determination. The measurement of these samples started 3 d after the irradiation.

The bromine concentration in drinking water was chosen in such a way that the intake of bromine from the water with the lowest bromine concentration was comparable to the bromine intake from the diet. The highest dose of bromine was not to influence the growth of the animals. At the termination of experiments, the whole daily intake of bromine may be estimated as follows: from the diets A and B 0.09 and 0.03 mg of Br, respectively, from the drinking water with the Br concentrations of 0.3, 1.5, and 3.0 mg. In our experimental conditions, no differences were found in the weight increments between the control and experimental group A. Differences in the weight increments were also not found by Krari et al. *(3)* for the dose of 5 µmol of NaBr/kg/d, which corresponds with our groups drinking water with the Br concentration of 10 µg/g (Tables 2 and 3). On the other hand, Van Leeuwen et al. *(4)* found a 17% decrease in the body weight of rats that were given a diet containing 19 g NaBr/kg diet for 2 wk. However, the intake of bromine from this diet exceeds the bromine intake of our most exposed groups by two orders of magnitude.

The results of bromine and iodine determination in the rat thyroid gland are given in Tables 2 and 3 for all four groups ($n \approx 6$ in each group) as the I/Br concentration ratios (weight and molar). The animals that were exposed for 15 d received diet A, whereas the animals of the second experimental series, which were exposed for 57 d also received diet A at the beginning, but during the last 30 d they received diet B. From the results given in the Tables 2 and 3, it follows that with increasing bro-

Table 1
γ Lines of the Short-Term Activated Rat Thyroid Gland (Irradiation Time 1 Min, Decay Time 4 or 15 Min, Measurement 10 Min), Spectroscopic Data According to *(2)*

Radionuclide	Half-life	Gamma line	
		Energy, keV	Intensity, %
^{24}Na	15.02 h	1368.5	100
		2753.9	99.9
^{38}Cl	37.24 min	1642.4	31.6
		2167.6	42.4
^{41}Ar	1.83 h	1293.7	99.1
^{42}K	12.36 h	1524.6	18.3
^{80}Br	17.4 min	616.2	6.7
^{128}I	25.0 min	442.9	16.0
		526.6	1.50
		743.6	0.147
		969.5	0.384

mide intake, the concentration of bromine in the thyroid gland increases with simultaneous decrease of the iodine content, which is no more distinctly seen in the decrease of the concentration ratio I/Br. This indicates that substitution of iodine by bromine in the thyronine molecules occurs. The difference in the I/Br values between the control groups (cf Tables 2 and 3) refers to the sensitivity and dynamics of the I/Br ratio response in the region of low bromide intake. No differences were found in the content of chlorine, sodium, and potassium in the thyroid glands between the control and experimental groups.

Assuming that in the thyroid gland hormone there exists a definite number of positions for the location of halogens (iodine, bromine) and

Table 2
Content of Bromine and Iodine in the Rat Thyroid Gland, Intake of Drinking Water for 15 D, Diet A (c_{Br}4.7 μg/g Br)*

Concentration of bromine in drinking water (μg g^{-1})	Concentration in the thyroid gland (μg g^{-1})		Concentration ratio I/Br	
	I	Br	Weight	Molar
0	2475±993	27.6± 3.2	89.7±35.4	56.5±22.3
10	2236±655	74.1±12.7	30.2±10.2	19.0± 6.4
50	2154±478	321.1±62.2	6.7± 1.9	4.2± 1.2
100	1985±495	589.1±41.5	3.4± 0.9	2.1± 0.6

*The values of concentration are given as arithmetic means ± SD (n = 6). Molar ratio I/Br = weight ratio I/Br multiplied by a factor of 79.904/126.9045.

Table 3
Content of Bromine and Iodine in the Rat Thyroid Gland, Intake of Drinking Water for 57 D*

Concentration of bromine in drinking water (μg g^{-1})	Concentration in the thyroid gland (μg g^{-1})		Concentration ratio I/Br	
	I	Br	Weight	Molar
0	2540±514	14.5± 2.0	175.2±42.8	110.3±26.9
10	2423±514	159.3±14.0	15.2± 3.5	9.6± 2.2
50	1796±247	284.5±84.3	6.3± 2.0	4.0± 1.3
100	2143±412	432.4±80.0	5.0± 1.3	3.1± 0.8

*The values of concentration are given as arithmetic means ± SD (n = 6). Molar ratio I/Br. Cf Table 2.

that this number is constant, the number of iodine atoms can be expressed by the relation:

$$N_I = k - N_{Br} \quad (1)$$

where k is the number of binding positions for halogens (I,Br) and N_{Br} is the number of bromine atoms.

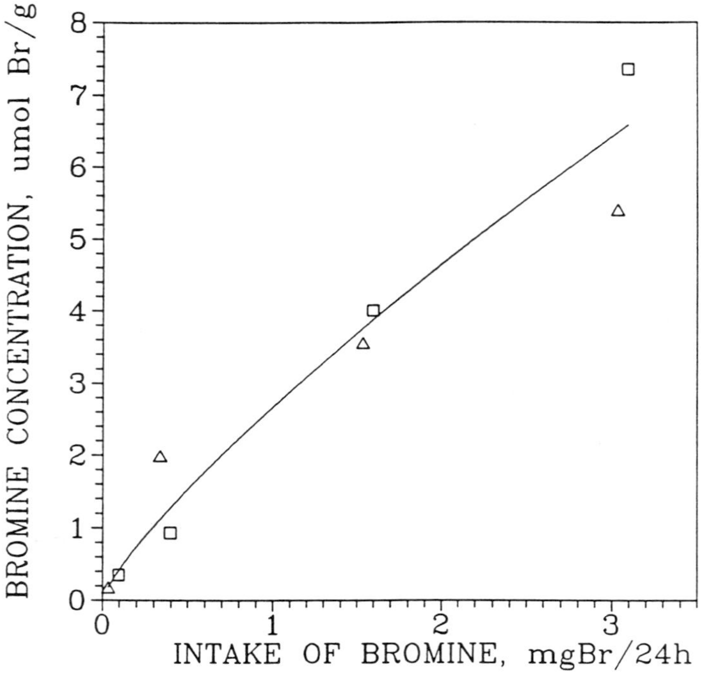

Fig. 1. Changes in the bromine concentration in the rat thyroid gland (μmol Br/g dry wt) as a function of the bromine intake (mg Br/24 h). □ Values from the first experimental series (Table 2). △ Values from the second experimental series (Table 3).

Then the following ratio applies:

$$I/Br = (k/N_{Br}) - 1 \tag{2}$$

From this equation, it follows that the ratio I/Br depends on the bromine supply and on the number of substitution-capable positions for the given halogens. This number, i.e., the concentration of iodine at the zero intake of bromine, corresponds to the constant k. Under our experimental conditions, the iodine concentration found in the thyroid dry wt of the control group of the second experimental series (Table 3) approaches the constant k. Thus, k = 2540 μg/g expressed in the molar concentration: c_I = 2540/126.9045 ≈ 20. Experimentally determined changes in the bromine concentration in the rat thyroid gland as a function of the bromine intake (Fig. 1) can be expressed by the relation:

$$c_{Br(max)} = 720[1 - \exp(-0.4x)] \tag{3}$$

where x is the bromine intake in mg/24 h. From the Eq. (3), it follows that $c_{Br(max)}$ = 720 μg/g, or when expressed in the molar, $c_{Br(max)}$ = 720/79.904 ≈ 9. The concentration $c_{Br(max)}$ is the limit value of bromine concentration at enormously high intake of bromine, but under condi-

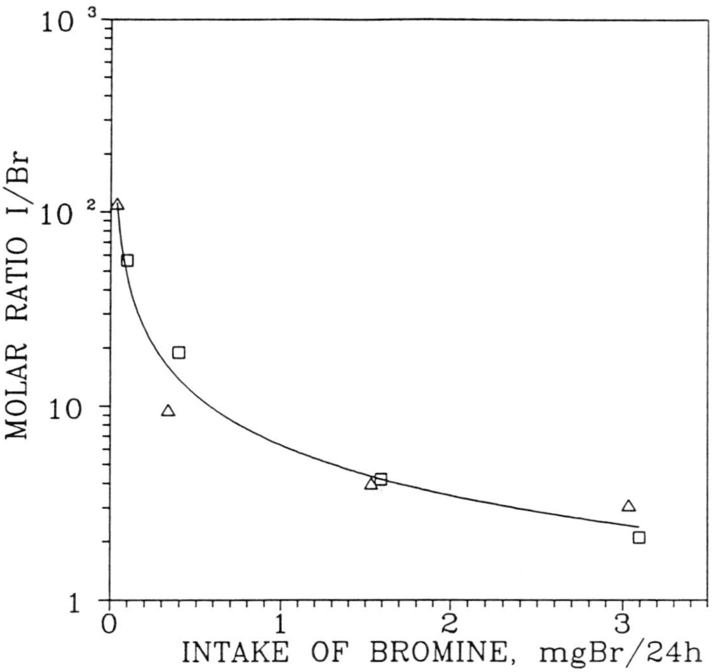

Fig. 2. Changes in the molar concentration ratio I/Br in the rat thyroid gland in dependence on the bromine intake. □ Values from the first experimental series (Table 2). △ Values from the second experimental series (Table 3).

tions of normal iodine intake from the diet. It means that under these conditions, it is not possible to reach lower I/Br ratio than $20/9 \approx 2.22$. The calculated I/Br ratios and experimentally found values are shown in Fig. 2. It follows from the dependence of the I/Br ratio on the bromide intake that the correct criterion for assessment of the studied halogen substitution is the I/Br ratio value, and not the concentration values of both halogens. The latter values can be influenced by a number of factors, e.g., the present state of animals, their nutrition, and so on (cf Table 3). If the I/Br ratio represents the number of substituted positions in thyronine molecules by bromine as mentioned, it will be necessary to investigate the chemical composition of the products of the bromine/iodine interaction in the thyroid gland and the physiological significance of this compounds.

CONCLUSIONS

In the course of increased bromine intake, the bromine concentration in the thyroid gland increases with simultaneous decrease of the iodine concentration. The concentration ratio I/Br in the rat thyroid gland reacts sensitively to the changes in the bromine supply, especially in the region of low bromine intake.

Under the condition of normal (natural) iodine intake the I/Br ratio depends on the number of binding positions for halogens (iodine and bromine) and on the intake of bromine; and even at enormously high intake of bromine, it is not possible to occupy all binding positions in the thyronine molecules reserved for halogens with bromine.

ACKNOWLEDGMENTS

The financial support from the Ministry of Environment of the Czech Republic is gratefully acknowledged. The authors thank Dr. J. Velický from the Institute of Experimental Medicine for preparation of rat thyroid gland.

REFERENCES

1. J. Wolff, *Physiol. Rev.* **44,** 45–90 (1964).
2. U. Reus and W. Westmeier, Atomic Data and Nuclear Data Tables **29,** 193–406 (1983).
3. N. Krari, S. Berre, and P. Allain, *Biol. Trace Element Res.* **32,** 275–279 (1992).
4. F. X. R. Van Leeuwen, R. Hanemaaijer, and J. G. Loeber, *Arch. Toxicol. Suppl.* **12,** 93–97 (1988).

Neutron Activation Analysis of Biological Samples at the Radiochemistry Division of IPEN-CNEN/SP

M. SAIKI,* M. B. A. VASCONCELLOS, V. A. MAIHARA, M. J. A. ARMELIN, D. I. T. FÁVARO, AND S. O. ROGERO

Instituto de Pesquisas Energéticas e Nucleares, IPEN-CNEN/SP Radiochemistry Division, CEP 05422-970, São Paulo, SP, Brazil

ABSTRACT

Neutron activation analysis is a very useful method for determination of a great number of elements in biological samples. At the Radiochemistry Division of the IPEN-CNEN/SP, this method is being extensively applied to study several materials, such as extracts from medicinal plants, human hair, snake venoms, human lungs, foodstuffs, and corn samples. Both instrumental neutron activation analysis (INAA) and radiochemical neutron activation analysis (RNAA) are used to analyze real samples, as well as biological standard reference materials to evaluate the accuracy and precision of the results.

Index Entries: Neutron activation analysis (NAA); medicinal plants; human hair; human lung; snake venoms; foodstuffs.

INTRODUCTION

With the increasing knowledge of the function of elements in living organisms, the determinations of trace elements in biological systems have become of great interest. In this respect, NAA plays an important role owing to its accuracy, precision, sensitivity, multielemental character, and for some elements, speed of analysis.

The purely instrumental mode of NAA based on measurement of γ activities of radionuclides produced by neutron irradiation is widely used for the determination of several elements in biological materials.

*Author to whom all correspondence and reprint requests should be addressed.

Biological Trace Element Research Editor: G. N. Schrauzer © 1994 by Humana Press Inc.

However, for the determination of some elements of interest, such as As, Cr, Hg, Sb, and Se, present at very low concentrations, radiochemical separation is required to eliminate interfering activities and to improve in this way sensitivity and accuracy of analysis.

This article describes the most recent investigations using INAA and RNAA of biological materials performed at the Radiochemistry Division of IPEN-CNEN/SP in collaboration with universities, research institutions, and private medical clinics. Extracts from Brazilian medicinal plants were analyzed as a contribution to studies of their toxicity, medicinal action, or active components *(1)*. Investigations concerning medicinal plants have been of great importance, since their use is widespread in Brazil.

Human head hair was studied as an indicator of nutritional or health status of subjects by determining toxic and essential elements of interest. Total Hg in hair was also determined in order to detect population groups suspected of contamination by this element, mainly by ingestion of contaminated fish *(2)*. Determination of trace elements in snake venoms was carried out to study a role of metals in their toxic action and to characterize snake species *(3)*. Lungs from smokers, nonsmokers, and one stillborn were analyzed to determine normal levels of elements in hilum lymph nodes and pulmonary tissues, as well as to examine correlation between trace elements in these tissues and the smoking habit of the subjects *(4)*. Results of this study are also reported in detail elsewhere in this volume and will not be discussed here *(5)*.

Toxic and other elements were analyzed in Brazilian foodstuffs, such as bread, milk powder, and rice (essential components of the Brazilian diet *[6,7]*), as well as in gelatine samples used by the population for cooking *(8)* and in corns obtained as a result of studies aimed at increasing a nutritional value of the protein contained in this foodstuff *(9)*.

EXPERIMENTAL

Preparation of Synthetic Standards of the Elements

Standard solutions were prepared by dissolving high-purity metals, oxides, or salts of elements with appropriate reagents. Aliquots of multielemental standard solutions were pipeted onto a small sheet of Whatman filter paper and dried in a dessicator at room temperature. In the case of the Hg standard for short irradiation of 1 h, standard solution of this element was pipeted together with thioacetamide solution according to Noguchi et al. *(10)* to avoid losses of Hg during irradiation. The sheets of filter paper were put into polyethylene bags for irradiation. Long irradiations of standard solutions of volatile elements such as Hg, Cr, As, and Se were carried out in the same conditions used for samples, in quartz ampules.

Sample Collection and Preparation

Dried extracts were obtained from the Brazilian medicinal plants: *Cordia Verbenacea DC* (Boraginaceae) with antiinflammatory properties; *Folidago Microglossa DC* (Compositae), known by the name Brazilian Arnica and used as an antiinflammatory and cicatrizing drug; and *Petiveria Alliacea* (Phytolaccaceae), known by the name "Guiné" used in popular medicine "to calm nerves" and as an anesthetic, diuretic, and analgesic drug, too. Fresh materials of these plants were extracted using a 70% alcoholic solution at room temperature. The dark green solution was filtrated and concentrated under reduced pressure to obtain aqueous viscous material that was dried afterward.

Hair samples were collected from patients of a medical clinic and from control group formed by friends, relatives, and students. Hair strands were cut close to the scalp of the occipital region, and the length did not exceed 5 cm. In the laboratory, hair filaments were cut in lengths, smaller than 2 mm, and then they were washed with a solution of nonionic detergent (Triton X 100), acetone, and water *(11)*.

Also the procedure recommended by the IAEA *(12)* was used for washing of hair collected for total Hg determination in two population groups in Brazil that could be at risk with respect to Hg contamination. The two population groups selected for this work were: Indian tribes that consume fish from rivers located in Xingu Park (Amazonic region) and the population that frequently consume fish from the Billing Dam in São Paulo, which is situated near chlor-alkali and other industries.

Snake venoms were obtained at the Instituto Butantan-SP by squeezing the glands and collecting the liquids in Petri dishes. Then the samples were freeze-dried. Special care was taken to avoid their contamination before irradiation, as well to prevent breathing the venoms in or to prevent their contact with the skin. Snake venoms were analyzed from the following species: *Bothrops jararacussu, Crotalus durissus terrificus,* and *Bothrops jararaca.*

Foodstuffs (bread, milk, rice, and gelatine) were collected in the supermarkets of São Paulo city. Corn samples provided by Agronomy School, Piracicaba-SP and by Brazilian Company for Agropecuary Research (EMBRAPA) were sampled by quartering process, rinsed with water and dried at 50°C for 72h.

Irradiation and Counting

Irradiations have been carried out at the IEA-R1 swimming pool-type reactor of IPEN-CNEN/SP in a thermal neutron flux that ranged from 10^{11} to 10^{13} $n/cm^2/s$. Time of irradiation varied from 5 min to 16 h. Samples and standards were counted in close geometry using two counting systems: EG & G Ortec Model GEM 20190 Ge detector or ENERTEC hyperpure Ge detector coupled to an EG & G Ortec 4096 channel pulse height analyzer and EG & G Ortec Model GMX20190 Ge detector coupled to an ADCAM 918A multichannel buffer. Analyses of γ-ray spectra were carried out by using appropriate computer programs.

Instrumental and Radiochemical Neutron Activation Analyses

INAA was applied to analyze several elements in medicinal plant extracts, human head hair, snake venoms, human lung samples, and foodstuffs. To determine As, Cr, Hg, Sb, and Se at very low concentrations, several radiochemical procedures were used after wet ashing of the irradiated samples.

Hg and Se in Brazilian rice grains were determined using distillation in HBr medium and subsequent precipitation of Se by sodium metabisulfite and Hg by thiocetamide *(6,7)*. Inorganic exchanger (HAP) was also used in order to eliminate interferences from ^{24}Na and ^{42}K radioisotopes *(6)*. For determination in gelatine, a radiochemical separation procedure based on solvent extraction of Cr(VI) into tribenzylamine/chloroform solution *(8)* was used.

RNAA using inorganic exchangers (TDO and HMD), anionic resin (Dowex 2X8), and extraction of diethyldithiocarbamates were used for analysis of biological reference materials *(13)*.

RESULTS AND DISCUSSION

In medicinal plant extracts, concentration of the elements Ca, Cl, Mg, and Na was in the percent range, the elements Al, Br, Mn, Rb, and Zn were present at the ppm level, and the elements Co, Cs, and Sb at the ppb level *(1)*. Concentrations of Cl found in extracts from aerial parts were almost the same as those determined in extracts from roots. This fact was not expected, since Cl was found in a higher concentration in leaves than in roots. Extracts analyzed in this article exhibited high concentrations of K and Mg. It is important to note that potassium salts actively participate in constipation and the cardiac rhythm, whereas Mg and Ca can neutralize the acidity and avoid side effects of stomach lesions.

Results of hair analyses from patients of a medical clinic and control group are presented in Table 1. A comparison of geometric means indicates no significant difference between the two groups for concentration of the elements As, Br, Ca, Cr, Cu, Fe, Hg, K, Mg, Mn, and Zn. Nevertheless, the highest concentrations of As, Br, Cl, Cu, K, Mn, and Se as well as the highest variability of results for Cu, Cd, and K were found in the patient group. These hair analyses were used by physicians from the orthomolecular medicine field, together with other clinical exams, to determine how the element variations may contribute to health problems of the patients.

Table 2 presents results obtained for Hg determination of the group of Indians living in the Xingu Park, the group of people living near the

Table 1
Elemental Concentrations in Hair Samples from Patient and Control Groups Obtained by INAA (in ppm Unless Otherwise Indicated)

Element	Patients			Controls		
	Range	X_G x : s_G	n	Range	X_G x : s_G	n
As, ppb	5.5 — 125	27.9 x : 2.0	22	10 — 40	21.8 x : 1.5	14
Br	1.0 — 15.6	3.3 x : 2.1	24	0.46 — 6.04	2.4 x : 2.2	14
Ca	87 — 1820	445 x : 2.5	21	140 — 1788	412 x : 1.9	14
Cd, ppb	0.1 — 590	63 x : 25	6	74 — 800	263 x : 2.4	7
Cl	40 — 2080	463 x : 3.3	24	62.5 — 1339	263.4 x : 2.9	14
Cr, ppb	46 — 256	95 x : 2.3	24	69 — 692	149.9 x : 1.8	13
Cu	0.4 — 168	15.8 x : 3.5	21	4.0 — 50	19.8 x : 2.3	14
Fe	6.5 — 39	14.3 x : 1.7	23	7.9 — 30.0	13.9 x : 1.4	14
Hg	0.17 — 4.37	0.93 x : 2.7	22	0.46 — 4.8	1.3 x : 2.3	14
K	0.5 — 200	6.2 x : 4.3	23	0.53 — 15.4	2.5 x : 2.4	12
Mg	18 — 231	81.3 x : 2.3	9	21 — 251	60.1 x : 2.3	12
Mn	0.13 — 1.9	0.43 x : 2.1	24	0.125 — 0.670	0.4 x : 2.2	14
Na	1.0 — 95	6.7 x : 3.7	24	2.1 — 7.0	3.3 x : 1.5	14
Sb, ppb	6.3 — 120	24.0 x : 2.2	24	11.3 — 125	32.6 x : 2.5	14
Se	0.20 — 150	0.5 x : 3.9	24	0.034 — 9.03	0.47 x : 3.7	13
Zn	102 — 224	160x : 1.2	24	109 — 264	158 x : 1.2	14

X_G x:S_G = Geometric mean and standard deviation.
n = Number of individuals.

Table 2
Summary of the Results Obtained for Mercury Contents in the Hair of the Population Groups Studied *(2)* (in ppm)

Population Group	X + - s	Median	X x : sG	Range
Controls (n = 25)	1.06 + - 0.55	0.96	0.93 x : 1.71	0.20 - 2.5
Indians from Xingu Park (n = 27)	18.5 + - 5.9	18.0	17.6 x : 1.38	6.9 - 34
Population from the Billing Dam (n = 28)	0.88 + - 0.68	0.74	0.71 x : 1.85	0.30 - 3.0

Billing Dam, and for the control group. A comparison of these results shows highly elevated Hg contents in the Indian group, with the Hg mean concentration about 20 times higher than the mean of the control group. Table 3 presents the concentration ranges of the elements found in snake venom analyses, and shows significantly higher Zn concentra-

Table 3
Concentration Range of Elements Obtained in Brazilian Snake Venoms by INAA

Element	*Bothrops jararacussu* (n = 4)	*Crotalus durissus terrificus* (n = 4)	*Bothrops jararaca* (n = 4)
Br, ppm	4.24 — 19.45	2.3 — 8.6	1.3 — 13.6
Ca, ppm	625 — 1661	325 — 441	919 — 1052
Cl, ppm	1184 — 3906	505 — 1256	368 — 768
Cs, ppb	101 — 319	17 — 66	7.8 — 34.6
K, %	1.00 — 1.15	< 0.3	0.04 — 2.62
Mg, ppm	< 970	3340 — 3830	< 970
Na, %	0.763 — 0.947	2.04 — 2.17	1.28 — 2.31
Rb, ppm	56.2 — 82.1	5 — 12	4.4 — 26.3
Sb, ppb	91 — 472	5 — 29	24 — 2296
Se, ppm	1.66 — 3.55	3 — 6	1.83 — 3.57
Zn, ppm	537 — 678	96 — 142	587 — 803

n = Number of venom pool samples analyzed for each specie.

tions in *Bothrops* genera venoms than in venoms of *Crotalus durissus terrificus*.

Results obtained by INAA method in some foodstuffs, as an example, are reported in Table 4. Corn grains studies showed that the tryptophan content in the improved corn samples (EMBRAPA: BR451) was twice as high as in normal samples, whereas varying element concentration changes were found mainly for the elements Br, Ca, and I.

Results obtained for gelatine, bread, rice, and milk powder samples by employing various radiochemical procedures are presented in Table 5. The levels of toxic elements Hg, As, and Se found in rice grains were below the values established by the national legislation *(7)*.

In order to evaluate the accuracy and precision of INAA and RNAA results, several biological reference materials were analyzed. These results are given in Table 6. By comparing our results with literature data, it can be concluded that these analyses performed are accurate and precise within the limits expected for trace element analyses.

ACKNOWLEDGMENT

The authors are grateful to FAPESP and CNPq from Brazil, and to the IAEA for financial support.

Table 4
Elemental Concentrations in Brazilian Foodstuffs Determined by INAA
(µg/g dry wt Unless Otherwise Indicated)

Element	Industrialized Bread	Milk Powder	Rice		Corn	
			Polished Grain (1)	Parboiled Grain (2)	ESALQ: UD-2 (normal)	EMBRAPA: BR 451 (white)
As	-	-	0.093 +- 0.014	0.310 +- 0.021	-	-
Br	10.9 +- 0.3	14.7 +- 0.4	0.43 +- 0.04	0.33 +- 0.02	15.2 +- 2.9	0.95 +- 0.06
Ca	-	10200 +- 1100	-	-	51 +- 3	83 +- 6
Cl	10400 +- 600	7600 +- 500	-	-	761 +- 77	716 +- 119
Co, µg/kg	-	-	25.4 +- 5.7	27.9 +- 4.3	-	-
Cr	0.30 +- 0.01	0.21 +- 0.03		-	-	-
Cu				-	13 +- 4	17 +- 4
Fe	75 +- 3	-	7.58 +- 0.83	12.6 +- 1.2	22 +- 3	21 +- 4
I, µg/kg	-	-	-	-	482 +- 69	210 +- 53
K	-	11500 +- 200	654 +- 41	1079 +- 58	3.3 +- 0.2	3.9 +- 0.2
Mg, µg/kg	-	1040 +- 25	-	-	1.03 +- 0.04	1.30 +- 0.10
Mn	10.2 +- 0.8	0.52 +- 0.04	-	-	6.6 +- 0.7	5.6 +- 0.4
Na	7700 +- 200	3230 +- 40	12.3 +- 1.6	16.2 +- 0.7	3.8 +- 0.5	2.7 +- 0.5
Rb	3.4 +- 0.4	24 +- 2	3.81 +- 0.22	11.0 +- 0.7	6.5 +- 1.1	7.8 +- 0.2
S	-	-	-	-	2.0 +- 0.3	2.0 +- 0.4
Sb	0.90 +- 0.01	1.07 +- 0.06	-	-	-	-
Sc, µg/kg	2.35 +- 0.56	-	2.16 +- 0.08	1.11 +- 0.08	-	-
U, µg/kg	-	-	-	-	23 +- 4	16 +- 3
Zn	14.9 +- 0.6	24 +- 3	16.4 +- 0.4	7.61 +- 0.66	17 +- 1	29 +- 1

(1) Long Fine 2, Cerejeira Brand; (2) Long Fine, Mingote Brand.

Table 5
Determination of Cr, Hg, Sb, and Se in Foofstuffs by RNAA Method *(6–8)*
Results in µg/kg of Dried Material

Sample	Element	Rad. Proc. *	Mean +- s
Rice - Polished Grains	Hg	1	14.70; 18.89 (a)
	Se	1	34.70; 18.67 (a)
Rice - Parboiled Grains	Hg	1	5.20; 2.67 (a)
	Se	1	< 24
Natural Gelatine	Cr	2	478 +- 30
Flavoured Gelatine	Cr	2	85 +- 10
Industrialized Bread	Sb	3	1200 (a)
Milk Powder	Sb	3	910 (a)

* Radiochemical procedure: 1. distillation + precipitation; 2. solvent extraction with tribenzylamine in chloroform; 3. retention of interferents in HAP.

(a). Individual results

Table 6
Elemental Concentrations in Reference Materials Obtained by INAA and RNAA (in μg/g dry wt Dried Material)

Element	Reference Materials	This Work	Literature Values (14 – 16)
Al	NIST Citrus Leaves	137 +- 16	92 +- 15
As	Bowen's Kale	0.128 +- 0.014 (*)	0.131 +- 0.045
	SHINR-HH Human Hair	0.56 +- 0.06	0.59 +- 0.06
	NIST Citrus Leaves	2.97 +- 0.16	3.1 +- 0.3
	NIST Citrus Leaves	3.17 +- 0.54 (*)	3.1 +- 0.3
Br	NIST Bovine Liver	9.03 +- 0.43	9
	Bowen's Kale	24.2 +- 2.2	24.9 +- 2.5
Ca	SHINR-HH Human Hair	856 +- 94	790 +- 72
	Bowen's Kale	40890 +- 1890	41060 +- 2217
Cl	NIST Bovine Liver	2769 +- 141	2800 +- 98
	Bowen's Kale	3571 +- 173	3560 +- 427
Co	NIST Citrus Leaves	0.0211 +- 0.0029	0.020
	SHINR-HH Human Hair	0.144 +- 0.007	0.135 +- 0.008
Cr	Mixed Human Diet	0.110 +- 0.040 (*)	0.150 +- 0.040
	Mixed Human Diet	0.117 +- 0.007 (*)	0.150 +- 0.040
	NIST Citrus Leaves	0.74 +- 0.07	0.8 +- 0.2
	NIST Citrus Leaves	0.813 +- 0.020 (*)	0.8 +- 0.2
	NIST Citrus Leaves	0.820 +- 0.150 (*)	0.8 +- 0.2
	SHINR-HH Human Hair	4.36 +- 0.75	4.77 +- 0.38
Cs	Bowen's Kale	0.085 +- 0.004	0.0763 +- 0.0058
	NIST Citrus Leaves	0.097 +- 0.003	0.098
Cu	NIES 10A Rice Flour	4.6 +- 1.2	3.5 +- 0.3
	SHINR-HH Human Hair	22.9 +- 7.1	23.0 +- 1.4
Fe	NIST Citrus Leaves	84 +- 2	90 +- 10
	NIST Bovine Liver	193 +- 13	194 +- 19
	SHINR-HH Human Hair	69.1 +- 6.2	71.2 +- 6.6
I	NIST Citrus Leaves	2.2 +- 0.5	1.84 +- 0.03
Hg	NIST Bovine Liver	0.0054 +- 0.0014 (*)	0.004 +- 0.002
	Mixed Human Diet	0.0055 +- 0.0004 (*)	0.0048 +- 0.0014
	NIST Citrus Leaves	0.084 +- 0.004 (*)	0.080 +- 0.020
	Bowen's Kale	0.136 +- 0.022 (*)	0.171 +- 0.028
	Bowen's Kale	0.169 +- 0.019 (*)	0.171 +- 0.028
	SHINR-HH Human Hair	2.16 +- 0.19	2.16 +- 0.21
	NIES No.5 Human Hair	4.25 +- 0.24	4.4 +- 0.4
K	NIST Citrus Leaves	18000 +- 9000	18200 +- 600
	NIST Bovine Liver	10200 +- 700	9960 +- 70
	Bowen's Kale	24140 +- 1480	24370 +- 1462
La	Bowen's Kale	87 +- 20	86.4 +- 13.0
Mg	NIST Citrus Leaves	6227 +- 177	5800 +- 302
	Bowen's Kale	1645 +- 49	1605 +- 176
Mn	SHINR-HH Human Hair	2.86 +- 0.14	2.94 +- 0.20
	NIST Bovine Liver	10.7 +- 0.8	9.9 +- 0.8
Na	NIES No.5 Human Hair	25 +- 2	26 +- 1
	Bowen's Kale	2127 +- 157	2360 +- 284
S	NIST Citrus Leaves	4.8 +- 1.3	4.07 +- 0.09
Sb	NIST Citrus Leaves	0.036 +- 0.006 (*)	0.040
	Bowens Kale	0.068 +- 0.006 (*)	0.0685 +- 0.0144
	Bowen's Kale	0.064 +- 0.008	0.0685 +- 0.0144
	SHINR-HH Human Hair	0.248 +- 0.028	0.210
Sc	NIST Citrus Leaves	0.0119 +- 0.0010	0.010
	Bowen's Kale	0.0152 +- 0.008	0.0948
Se	Bowen's Kale	0.123 +- 0.020 (*)	0.134 +- 0.020
	Bowen's Kale	0.147 +- 0.025	0.134 +- 0.020
	NIST Bovine Liver	0.638 +- 0.075 (*)	0.71 +- 0.07
	NIST Bovine Liver	0.72 +- 0.07	0.71 +- 0.07
Th	Bowen's Kale	0.0122 +- 0.0022	0.0104
Zn	NIST Citrus Leaves	30 +- 2	29 +- 2
	Bowen's Kale	32.9 +- 1.5	32.3 +- 2.7

Results in μg/g of dried material

* Results obtained by RNAA method.

REFERENCES

1. M. Saiki, M. B. A. Vasconcellos, and J. A. A. Sertié, *Biol. Trace Element Res.* **26–27,** 743–750 (1990).
2. M. B. A. Vasconcellos, M. Saiki, G. Palleti, R. M. M. Pinheiro, R. G. Baruzzi, and R. Spindel, *J. Radioanl. Nucl. Chem., Articles,* **779,** 369–376 (1994).
3. M. Saiki, M. B. A. Vasconcellos, J. R. Rogero, and M. C. G. Cruz, *J. Radioanal. Nucl. Chem., Articles* **151,** 271–276 (1991).
4. S. O. Rogero, Master's dissertation, Instituto de Pesquisas Energéticas e Nucleares, São Paulo, 1992.
5. S. O. Rogero and M. Saiki, this vol.
6. V. A. Maihara and M. B. A. Vasconcellos, Publicaçäo IPEN 220 São Paulo, Oct/1988.
7. V. A. Maihara and M. B. A. Vasconcellos, *J. Radioanal. Nucl. Chem., Articles* **132,** 329–337 (1989).
8. M. Saiki, *Anal. Sci.* **Supplement 7,** 825–827 (1991).
9. M. J. A. Armelin, V. A. Maihara, M. B. A. Vasconcellos, D. I. T. Favaro, and V. F. Nascimento, *J. Radioanal. Nucl. Chem., Lett.* **164,** 265–274 (1992).
10. K. Noguchi, M. Shimizu, and K. Moriwaki, *Radioisotopes* **29,** 221 (1980).
11. B. L. Smith, Doctor's Data Inc., Chicago, IL.
12. Reference Methods for Marine Pollution Studies, n. 46 (draft) October 1987.
13. M. B. A. Vasconcellos, V. A. Maihara, D. I. T. Favaro, M. J. A. Armelin, E. C. Toro and R. Ogris, *J. Radioanal. Nucl. Chem., Lett.* **153,** 185–199 (1991).
14. E. Cortes Toro, R. M. Parr and S. A. Clements, IAEA/RL/128 (Rev. 1) Vienna, 1990.
15. Y. Muramatsu and R. M. Parr, IAEA/RL/128, Vienna, 1985.
16. R. M. Parr. *J. Radioanal. Nucl. Chem.* **123,** 259 (1988).

Quality Assurance and Comparison with Other Methods

Some Improvements in the Quality of NAA Procedures and the Reliability of the Results

A. R. BYRNE

Department of Nuclear Chemistry, "J. Stefan" Institute, 61000 Ljubljana, Slovenia

ABSTRACT

After considering the need for quality control in NAA, the concept of quality in NAA procedures themselves is discussed, and some important factors identified. Two approaches to improve quality are then described in more detail. The first concerns the unique ability of NAA using different isotopic reactions and different modes (INAA/RNAA) to provide independent data sets in the same laboratory, thus allowing internal validation or crosschecking. The second discusses the need for chemical yield measurements in RNAA and the advantages of the radioisotopic tracer technique. Some recent advances and further possibilities for this use of tracers are listed.

Index Entries: Neutron activation analysis; radiochemical NAA; quality control; independent data; crosschecking; chemical yields; radioisotopic tracers.

INTRODUCTION

An important practical point to be considered in relation to the quality of results from neutron activation analysis (NAA) is any *a priori* decision taken before analysis on the degree of quality required or affordable for reasons of cost. Because of advances in the sensitivity, accuracy, and multielement character of competing techniques, there is increasing commercial pressure on NAA laboratories to provide determinations in a suboptimum manner as concerns the quality of the procedure and the quality-assurance (QA) measures. Are we justified in cutting corners, e.g.

by omitting chemical yield measurements or making fewer (or no) quality-control (QC) analyses, for economic reasons? Perhaps the best answer to this is to stress rather that the best field of application of NAA and especially radiochemical NAA (RNAA) is in the "quality" market, where all the inherent advantages of the method may be realized—at a price. This point and some related questions were recently discussed by de Goeij and Woittiez *(1)* Dams *(2)*, and de Bruin *(3)*.

It is common to draw attention to the gap between the intrinsic advantages of NAA and its actual performance in trace analysis, as illustrated by the results of some intercomparisons, such as those organized by the IAEA. Perhaps it is now better appreciated that part, at least, of the problem lay in the nature of the intercomparisons themselves, such as the different backgrounds and experiences of the participants, the lack of commercial or competitive stimuli, the treatment of systematic errors only by statistical tools, and the absence of an in-built system of QC in the intercomparison. This is by no means to decry the usefulness of such exercises, which indeed lead to a timely and widespread reappraisal of the magnitude and sources of errors and their detection in NAA.

The question of the quality of data generated by NAA and the QC measures necessary to achieve acceptable results and exploit the inherent QA aspects of the method, has been under vigorous discussion. An IAEA Advisory Group in 1984 made a series of recommendations on QA in biomedical NAA *(4)*, and a detailed discussion of the points at issue can be found in the full report *(5)*. The general aspects of QA in NAA were also recently discussed by Heydorn *(6)*, a tireless worker in this field.

It is clear that in the more demanding types of trace determinations, exemplified by reference material (RM) certification, NAA has proven to be an essential contributor and has often been the dominant technique. The reasons for this and suitability of NAA for RM homogeneity studies and certification have recently been considered by Becker *(7)* and the present author *(8)*.

In view of space considerations, only two or three aspects relating to quality can be considered here. One of these is the exploitation of the unique possibilities offered in NAA for internal self-validation of results by the use of independent isotopic reactions and/or the use of both instrumental and radiochemical modes, in order to crosscheck results. Another concerns determination of the chemical yield in RNAA, which in quality analysis should be done for every sample aliquot, where the use of radioisotopic tracers is especially suitable. Some additional possibilities in this respect are discussed.

QUALITY IN RNAA PROCEDURES

The inherent quality of an NAA and especially an RNAA procedure is a characteristic that should be distinguished from QC or the QA of the whole measurement process. A poor procedure will always give poor results; the checks on quality applied will merely confirm this to be so more definitely, and will not by themselves improve the procedure or the results, except in so far as they better define their uncertainty and bias. Hence, QC must be used in a positive feedback manner to improve the procedure itself.

Some of the elements of quality in an RNAA procedure may be listed (a more comprehensive discussion of many of these can be found in de Goeij *(9)*):

1. Early achievement of radiochemical equilibrium between added carrier/radioisotopic tracer and the induced radionuclide—this necessarily also demands total, loss-free dissolution of the sample, since in a heterogeneous system containing solid particles, equilibrium will not be achieved. It also demands prevention of early (pre-equilibration) volatilization losses and incorporation of a redox step to ensure identity of chemical species for multivalent elements. Thus, the use of closed dissolution systems, e.g., bombs and Schöniger flasks, is preferable.

   As concerns total dissolution, this is not easy to obtain with some matrices, even on prolonged HF treatment, as shown by Greenberg et al. *(10)*. It is also advisable to add boric acid or borate to such solutions after use of HF to ensure solubility of fluorides and avoid negative effects of F^- on the solution chemistry of other elements.
2. Specific separations, leading to good radiochemical purity.
3. Appropriate and correct measurement conditions—2 and 3 are self-evident.
4. Individual chemical yield measurements in each analysis—this point is discussed more fully in Radiochemical Yield Determination below.
5. Robustness of the procedure—by robustness is meant that the procedure should not be sensitive to small manipulative errors, such as addition of reagents, adjustment of pH, transfer of solution, and so forth, and also be operator-independent; that is, good results may be obtained by anyone following the procedure. This implies that it should be manipulatively straightforward and not too long and complicated. In other words, it should be fairly "foolproof." Use of chemical yield measurements makes an important contribution to the robustness of a procedure.

INTERNAL SELF-VALIDATION OF NAA DATA

It is mandatory in certification work or other cooperative quality analyses to obtain results by at least two independent methods *(11)*. The fact that NAA is a nuclear-based method, totally different from the electronic nature of other chemical analyses, has of course been an additional reason, apart from its inherent QA features, why it is in demand in this type of work. What is less appreciated, however, is the capability of NAA in its various forms and modes to supply essentially independent sets of analytical data within the same laboratory—what may be termed its capacity for internal self-validation (or crosschecking), as discussed elsewhere *(8)*.

The prime source of this independence lies in the isotopic character of NAA and the possibility of using different isotopic nuclear reactions. NAA is not an elemental technique of analysis, but an isotopic one. Also, if we examine the two major modes of NAA, by which is meant nondestructive (instrumental) NAA (INAA) and RNAA, it is apparent that the techniques are methodologically independent, especially if we use different isotopic reactions for the two modes. For example, INAA of antimony via the reaction $^{123}Sb(n,\gamma)$ $^{124}Sb(t_{1/2}$ = 60 d) contrasted to RNAA using the reaction $^{121}Sb(n,\gamma)$ $^{122}Sb(t_{1/2}$ = 2.7 d) with sample dissolution, radiochemical separation of antimony, and a measurement of the yield (thus falling into Heydorn's category *(12)* of "comprehensive" NAA), which can be easily be determined using the radioisotopic tracer $^{125}Sb(t_{1/2}$ = 2.77 yr). This independent approach from two routes was used in a study on Sb in IAEA Soil-5, where the internal verification approach was first utilized *(13)*. It can also be argued that INAA and RNAA are independent even using the same isotopic reaction, and this is supported by NIST's recent certification of vanadium in apple and peach leaves based on INAA and RNAA, both using the ^{52}V nuclide.

Naturally, the possibilities for independent isotopic routes to analytical information are greatly widened by using other forms of activation analysis, such as photon and charged particle AA, and neutron-induced prompt γ activation. However, these alternatives are less frequently available in one laboratory.

It is evident that the two modes of INAA and RNAA are essentially independent, but it is not so well understood that two RNAA approaches can also be strictly methodologically distinct. For example, consider the determination of tin by the isotopic reactions $^{120}Sn(n,\gamma)$ $^{121}Sn(\beta^-, t_{1/2}$ = 27 h) and $^{116}Sn(n,\gamma)$ $^{117m}Sn(t_{1/2}$ = 14 d). After irradiation with two separate standards from different stoichiometric tin compounds, the samples may be dissolved by distinct techniques, followed by different radiochemical separations. ^{121}Sn can be solvent extracted with *N*-benzoylphenylhydroxylamine and the β particles counted in a liquid scintillation detector *(14)*, whereas ^{117m}Sn is counted by γ spectrometry. Thus, every phase of the process is different, and hence, the resulting data are independent.

It appears that the independence of the information provided by NAA methods is unique. In other isotopic techniques, such as ICP-MS, the different isotopic signals do not give methodologically independent data.

RADIOCHEMICAL YIELD DETERMINATION

Determination of individual chemical yields in RNAA improves the quality of the procedure and the data by eliminating one of the most important sources of variability—the nonconstant recovery. Heydorn *(15)* has given a detailed and exhaustive discussion of the yield correction and its influence on precision and accuracy. In spite of the fact that the radioisotopic technique for yield determination has undeniable advantages, its use is still relatively uncommon. When the radiotracer is measured simultaneously with the indicator nuclide, the procedure is not only simplified by one stage and efficient, but also increased in overall accuracy by compensating for differences in pile-up, dead-time, or counting geometry, the yield becoming an overall yield.

The desirable properties of radiotracers for yield correction include a lower γ energy than the indicator peak (to avoid deteriorating the signal/background ratio), a longer half-life, and nonproduction (or negligible production) in the irradiated sample. These and some other factors have recently been considered by Schelhorn et al. *(16)*. With the advent of low-energy photon detectors (LEPDs) and well-type Ge detectors, which are more sensitive to lower-energy γ-rays (i.e., ca. 20–60 keV) than conventional Ge detectors, the range of radiotracers useful for chemical yield determinations has expanded. LEPDs also minimize Compton effects from higher-energy γ-rays on lower ones. Some newer examples of useful isotopic tracers will now be mentioned.

^{231}Pa for Thorium via ^{233}Pa

Determination of Th by RNAA via the indicator nuclide ^{233}Pa ($t_{1/2}$ = 27 d) has been troublesome owing to the absence of Pa carrier and the lack of a suitable Pa radioisotopic tracer. Use of ^{231}Pa ($t_{1/2}$ = 33000 yr) can solve this problem *(17)* (suitable precautions are of course necessary when using α-emitting tracers). Either the 27-keV or weaker 283.6-keV γ-rays may be measured with an LEPD or well-type detector, when quantifying the 311-keV line of ^{231}Pa.

^{235}U for Uranium via ^{239}U

Milligram quantities of natural uranium emit an easily measurable γ-ray at 185.6 keV, which can be used as radiotracer for the indicator nuclide produced by ^{238}U(n,γ) ^{239}U ($t_{1/2}$ = 25 min, E_γ = 75 keV). Since the carrier solution is naturally radioactive, this represents one of the simplest radiotracer methods and is in routine use in our laboratory *(18)*. When the uranium content is in the picogram range, the amount of carrier is reduced to minimize Compton effects on the background under the peak, and a second longer count made of the ^{235}U peak with good statistics *(19)*.

^{238}Np for Uranium via ^{239}Np

The problems with ^{239}Np as an indicator nuclide for U are similar to those of ^{233}Pa for Th. Again they may be overcome by the use of a long-lived α-active nuclide, in this case ^{237}Np ($t_{1/2}$ = 2 × 10^6 yr), which is, however, now coirradiated with the sample to produce ^{238}Np ($t_{1/2}$ = 2.2 d, Eγ = 984, 1026 and 1028.5 keV). Thus, the irradiated solution acts as both carrier (^{237}Np) and tracer (^{238}Np). This technique has been used by Niese and Niese *(20)*, and in our laboratory. By using both approaches with ^{238}Np and ^{239}U, we were able to crosscheck uranium results *(18)*. Note also that, contrarily, ^{239}Np can be used as a radiotracer to monitor recovery of ^{238}Np in RNAA determination of ^{237}Np *(21)*.

^{103}Pd for Palladium via ^{109}Pd

^{103}Pd($t_{1/2}$ = 17 d), produced with low specific activity by long irradiation of a Pd target, has weak γ-rays at 357 keV (0.02%) and at 39.9 KeV (0.07%). This means that no ^{103}Pd will be detectable from Pd in an irradiated normal sample. Thus, aged irradiated Pd, containing ^{103}Pd, in relatively large carrier amounts (ca. 1 mg), can be used as a radiotracer for the indicator nuclide ^{109}Pd ($t_{1/2}$ = 13.4 h, Eγ = 88 keV). Most useful Pd separations, such as precipitation of PdI_2 or extraction (or precipitation) of Pd-dimethylglyoxime, require rather large amounts of carrier.

Purification of ^{103}Pd from irradiated Pd containing large excesses of ^{111}Ag, ^{192}Ir, and ^{198}Au can be performed by precipitation of AgCl, followed by anion exchange in chloride medium, where Ag and Ir(III), [maintained free of Ir(IV) with hydroxylamine and hydrazine], are eluted with 6*M* HCl and Pd stripped with 12*M* HCl, whereas Au remains on the column (Byrne, unpublished).

^{67}Cu for Copper via ^{64}Cu

^{67}Cu ($t_{1/2}$ = 62 h), the longest-lived Cu isotope, is obtained by a variety of nuclear reactions, mainly via high-energy photon bombardment *(22)* of zinc via ^{68}Zn(γ,p) ^{67}Cu, or by protons via ^{68}Zn(p,2p) ^{67}Cu and ^{70}Zn(p,α) ^{67}Cu, and has useful γ-rays at 184.6 (48.6%), 93.3 (16%), and 91.3 keV (7%). It can also be produced in a reactor via ^{67}Zn(*n*,p) ^{67}Cu, but large amounts of ^{64}Cu are coproduced from ^{64}Zn, and thus isotopically enriched ^{67}Zn (costing around $40/mg!) is normally used. However, this author recently showed *(23)* that Zn of natural isotopic composition can be activated to produce useful tracer quantities (0.1–1 μCi) of ^{67}Cu containing, after 10 d of decay, practically negligible quantities of ^{64}Cu. The details of the one-step anion-exchange separation of ^{67}Cu from the irradiated ZnO target were given in a poster at this meeting *(24)*. ^{67}Cu represents an excellent yield radiotracer for copper when ^{64}Cu is the indicator isotope, since it has a longer half-life and lower γ energies.

^{108m}Ag for Silver via ^{110m}Ag

Schelhorn et al. *(16)* mentioned the use of ^{111}Ag ($t_{1/2}$ = 7.5 d, obtained from ^{110}Pd[*n*,γ] ^{111}Pd β-^{111}Ag), as a yield monitor for Ag, with its main γ-ray at 342 keV. A longer-lived alternative is ^{108m}Ag ($t_{1/2}$ = 127 yr) obtained simply by several years of cooling of irradiated silver, and having γ-lines at 76, 434, 614, and 723 keV.

^{113m}Sn for Tin via ^{117m}Sn

The relative proportions of ^{113m}Sn ($t_{1/2}$ = 115 d) and ^{117m}Sn ($t_{1/2}$ = 14 d) produced on reactor irradiation depend very much on the neutron spectrum. In the relatively poorly thermalized neutron flux of our TRIGA Mk II reactor, the ratio of ^{117m}Sn/^{113m}Sn is > 10, and this allows aged ^{113m}Sn, free of ^{117m}Sn, to be used as a yield tracer in RNAA of tin using ^{117m}Sn, with a small correction for the ^{113m}Sn produced in the sample easily derived from the ^{117m}Sn/^{113m}Sn ratio in the standard.

An alternative is to use even older irradiated Sn (purified from ^{125}Sb) containing ^{121m}Sn ($t_{1/2}$ = 55 yr) and its low-energy γ-ray (E_γ = 37.1 keV, 1.9%). A third, apparently unexplored possibility is ^{126}Sn ($t_{1/2} = 1 \times 10^5$ yr), a fission product of uranium, with only low-energy γ-rays (E_γ = 87.6 keV, 37%).

^{204}Tl for Thallium via ^{202}Tl

The best Tl radiotracer for RNAA of Tl using fast neutrons via ^{203}Tl (*n*, 2*n*) ^{202}Tl ($t_{1/2}$ = 12.2 d) is ^{201}Tl ($t_{1/2}$ = 3.05 d), with useful γ-rays at 167 and 135 keV. However, it is produced by charged particle bombardment of lead or mercury, and is not a feasible option for those without access to its production.

^{204}Tl ($t_{1/2}$ = 3.78 yr) is a β^- emitter, but also emits weak Hg X-rays, notably at 68.9, 70.8, and 80.1 keV. Thus, aged irradiated thallium in carrier quantities could be used as a radioisotopic tracer for ^{202}Tl. This indicator nuclide enables a sensitivity in the region of 1 ppm for Tl determination by fast-neutron AA *(25–27)*.

^{81m}Se for Selenium via ^{75}Se

^{81m}Se ($t_{1/2}$ = 57 min, E_γ = 103 keV) has been used for determination of the chemical yield of Se in RNAA employing ^{75}Se ($t_{1/2}$ = 120 d) by the reactivation technique, e.g., by Polkowska-Motrenko et al. *(28)*. However, as a simpler alternative to the reactivation method, we have recently used ^{81m}Se as a radioisotopic tracer *(29)*, freshly prepared before use by irradiation of isotopically highly enriched ^{80}Se (Oak Ridge), thus avoiding formation of ^{75}Se and the necessary correction. Because of the volatility of selenium and its multivalent character, in our experience, chemical yields can typically vary by 10–15% in RNAA procedures, so that individual yield measurements using ^{81m}Se provide a worthwhile improvement in quality.

Some Other Radiotracers for NAA

^{57}Co and ^{109}Cd are relatively cheap, long-lived, and have ideal low-energy γ-rays (122 and 88 keV, respectively). In view of the fact that in trace analysis Co and Cd are considered "difficult" elements, it is surprising that they are not more frequently used in RNAA. ^{57}Co is also useful for Ni determination *(30,31)* by fast-neutron activation via ^{58}Ni (n,p) ^{58}Co.

^{77}As ($t_{1/2}$ = 39 h, E_γ = 239 keV) is a useful alternative to cyclotron-produced ^{74}As, since it may be easily produced from reactor irradiation of Ge, and used for yield determination of As in RNAA via ^{76}As *(32,33)*.

^{181}W ($t_{1/2}$ = 121 d) was reported by Schelhorn et al.*(16)* to be a useful yield monitor for the indicator isotope ^{187}W, using its Ta $K_{\alpha 2}$ and $K_{\beta 1}$ X-rays of 56.3 and 65.1 keV, which are not overlapped by possible Re X-rays from ^{187}W and W X-rays induced in the carrier. A way of avoiding these possible difficulties, especially for those without an LEPD, is to use the 125-keV γ-ray (0.02%) of ^{125}W (75 d) also present in aged highly irradiated tungsten.

Finally, we note that sadly the "problem" elements Mo and Cr do not have suitable radiotracers for use in RNAA, and particularly for Cr, this undoubtedly contributes to continuing analytical difficulties in its determination. The production of ^{48}Cr ($t_{1/2}$ = 21.6 h, E_γ = 308 [100%], 112 [99%] from ^{46}Ti $(\alpha,2n)$ ^{48}Cr might be worth considering.

ACKNOWLEDGMENTS

The assistance of the Ministry of Science and Technology of Slovenia is gratefully acknowledged.

REFERENCES

1. J. J. M. de Goeij and R. W. Woittiez, *J. Radioanal. Nucl. Chem., Articles* **168,** 429 (1993).
2. R. Dams, *Fresenius J. Anal. Chem., Articles* **337,** 492 (1990).
3. M. de Bruin, *J. Radioanal. Nucl. Chem., Articles* **160,** 31 (1992).
4. IAEA Advisory Group, *Anal. Chim. Acta* **163,** 1 (1984).
5. *Quality Assurance in Biomedical Neutron Activation Analysis,* IAEA TECDOC-323, IAEA Vienna, Nov. 1984.
6. K. Heydorn, *J. Radioanal. Nucl. Chem., Articles* **151,** 139 (1991).
7. D. A. Becker, *Fresenius J. Anal. Chem.* **345,** 298 (1993).
8. A. R. Byrne, *Fresenius J. Anal. Chem.* **345,** 144 (1993).
9. J. J. M. de Goeij, in *Quality Assurance in Biomedical Neutron Activation Analysis,* IAEA TECDOC-323, IAEA Vienna, Nov. 1984, pp. 153–161.
10. R. Greenberg, H. M. Kingston, R. L. Walters, Jr., and K. W. Pratt, *Fresenius J. Anal. Chem.* **338,** 394 (1991).
11. H. Merchandise and E. Colinet, *Fresenius Z. Anal. Chem.* **316,** 669 (1983).
12. K. Heydorn, *Neutron Activation Analysis for Clinical Trace Element Research,* CRC Press, Boca Raton, FL, 1984.
13. L. Kosta and A. R. Byrne, *J. Radioanal. Chem.* **69,** 117 (1982).
14. A. R. Byrne, *J. Radioanal. Chem.* **27,** 591 (1977).
15. K. Heydorn, in *Neutron Activation Analysis for Clinical Trace Element Research,* CRC Press, Boca Raton, FL, 1984, pp. 142–150.
16. H. Schelhorn, M. Geisler, and A. Rämmler, *J. Radioanal. Nucl. Chem., Articles* **168,** 265 (1993).

17. L. Benedik and A. R. Byrne, *J. Radioanal. Nucl. Chem., Articles,* in press (1994).
18. A. R. Byrne and L. Benedik, *Talanta* **35,** 161 (1988).
19. A. R. Byrne and L. Benedik, *Sci. Total Environ.* **107,** 143 (1991).
20. U. Niese and S. Niese, *J. Radioanal. Nucl. Chem.* **91,** 17 (1985).
21. A. R. Byrne, *J. Environ. Radioactivity* **4,** 133 (1986).
22. S. Muzadek, L. F. Mausner, and S. C. Srivastava, *Appl. Rad. Isotopes* **37,** 29 (1986).
23. A. R. Byrne, in *IJS Report* DP-6693, March 1993, J. Stefan Institute, Ljubljana, Slovenia.
24. A. R. Byrne and M. Dermelj, *Biol. Trace Element Res.* This issue.
25. A. de Ruck and R. Dams, *J. Radioanal. Nucl. Chem. Lett.* **94,** 97 (1985).
26. I. M. Cohon, S. M. Resnifzky, and G. M. Bars, *J. Radioanal. Nucl. Chem.* **72,** 451 (1982).
27. A. Fajgelj and A. R. Byrne, *J. Radioanal. Nucl. Chem., Articles* in press (1994).
28. H. Polkovska-Motrenko, M. Dermelj, A. R. Byrne, A. Fajgelj, P. Stegnar, and L. Kosta, *Radiochem. Radioanal. Lett.* **52,** 99 (1982).
29. V. Stibilj, M. Dermelj, and A. R. Byrne, *J. Radioanal. Nucl. Chem., Articles* **182,** 317 (1994).
30. A. R. Byrne and I. Krašovec, *Fresenius J. Anal. Chem.* **332,** 666 (1988).
31. J. Kučera and A. R. Byrne, *J. Radioanal. Nucl. Chem., Articles* **168,** 201 (1993).
32. H. Schelhorn and M. Geisler, *J. Radioanal. Nucl. Chem.* **83,** 5 (1984).
33. A. R. Byrne, *Vestnik Slov. Kem. Druš.* **32,** 311 (1985).

RNAA as Compared to ICP-MS for the Analysis of Normal Human Serum

R. DAMS

Lab. Analytical Chemistry, Institute for Nuclear Sciences, University Gent, Proeftuinstraat 86, B-9000 Gent, Belgium

ABSTRACT

The merits of radiochemical neutron activation analysis (RNAA) and inductively coupled plasma mass spectrometry (ICP-MS) are critically discussed for the determination of trace and ultratrace elements in normal human serum. For RNAA, two semiautomated procedures, allowing the determination of up to 18 elements, are briefly described. ICP-MS has a series of interesting features for the determination of trace elements. Matrix and spectral interferences can mostly be avoided or corrected for. After a simple 5- or 10-fold dilution and addition of an internal standard, more than 20 elements can be measured precisely and accurately.

Index Entries: Human serum; radiochemical neutron activation analysis; inductively coupled plasma mass spectrometry.

INTRODUCTION

Comparing two analytical methods is always very difficult. In fact, both have their merits and their shortcomings. One is more suited for solid samples, and the other for liquid ones. A method may be very sensitive for certain components or elements, and the other may be less sensitive to spectral or matrix interferences, to contamination, to blanks, and so forth. Are chemical separations needed, or is the method completely instrumental? What is the speed of the analysis, the costs, the attainable precision or accuracy, the multielement character, the possibility for speciation studies, and so on? It all depends on the objectives put forward by the analyst and above all on the nature of the sample.

Biological Trace Element Research Editor: G. N. Schrauzer © 1994 by Humana Press Inc.

The trace element content of human serum has attracted the interest of many analytical and clinical scientists for a number of reasons. It is of vital importance for the human organism, since it is involved in all metabolic activities, and in addition, it is easily accessible. Although present in minute amounts, trace elements ensure vital biochemical functions. On the other hand, in excessive concentrations, they interfere with vital processes and exert deleterious effects. The concentrations of the metallic trace elements of potential interest vary from 10 µg/mL to <0.1 ng/mL.

ANALYSIS OF NORMAL HUMAN SERUM

From the matrix composition of normal serum, it appears that it is an aqueous medium, easily diluted, but also not too hard to be converted into a solid by freeze-drying. It contains about 7% of organic material, soluble in water, and in addition some salts, primarily chlorides of Na, K, and Ca equivalent to about 9 g NaCl/L. Also the presence of S, and of more than traces of Br and P is important. As a matter of fact, the sampling itself and the preanalysis treatment must be performed under meticiously controlled conditions to avoid contamination or losses. The main requirements to develop sampling and sample handling bear on clean room conditions (Class 100), cleaning of laboratory ware, purity of the chemicals, and last but not least GLP. GLP means in this case "developing a systematic attitude towards dealing with and evaluating adventitious additions or losses." The less elaborate the sample handling is, the lesser the ultratrace elements content will be jeopardized *(1–3)*.

Up to about 5 yr ago, RNAA and INAA together with GF-AAS (and in specific cases ICP-AES, Voltametry, PIXE, TR XRF, or IDMS) were about the only techniques that had been employed, at least for the ultratrace (<10 ng/mL) elements. GF-AAS is hampered by the occurrence of matrix effects and has a monoelement character. INAA yields mostly insufficiently low detection limits, and thus, elaborate radiochemical separations are involved (RNAA) as well as often long waiting times. About 5 yr ago, ICP-MS became a very strong competitor to RNAA, not only for ultratrace elements in serum, but in more general terms, for analysis of body fluids and tissues *(4)*.

RADIOCHEMICAL AND INSTRUMENTAL NEUTRON ACTIVATION ANALYSIS (RNAA-INAA)

The principles of neutron activation analysis and its advantages are well known. It is not necessary to illustrate once more its inherent sensitivity for a large number of elements, its freedom of contamination after irradiation, its high specificity when γ-spectrometry is applied, and its totally independent principle as a nuclear-based property. Matrix or other interferences can be corrected for. The addition of carriers allows one to work in more controlled conditions and to determine the radiochemical yield.

Procedures

As an illustration, two semiautomated procedures for RNAA developed by the group of R. Cornelis *(5,6)* will briefly be explained. These methods are being designed according to the irradiation period required to induce measurable amounts of radioisotopes of the elements of interest. The first procedure *(5)* largely covers the elements determinable by shorter-lived radioisotopes. About 500 mg lyophilized serum, originating from about 6 mL serum, are irradiated in a thermal flux of 4.3×10^{12} $n/cm^2/s$ for 7 h in the Thetis reactor. After about 18 h of decay, the sample is burned in a Trace-O-Mat apparatus (O_2 atmosphere, two halogen lamps). The elements of interest are collected at the bottom of the apparatus by refluxing with HBr, and consequently separated from the matrix activity (Na, K, Br, P) on the organic anion exchanger DOWEX 1X-8, 100–200 mesh. Elutions with 10 mol/L HCl and 3 mol/L HCl + 0.2% H_2O_2 allow separations in three groups and determinations of:

1. Rb and Cs (after decay of the matrix);
2. As, Cu, Fe, and Se; and
3. Au, Cd, Hg, Fe, Mo, Sb, Zn (and some traces of Se).

The third group elements are measured on the ion exchanger, which is therefore transferred to a polyethylene-counting vial and homogenized. The concentrations are calculated using the k_0-standardization method. Mostly 11 out of 12 elements can be determined, for Cd the content is often below the detection limit.

The second procedure *(6)* covers a group of elements to be determined through long-lived radioisotopes. After 10 d of irradiation of about 150 mg dry-ashed lyophilized serum in high-purity Suprasil quartz at a much higher neutron flux ($10^{14}n/cm^2/s$) in the BR2 reactor, the sample is dissolved in a $HNO_3/HClO_4$ mixture. The elements are separated into four balanced groups, using inorganic ion-retention media, namely AAO (acid Al_2O_3), HAP (hydrated Sb_2O_5), and HMD (hydrated MnO_2). In the second-generation biological reference material (lyophilized human serum) prepared by Versieck et al. *(8)*, the concentrations of the 13 elements (Sc, Cr, Fe, Co, Zn, Se, Rb, Ga, Ag, Sn, Sb, Cs, Hg) ranging from 1 ng/g (Cr) to 26 µg/g (Fe) in the lyophilized material, or about 11 times lower in the original serum could be determined. Together both lengthy procedures allow the detection of 16–18 elements in this material. To this list may be added the elements V, Mn, and I, which can be determined from their short-lived radioisotopes after a fast single-element radiochemical separation *(3,7)*.

INDUCTIVELY COUPLED PLASMA-MASS SPECTROMETRY (ICP-MS)

Principles

Mass spectrometry when coupled to an ICP has a series of interesting features for the determination of trace and ultratrace elements. It is nearly the only universal quasisimultaneous multielement method. Preferentially, liquid samples, but in principle also solid samples can be analyzed (with an alternative sample introduction system, such as ETV or LA). Excellent detection limits, on the order of 0.5–20 ng/L can be reached. In principle, simple line spectra are obtained rather than the line-rich spectra of emission techniques. There is a linear dynamic range of 5 orders of magnitude, and a high precision is possible, because an internal standard can be added. Since isotopic-ratio information is obtained, spectral interferences can be traced down and often corrected for, and finally when isotope dilution is applied, the precision can be markedly increased. A high sample throughput is possible.

The most important disadvantages are the occurrence of matrix effects owing to clogging of the entrance aperture of the sampling cone, or of the pneumatic nebulizer, signal suppression as a result of ionization suppression, and space charge effects. Heavy matrices are thus to be avoided by dilution, or must be corrected for by addition of an internal standard, by matrix simulation, or by standard addition.

The matrix effect is primarily caused by easily ionized ions, such as sodium, potassium, calcium, and so forth. It was shown experimentally that the ^{115}In signal is suppressed by as much as a factor of three by NaCl contents encountered in human serum. In addition, this suppression is strongly mass-dependent *(9)*. Corrections for signal instabilities and for such suppressions are generally performed by the addition of an internal standard and using the experimental signal ratio (analyte/internal standard). Since the matrix effect is mass-dependent, it is obvious that one internal standard cannot efficiently be used for all elements. Several such standards must be used over the entire mass range *(10,11)*.

A second shortcoming of ICP-MS is its limited resolution. Most units for ICP-MS are equipped with a quadrupole mass spectrometer, instead of a double-focusing magnetic spectrometer. A quadrupole allows only a resolution of 1.0–0.5 amu. Spectral overlaps resulting from interferences by isobaric ions, polyatomic ions, and doubly charged ions thus cannot be resolved. Polyatomic ions formed by the ICP matrix elements Ar, H, N, and O often interfere with the detection of ions with atomic masses ranging from 27–81. In addition to these, molecular ions formed with elements from the acid or the sample matrix, such as S, Cl, Na, C, and N, must be taken into account. Many interferences can however, be, avoided by a careful selection of the isotope measured, and for unexpected interferences, the measurement of the experimental isotopic ratios is a powerful weapon in the hands of the analyst.

DETERMINATION OF TRACE ELEMENTS

As an example of the determination of an element that has been spectrally interfered with, Zn can be mentioned. It has four stable isotopes with isotopic abundances ranging from 48 to 4%. All of them are interfered with by SO_2^+ or HSO_2^+ ions. The interference factors or the apparent Zn concentrations for a sulfur content of 1 g/L can be determined experimentally. With a sulfur content of 0.9 g/L of serum, the interference is only about 0.4% for the ^{68}Zn isotope and 65% for the ^{64}Zn. This correction can thus be calculated or experimentally determined by using a simulated serum solution as blank. To match the S content, cystein is added. The results for the second-generation biological reference material proved that such correction is very well feasible, although as expected, the ^{68}Zn measurements give the most precise and accurate results *(12)*.

Other simularly interfered with elements are ^{57}Fe ($^{40}Ca^{17}O$), ^{65}Cu ($^{33}S^{16}O_2$; $^{32}S^{33}S$), and ^{59}Co ($^{36}Ar^{23}Na$; $^{43}Ca^{16}O$). Table 1 shows some results for the second-generation reference material. Concentrations are expressed in the reconstituted human serum *(12)*. The agreement with the certified values is generally excellent. Typically, addition of an internal standard and a dilution by a factor of 5 or 10 with 0.14 mol/L HNO_3 are applied as the only sample treatment. Mostly external calibration is applied. The standards are also prepared in 0.14 mol/L HNO_3. For cobalt, a very large correction (±50%) must be applied, resulting in a less accurate value. For Sr, also in the ng/mL level, comparison is only possible with two RNAA results (31 ng/mL using the long-lived ^{85}Sr [$t_{1/2}$ = 64.8 d] and 22.2 ng/mL using the short-lived ^{87m}Sr [$t_{1/2}$ = 2.8 h]) *(13,14)*. ICP-MS (25.5 ± 0.6 ng/mL) using ^{88}Sr (82.6%) compares again favorably with RNAA. The determination of Br presents some problems. Since the two stable isotopes of Br (^{79}Br and ^{81}Br) are situated on both sides of a large $^{40}Ar_2$ peak, the peak integration range and the mass calibration had to be adjusted and the resolution optimized (ΔM at 5% of maximum equals 0.61 U), at the expense of sensitivity. The results for both peaks (4.30 ± 0.05 μg/mL) agree with the certified value (4.44 ± 0.35 μg/mL) *(15)*.

DETERMINATION OF ULTRATRACE ELEMENTS

To determine the ultratrace elements (ng/mL), several aspects must be taken into account, such as dilution factor, scanning or counting mode, blanks, and memory effects. The measurement is only quasisimultaneous, since the different channels, corresponding to a specific mass, must be scanned. In the scanning mode, 120 sweeps, with a dwell time of 250 μs/channel and 2048 channels divided over 250 mass units, can be performed in about 1 min (0.25 s/mass). With peak jumping, a dwell time of 25 ms can be set, and the detection limit can be lowered by about a factor of 10, if the detection limit is not determined by the blank. In the present work intermediate conditions are used. The mass range is divided in about four or five parts consisting of 20–30 mass units. Each mass consists of about 20 channels to allow for a correct integration of the peaks over 0.8 amu. The dwell time is 320 μs, and 400 or 200 sweeps are performed over 512 or 1024 channels. A measurement of two to three elements together with the internal standard thus takes about 60 s. In between two measurements, the sample introduction system is rinsed for 2 min with 0.14 mol/L HNO_3 to avoid memory effects.

Table 1
Comparison of the Results of Seven Trace-Elements in Biological Reference Material Human Serum by ICP-MS with the Certified Values*

Element	Nuclide	ICP-MS		Certified
		10-fold diluted	5-fold diluted	
Fe (μg/g)	^{57}Fe	26.0 ± 3.6	24.3 ± 3.9	25.9 ± 1.5
Co (ng/g)	^{59}Co		2.6 ± 0.6	3.6 ± 0.6
Cu (μg/g)	^{65}Cu	11.4 ± 0.5	11.5 ± 0.6	11.1 ± 0.4
Zn (μg/g)	^{66}Zn	10.1 ± 0.2	10.3 ± 0.2	9.6 ± 0.2
Rb (μg/g)	^{85}Rb	1.74 ± 0.02	1.76 ± 0.01	1.85 ± 0.33
Br (μg/g)	^{79,81}Br	4.30 ± 0.05	-	4.44 ± 0.35
Cs (ng/g)	^{133}Cs		10.1 ± 0.6	10.0 ± 2.3

*Values expressed in lyophilized material *(12)*.

The elevation of the blank level can have several origins. The milli-Q water (resistivity 18 M Ωcm) and the HNO_3, prepared by subboiling distillation, are significant sources for B, Sn, Ba, and Pb, whereas for Hg, the blank originates from impurities present in the argon gas. For the other elements, the blank signals observed are the result of the continuous background present over the whole mass range. Table 2 gives detection limits (DL) for 11 elements based on 3σ of the blank or the background, and on a dilution by a factor of 5 of the original serum *(11)*.

Table 3 gives the results as obtained for these 11 elements, together with results for Mg and I. When comparison with certified values is possible (Mo, Cd, Cs, and Hg), there is always agreement. For several elements (Li, B, Mg, Sn, Sb), only comparison with one or two literature values can be made. For Sb, only a comparison with one NAA value is possible. There is no agreement. For the elements I, Ba, Pb, and Bi, no other values for this reference material are available. For Li, Mg, and Pb, the accuracy could be tested by analysis of NIST SRM 909, and the accuracy could be tested for Li and Pb at a 1000 times higher concentration. In total, 7 + 13 trace and ultratrace elements were determined, whereas before the contents of Na, Cl, Ca, and S had been determined after a dilution ranging from a factor of 10 for Ca to 5000 for Na *(11,16–18)*.

An interesting result is the determination of Sn. It has 10 stable isotopes, but 8 of them are isobarically overlapped by Cd, In, or Te. Only ^{118}Sn (24%) and ^{120}Sn (32%) can be used. After a fivefold dilution, detection is possible since the blank is about 10 times lower than the sample content. The standard deviations on several replicate analyses on the lyophilized serum and on the original serum are on the order of 10%. Comparison is possible with two RNAA values using, respectively, the long-lived ($t_{1/2}$ = 115 d) ^{113}Sn (10 ± 2.6 ng/g) in the earlier-described multielement procedure *(6)* and both ^{113}Sn (7.96 ± 0.97 ng/g) and the shorter-lived ($t_{1/2}$ = 13.6 d) ^{117m}Sn (8.38 ± 0.48 ng/g) after a single-element separation based on extraction of SnI_4 *(19)*. Another author

Table 2
Blank Values and Detection Limits by ICP-MS as Compared to Concentrations in Human Serum *(11)*

Element (nuclide)	Blank value ($\mu g\ L^{-1}$)	Detection limit * ($\mu g\ L^{-1}$)	Concentration in human serum ($\mu g\ L^{-1}$)
Li (^{7}Li)	0.02 - 0.04	0.05	1
B (^{10}B)	1 - 2	0.5	5 - 50
Mo (^{98}Mo)	0.01 - 0.03	0.04	0.3 - 1.2
Cd (^{114}Cd)	0.01 - 0.02	0.02	0.1 - 0.2
Sn (^{120}Sn)	0.02 - 0.06	0.05	< 5
Sb (^{121}Sb)	0.01 - 0.02	0.03	< 0.5
Cs (^{133}Cs)	0.005 - 0.015	0.03	1
Ba (^{138}Ba)	0.02 - 0.05	0.04	? (< 10)
Hg (^{202}Hg)	0.04 - 0.08	0.2	0.1 - 5
Pb (^{208}Pb)	0.08 - 0.15	0.1	0.02 - 4
Bi (^{209}Bi)	0.001 - 0.003	0.007	< 0.1

*Taking into account a fivefold dilution of the serum sample.

obtained 7.6 ± 1.4 ng/g by ICP-MS *(20)*. The Ba value could not be compared to a literature value. Scarce literature values for Ba in serum are mostly of 50 to 2000 times higher, although determined by NAA.

A detection without any preliminary chemical separation has of course a number of limitations. Owing to spectral overlaps, the following elements cannot be determined in this way: Mn, V, Cr, As, and Se. By a simple anion-exchange separation of SO_4^{2-} and Cl^-, it was possible to determine accurately Se and As *(21)*. An alternative approach for As and Se is a mathematical correction based on the isotopic-ratio measurement *(22)*. The separation on anion exchange simultaneously removes the most important interferences with ^{65}Cu, ^{64}Zn, ^{66}Zn, ^{51}V, and ^{50}Cr. Unfortunately, V and Cr still cannot be determined, because their contents are so low in normal human serum that the required dilution brings the concentrations below the detection limits.

SUMMARY

If one tries to compare RNAA and ICP-MS, it appears that both allow a long list of elements to be determined in normal human serum, 21 and 22, respectively, with Sb and Cd being very close to the detection limit (Table 4). In addition, the minor elements Na, Cl, and Ca, and by ICP-MS, also S can be determined. Missing from this list of 27 elements

Table 3
Comparison of Results for Ultratrace Elements in Biological Reference Material Human Serum by ICP-MS*

Second generation biological reference material		
Isotope measured	Result obtained	Certified or literature (lit) value
^{7}Li	15.04 ± 0.39 ng/g	19.3 ; 14.8 (lit)
^{10}B	227 ± 18 ng/g	222 (lit)
^{24}Mg (100 x dil.)	188.1 ± 8.7 μg/g	180 ± 10 (lit)
^{98}Mo	7.51 ± 0.77 ng/g	7.5 ± 0.8
^{111}Cd	2.23 ± 0.61 ng/g	2.0 (1.7 - 2.5)
^{118}Sn	9.92 ± 0.83 ng/g	10 ; 7.6 ; 8.0 ; 8.4 (lit)
^{121}Sb	0.87 ± 0.29 ng/g	0.25 ± 0.15 (lit)
^{127}I	0.710 ± 0.048 μg/g	-
^{133}Cs	10.20 ± 0.79 ng/g	10.0 ± 2.3
^{138}Ba	11.1 ± 3.1 ng/g	-
^{200}Hg	6.49 ± 0.89 ng/g	6.6 ± 0.4
^{208}Pb	46.3 ± 4.6 ng/g	-
^{209}Bi	0.69 ± 0.11 ng/g	-

*Values expressed in lyophilized material *(11)*.

are unfortunately Al, Be, and Ni. Up to now the detection limits have been insufficient for their detection in normal serum.

When going back to the originally mentioned requirements for an analytical technique, it is obvious that ICP-MS is best suited for treating a liquid serum sample. Both techniques are highly sensitive and largely multielemental. Contamination and blanks are sometimes limiting factors in ICP-MS. Precision and accuracy can be excellent for both techniques. Purely instrumental analysis, that is, without preliminary chemical separations, is possible for all elements detected by ICP-MS, with only a dilution before measurement, except for As and Se, where interferences must be removed or corrected for. Purely instrumental NAA is only feasible for about six elements (Co, Zn, Se, Br, Rb, Cs). For the other elements, multielement radiochemical procedures or even single-element procedures (V, Mn, I) must be applied. Speciation is possible by ICP-MS, when it is used as a detector after a chromatographic separation. In terms of speed and cost of the analysis and availability of the equipment needed, ICP-MS is obviously largely favored.

Table 4
Comparison of RNAA and ICP-MS for Analysis of Normal Human Serum

Elem.	Conc. Ref.serum ($ng.g^{-1}$)	RNAA ($n\ cm^{-2}\ s^{-1}$) 7h; $4x10^{12}$	13d; 10^{14}	ICP-MS No corr.	Corr. interf.
Li	(15)			+	
B	(225)			+	
Mg	(180 000)			+	
Sc	(0.018)		+		
V	(0.35)	+ (*)			
Cr	0.76		+		
Mn	7.7	+ (*)			
Fe	25 900	+	+		+
Co	3.6		+ (**)		+
Cu	11 100	+			+
Zn	9 600	+	+ (**)		+
As	19.6	+			+ (***)
Se	1 050	+	+ (**)		+ (***)
Br	48 800	+ (**)		+	
Rb	1 850	+	+ (**)	+	
Sr	(260)	+		+	
Mo	7.5	+		+	
Ag	(0.93)		+		
Cd	2.0	(+)		+	
Sn	(9)		+	+	
Sb	(0.25-0.8)	(+)		(+)	
I	(680)	+ (*)		+	
Cs	10.0	+ (**)	+ (**)	+	
Ba	(12)			+	
Hg	6.6	+	+	+	
Pb	(46)			+	
Bi	(0.7)			+	
TOTAL		**19 (+2)**		**21 (+1)**	

( ) No certified value; (*) separate monoelement procedure; (**) purely instrumental (INAA) possible; (***) chemical separation needed.

REFERENCES

1. G. V. Iyengar, *Elemental Analysis of Biological Systems. Vol. 1. Biomedical, Environmental, Compositional and Methodological Aspects of Trace Elements*, CRC Boca Raton, FL, 1989
2. R. Cornelis, *J. Trace Elem. Electrolytes Health Dis.* ***6,*** 129 (1992)
3. J. Versieck and R. Cornelis, *Trace Elements in Human Plasma or Serum*, CRC Boca Raton, FL, 1988.
4. H. Vanhoe, C. Vandecasteele, J. Versieck, and R. Dams *Plasma Source Mass Spectrometry*, Proc. 3rd Surrey Conference 1989. K. E. Jarvis and A. L. Gray, eds., Royal Society of Chemistry, 66 (1990).
5. D. Van Renterghem, R. Cornelis, and R. Vanholder, *Anal. Chim. Acta* ***257,*** 1 (1992).
6. Lin Xilei, D. Van Renterghem, R. Cornelis, and L. Mees, *Anal. Chim. Acta* ***211,*** 231 (1988).
7. J. Versieck, J. Hoste, F. Barbier, L. Vanballenberghe, J. De Rudder, and R. Cornelis, *Clin. Chim. Acta* ***135,*** 87 (1987).
8. J. Versieck, L. Vanballenberghe, A. De Kesel, J. Hoste, B. Wallaeys, J. Vandenhaute, N. Baeck, and F. W. Sunderman, Jr., *Anal. Chim. Acta* ***204,*** 63 (1988).
9. C. Vandecasteele, M. Nagels, H. Vanhoe, and R. Dams, *Anal. Chim. Acta* ***211,*** 91 (1988).
10. F. Vanhaecke, H. Vanhoe, R. Dams, and C. Vandecasteele, *Talanta* ***39,*** 737 (1992).
11. H. Vanhoe, Inductief gekoppeld Plasma Massaspectrometrie voor de bepaling van spoor- en ultraspoorelementen in menselijk serum, Ph.D. Thesis, University Gent, Belgium, 1992.
12. H. Vanhoe, C. Vandecasteele, J. Versieck, and R. Dams, *Anal. Chem.* ***61,*** 1851 (1989).
13. J. Versieck, L. Vanballenberghe, A. Wittoek, and H. Vanhoe, *J. Radioanal. Nucl. Chem.* ***168*** 243 (1993).
14. C. Vandecasteele, H. Vanhoe, R. Dams, L. Vanballenberghe, A. Wittoek, and J. Versieck, *Talanta* ***37,*** 813 (1990).
15. C. Vandecasteele, H. Vanhoe, R. Dams, and J. Versieck, *Anal. Lett.* ***23,*** 1827 (1990).
16. H. Vanhoe, C. Vandecasteele, J. Versieck, and R. Dams, Microchim. *Acta* ***III,*** 373 (1989).
17. H. Vanhoe, F. Allemeersch, J. Versieck, and R. Dams, *The Analyst* in press.
18. H. Vanhoe, C. Vandecasteele, J. Versieck, and R. Dams, *Anal. Chim. Acta* ***244,*** 259 (1991).
19. J. Versieck and L. Vanballenberghe, *Anal. Chem.* ***63,*** 1143 (1991).
20. H. Baumann, private communication (1989).
21. J. Goossens and R. Dams, *J. Anal. Atomic Spectrom.* ***7,*** 1167 (1992).
22. J. Goossens, L. Moens, and R. Dams, *Talanta* **41,** 187 (1994).

Intercomparison of Techniques Available at INETI in the Analysis of Two IAEA Candidate Research Materials

M. C. FREITAS,*[,1] M. H. AFONSO,[2] C. ALMEIDA,[1] L. C. ALVES,[3] M. F. ARAÚJO,[1] M. A. BARREIROS,[4] J. SEABRA E BARROS,[2] M. BORDALO COSTA,[4] M. A. GOUVEIA,[1] AND M. A. REIS[3]

[1]Departamento de Química (DCTN/SQ), 2686 Sacavém Codex; [2]Departamento Central de Estudos e Análises Industriais (DCEAI); [3]Departamento de Física (DCTN/SF) and [4]Departamento de Estudos de Impacte Industrial (DEII), Instituto Nacional de Engenharia e Tecnologia Industrial (INETI), Portugal

ABSTRACT

Under contract with the IAEA, the epiphytic lichen *Evernia prunastri* was collected to prepare a multielement lichen reference material for quality assurance of environmental studies. An intercomparison run on trace and minor elements in this candidate research material (IAEA-336) was organized in which six analytical groups of the National Institute of Engineering and Industrial Technology (INETI) took part. INAA, PIXE, XRF, AAS, and ICP-ES were applied. The results obtained by different methods are compared, and their complementarity is discussed. As a quality control, the IAEA cabbage research material (IAEA-359) was analyzed. The results agree quite well with the estimated values given by the IAEA.

Index Entries: Nuclear techniques; atomic methods; biological materials; minor and trace elements; IAEA materials.

*Author to whom all correspondence and reprint requests should be addressed.

Biological Trace Element Research Editor: G. N. Schrauzer © 1994 by Humana Press Inc.

INTRODUCTION

Multielement plant analyses are routinely performed for monitoring purposes, mapping, and risk assessment. In such applications, relevant information is commonly obtained by studying an organism used for monitoring (the monitoring organism). In recent years, a variety of different biological materials, including lichens, have been suggested for use as biomonitors for environmental pollution. However, the analytical quality control in this field is hampered by the nonexistence of suitable reference materials. A lichen research material (IAEA-336) was prepared at INETI under contract with the IAEA. It was analyzed by different groups of INETI, as part of an IAEA intercomparison run. The cabbage IAEA-359 was also analyzed as a quality-control sample.

INETI has acquired experience and capacity in different analytical methods, namely instrumental neutron activation analysis (INAA) (both comparative and k_0-based methods), proton-induced X-ray emission (PIXE) (both thick and thin targets), energy-dispersive X-ray fluorescence (XRF), atomic absorption spectrometry (AAS), and inductively coupled plasma emission spectrometry (ICP/ES). Some of these techniques have been already used in intercomparison runs of candidate reference materials organized by international groups (e.g., *1,2*), in addition to their application in the determination of elemental concentrations in geological, environmental, biological, and high-purity materials. They were implemented by the six analytical groups of four departments of INETI (chemistry, physics, industrial analysis, and environment). In this article, the methods used are shortly described, and the results shown and compared. Complementary features of the methods are discussed.

EXPERIMENTAL

The candidate reference materials IAEA-336 and IAEA-359 were carefully homogenized. Fractions were separately oven-dried for 2 h at 100°C (IAEA-336) and 72 h at 80°C (IAEA-359).

INAA

Six pellets (diameter: 1.3 cm; height: 0.3 cm) of 500 mg were made for each material using a pressure of 6 t/cm^2 and put into polyethylene containers for irradiation.

k_0 Method

Short irradiation (1-min irradiation time) was carried out in a pneumatic system of the INETI's Portuguese Research Reactor (RPI) at a thermal neutron flux of 2.7×10^{12}/cm^2/s; long irradiation (7 h for IAEA-336 and 4 h for IAEA-359) was carried out in the core of the RPI at a thermal neutron flux of 3.6×10^{12} (IAEA-359) and 2.2×10^{13} (IAEA-336)/cm^2/s. Wires of an 0.1% Au-Al alloy of 1 mm diameter (CBNM) and alloy of Mo 125 μm diameter (Goodfellow) were coirradiated as comparators.

All measurements were carried out with an HPGe and an LEPD. The nuclides, the γ-ray energies, and the decay times used in the determination of the elemental concentrations are given in Table 1. The spectra were processed using SEQAL computer code. The calculations were made by using the k_0-standardization method's computer codes SOLANG and SINGCOMP. The Gent k_0-factors were used.

Comparative Method

Two IGGE standards were employed as reference samples, namely GSS-4 and GSS-5. After oven-drying for 24 h at 110°C, they were mixed with cellulose (also oven-dried at the same conditions), and pellets similar to the samples were made. The certified values of element contents were taken from Govindaraju *(3)*. The irradiations were done in the core of the RPI at a thermal neutron flux of $3.2 \times 10^{12}/cm^2/s$.

All measurements were carried out with a Ge(Li) and an LEPD. The nuclides, the γ-ray energies, and the decay times used in the determination of the elemental concentrations are given in Table 1. The spectra were processed using the programs GELI and DAISY (GELIAN and OLIVE program adaptations).

PIXE

Thick Target (TT)

Six pellets were made out of each sample using a pressure of 6 t/cm^2. The TTPIXE analyses were carried out at an INETI's Van de Graaff accelerator. The samples were irradiated in vacuum (10^{-5} Torr) with a 1.9-MeV proton beam (20 mm^2 cross-section) and a 750 nA/cm^2 maximum current density. The total irradiation time was 30 min/pellet. A 200-eV resolution Si(Li) detector was placed at 90° relative to the proton beam direction.

The spectra analyses were performed with AXIL *(4)* Computer Code V3.0. The quantitative analysis was made using DATTPIXE *(5)*. The values reported in this work were corrected for radiation damage effects during irradiation.

Thin Target

The major elements (Al, Si, K, Ca) were determined without preconcentration. Samples of 200 mg were digested in a Teflon^{tm} bomb with HNO_3 and 100 μg of yttrium as an internal standard. A drop of the resulting solution was pipeted onto a mylar foil for PIXE measurements *(6)*.

In order to improve detection limits of some elements, preconcentration was employed. About 200 mg of the samples were digested with HNO_3 in a Teflon^{tm} bomb and diluted to 50 mL. Aliquots of 15 mL were reacted with ammonium pyrrolidine dithiocarbamate (APDC) at pH = 3 with 100 μg of palladium as a coprecipitant and internal standard *(7)*. The palladium dithiocarbamate precipitates were filtered through Nucle-

Table 1
Nuclides, γ-Ray Energies and Decay Times Used for INAA Measurements

Element	Radionuclide	Energy of gamma-ray used (keV)	Cooling time
Na	^{24}Na	1368	4-7 d
Mg	^{27}Mg	844, 1014	15 min.
Al	^{28}Al	1779	5 min.
Cl	^{38}Cl	1642	15 min.
K	^{42}K	1525	4-7 d
Ca	^{47}Ca	1297	4-7 d
Fe	^{59}Fe	1099	30 d
Sc	^{46}Sc	889	30 d
V	^{52}V	1434	5 min.
Ti	^{51}Ti	320	5 min.
Cr	^{51}Cr	320	30 d
Mn	^{56}Mn	847, 1810	2 h
Co	^{60}Co	1173, 1332	30 d
Zn	^{65}Zn	1115	30 d
As	^{76}As	559	4-7 d
Se	^{75}Se	264	30 d
Br	^{82}Br	776	4-7 d
Rb	^{86}Rb	1077	30 d
Sr	^{85}Sr	514	30 d
Ag	^{110m}Ag	658	30 d
Sb	^{124}Sb	602, 1691	30 d
Cs	^{134}Cs	796	30 d
Ba	^{131}Ba	496	4-7 d
La	^{140}La	1596	4-7 d; 30 d
Ce	^{141}Ce	145	30 d
Nd	^{147}Nd	91	4-7 d
Sm	^{153}Sm	103	4-7 d
Eu	^{152}Eu	121, 244, 344, 1408	30 d
Gd	^{153}Gd	103	30 d
Tb	^{160}Tb	879	30 d
Tm	^{170}Tm	84	30 d
Yb	^{169}Yb	63, 130, 177	30 d
Lu	^{177}Lu	208	4-7 d
Hf	^{181}Hf	482	30 d
Ta	^{182}Ta	67, 100, 152, 1221	30 d
Hg	^{203}Hg	279	30 d
Th	^{233}Pa	312	30 d
U	^{239}Np	106,277	4-7 d

pore filters to form the targets to be analyzed by the PIXE technique. The experimental conditions in the Van de Graaff accelerator were as follows: 1.7 MeV energy of proton beam, 20 nA current, 30 min counting time, 30 μC collected charge, 25-μm thick aluminum absorber between the sample and the detector for the preconcentrated targets. The Al and the other minor elements were measured without absorber.

The characteristic X-ray spectra were measured with an Si (Li) detector, and the peak areas were evaluated using the AXIL program *(4)*. The experimental efficiencies (calibration factors) were determined using samples prepared from standard solutions with known concentration of each element and of internal standard (Y, Pd) using the same chemical procedure as for the unknown samples, and submitting the targets to the same experimental conditions. Blanks were studied in order to detect possible contamination during the chemical procedure.

XRF

Six subsamples (1–2 g each) were pressed into pellets, with a diameter of 3.2 cm and a sample thickness of about 300 mg/cm^2 for the lichen material and of 100–150 mg/cm^2 for the cabbage material, using a pressure for 10 min. The pellets were measured using an energy-dispersive X-ray fluorescence spectrometer (Kevex Delta XRF Analyst System) equipped with adequate secondary targets and filters.

The primary photon beam is produced by a high-power rhodium anode X-ray tube. The characteristic X-rays emitted by the samples were collimated at 90° and measured with an Si(Li) detector (resolution of 165 eV at 5.9 keV and an active area of 30 mm^2). The operating conditions used for analysis are shown in Table 2.

Peak area determinations were routinely performed by using a Gaussian deconvolution that assumes a theoretical relationship among the X-ray line intensities within each element line series *(8)*. Elemental concentrations were determined through the MARS computer program, which corrects for light element absorption based on the mean atomic number of the sample. The mean atomic number is determined from the Compton/Rayleigh scatter intensity ratio *(9)*. Pellets made of powders of pure compounds ($C_6H_{10}O_5$, H_3BO_3, $NaNO_3$, and NaCl) were used to establish the correlations between the ratios of the intensities of the incoherently and coherently scattered X-rays of the Ag secondary target by the sample, with the atomic number (Z).

AAS and ICP-ES

Minor and trace elements were determined by atomic absorption spectrometry (AAS) and inductively coupled plasma emission spectrometry (ICP-ES). The analytical procedure is schematized in a diagram shown in Fig. 1.

Standard solutions of each element were prepared from Johnson Matthey spectrographically standardized metals or suprapure salts. Alternatively, 1.00 g/L spectrosol-grade (BDH) standard solutions was

Table 2
Operating Conditions Used for Analysis of IAEA-336 and IAEA-359 by XRF

		Secondary targets		
Exc. mode	primary	Ge	Ag	Ti
KV	4	15	35	15
mA	0.2	3.0	1.0	1.3
Atm.	vacuum	vacuum	vacuum	vacuum
Time (sec.)	1000	1000	1000	1000
Det. elements	Mg	Mn, Fe, Cu, Zn	Br, Rb, Sr	P, S, Cl, K, Ca

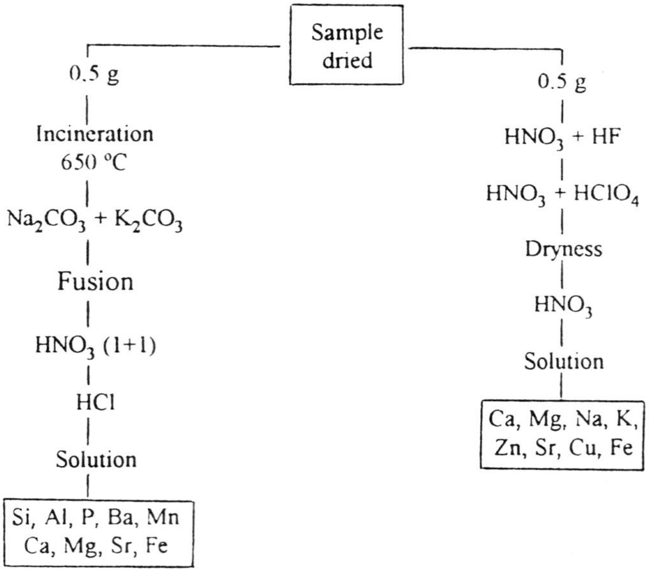

Fig. 1. Analytical diagram for AAS and ICP-ES.

used. Calibration solutions were prepared by dilution of stock solution, and for ICP-ES analysis, some multielement standards were prepared. All volumetric glassware were class by AS. The test solutions were prepared by using platinum crucibles. The atomic absorption spectrometer used was UNICAM PU 9100 X, with hollow cathode lamps of each element. The spectrometer for ICP-ES was UNICAM PU 7000, sequential, with grid nebulizer, a fixed quartz torch, and a 40.68 MHz free-running generator. The elements Ca, Mg, Na, K, Cu, Fe, Sr, and Zn were determined in the solution obtained by acid treatment with HNO_3, HF, and $HClO_4$, evaporation to dryness and HNO_3 dissolution. For determination of the elements Si, Al, P, Ba, and Mn, sample fusion with a mixture of potassium carbonate and sodium carbonate after incineration at 650°C was carried out. The elements Ca, Mg, Sr, and Fe were determined in both solutions and the results compared. For each procedure, a blank solution was always prepared. The calibration solutions were prepared with the same volume of the blank solution as in the test solution.

AAS was used for sodium and potassium determination. Operating conditions were optimized for maximum absorbance while aspirating the standard solutions, using the most sensitive resonance line for each element. The spectrometer was operated in absorbance mode, and an integration time of 3 s was used. The results were the means of three measurements. The linear regression was applied with a calculator for AAS data. ICP-ES was used for determination of the elements Ca, Mg, Cu, Fe, Sr, Si, Al, P, Ba, Mn, and Zn, using 1.0 kW power, 13 L/min argon coolant, and other conditions as recommended by the manufacturer. The integration time (3–10 s) and number of measurements (3–6) were dependent on the sensitivity of the elements and stability of the results. The most sensitive lines free of interference were used: Ca—393.366 nm, Mg—279.553 nm, Cu—324.754 nm, Fe—259.940 nm, Sr—421.552 nm, Si—251.611 nm, Al—308.215 nm, P—214.910 nm, Ba—455.403 nm, Mn—257.610 nm, and Zn—213.856 nm. The results were obtained using calibration curves for each element with at least three points. For ICP-ES, the concentration value was directly obtained from the straight-line calibration of the computer program.

RESULTS

Tables 3 and 4 show the results obtained by the six INETI analysis groups using INAA, PIXE, XRF, AAS, and ICP-ES for the lichen IAEA-336 and the cabbage IAEA-359, respectively. In Table 4, the IAEA estimates are also presented (no uncertainties are assigned to them).

The values in Table 3 show that, for IAEA-336, INAA provided results on 38 elements out of a total of 44; 21 of the determined concentrations are at the ng/g level. The following heavy metals, relevant for pollution studies, were determined with this technique: As, Co, Cr, Fe, Hg, Mn, Sb, Se, Ti, V, and Zn. The other relevant heavy metals, Cu, Ni, and Pb, could be determined by PIXE (Cu also by XRF, and ICP-ES). The minor elements P, S, and Si were also determined by PIXE (S and P also by XRF, and P and Si by ICP-ES). The detection limits of the elements determined by PIXE, XRF and AAS/ICP-ES were not below the μg/g level. It should be stressed that the IAEA values in Table 4 are only estimates. Nevertheless, it can be concluded that 23 values out of 34 results obtained in this work agree within ±10% with the IAEA estimates. The IAEA value for Mn seems to be overestimated.

Table 3
Elemental Concentration in μg/g for the Lichen IAEA-336**

		Group 1 INAA k_0 method	Group 2 INAA comp. method	Group 3 PIXE thick target	Group 4 PIXE thin target	Group 5 XRF	Group 6 AAS/ICP-ES
ng/g	Ag	29.3±2.0	–	–	–	–	–
	Al	631±20	–	–	650±28	–	711.3±2.6
ng/g	As	568±16	636.7±6.2	–	–	–	–
	Ba	6.82±0.18	5.92±0.18	–	–	–	4.50±0.50
	Br	15.49±0.32	12.90±0.11	3.40±0.24	–	18.00±0.32	–
	Ca	3375±155	–	2888.3±8.7	3298±84	3880±20	2405±24
	Ce	1.269±0.035	1.146±0.022	–	–	–	–
	Cl	1696±31	–	2165±22	–	2200±45	–
ng/g	Co	332.6±3.5	292.2±5.2	–	–	–	–
	Cr	1.338±0.012	1.0911±0.0075	–	1.63±0.13	–	–
ng/g	Cs	140.0±6.6	113.3±1.7	–	–	–	–
	Cu	–	–	3.83±0.46	6.01±0.63	6.80±0.37	5.17±0.54
ng/g	Eu	18.75±0.50	24.00±0.47	–	–	–	–
	Fe	494.1±4.6	441.4±3.6	418±15	490.0±7.2	664.0±2.7	405.2±3.3
ng/g	Gd	891±29	–	–	–	–	–
ng/g	Hf	58.72±0.82	58.56±0.58	–	–	–	–
ng/g	Hg	247.7±6.4	192±11	–	–	–	–
	K	1859±33	1793±10	2005±11	2430±69	2020±73	1680±38*
ng/g	La	650±35	620.0±8.5	–	–	–	–
ng/g	Lu	7.26±0.75	6.77±0.14	–	–	–	–
	Mg	585±49	–	–	–	251±28	624.9±7.8
	Mn	53.1±1.6	61.44±0.47	63.2±1.6	67.1±2.1	–	65.33±0.61

However, several discrepancies between the results of methods employed or between the results of this work and the IAEA estimates were also detected, which would need further investigation. Such examples are the results by PIXE (thick target) for Br, Cu, and Pb, the results by PIXE (thin target) for Cr, Pb, Si, and Ti, the results by XRF for Fe and

Table 3 *(continued)*

		Group 1 INAA k_0 method	Group 2 INAA comp. method	Group 3 PIXE thick target	Group 4 PIXE thin target	Group 5 XRF	6 AAS/ICP-ES
	Na	361.8±6.1	389.7±2.9	–	–	–	289.8±2.9*
ng/g	Nd	490±26	543±16	–	–	–	–
	Ni	–	–	–	2.95±0.26	–	–
	P	–	–	675±45	–	590±44	572.5±2.5
	Pb	–	–	7.08±0.39	4.13±0.47	–	–
	Rb	1.926±0.034	1.634±0.019	1.97±0.27	–	2.20±0.20	–
	S	–	–	836±28	–	871±26	–
ng/g	Sb	89.9±2.0	78.6±5.3	–	–	–	–
ng/g	Sc	205.1±2.0	170.0±1.7	–	–	–	–
ng/g	Se	261.0±7.2	268.9±8.4	–	–	–	–
	Si	–	–	1897±134	285±29	–	1877±23
ng/g	Sm	101.4±9.2	111.2±3.3	–	–	–	–
	Sr	11.20±0.70	8.58±0.18	7.47±0.38	–	11.00±0.32	8.64±0.17
ng/g	Ta	15.25±0.34	16.44±0.47	–	–	–	–
ng/g	Tb	15.03±0.76	15.26±0.66	–	–	–	–
ng/g	Th	166.1±2.0	144.4±2.4	–	–	–	–
	Ti	56.8±4.2	–	51.67±0.99	37.4±1.5	–	–
ng/g	Tm	5.75±0.65	–	–	–	–	–
ng/g	U	47.68±0.45	–	–	–	–	–
	V	1.436±0.072	–	–	–	–	–
ng/g	Yb	23.8±1.0	45.22±0.32	–	–	–	–
	Zn	36.14±0.35	28.10±0.23	31.83±0.70	36.1±1.5	36.60±0.40	32.60±0.51

*AAS
**The uncertainties are standard errors.

Mg, the results by AAS/ICP-ES for Al, Ba, Ca, K, and Na in the lichen IAEA-336 material (cf Table 3) and the results by PIXE (thin target) for K and Pb, and the results by XRF for Br, Rb, and Sr in the cabbage IAEA-359 material (cf Table 4).

Table 4
Elemental Concentration in μg/g for the Cabbage IAEA-359**

		Group 1 INAA k_0 method	Group 2 INAA comp. meth.	Group 3 PIXE thick target	Group 4 PIXE thin target	Group 5 XRF	Group 6 AAS/ICP-ES,	IAEA estimated
	Al	127±21	--	--	457±15	--	169.0±2.9	--
ng/g	As	72.8±6.3	--	--	--	--	--	--
	Ba	10.49±0.46	8.42±0.17	--	--	--	9.5±1.5	--
	Br	7.81±0.27	6.37±0.11	7.00±0.52	--	4.67±0.19	--	--
%	Ca	2.08±0.10	--	2.245±0.016	2.05±0.11	1.970±0.015	1.895±0.015	(2.0)
ng/g	Ce	220±14	--	--	--	--	--	--
%	Cl	0.818±0.026	--	1.0633±0.0080	--	0.8533±0.0067	--	--
ng/g	Co	184±11	110.0±1.7	--	--	--	--	--
	Cr	1.739±0.099	1.417±0.032	--	1.98±0.12	--	--	(2.0)
ng/g	Cs	27.8±2.1	27.78±0.76	--	--	--	--	--
	Cu	--	--	6.33±0.42	6.02±0.20	5.67±0.33	5.50±0.50	--
	Eu	--	4.11±0.11	--	--	--	--	--
	Fe	188±10	155.9±3.2	168.0±2.5	166±16	176.0±3.5	--	(170)
ng/g	Hf	15.95±0.95	14.11±0.35	--	--	--	--	--
ng/g	Hg	36.0±3.4	--	--	--	--	--	--
%	K	3.43±0.092	3.130±0.030	3.897±0.032	4.67±0.28	3.547±0.024	2.917±0.059*	(3.0)
ng/g	La	92.8±3.1	120.0±9.1	--	--	--	--	--

%	Mg	0.2040±0.0099	--	--	--	0.2200±0.0044	0.2074±0.0038	(0.2)
	Mn	27.11±0.92	30.24±0.41	35.33±0.56	37.8±2.2	36.3±1.2	33.00±0.66	(40)
	Na	640.0±8.9	682.1±6.3	--	--	--	524.3±3.7*	(600)
	Ni	--	--	--	2.08±0.42	--	--	(1.0)
%	P	--	--	0.6713±0.0072	--	0.5233±0.0088	0.5410±0.0030	--
	Pb	--	--	--	3.37±0.79	--	--	(1.5)
	Rb	7.21±0.10	5.97±0.11	5.83±0.40	--	4.67±0.33	--	--
%	S	--	--	1.848±0.017	--	1.493±0.014	--	--
ng/g	Sb	24.63±0.72	--	--	--	--	--	--
ng/g	Sc	33.4±1.8	24.11±0.35	--	--	--	--	--
ng/g	Se	157±11	--	--	--	--	--	--
	Si	--	--	894±74	969±53	--	768±16	--
ng/g	Sm	10.49±0.47	--	--	--	--	--	--
	Sr	53.3±1.1	--	49.5±1.1	60±13	34.3±1.8	48.60±0.40	--
ng/g	Th	30.7±1.4	26.89±0.70	--	--	--	--	--
	Ti	--	--	13.33±0.56	15.93±0.84	--	--	--
	Zn	42.7±2.2	36.20±0.72	42.83±0.83	40.8±1.6	46.67±0.88	35.33±0.88	(40)

*AAS
**The uncertainties are standard errors.

CONCLUSION

From this work it is obvious that atomic analysis methods produce valuable results, although they cannot provide as much information on elemental composition of biomonitors of environmental pollution as do the nuclear methods. The INAA method, because of very low detection limits, is capable of providing information on elements, such as Hg, Sb, Sc, and Se, usually present at very low concentrations in biomonitors. However, determination of these elements is important in environmental studies. In spite of its very low detection limits for a considerable amount of elements, the INAA method did not provide information on Pb, Cu, Ni, P, S, and Si. Except for Pb, these elements might not be very toxic, but knowledge of their concentration levels can be very important for environmental studies. Thick-target PIXE appeared to be a very promising technique, although its methodology is still in development. XRF can be considered as a technique to be relied on because of the fact that commercial systems are available in the market, which is not the case for INAA or PIXE.

REFERENCES

1. M. C. Freitas, F. M. Catarino, C. Branquinho, and C. Máguas, *J. Radioanal. Nucl. Chem., Articles* **169,** 47 (1993).
2. M. C. Freitas, *Fresenius J. Anal. Chem.* **345,** 304 (1993).
3. K. Govindaraju, *Geostand. News. Spec. Issue* **13,** 1 (1989).
4. P. Van Espen, K. Janssens, and I. Nobels, *Chemolab.* **1,** 109 (1986).
5. M. A. Reis and L. C. Alves, *Nucl. Instr. Methods* **B68,** 300 (1992).
6. T. Badica, C. Ciortea et al. *Nucl. Instr. Methods* **B3,** 288 (1984).
7. P. Marijanovic, J. Makjanic, and V. Valkovic., *J. Radioanal. Nucl. Chem.* **81,** 353 (1984).
8. T. P. Schreiber and A. M. Wims, *X-Ray Spectrom.* **11,** 42 (1982).
9. T. O'Reilly and B. W. King, *Adv. X-Ray Anal.* **30,** 165 (1986).

Comparative Study of Methods for Determining Lanthanide Elements in Biological Materials by Using NAA, HPLC Postcolumn Reaction, and ICP-MS

MOMOKO CHIBA,*[,1] ATSUKO SHINOHARA,[1] MITIKO SAIKI,[2] AND YUTAKA INABA[1]

[1]*Juntendo University School of Medicine, Tokyo 113, Japan; and* [2]*Nuclear and Energy Research Institute, São Paulo, Brazil*

ABSTRACT

Mice were injected intravenously with one of the following elements: Y, La, Ce, Nd, Sm, Eu, Gd, Tb, Dy, Ho, Er, or Yb, at a dose of 25 mg/kg body wt. After 20 h, the various organs were taken out, and the element concentrations were determined by NAA, HPLC postcolumn reaction, or ICP-MS. The results by these three methods were within the acceptable range. About 85% of the amount administered of each element was found in liver, lung, and spleen.

Index Entries: Lanthanide elements; NAA; ICP-MS; HPLC postcolumn reaction; mouse; iv injection; distribution; Arsenazo-III.

INTRODUCTION

The use of lanthanide elements has been increasing in the fields of both traditional and ultramodern industries. There have been no reports on the health effects in relation to environmental contamination owing to disposal of used products containing such rare elements. The possibility that general residential areas may be influenced by rare elements cannot

*Author to whom all correspondence and reprint requests should be addressed.

Biological Trace Element Research Editor: G. N. Schrauzer © 1994 by Humana Press Inc.

be denied, because the types and amounts of the products containing rare elements are increasing and old or used ones may be discharged into the common environment. There is a concern that leaching by rain, especially acid rain, may enlarge the contaminated area.

We succeeded in determining Tb, Eu, and Dy using 2,6-pyridinedicarboxylic acid as a fluorometric ligand *(1,2)*. In this article, we introduce a colorimetric method of the HPLC postcolumn reaction as a determination technique for the lanthanide elements. The results were comparable to the values determined by neutron activation analysis (NAA) and inductively coupled plasma-mass spectrometry (ICP-MS).

In addition, we describe the distribution of the lanthanide elements after 20 h of iv injection in mice.

MATERIALS AND METHODS

Chemicals Administered to Mice

The chloride compounds of yttrium (Y), lanthanum (La), cerium (Ce), neodymium (Nd), samarium (Sm), europium (Eu), gadolinium (Gd), terbium (Tb), dysprosium (Dy), holmium (Ho), erbium (Er), and ytterbium (Yb) were used. Each compound was dissolved in a 5% glucose solution.

Animal Treatment

Five-week-old mice were divided into 12 groups, which consisted of five mice each, and another group served as a control. Each mouse received a single iv injection with one of the elements at the dose of 25 mg element/kg body wt. At 20 h after the injection, the animals were sacrificed using ether anesthesia, and organ specimens were removed after perfusion by the 5% glucose solution.

Reagents

Nitric acid (HNO_3) and perchloric acid ($HClO_4$) of analytical grade or reagents for atomic absorption were used. Formic acid, 2-hydroxybutyric acid, and 2,7-*bis*-(2-arsenophenylazo)-1,8-dihydroxynaphthalene-3,6-disulfonic acid (Arsenazo-III) were of the reagent grade commercially available.

Analytical Procedure

NAA

The lyophilized liver samples were powdered by grinding. Then samples of about 100 mg each were weighed in polyethylene bags and sealed by heating. Irradiations were carried out with appropriate standard samples in an IEA-RI nuclear research reactor. The detailed condi-

Table 1
Experimental Conditions of NAA

Element	Irradiation time	Neutron flux	Decay time	Radioisotope measured	half life	gamma ray energy(keV)
La	30 min	(a)	1d	^{140}La	40.27h	1597
Ce	30 min	(b)	4d	^{141}Ce	32.5d	145.4
Nd	8 h	(c)	6d	^{147}Nd	11.1d	91
Sm	30 min	(a)	1d	^{153}Sm	47.1h	69 and 103
Eu	3 min	(a)	1h	^{152m}Eu	9.35h	841
Gd	30 min	(a)	4h	^{159}Gd	18h	363
Tb	30 min	(b)	11d	^{160}Tb	73d	298
Dy	30 min	(a)	19h	^{165}Dy	2.36h	634
Ho	10 min	(a)	20h	^{166}Ho	26.9h	80 and 1378
Er	15 min	(a)	1d	^{171}Er	7.8h	308
Yb	30 min	(a)	15h	^{175}Yb	101h	396

(a) $2.67 \times 10^{11} n/cm^2/s$.
(b) $1.54 \times 10^{12} n/cm^2/s$.
(c) $10^{12} n/cm^2/s$.

tions are listed in Table 1. The radioactivities were counted by an ENERTEC hyperpure Ge detector coupled to an EG & G Ortec 4096 channel pulse height analyzer connected to a Monydata PC 200 Plus microcomputer.

HPLC Postcolumn Reaction

An aliquot of the liver samples, 20–100 mg, was wet-ashed using HNO_3 and $HClO_4$. After heating at 100–120°C for 1–2 h and then at 200°C for about 1 h, the residue was dissolved in 0.5% HNO_3. This solution was passed through an ion-exchange column (Shim-pack IC-CI) installed with HPLC (Shimadzu LC 6A); 0.05–0.2 mol/L 2-hydroxybutyric acid (pH 4.4) was used as a mobile phase. The eluents were reacted with Arsenazo-III to develop color. Optical density at 655 nm was measured for all lanthanide elements, with 2.5 ng as the detection limit.

ICP-MS

0.5% HNO_3 solution after digesting liver samples was applied to ICP-MS (Yokogawa PMS-200). The detection limits were 0.02 ng element/mL.

RESULTS

The Element Concentrations in Liver

Table 2 shows the element concentrations in liver, determined by HPLC-colorimetric and ICP-MS methods as a mean value and SD (in µg/g, wet wt). Table 3 shows the results obtained by HPLC-colorimetric and NAA methods. In order to compare the results obtained by these three methods, element ratios were calculated as shown in Table 4. The ratios for HPLC/ICP-MS ranged from 0.91–1.17, whereas the ratios for NAA/ICP-MS were from 0.76–1.33. As for the CV(%), values were relatively small for HPLC/ICP-MS, below 10%, whereas for NAA/ICP-MS, they were relatively large.

The Element Concentrations Determined in Various Organs of Mice

Figure 1A shows the Tb or Yb concentrations in various organs of mice. In control mice, all lanthanide elements were under the detection limit. Figure 1B shows the distribution ratios of the total amounts administered. More than 85% of the dose administered was found in liver, lung, and spleen. Figure 2 shows the element concentrations in spleen, lung, and liver.

DISCUSSION

In the papers reported previously, analytical methods applied to investigate the distribution of the lanthanide elements administered, used mainly radioisotopes *(3–5)*. Since analytical instruments and techniques have been markedly improved recently, we tried to determine the lanthanide elements in biological materials using HPLC equipped with fluorometric and colorimetric detectors, and ICP-MS. In addition, some of the samples were analyzed by NAA in order to check results of the previously mentioned methods.

The injection dose chosen of 25 mg element/kg was extremely high. The reason was that neither the distribution ratios in individual organs nor the detection limits of the methods applied were known. It was assumed that the element concentrations in organs, such as thymus, submandibular gland, or sexual organs, could be very low and close to the detection limits. Therefore, such a relatively high dose was chosen.

The detection limit of the HPLC postcolumn colorimetric determination using Arsenazo-III was 2.5 ng which allowed determination of all lanthanide element included in this study. In the case of using the gradient mode for the mobile phase, each element could be separated, although it took a little longer and the detection limits were somewhat higher.

Table 2
Concentrations of Lanthanide Elements Determined in the Liver of Mice by HPLC Postcolumn Reaction and ICP-MS Methods

Element	Concentrations (μg/g wet wt)	
	HPLC (Arsenazo-III)	ICP-MS
Y	138.7 ± 15.5	141.5 ± 12.8
La	151.5 ± 17.7	164.1 ± 19.6
Ce	155.8 ± 15.1	161.3 ± 12.3
Nd	129.3 ± 16.9	141.6 ± 19.3
Sm	129.4 ± 28.4	134.6 ± 31.5
Eu	149.8 ± 25.4	147.7 ± 21.7
Gd	126.8 ± 7.4	108.4 ± 7.5
Tb	158.2 ± 11.8	156.8 ± 10.9
Dy	149.3 ± 19.9	141.2 ± 19.1
Ho	186.4 ± 15.8	187.1 ± 13.0
Er	191.7 ± 31.5	190.0 ± 31.0
Yb	154.3 ± 18.9	163.4 ± 21.8

Mean ± SD, n = 4 ~ 5, 20 h after iv injection of 25 mg element/kg.

Table 3
Concentrations of Lanthanide Elements Determined in the Liver of Mice by HPLC Postcolumn Reaction and INAA

Element	Concentation (μg/g dry wt)	
	HPLC (Arsenazo-III)	INAA
La	950 ± 162	1039 ± 154
Ce	848 ± 131	753 ± 131
Nd	764 ± 255	598 ± 211
Sm	718 ± 146	556 ± 102
Eu*	212 ± 47	247 ± 55
Gd	757 ± 68	676 ± 75
Tb	583 ± 23	695 ± 66
Dy*	243 ± 30	254 ± 31
Ho	1087 ± 380	790 ± 126
Er	986 ± 106	1049 ± 197
Yb	937 ± 19	1170 ± 105

Mean ± SD, n = 4 ~ 5, 20 h after iv injection of 25 mg element/kg (*10 mg element/kg)

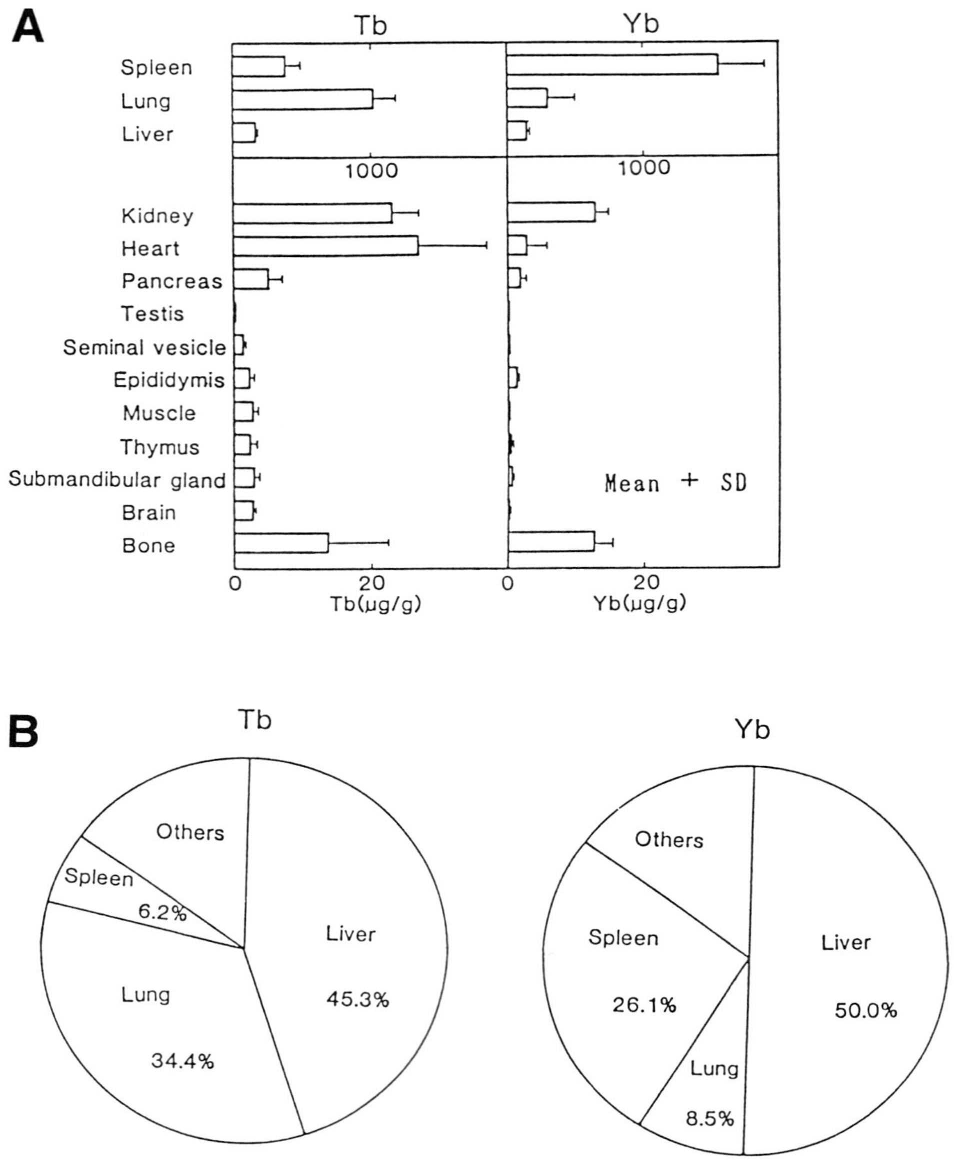

Fig. 1. The concentrations of Tb and Yb in various organs of mice injected with Tb or Yb (A), and the distribution ratios of the amounts administered (B).

Element recoveries by HPLC postcolumn colorimetric method were also studied, and values from 98.2% (Nd) to 105.0% (Ce) were found.

Lyophilized samples for NAA and wet-ashed samples for HPLC and ICP-MS were prepared separately from the same liver specimens. The sample solutions after wet-ashing were divided into two portions, which were then applied to HPLC and to ICP-MS. This way of sample prepa-

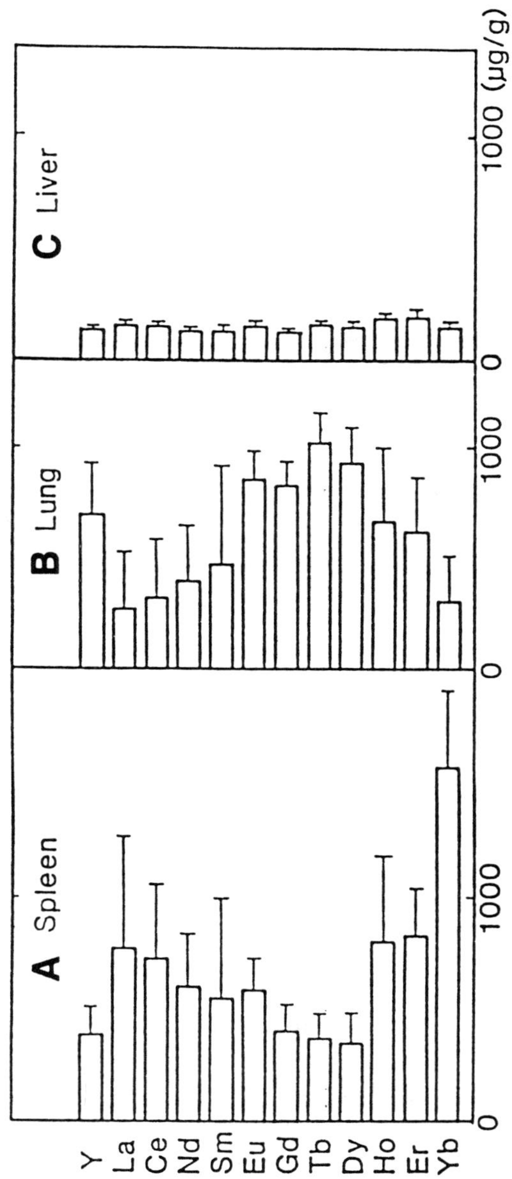

Fig. 2. The concentrations of the elements determined in the spleen, lung, and liver of mice injected with respective lanthanide elements.

Table 4
Comparison of the Results Obtained by the Three Methods Studied

Element	HPLC/ICP-MS		INAA/ICP-MS	
	MEAN	CV(%)	MEAN	CV(%)
Y	0.98	5.1	-	-
La	0.93	9.1	1.17	6.9
Ce	0.97	7.3	0.86	13.8
Nd	0.91	3.5	0.78	2.9
Sm	0.97	4.5	0.76	13.5
Eu	1.01	2.6	0.77*	14.3
Gd	1.17	2.0	1.15	4.3
Tb	1.01	2.2	0.82	3.9
Dy	1.06	2.6	0.83*	12.8
Ho	1.00	2.5	0.78	11.1
Er	1.01	1.6	1.02	20.4
Yb	0.94	3.7	1.33	3.7

*Estimated value.
$n = 4 \sim 5$.

ration may be considered as a reason why better agreement was found between the results by HPLC postcolumn reaction and ICP-MS than agreement between the results of either of these methods and the NAA results, because differences in the element distribution inside the liver may occur.

SUMMARY

In order to be able to investigate health effects of lanthanide elements, three methods of their determination were examined. Twelve of the lanthanide elements, Y, La, Ce, Nd, Sm, Eu, Gd, Tb, Dy, Ho, Er, and Yb, were determined in various organ samples of mice by NAA, HPLC postcolumn reaction, and ICP-MS. The results obtained by the three methods agreed well within acceptable limits. The lanthanide elements injected intravenously were found mainly in the liver, lung, and spleen, whereas much lower concentrations were detected in other organs of mice.

REFERENCES

1. A. Shinohara and M. Chiba, Anal. *Sci.* **7,** suppl. 775 (1991).
2. A. Shinohara, M. Chiba, and M. Kikuchi, *J. Anal. Toxicol.* **13,** 135 (1989).
3. B. Ewaldsson and G. Magnusson, *Acta Radiol.* **2,** 65 (1964).
4. B. Ewaldsson and G. Magnusson, *Acta Radiol.* **2,** 121 (1964).
5. P. B. Hammond and R. P. Beliles, in *Casarett and Doull's Toxicology,* J. Doull, C. D. Klaassen, and M. O. Amdur, eds., Macmillan, New York, 1980, pp. 409–469.

The IRMM—International Measurement Evaluation Program (IMEP)

IMEP-3—Trace Elements in Synthetic and Natural Water

A. LAMBERTY,*[,1] G. LAPITAJS,[2] L. VAN NEVEL,[3] A. GÖTZ,[4] J. R. MOODY,[5] D. E. ERDMANN,[6] AND P. DE BIEVRE[1]

[1]Institute for Reference Materials and Measurements (IRMM), B-2440 GEEL, Belgium; [2]University of Saarbrücken, FRG; [3]IRMM/University of Antwerpen, Belgium; [4]Visiting Scientist at IRMM, Belgium; [5]NIST, Gaithersburg, MD 20899; and [6]US Geological Survey, Denver, CO

ABSTRACT

The aim of the IRMM—International Measurement Evaluation Program (IMEP) is to test out a possible realization of international measurement comparability for field laboratories through traceability of their measurements to the SI Unit for amount of substance: the mole. In IMEP-3, 10 different trace elements, B, Ca, Cd, Cu, Fe, K, Li, Pb, Rb, and Zn, were determined in a synthetic and a natural water by participating laboratories using their routine methods and graphically compared (in coded form) to certified values, established by IRMM and NIST using an isotope-specific method (Isotope Dilution Mass Spectrometry, Neutron Activation Analysis). The number of participants was 70; 64 laboratories have reported results. The results show a spread of more than 50% asymmetrically around the certified value. The Youden graphs allow evaluation of the overall performance of the laboratories in the IMEP-3 round.

Index Entries: International Measurement Evaluation Program (IMEP); certification by isotope-specific methods; isotope dilution

*Author to whom all correspondence and reprint requests should be addressed.

Biological Trace Element Research Editor: G. N. Schrauzer © 1994 by Humana Press Inc.

mass spectrometry (IDMS); neutron activation analysis (NAA); water analysis; trace elements in water; water pollution; comparability and traceability of analytical results.

INTRODUCTION

Good analytical measurement capability for water analysis is important because of the environmental importance—water quality control in drinking water (EC directive: max. 5 mg/L Cd, max 50 mg/L Pb), in groundwater, in river water, and in treated waste water—and because of the political importance—border crossing pollution of rivers and seas, and scarcity of drinking water. Since measurements determine the official quality and use of the water, it is important to know the true analytical capability of laboratories, and what is the "state of the practice" of water analysis compared to the "state of the art."

It is the intent of IRMM's International Measurement Evaluation Program, IMEP, to provide "state-of-the-practice" pictures of the assay of toxic or life-essential elements in a given matrix and compare them to a certified value obtained using an isotope-specific method, with a provable total uncertainty, considerably smaller than the interlaboratory spread. The certified value is traceable to the SI unit for amount of substance: the mole, in the shortest possible, transparent way. Thus, field laboratories are given the opportunity to realize the comparability of their measurements across borders because their measurements are traceable to a common basis.

Further characteristics of the IMEP program are:

- Participants to work under normal working conditions with choice of measurement methods, procedures, and instrumentation at their discretion;
- Officially guaranteed coded participation—individual results are only decoded to the participant concerned;
- Reporting of results in graphical display; and
- Conclusions from graphical display to be drawn by participants for themselves—assistance is only given if requested and then only **after** the measurement round.

METHODS

Preparation of the Water Materials

Synthetic water A was prepared at the facilities of the US Geological Survey (USGS), Branch of Quality Assurance, Golden, CO under the responsibility of NIST/USGS. Only high-purity reagents were used; the containers were acid-cleaned before use. It was prepared from distilled water and sufficient nitric acid to make the solution approx 0.5 mol/L

HNO_3. Solutions containing known amounts of elements to be determined were added to the acidified water solution. After mixing thoroughly, the solution was filtered through 0.2-μm filters, UV sterilized, and transferred to cleaned 500-mL polyethylene bottles. For the IMEP-3 round, several of these bottles were pooled and repacked into 50-mL acid-cleaned polyethylene bottles. It can now be revealed that water A is identical to what was later distributed as NIST-SRM 1643c *(1)*.

Natural water B was prepared from the Denver Federal Center tap water. The water was acidified to pH 1.5–2 with nitric acid and then "spiked" with constituents needed to attain the desired concentration levels. After mixing, the water was UV-sterilized, filtered, and finally bottled into acid-washed 1-L polypropylene bottles. For the IMEP-3 round, some of these bottles were pooled and repacked into 50-mL acid-cleaned polyethylene bottles. It can now be revealed that water B is identical to USGS T-107 *(2)*.

Establishment of Certified and Assigned Values

The certification of the synthetic water A was done by IRMM and NIST using isotope-specific methods: isotope dilution thermoionic mass spectrometry (IDThMS), isotope dilution inductively coupled plasma (ID-ICP), instrumental neutron activation analysis (INAA), and radiochemical neutron activation analysis (RNAA). The certified values are summarized in Table 1.

For the natural water B, "assigned" values were established for all elements based on the results obtained in a round-robin testing program under the responsibility of USGS based on an educated interpretation of data *(2)*. The results are shown in Table 2. These are not certified values in the sense of the definition used in the IMEP program. These certified and assigned values serve to compare the results obtained under routine conditions by the participating laboratories in the IMEP-3 round.

Description of IMEP-3

Seventy participants requested participation, and 64 returned results to IRMM. Participation was international; laboratories from 13 countries took part (+IRMM); most of them were from Austria and Germany. The methods used for the determination were mainly atomic absorption and emission (cf Table 3) spectroscopic methods.

"State of the Practice" in Graphical Display

The results obtained from the participants were grouped in different ways, as illustrated for Pb in:

Figs. 1 and 2: coded laboratories grouped with ascending values of results making a distinction between laboratories that

Table 1
Certified Values: Synthetic Water A

Element	Concentration [μmol kg^{-1}]*		Concentration [ng mL^{-1}]*
B	10.83±0.13	(1.2%)	119.0±1.4
Ca	904±34	(3.8%)	36800±1400
Cd	0.107±0.009	(8.4%)	12.2±1.0
Cu	0.345±0.043	(12%)	22.3±2.8
Fe	1.884±0.053	(2.8%)	106.9±3.0
K	58**		2300
Li	2.34±0.14	(6.0%)	16.5±1.0
Pb	0.168±0.004	(2.4%)	35.3±0.9
Rb	0.13±0.002	(1.5%)	11.4±0.2
Zn	1.112±0.014	(1.3%)	73.9 ±0.9

*Uncertainties indicated here are 2 s.
**No certified value.

declare themselves as "more" or "less experienced" in these determinations;

Figs. 3 and 4: coded laboratories grouped in order to compare different methods used;

Figs. 5 and 6: results obtained for synthetic water A compared with results obtained for synthetic water A divided by results obtained for natural water B (A/B); and

Fig. 7: Youden plot to show the overall performance of the IMEP-3 round.

RESULTS AND DISCUSSION

The results for all elements show a spread of at least ±50%. The graphs indicate, on one hand, what is the best possible accuracy obtain-

Table 2
Assigned Values: Natural Water B

Element	Concentration [μmol kg^{-1}]*		Concentration [ng mL^{-1}]*
B	12.04±0.65	(5.4%)	130±7
Ca	292.4±5.0	(1.7%)	11.7±0.2
Cd	0.127±0.005	(3.9%)	14.3±0.5
Cu	0.473±0.010	(2.1%)	30.0±0.6
Fe	0.933±0.036	(3.9%)	52±2
K	21.5±1.0	(4.7%)	840±40
Li	27.8±1.0	(3.6%)	193±7
Pb	0.126±0.005	(4.0%)	26±1
Rb	**		
Zn	1.161±0.038	(3.3%)	75.8±2.5

*Uncertainties indicated here are 95% confidence interval.
**No assigned value.

able (state of the art) on the certified value and what is obtained in practice under normal, routine conditions (state of the practice). They give an excellent basis to judge the attainable quality of implementation of official regulatory levels.

By elimination of outliers, the spread of the results reduces to ±30% for most elements. This is a common practice of some reference material producers. However, if important decisions are to be based on measurement results of regulatory laboratories, there are no "outliers": each of the outlying laboratories may be a very official laboratory!

There is no correlation between the self-assessment of the laboratories declaring themselves as "more" or "less experienced" and the reality. The same laboratory may show good performance for one element and bad performance for another. Quality of measurement results depends on the person who is doing the analysis. This was also observed in IMEP-2: Cd in polyethylene *(3)*. Therefore, it is better to impose to a laboratory good quality than a specific method or procedure.

Table 3
Methods Used by Participants (64 Laboratories)

Method	Number of participants
Flame - AAS	29
GF - AAS	54
Flame AES	13
ICP - AES	34
ICP - MS	9
Photometry	9
Ion chromatography	2
Anodic stripping voltametry	1

For all elements, there are far more outliers at the high than on the low concentration side: this indicates contamination problems, even though the concentration levels are quite high (the Cd concentration is higher than allowed in drinking water; *see* EC Directive for drinking water). In the way they have been applied, all methods produced results of the same quality. There was no method that was found to be superior. The best method (IDMS included!) can give wrong results if not correctly applied. On the other hand, "routine methods" can give adequate quality if sufficient attention is given to quality control.

The large spread is not the result of systematic errors that "could have been corrected if we had had the appropriate RM." There is no significant reduction of the spread of the results if results obtained for synthetic water A are divided by the results obtained for natural water B; one would expect that calibration errors cancel out since water A was, unknowingly, a Reference Material NIST-SRM 1643c.

The Youden graphs show the overall performance of the laboratories for each element. They allow one to distinguish the results of laboratories that are too high for both water A and B (upper right quadrant), too low for both water A and B (lower left quadrant), too high for water A and too low for water B (lower right quadrant), and too low for water A and too high for water B (upper left quadrant). Results on the diagonal show a constant deviation (systematic error) from the certified value.

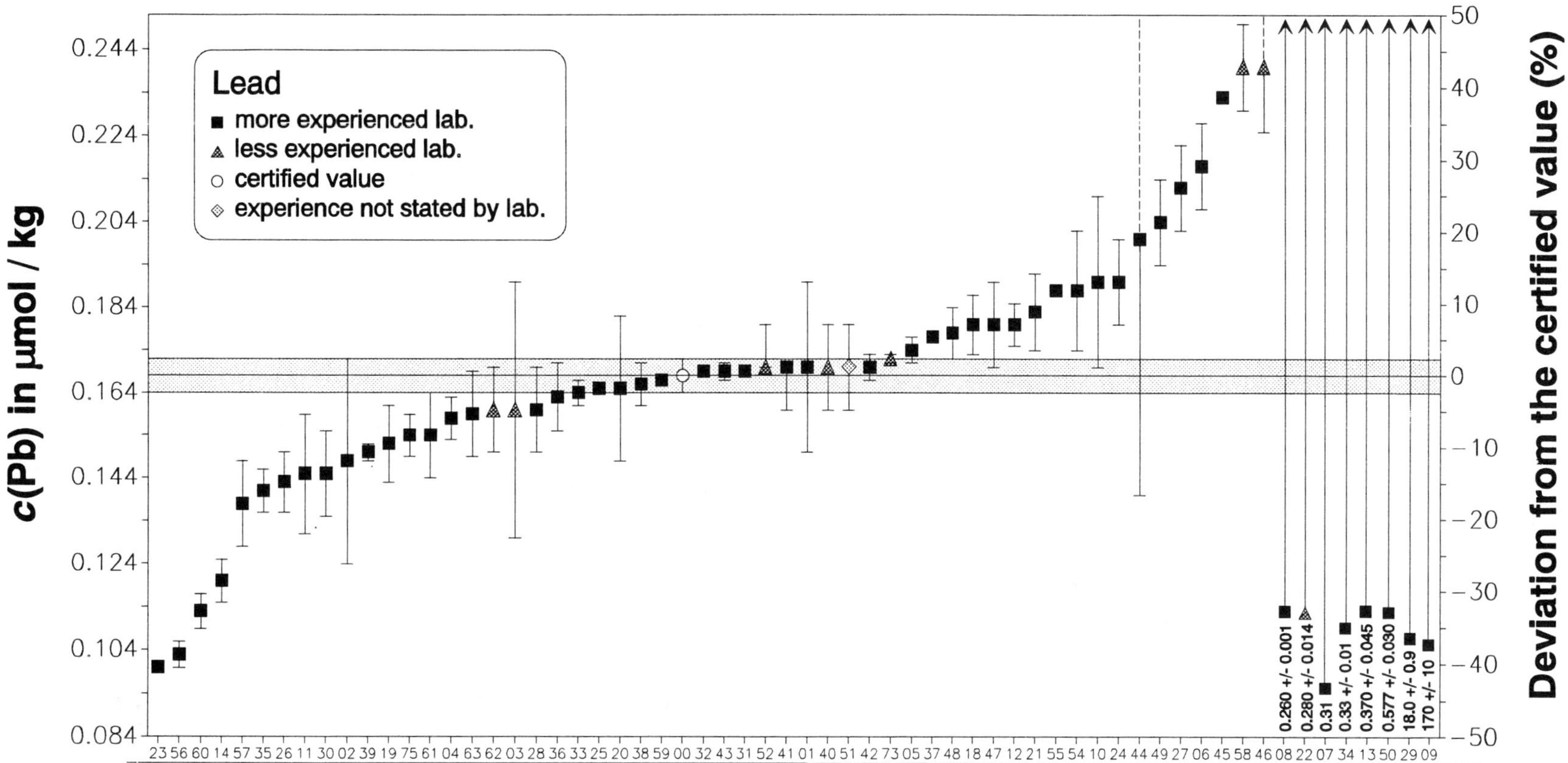

Fig. 1. Coded laboratories arranged by ascending values in making a distinction between laboratories that declare themselves "more" or "less" experienced in the determination of IMEP-3: Trace elements in synthetic water A. Certified value: 0.168(4) µmol/kg, 35.3(9) ng/mL.

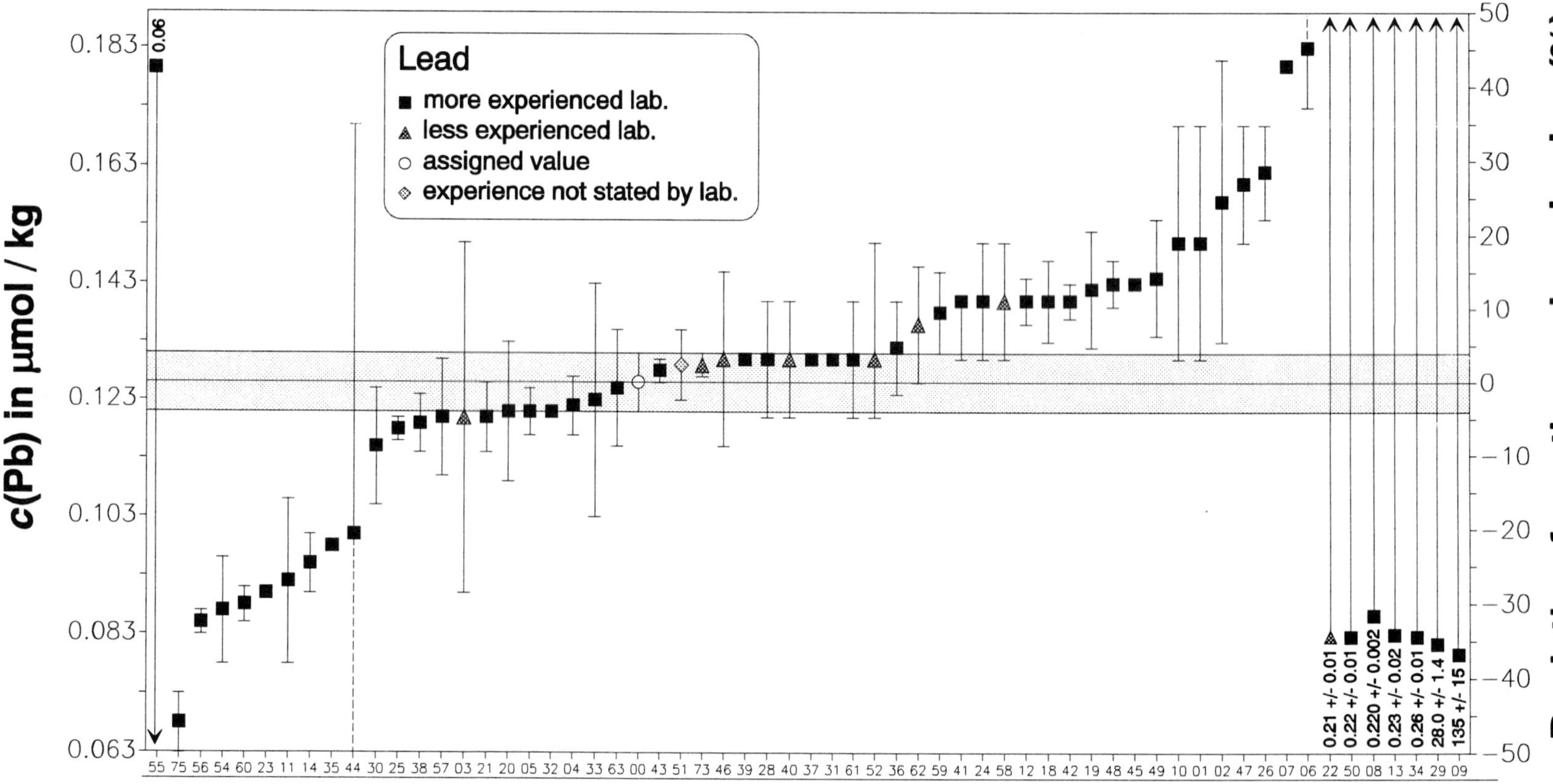

Fig. 2. Coded laboratories arranged by ascending values in making a distinction between laboratories that declare themselves "more" or "less" experienced in the determination of IMEP-3: Trace elements in natural water B. Certified value: 0.126(5) µmol/kg, 26(1) ng/mL.

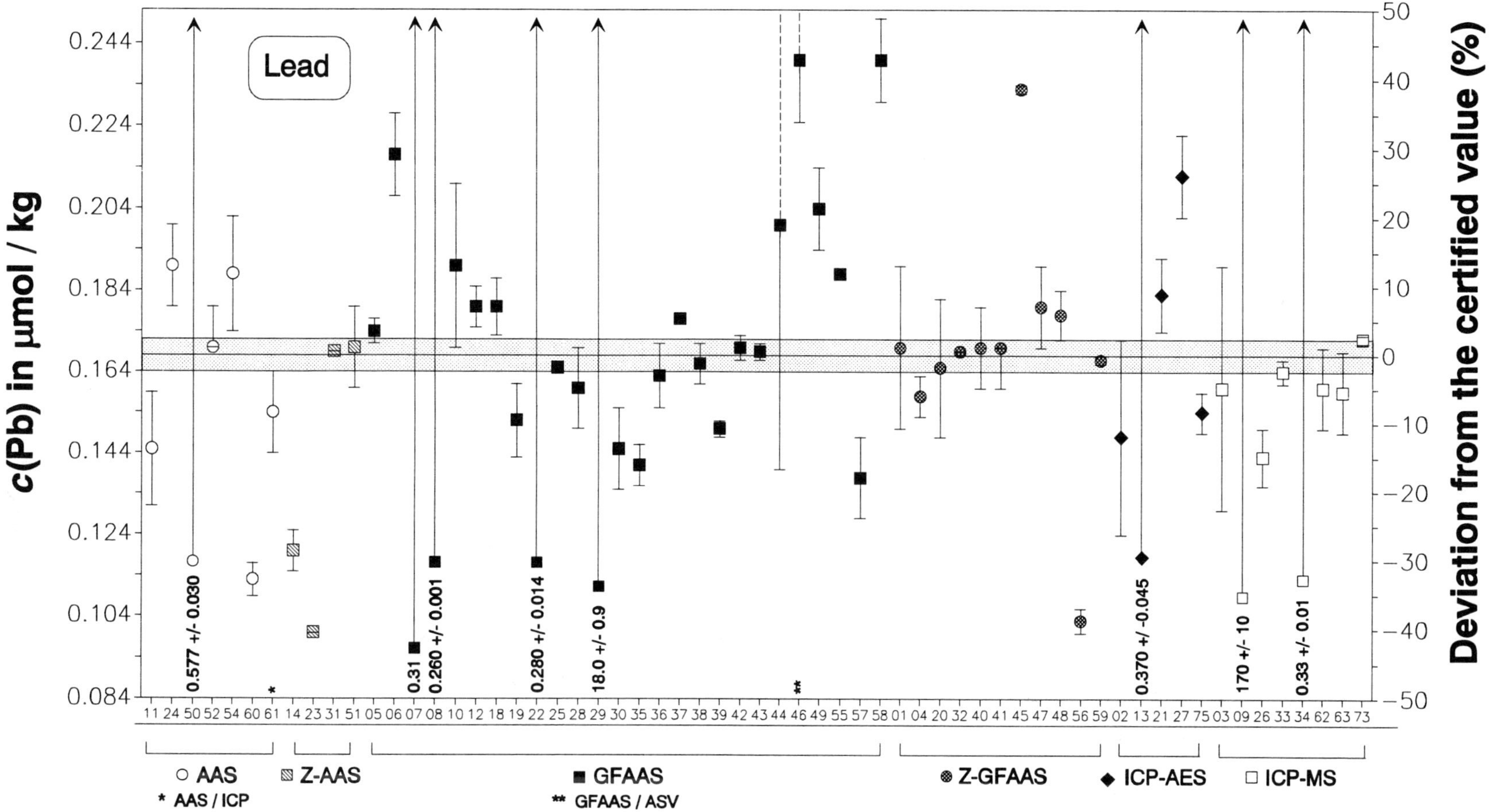

Fig. 3. Coded laboratories arranged in order to compare different methods used in the determination of IMEP-3: Trace elements in synthetic water A. Certified value: 0.168(4) μmol/kg, 35.3(9) ng/mL.

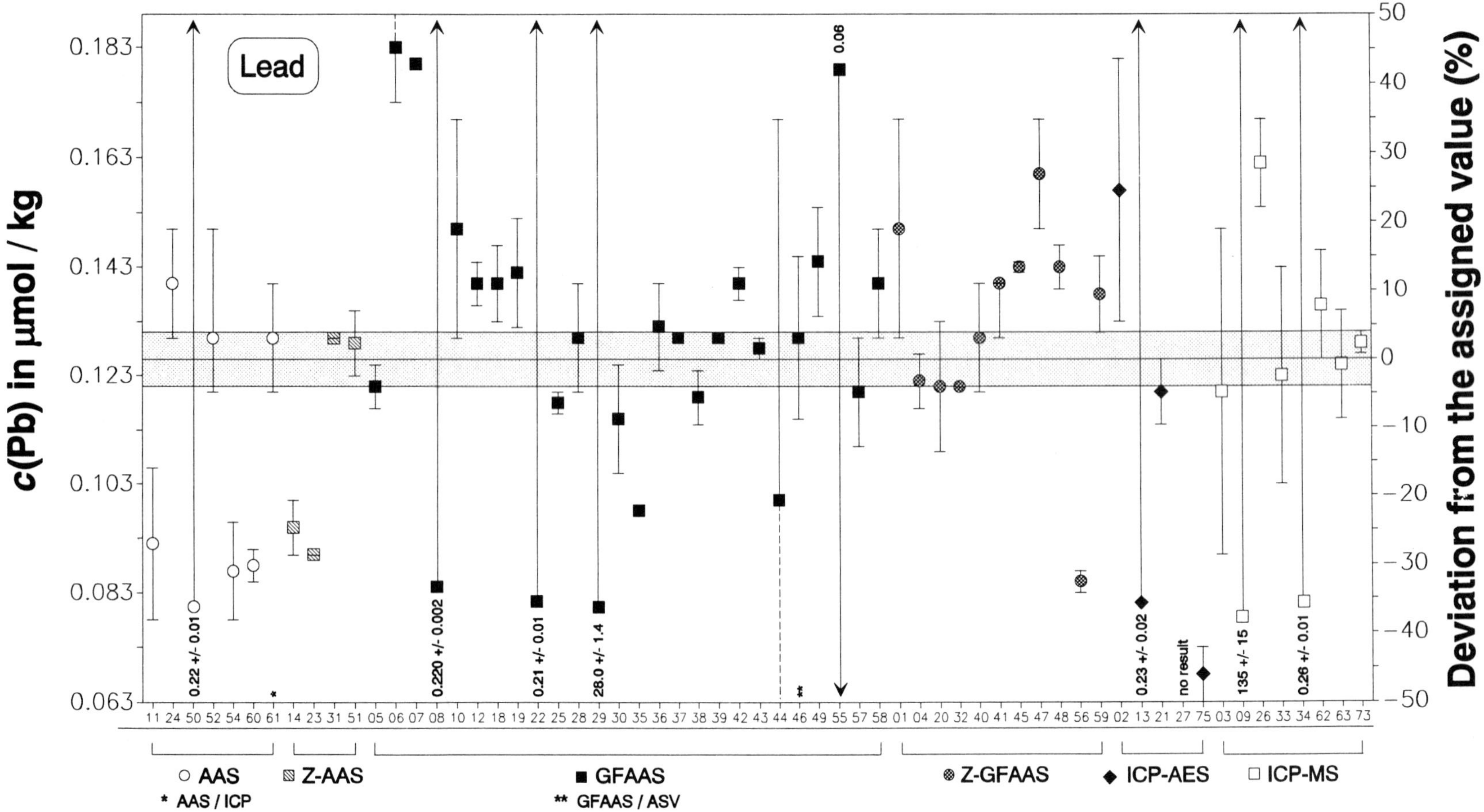

Fig. 4. Coded laboratories arranged in order to compare different methods used in the determination of IMEP-3: Trace elements in natural water B. Certified value: 0.126(5) µmol/kg, 26(1) ng/mL.

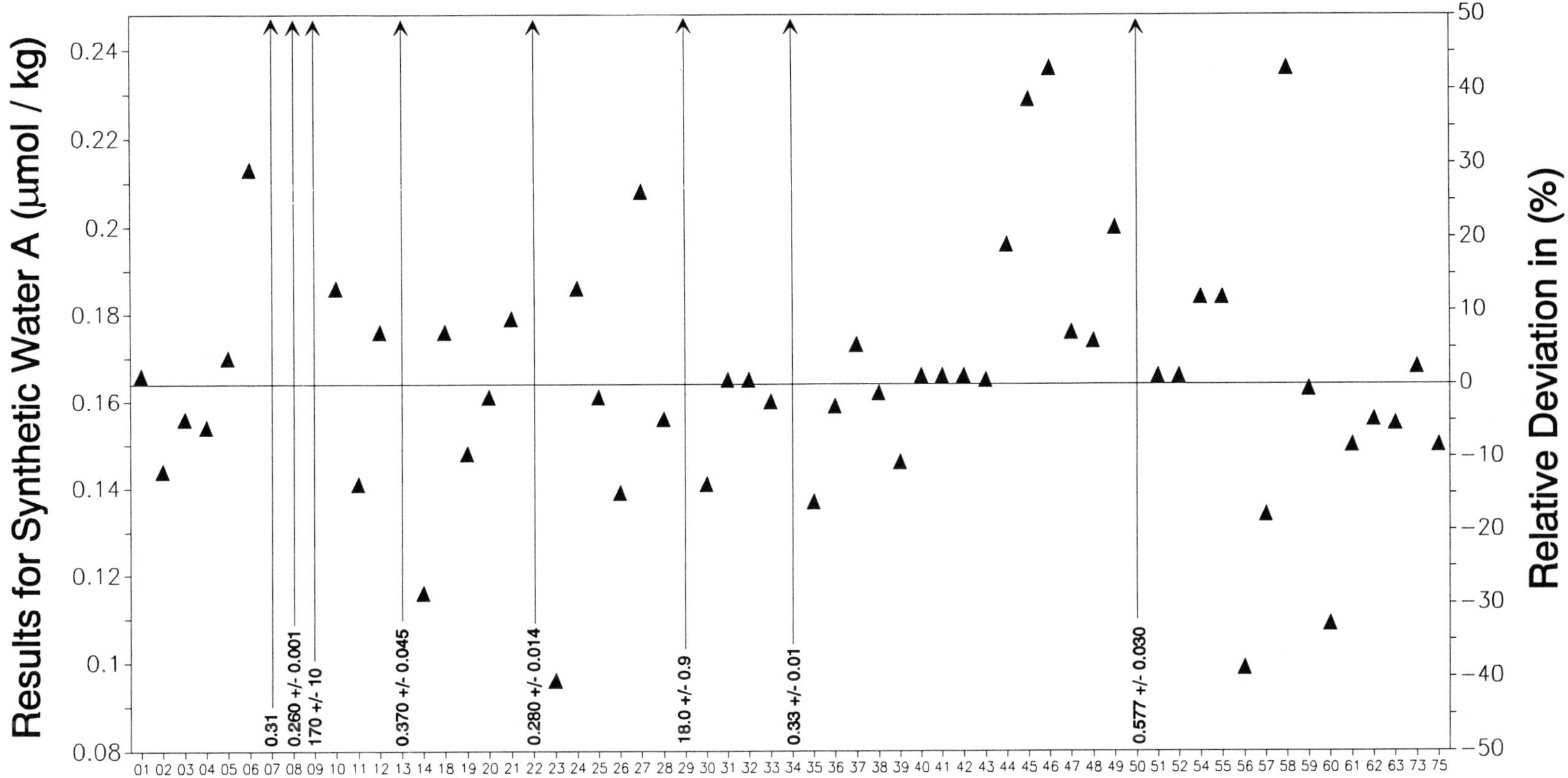

Fig. 5. Results obtained for synthetic water A in the determination of IMEP-3: Lead.

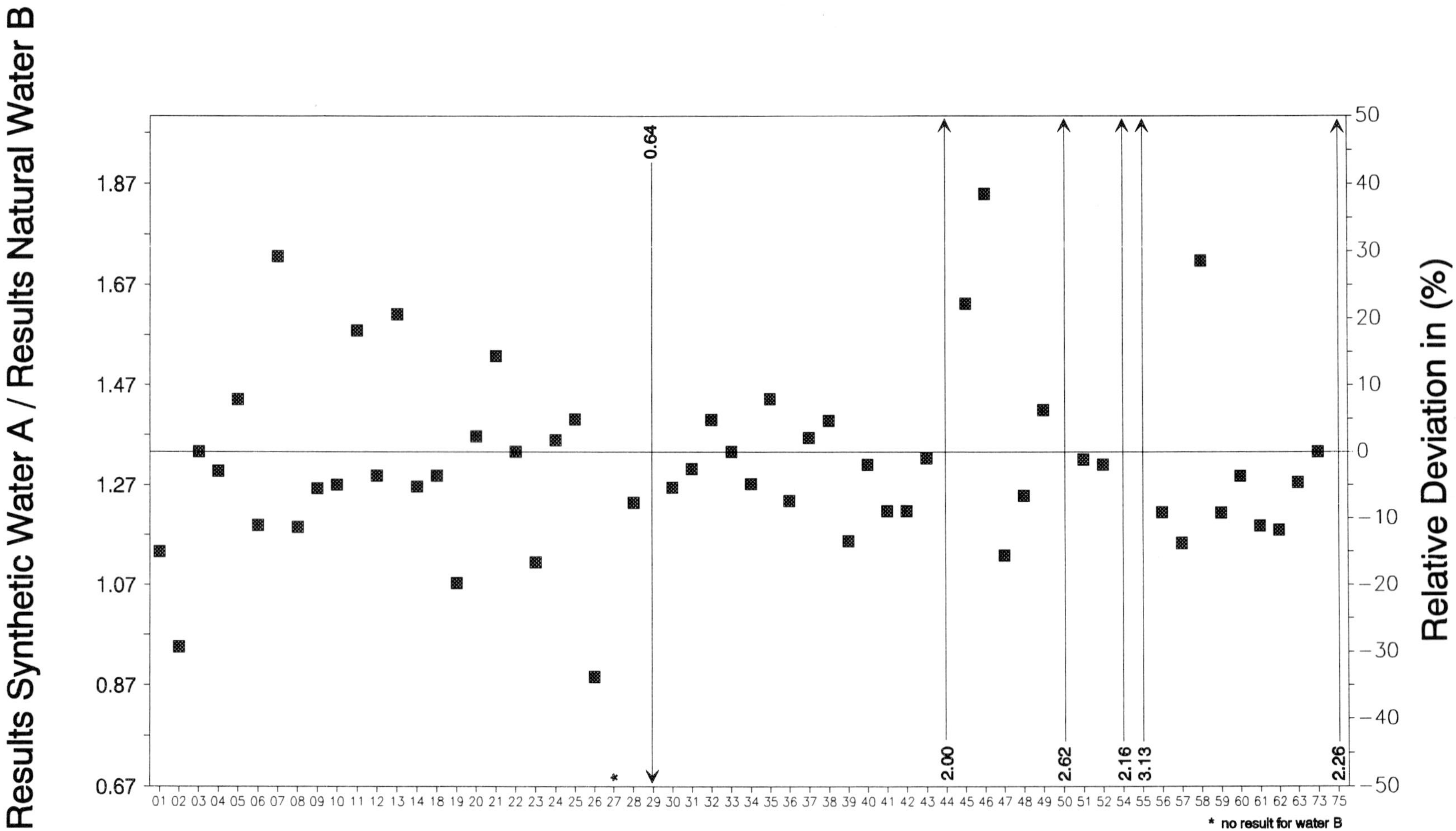

Fig. 6. Results obtained for synthetic water A divided by results obtained for natural water B (A/B) in the determination of IMEP-3: Lead.

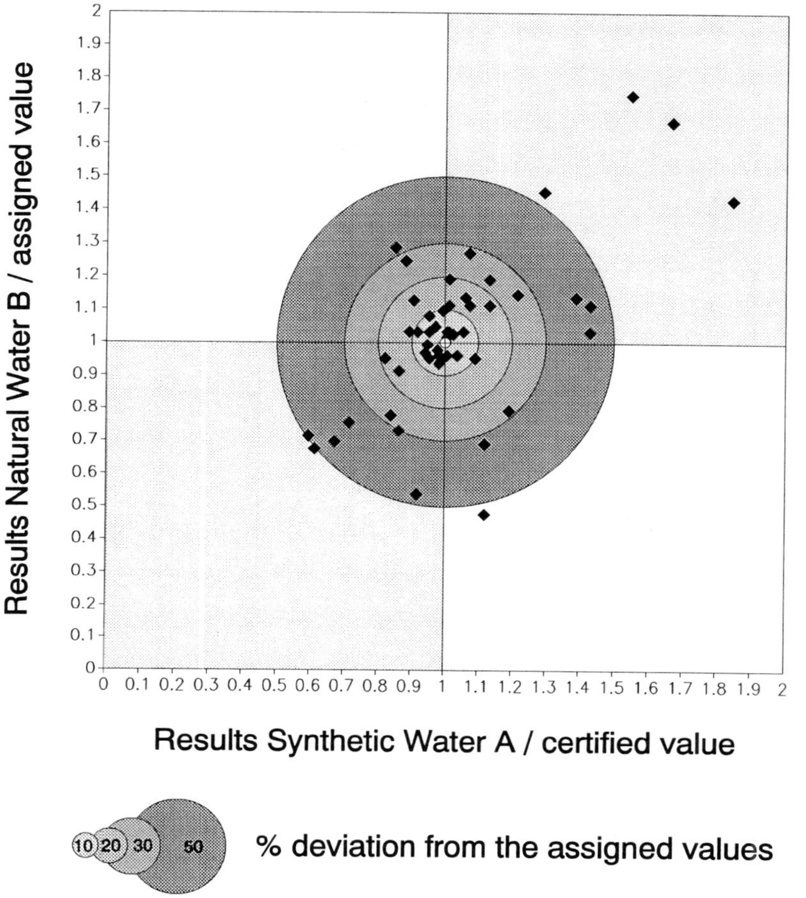

Fig. 7. Youden plot to show the overall performance of the IMEP-3: Lead round.

REFERENCES

1. NIST Certificate SRM 1643c, Trace elements in water, NIST, Gaithersburg, MD 20899, June 24, 1991.
2. Report of the US Geological Survey's Analytical Evaluation Program—Standard Reference Water Samples T-107 (trace constituents), M-110 (major constituents), N-23 (nutrients), P-13 (precipitation snowmelt) and Hg-5 (mercury), Denver, CO, August 1989.
3. A. Lamberty, P. De Bievre, and A. Götz, *Fresenius J. Anal. Chem.* **345,** 310 (1993).

Validation of the Cumulative or Replicate NAA Method for the Determination of Trace Elements in Biological Materials

M. F. REIS,*[,1] A. A. S. C. MACHADO,[2] M. ABDULLA,[3] J. CANÁRIO,[4] AND J. C. SILVA[1]

[1]*INETI - ICEN - DEEN, Sacavém, Portugal;* [2]*Laquipai, Faculty of Science, University of Oporto, Portugal;* [3]*Department of Clinical Pharmacology and Therapeutics, Hamdard University, New Delhi, India; and* [4]*Faculty of Science, University of Lisbon, Portugal*

ABSTRACT

To make the best use of time and facilities, a neutron activation system, fully automatic, including spectrum and data processing, to be used with short-lived nuclides, has been recently developed at the Portuguese Research Reactor (in Portuguese "Reactor Português de Investigação," RPI). Using this system, the cumulative neutron activation analysis method is now being implemented. This article summarizes the experimental procedures used to validate the cumulative NAA method for the determination of selenium in biological samples, emphasizing the determination of the essential characteristics of precision, accuracy, and limits of detection and of quantification of the method. The article also examines how detection limits and precision are improved when the samples are analyzed by this method compared to the cyclic activation measurements in use at RPI for the determination of selenium. The improvement is demonstrated for the measurement of selenium in several reference materials when the result obtained by adding up seven spectra of separate aliquots of the same sample is compared to the result from a cumulative spectrum of seven (whenever possible) consecutive cycles of a single sample.

*Author to whom all correspondence and reprint requests should be addressed.

Index Entries: Validation of analytical methods; cumulative or replicate neutron activation analysis; human hair, blood, and diet samples; selenium; short-lived nuclides.

INTRODUCTION

Neutron activation analysis with short-lived nuclides has a great potential for the determination of a number of elements in a variety of sample matrices. The disadvantages are the fact that a number of elements have no short-lived nuclides and the need for rather sophisticated irradiation and measurement equipment, if very short-lived nuclides are involved. The main advantages are: the possibility of repeating the analysis with the same sample in very few days or, sometimes, within a few hours or even minutes; the significant reduction of experimental time; the attainment of fast results; and, generally, better sensitivities and limits of detection than those offered by the corresponding long-lived nuclides under standardized experimental conditions *(1–4)*.

The problems associated with the application of neutron activation analysis with short-lived nuclides to some samples of nutritional and clinical origin (particularly composite diets, foodstuffs, and blood samples) are the result of background activity induced in the matrices. Signal-to-background ratio can be significantly improved by using methods based on the principles of cyclic activation *(1,5–7)*. Usually, in cyclic activation, the sample is irradiated for a short period of time (relative to the half-life of the isotope of interest), rapidly transferred to the detector for counting (again for a short period of time), and the entire process repeated for a number of cycles. The activities detected at each counting period are summed to produce a cumulative spectrum over the total experimental time *(5)*. However, the growth of the activity from radionuclides with longer half-lives restricts the number of cycles that may be made before high background-to-signal ratio occurs *(8,9)*. Guinn *(10)* has shown that cumulative or replicate analysis, consisting of analyzing n aliquots of the same sample and accumulating the n spectra as a single cumulative spectrum improves considerably the precision and detection limits of single-sample analysis, even when n cyclic measurements of the single sample are performed. These procedures can be applied using the facilities available for cyclic instrumental NAA (CINAA) and do not require any additional expensive equipment.

To be able to use this method at the RPI, some improvements have been done at the recently developed neutron activation system *(4)* in use at RPI since 1990. The system is now fully automated, including spectrum and data-processing, and can be used routinely for the analysis of large amounts of samples. A detailed description will be given elsewhere *(11)*.

When a new or modified analytical method is being considered for use in routine analysis of a large number of samples, it has to be subjected to a validation study, in order to demonstrate that the method is appropriate for the analysis of the matrices to which it will be applied and to gather preliminary information on precision, accuracy, and detection limits of the method itself *(12)*. The present study describes the experimental procedures to validate the cumulative or replicate NAA method at the RPI, since the method is intended to be used in routine determination of selenium (and, in the near future, of other trace elements producing short-lived nuclides) in complement with and/or alternative to cyclic activation analysis. This validation study is part of a quality-assurance program of trace element analysis in progress at RPI Laboratory for the application of INAA methods with short-lived nuclides. To demonstrate the better performance of the cumulative NAA method for the determination of selenium compared to cyclic activation analysis, results from measurements of selenium in several reference materials using both methods are also presented in this article.

MATERIAL AND METHODS

Standards

Selenium and multielemental comparator standards were prepared by depositing microliter portions, containing microgram quantities, of ultrapure (Alpha Ventron Corp., Alpha Johnson Matthey Gmbh, Karlsruhe, Germany) solutions on a cellulose support, wetting them with ultrapure water to achieve homogeneity, and evaporating them to dryness at low heat under an infrared lamp. The standards were heat-sealed in small polyethylene vials (1.2 mL) and placed in medium-sized irradiation vials (7 mL), together with a second small vial, used as spacer to keep the sample in place and to prevent displacement of the sealed cap. The irradiation vial was then heat-sealed according to the procedure described elsewhere *(4,13,14)*.

Reference Materials

Samples of reference materials from NIST and IAEA, dried according to the procedures recommended by the issuing agencies, were weighed into small polyethylene vials and prepared for irradiation following the procedure described in the preceding section.

Sample Preparation

Following the scheme shown in Fig. 1, three series of samples were prepared (each series corresponding to a different type of matrix expected to be analyzed by the cumulative method). The samples included homogenized and lyophilized composite daily diets, lyophilized human whole-blood samples, and samples of human hair, washed according to the protocol recommended by IAEA *(15)*. Preparation of samples involved spiking substrates with known amounts of selenium, selected in order to cover the range of the element concentrations normally expected to be encountered in the corresponding type of matrix. Five different concentration levels were obtained, the substrate itself being used as the lowest level. For each substrate, seven replicates of each concen-

Material	Mass of selenium added to the samples*							
	0 μg (Blank)	0 μg (Base)	0.1 μg	0.5 μg	1 μg	2.5 μg	5 μg	10 μg
Cellulose	PBRA#		P0-1#	P0-5#	P001#	P2-5#	P005#	P010#
Blood		SDES#	S0-1#	S0-5#	S001#	S2-5#		
Hair		CDES#	C0-1#	C0-5#	C001#	C2-5#		
Diet		DDES#	D0-1#	D0-5#	D001#	D2-5#		

Fig. 1. Scheme of the preparation of spiked samples.

*The symbol (□) represents one replicate. The string of characters on each set of replicates is the replicate code where # can assume the value 1 to 7.

tration were prepared in a total of 35 replicates/sample. The "spikes" were microliter portions of ultrapure solutions, prepared at the same time for the standards (cellulose support) and containing known amounts of selenium, which were deposited into small vials containing substrate and evaporated to dryness at low heat under an infrared lamp. No special precautions were taken to ensure homogeneity of the spiked samples, since the inhomogeneity of the sample material is not a serious concern when using the cumulative NAA method *(10)*.

Irradiations

All samples and standards were irradiated in the RPI at a thermal neutron flux of 1.9×10^{12} $n{\cdot}cm^{-2}{\cdot}s^{-1}$ with 1 and 3% of epithermal and fast neutrons respectively. The neutron flux remains reasonably constant during sequential irradiations of large sets of samples meaning that the samples and comparator standards can be irradiated separately. To safeguard against unexpected changes of stability of the system during the analysis of a large set of samples, the same set of standards was irradiated after irradiation of a small number of samples (usually five or seven). The time between irradiation of standards was selected in such a way to allow them completely to decay before the next irradiation. The CINAA and cumulative NAA methods involved the following timing parameters: an irradiation time of 10 s, a decay time of 5 s, and a counting time of 10 s. The number of cycles was planned to be seven. However, because of growing background activity, the maximum number of cycles used was restricted to four. For certain reference materials, such as Oyster Tissue, NIST-SRM-1566a, and Total Diet, NIST-SRM-1548, that number could not even be higher than two. The cumulative NAA method was applied with 7 replicates/sample.

Transfer and Counting Systems

Routing of the samples and acquisition of the spectra are controlled by the fully automatic neutron activation system in use at the RPI *(4)*. The γ-ray spectrometer is a CANBERRA HPGe coaxial detector, model GC3518 (relative efficiency at 1.33 MeV, Co-60, 35%) coupled to an MCA, which is a board level product (CANBERRA AccuSpec/A) with emulation software. Spectral data are dumped on floppy disks for off-line spectral analysis by SAMPO-90, a γ-ray spectrum analysis program from CANBERRA. The 162-keV γ-ray of ^{77m}Se (17.4 s) was used to determine the selenium content.

RESULTS AND DISCUSSION

For the validation of the method, all the replicates for all the samples and concentrations, as well as several standard reference materials and standards of selenium were processed, in random order, seven times, on seven different days and by the same analyst. This procedure aimed at subjecting the method to all factors likely to affect it without introducing any between-analyst variance. For each sample and each concentration level, the mean concentration and standard deviation were calculated based on the seven runs performed.

Precision of Cumulative NAA Method for Diet, Blood, and Hair Samples

From Fig. 2 it can be seen that the standard deviation varies with the selenium level, over the ranges of concentration studied, for all the matrices used in this work. In the case of diet, the function seems to be linear over the whole range, which means that the relative standard deviation (*see* Fig. 3) can be considered constant around a mean value. However, more results would be needed to establish this relationship unambiguously.

Concerning blood and particularly hair samples, the relative standard deviation presents the expected variation at the low concentration end. When higher concentration levels are considered, defined trends for constant values are evident. The plots of standard deviation vs mean concentration can be used to interpolate the value of the standard deviation for any level of analyte that has not been covered in this work. They are particularly useful for the establishment of the limits of detection and quantification, as will be demonstrated in the next section.

As normally observed, the relative standard deviation for blood and hair samples becomes nearly constant for the ranges of higher concentrations. For the lower concentration ranges, the absolute standard deviation is constant, so that the relative deviation appears to double for each half decrease in analyte concentration. The course of this variation is in agreement with that of the Horwitz curve *(16)* calculated for a wide variety of analytes and different instrumental methods of analysis.

Limits of Detection and Quantification

The limit of detection is defined by the American Chemical Society (ACS) *(17,18)* as the blank signal plus at least three times the standard deviation of the blank value. In analytical methods that use a spectral peak and that have no need of using blank samples, the blank may be estimated from the height of the background to the actual or expected peak. Its standard deviation may be based on the evaluation of the counting statistics *(19)*. If, as is the case, the regression line "standard deviation vs analyte concentration" is available for a particular matrix (*see* Fig. 2), the limit of detection may be calculated directly as three times the intercept (s_0) of this line *(20)*.

The ACS guidelines define the limit of quantification as the blank signal plus ten times the standard deviation of the blank. Based on the plot of standard deviation vs analyte concentration, this parameter can be simply calculated by multiplying the intercept of the line by ten. Table 1 shows the detection and quantification limits for the diet, blood, and hair samples studied by the cumulative NAA method in this work.

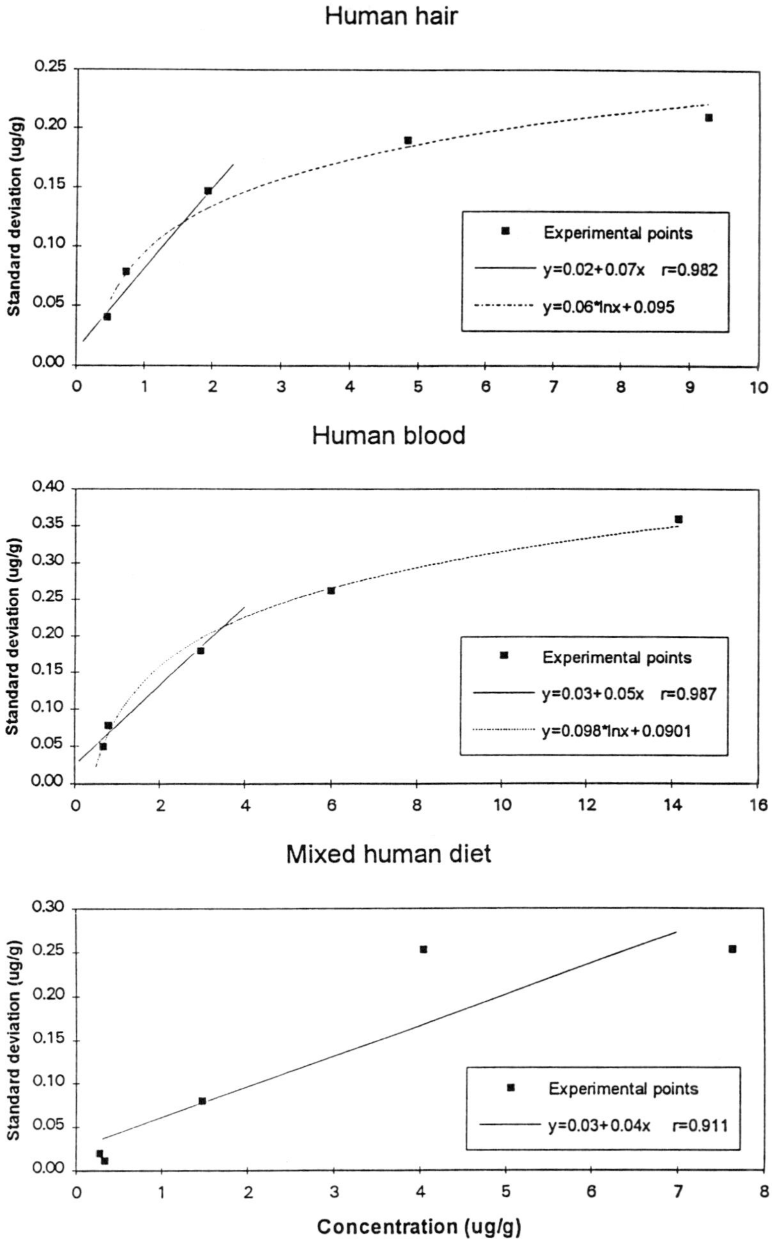

Fig. 2. Variation of standard deviation with concentration for biological matrices analyzed by cumulative NAA method.

Accuracy

To validate an analytical method, external quality assessment is mandatory, since this is a sure means of detecting problems of bias that are difficult to identify by internal control measures. One of the most powerful and convincing ways to establish the reliability of an analytical

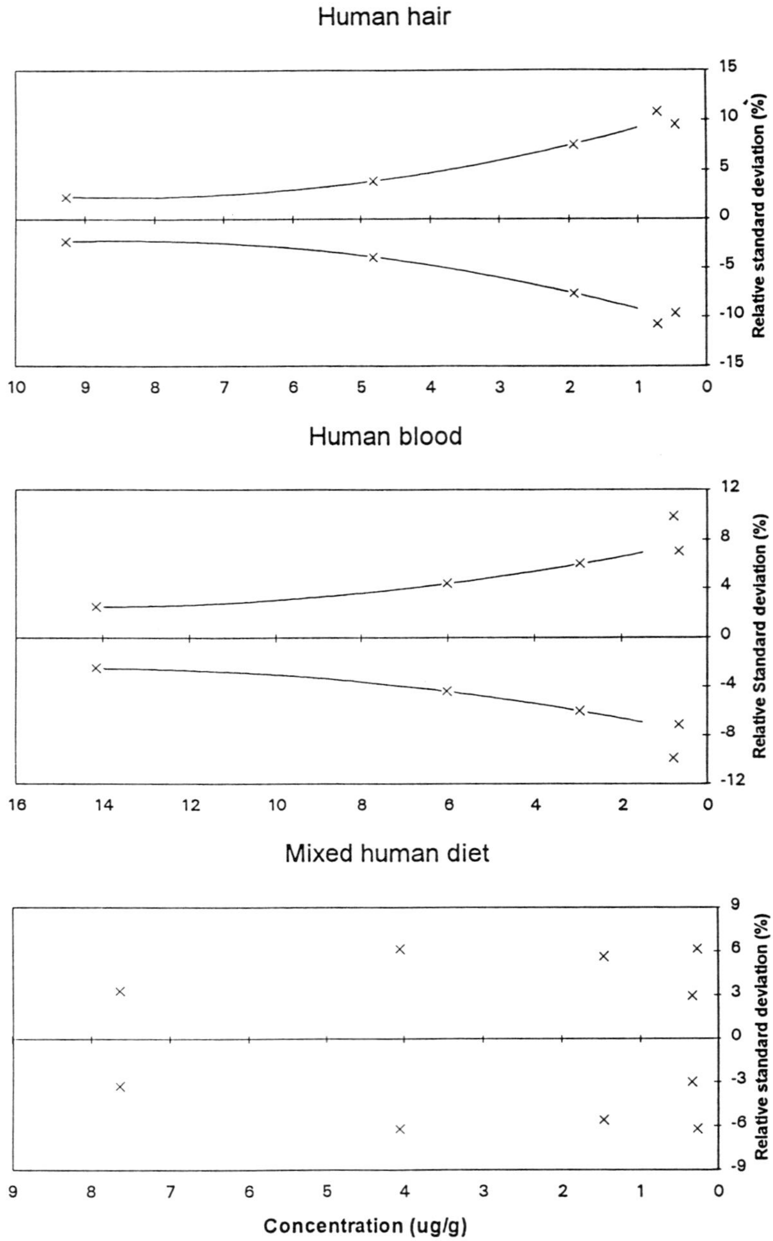

Fig. 3. Variation of precision with concentration for biological matrices analyzed by cumulative NAA method.

method is by the use of suitable standard reference materials (SRM) *(19)*. In this work, a number of reference materials (nine) were analyzed for selenium, and the concentration values, based on the averages of seven determinations, were compared with certified values, whenever possible. As can be seen from Table 2 and Fig. 4, the agreement is generally good.

Table 1
Limit of Detection and Limit of Quantification of Selenium for Biological Materials Analyzed by Cumulative NAA Method

Sample material	Intercept s_0 (μg/g)	LOD (μg/g)	LOQ (μg/g)
Mixed human diet	0.026	0.078	0.26
Human blood	0.025	0.075	0.25
Human hair	0.018	0.054	0.18

Slight differences, from unity and zero respectively, shown by the slope and intercept of the "best" line fitted to the experimental points, merely represent a random variation. To confirm this a "confidence band" around the least-squares straight line was calculated. The theoretical line of unity slope and zero intercept lies within this confidence band meaning that, with the selected level of confidence (95%), the experimental slope and intercept are not significantly different from unity and zero, respectively. Since the theoretical line is well centered in the band, the method can be considered unbiased.

Improvement in Precision and Detection Limits

The selenium contents of a number of SRMs, determined by CINAA and cumulative NAA methods, are given in Table 3. Each concentration value is the average of three independent determinations by CINAA, obtained with the indicated number of cycles, and of four determinations by the cumulative NAA method, with seven replicates for each material. Detection and quantification limits were calculated according to the methods described by Currie *(21)*.

From Table 3, it can be seen that for most of the SRMs studied, the detection limits and precision of measurements improve in CINAA with an increasing number of cycles, as is often the case. However, it is also evident that the improvement is much more significant when the cumulative NAA method is used for the same material. As predicted by Guinn *(10)*, for most of the SRMs, precision and detection limits are improved by approximately a factor of $n^{1/2}$, where n is the number of replicates used.

CONCLUSIONS

The work performed to validate the cumulative NAA method at RPI Laboratory, corresponds to a need for available realistic information on the main characteristics of this analytical method for the determination of selenium in three different biological matrices already being analyzed in our laboratory. The analysis of these samples is part of a nutritional study in progress in a well-defined Portuguese population group of rural

Table 2
Selenium content* of Reference Materials Analyzed
by Cumulative NAA Method

Reference Material		This work (μg/g)	Recommended value (μg/g)
Mixed Human Diet	IAEA-H-9	n.d.	0.11 ± 0.01
Non-Fat Milk Powder	NIST-SRM-1549	n.d.	0.11 ± 0.01
Total Diet	NIST-SRM-1548	0.24 ± 0.02	0.25
Animal Muscle	IAEA-H-4	0.33 ± 0.01	0.28 ± 0.03
Rice Flour	NIST-SRM-1568	0.29 ± 0.01	0.4 ± 0.1
Bovine Liver	NIST-SRM-1577a	0.71 ± 0.04	0.71 ± 0.07
Wheat Flour	NIST-SRN-1567a	1.07 ± 0.1	1.1 ± 0.2
Fish Flesh	IAEA-MA-A-2	1.20 ± 0.04	1.7 ± 0.3
Oyster tissue	NIST-SRM-1566a	2.04 ± 0.13	2.21 ± 0.24
Albacore Tuna	NIST-RM-50	4.3 ± 0.28	3.6 ± 0.4

*Content based on four determinations of seven replicates per reference material. n.d.: not detected.

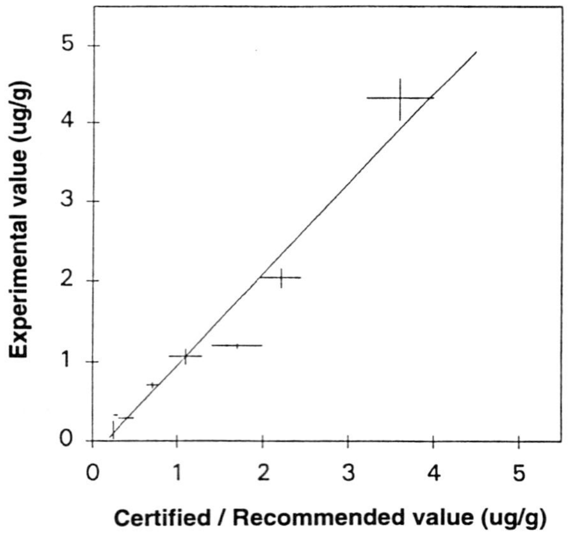

Fig. 4. Selenium content in SRM.

workers, where the selenium content in the diets is known to be very low *(13)*. After validation, one can establish whether our data, derived from samples of the type of the matrices studied, have true quantitative value or whether they are mainly qualitative. The results of this study are also very useful in validating the working range of the analytical method confirming the reliability of the method in the determination of selenium for the matrices involved. At present, similar work is under way to produce the same kind of results for other trace elements of nutritional impor-

Table 3
Precision, Accuracy, Detection, and Quantification Limits for Selenium in Biological Reference Materials Analyzed by CINAA* and Cumulative or Replicate NAA Method

Reference Material		Reported value (µg/g)	# cycles	Mean ± SD (µg/g)	%SD (%)	LOD (µg/g)	LOQ (µg/g)
Oyster tissue	NIST-SRM-1566a	2.21 ± 0.24	1	2.24 ± 0.30	13.4	0.239	1.010
			2	2.25 ± 0.16	7.2	0.168	0.629
			Rep**	2.04 ± 0.12	5.7	0.105	0.361
Rice Flour	NIST-SRM-1568	0.4 ± 0.1	1	0.32 ± 0.02	5.6	0.084	0.423
			2	0.33 ± 0.01	1.6	0.053	0.231
			3	0.36 ± 0.02	5.3	0.048	0.190
			4	0.35 ± 0.02	5.6	0.038	0.147
			Rep	0.29 ± 0.01	4.8	0.026	0.097
Bovine Liver	NIST-SRM-1577a	0.71 ± 0.07	1	0.80 ± 0.06	7.3	0.129	0.643
			2	0.74 ± 0.05	7.3	0.088	0.302
			3	0.75 ± 0.07	8.7	0.076	0.299
			4	0.77 ± 0.05	6.8	0.071	0.263
			Rep	0.71 ± 0.04	5.2	0.049	0.178
Albacore Tuna	NIST-RM-50	3.6 ± 0.4	1	4.60 ± 0.16	3.6	0.158	0.792
			2	4.36 ± 0.11	2.6	0.123	0.499
			3	4.49 ± 0.11	2.4	0.103	0.395
			Rep	4.32 ± 0.28	6.5	0.094	0.320
Fish Flesh	IAEA-MA-A-2	1.7 ± 0.3	1	1.26 ± 0.15	12.3	0.165	0.841
			2	1.16 ± 0.01	0.8	0.122	0.507
			3	1.21 ± 0.06	5.1	0.100	0.393
			4	1.21 ± 0.07	5.7	0.089	0.335
			Rep	1.20 ± 0.04	3.2	0.066	0.240
Animal Muscle	IAEA-H-4	0.28 ± 0.03	1	0.31 ± 0.12	10.4	0.104	0.582
			2	0.31 ± 0.04	12.7	0.076	0.335
			3	0.30 ± 0.04	11.7	0.064	0.264
			Rep	0.33 ± 0.01	3.3	0.040	0.15[illegible]
Total Diet	NIST-SRM-1548	0.25	1	0.30 ± 0.02	7.1	0.089	0.569
			2	0.28 ± 0.03	9.9	0.086	0.363
			Rep	0.24 ± 0.02	9.5	0.049	0.178

*Results obtained as average of three determinations with the indicated number of cycles.
**Rep: Results obtained by cumulative or replicate NAA method.

tance: copper, zinc, and manganese *(22)*. The results will be published very soon.

REFERENCES

1. K. N. DeSilva and A. Chatt, *J. Trace Microprobe Tech.* **1,** 307–337 (1982–83).
2. A. Chatt and J. Holzebecher, *Trans. Am. Nucl. Soc.* **62,** 220–222 (1990).
3. F. Grass, *Trans. Am. Nucl. Soc.* **62,** 220 (1990).

4. M. F. Reis and E. Martinho, *J. Radioanal. Nucl. Chem., Articles* **167,** 39–44 (1993).
5. N. M. Spyrou, *J. Radioanal. Chem.* **61,** 211–242 (1981).
6. A. Chatt, K. N. DeSilva, J. Holzbecher, D. C. Stuart, R. E. Tout, and D. E. Ryan, *Can. J. Chem.* **59,** 1660–1664 (1981).
7. L. S. McDowell, P. R. Giffen, and A. Chatt, *J. Radioanal. Nucl. Chem.* **100,** 519–529 (1987).
8. S. J. Parry, *J. Radioanal. Chem.* **75,** 253–258 (1982).
9. G. Erdtmann and H. Petri, in *Treatise on Analytical Chemistry*, part I, vol. **14,** 2nd ed. Wiley, (1986).
10. P. Guinn, *Radiochem. Radioanal. Lett.* **44,** 133–138 (1980).
11. M. F. Reis, A. A. S. C. Machado, and M. Abdulla, to be published.
12. J. P. Dux, in *Handbook of Quality Assurance for the Analytical Chemistry Laboratory*, Van Nostrand Reinhold, 1986.
13. M. F. Reis, J. Holzbecher, E. Martinho, and A. Chatt, *Biol. Trace Element Res.*, **26/27,** 629–635 (1990).
14. M. F. Reis, *Report LNETI/DEEN-R-91/22* (1991).
15. Yu. S. Ryabukhin, *Report IAEA/RL/50*, Vienna, Austria (1978).
16. W. Horwitz, L. Kamps, and K. W. Boyer, *J. Assoc. Off. Chem.* **63,** 1344–1354 (1980).
17. American Chemical Society, *Anal. Chem.* **52,** 2242–2249 (1980).
18. American Chemical Society, *Anal. Chem.* **55,** 2210–2218 (1983).
19. R. M. Parr, *Nutr. Res.*, Suppl. **1,** 5–11 (1985).
20. A. A. S. C. Machado, in *Controlo de Qualidade em Química Analítica*, MSc Course, Faculdade de Ciências, Porto (1991).
21. L. A. Currie, *Anal. Chem.* **40,** 586–593 (1968).
22. M. Abdulla, PhD. dissertation, Lund University, Sweden (1985).

SUM and MEAN

Standard Programs for Activation Analysis

RICHARD M. LINDSTROM

Inorganic Analytical Research Division, National Institute of Standards and Technology, Gaithersburg, MD

ABSTRACT

Two computer programs in use for over a decade in the Nuclear Methods Group at NIST illustrate the utility of standard software: programs widely available and widely used, in which (ideally) well-tested public algorithms produce results that are well understood, and thereby capable of comparison, within the community of users. SUM interactively computes the position, net area, and uncertainty of the area of spectral peaks, and can give better results than automatic peak search programs when peaks are very small, very large, or unusually shaped. MEAN combines unequal measurements of a single quantity, tests for consistency, and obtains the weighted mean and six measures of its uncertainty.

Index Entries: Peak integration, γ-ray spectrometry, weighted mean.

INTRODUCTION

We define a standard program as one that is widely used because it is reasonably easy to apply in a number of contexts. In such a program, ideally, well-tested and publicly known algorithms are employed to produce results that are well understood within the community of users, and thereby capable of easy comparison and discussion. For instance, users of commercial computer-based γ-ray spectrometry systems employ widely sold programs for such common operations as peak finding and integration: if two workers use the same manufacturer's counting system, then they are probably processing their data in much the same way (always excepting user choices in the manner in which programs are used *(1)*) and their results can thus be readily compared. Another example is the CLSQ program *(2)*, which was used for years to fit decay curves at Brookhaven. More recent examples are the spreadsheet programs in use on personal computers, where (one believes) the functions have all been properly coded in correct, efficient, and robust algorithms.

Two well-tested programs have been in use for over a decade in the Nuclear Methods Group at NIST: SUM, to integrate difficult peaks in a γ spectrum, and MEAN, to compute the best average value from repeated measurement of a quantity. These programs, both of which were written to mechanize operations that had been performed with a calculator, illustrate the utility of standard software as tools to help perform accurate measurements.

SUM

The most basic operation in γ-ray spectrometry is identifying significant peaks in the spectrum, and determining the position, net area, and uncertainty of the area of each peak. The SUM program was originally written to perform the integration function because the commercial peak search in use in our laboratory (an early version of Nuclear Data's PEAK) calculated the uncertainty of the area incorrectly. SUM uses simple standard algorithms based on correct error propagation *(3,4)*. Net areas are calculated as the total peak area above a trapezoidal baseline (Fig. 1). The user enters the channel numbers through the keyboard or mouse for each peak and for its baseline regions. Different versions of the program allow for the baseline to be adjacent to the peak region or detached from it.

The SUM program often gives better results than most automatic peak search programs in at least three cases: when counting statistics are poor, when they are very good, or when peak shapes are unusual. In these cases, judicious use of SUM can eliminate certain non-Poisson sources of error inherent in the peak search process. These cases will be considered in turn.

When there are very few counts per channel, the eye of an experienced analyst, guided by judgment based on counting a standard with good statistics, can be better than an automatic program for detecting peaks and judging where their boundaries occur *(1,5)*. For weak peaks, integration about a known energy, which is always the case in conventional activation analysis (but not always with prompt γ-rays), is more precise and can be more accurate than allowing the energy to be a free parameter *(6)*. The simplest peak search programs define the boundaries of a peak as the channels with the lowest number of counts on either side of the peak maximum. This definition automatically biases the baseline low and the net area high *(5,7)*.

More sophisticated methods (including fits to analytical functions), involving smoothing of the baseline, improve the accuracy, but whenever random variations in the baseline can affect the choice of peak boundaries, the peak search process itself introduces bias. In critical cases, repeated manual integrations using different selections of peak and base-

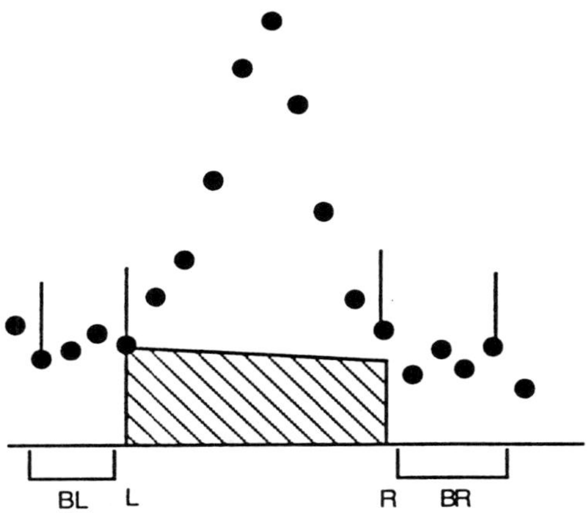

Fig. 1. SUM adds the contents of the peak channels between L and R, and subtracts the baseline trapezoid determined by averaging over BL and BR channels on left and right. The baseline regions need not be adjacent to the peak.

line channels may give a better idea of the real imprecision than does Poisson statistics using only one choice of channels. It is not widely appreciated that the net area reported by some commonly used peak search programs depends on the user-selected "sensitivity" and other adjustable parameters *(1,8)*. This may help to explain why intercomparisons of peak integration algorithms (e.g., *9*) often are not fully conclusive: more skilled users make better choices of these parameters. Prompt-γ activation analysis, for example, often gives spectra where peaks are numerous and small, and therefore benefit from the use of SUM for interpretation.

A particularly good recent example from our laboratory of the shortcoming of a commonly used peak search program is the determination of iodine using the 443-keV peak of 25-min ^{128}I in the presence of the activation products of other halogens and manganese. For some samples, this was a small peak on a large, locally smooth baseline. Using SUM, the precision of the analysis was improved by nearly a factor of two by showing experimentally (Fig. 2) that the optimum number of baseline channels was much greater than the automatic program chose to use.

Manual selection of peak boundaries may also improve precision in the opposite situation: when there is a very large number of counts in the peak of interest. In a high-precision nuclear safeguards application, Greenberg and Carpenter *(10)* showed that the peak search process introduces excess non-Poisson variability into the precision of repeated counts of the same sample. Using the SUM program with nonadjacent baselines

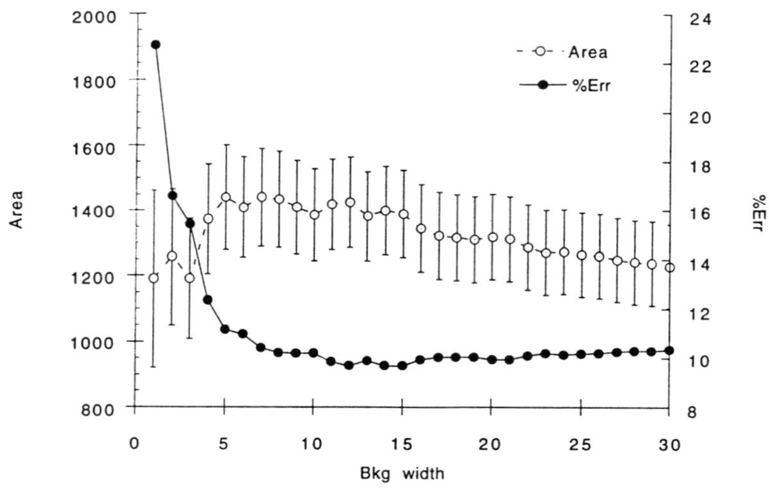

Fig. 2. Use of SUM to find the optimum (symmetric) baseline width for the 443-keV peak of ^{128}I on a baseline dominated by ^{82}Br. For this case a commercial peak search program (using two baseline channels on each side of the peak) gave 1258 counts ± 13.7% (1σ); SUM (using 10 channels) gave 1471.4 ± 9.2%. The conversion gain is 0.38 keV/channel.

improved the precision by more than a factor of two (to about 0.05%) compared to a Gaussian fit.

For peaks with unique shapes, such as the heavily Doppler-shifted boron capture γ-ray, no automatic peak fitting method is reliable, so that a routine like SUM is required. SUM has been adapted to analyze another common case, when the peak is in fact absent and a detection limit is sought. A variant called DETLIM calculates the Currie detectability parameters L_C, L_D, and L_Q, *(11)* and the 95% confidence limit, using boundaries chosen with SUM based on comparison spectra where the peak is visible.

The outputs of SUM, written both to printed reports and to binary files, are in the same form as those from the Canberra Nuclear Data PEAK program, so that SUM can be used in the same manner in automatic command procedures. In addition, the results are written to a tab-delimited text file that can be exported to a personal computer and read directly by a spreadsheet program. The original version was written in PDP-11 Fortran for the Midas operating system in 1978, and later translated to VAX/VMS Fortran, using the Canberra Nuclear Data CAM data structures. The source code is about 400 lines. Definitions of terms and algorithms are documented in a five-page user's guide. SUM has been used in numerous applications at NIST and elsewhere *(10,12–17)*.

MEAN

A fundamental operation in all of measurement science is to combine several measurements of a single quantity, in general with different uncertainties, to obtain the best value for the average and for its uncertainty. Before combining the individual measurements, we usually want to test them for consistency and for the presence of unsuspected sources of variability *(18)*, and perhaps reject outliers and re-average the remainder. The appropriate algorithm to perform the combination is a weighted mean, where the weight assigned to each point is inversely proportional to its variance. The MEAN program performs this function.

Most of the algorithms used in the program were taken from a standard textbook *(19)* and first implemented on a programmable calculator. The program was later expanded and translated to PDP-11 Fortran for the Midas operating system, to VAX/VMS, to MS-DOS Fortran *(20)*, and to the Excel spreadsheet for Macintosh. The same set of test data has been run for each version of the program to ensure that the outputs are identical.

In addition to weighted and unweighted means, the program displays the residual for each input point in native units and in standard deviations, computes six measures of uncertainty of the means, displays chi-squared and its probability, and offers the option of refitting the data set with points omitted or added, or with an additional component of error added in quadrature to each point. The results of the analysis of a standard data set are given in Fig. 3. The normalized residuals efficiently highlight outliers. Comparing the observed standard deviation of the mean with that expected *a priori* from the input precision (essentially an *F* test) helps detect unexpected sources of error *(21)*. The computation is carried out to as many significant figures as the computer allows (17 decimal digits for the PDP-11, 16 for the VAX, 10 for the HP15C calculator, and 15 for Excel). MEAN passes robustness tests *(22,23)* which have been proposed to determine when round-off becomes noticeable; this is generally important for precise numbers where differences are small.

The definition of each item displayed on the output page, and the algorithm by which it is calculated, are documented in a three-page write-up made available to all users of the program. The program has been used for more than 10 yr in Reports of Analysis for certification of Standard Reference Materials.

SUMMARY

The SUM and MEAN programs have been in regular use for over a decade in the Nuclear Methods Group at NIST. These well-tested standard programs provide uniform definitions of terms and common algorithms for comparing measurements. Both are useful for exploratory data analysis, as tools for adjusting measurement processes and interpreting

```
Standard Test Data

                         Standard     % standard    Normalized  Relative
I        X(I)            deviation     deviation      residual    weight

1     10.0000             2.00000       20.0000         0.288     0.3462
2     11.0000             2.00000       18.1818         0.788     0.3462
3     8.00000             3.00000       37.5000        -0.474     0.1538
4     6.00000             3.00000       50.0000        -1.141     0.1538

Weighted mean          =    9.42307692       unweighted   8.75000000
Weighted s             =    2.02985605       (  21.5413 percent)
s about wtd mean       =    2.34961915       (  24.9347 percent)
s about unwtd mean     =    2.21735578       (  25.3412 percent)
Observed s.d. of mean (s/sqrt(n)) = 1.10867789 ( 12.6706 percent)
A priori s.d. of mean             = 1.17669681 ( 12.4874 percent)
ts/sqrt(n) [at P=.05; t=  3.182]  =    3.528     ( 40.3179 percent)
Reduced chi squared    =    0.744 for    3 degrees of freedom
Prob of exceeding      =    52.6      percent in a normal population
```

Fig. 3. Output from the VAX/VMS version of the MEAN program for input data 10 ± 2, 11 ± 2, 8 ± 3, and 6 ± 3, where "±" denotes the standard deviation of the values.

their results. They have also proven to be useful teaching tools for new workers.

ACKNOWLEDGMENTS

I thank L. A. Currie, R. F. Fleming, R. R. Greenberg, and other coworkers for useful discussions over the years. Source and executable copies of these programs are available on request, without charge and without warranty. Certain commercial equipment is identified in this article in order to specify the procedures in adequate detail. This identification does not imply recommendation or endorsement by the author or by the National Institute of Standards and Technology, nor does it imply that the equipment identified is necessarily the best available for the purpose. Contributions of the National Institute of Standards and Technology are not subject to copyright.

REFERENCES

1. L. A. Currie, *J. Res. NBS* **90,** 409–419 (1984).
2. J. B. Cumming, in *Applications of Computers to Nuclear and Radiochemistry*, G. D. O'Kelley, ed., NAS-NRC, Washington, DC, 1962, pp. 25–32.
3. H. P. Yule, *Anal. Chem.* **40,** 1480–1486 (1968).
4. R. F. Fleming (1978), unpublished lecture notes.
5. G. W. Phillips, *Nucl. Instr. Methods* **153,** 449–455 (1978).
6. J. S. Yadav, J. Brückner, and J. R. Arnold, *Nucl. Instr. Methods.* ***A277,*** 591–598 (1989).

7. T. P. Kohman, *J. Chem. Ed.* **47,** 657–658 (1970).
8. R. R. Greenberg, personal communication (1990).
9. R. M. Parr, H. Houtermans, and K. Schaerf, in *Computers in Activation Analysis and Gamma-Ray Spectroscopy (CONF-780421),* B. S. Carpenter, M. D. D'Agostino, and H. P. Yule, eds., US DOE, Springfield, VA, 1979, p. 544–562.
10. R. R. Greenberg and B. S. Carpenter, *J. Radioanal. Nucl. Chem.* **111,** 177–197 (1987).
11. L. A. Currie, *Anal. Chem.* **40,** 586–593 (1968).
12. D. L. Anderson, W. A. Cunningham, and E. A. Mackey, *Biol. Trace Element Res.* **27,** 613–622 (1990).
13. R. M. Lindstrom, G. J. Lutz, and B. R. Norman, *J. Trace Microprobe Tech.* **9,** 21–32 (1991).
14. E. A. Mackey, G. E. Gordon, R. M. Lindstrom, and D. L. Anderson, *Anal. Chem.* **63,** 288–292 (1991).
15. E. A. Mackey, G. E. Gordon, R. M. Lindstrom, and D. L. Anderson, *Anal. Chem.* **64,** 2366–2371 (1992).
16. S. O. Yusuf, Ph. D. thesis, University of Michigan (1993).
17. S. O. Yusuf and R. F. Fleming, *J. Radioanal. Nucl. Chem.* **179** 105–112 (1994).
18. L. A. Currie, *Nucl. Instr. Methods* **100,** 387–395 (1972).
19. P. R. Bevington, *Data Reduction and Error Analysis for the Physical Sciences,* McGraw-Hill, New York, 1969.
20. G. P. Lamaze, personal communication (1992).
21. K. Heydorn, in *Accuracy in Trace Analysis: Sampling, Sample Handling, Analysis (NBS Spec. Publ. 422),* P. D. LaFleur, ed., US GPO, Washington, DC, 1976, v. 1, pp. 127–139.
22. P. M. Wanek, R. E. Whipple, T. E. Fickies, and P. M. Grant, *Anal. Chem.* **54,** 1877–1878 (1982).
23. H. E. Solberg, *Anal. Chem.* **55,** 1611 (1983).

A PC-Based Computer Program for Quality Assessment of INAA Results of Emission and Air Particulate Samples

JIŘÍ FALTEJSEK*,[1,2] AND IVAN OBRUSNÍK[3]

[1]*Czech Ecological Institute, NAA Laboratory, Řež, Czech Republic; and* [2]*Present address: Nuclear Physics Institute, Academy of Sciences of Czech Republic, 25068 Řež, Czech Republic;* [3]*Czech Hydrometeorological Institute, Prague, Czech Republic*

ABSTRACT

A PC-based computer program for quality assessment (QAS) of results obtained by neutron activation analysis (NAA) is described. Program QAS is designed for the quality assurance system based on a routine use of certified reference materials (CRMs) with similar matrix composition as the samples analyzed together with each sample batch. The results obtained for CRMs samples are then compared with certified (true) values and with the results obtained for particular CRM in previous analysis runs. Five various tests are applied to check the quality of NAA results (for CRM and, consequently, for real samples) obtained for a particular analysis run. Control charts for elemental concentrations found in CRM and ranges of results for duplicate aliquots of CRM are displayed to visualize possible discrepancies. The program accepts files transported from a γ-ray spectrometric system ND 683 (based on a PDP 11 computer), extracts and stores reference material results, and prepares a work spreadsheet for each set of results to avoid typist and other kinds of errors.

Index Entries: Quality assessment; quality assurance; control charts; neutron activation analysis; reference materials; computer program.

*Author to whom all correspondence and reprint requests should be addressed.

Biological Trace Element Research Editor: G. N. Schrauzer © 1994 by Humana Press Inc.

INTRODUCTION

Measurement data of an adequate quality are necessary if environmental (or any other) problems are to be properly understood and effective action is to be taken. Quality-assurance programs involving all steps of an analytical procedure are necessary to ensure a good quality of data. The quality assurance (QA) system includes separate, but coordinated activities of quality control and quality assessment *(1–6)*. QA is a system of activities, the purpose of which is to provide the user of analytical results with the assurance that the presented results have defined standards of quality with a stated level of confidence. The quality control (QC) system of activities is designed to control the quality of analyses to fulfill the demands of the result user. It is a subset of the whole QA system.

The quality assessment (QAS) is an overall system of activities to provide assurance that the QC system works properly and effectively. QAS techniques are used to monitor the precision and accuracy of a system operating in the state of statistical control. There are some QAS techniques that may be applied only rarely (e.g., participation in interlaboratory tests), but others can be routinely used (e.g., analysis of control samples, reference materials, and so forth).

NAA PROCEDURE SUPPORTED BY QAS

In this work, a part of the QAS system routinely used for NAA of environmental samples is presented. NAA is usually carried out in batches consisting of analyzed samples, standards, flux monitors, and additional samples, such as blanks and control samples *(7–9)*. Irradiated samples are measured under various conditions (e.g., with various decay and cooling time). Thus, for each sample batch, we obtain several sets of results and each set should be checked by QAS. The final results of element concentration are calculated from these particular results given for various γ-lines by a selection of analytical lines or, in some cases (e.g., ^{60}Co), as weighted average of more such lines. Results of QAS together with detection limit calculated for individual γ-line determination help to make a correct selection for the final value of elemental concentration.

Two duplicate aliquots of suitable certified reference materials (CRM) are used as control samples in our QAS system *(10)*. These samples are handled and measured as ordinary samples. In this way, the following procedure can be applied:

- Sample preparation for irradiation;
- Irradiation;
- Preparation for radioactivity counting;
- Counting; and
- Result evaluation.

The results of CRM analyses are evaluated and compared with CRM certified values, their uncertainties, and also with the control limits of QAS by the program. Only if the QAS results satisfy the relevant criteria for QAS, the results obtained for real samples analyzed in the same batch can be reported to customers.

A matrix-based reference material used as a quality-control sample in each batch makes it possible to check the batch-to-batch repeatability, the accuracy of the results for the analyzed batch, and a long-term stability of the analytical system.

TESTS USED FOR QAS IN NAA

The five following tests are carried out in our QAS program: combined QAS, repeatability, test based on the use of robust statistics, accuracy, and repeatability within batches. Four tests are used to check mean elemental concentrations of CRM analyses, and the last one evaluates the range of CRM replicates. The tests are based on a strategy of control charts with the use of control and warning limits. The QAS program also contains a graphic presentation of these tests for the comfort of an analyst.

Combined QAS Test

The main QAS test combines a check of repeatability and accuracy. It has been derived from the test used in the Health and Environmental Chemistry Group at the Los Alamos National Laboratory *(11)*. In this test, a difference is compared between a particular result of CRM analysis together with its uncertainty, and the reference value propagated with the standard deviation of our laboratory mean according to Eqs. (1) to (3) *(see 10,11)*.

$$CL_{QAS} = RV \pm 3\sigma \tag{1}$$

$$WL_{QAS} = RV \pm 2\sigma \tag{2}$$

where CL_{QAS} are control limits (upper and lower), WL_{QAS} are warning limit (upper and lower), and:

$$\sigma = (s^2 + s_{RV}^2)^{1/2} \tag{3}$$

RV is the reference value (certified, information, or consensus value of the concentration of the element in the CRM), s is standard deviation computed from all previous results, and S_{RV} is an uncertainty related to the RV.

Possible results of this test may be:

1. A current result is within the warning limits—**system is under statistical control** ("O.K." in the report);
2. A current result lies between warning and control limits ("Warning!" in the report). Results of this batch can be

reported to the customer provided that all previous results of CRM analyses were okay. In any case, a detailed investigation of the analytical procedure is needed to elucidate the reason for this deviation;

3. A current result lies outside control limits—**system is out of control** ("OUT!" in the report). Such results cannot be reported to the customer. The detailed examination and correction of the analytical procedure are necessary; and
4. **There is not enough data to construct limits (No data)** (<7 data stored in the database and no reference value is given).

If no reference value is available, arithmetic mean of all previous results is used as the reference value. In this case, the test has the same meaning as the test of repeatability, the next test performed by the QAS program.

REPEATABILITY TEST

A current result is compared with the control and warning limits constructed as 2 and 3 SDs, respectively, of the laboratory mean *(12,13)*. Possible outputs are the same as in the first test. Both tests described above require at least more than 7, preferably about 20 results of CRM analyses to be carried out before their use for QAS purposes.

ROBUST STATISTICS

The empirical distribution of results of CRM control samples does not necessarily conform with normal distribution, since outliers can occur. In such a case, robust statistic *(14)* is a more appropriate approach. Moreover, this test can be used to define tolerance bounds from four control results of CRM analyses only. Median of these results can be calculated, and the warning limits are derived as the first and third quartiles. Then, the control limits are defined as:

$$UCL_{ROBUST} = UWL_{ROBUST} + 1.5h \tag{4}$$

$$LCL_{ROBUST} = \max\{0, LWL_{ROBUST} - 1.5h\} \tag{5}$$

where:

$$h = UWL_{ROBUST} - LWL_{ROBUST} \tag{6}$$

UCL_{ROBUST} and LCL_{ROBUST} are upper and lower control limits, respectively, and UWL_{ROBUST} and LWL_{ROBUST} are upper and lower warning limits (third and first quartiles), respectively.

The graphical presentation of this test in the form of boxplot together with the control chart is also done by the program for the user comfort.

ACCURACY TEST

This test is designed to test the accuracy of the results obtained for a control CRM sample using relation (7).

$$|RV - RES| \leq (s_{RV} + s_{RES}) \tag{7}$$

where RV and RES are the reference value and a current result, and S_{RV} and S_{RES} are related reference value uncertainty and standard error of the result (counting error of radioactivity measurement).

In the case that the current result is below the determination limit, the accuracy test is fulfilled if:

$$DL \geq RV - s_{RV} \tag{8}$$

where DL is detection limit.

TEST OF REPEATABILITY WITHIN BATCHES USING RANGE

The repeatability within batches is tested by an evaluation of the range of results obtained in the tested batch and the ranges obtained for previous results. For the case of duplicates, the control and range limits are defined (12) as:

$$CL_{Range} = 3.267\bar{R} \tag{9}$$

$$WL_{Range} = 2.512\bar{R} \tag{10}$$

where $\bar{R}$ is the mean range. In this test, both lower (control and warning) limits are set to zero.

QAS PROGRAM DESCRIPTION

The QAS program has been written in FOXPRO v.2.1, and its graphic part in Turbo Pascal for IBM PC-compatible computers. The program consists of several parts designed to create a working data base from the tested data, set up environment (selection of output, data bases, condition for the data selection), select relevant stored data, and compute the test limits. Other parts of the program are designed for presenting test results (using screen or a printer), making control charts and storing tested results in data bases. Moreover, a Base Manager is incorporated into the QAS program, but access to this part of the program is permitted only for a QA supervisor, and it is secured by a password. Three data bases are necessary for the construction of the limits for all tests. The first data base stores information on CRMs used for QAS (CRM data base). It contains reference values of element concentration together with the certification value status (certified, information, consensus, and so on) and uncertainties of the stored values. The results of all CRM analyses are stored in two other data bases organized according to elements and analytical γ-lines of particular radionuclides. The first of these result data bases contains the results of each particular analysis in separated records, including the flag for an aerosol filter selection (Nuclepore, Synpor, and so forth) in the case of NAA of aerosols and a sample identification. The second result data base stores information about the mean values and ranges of CRM control analyses for a particular series of analyses described by unique date (e.g., the end of irradiation).

Creating and Handling of Working Data Base

The QAS program cannot handle the original measured data files, because the results of NAA are evaluated on a PDP 11/73 computer with the different file and diskette formats. Therefore, the results are transferred in the form of an ASCII file from the PDP computer to PC and a simple translating interface is used for the data input into a working file. The data input from a keyboard, browsing, and corrections can also be performed. A work spreadsheet file in the format of Lotus 123 (WK1) can be created.

Data Tests and Report Output

The QAS program computes the limits for all elements, and the analytical γ-lines for all tests and stores these values and test results obtained in the working data base. A short summary or full protocol can be directed to the screen, printer, or a text file. The examples of such reports are shown in Fig. 1 and 2.

QAS Control Chart

The QAS control chart is a comprehensive presentation of the QAS tests. The QAS program creates a control chart together with a boxplot that characterizes parameters of the robust test and a simple histogram to be displayed for better visualization of the distribution of stored results (cf Fig. 3). The control chart of ranges can also be displayed (cf Fig. 4). An inspection of the data obtained for a particular batch can be performed by marking with cursor (the data are displayed in the bottom of the control chart). The last value is placed at the right side of the chart. Selected results can be zoomed to the full screen. The limits for QAS test are displayed by the full line (control limits red, warning yellow). The limits for repeatability are presented as dashed lines, and mean value as a dotted line. The uncertainty interval of the reference value has another background color than the rest of the chart.

Visualization of the data is very useful not only for quality assessment, but also for the whole QA system in general, especially when some problems appear. For this reason, a simultaneous display of more analytical lines (even for various elements) can be performed by the program.

Summary of QAS for RM 1648						
Analytical line	QAS	Repeat.	Precision	Robust	Accuracy	
Na 1368.00	O.K.	O.K.	O.K.	O.K.	O.K.	(CONS)

Fig. 1. Example of the short summary QAS protocol.

```
Reference material :SRM 1648   Element : Na
Reference value :  4600.0000 ±  700.0000 (CONS)

Line : 1368.00    ( 12 results in archive)
Control limits for  tests :
QAS test        4600.0000 (  16.6126%)  1772.3530   - 6357.4378
Repeatability   4064.8954 (   7.5414%)  3145.2484   - 4984.5424
Robust          4069.3900               3312.1588   - 4686.2688

Tested batch :                            QAS    Repeat.  Robust   Accuracy
Batch mean :                 4076.0867     O.K.    O.K.     O.K.     O.K.
Batch range :  418.3700  Range test :  O.K.  (ctrl. lim.:  564.4586 )

Individual results                         QAS    Repeat.  Robust   Accuracy
 VAN509DAT  4308.3100± 127.50400   O.K.    O.K.    Warning!  O.K.
 VAN535DAT  4030.0100± 107.58100   O.K.    O.K.    O.K.      O.K.
 VAN539DAT  3889.9400± 112.22000   O.K.    O.K.    O.K.      O.K.
```

Fig. 2. Example of the full QAS protocol.

Data Base Manager

Generally, all results should be stored and used for the calculation of particular limits. However, under certain circumstances, it might be desirable to make changes in the CRM or result data bases. This can be done using the data base manager.

SPECIFIC PROBLEMS RELATED TO INSTRUMENTAL NAA OF ENVIRONMENTAL SAMPLES

A specific problem occurs in the case of instrumental NAA (INAA) of environmental samples, e.g., aerosols, when a blank originated in the filter or in a polyethylene bag used for sample wrapping is so high that

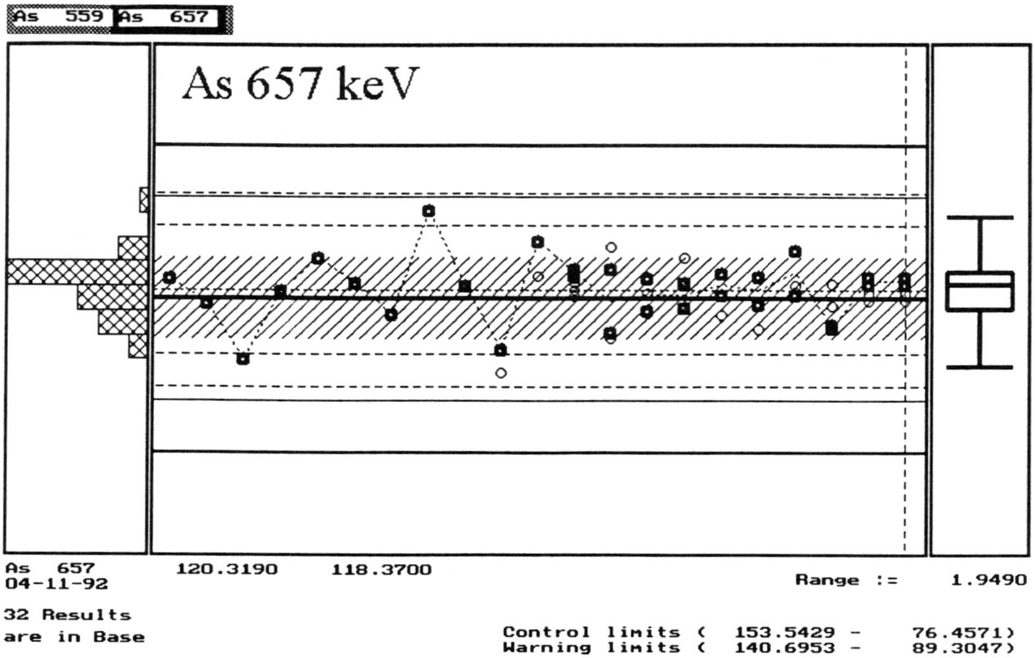

Fig. 3. Example of control chart of As determination in the SRM NIST 1648.

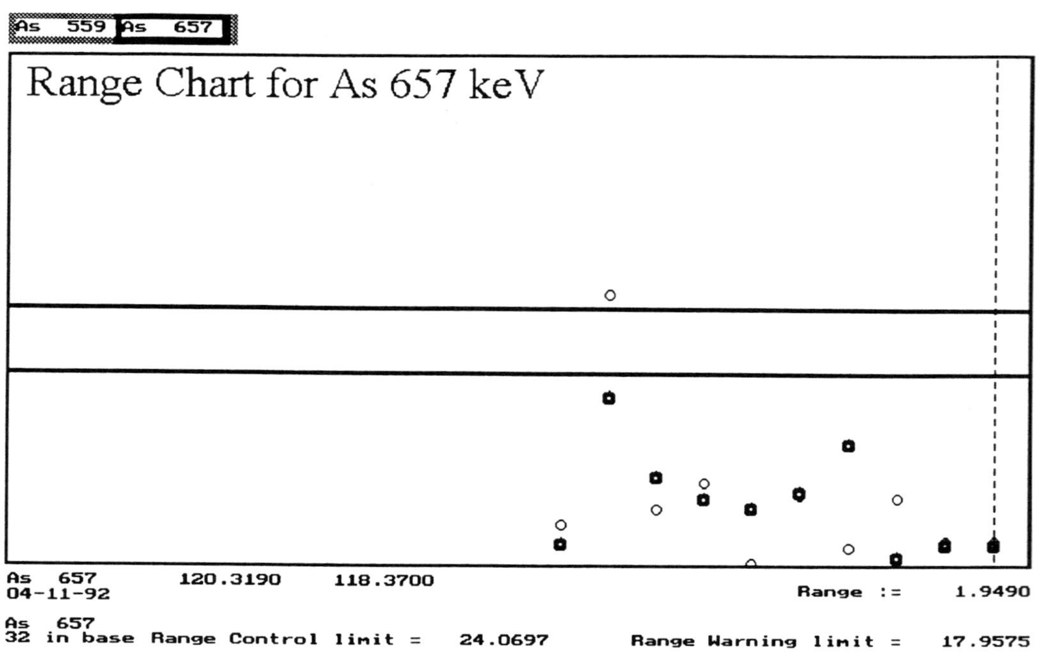

Fig. 4. Example of the control range chart for As in the SRM NIST 1648.

a blank subtraction is necessary. In the routinely used INAA scheme, several measurement groups for different conditions are collected. The QAS program creates a work spreadsheet for each group where the results of analysis for each sample (including control samples) and analyzed γ-lines are organized in the table. The final element concentration is computed from particular results spreadsheets with respect to detection limits and QAS test results, and then the new summary spreadsheet table is created.

Elemental concentrations found in blanks (empty polyethylene and clear filter) for current sample set are in the table, and therefore, correction for a blank contribution can be done. Blank values usually vary, and therefore, it is better to use average values from a large set of data. For this purpose, a computer program was designed to extract results of blank samples from a summary spreadsheet, store these values to data base, and compute average values for elemental concentrations in the blanks. Output of calculated averages and their uncertainties is carried out in the spreadsheet with the same organization as the summary table from which data have been extracted without any manual handling to avoid the possibility of a typist error.

The program for blank evaluation gives an opportunity for testing stored results for outliers. Elemental concentrations or their logarithms may be tested according to Grubbs *(15)*, Dean and Dixon *(16)*, Huber *(17)*, Kolmogorov et al. *(18)*, and Shapiro-Wilk *(19)*, as well as by the tests for skeweness and kurtosis *(18)*. Detected outliers can be interactively rejected from the set of blank concentration while the average value is computed. The way to detect outliers is described in detail in the program for the evaluation of round-robin analyses *(20)*.

SUMMARY

The programs described are routinely used for the management and assessment of the quality of analytical results, especially for NAA of air particulate samples. For this purpose, an NIST SRM 1648 Urban Air Particulate is the most frequently used RM. The results of QAS tests can demonstrate that the analytical system works under statistical control.

The data processing in spreadsheets enables the analyst to control data flow and to decrease the amount of tedious work during processing large sets of samples. Moreover, checking the element concentrations in blank samples may help one to discover or solve problems related to the quality of analytical results and procedures used.

The QAS program can be used for INAA of various materials. Its application to NAA with radiochemical separation is limited because of the problems related to the determination of a chemical yield of separation, the necessity of splitting of data bases according various separation methods, and so on. However, the program has been used in a routine basis in our NAA laboratory with rather encouraging results.

REFERENCES

1. American Chemical Society Subcommittee on Environmental Analytical Chemistry, *Anal. Chem.* **52,** 2242–2249 (1980).
2. J. K. Taylor, *Environmental Sampling For Hazardous Wastes,* G. E. Schwartzer and J. A. Santolucio, eds., ACS Symposium Series 267, 1984, p. 105.
3. I. Obrusník, *Chem. listy* ***81,*** 1256 (1987).
4. R. A. Nadkarni, *Anal. Chem.* ***63,*** 675A–681A (1991).
5. R. J. Mesley, D. W. Pocklington, and R. F. Walker, *Analyst* ***116,*** 975 (1991).
6. J. K. Taylor, *Am. Lab.* October, 67 (1985).
7. I. Obrusník, J. Horáková, and L. Kolář, Report NRI Řež 8303CH (1987).
8. I. Obrusník, J. Horáková, and L. Kolář, Report NRI Řež 8992CH (1989).
9. J. Kučera, J. Faltejsek, J. Horáková, Report NRI Řež 9574CH (1991).
10. L. Kolář and I. Obrusník, Report NRI Řež 8439CH (1988).
11. M. A. Gautier, E. S. Gladney, and B. T. O'Malley, *Quality Assurance for Health and Environmental Chemistry,* LA 10813 MS, Los Alamos, 1986.
12. S. L. Inhorn, *Quality Assurance Practices For Health Laboratories,* Byrd Pie-Press, Springfield, VA, 1978.
13. J. M. Juran, *Quality Control Handbook,* 3rd ed., McGraw Hill, New York, 1974.
14. C. Reimann and F. Würtzer, *Mikrochim. Acta* ***1986(II),*** 31–42 (1986).
15. K. Eckschlager, and I. Horsák, *Vyhodnocování analytických výsledků a metod,* SNTL, Praha, 1980.
16. R. Dybczynski, *Analytica Chimica Acta* ***117,*** 53–70 (1980).
17. P. L. Davies, *Fresenius Z. Anal. Chem.* **331,** 513–519 (1988).
18. L. Sachs, *Applied Statistics, A Handbook of Techniques,* 5th ed., Springer-Verlag, New York, 1978.
19. S. S. Shapiro, and M. B. Wilk, *Biometrika,* **52,** 591–611 (1965).
20. J. Kučera and J. Faltejsek, *IPECA—Interactive Programme for Evaluation of Circular Analyses, Fresenius J. Anal. Chem.*, in press.

Accurate Determination of Cobalt Traces in Several Biological Reference Materials

RAJMUND DYBCZYŃSKI* AND BOŻENA DANKO

Department of Analytical Chemistry, Institute of Nuclear Chemistry & Technology, 03-195 Warszawa, ul. Dorodna 16, Poland

ABSTRACT

A newly devised, very accurate ("definitive") method for the determination of trace amounts of cobalt in biological materials was validated by the analysis of several certified reference materials. The method is based on a combination of neutron activation and selective and quantitative postirradiation isolation of radiocobalt from practically all other radionuclides by ion-exchange and extraction chromatography followed by γ-ray spectrometric measurement. The significance of criteria that should be fulfilled in order to accept a given result as obtained by the "definitive method" is emphasized. In view of the demonstrated very good accuracy of the method, it is suggested that our values for cobalt content in those reference materials in which it was originally not certified (SRM 1570 spinach, SRM 1571 orchard leaves, SRM 1577 bovine liver, and Czechoslovak bovine liver 12-02-01) might be used as provisional certified values.

Index Entries: Determination of cobalt; RNAA; definitive methods; biological materials; certified reference materials.

INTRODUCTION

Cobalt is one of the elements essential for life, but on the other hand, excessive occupational exposures to cobalt may result in cancer *(1)*. At the same time from an analytical point of view, cobalt belongs to the so-called difficult elements *(2,3)*. Accurate determination of trace amounts of cobalt in biological materials is in fact a difficult task. Such conclusions

*Author to whom all correspondence and reprint requests should be addressed.

Biological Trace Element Research Editor: G. N. Schrauzer © 1994 by Humana Press Inc.

can be drawn from results differing by several orders of magnitude obtained by various investigators participating in international intercomparisons (3–8), as well as from wide range of data reported for the so-called normal level of Co in serum as compiled by Versieck and Cornelis *(9)*.

Further proof for the existence of serious, although not clearly identified, analytical difficulties in the determination of trace amounts of cobalt in biological materials is the fact that cobalt could not be certified in a majority of certified reference materials (CRMs) issued by NBS/NIST *(10–14)*, IAEA *(15)*, BCR *(16,17)* and other institutions *(18)*.

Recently, a very accurate ("definitive") method for the determination of cobalt traces in biological materials was devised in our laboratory *(19,20)*. Results of cobalt determination in several CRMs in which it was not certified by the Issuing Organizations are presented in this article.

EXPERIMENTAL

Sample Preparation and Irradiation

One hundred to 200-mg samples of biological materials, most of which were CRMs, were weighed on an analytical balance into small handmade Al foil bags of very low Co content. Standards (200 ng Co) were prepared by evaporating an aliquot (50 μL) of freshly prepared standard Co solution in dilute HNO_3 on a small strip of Al foil and placed in an Al foil bag analogous to those used for the samples. The irradiation package, which consisted of four to eight standards, some of which served also as irradiation monitors to account for possible flux gradient, six to eight samples, and a blank (empty Al bag) wrapped in Al foil, was irradiated at a flux of $1–3 \times 10^{13}$ n/cm^2/s for 50–100 h, and at a later stage of work, also at the flux of $6–9 \times 10^{13}$ n/cm^2/s for 100–200 h, and cooled for at least 3 wk before processing.

Radiochemical Separation

Samples were removed from the Al foil bags, and the interior of the unfolded bag was washed with concentrated HNO_3. The washings were added to the sample. The separation scheme is shown in Fig. 1. Details on the ion-exchange and extraction chromatography systems used in this work can be found in earlier papers *(19,20)*. A residual blank (washings from the empty Al foil bag) and one of the standards were run through the same separation procedure as the samples.

γ-Ray Spectrometric Measurements

γ-Ray spectrometric measurements of cobalt fractions from the samples, residual blank, and the standards were performed with the aid of a 55 cm^3 Ge(Li) detector (ORTEC), efficiency ca. 10% with respect to 3″ × 3″ NaI (Tl) detector, ORTEC analog line, and Plurimat 20 (Intertechnique)

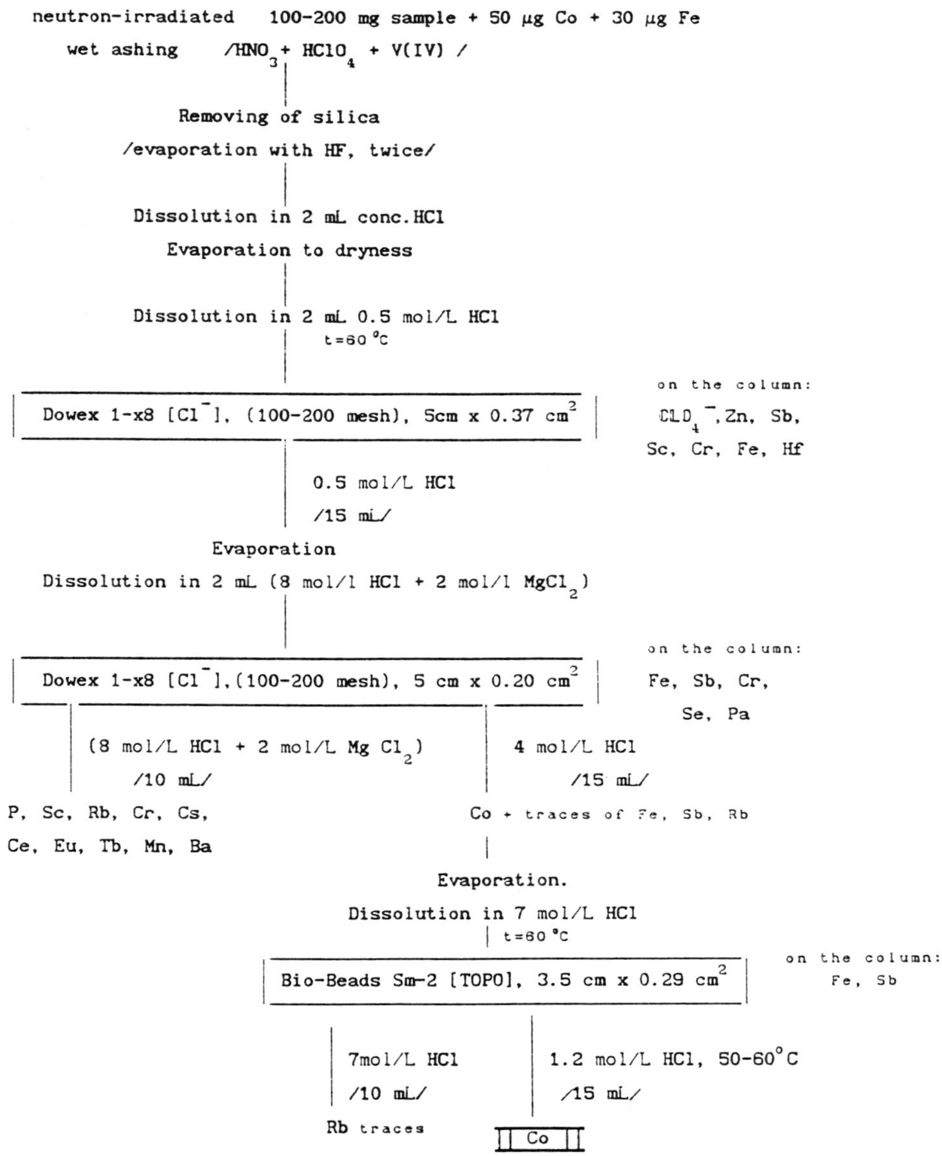

Fig. 1. Radiochemical separation scheme for the quantitative isolation of cobalt from irradiated biological materials *(19,20)*.

pulse height analyzer or IN-45 PHA (Intertechnique). Energy resolution was 3.1 keV for 1332.4 keV ^{60}Co line, with peak-to-Compton ratio: 33:1.

Some measurements were done with an HPGe detector (ORTEC), volume 212 cm^3, relative efficiency 47%, ORTEC analog line and SWAN-3 PHA of the plug-in card type with appropriate software produced by the Institute of Nuclear Problems, Swierk, Poland. Energy resolution was 1.81 keV for 1332.4-keV ^{60}Co line, with peak-to-Compton ratio: 72:1.

RESULTS AND DISCUSSION

Criteria Assuring the Quality of Results

As was pointed out in our earlier works *(20,21)*, the "definitive method" should be equipped with certain warning mechanisms fitted into the procedure to safeguard against gross errors, which could otherwise occur unnoticed by the analyst. The following criteria *(20)* were adopted in this work:

1. There are no indications that something might go wrong during the analysis;
2. Visual control of the correctness of the separation procedure: in the retention stage, both on ion-exchange and on extraction chromatography columns, the blue cobalt band should not travel more than ½–⅔ of the bed length;
3. Small residual blank: the results were accepted if the correction for residual blank determined in a given run was below 10% of the cobalt contents in the analyzed sample;
4. Agreement of standards: The results for a given series of samples were accepted if the normalized count rates of the two standards, one of which was processed in the same way as the samples and the other one was measured directly after dissolution, did not differ by more than 5% (after correcting for flux gradient if any); and
5. The result for the CRM irradiated and analyzed together with the samples should be in agreement with the "recommended" or "certified" value. The criterion adopted here was that the interval $X \pm S$, where X is the result and S standard deviation of single determination known from previous experience with that type of sample, should overlap with a confidence interval of the certified value.

Only when all of the above criteria were simultaneously fulfilled was the result acknowledged as being obtained by the definitive method.

Validation of the Method

The separation scheme as shown in Fig. 1 was thoroughly tested by tracer experiments, and it was demonstrated *(19,20)* that the procedure assures truly quantitative recovery. At the same time, radiochemical purity of cobalt fractions isolated from real samples was very high and practically no other activities, except photopeaks present in the natural background, could be detected (cf Fig. 2).

Certified or recommended value established in the course of an international intercomparison employing a variety of analytical techniques should have in most cases (i.e., in the absence of predominating systematic errors) an equal chance of being higher or lower than the unknown "true value." Therefore, when analyzing several CRMs by a very accurate method that should yield in each case results close to the

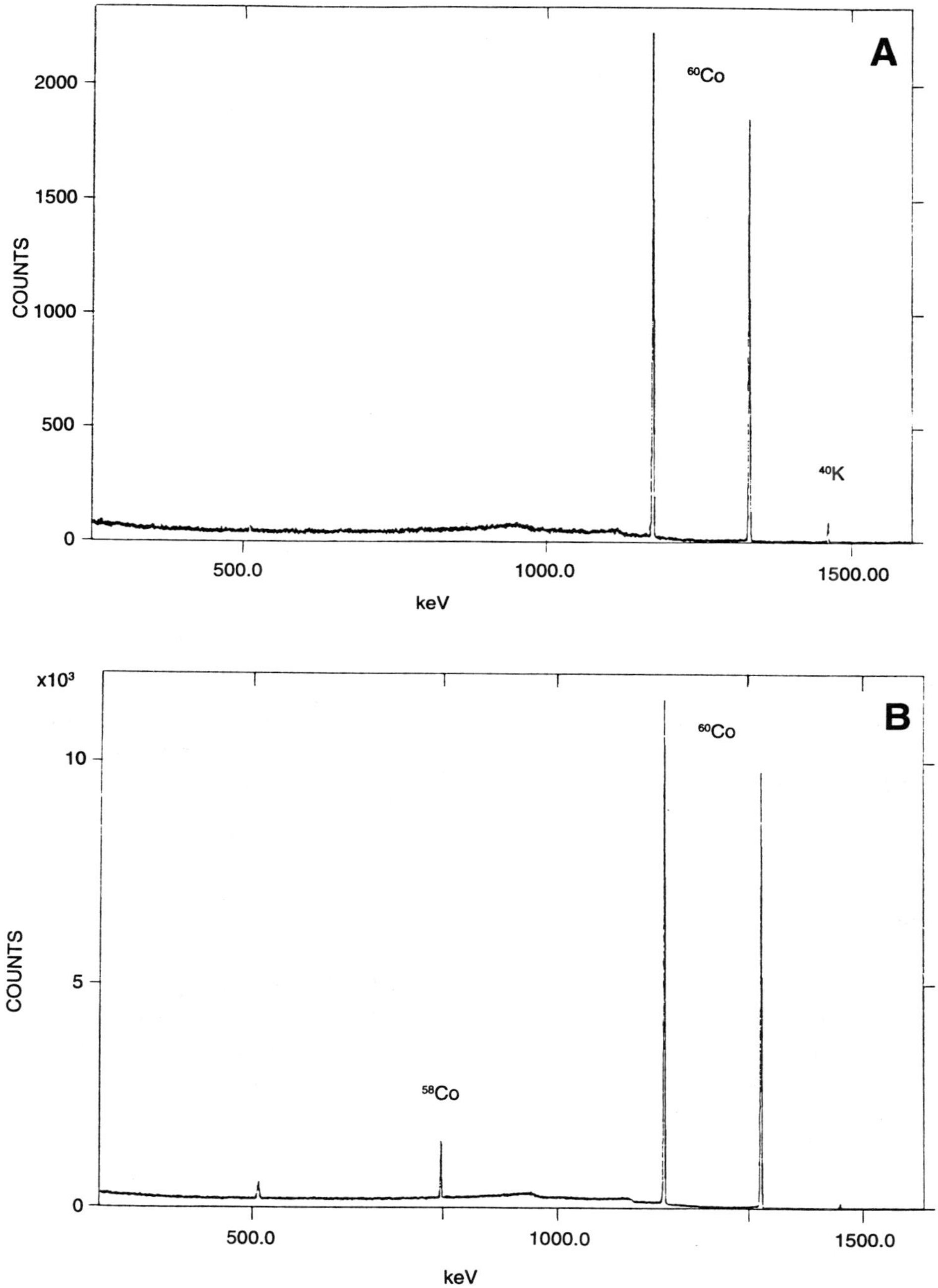

Fig. 2. γ-Ray spectra of cobalt fraction from: **(A)** Bowen's kale, t = 2500 s, and **(B)** IAEA V-10 Hay Powder t = 3000 s.

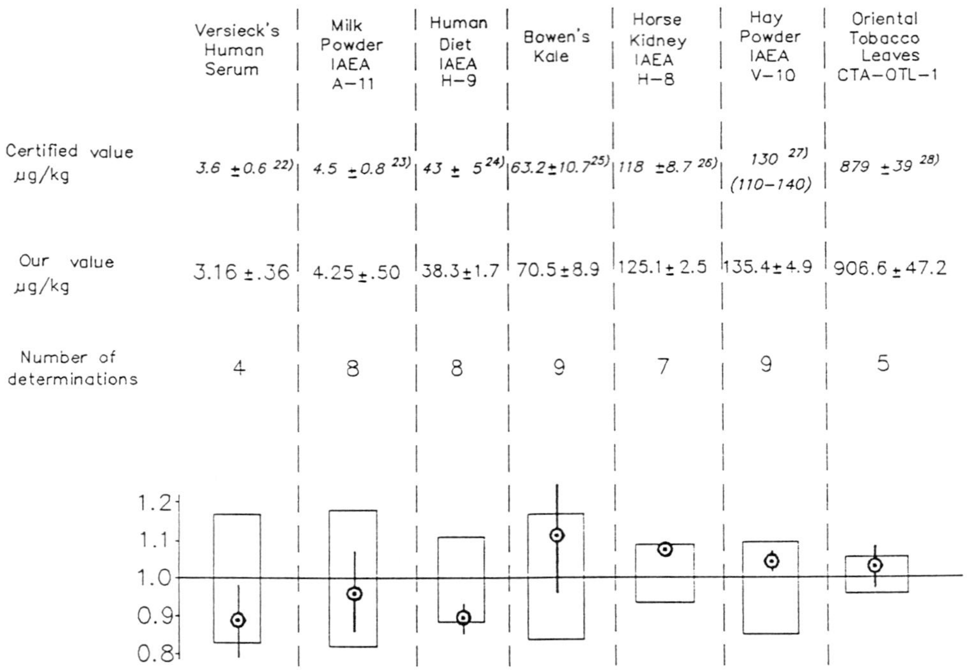

Fig. 3. Comparison of our results obtained by definitive method with the certified values. The diagram shows the ratio of the mean of our results and its confidence limits to the certified value. Value 1—normalized certified value shown together with its confidence interval.

"true values," both positive and negative deviations from the certified or recommended value are to be expected. In Fig. 3, the results obtained in our previous work (20) by the present method for several CRMs of both animal and plant origins are shown in the form: mean ± confidence interval (showing also the number of determinations) and compared with the certified values and their confidence limits. As can be seen in each case, our value is within the confidence limits of the certified or recommended value. Both positive and negative deviations from the certified values are observed as expected for the truly unbiased method.

Determination of Cobalt Concentration in Some CRMs in Which Co Was Not Certified

After having proven that the method gives highly reliable results, it was decided to use it for the determination of cobalt contents in some CRMs in which it was originally not certified and to compare the outcome with some literature data. The results are shown in Table 1. One of the conclusions that can be drawn from inspection of data presented in Table 1 is that so-called consensus values, e.g., those from the compilation by Gladney et al. *(29)*, should not be indiscriminately used in lieu of

Table 1
Results of Cobalt Determination in Some CRMs in Which It Was Not Certified

Material	Information value for Co µg/kg	Results of individual determinations x_i, µg/kg	Arithmetic mean and its confidence limits $\bar{x} \pm t_{0.05} s\, n^{-1/2}$	Relative standard deviation s_r	Gladney's compilation value *) µg/kg[29]
NBS SRM 1570 Spinach	(1500)	1268			
		1050			
		1298			
		1207			
		1298			
		1226			
		1254			
		1354	1244 ± 76	0.073	1560±76
Czecho-Slovak Bovine Liver 12-02-01	(374)	341.6			
		341.4			
		339.8	341.0 ± 2.5	0.003	-
NBS SRM 1577 Bovine Liver	(180)	239.6			
		242.7			
		250.4			
		241.9			
		252.1			
		241.5	244.7 ± 5.5	0.021	230±40
NBS SRM 1571 Orchard Leaves	(200)	140.0			
		117.1			
		121.8			
		136.4			
		134.1			
		141.0			
		140.1			
		142.9			
		145.0			
		146.1	136.5 ± 6.9	0.071	160±37

*Original data by Gladney et al. [29] given in the form: mean ± SD (*n*) were recalculated into: mean ± confidence interval to allow direct comparison with our results.

Table 2
Significance of Qualifying Criteria in Assuring Quality of Results

MATERIAL	INFORMATION OR "CONSENSUS" VALUE FOR Co μg/kg	MEAN ± CONF. INT. μg/kg	REL. SD
NBS SRM 1570 SPINACH	(1500)[11] 1560±76[29]	A 1244 ± 76	A 0.073
		B 1163 ± 111	B 0.151
NBS SRM 1571 ORCHARD LEAVES	(180)[11] 160±37[29]	A 136.5 ± 6.9	A 0.071
		B 128.4 ± 10.4	B 0.127

(A) Calculated from determinations in which all criteria required for "definitive method" were fulfilled.
(B) Calculated from determinations in which not all qualifying criteria were simultaneously fulfilled.

certified values. For instance, both NBS "information value" and Gladney's consensus value (which incidentally is based mostly on INAA results) for SRM 1570 spinach are certainly too high. Recent accurate measurements by Becker also by INAA *(30)* gave 1350 and 1340 μg/kg, respectively. Our single measurement by INAA on a sample that has been cooled for 1.5 yr gave the 1353 μg/kg. The results by our definitive method are even lower (cf Table 1), and we believe that the latter (1244 ± 76 μg/kg) is the best approximation of the true value. A similar situation seems to exist with SRM 1571 orchard leaves, although the confidence limits of Gladney's "consensus value" are so wide that it is impossible to draw meaningful conclusions. Our single INAA measurement after 1.5 yr of cooling gave the result 155.7 μg/kg. A recent result of Kucera et al. *(31)* also by INAA was 154 ± 20 μg/kg. Again we believe that our result by the definitive method (136.5 ± 6.9 μg/kg) is at present the best approximation of the true value.

It is perhaps worth noting that our single INAA measurements after 1.5 yr of cooling for other materials (39.2 μg/kg for IAEA H-8 and 67.9 μg/kg for Bowen's kale) gave the values lying within the confidence intervals of our results by the definitive method or slightly above the confidence interval (143 μg/kg for IAEA V-10 Hay Powder) (cf Fig. 3).

It looks like certain neutron-irradiated materials may contain something other than ^{60}Co radioactivities contributing counts to the region of 1332 keV that leads to positive systematic errors when the analysis is made by instrumental γ-ray spectrometry. Finally, one may note very good agreement of our result for SRM 1577 bovine liver (244.7 ± 5.5 μg/kg, cf. Table 1) with the recent results of Woittiez et al. (244 ± 13 μg/kg) *(32)* by radiochemical NAA, although the latter has a considerably wider confidence interval.

In Table 2, there is an illustration of the significance of qualifying criteria on accuracy and precision of results. The data obtained under conditions where all the criteria were simultaneously fulfilled and the results were considered as having been obtained by the "definitive method" are listed under "A category." Those shown under "B category" were obtained by the same procedure, but for different reasons, one or more criteria were not met and hence these results were not considered as having been obtained by the "definite method." One can easily note much better precision of results in "A category," which must also be more accurate, since both series of determinations were done by the same method.

CONCLUDING REMARKS

As was demonstrated above, the method presented in this article is highly reliable and free from systematic errors. It has a very favorable detection limit of 0.008 ng Co, i.e., 0.04 μg/kg for 200 mg samples *(20)*. Because of the use of Al foil of high purity as a canning material, the residual blank that should be determined in each series of determinations was small ($x \pm s = 0.083 \pm 0.048$ ng Co/sample) *(20)*.

Theoretically, positive systematic errors may occur when analyzing samples with very high Ni, Cu, and Fe contents owing to threshold or second-order reactions: $^{60}Ni(n,p)^{60}Co$; $^{63}Cu(n,\alpha)^{60}Co$; and $^{58}Fe(n,\gamma)^{59}Fe \text{—} \beta^- \rightarrow ^{59}Co(n,\gamma)^{60}Co$.

The magnitude of possible errors was evaluated experimentally *(20)*, the interfering factors ($IF_{Co(El)}$) being as follows: $IF_{Co(Cu)} = 2.2 \times 10^{-6}$ g Co/g Cu, $IF_{Co(Ni)} = 4.1 \times 10^{-6}$ g Co/g Ni, and $IF_{Co(Fe)} = 1.0 \times 10^{-6}$ g Co/g Fe. Knowing the concentration of an interfering element in a sample, C_{El} the apparent cobalt concentration owing to interfering reaction, $C_{appCo(El)}$ can be calculated from the relationship (concentrations for instance in μg/kg):

$$C_{appCo(El)} = C_{El} \times IF_{Co(El)} \qquad (1)$$

For instance, for SRM 1577 bovine liver, which contains relatively very high concentrations of metallic elements (C_{Cu} = 193 mg/kg; C_{Ni} = 160 μg/kg C_{Fe} = 268 mg/kg) *(29)* the total apparent concentration of cobalt formed as a result of interfering reactions as calculated from eq. (1) amounts to 0.69 μg/kg.

So, possible uncertainty resulting from nuclear interferences can reach 0.28% of the cobalt content, nearly an order of magnitude less than the analytical uncertainty (cf Table 1). For most of other samples, the potential errors owing to interfering reactions are even much lower.

In view of proven good accuracy of our method, it is suggested that cobalt concentrations established for CRMs in which they were originally not certified (cf Table 1) might be used by the scientific community as provisional certified values.

ACKNOWLEDGMENTS

This work was in part supported by the State Scientific Research Committee (KBN) (grant No. 2 0675 91 01) and the Maria Skłodowska-Curie Joint Fund II (grant PAA/NIST 90-045), as well as the International Atomic Energy Agency (contract No. 5955/RB). The authors wish to thank Professor J. Versieck for supplying samples of his Human Serum CRM.

REFERENCES

1. *Metals and Their Compounds in the Environment* E. Merian, ed., VCH Verlagsgesellschaft, Weinheim, 1991, p. 503.
2. R. M. Parr, in *Trace Element Metabolism in Man and Animals—3, Proc. 3rd Intern. Symposium, Freising—Weihenstephan, 1977,* Technical University, Munich, 1978, p. 622.
3. *Quality Assurance in Biomedical Neutron Activation Analysis*, IAEA-TECDOC-323, IAEA, Vienna, 1984.
4. R. M. Parr, in *Trace Element Analytical Chemistry in Medicine and Biology*, P. Brätter, and P. Schramel, eds., W. de Gruyter, New York, 1980, p. 631.
5. L. Pszonicki, A. N. Hanna, and O. Suschny, Report on Intercomparison V-9 of the Determination of Trace Elements in Cotton Cellulose, IAEA/RL/97, IAEA, Vienna, 1983.
6. E. Zeiller, V. Strachnov, and R. Dekner, Intercomparison study IAEA-155, on the determination of inorganic constituents in whey powder, IAEA/AL/034, IAEA, Vienna, Nov., 1990.
7. R. Dybczyński, A. Veglia, and O. Suschny, in *Trace Element Analytical Chemistry in Medicine and Biology*, P. Brätter and P. Schramel, eds., W de Gruyter, New York, 1980, p. 657.
8. L. Pszonicki, A. Veglia, R. Dybczyński, and O. Suschny, Report on intercomparison V-8 of the determination of trace elements in rye flour, IAEA/RL/91, IAEA, Vienna, Jan., 1982.
9. J. Versieck, and R. Cornelis, *Anal. Chim. Acta* **116**, 217 (1980).
10. E. S. Gladney, *Anal. Chim. Acta* **118,** 385 (1980).

11. C. H. Hudson, ed., *NBS Standard Reference Materials Catalog 1984–85*, NBS Special Publication 260, US Government Printing Office, Washington, DC, Feb., 1984.
12. N. M. Trahey, ed., *NIST Standard Reference Materials Catalog* 1992–93, NIST Special Publication 260, US Government Printing Office, Washington, DC, Feb., 1992.
13. National Institute of Standards & Technology, Certificate of Analysis, Standard Reference Material 1515, Apple Leaves, Gaithersburg, MD, Jan. 22, 1993.
14. National Institute of Standards & Technology, Certificate of Analysis, Standard Reference Material 1547, Peach Leaves, Gaithersburg, MD, Jan. 22, 1993.
15. International Atomic Energy Agency, AQCS Programme 1992/93, IAEA, Vienna, Apr. 1992.
16. J. J. Bellardo and P. J. Wagstaffe, *Fresenius Z. Anal. Chem.*, **332**, 533 (1988).
17. B. Griepink, H. Muntau, and E. Colinet, *Fresenius Z. Anal. Chem.* **315**, 193 (1983).
18. J. Kučera, P. Mader, D. Miholová, J. Cibulka, M. Poláková, and D. Kordík, *Fresenius J. Anal. Chem. 338*, 66 (1990).
19. R. Dybczyński, B. Danko, and H. Maleszewska, *J. Anal. Chem. (Moscow)*, **49,** No. 1, (1994).
20. R. Dybczyński and B. Danko, *J. Radioanal. Nucl. Chem.*, **181,** 43 (1994).
21. R. Dybczyński, M. Wasek, and H. Maleszewska, *J. Radioanal. Nucl. Chem. Articles*, **130,** 365 (1989).
22. J. Versieck, L. Vanballenberghe, A. De Kesel, J. Hoste, B. Wallaeys, J. Vandenhaute, N. Baeck, H. Steyaert, A. R. Byrne, F. W. Sunderman, Jr., *Anal. Chim. Acta* **204**, 63 (1988).
23. A. R. Byrne, C. Camara-Rica, R. Cornelis, J. J. M. De Goeij, G. V. Iyengar, G. Knapp, R. M. Parr, and M. Stoeppler, *Fresenius Z. Anal. Chem* **326,** 723 (1987).
24. R. M. Parr, *J. Radioanal. Nucl. Chem.* **123,** 259 (1988).
25. H. J. M. Bowen, in *Biological Reference Materials*, W. W. Wolf, ed., Wiley, New York, 1985, pp. 3–17.
26. R. M. Parr, Progress report No. 1, Intercomparison of Cadmium and Other Elements in IAEA Horse Kidney (H-8), IAEA, Vienna, Jan., 1985.
27. L. Pszonicki and A. N. Hanna, Report of the Intercomparison IAEA (V-10) of the Determination of Trace Elements in Hay Powder, IAEA/RL/123, IAEA Vienna, July 1985.
28. R. Dybczyński, H. Polkowska-Motrenko, Z. Samczyński, and Z. Szopa, *Fresenius J. Anal. Chem.* **345,** 99 (1993).
29. E. S. Gladney, B. T. O'Malley, I. Roelandts, and T. E. Gills, Standard Reference Materials: Compilation of Elemental Concentration Data for NBS Clinical, Biological, Geological and Environmental Standard Reference Materials, NBS Special publication 260–111, US Government Printing Office, Washington, DC, 1987.
30. D. A. Becker, private information (1993).
31. J. Kučera, L. Soukal, and J. Horákova, *Fresenius J. Anal. Chem.* **345**, 188 (1993).
32. J. R. W. Woittiez, and M. De La Cruz Tangonan, *J. Radioanal. Nucl. Chem. Articles* **158**, 313 (1992).

NAA Methodology for the Certification of Eight Trace Elements in a Plankton Reference Material

A. P. GRIMANIS* AND MARIA VASSILAKI-GRIMANI

Radioanalytical Laboratory, NCRS Demokritos, 153 10 Ag. Paraskevi Attiki, Athens, Greece

ABSTRACT

Neutron activation analysis (NAA) methodology was applied for the certification of eight trace elements (As, Co, Cr, Cu, Hg, Mn, Se, and Zn) in a plankton reference material (CRM 414) prepared by the BCR of the Commission of the European Communities. Average concentrations (μg/g dry wt) and standard deviations found in plankton for the eight trace elements were as follows: As 6.29 ± 0.19, Co 1.37 ± 0.05, Cr 23.7 ± 0.56, Cu 27.1 ± 0.88, Hg 0.29 ± 0.02, Mn 285 ± 4.1, Se 1.62 ± 0.10, and Zn 114 ± 3.4. All values found were to be in very good agreement (0.5%–8.1%) with certified values.

Index Entries: INAA; RNAA; trace elements; CRM certification; plankton.

INTRODUCTION

Phytoplankton and zooplankton play a major role in the biogeochemical cycles of trace elements not only because they can concentrate large quantities of heavy metals from water and pass it along the food chain, but also because they can redistribute trace elements by sinking of planktonic skeletal structures, organic detritus and fecal pellets, moulting of crustacean exoskeletons, and vertical migrations *(1)*. Furthermore, plankton may be used as an indicator for water pollution and for comparing the levels of contamination of different aquatic environments *(2)*.

*Author to whom all correspondence and reprint requests should be addressed.

Biological Trace Element Research Editor: G. N. Schrauzer © 1994 by Humana Press Inc.

The plankton was collected by the CNR Institute for Limnology (Pallanza, Italy) in April and May 1988 from several ponds situated in the vicinity of the river Po. The plankton reference material (CRM 414) was prepared by the Community Bureau of Reference (BCR) of the Commission of the European Communities (CEC) *(2)*. Among the 16 laboratories from 8 member states and the Joint Research Centre of Ispra of the CEC, which participated in this certification campaign, was our Radioanalytical Laboratory of the Demokritos Centre, Greece *(2,3)*. INAA was applied for the determination of Co, Cr, Mn, Se, and Zn in plankton. Arsenic, Cu, and Hg were estimated in plankton by RNAA.

EXPERIMENTAL

Bottles of Plankton

Two bottles of plankton (CRM 414) were sent to us for analysis (BCR Project RM 370). Prior to analysis, each bottle of plankton was shaken for 2 min for proper mixing. Portions from each of the two bottles of plankton were taken for analysis. To check the reproducibility of the NAA methodology used, as well as the homogeneity of the plankton samples, the analysis of plankton samples was performed five times.

Preparation of Samples and Standards for Irradiations

Proper liquid standards of As, Co, Cr, Cu, Hg, Mn, Se, and Zn were prepared. The concentration of these elements in standard solutions was as follows:

Mixed standard solution of Co, Cr, Se, and Zn: Co 5 μg/mL, Cr 20 μg/mL, Se 5 μg/mL, and Zn 100 μg/mL;
Mixed standard solution of As and Cu: As 20 μg/mL and Cu 40 μg/mL;
Mercury standard solution: 1 μg Hg/mL; and
Manganese standard solution: 100 μg Mn/mL.

Plankton samples weighing about 200 mg each were placed in polyethylene tubes (10 × 30 mm) for short irradiation or in quartz tubes (10 × 30 mm) for long irradiations. Aliquots of 0.5 mL of each of the standard solutions were pipeted into identical polyethylene or quartz tubes. The tubes were heat-sealed.

Irradiation

All neutron irradiations were carried out in the swimming pool-type reactor of the NCSR "Demokritos" using the rotation system. The thermal neutron flux as well as the irradiation, the decay, and the counting time were varied depending on the trace elements under investigation, and are given in Table 1.

Table 1
Experimental Parameters Used for the Determination of Eight Trace Elements in Plankton by NAA Methodology

Thermal neutron flux $n.cm^{-2}.sec^{-1}$	Irrad. time	Cooling time	Methodol.	Counting time	Elements determined
$2.78x10^{12}$	210 sec	2.5 h	INAA	180 sec	Mn
$2.78x10^{13}$	4 hours	21 d	INAA	30 min	Co, Cr, Se, Zn
$2.78x10^{13}$	7 hours	7 d	RNAA	60 min	Hg
$2.78x10^{13}$	5 min	20 h	RNAA	1000 sec	As, Cu

γ-Ray Spectroscopy

A germanium detector (ORTEC) was used. The output signals were passed into a preamplifier (ORTEC) and spectroscopy amplifier (ORTEC). The Ge detector system was connected with a 4096 channel pulse height analyzer (Canberra model 35/plus). System resolution was 1.69 keV, and the peak Compton ratio was 59.1 for the 1332 keV of ^{60}Co. The efficiency of the Ge detector was 20%. The calibration of the multichannel analyzer was done with a Europium-152 standard source. The analyzer was coupled with an IBM AT Computer. The data were processed using the Canberra spectran AT V 3.0 software.

NAA METHODOLOGY

INAA (Mn, Co, Cr, Se, and Zn)

For the determination of Co, Cr, Mn, Se, and Zn, instrumental NAA was used. The experimental parameters are given in Table 1. After counting, the areas under the photopeaks of ^{56}Mn at 847 keV, of ^{75}Se at 264 keV, of ^{51}Cr at 320 keV, of ^{65}Zn at 1115 keV, and of ^{60}Co at 1173 keV or 1332 keV for samples and standards were compared.

RNAA (Hg, As, and Cu)

Mercury

For the determination of Hg, a radiochemical NAA method developed at our radioanalytical laboratory has been applied *(4)*. The radio-

Table 2
Average Concentrations (μg/g dry wt) and Standard Deviations of As, Co, Cr, Cu, Hg, Mn, Se and Zn Found in a Plankton Reference Material (CRM 414)

Element	Present work Average	Present work St. deviation	BCR Certified value
As	6.29	0.19	6.82
Co	1.37	0.05	1.43*
Cr	23.7	0.56	23.8
Cu	27.1	0.88	29.5
Hg	0.29	0.02	0.276
Mn	285	4.1	299
Se	1.62	0.10	1.75
Zn	114	3.4	112

*Indicative value.

chemical procedure is based on the digestion of irradiated samples with HNO_3–H_2SO_4 in the presence of an Hg carrier. The resulting solution was made 10 mol/L H_2SO_4–0.025 mol/L KI. Mercury was extracted as HgI_2 into toluene and back-extracted into 0.034 mol/L EDTA in 5% aqueous ammonia solution. The aqueous phase was drawn off into a 25-mL volumetric flask and diluted to volume with 0.5 mol/L HNO_3. The ^{197}Hg was counted on the Ge detector. The mercury standard was processed in the same manner. The chemical yield of Hg was estimated by a reactivation procedure.

Arsenic and Copper

The simultaneous determination of As and Cu was made by a modified radiochemical NAA method based on solvent extraction as developed at our radioanalytical laboratory *(5)*. The modification consists of using toluene as a solvent instead of benzene, which is equally active, but not toxic like benzene. The radiochemical procedure is based on the digestion of irradiated plankton samples with $HClO_4$–HNO_3 in the presence of Na, As, and Cu carriers. The resulting solution was made 7 mol/L $HClO_4$–1 mol/L HBr. The As was extracted as $AsBr_3$ into toluene and back-extracted into water. Arsenic was precipitated as metallic As with sodium hypophosphite, dissolved in HNO_3, transferred to a 5-mL volumetric flask, and the ^{76}As was counted. The As standard was processed in the same manner.

The Cu in the aqueous phase was extracted as copper cupferronate into chloroform and back-extracted into concentrated ammonia solution.

The Cu was precipitated as sulfide with 3% aqueous thioacetamide solution, and the precipitate was dissolved in hot HNO_3. The resulting solution was transferred to a 5-mL volumetric flask diluted to volume with distilled water, and the ^{64}Cu was counted on the Ge detector. The Cu standard was processed in an identical manner. The chemical yields of As and Cu were estimated by reactivation procedures.

RESULTS AND DISCUSSION

Average concentrations (μg/g dry wt) and standard deviations of the eight trace elements (As, Co, Cr, Cu, Hg, Mn, Se, and Zn) found in the plankton reference material are presented in Table 2. In the same table, BCR certified values *(2)* are also included for comparison. Values found and certified values are in very good agreement (0.5–8.1%).

CONCLUSION

NAA, because of its precision and accuracy, can successfully be applied as a reference method for the certification of several biological, geological, botanical, and environmental materials.

REFERENCES

1. J. H. Martin and G. A. Knauer, *Geochim. Cosmochim. Acta* **37,** 1639 (1973).
2. P. Quevauviller, K. Vercoutere, H. Muntau, and B. Griepink, *Fresenius J. Anal. Chem.* **345,** 12 (1993).
3. The BCR programme on methodology and chemical analysis. Projects and results 1988–1992, p. 206 (1992) Report EUR 14800 CEC, Brussels.
4. A. P. Grimanis and G. D. Kanias, *J. Radioanal. Chem.* **72,** 587–596 (1982).
5. A. P. Grimanis, in *Modern Trends in Activation Analysis*, J. R. DeVoe and P. D. LaFleur, ed., Nat. Bureau Stand. (US) Spec. Publication 312, vol. 1, 1969, pp. 197–202.

Interlaboratory Analysis of IRM NSC-21 Compost Vitahum

PAVEL MADER,*[,1] JIŘINA SZÁKOVÁ,[1] AND JAN KUČERA[2,3]

[1]University of Agriculture, Faculty of Agronomy, 165 21 Prague 6, Czech Republic; [2]Czech Ecological Institute, NAA Laboratory, Řež near Prague, Czech Republic; and [3]Present Address: Nuclear Physics Institute, Academy of Sciences of Czech Republic, 250 68 Řež near Prague, Czech Republic

ABSTRACT

The preparation of internal reference material NSC-21 Compost Vitahum, including its stabilization, homogeneity testing, and evaluation of an interlaboratory comparison in which 19 laboratories took part using several analytical methods is described. From the results of the intercomparison, consensus values were derived for the contents of Cd, Cu, Ni, Pb, and Zn, whereas for As, Co, Cr, Fe, Hg and Mn, information values were derived. Bimodal distribution of the As and Cr results was obtained, presumably because of incomplete dissolution of the elements from the matrix if insufficiently aggressive decomposition procedures were employed. Some problems encountered were elucidated by application of neutron activation analysis (NAA) in its nondestructive and destructive (radiochemical) modes.

Index Entries: Internal reference material; industrial compost; interlaboratory comparison; consensus value; information values; decomposition yield.

INTRODUCTION

Reference materials (RMs) of various classifications (internal IRMs, certified CRMs, or standard SRMs) play a very important role in quality assurance (QA) of chemical analysis *(1)*. RMs are useful tools in the

*Author to whom all correspondence and reprint requests should be addressed.

Biological Trace Element Research Editor: G. N. Schrauzer © 1994 by Humana Press Inc.

development and optimization of analytical methods, and their role, especially of CRMs and SRMs, is indispensable in quality assessment for proving accuracy.

Analyses of industrial composts represent an important task for many agricultural and hygienic laboratories, because maximum admissible levels of heavy metals and other toxic elements are set for both composts and their components, and thus contents of the above elements should periodically be checked in these materials. QA of compost analysis is hampered by lack of an appropriate RM, because RMs of soils and/or sewage sludges do not provide good matrix and analyte level matches. Therefore, an IRM of industrial compost denoted NSC-21 was prepared at the University of Agriculture, Prague, and an anonymous interlaboratory comparison was organized in which 11 Czech and 8 laboratories from abroad participated on a voluntary basis.

METHODS

Approximately 3 kg of the industrial compost "Vitahum" were freeze-dried, ground, manually homogenized in a plastic bag, and sterilized with γ-rays of ^{60}Co by a dose of 25 kGy. Homogeneity of the material was tested by determination of 17 elements (As, Br, Ce, Co, Cr, Cs, Eu, Fe, Hf, K, La, Na, Sc, Sm, Ta, Th, Zn) by instrumental neutron activation analysis (INAA). The relative standard deviation (RSD) owing to inhomogeneity calculated according to *(2)* amounted to 3–8% for sample weights of 100 mg. The highest values (RSD 7–8%) were observed for the elements Ce, Co, Cr, La, Sm, and Th, which apparently originate from the mineral fraction of the compost. Mercury homogeneity was tested separately for aliquots weighing about 50 mg, which is a typical sample weight when Hg is determined by a single-purpose atomic absorption spectrometer TMA-254 (Tesla, Czech Republic) or the latest version AMA-254 (Altec, Czech Republic). Using the latter device, the RSD of Hg determination was below 1%. The material was further characterized by classical silicate analysis, which yielded the following results *(3)*: SiO_2 46.11%, Al_2O_3 + TiO_2 4.8%, all Fe as Fe_2O_3 2.05%, MnO 0.07%, MgO 0.95%, CaO 2.81%, Na_2O 0.71%, K_2O 2.51%, P_2O_5 1.46%, loss on ignition at 1000°C 37.21%; total 98.69%.

The participants received 25-g aliquots of the material and were asked to determine the elements of interest, namely As, Cd, Co, Cr, Cu, Hg, Mn, Ni, Pb, and Zn, employing those analytical techniques they believed they had experience in for routine analysis of this kind.

Modified statistical evaluation according to *(4)* of the results obtained was applied. Laboratory means and their standard deviations were calculated for each laboratory set, which usually consisted of three to six replicates. However, from some laboratories, only laboratory means were obtained. The normality of distribution of the means was checked by the Kolmogorov-Smirnov test. Outliers were identified as described earlier *(4)* and rejected only if analytical reasons were also found for their elimination. Since the variation between the laboratory sets was significantly higher than the variation within the sets, the overall mean was calculated as the mean of laboratory means. For all statistical tests, the significance level $\alpha = 0.05$ was employed.

RESULTS AND DISCUSSION

In total, 125 laboratory sets of individual results (exceptionally only means) were obtained. In some laboratories, more then one method of either decomposition or measurement was employed. The following numbers of laboratory means were reported: Cd 21, Pb 17, Cr 16, Cu 15, As 14, Zn 13, Hg 10, Ni 8, Co 8, Mn 7, Fe 5. These results were subjected to combined statistical-analytical evaluation. For many other elements (Al, Ba, Br, Ca, Ce, Cl, Cs, Eu, K, La, Mg, Mo, Na, P, Rb, Sb, Sc, Sm, Sr, Th, Ti, V, W) less than three laboratory sets were obtained, and no attempt was made to critically evaluate the contents of these elements.

Only one laboratory provided results by nondestructive analysis (INAA); all remaining laboratories employed analytical techniques that involved sample decomposition. Of these, 36 laboratory sets were obtained using dry-ashing in both classical (in air under atmospheric pressure in open system) and modern (oxygen at atmospheric or increased pressure) versions. In this number, the majority of Czech and Slovak laboratories are involved in which dry-ashing in a stream of oxygen was carried out *in situ* and Hg trapped on an Au amalgamator in the single-purpose atomic absorption spectrometer TMA-254 or AMA-254 for Hg determination. Wet-ashing under atmospheric pressure and pressurized wet-ashing in various mixtures of acids (including microwave decomposition), with and without HF, were employed for obtaining 34 and 34 laboratory sets, respectively. Ten laboratory sets resulted from a combination of some of the above mentioned decomposition methods. One laboratory set was obtained after alkaline fusion with $Li_2B_4O_7$ and one after leaching with 2 $mol{\cdot}L^{-1}$ HNO_3.

From all the measurement techniques, AAS in various modes (i.e., flame, flameless, cold vapor, or hydride generation techniques) was most frequently used (60 laboratory sets), whereas other techniques provided the following numbers of laboratory sets: NAA (both INAA and radiochemical mode) 47, ICP-AES 22, electrochemical methods 14, ICP-MS 10, spectrophotometry 1. In some laboratories, the accuracy of their results was checked by concurrent analyses of various CRMs (soil, sewage sludge, plant material, bovine liver, river sediment).

The results of the intercomparison (if at least six laboratory means were obtained) are presented anonymously, using codes A–S assigned to participating laboratories, in Fig. 1. The laboratory means and their standard deviations are plotted in ascending order together with the derived consensus and/or information values and their 95% confidence intervals (thick and thin lines, respectively). Accepted laboratory means are marked by crosses, whereas rejected outliers are marked as points.

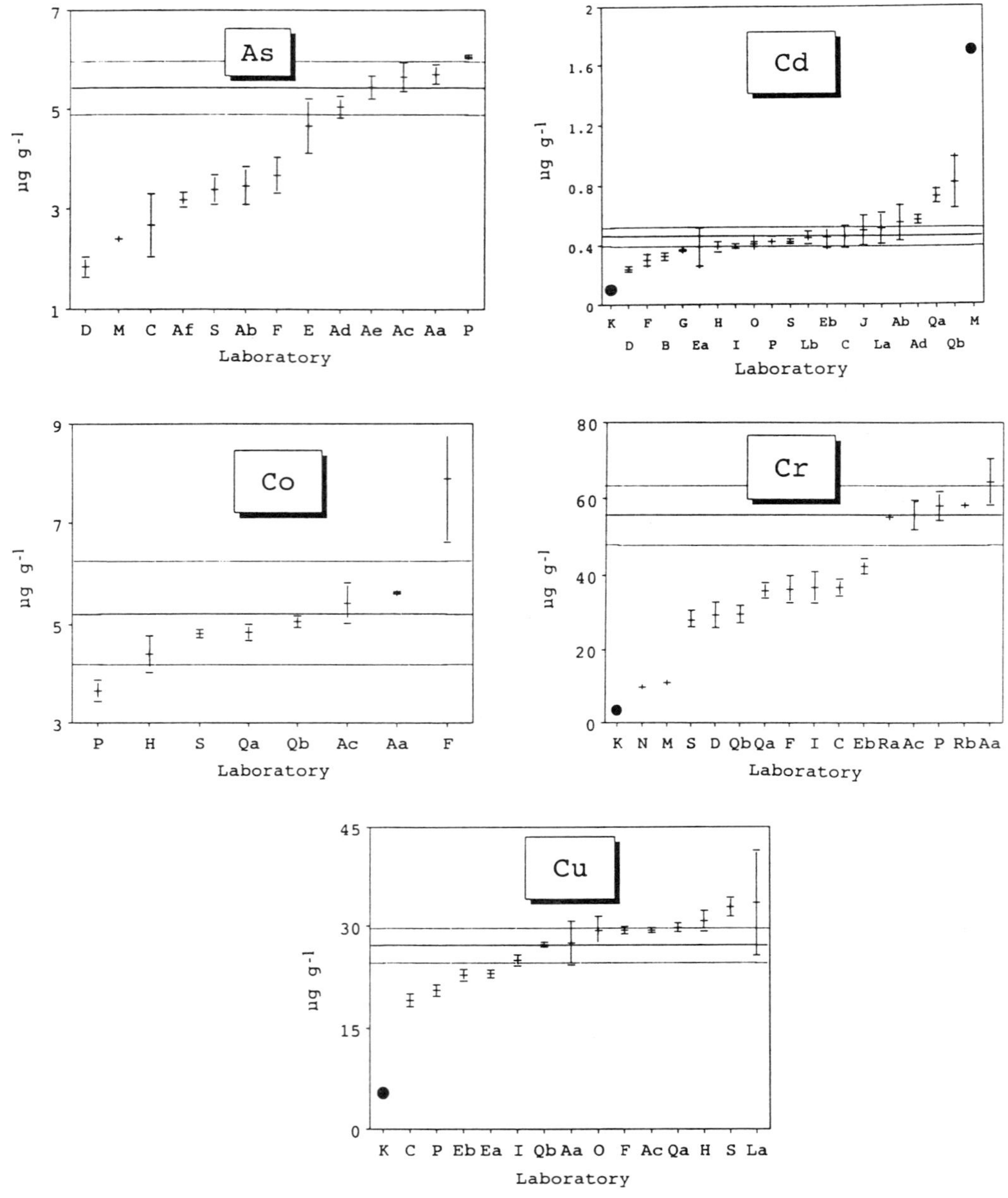

Fig. 1. Results of interlaboratory analysis of IRM NSC-21 Compost Vitahum for selected elements.

All results obtained in laboratory K were identified as outliers and were rejected. These values were significantly lower compared to results from other laboratories (up to one order of magnitude), because the decomposition method employed (dry-ashing of a 5-g sample in a muffle furnace at 500°C for 5 h followed by leaching of the resulting ash by

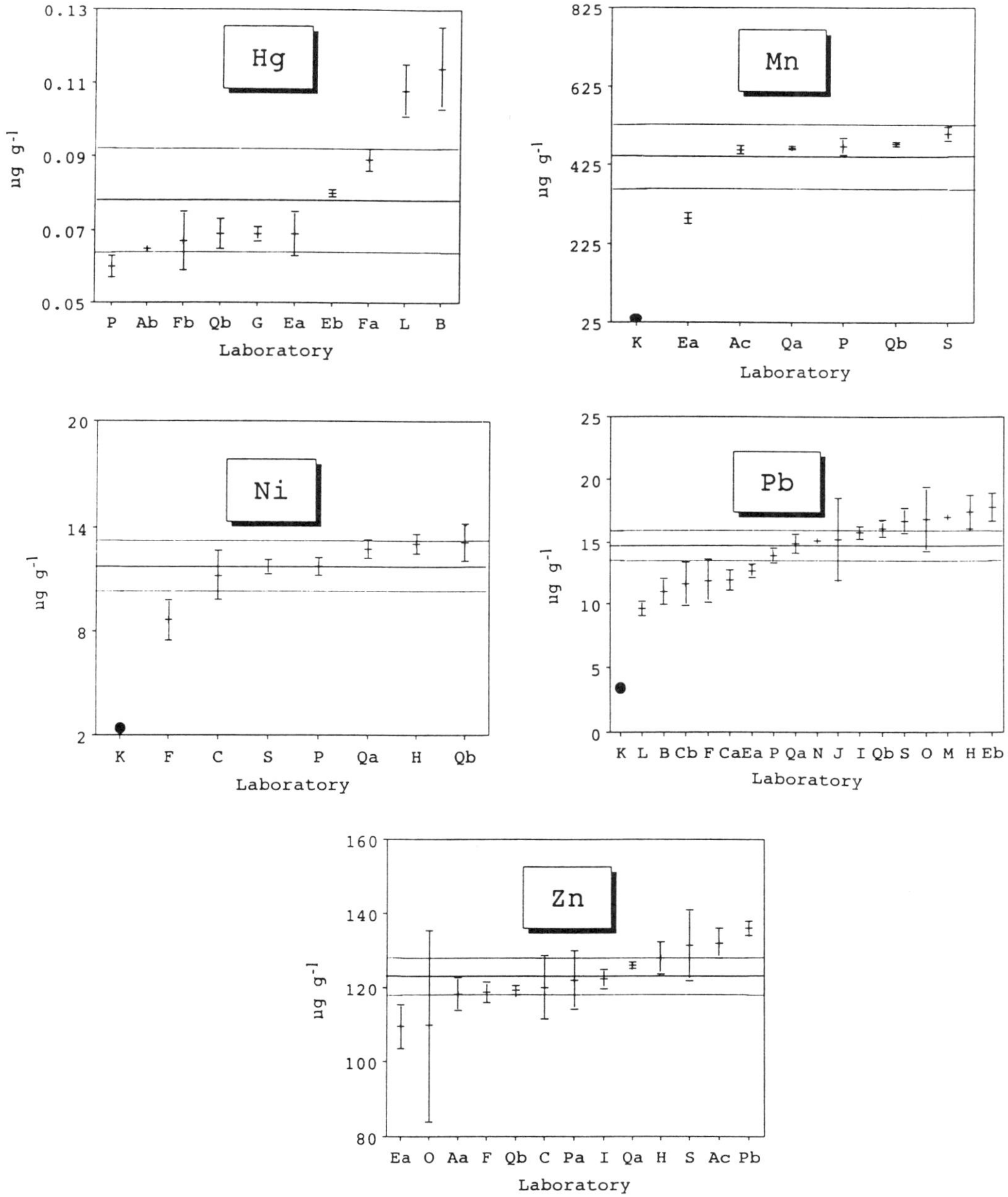

Fig. 1 *(continued).*

dilute HNO_3) should be considered insufficient for complete destruction of organic matter owing to a short ashing time *(5,6)*. The cadmium content determined by laboratory M was rejected as well, because it was identified as outlier and generally insufficient analytical information was provided concerning this value.

Statistical evaluation of the remaining intercomparison results revealed (cf Fig. 1) that two types of distribution of the results can be distinguished: (1) approximately normal, rather narrow distribution of the laboratory means for Cd, Cu, Ni, Pb and Zn; and (2) skewed distribution and/or higher variability of the laboratory means for As, Co, Cr, Fe, Hg and Mn with distinct bimodality for the As and Cr values. For the former group of elements, the overall mean of the accepted laboratory means was assigned the status of a consensus value, because good agreement was reached among results obtained by various decomposition and measurement techniques, and the relative width of the 95% confidence interval (CI) was satisfactorily low, not exceeding 15%. The consensus values are considered to be a very good estimate of the true total analyte contents.

For the latter group of elements, information values were derived either as the overall mean of the accepted laboratory means of higher variability, if analytical reasons of the increased variability were not identified (for the elements Co, Fe, Hg, and Mn) or as the overall mean of laboratory means selected according to analytical considerations (for As and Cr). The relative width of 95% CI is usually higher compared to that of consensus values owing to a high variability of results (Hg) or to a small number of results obtained (Co, Fe, Mn). The information values are considered to be a reasonably good estimate of the true value of the total analyte contents. They would need, however, further improvements concerning mainly their precision. The established consensus and information values are summarized in Table 1.

In the case of As, significantly lower results of RNAA (denoted RNAA[1] in Table 2) with proven accuracy for many types of biological materials *(7)* were found compared to INAA results. The accuracy of INAA results has also been proven by concurrent analysis of US NIST SRM-2704 Buffalo River Sediment. Since the chemical yield of As separation in RNAA was checked using the ^{77}As radiotracer *(7)*, the decomposition procedure employed had to be suspected to provide low decomposition yields of As from the NSC-21 matrix. Therefore, various decomposition procedures were studied that were followed by the same radiochemical separation of the elements As, Cd, Cu, Mo, and Sb with controlled chemical yields for "critical" elements As and Cd *(7)* on analysis of about 150-mg samples of NSC-21. The results of this study are shown in Table 2. It can be inferred from Table 2 that the temperature in wet-ashing was critical for complete As dissolution from the matrix. The composition of acid mixture used in wet-ashing (including HF) did not affect As dissolution in this case. The highest As decomposition yield was obtained when temperature of the boiling point (and decomposition) of H_2SO_4 was reached. This seems to suggest a high resistance of some As compounds to wet-ashing, which was also mentioned by Byrne *(8)*. For this reason, only two sets (Aa, Ac) of INAA results, the RNAA[c] and RNAA[d] (cf. Table 2) results (Ad, Ae), and two sets of HG-AAS results (E, P) obtained after wet-ashing in a mixture containing H_2SO_4 at 250°C and pressurized wet-ashing in a High-Pressure Asher (Paar, Austria) at 300°C, were used for deriving the As information value.

Table 1
Consensus and Information Values of Element Contents in IRM NSC-21 (µg/g, dry wt[a])

Element	N[b]	Consensus value Mean ± 95% CI[c]	Element	N[b]	Information value Mean ± 95% CI[c]
Cd	19	0.457 ± 0.067	As	6	5.43 ± 0.53
Cu	14	27.2 ± 2.6	Co	8	5.22 ± 1.03
Ni	7	11.8 ± 1.50	Cr	6	55.7 ± 7.7
Pb	16	14.7 ± 1.2	Fe, %	6	1.00 ± 0.23
Zn	13	123 ± 5	Hg	10	0.078 ± 0.014
			Mn	6	446 ± 82

[a]Unless otherwise indicated.
[b]Number of accepted laboratory means.
[c]Confidence interval.

Table 2
Results for As, Cd, Cu, Mo, and Sb Obtained by INAA and RNAA Using Various Decomposition Procedures (µg/g dry wt, Mean ± SD)

Procedure	N	As	Cd	Cu	Mo	Sb
RNAA - a	4	3.46 ± 0.38	0.45 ± 0.02	27.5 ± 3.2	3.67 ± 0.11	0.58 ± 0.06
RNAA - b	3	3.18 ± 0.15	0.44 ± 0.05	24.6 ± 0.4	3.81 ± 0.08	0.51 ± 0.02
RNAA - c	3	5.05 ± 0.21	0.47 ± 0.03	29.4 ± 0.4	3.79 ± 0.05	0.32 ± 0.03
RNAA - d	3	5.44 ± 0.23	0.48 ± 0.02	29.4 ± 1.0	3.86 ± 0.11	0.67 ± 0.06
INAA	20	5.65 ± 0.29	< 5	-	3.5 ± 0.8	0.62 ± 0.06
INAA	5	5.70 ± 0.20	< 5	-	3.6 ± 0.4	0.61 ± 0.05

[a]Decomposition by heating at 150°C in Teflon[tm] bomb in 3 mL of conc. H_2SO_4 + 4 mL of conc. HNO_3 + 3 mL of conc. HF + 1 mL of 30% H_2O_2 for 4 h, and then heating with 3 mL of conc. HNO_3 at 160°C for 1 h.

[b]Decomposition by heating at 160°C in 5 mL of conc. HNO_3 + 3 mL of conc. $HClO_4$ + 3 mL of conc. HF in Teflon[tm] crucible until white fumes of $HClO_4$, and then additional heating with 3 mL of conc. HNO_3 for about 1 h.

[c]Decomposition in 5 mL of conc. HNO_3 + 3 mL of conc. $HClO_4$ in glass decomposition flask at 200°C until white fumes of $HClO_4$ appear.

[d]As in the case of (a), and then additional heating with 3 mL of conc. HNO_3 in silica glass Kjeldahl flask over gas burner flame until white fumes of H_2SO_4 at about 340°C.

It can also be inferred from Table 2 that complete dissolution of Sb requires both high temperature and the presence of HF on wet-ashing (apparently a part of Sb is contained in the mineral fraction of the matrix), whereas complete dissolution of Cd, Cu, and Mo can be achieved by wet-ashing at 160°C even in the absence of HF in the decomposition mixtures studied (most likely, these elements are bound to the organic fraction of the matrix).

A bimodal distribution of the Cr results is also obvious from Fig. 1. Significantly higher values were determined after more aggressive decompositions (presence of HF on wet-ashing, on dissolution with HF of dry-ashing residues, alkaline fusion) and by INAA. This again seems to suggest that a considerable part of Cr is contained in the mineral fraction of the matrix, which is difficult to decompose. This problem has already been dealt with by Byrne *(9)*, who concluded that lower Cr values obtained by AAS in comparison with INAA values in intercomparisons on soil and marine sediment resulted from incomplete Cr dissolution from the matrix, and by Becker *(10)*, who suggested nondestructive analytical techniques, such as INAA, to be used when biological materials containing a mineral fraction are to be certified for content of elements that are difficult to dissolve, such as Co, Cr, Fe, Hf, and Th, and for volatile elements. Therefore, we have derived the information value of the total Cr content only from two INAA sets (Aa, Ac) and from those results that were obtained using sufficiently aggressive decomposition procedures (Eb, P, Ra, Rb). On the other hand, the remaining lower Cr values were also quite consistent, which seems to suggest that a value of about 33 μg kg^{-1} represents an easily leachable (presumably from the organic fraction of plant material) fraction of Cr.

Rather disparate results were obtained in this intercomparison for Hg (cf. Fig. 1). Homogeneity of Hg distribution in sample aliquots of about 50 mg used in the majority of laboratories employing the TMA-254 or AMA-254 devices for the element determination was satisfactory. No significant differences were found between the results obtained by the above instruments and other analytical techniques (CV-AAS, RNAA). The Hg level was the lowest among the elements for which consensus or information values were derived, but still well above the limit of determination of all analytical techniques employed. Therefore, the overall mean was calculated from all laboratory means and assigned the status of an information value. The same was done with the Co, Fe, and Mn results, except for the results from laboratory K.

SUMMARY

This interlaboratory comparison revealed that the accurate determination of the elements Cd, Cu, Ni, Pb, and Zn in industrial compost is feasible by the currently used decomposition and measurement techniques in most laboratories that took part in this intercomparison. However, the opposite is true for several other important trace elements, namely As, Cr, and Hg. For the accurate determination of total As and Cr contents in industrial compost and similar matrices, a sufficiently aggressive decomposition procedure must be chosen (and tested) to guarantee that complete dissolution of the elements from the matrix is achieved. In this respect, temperature on wet-ashing and the presence of HF in decomposition mixtures are the crucial factors for the accurate determination of As and Cr, respectively. A very useful role of NAA in checking accuracy of other analytical methods, and in optimizing analytical procedures for elements difficult to determine in complex matrices by comparing the results of its nondestructive and destructive modes (a part of a self-validation principle *(8)* of the method) was again demonstrated in this work.

ACKNOWLEDGMENTS

The authors acknowledge the collaboration of the following workers and laboratories: Dr. J. Čečka, Bioanalytika, Hradec Králové, Czech Republic; Dr. O. Čelechovská, Dr. E. Šucman, University School of Veterinary Medicine, Brno, Czech Republic; Dr. E. Čurdová, National Institute of Public Health, Prague, Czech Republic; Dr. F. Frantík, Research Institute for Brewery, Prague, Czech Republic; Dr. W. Gössler, Karl-Franzens Universität, Graz, Austria; Dr. S. K. Gupta, Soil Chemistry and Soil Testing Laboratory, Lieberfeld-Bern, Switzerland; Dr. Kňourková, Central Control and Testing Agricultural Institute, Prague, Czech Republic; Dr. P. Kokkonen, University of Oulu, Kajaani, Finland; Dr. S. Krýsl, District Hygiene Centre, Klatovy, Czech Republic; Dr. M.-L. Linn, Resource Consultants, Brentwood, TN; Dr. M. Matúška, Central Control and Testing Agricultural Institute, Bratislava, Slovak Republic; Dr. P. Mikulica, Agrochemical Enterprise, Mstětice, Czech Republic; Dr. C. Placko-Moore, NET Gulf Coast, Inc., Carroliton, TX; Dr. J. Sippola, Agricultural Research Centre, Jokioinen, Finland; Dr. M. Stoeppler, Research Center (KFA), Juelich, Germany; Dr. L. Škvorová, District Hygiene Centre, Příbram, Czech Republic.

REFERENCES

1. J. K. Taylor, *Quality Assurance of Chemical Measurements*, Lewis Publishers, Chelsea, MI, 1987.
2. J. Kučera and L. Soukal, *J. Radioanal. Nucl. Chem., Articles* **121,** 245 (1988).
3. R. Přibil, personal communication.
4. J. Kučera, P. Mader, D. Miholová, J. Cibulka, M. Poláková, and D. Kordík, *Fresenius J. Anal. Chem.* **338,** 66 (1990).
5. P. Mader, J. Kučera, J. Cibulka, and D. Miholová, *Chem. listy* **83,** 765 (1989).
6. P. Mader, J. Száková, and E. Čurdová, *Talanta*, in press.
7. J. Kučera and L. Soukal, *J. Radioanal. Nucl. Chem., Articles* **168,** 185 (1993).
8. A. R. Byrne, *Fresenius Z. Anal. Chem.* **326,** 733 (1987).
9. A. R. Byrne, *Fresenius J. Anal. Chem.* **345,** 144 (1993).
10. D. A. Becker, *Fresenius J. Anal. Chem.* **345,** 298 (1993).

Nuclear and Conventional Methods for Soil Determination in Sugar Cane Industry

Validity of Sampling Procedure

ELISABETE A. DE NADAI FERNANDES*
AND MÁRCIO ARRUDA BACCHI

Centro de Energia Nuclear na Agricultura, P.O. Box 96 13400-970 Piracicaba SP Brazil

ABSTRACT

Scandium and ash methods' performances were compared in terms of soil content assessment in sugar cane loads, emphasizing the common sampling drawbacks. Both methods are adequate for such determination in controlled conditions. The scandium has demonstrated better analytical characteristics, since it is free from interferences of cane matrix, which decreases the accuracy of the ash method in normal mill conditions.

Index Entries: Sugar cane; impurities; soil; cane consignments; INAA; ash content; scandium; sampling.

INTRODUCTION

As a consequence of the conversion of the Brazilian agroindustry from manual to mechanical loading in the beginning of the 1970s, the soil level in cane received at the sugar mill has shown a relatively sharp rise with a gradual upward trend since then. The well-known drawbacks of soil on factory performance and costs *(1,2)* have prompted an investigation to develop a reliable method for assessing its content in the cane supply. Among conventional procedures for industrial daily control,

*Author to whom all correspondence and reprint requests should be addressed.

Biological Trace Element Research Editor: G. N. Schrauzer © 1994 by Humana Press Inc.

emphasis may be given to the ash and separation methods, both showing particular variations from one mill to another. The ash method estimates the soil content by measuring the residual mass after incineration *(3)*, whereas the separation method directly determines the soil amount manually withdrawn from the cane *(4)*. Nuclear methods involving the determination of elements present in much higher concentration in soil than in cane such as aluminum *(5)*, silicon and aluminum *(6,7)*, and scandium *(8)*, have been under consideration.

In the last two decades, attempts have been directed to at defining a sugar cane quality-control system, either for payment of growers or industrial improvements. This system is based on determinations of fiber, sugar, and soil contents in delivered canes. Usually, the Brazilian mills sample the sugar cane with a horizontal probe of about 20 cm in diameter, perforating the load at three consecutive positions, the final sample being a composite of total extracted material. Since mineral impurities in a load are unevenly distributed over the stalks, much effort has to be spent to overcome difficulties of sample representation to provide accurate soil level evaluation. This study has been carried out to examine the scandium and ash methods' performance in terms of soil content assessment in cane loads, stressing the common sampling problems.

MATERIALS AND METHODS

Field tests were conducted at the Usina Santa Bárbara (city of Santa Bárbara D'Oeste, state of São Paulo) during the 1992 crop season. Three loads of the same sugar cane variety were harvested in the area selected for study. Cane was sampled from each load by doing 18 cores with the horizontal probe, comprising both sides of the truck (Fig. 1). The material from each core weighing about 15 kg was collected separately, manually homogenized, dried at 60°C, and thereafter ground in a Wiley mill. Clean cane samples were obtained by randomly taking 10 cane stalks from each load, which were carefully washed to remove the adhered soil. From several points in the field where the cane was grown, a soil sample was taken, dried at 100°C, and pulverized in a mortar. The washed canes stalks were disintegrated in a laboratory shredder before being submitted to the procedure followed for other cane materials. All samples were directed to both instrumental neutron activation and ash analysis.

Samples of about 200 mg of canes and 100 mg of soil were inserted into special polyethylene vials and irradiated for 8 h at a thermal neutron flux of about 10^{12} n/cm^2/s, in the research nuclear reactor IEA-R1 of Instituto de Pesquisas Energéticas e Nucleares (IPEN/CNEN/SP). After suitable cooling times, the induced radioactivity was measured by using an HPGe semiconductor detector (10% relative efficiency and 1.85 keV resolution at the ^{60}Co photopeak) connected to a computerized multichannel buffer. Standard reference materials V 10, SRM 1515, SRM 1572, and Soil 7 were coirradiated with the samples.

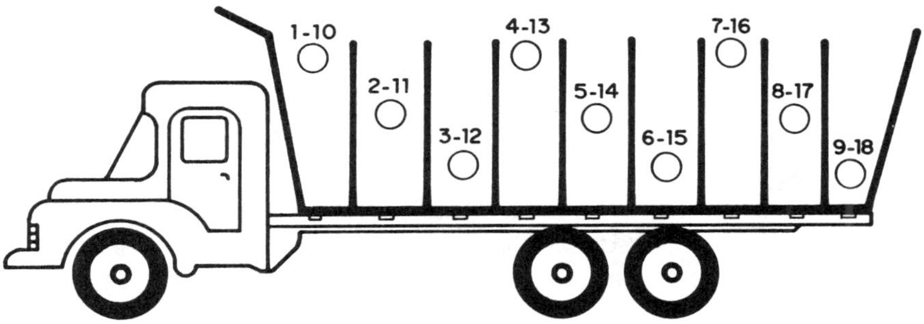

Fig. 1. Points of core samplings, 1–9 in the left side, and 10–18 in the right. Combinations (1-2-3), (4-5-6), (7-8-9), (10-11-12), (13-14-15), and (16-17-18) are valid for quality-control system.

The ash content of canes was determined following the analytical procedure described elsewhere *(3)*. In brief, it consists of incinerating the sample to constant mass in a muffle furnace at 850°C, using fused silica basin. Samples of 50 g were used, requiring a total incinerating time of about 120 min. Tests were also undertaken to check the response of the method to varying amounts (0–10%) of soil added to subsamples of clean shredded cane.

RESULTS AND DISCUSSION

Results of the soil addition curve showed a high linear correlation between the incineration residues of spiked cane samples and added soil contents (r^2 = 0.99997). Such a good relationship has been possible because the same cane and soil were used in sample preparation, being changed only in their proportions. However, this is not valid for real conditions, when individual ash contents of soil and cane itself must be well known. Otherwise, errors in these parameters would cause strong deviations in the relationship. The ash values recovered by the correlation equation are 1.10% for the clean cane and 88.6% for the soil, in agreement with respective results of their determinations (1.10 and 88.9%).

The mineral content of soils is assumed to have small variation among different types, the mass reduction after soil calcination being especially the result of losses of residual moisture, organic matter, and volatile elements. Thus, an average for all soils of the mill can be used, and for more accurate analysis, the determination should be done for a specific field area or, at least, for each soil type. On the other hand, the cane ash content usually varies between 0.4 and 1.2% *(3)*, depending on factors, such as soil fertility, sugar cane variety, maturation stage, and proportions of tops and trash. For routine applications, it is not feasible to perform individual determinations of cane ash content, causing distortions in the method response and reducing its precision and accuracy.

The ash amount of the cane itself is a straight function of its mineral element contents, potassium being one of the highest. In this test, the potassium determined by INAA showed an average concentration of 0.511 ± 0.011% for the three loads, whereas for the 54 industrial cane samples, it ranged from 0.4–0.9%. The amount of potassium resulting from the soil present in the samples, even in the high contaminated ones, would not exert a significant contribution to the total content, taking into account its low concentration of 0.055%. Thus, it can be concluded that the great variation observed among potassium values and, consequently, in the ash contents is the result of different relative proportions of cane components in samples.

A Brazilian mill usually determines its own average clean cane ash for deduction in all industrial cane ash calculations, which leads to considerable errors mainly for samples with less amount of soil. Since the individual determinations of ash in clean canes show far great differences than a minimum and a maximum of 0.4 and 1.2%, it is not worth averaging such values to be used for each deduction. In some cases, the uncertainty in the discount of the clean cane blank may be even greater than the real soil contribution.

In the scandium method, the cane blank is negligible, even in the worst condition of analyzing canes coming from sandy soils. The Sc content of Brazilian soils cultivated with sugar cane has varied from 4–66 ppm *(8)*, whereas in sugar cane plants, it has been on the order of 3 ppb *(8,9)*, no matter what soil they were grown in. It means that the soil poorest in Sc is at least a thousand times as concentrated in this element as canes. Being only a function of Sc in soil, the accuracy of the method can be assessed providing that the soil concentration is known. In this way, whenever Sc is detected in an industrial cane sample in levels above that of sugar cane itself, one actually measures the Sc from the soil existing in such material.

For practical reasons, a soil inventory in a mill area where the method should be applied is helpful in establishing an Sc data bank, based on the fact that Sc concentration is related more to soil parent material than to environmental influences. Like scandium, the elements iron, thorium, and hafnium have also shown good characteristics to trace soil in cane *(9)*, but their limits of detection by the established INAA procedure were too close to the measuring range.

Data obtained by the ash and scandium methods were compared (Fig. 2), presenting a good linear correlation (r^2 = 0.9529). Presuming an ideal situation, in which there is a perfect correlation of data from both methods, a resultant equation with linear coefficient equal to 1 and zero as interception value is expected. The linear coefficient did not differ statistically from unity, but the obtained interception value was 0.301. This discrepancy may be explained by the blank problem of the ash method, which is that even in such a controlled condition, occurred considerable uncertainties. In the same way, the correlation coefficient was different

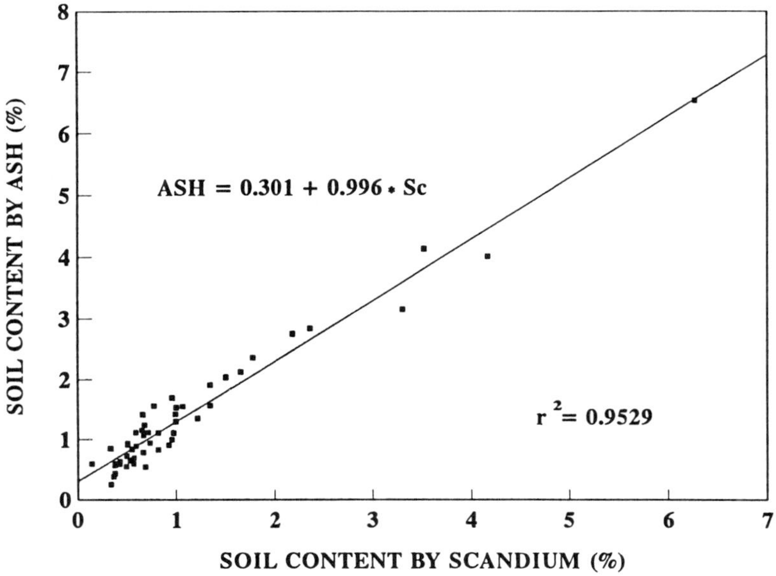

Fig. 2. Relationship between soil content of the 54 core samples, in percent of dry mass, obtained by ash and scandium methods, showing the correlation coefficient and the equation.

from the ideal, partially because of the referred blank problem and also because of random errors in both methods determinations.

In accordance with the Brazilian system of sorting samples for analysis of canes, six combinations of three cores were available in each load (Fig. 1). The mean soil contents calculated by the scandium method for each valid combination had their errors determined in relation to the true soil content, obtained by averaging the results of the 18 samples of each load (Fig. 3). These estimates can be accepted as the real soil level, since they are represented by a mass of about 300 kg, somewhat significant for the total load of 14 t. A high variation among distinct core combinations can be seen, demonstrating the erratic distribution of soil in loads without any relationship between the results obtained and the position of sampling. The large range of soil contents found in the three loads (0.606, 2.06, and 1.18%) should be pointed out, even though the same field conditions were available at the time of harvesting, with higher deviations among combinations as the soil amount is increased.

The scandium method seems potentially better than the ash method for soil assessment in sugar cane loads, achieving more accurate and precise results, mainly in the presence of minimum amounts of soil. Nevertheless, the challenge of the sampling procedure must be solved before envisaging method changes, since limitations imposed by uncertainties of sample composition are larger than those of the analysis itself, whichever the chosen analytical procedure.

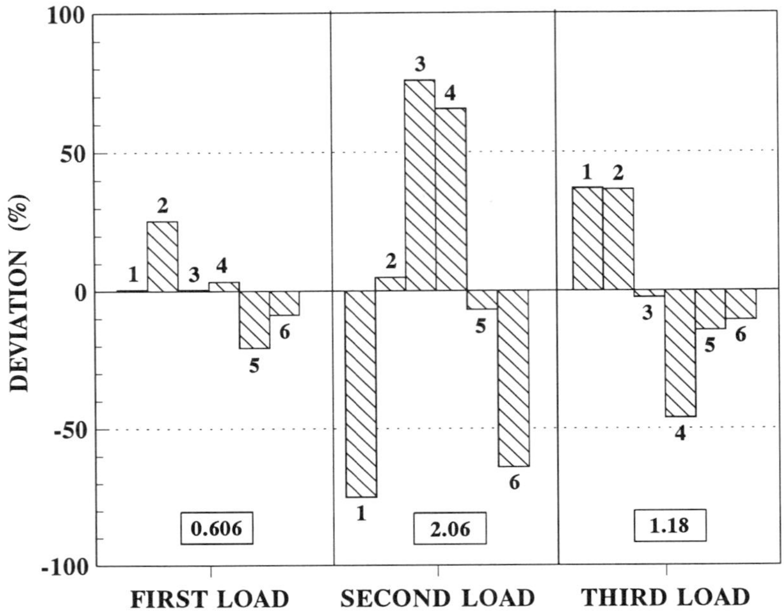

Fig. 3. Percent error of each valid sampling combination, in relation to the average of the 18 samples of each load, depicted inside boxes. Results obtained by scandium method.

ACKNOWLEDGMENTS

The authors wish to acknowledge the assistance of various members of the Usina Santa Bárbara staff who have given their time freely. This research was supported by CNPq—Conselho Nacional de Desenvolvimento Científico e Tecnológico, and FAPESP—Fundação de Amparo à Pesquisa do Estado de São Paulo.

REFERENCES

1. G. Trabanelli, G. Mantovani, and F. Zucchi, *Sugar Technol. Rev.* **14,** 1–27 (1988).
2. J. H. M. Smits and R. L. Blunt, *Proc. SASTA* **50,** 1–4 (1976).
3. M. A. Brokensha and P. Mellet, *Proc. SASTA* **51,** 97–100 (1977).
4. A. C. Fernandes and E. R. Oliveira, *Proc. ISSCT* **16,** 1963–1973 (1977).
5. V. F. Nascimento Filho, E. A. N. Fernandes, E. S. B. Ferraz, and L. C. R. Pessenda, *Proc. Reunião Annual SBPC* **38(7),** 7, 8 (1986).
6. J. A. Aylmer and R. J. Holmes, *Int. Sugar J.* **83,** 362–364, 996 (1981).
7. J. G. Loughran, G. Cowan, and V. Mason, *Proc. ISSCT* **18,** 942–951 (1983).
8. E. A. N. Fernandes, *J. Radioanal. Nucl. Chem.* **168,** 1, 41–46 (1993).
9. M. A. Bacchi, Thesis, CENA/USP, Piracicaba, p. 80 (1993).

Miscellaneous

Programs of the IAEA Utilizing Nuclear Analytical Techniques in the Life Sciences

R. M. PARR* AND R. ZEISLER

International Atomic Energy Agency, P.O. Box 100, A-1400 Vienna, Austria

ABSTRACT

Nuclear analytical techniques have an important role in the IAEA in life sciences programs relating to human health. A major effort has recently been devoted to the determination of dietary intakes of trace elements, including studies in some countries affected by the Chernobyl accident. The same accident stimulated interest in using ^{129}I (as determined by NAA and other techniques) to trace the distribution of fission products in the environment. Other research topics concern solid wastes, air pollution, exposure to mercury in selected human populations, and osteoporosis (including NAA studies of trace elements in human bone). Much of the IAEA's work is concerned with analytical quality control; new reference materials in preparation include lichen (for trace elements) and human hair (for total and methyl mercury).

Index Entries: Aerosols; air pollution; analytical quality assurance; biomonitors; bone; Chernobyl; diets; environment; foodstuffs; hair; iodine-129; lichen; mercury; methyl mercury; neutron activation analysis; nutrition; occupational health; osteoporosis; PIXE; radioimmunoassay; solid wastes; XRF.

INTRODUCTION

The International Atomic Energy Agency (IAEA), which came into being in 1957, is an independent intergovernmental organization within the United Nations system. Together with its member states (presently 114 in number), the IAEA is called upon, by its statute, "to accelerate and

*Author to whom all correspondence and reprint requests should be addressed.

Biological Trace Element Research Editor: G. N. Schrauzer © 1994 by Humana Press Inc.

enlarge the contribution of atomic energy to peace, health and prosperity throughout the world." The mention of "health" in this statute has, for many years, provided the justification for implementing a program on applications of nuclear and nuclear-related analytical techniques in relation to human health. Within this area, a variety of topics are currently being supported that make use of nuclear and nuclear-related analytical techniques, such as neutron activation analysis (NAA), energy-dispersive X-ray fluorescence (ED-XRF), particle-induced X-ray emission (PIXE), and, more recently, inductively couple plasma mass spectrometry (ICP-MS).

The main beneficiaries of the IAEA's programs are institutes and individual scientists in all member states, but particularly in developing countries. The main mechanisms of support are technical cooperation (with a budget of about US$40 million p.a.) and coordinated research (about $4 million p.a.). Since Coordinated Research Programs (CRPs) provide the best examples of the current applications of nuclear analytical techniques in the life sciences, most of what follows in this article is concerned with these programs.

The need for more work in this area was highlighted recently by two United Nations Conferences that took place in 1992. The United Nations Conference on Environment and Development, held in Brazil, identified a number of environmental topics of high priority under its Agenda 21. Nuclear analytical techniques are potentially able to address many of these needs. Similarly, the International Conference on Nutrition, organized by FAO and IAEA in Italy, identified micronutrient malnutrition as a high-priority topic. Again, nuclear analytical techniques have many useful applications in this area. Partly as a consequence of these UN initiatives, applications of nuclear analytical techniques in health-related environmental studies and in human nutrition have come to be recognized as high-priority subjects both within the IAEA's own program and by many of its member states.

COORDINATED RESEARCH

As explained in the review of the IAEA's programs presented at the previous conference in this series *(1)*, the IAEA is probably unique among UN organizations in promoting technology transfer to developing countries by means of a wide-ranging research contract program. A list of recent, current and possible future Coordinated Research Programs (CRPs) involving applications of nuclear analytical techniques in the life sciences is given in Table 1. The CRPs with the greatest impact and other related projects are described briefly in the following paragraphs.

Dietary Intakes of Trace Elements

Global Studies

Nuclear analytical techniques have been found to be particularly appropriate for determining trace and other elements in human food-

Table 1
Recent, Current, and Future CRPs Using Nuclear Analytical Techniques in the Life Sciences

Years	No.[a]	Title
1984-1989	14	The significance of hair mineral analysis as a means for assessing internal body burdens of environmental pollutants
1984-1990	14	Human daily dietary intakes of nutritionally important trace elements as measured by nuclear and other techniques
1985-1990	11	Nuclear techniques for toxic elements in foodstuffs (in part of the Asia & Pacific Region)
1987-1992	20	Use of nuclear and nuclear-related techniques in the study of environmental pollution associated with solid wastes
1987-1992	10	Nuclear analytical techniques for the analysis of trace elements in agroindustrial products and foodstuffs (Latin America Region)
1990-1994	10	Assessment of environmental exposure to mercury in selected human populations as studied by nuclear and other techniques
1992-1996	9	Reference materials for micro-analytical nuclear techniques
1992-1997	19	Applied research on air pollution using nuclear-related analytical techniques
1993-1998	8	Comparative international studies of osteoporosis using isotope techniques
1994-1999	-	Assessment of environmental pollutants using radioimmunoassay and other related techniques
1994-1999	-	Compilation of anatomical, physiological and metabolic characteristics for a Reference Asian Man: dietary intake studies
1995-2000	-	Workplace monitoring and occupational health studies using nuclear and related analytical techniques
1995-2000	-	Secondary (regional) reference materials for environmental studies
1995-2000	-	Environmental biomonitoring and specimen banking for developing countries

[a]No. of participating countries.

stuffs and total diets. Another paper in these proceedings describes a CRP on toxic elements in foodstuffs, which was recently implemented in 11 countries in the Asia and Pacific region *(2)*. In another related CRP, a major effort has been devoted to a study of dietary intakes of trace elements in 25 study groups in 16 countries. Neutron activation analysis (NAA) served as the reference analytical method for 14 of the elements of interest and as quality-control check on a further four elements. Although the final report on this CRP has not yet been published, much of the data generated by it have already been used in the preparation of working documents for the WHO/FAO/IAEA Expert Consultation on Trace Elements in Human Nutrition *(3)*, and as input to a data base on dietary intakes of trace elements in 47 different countries *(4)*. Figure 1 exemplifies some of the output of these studies. It shows the distribution of dietary intakes of zinc in various countries expressed relative to the new WHO/FAO/IAEA basal requirements (the plural is used here because the requirements vary according to age and sex). Conclusions that may be drawn from these evaluations are that (1) only very few data on dietary intakes of trace elements are so far available from developing countries, and (2) in the few cases where such data are available, they indicate that dietary intakes of zinc in most developing countries are, at best, only marginally sufficient (and in some cases are actually **deficient** as judged by comparison with the basal requirements). Partly as a result of this kind of investigation, the possibility that zinc malnutrition may be relatively widespread is now coming to the attention of international bodies responsible for nutrition. It may be anticipated that, during the next few years, there will be a rapid growth in the number of studies of zinc nutrition in developing countries. Nuclear techniques are potentially able to play an important role in this work.

Chernobyl

The experience gained in the CRP just mentioned was of great value in connection with a study of trace elements in foods and diets of some population groups living in three regions of the former USSR (Novozybkov, Russian Federation, Ovruch, Ukraine and Bragin, Belarus) affected directly by the Chernobyl accident. This study was part of the International Chernobyl project *(5)*, carried out mainly during the period 1990–1991. The main focus of the whole project, of course, had to do with issues concerning the radiation doses delivered to people in the affected areas. However, there were also some important issues of nutritional significance. Local authorities were concerned about the possibility of deficiencies of essential trace elements in the diets (particularly of iodine, iron, and selenium). Another concern was that the environment, including local foodstuffs, may have been contaminated with lead and cadmium arising from the application of these elements to the open Chernobyl reactor immediately after the accident.

An international team of scientists assembled by the IAEA collected and analyzed more than 600 samples for more than 50 different analytes

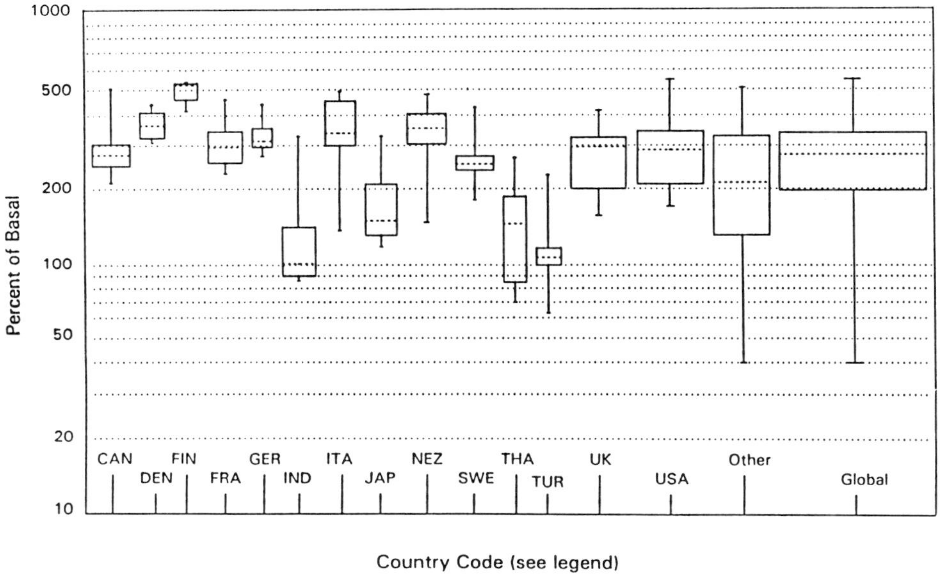

Fig. 1. Dietary intakes of zinc: graphical overview of studies conducted in various countries *(4)*. Intake values are expressed as a fraction (percent) of the proposed minimum mean basal population intake of zinc defined by WHO *(3)* and are presented separately for each country in the form of box plots. The right-most box shows global data (i.e., the complete data set). The countries presented individually are as follows: Canada, Denmark, Finland, France, Germany, India, Italy, Japan, New Zealand, Sweden, Thailand, Turkey, UK, and USA. Countries in the category "other" are Australia, Belgium, Brazil, China (incl. Taiwan), Iran, Malawi, Morocco, Myanmar, Netherlands, Nigeria, Philippines, Poland, Spain, Sudan, Switzerland, former USSR, and former Yugoslavia.

(trace elements, minor elements, and organic microcontaminants). Many of the analyses were performed by NAA. Table 2 illustrates some of the results obtained. With regard to trace elements, the main conclusions were:

1. Iodine intakes and excretion were at the low end of the acceptable range, but apparently not low enough to be causing goiter or other thyroid abnormalities;
2. Dietary intakes of essential elements (Ca, Cu, Fe, K, Mg, Mn, Na, P, Se, and Zn) were within the normal ranges observed in other countries; and
3. Dietary intakes of toxic elements (Cd, Hg, and Pb) were low in comparison with many other countries, and were also well below the maximum tolerable intakes specified by WHO and FAO. Levels of lead in blood and environmental biomonitors (e.g., lichen) were also investigated and found to be well within the normal range.

Table 2
Dietary Intakes of Minor and Trace Elements in Regions of the Former USSR Immediately Affected by the Chernobyl Nuclear Power Plant Accident

Element	Unit	Chernobyl[(a)] Mean±SE	(n)	Comparative intake values[(b)]
B	mg/d	2.0±0.2	(32)	1.6 (1.3-3.7)
Ca	mg/d	863±133	(34)	760 (210-1650)
Cd	μg/d	8.4±1.4	(34)	14 (8-200)
Cu	mg/d	0.82±0.1	(32)	1.5 (0.6-5.8)
Fe	mg/d	15.4±1.5	(34)	13 (5.1-47)
Hg	μg/d	3.4±0.5	(32)	4.6 (0.7-76)
I	μg/d	81±12	(34)	190 (50-1050)
K	mg/d	3170±328	(34)	2670 (500-4500)
Mg	mg/d	270±22	(34)	300 (120-680)
Mn	mg/d	3.3±0.4	(34)	3.1 (1.6-8.8)
Na	mg/d	4760±480	(34)	3510 (1570-6900)
P	mg/d	1326±135	(34)	1300 (470-2010)
Pb	μg/d	49±13	(34)	54 (7-510)
Se	μg/d	63±10	(34)	60 (8-1340)
Zn	mg/d	9.9±0.8	(34)	10 (4.2-19)

(a)Mean intake ± SE (number of measurement for all three regions combined)—*see text*.

(b)Median (range): from a global survey of dietary intakes *(3)*.

Health-Related Environmental Studies

As shown in Table 1, the IAEA has also been supporting a number of CRPs dealing with health-related environmental applications of nuclear analytical techniques. The main topics of current interest concern the use of nuclear analytical techniques (1) for assessing exposure to mercury in selected human populations, and (2) for air pollution studies.

In the program on mercury pollution, interest centers on using human hair as a bioindicator of exposure to inorganic and methyl mercury. WHO has tentatively set an action level of 5 μg/g of total mercury in hair; i.e., if levels of 5 μg/g or higher are observed, then this suggests that further studies, including measurements of methyl mercury, are advised. Preliminary findings from this CRP already indicate that this level is exceeded in individuals from some populations in Brazil, Chile, China, Malaysia, and other countries.

In the program on air pollution, nuclear analytical techniques are being used to study trace elements on air-filter samples collected, in the participating countries, in urban areas and remote locations. Such studies represent an almost ideal application of nuclear analytical techniques, since there are no other competing methods that can provide multielement analytical data for so many elements (generally more than 20) at the required level of sensitivity and by purely instrumental means. The data generated by these measurements are amenable to evaluation by sophisticated statistical techniques. Thereby, important information can be obtained on pollution sources that may be having an impact on human health and on the long-range transport of air pollutants. This CRP also has a component concerned with the use of biomonitors of environmental pollution, such as lichens and mosses.

Future Developments

Some of the new CRPs that are expected to be implemented during the next 2–3 yr are identified in Table 1. The program on osteoporosis will mainly be concerned with measurements of bone density in selected human populations using dual-energy photon absorptiometry (DEXA). However, since trace elements are intimately concerned with bone metabolism, it is also planned to include some studies of the trace element composition of human bone collected as biopsy and autopsy specimens. The other CRPs mentioned are also concerned with trace elements (essential and/or toxic), except for the program on radioimmunoassay. The latter will make use of radiolabeled immunoassays for the determination of organic microcontaminants (both "naturally occurring," e.g., aflatoxin, and as pesticide residues).

ANALYTICAL QUALITY-CONTROL SERVICES

For more than 30 yr the IAEA has been operating a program called Analytical Quality Control Services (AQCS), concerned mainly with intercomparison and reference materials for radionuclides, trace elements, and pesticide residues. In addition to preparing and distributing its own analytical quality-control materials, the IAEA also maintains a data base of biological and environmental reference materials for trace elements (and some other analytes) that are available internationally from any supplier *(6)*. This data base is currently being updated in a collaborative project with UNEP. Many of the materials in the data base have been certified with the aid of NAA and/or are suitable for the quality control of measurements made by nuclear analytical techniques.

New materials awaiting certification and/or in preparation under the IAEA's own program of AQCS include a lichen research material (for

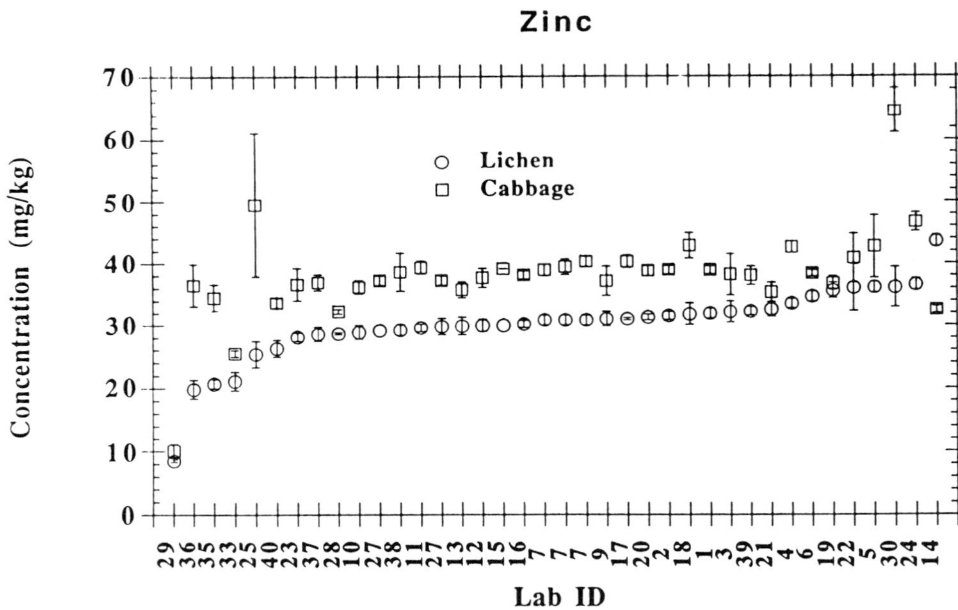

Fig. 2. Results of an interlaboratory comparison run on Candidate Lichen Research Material (IAEA-336). Laboratory means and their respective confidence intervals are displayed for the trace element zinc. The data are arranged in order of increasing numerical value. Also displayed are corresponding data for a quality-control material, IAEA-359 cabbage.

trace elements), a soil reference material (for fission products, including ^{129}I), and human hair (for total and methyl mercury).

IAEA-336 Lichen

Lichen is attracting increasing attention as a biomonitor of environmental pollution, and the use of such biomonitors is already being promoted in a number of different IAEA projects. Since no suitable certified reference material is yet available for such studies, the IAEA has initiated a project to satisfy this need. With the cooperation of the Portuguese National Institute of Energy and Industrial Technology (INETI) in Sacavem, Portugal, a 20-kg batch of the epyphytic lichen *Evernia Prunasti*, IAEA-336 lichen, has recently been prepared and distributed to a network of collaborating analytical laboratories having access to a variety of nuclear and other analytical techniques. The final stage of certification is now in progress based on 40 sets of results. Figure 2 illustrates the data obtained for one of the elements of primary interest, zinc. Also included in this plot are similar results for IAEA-359 cabbage, which was used as a quality-control standard. It is expected that IAEA-336 lichen can eventually be certified for more than 25 minor and trace elements, and that it will become available under the IAEA's AQCS Program early in 1994.

Human Hair

A human hair reference material is currently being prepared for use, initially, in connection with the CRP on mercury pollution described above. Later, it will be made available under the IAEA's AQCS Program. The analytes of primary interest are total mercury and methyl mercury; however, it is anticipated that the material will also be certified for a wide variety of minor and trace elements. The starting material was a 10-kg batch of human hair provided by a collaborating scientist in India. As of the time of writing this article, the material still had to be homogenized and distributed for intercomparison. It was planned to use the technique of cryogenic homogenization in a titanium rod mill.

The level of total mercury in the starting material is <0.5 μg/g. For the purposes of analytical quality control, it is also desirable to have a second-reference material with a much higher level of mercury (preferably in the range 10–20 μg/g). Such a material is not readily available in a sufficient amount from any known source. It is therefore planned to use part of the batch of the original starting material and to spike this artificially with methyl mercury in a controlled manner. Preliminary data indicate that the spike may be expected to behave analytically in a way that is very similar to that of naturally incorporated methyl mercury. It is expected that this intercomparison material (actually two samples with low and high levels of methyl mercury) will become available during the course of 1994.

IAEA-375 Soil

The Chernobyl accident has stimulated interest in using ^{129}I to trace the distribution of fission products in the environment. One possible indirect application of this research is to obtain better estimates of the thyroid radiation doses delivered to people living in some of the affected regions.

The Agency's laboratory has recently established an NAA procedure for determining ^{129}I in biological and other environmental materials, and has applied it to a variety of samples, including some from the Chernobyl region. This method involves both pre- and postirradiation separations of iodine *(7)*. An effort is presently under way to establish a reference material for ^{129}I, IAEA-375 soil, which was prepared from material obtained from one of the contaminated areas near Chernobyl. This material is presently in the final stages of certification. It will become available under the IAEA's AQCS Program early in 1994.

PUBLICATIONS

Many of the IAEA's Coordinated Research Programs and other activities give rise to technical reports, proceedings, and other kinds of publications. Some recent publications of relevance to the subject of this conference are listed in the references *(4–6,8–12)*. Those indicated by an asterisk are available cost-free on request.

CONCLUSIONS

It appears likely that nuclear analytical techniques will continue, at least for several more years, to have an important role in programs of the IAEA relating to the life sciences. Relevant applications of these techniques will continue to be promoted by the IAEA mainly within the framework of coordinated research and for the certification of appropriate analytical reference materials.

ACKNOWLEDGMENTS

Some of the information presented in this article is drawn partially from the programs of other present and former colleagues in the IAEA. The authors would like to acknowledge in particular their collaboration with E. Cortes Toro, R. Schelenz, S. Stone, V. Valkovic, and H. Vera Ruiz. In addition, the authors would like to thank all the scientists in many different countries who have participated in the Coordinated Research Programs and other projects referred to in this article.

REFERENCES

1. R. M. Parr and E. Cortes Toro, *Biol. Trace Element Res.* 27/28, 671–681 (1990).
2. E. Cortes Toro, H. A. Das, J. J. Fardy et al., these proceedings.
3. World Health Organization, Report of the Expert Consultation on Trace Elements in Human Nutrition and Health. WHO, Geneva (in press, due 1994).
4. R. M. Parr, H. Crawley, M. Abdulla, G. V. Iyengar, and J. Kumpulainen, Human Dietary Intakes of Trace Elements: A Global Literature Survey Mainly for the Period 1970–1991: I. Data Listing and Sources of Information. Report NAHRES-12, IAEA, Vienna pp. 109 (1992)*.
5. International Atomic Energy Agency, The International Chernobyl Project Technical Report. IAEA, Vienna (1991).
6. E. Cortes Toro, R. M. Parr, and S. A. Clements, Biological and Environmental Reference Materials for Trace Elements, Nuclides and Organic Microcontaminants: A Survey. Report IAEA/RL/128 (Rev. 1), IAEA, Vienna (1990)*.
7. V. Karg and T. Schönfeld, Determination of ^{129}I in environmental samples from Austria by means of neutron activation analysis. Report OEFZS-4378, Austrian Research Centre, Seibersdorf (1986).
8. International Atomic Energy Agency, Practical Aspects of Operating a Neutron Activation Analysis Laboratory. Report IAEA-TECDOC-564, IAEA, Vienna (1990)*.
9. International Atomic Energy Agency, Guidelines for Leaching Studies on Coal Fly Ash and Other Solid Wastes with Special Reference to the Use of Radioanalytical Techniques. Report IAEA-TECDOC-616, IAEA, Vienna (1991)*.

10. International Atomic Energy Agency, Sampling and Analytical Methodologies for Instrumental Neutron Activation Analysis of Airborne Particulate Matter, Report IAEA-TCS-4, IAEA, Vienna (1992)*.
11. International Atomic Energy Agency, Applications of Isotopes and Radiation in Conservation of the Environment (Proc. Symposium, Karlsruhe, Germany), IAEA-SM-325, Vienna (1992).
12. International Atomic Energy Agency, Sampling and Analytical Methodologies for Energy Dispersive X-Ray Analysis of Airborne Particulate Matter. Report NAHRES-16, IAEA, Vienna (1993)*.

*Available cost-free on request.

Evaluation of the Washing System Efficiency in Sugar Cane Mills by Neutron Activation Analysis

MÁRCIO ARRUDA BACCHI AND
ELISABETE A. DE NADAI FERNANDES*

*Centro de Energia Nuclear na Agricultura, P.O. Box 96
13400-970 Piracicaba SP Brazil*

ABSTRACT

The effectiveness of industrial washing operation in minimizing soil in sugar cane delivered to mills is evaluated. Instrumental neutron activation analysis was the nuclear analytical technique chosen for this work, with Fe, Hf, Sc, and Th selected as soil tracers. On the basis of differences between elemental concentration of integral and prepared cane, that is before and after washing, the feasibility of the method for evaluation of the washing efficiency is demonstrated.

Index Entries: Sugar cane washing; INAA; soil tracers; sugar mill; trace elements.

INTRODUCTION

Sugar cane plays an important role in the Brazilian economic scenario. From a cultivated area of about 5 million hectares, every year 240 million tons of cane are manually cut and mechanically loaded for production of sugar and alcohol. The mechanization has induced extraneous matter to be carried to mills together with the stalks. Among different categories of extraneous matter, tops and leaves can be useful in raising bagasse for energy generation, whereas sand and soil can damage the industrial system. Being influenced by the harvesting system, soil type, and weather conditions, the dirt level in cane has reached unacceptable levels around 5% *(1)*. In addition to the policy of encouragement of har-

*Author to whom all correspondence and reprint requests should be addressed.

Biological Trace Element Research Editor: G. N. Schrauzer © 1994 by Humana Press Inc.

vesting green cane because of ecological claims, growers have been still burning cane while standing in the field. This operation results in cane exudation, promoting even more adherence of soil onto stalks.

High priority has been given by sugar cane producers to research an approach for reducing dirt levels in harvested cane, improving field conditions so that soil intake is depressed, and implementing industrial washing process on the cane delivered to the mill. For this, huge amounts of water have to be promptly available followed by a special treatment system to reduce the organic and inorganic materials that are incorporated into fed water.

Nowadays, the usual washing system in Brazil consists of the application of recycled water by means of aspersion at a rate of 3–5 m^3/t of cane over feeder tables. Sugar cane mills have been gradually replacing their feeder tables by recently developed designs with higher slopes, enhancing the efficiency of washing and using much less water. The evaluation of such improvements relies on accurately measured dirt reductions in the washed material.

Among existing methods for soil assessment in cane, preference must be given to those applying activation analysis (AA) of elements present to a much higher extent in soils than in canes, such as Si and Al *(2,3)*, Fe *(4)*, and Sc *(5)*. Accurate results can be provided by AA as this method is interference-free from cane composition. Conversely, the high technology dependence and costs are problems that can reduce their application on a routine basis.

This research was concerned with the use of instrumental neutron activation analysis (INAA) for the assessment of efficiency of the cane-washing system installed in mills, taking into account alterations in elemental concentration of industrial cane owing to the action of washing.

MATERIALS AND METHODS

Sampling was performed at the Usina Santa Bárbara, a typical mill of medium size, located in the city of Santa Bárbara D'Oeste, state of São Paulo, processing about 1.5 million t of cane/y. The cane was harvested after burning, with manual cutting and mechanical loading, the discharge of trucks occurring on a feeder table of 45° slope, with dimensions of about 8 m length and 10 m width. The washing operation has been made by an aspersion system, at a rate of water of about 3 m^3/t of cane.

Three different sampling times were accounted for in a first trial during the 1990/91 crushing season. At each time, three cane consignments from the same field area were mechanically sampled with an horizontal probe according to the Quality Payment System. Briefly described, it consists of each consignment material being composed of three core samples of about 15 kg, which were disintegrated and well homogenized before taking representative subsamples for analysis. After passing through the washing system, the cane was conveyed upward into the shredder, from whose elevator was taken material at intervals of 3 min, the composite sample representing cane after washing. About 15 kg of cane were randomly taken from each consignment, washed in a laboratory, and considered as clean cane, with no soil contamination. These whole stalks were shredded, homogenized, and like all other cane materials, dried at 60°C to constant mass and powdered. Soil samples from the field areas where the canes were harvested have also been collected, dried at 100°C, and ground in a mortar.

In the 1991/92 season, a second trial was performed with three consignments from another selected field area by doing 18 holes in each one, with sampled material collected separately for analysis. Following a procedure identical to the first trial, shredded cane, clean cane, and soil were also provided.

About 250 mg of cane and 100 mg of soil were placed into high-purity polyethylene vials and irradiated in the IEA-R1 research reactor of Instituto de Pesquisas Energéticas e Nucleares, IPEN/SP, with a thermal neutron flux of about 5.10^{12} $n{\cdot}cm^{-2}{\cdot}s^{-1}$ for 7 h, together with soil and plant reference materials. The induced radioactivity was measured with a hyperpure germanium detector after decay times of 3, 6, 15, and 30 d.

RESULTS AND DISCUSSION

The elements determined in samples showed much higher concentration in soil than in clean cane, except K and Rb. Average concentrations of Fe, Hf, Sc, and Th showed the largest differences, since they are present in levels at least a thousand times higher in soils than in cane, whereas smaller contrasts were seen for Ce, Co, Eu, and La (Fig. 1). Elements predominating in soil over cane matrix can then be considered as soil tracers, their usefulness being proportional to the magnitude of such ratio.

An increase of elemental concentration in industrial canes can easily be noted in comparison with the clean cane values (Fig. 2), such expected occurrence being justified by the presence of soil in samples of the consignments. The concentrations in shredded cane have decreased compared with consignment canes, in a variable pattern for each element, by the action of the industrial washing that effectively removed part of the dirt. The reduction behavior has been also observed for K and Rb, to a lesser degree, even though they originated mainly from cane. This can be explained by cane juice losses during washing and, therefore, by the losses of mineral elements.

Ce, Co, Eu, Fe, Hf, La, Sc, and Th mean concentrations in the cane samples of the three consignments of the first trial, and in the 18 holes of each one of the second were related to the respective concentrations in shredded cane, subtracting the blanks owing to clean cane. Such a relationship considers the reductions of concentration as a decrease of the dirt content, thus being an estimation of the washing efficiency. In general, similar efficiency values were obtained for all those elements, with deviations accounted for mainly by distinctive behavior of the soil elements during washing, as well as for sampling uncertainties (Fig. 3).

Sc, Fe, Th, and Hf predominate in such a high level in soils that the efficiency can be calculated only by the differences in concentration

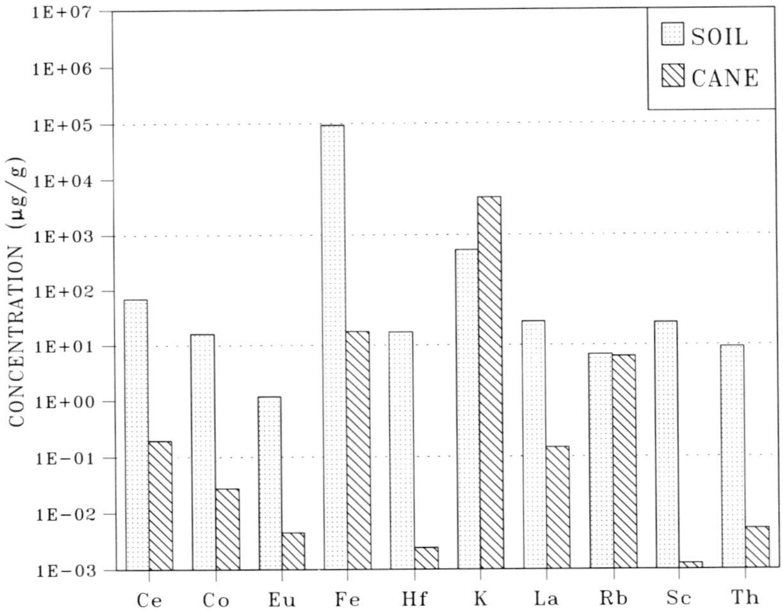

Fig. 1. Average elemental concentration in soils and clean canes of the two trials.

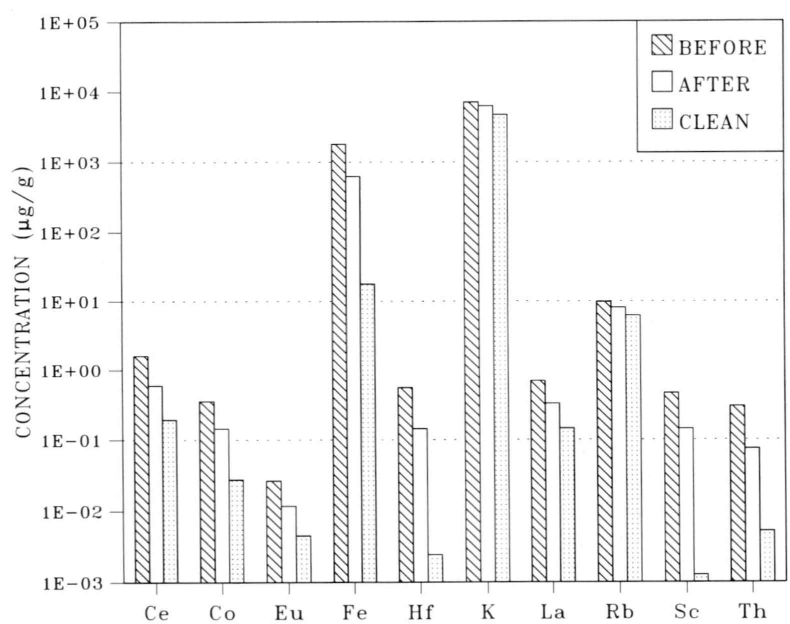

Fig. 2. Average elemental concentration in industrial cane, before and after washing, and in clean cane for both trials.

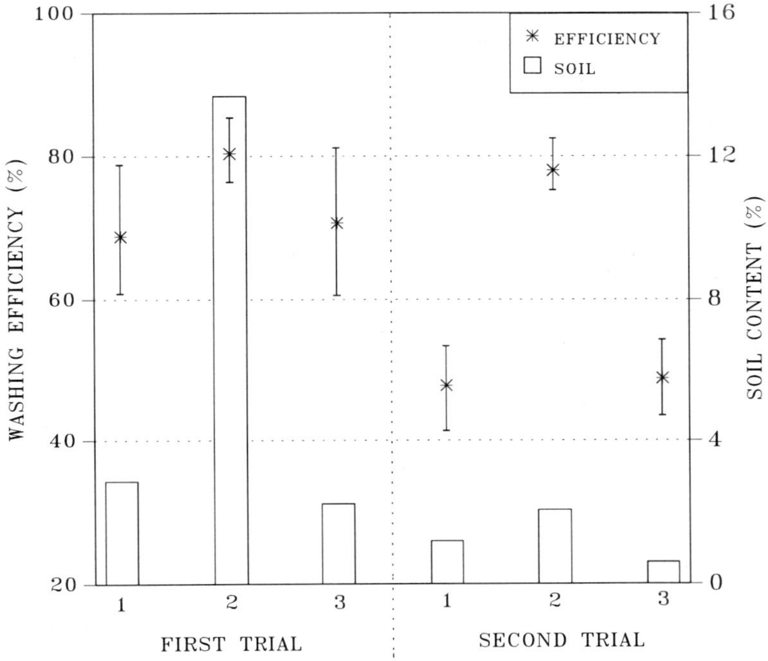

Fig. 3. Mean washing efficiency for each sampling calculated from Ce, Co, Eu, Fe, Hf, La, Sc, and Th concentrations, showing the upper and lower limits, and the soil contents in consignments determined by Sc.

Table 1
Washing Efficiency Average Values (%) Calculated from Fe, Hf, Sc, and Th Concentrations, for the Three Sampling Dates of First Trial and for the Three Consignments of the Second One[a]

	FIRST TRIAL			SECOND TRIAL		
SAMPLE NUMBER	1	2	3	1	2	3
WITHOUT SUBTRACTION	68.2 ± 6.0	79.2 ± 3.1	75.6 ± 4.4	43.6 ± 3.4	78.3 ± 1.6	51.5 ± 1.4
WITH SUBTRACTION	69.2 ± 5.9	79.8 ± 3.0	76.3 ± 4.7	43.7 ± 3.5	78.4 ± 1.6	51.6 ± 1.4

[a]Results obtained with clean cane blank subtraction or without such correction.

between industrial cane before and after washing. Values thus obtained were compared with those of subtracted blanks, observing differences of < 1% in calculated efficiencies as shown in Table 1. Consequently, it becomes unnecessary to analyze clean cane in routine applications, which reinforces the method practicability.

Soil composition is not used for efficiency estimates if all the cane milled during sampling time comes from the same field area, assuring the uniformity of dirt composition. Nevertheless, under normal operational conditions, when the industry receives cane from different field areas, sampling can be a serious problem in determining the efficiency, being essential to consider the cane mass and respective soil composition of each provenience.

An association between soil level in consignment canes, inferred through scandium content, and the average washing efficiency was also seen, with a tendency toward better efficiency as the dirt levels increase (Fig. 3). Such behavior confirms the expectation of the suitability of washing systems for a high dirt content in canes. Anyway, the drawbacks of the sampling representation in terms of soil assessment of cane consignments must be emphasized. The soil values found in the second trial can be regarded as the true soil contents in the cane load, because they represent the average of 18 sampling points well allotted in the consignment, against only 3 in the first. Therefore, much more accuracy is expected from the relationship between soil % cane and washing efficiency, which was really observed. Since this method can discriminate low levels of soil in industrial cane samples accurately, it may assist in providing a basis for evaluation programs of the effectiveness of industrial washing installations in sugar cane mills and also the improvements in washing system designs.

ACKNOWLEDGMENTS

The authors wish to thank the staff of Usina Santa Bárbara for assisting with the practical work involved. Thanks are also owed to FAPESP—Fundação de Amparo à Pesquisa do Estado de São Paulo, and CNPq—Conselho Nacional de Desenvolvimento Científico e Technológico for supporting the research.

REFERENCES

1. O. A. Braunbeck and A. D. Banchi, *Bol. Téc. Copersucar*, **43**, 35–42 (1988).
2. J. A. Aylmer and R. J. Holmes, *Int. Sug. J.* **83,** 996, 362–364 (1981).
3. J. G. Loughran, G. Cowan, and V. Mason, *Proc. ISSCT* **18**, 942–951 (1983).
4. E. A. N. Fernandes, M. A. Bacchi, and E. S. B. Ferraz, *Int. Sug. J.* **95,** 1129, 33–36 (1993).
5. E. A. N. Fernandes, *J. Radioanal. Nucl. Chem.*, **168,** 1, 41–46 (1993).

Determination of Copper by Isotopic Dilution

E. S. Faquim and C. S. Munita*

Radiochemistry Division, Instituto de Pesquisas Energéticas e Nucleares, Comissão Nacional de Energia Nuclear, C.P. 11049, CEP 05422-970, São Paulo, Brasil

ABSTRACT

A rapid and selective method was used for the determination of copper by isotopic dilution employing substoichiometric extraction with dithizone in carbon tetrachloride. The appropriate pH range for the substoichiometric extraction was 2–7. In the analysis, even a large excess of elements forming extractable complexes with dithizone does not interfere. The accuracy and precision of the method were evaluated. The method has been applied to analysis of reference materials, wheat flour, wine, and beer.

Index Entries: Copper; isotopic dilution; substoichiometry; copper dithizonate; application of tracers; environmental samples.

INTRODUCTION

Copper is one of the elements that has biological interest at trace and ultratrace levels. Analytical difficulties often seem to increase drastically when analyzing materials with very low contents of trace elements *(1)*. Thus, the availability of sensitive and selective methods is of great importance for biological and environmental samples. Neutron activation analysis with inherent sensitivity is the obvious method of choice. However, the ^{64}Cu isotope can also be produced by the reaction ^{64}Zn $(n, p)^{64}Cu$, which has a cross-section of 390 mb and ^{64}Zn has 48.9% natural abundance. Since most biological samples contain amounts of Zn

*Author to whom all correspondence and reprint requests should be addressed.

 © 1994 by Humana Press Inc.

comparable to those of copper, this reaction may cause some interference in the determination of copper. In this context, substoichiometric isotopic dilution analysis proposed by Suzuki *(2)* and Ruzicka and Stary *(3)* is a very attractive analytical method to determine copper in these kinds of samples.

The method offers an accurate and precise determination of trace and ultratrace amounts of elements by measurements of radioactivity alone, without corrections for chemical yield. The advantages of substoichiometric separation are that it is highly selective, quick, and reproducible. The principles of the method were treated in several papers (4–7). This article describes a simple procedure for the determination of copper in biological and environmental samples by isotope dilution analysis involving the substoichiometric extraction technique with dithizone into carbon tetrachloride.

MATERIALS AND METHODS

Counting Apparatus

A nuclear Chicago single-channel analyzer coupled to a 3.7 × 6.2 cm well-type NaI (Tl) crystal was used for the measurements of the activity.

Reagents

All the chemicals, reagents, and solvents used were of A.R. grade. The acids and ammonium hydroxide were purified by distillation.

Radio-Copper Solution

^{64}Cu was produced by the neutron irradiation of metallic copper for 30 min or 8 h at a neutron flux of 10^{12} n/cm^2/s. The irradiated target was dissolved in nitric acid, and excess acid was eliminated by evaporation. The radioactive solution was transferred and diluted to give a concentration of 1 mg/mL.

Dithizone Solution

The solution was prepared dissolving dithizone in CCl_4 to give a concentration of 10^{-3} mol/L. The solution was transferred into an amber flask and kept in the refrigerator, since it decomposes rapidly with formation of carbodiazone if exposed to strong light and subjected to relatively high temperatures.

Buffer Solution

Buffer solution was prepared by the dissolution of 15 g of potassium iodide, 13 g of ammonium acetate in 100 mL of water, and the addition of glacial acetic acid to attain a pH value of 4.8 *(8)*. It was purified by shaking with a concentrated solution of dithizone in CCl_4.

Interfering Elements Solutions

Appropriate amounts of the elements; salts, oxides, or metals were dissolved in acid or deionized distilled water. The excess acid was eliminated by evaporation. The solutions were transferred to give a concentration of 1 mg/mL. The procedure was applied to prepare As(III), Au(III), Bi(III), Cd(II), Co(II), Hg(II), In(III), Mn(II), Ni(II), Pb(II), Pd(II), Pt(II), Sb(III), Se(II), Sn(IV), and Zn(II).

Sodium Thiosulfate Solution 2%

An amount of sodium thiosulfate was dissolved in distilled water.

COPPER EXTRACTION

The high selectivity and simplicity of the method that give optimum conditions for the determination of copper were studied. The influence of hydrochloric, sulfuric, and nitric acids was examinated because of the possibility of using these acids in the digestion process of samples.

EQUILIBRIUM TIME

A series of solutions, pH 5, with the same amounts of ^{64}Cu were shaken with dithizone in CCl_4 for different times. The γ activity of the extracted copper was measured and plotted against shaking time. The equilibrium was reached after 2 min of shaking.

EFFECT OF ACIDS

The influence was verified by the addition of an aqueous solution to the tracer dissolved in HCl, HNO_3, or H_2SO_4 separately. The excess of acid was eliminated by evaporation, and the pH was adjusted to 4.8 with buffer solution. The solution was shaken for 2 min with the dithizone solution. The phases were separated, and the activity of the organic aliquot was measured. The results showed that copper is quantitatively extracted independently on the acid used.

SUBSTOICHIOMETRIC EXTRACTION

A known amount of ^{64}Cu and a substoichiometric amount of dithizone in CCl_4 solution were added to an aqueous solution. A series of preliminary experiments were carried out to ascertain the best conditions of

Table 1
Effect of pH on the Substoichiometric Extraction of Copper[a]

pH	Percentage of extraction(%)
0.5	67.8
1.0	84.5
1.5	87.1
2.0	89.4
3.0	89.0
4.0	90.9
5.0	93.3
6.0	95.4
7.0	95.3
8.0	64.9

[a]Dithizone concentration 10^{-3} mol/L, 6.3×10^{-3} mol/L ^{64}Cu. Shaking time: 3 min.

pH range and influence of masking agent for finding the reproducibility of the substoichiometric extraction.

pH RANGE

A series of solutions with known amounts of radio-copper with a constant, but substoichiometric amount of dithizone in CCl_4 were prepared and investigated at different pHs. The dependence of the substoichiometric extraction of Cu with dithizone on the pH can be seen in Table 1. Copper could be extracted substoichiometrically from pH 2 to 7, which is in agreement with the predicted value *(9)*.

REPRODUCIBILITY OF SUBSTOICHIOMETRIC EXTRACTION

The reproducibility of the substoichiometric extraction of copper with dithizone into CCl_4 was examined in solutions containing different

Table 2
Substoichiometric Extraction of Cu(II) with Dithizone into CCl_4[a]

$^{64}Cu(II)$ (ug)	Activity of 1mL aliquot of organic phase (cpm)
8.09	4003
12.13	5967
16.18	7720
24.27	8391
32.36	8550
36.40	8835
40.45	8841
50.56	9016

Dithizone concentration 10^{-4} mol/L, shaking time: 3 min.

amounts of radioactive Cu(II). A graph was plotted of the activity of an aliquot of the organic phase vs the amount of Cu(II) taken. As shown in Table 2, the activity of the organic phase increased with the growing amount of Cu(II); after the point, which corresponds to a copper–dithizone ratio of 1:2, the amount of copper in the organic phase remained constant. The aqueous solution containing excess labeled copper was treated repeatedly with dithizone in CCl_4. In each case, the activity of the organic phase was noted to be the same as that of the other extracts. This shows that the same amount of copper was extracted in each extraction.

INTERFERENCES OF OTHER ELEMENTS IN THE SUBSTOICHIOMETRIC EXTRACTION OF Cu(II)

The study of the selectivity of substoichiometric separation of Cu(II) was carried out by means of a series of solutions containing known amounts of nonactive copper and active copper and substoichiometrically extracted with dithizone in CCl_4, in the presence of As(III), Au(III), Bi(III), Co(II), Cd(II), Hg(II), In(III), Mn(II), Ni(II), Pb(II), Pd(II), Pt(II),

Table 3
Interference with Extraction of Copper Caused by Addition of Nonactive Elements

Copper taken (μg)	Active copper (μg)	Other cations present (μg)	Copper found (μg)*
1.0	1.0	–	1.17±0.02
		25.0μg Hg	1.01±0.02
1.0	1.0	–	1.25±0.03
		1.01μg Au	1.44±0.04
		0.95μg Pd	3.12±0.08
		1.04μg Pt	1.33±0.03
1.0	1.0	–	1.22±0.03
		93.2μg Bi	1.10±0.03
		98.5μg In	1.40±0.03
		100.8μg Mn	1.31±0.03
		98.7μg Se	1.38±0.03
1.0	1.0	–	1.01±0.03
		100.0μg As	1.12±0.04
		100.0μg Cd	1.05±0.03
		100.0μg Co	1.01±0.03
		100.0μg Ni	1.05±0.03
		100.0μg Pb	1.00±0.03
		100.0μg Sb	1.10±0.03
		100.0μg Sn	1.08±0.03
		100.0μg Zn	1.09±0.03

*Arithmetic mean, $n = 4$.

Sb(III), Se(II), Sn(IV), and Zn(II). As shown in Table 3 even a large excess of metals that form extractable complexes with dithizone when iodide is present in the aqueous phase will not inhibit the extraction. It can be seen that only Au, Pd, and Pt interfere.

Table 4
Results of Copper Determination

Material	Certified value (μg/g)	Experimental value (μg/g)*	RSD	Deviation from the certified value %
SRM 1567a	2.1±0.2	1.79±0.16	0.057	-14.6
SRM 1568	2.2±0.3	2.04±0.22	0.043	-7.4
SRM MA-A-2 621	4.0±0.1	4.83±0.6	0.078	20.8
SRM 1572	16.5±1.0	17.0±0.6	0.022	3.0
Wheat flour		1.00±0.09	0.053	
Wine (μg/mL)		0.17±0.01	0.051	
Beer (μg/mL)		0.28±0.02	0.063	

*Arithmetic mean (n = 4) and its confidence limits ($\bar{x} \pm t_{0.05}\, sn^{-1/2}$).

APPLICATIONS

The method described has been applied to the determination of copper in reference materials (RMs): citrus leaves: NBS-SRM-1472; wheat flour: NBS-SRM-1567a; rice flour: NBS-SRM-1568; fish homogenate: MA-A-2 621, wine, beer, and wheat flour.

Sample Preparation

RMs

An amount of 200 mg of sample was added to 2–3 mL of HNO_3 and a known amount of ^{64}Cu in a microwave acid digestion bomb. The sample was digested in a microwave furnace at medium power for 2 min and cooled for 1 h. The sample solution was evaporated to eliminate the excess acid.

Wine and Beer

The procedure used was described by Kumar et al. *(10)*. An aliquot of 50 mL of wine or beer was evaporated to dryness, and the residue was dissolved in 100 mL of HNO_3.

Substoichiometric Determination of Copper

The appropriate amount of dilute HNO_3 or NH_4OH was added to the sample and standard solutions to adjust the pH to 4.0–5.0. After adjusting the acidity, 2 mL of buffer solution and 0.5 mL of sodium thiosulfate solution 2% were added. The solutions were simultaneously extracted with equal substoichiometric portions of dithizone in CCl_4. The activity of the organic layer was measured. Table 4 shows the results obtained.

DISCUSSION

As shown above, Cu(II) could be extracted with a substoichiometric amount of dithizone in CCl_4 solution between pH 2 and 7. Above and below this pH range, the extraction is not quantitative. To examine the influences of other elements, the substoichiometric extraction with 1×10^{-5} mol/L dithizone was applied to the synthetic samples solutions containing a small amount of copper and large amounts of other elements. As presented in Table 3, the substoichiometric determination of copper is possible together with other elements that have higher extraction constants than copper. The logarithm of the extraction constant of mercury with dithizone is 26, 85. However, the results showed that mercury does not interfere with the extraction of copper. This occurs because a buffer solution containing iodide was added to the aqueous solution to form an HgI_4^{2-} complex, which keeps mercury in the aqueous phase. The thiosulfate present prevents oxidation of the dithizone by free iodine.

This method is found to be reasonably selective for Cu(II). The elements Au, Pt, and Pd are extracted, but they may be left out of consideration because of their rarity in these kinds of samples.

The precision and accuracy of the method were tested with reference materials. From Table 4, the good precision of the method is shown. Even for materials with the lowest copper contents, relative standard deviation does not exceed 8%. The precision of substoichiometric analysis has mainly two components: one associated with substoichiometric separation and the other associated with radioactivity measurements. Since the latter component can be improved by use of a radioisotope with a high specific activity, the precision depends primarily on the former component.

The detection limit was found to be < 13 ng, which was obtained with radio-copper produced by irradiation of Cu(II) in a thermal neutron flux of 10^{12} n/cm^2/s for 8 h applying the substoichiometric radiochemical separation procedure.

ACKNOWLEDGMENT

The authors wish to acknowledge financial support from CNPq.

REFERENCES

1. R. Dybczynski, M. Wasek and H. Maleszewska, *J. Radioanal. Nuclear Chem.* **130,** 365–388 (1989).
2. N. Suzuki, in *Proceedings of 2nd Conf. Radioisotopes*; Japan Atomic Industrial Forum, Tokyo, 1958, p. 151.
3. J. Ruzicka and J. Stary, *Substoichiometry in Radiochemical Analysis*, Pergamon, London, 1968.
4. K. Kudo and N. Suzuki, *J. Radioanal. Chem.* **26,** 327–373 (1975).
5. P. C. Reddy, B. Polalah and R. Rengamanner, *Radioisotopes* **39,** 496–502 (1990).
6. S. S. Shetty and Z. R. Turel, *J. Radioanal. Nuclear Chem.* **145,** 223–229 (1990).
7. N. Suzuki, *J. Radioanal. Nuclear Chem., Articles* **124,** 197–210 (1988).
8. J. Ruzicka and J. Stary, *Talanta* **9,** 617–621 (1962).
9. K. Kudo and K. Kobayashi, *J. Radioanal. Chem.* **53,** 163 (1979).
10. B. Kumar, H. B. Singh, M. Katyal and R. L. Sharma. *Mikrochim. Acta* **III,** 79–87 (1991).

Simulation and Optimization of Cyclic Activation Analysis of Short-Lived Isotopes with 14MeV Neutron Generator

R. KHELIFI,* Z. IDIRI, AND S. TOBBECHE

Centre de Develop des Techniques Nucleaires, 2, Bd. Frantz Fanon, Algiers, Algeria

ABSTRACT

A program of simulation and optimization is developed for the case of cyclic activation analysis of short-lived isotopes with 14-MeV neutrons. The background line under the photopeaks of interest is simulated using Zikovsky's model. The reliability of the program is checked on real conditions with a geological reference sample "Soil 5" provided by the IAEA. Optimum experimental conditions (timing parameters, number of cycles) are determined, and corresponding detection limits calculated. A systematic study of short-lived isotopes with half-lives lower than 5 min is done for Soil 5, and the results are discussed.

Index Entries: Optimization; neutron activation; cyclic analysis.

INTRODUCTION

Simulation and optimization in activation analysis are practical concepts that permit the analysts to make an *a priori* decision on how an element could be detected or not without making trials that are time- and material-consuming. Generally, the quantity that is calculated is the limit of detection. The values of these limits given in the literature *(1)* are often not realistic, since they are calculated in the absence of a background.

Guinn et al. *(2,3)* have worked on a simulation and optimization of thermal neutron activation analysis where background is taken as being

*Author to whom all correspondence and reprint requests should be addressed.

the result of two components: natural background and Compton scattering of high-energy γ-rays. However, the optimum timing parameters (irradiation decay and counting time) are determined empirically. Spyrou et al. *(4,5)* have developed a program for a simulation and optimization applied to cyclic activation analysis with thermal neutrons. In this article, the simulation of the background line for a given matrix is carried out using Zikovsky's model *(6)*. The different detector responses (resolution, Compton-to-peak ratio, efficiency) are determined experimentally. The background line is then calculated using the different parameters of the analysis: timing parameters, mass of the irradiated sample, elemental composition, and neutron flux. The optimization of the different timing parameters and the number of cycles is done by minimizing the background-to-signal ratio. The search for the minimum is carried out using a direct search method (the flexible polyeder method) *(7)*, with the constraint that the total experimental time is fixed.

SIMULATION

Simulation Model

According to Zikovsky's model *(6)*, which is used in this work, the photopeak is represented by a triangle and the Compton continuum by a rectangle. Unlike the procedure of Al-Mugrabi and Spyrou *(5)*, who used photofractions, the Compton continuum was calculated using a peak-to-Compton ratio function (*see* Fig. 1), which is a fit of experimentally determined points. Photopeak areas were calculated by means of functions representing absolute detector efficiency (*see* Fig. 2) and energy resolution (*see* Fig. 3)

The simulated background line is computed according to the equation:

$$\mathrm{B} = \sum_{j=1}^{k} 3\Gamma(N_{cc}/N_c)_j M_e \varepsilon_j c_j \sigma_j \varnothing I_j S_j D_j C_j / E_c + tc\ BN \tag{1}$$

where j = element index, k = number of Compton contributions, BN = natural background spectrum per unit time, $(N_{cc}/N_c)_j$ and ε_j = peak-to-Compton ratio and absolute detector efficiency at the γ line of interest, respectively, I_j = γ line intensity, Γ = detector resolution in keV, E_c = Compton front energy, c_j = j-th element concentration, M_e = sample mass, σ_j = neutron cross section, and Φ = neutron flux. S_j, D_j and C_j - are saturation, decay, and counting factors, respectively.

$$S_j = 1 - \exp(-\lambda_j ti) \tag{2}$$

$$D_j = \exp(-\lambda_j td) \tag{3}$$

$$C_j = \{[1 - \exp(-\lambda_j tc)]/\lambda_j\} \tag{4}$$

where, λ_j = decay constants, and t_i, t_d, t_c are irradiation, decay, and counting time, respectively.

The program developed in this work is written in Fortran 77 and calculates the background line by summing the Compton contributions of

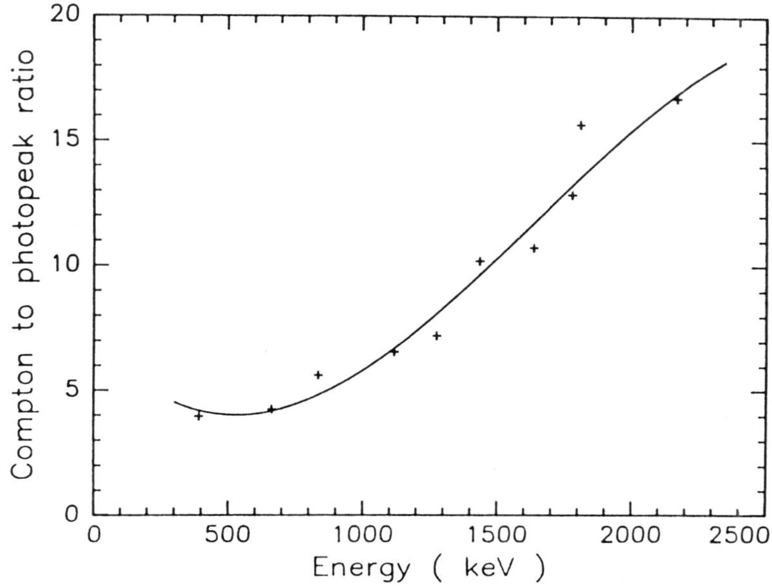

Fig. 1. Compton-to-photopeak ratio vs energy.

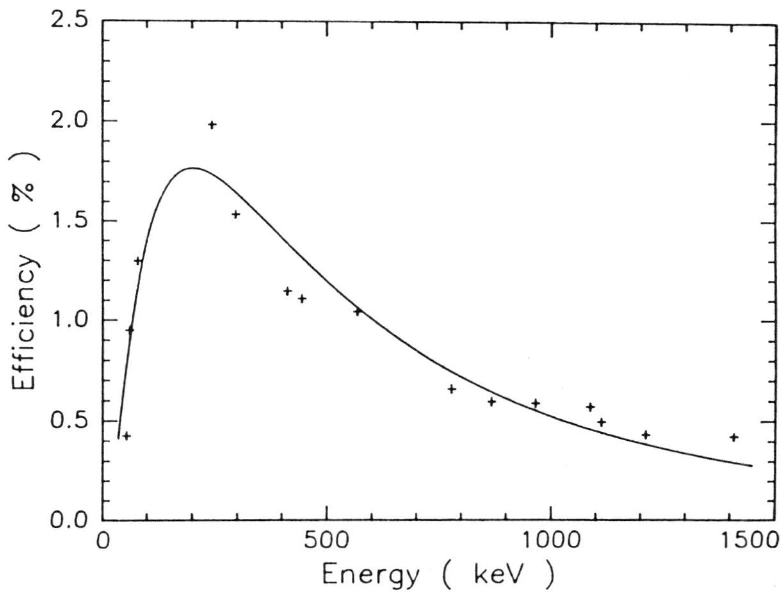

Fig. 2. Detector efficiency vs energy.

different γ-ray lines involved in the spectrum. It uses a nuclear data file supplied by the user together with experimental data (mass sample, matrice composition, detector responses, and timing parameters). The output file gives peak areas and the background line.

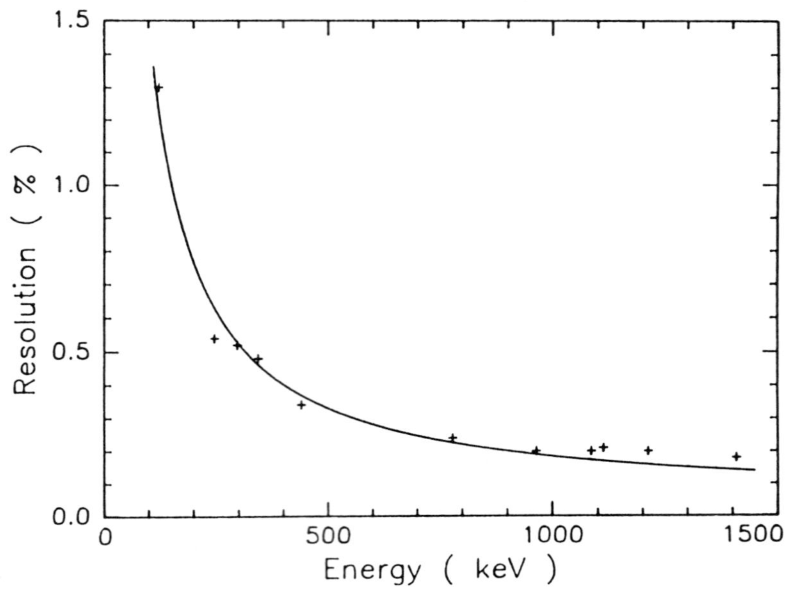

Fig. 3. Detector resolution vs energy.

Reliability of the Model

In order to check the reliability of the present model, an IAEA reference material Soil 5 was irradiated and counted two times. The neutron generator is a SAMES J25 operating at 150 kV with a yield of 10^{11} n/s. The detection system is mainly composed of an ORTEC HPGe detector with 2 keV resolution and relative efficiency of 20% and an ORTEC multichannel analyzer with 1024 channels.

The mass of the Soil 5 sample was 9.215 g and the average flux inside the sample 2.2 10^7 $n/\text{cm}^2/\text{s}$. Simulated and measured values of background in number of counts per keV are given in Tables 1 and 2. In these two countings, a mean value of simulated-to-measured background ratio of 1.1 has been observed for energies <1779 keV. For higher energies, the ratio is worse, because the peak-to-Compton ratio is not well known in this energy region. The fit function used does not cover this region, and counting statistics are poor in this energy region.

OPTIMIZATION

Objective Function

The optimization program used for cyclic activation analysis is based on a variant of the simplex method. It minimizes the detection

Table 1
Measured and Simulated Background Values for Certified Reference Material Soil 5, with Neutron Flux $\Phi = 2.6\ 10^7\ n/cm^2/s$ and $t_i = 1200$ s, $t_d = 600$ s, and $t_c = 1200$ s

Reactions	$T_{1/2}$ [s]	Energy [keV]	Measured background	Simulated background
$^{85}Rb(n,2n)^{84m}Rb$	1220	248	1121	1131
$^{88}Sr(n,2n)^{87m}Sr$	10100	388	634	623
$^{140}Ce(n,2n)^{139m}Ce$	56	754	379	555
$^{27}Al(n,p)^{27}Mg$	568	844	255	297
$^{27}Al(n,p)^{27}Mg$	568	1014	173	150
$^{29}Si(n.p)^{29}Al$	394	1273	95	96
$^{27}Al(n,\alpha)^{24}Na$	53900	1369	84	88
$^{28}Si(n,p)^{28}Al$	135	1779	33	13
$^{56}Fe(n.p)^{56}Mn$	9290	1811	27	5
$^{39}K(n,2n)^{38}K$	458	2168	17	3

Table 2
Measured and Simulated Background Values for Certified Reference Material Soil 5, with Neutron Flux $\Phi = 2.6\ 10^7\ n/cm^2/s$ and $t_i = 1200$ s, $t_d = 1800$ s, and $t_c = 7200$ s

Reactions	$T_{1/2}$ [s]	Energy [keV]	Measured background	Simulated background
$^{88}Sr(n,2n)^{87m}Sr$	10100	388	752	940
$^{56}Fe(n.p)^{56}Mn$	9290	847	315	349
$^{27}Al(n,p)^{27}Mg$	568	1014	305	211
$^{29}Si(n.p)^{29}Al$	394	1273	126	117
$^{41}K(n,p)^{41}Ar$	6590	1294	117	105
$^{27}Al(n,\alpha)^{24}Na$	53900	1369	82	74
$^{56}Fe(n.p)^{56}Mn$	9290	1811	93	23
$^{39}K(n,2n)^{38}K$	458	2168	83	19

limit (L_d), and gives the optimum timing parameters (t_i, t_d, t_c, t_r) and numbers of cycles (n) for analysis. L_d is proportional to the background fluctuations-to-signal ratio, and for a mass sample M_e, L_d is given in terms of concentration by:

$$L_d = 3(Bc)^{1/2}/M_e\, Nc \tag{5}$$

where Nc = the photopeak area of interest, and Bc = the cumulative Compton number of counts under the photopeak of interest:

$$Bc = \sum_{j=1}^{k} Bc_j \tag{6}$$

Bc_j is the contribution of the j-th isotope defined as follows:

$$Bc_j = N_j(N_{cc}/N_c)_j \varepsilon_j I_j \Phi \sigma_j [1 - \exp(-\lambda_j ti)] \exp(-\lambda_j td)$$
$$[1 - \exp(-\lambda_j t_c)]/\lambda_j$$
$$\left[\frac{n}{[1 - \exp(-\lambda_j t)]} - \frac{\exp(-\lambda_j t)[1 - \exp(-n\lambda_j t)]}{[1 - \exp(-\lambda_j t)]^2}\right] \tag{7}$$

where k = number of isotopes contributing to a Compton background, N_j = number of target nuclei, n = number of cycles; and n_j (N_{cc}/N_c), ε_j, I_j, Φ, and σ_j are defined in the simulation part.

The activation cycle is closed by the return of the sample to the irradiation position (t_r) so that one cycle time or period (t) is defined by the sum $t = ti + td + tr + tc$.

Calculation Procedure

The total experiment time is set constant and fixed by the user. Irradiation and counting times are set equal, because the detection limit function varies through saturation and counting factors, which have similar form.

Finally, the number of parameters used for optimization is reduced to 3, $ti = tc$, td, tr.

The number of cycles is deduced from the constraint of total experimental time t_e, which is fixed by the user:

$$nt = t_e \tag{8}$$

The optimization is carried out by introducing total experimental time and initial values for irradiation, counting, decay, and transfer times. The optimization program uses the simulation background line output file. The minimum values of decay and transfer times are set to 3 s, according to the minimum transfer time of our pneumatic transfer system around 2 s.

A systematic study of detection limits is carried out for the reference material Soil 5. Only nuclear reactions producing radionuclides with half-lives <300 s are considered. The optimum values of timing parameters and detection limits are given in Table 3. In this table, decay and

Table 3
Calculated Optimum Values for Timing Parameters, Number of Cycles, and Corresponding Detection Limits in Soil 5

Reactions	Eγ [keV]	$t_{1/2}$ [s]	t_i=t_c [s]	No. of cycles	L_D*	Certified values
$^{128}Xe(n,2n)^{127m}Xe$	125	69.2	63	19	13.6 %	
$^{76}Ge(n,2n)^{75m}Ge$	140	47.7	46	24	3.32 %	2.3
$^{75}As(n,p)^{75m}Ge$	140	47.7	51	22	17.0 %	93.9
$^{45}Sc(n,\gamma)^{46m}Sc$	142	18.7	20	26	3.77 %	14.8
$^{46}Ti(n,p)^{46m}Sc$	142	18.7	23	35	29.43 %	0.47 %
$^{78}Se(n,2n)^{77m}Se$	162	17.5	27	20	1.90 %	1.3
$^{82}Kr(n,2n)^{81m}Kr$	190	13.3	17	30	19.33 %	
$^{19}F(n,p)^{19}O$	197	27.1	28	19	2.80 %	628
$^{232}Th(n,f)^{136}I$	197	45	49	23	23.80 %	11.3
$^{238}U(n,f)^{136}I$	197	45	49	23	3.68 %	3.15
$^{79}Br(n,n')^{79m}Br$	207	4.86	7	30	3.60 %	5.4
$^{180}Hf(n,2n)^{179m}Hf$	214	18.7	19	27	2.37 %	6.3
$^{108}Pd(n,2n)^{107m}Pd$	215	21.3	25	22	6.01 %	
$^{197}Au(n,n')^{197m}Au$	279	7.80	14	18	64	
$^{26}Mg(n,\alpha)^{23}Ne$	440	37.2	41	14	9060	1.5 %
$^{208}Pb(n,2n)^{207m}Pb$	570	0.81	2	22	6450	129
$^{90}Zr(n,2n)^{89m}Zr$	588	251	196	6	1800	221
$^{138}Ba(n,2n)^{137m}Ba$	662	153	167	7	167	562
$^{140}Ce(n,2n)^{139}Ce$	754	56.4	49	17	62	59.7
$^{89}Y(n,n')^{89m}Y$	909	16	19	27	474	21
$^{71}Ga(n,\alpha)^{68}Cu$	1078	31	32	18	1.08 %	18.4
$^{52}Cr(n,p)^{52}V$	1434	225	156	8	1550	28.9
$^{55}Mn(n,\alpha)^{52}V$	1434	225	156	8	130	852
$^{40}Ar(n,p)^{40}Cl$	1461	81	63	14	7400	
$^{23}Na(n,\alpha)^{20}F$	1634	11	15	35	813	1.92 %
$^{28}Si(n,p)^{28}Al$	1779	135	80	11	22.8	26 %
$^{31}P(n,\alpha)^{28}Al$	1779	135	80	11	35.1	0.11 %
$^{11}B(n,p)^{11}Be$	2125	13.8	16	32	1400	63

*in μg/g except as noted.

transfer times are set to 3 s, since the calculated optimum values for all elements are equal to zero.

CONCLUSION

The reliability of the simulation model used in this work has been checked, and a good agreement between measured and simulated background has been observed for energies lower than 1779 keV. This is almost sufficient for activation analysis with 14-MeV neutrons, since most γ lines used for analysis are located in the region below the precedent "critical" energy.

It can be seen from the results of optimization obtained (Table 3) that only minor and major elements (Mg, Mn, Ba, Si, P, Na) can be determined in Soil 5 using 14-MeV cyclic activation analysis with a neutron generator like SAMES J25 with an output of 10^{11} n/s. The optimization program is under development for introducing bremsstrahlung, and single- and double-escape peaks components. The display of the simulated γ-ray spectra is in progress.

REFERENCES

1. R. Pepelnik, *J. Radioanal. Chem.* **112,** 435 (1987).
2. V. P. Guinn, E. Garzanov, and E. Cortes Toro, *J. Radioanal. Chem.,* **43,** 599 (1978).
3. V. P. Guinn, J. Leslie, and L. Nakazawa, J. *Radioanal. Chem.* **70,** 513 (1982).
4. N. M. Spyrou, *J. Radioanal. Chem.* **61,** 211 (1981).
5. M. A. Al-Mugrabi and N. M. Spyrou, *J. Radioanal. Nucl. Chem.* **110,** 67 (1987).
6. L. Zikovsky and E. A. Schweikert, *Nucl. Instr. and Methods,* **155,** 279 (1978).
7. R. O'Neil, *Appl. Statist.* **20,** 338 (1971).

Measurement of Uranium Content and Its Distribution in Some Crustacea of *Mysidacea* Species by the Fission Track Method

IULIA I. GEORGESCU*,[1] AND ANA DANIS[2]

[1]*Polytechnic Institute, Faculty of Chemical Technology, Bucharest, Romania; and* [2]*Institute of Physics and Nuclear Engineering, Section 6, Bucharest, Romania*

ABSTRACT

Uranium distributions and contents in *Mysidacea* species were investigated in order to be subsequently correlated with some specific properties of these crustacea. The fission track micromapping technique was used in uranium distribution and content measurements. The investigated biological samples were: (1) *Javanisomysis gutzui* Bacescu, 1992 and (2) *Mesopodopsis slabberi* Van Beneden, 1861. The determined contents varied in the range of natural background values of uranium.

Index Entries: Crustacea of *Mysidacea* species; uranium; biological samples; fission track micromapping technique.

INTRODUCTION

Some of *Mysidacea* species crustacea show scintillation properties that could be correlated with their uranium content chemically bound to the organic matter, as well as the presence of uranium inclusions. In this work, only the uranium distribution and its content were investigated by the fission track micromapping technique.

The investigated biological samples were: (1) *Javanisomysis gutzui* Bacescu, 1992 (Fig. 1) and (2) *Mesopodopsis slabberi* Van Beneden, 1861 (Fig. 2). These crustacea specimens were sampled: (1) in 1991, in Nord

*Author to whom all correspondence and reprint requests should be addressed.

 © 1994 by Humana Press Inc.

Fig. 1. *Javanisomysis gutzui* Bacescu, 1992. Real size: 3.5 mm.

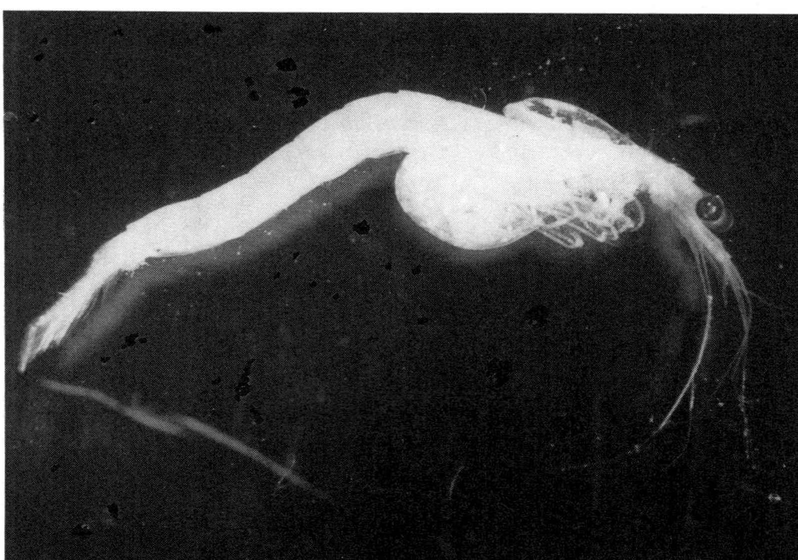

Fig. 2. *Mesopodopsis slabberi* Van Beneden, 1861. Real size: 7 mm.

West Sea, 35 km offshore Djakarta near Pari island, 0.5 m depth, and (2) in 1949, from the Black Sea shore at the mouth of the Saint George Branch of Danube River delta, 30–50 m offshore, 2–3 m depth. Immediately after sampling, the crustacea were put in ethyl alcohol, 72%.

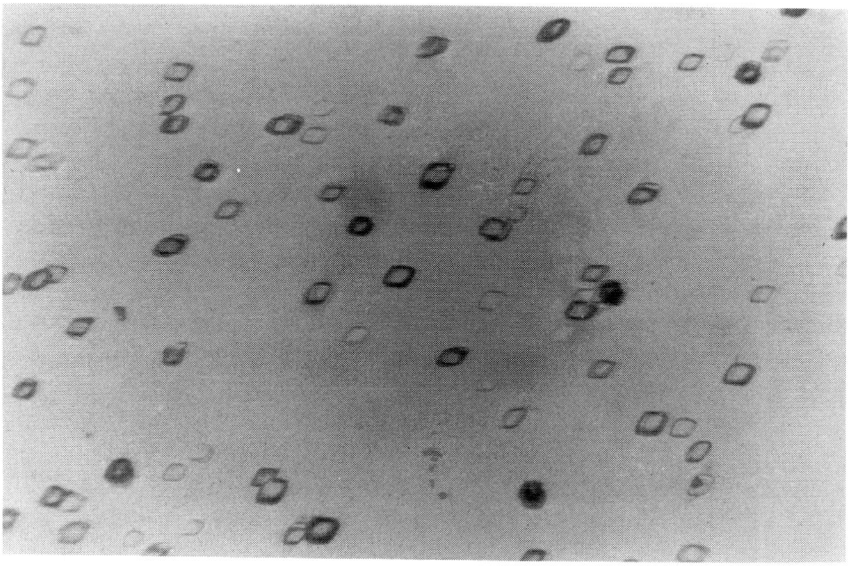

Fig. 3. Uranium track micromapping details in a muscovite detector for *Javanisomysis gutzui* Bacescu, 1992 (x600).

URANIUM TRACK MICROMAPPING METHODS

The track micromapping technique described by Fleischer et al. *(1)* was used in the following experimental conditions:

- The mica-muscovite as track detector;
- Thermal neutron fluences of ≈ 1.5 10^{15} neutrons/cm^2; and,
- The etching conditions for the muscovite detector: for the fossil fission track—HF ≥ 38%, 54 h at room temperature (pre-etching), and for the induced fission tracks—HF ≥ 38%, 6 h at room temperature *(2)*.

The uranium track micromappings obtained for the two investigated crustacea species are shown in Figs. 3 and 4.

In order to determine the uranium distributions and its contents, these micromappings were studied by optical microscopy. A stereomicroscope Wild M7S was used in the whole study, and a binocular Zeiss-Jena microscope for details and uranium content determinations. The SL-1, IAEA, Vienna Seibersdorf Laboratory certified reference material was used for calibration in the content measurements.

RESULT AND DISCUSSIONS

The studies of these micromappings led us to the following conclusions: For all the investigated crustacea samples, a quite uniform uranium distribution was found, and any uranium inclusions were not

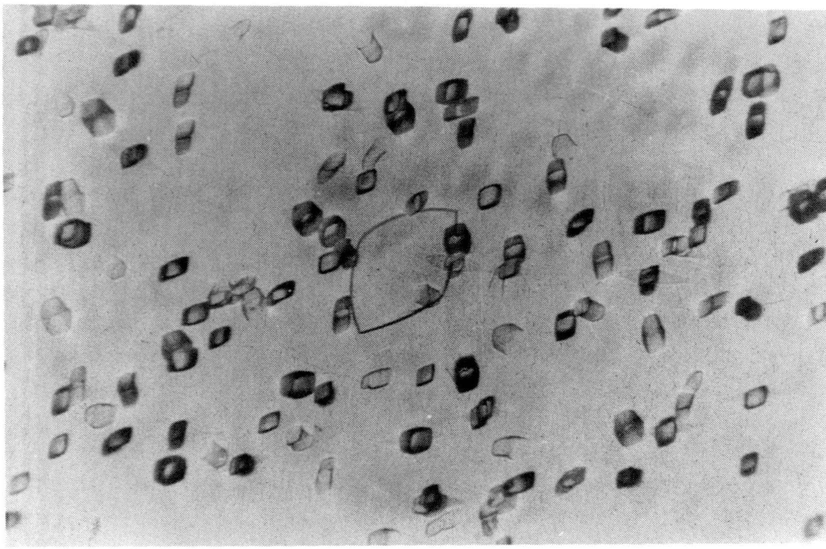

Fig. 4. Uranium track micromapping details in a muscovite detector for *Mesopodopsis slabberi* Van Beneden, 1861 (x600). The large track is the fossil uranium fission track, and the small tracks are induced fission tracks.

observed; and The uranium contents were found in the concentration ranges: (1.19 ± 0.18) μg/g ÷ (3.38 ± 0.51) μg/g for the *Javanisomymys gutzui* Bacescu, 1992 and (1.90 ± 0.29) μg/g ÷ (4.65 ± 0.70) μg/g for the *Mesopodopsis slabberi* Van Beneden, 1861.

The increased uranium concentrations in the Black Sea crustacea could be explained by the fact that these crustacea live on the sea floor, whereas the Java crustacea live as swimmers. It must be pointed out that our results are in agreement with those obtained for the other biological samples of the Black Sea Romanian shore *(3)*.

ACKNOWLEDGMENTS

The authors are indebted to Academician Professor Dr. Mihai Bacescu of the Natural History Museum of Bucharest, who gave us the samples and provide useful discussions.

REFERENCES

1. R. L. Fleischer, P. B. Price, and R. M. Walker, *Nuclear Tracks in Solids*, University of California Press, New York, 1975.
2. A. Danis, U and Th content measurements. Homologated Methods, IPNE, 433/1982.
3. I. I. Georgescu, M. Ciubotariu, and A. Danis, *Rapp. Comm. Int. Mer Médit.* **33,** 274 (1992).

True-Coincidence Correction When Using an LEPD for the Determination of the Lanthanides in the Environment via k_0-Based INAA

M. C. FREITAS*[,1] AND F. DE CORTE[†,2]

[1]*Instituto Nacional de Engenharia e Tecnologia Industrial (INETI), Instituto de Ciências e Engenharia Nucleares, Departamento de Química, P-2685 Sacavém, Portugal; and* [2]*Laboratory of Analytical Chemistry, Institute for Nuclear Sciences (INW), University of Gent, Proeftuinstraat 86, B-9000, Belgium*

ABSTRACT

As part of a recent study on the environmental effects caused by the operation of a coal-fired power station at Sines, Portugal, k_0-based instrumental neutron activation analysis (INAA) was used for the determination of the lanthanides (and also of tantalum and uranium) in plant leaves and lichens. In view of the accuracy and sensitivity of the determinations, it was advantageous to make use of a low-energy photon detector (LEPD). To begin with, in the present article, a survey is given of the former developments leading to user-friendly procedures for detection efficiency calibration of the LEPD and for correction for true-coincidence (cascade summing) effects. As a continuation of this, computer coincidence correction factors are now tabulated for the relevant low-energetic γ-rays of the analytically interesting lanthanide, tantalum, and uranium radionuclides. Also the 140.5-keV line of $^{99}Mo/^{99m}Tc$ is included, molybdenum being the comparator chosen when counting using an LEPD.

†Research Director of the Belgian National Fund for Scientific Research.

*Author to whom all correspondence and reprint requests should be addressed.

Biological Trace Element Research Editor: G. N. Schrauzer © 1994 by Humana Press Inc.

Index Entries: Activation analysis; true-coincidence effects; summing effects; low-energy photon detector; neutron activation analysis; detection efficiency; peak-to-total ratio.

INTRODUCTION

The desirability of extending the k_0-method to counting with an LEPD became especially clear from a number of recent applications, such as the panoramic analysis of silica, the determination of U in materials for fission-track and thermoluminescence dating, and the determination of lanthanides in environmental samples *(1)*. As part of a recent study of the environmental effects caused by the operation of a coal-fired power station at Sines, Portugal, k_0-based instrumental neutron activation analysis (INAA) with counting using an LEPD was used for the determination of the lanthanides, tantalum, and uranium in plant leaves and lichens. One of the parameters to be considered is the correction factor (denoted in the present work as COI) accounting for true-coincidence (cascade summing) effects. For the full-energy peak area N_p of a measured γ-line A (underlined), the following holds *(2,3)*:

$$N_{p,\underline{A}}(\text{corrected}) = N_{p,\underline{A}}(\text{observed})/\text{COI}_{\underline{A}} \quad (1)$$

This problem was already solved for counting with a large Ge detector, where γ-γ and γ-KX coincidence summing was taken into account. The extension to an LEPD necessitates that one consider the LX lines originating from internal conversion or electron capture. Previous publications *(2,4,5)* dealt with the determination of the full-energy peak detection efficiency and the total efficiency *(4)*, the fundamental principles of calculating COI factors *(2)*, and the computer codes developed *(5)*. In this article, a survey is given of these developments. Next, COI factors are computed and presented for the relevant low-energetic γ-rays *(6)* of the analytically interesting lanthanide, tantalum, and uranium radionuclides, and also for the 140.5-keV line of ^{99}Mo/^{99m}Tc (molybdenum being the comparator chosen when counting using an LEPD); the tabulated values shown in this work were obtained for point sources counted using the LEPD of INETI, Sacavém at useful source-detector distances.

SURVEY OF DEVELOPMENTS REGARDING THE DETECTION EFFICIENCY AND THE TRUE-COINCIDENCE CORRECTION

The use of an LEPD in k_0-standardized NAA requires the knowledge of the same type of detector characteristics as the use of a large Ge detector: the full-energy peak detection efficiency ε_p and the total detection efficiency ε_t, obtained as $\varepsilon_t = \varepsilon_p/(P/T)$, where P/T is the peak-to-total ratio. The parameter ε_p appears in the basic equation of the k_0-method for concentration calculation, and also in the correction for summing in by γ-γ coincidence. ε_t appears in the correction for summing out.

The efficiency calibration has been studied by using two different LEPDs at the INW, Gent and the INETI, Sacavém *(4)*. To determine the "reference" ε_p curve, use was made of absolutely calibrated and home-made point sources measured at a large distance from the detector, and polynomials were fitted to the experimental points. The accuracy was estimated at ~1–2% for higher energies (>11.1 keV) and ~5% for lower energies (<11.1 keV). The reference curve is the basis for calculation of the efficiency for actual geometries where extended samples are placed at any distance from the LEPD. In this calculation, use was made of the concept of effective solid angles as calculated with the SOLANG program *(7)*. The accuracy of the efficiency conversion, experimentally checked by measuring some coincidence-free γ-rays of point and cylinder sources on top of the LEPD and at the reference distance, was found to be better than 5%. The P/T ratio was experimentally determined using coincidence-free sources, and an interesting conclusion that came out is that P/T is not significantly dependent on the source-LEPD separation. In order to calculate coincidence correction factors when measuring using an LEPD, algorithms were developed *(2)* for the description of the summing-out probabilities taking into account γ-LX in addition to γ-γ and γ-KX true coincidences, where KX and LX can originate from internal conversion and electron capture; cascades with up to 5 components were considered. Three main computer codes were developed *(5)*: CADEIA, which detects both summing-out and summing-in cascades for any given γ in a decay scheme; REJECT, which classifies the cascades detected by CADEIA as relevant or irrelevant, following the criteria of De Corte *(3)*; and COINLEPD, which calculates the COI factors taking into account the cascades considered as relevant by REJECT. An up-to-date file of relevant nuclear data was created at the INETI *(8)*, including γ energies, absolute γ intensities, branching ratios (total; K; L_1; L_2; L_3), internal conversion coefficients, electron capture intensities, (K; L_1; L_2; L_3), electron capture probabilities, and weighted averages of the energies of KX- and LX-rays emitted after a primary K, L_1, L_2, or L_3 vacancy. An experimental check of the computed COI factors was performed for the 106.1-keV line of ^{239}Np and the 208.4-keV line of ^{177}Lu *(2)*. At the INETI/Sacavém, point sources were measured at distances of 2.09 and 16.28 cm from the LEPD (at the latter distance, coincidence effects are totally negligible). ^{239}Np was also measured as a 0.4-mL solution (6*M* HNO_3) in a small cylindrical polyethylene tube positioned at 2.09 cm from the LEPD. The ratios of the detection efficiencies from a point at 16.28 cm to a point or a cylinder at 2.09 cm were calculated with the program SOLANG *(7)*. The experimentally obtained COI values (2 s uncertainties) agree quite well with the calculated ones:

Table 1
Summing Effect Correction Factors (COI Factors) for the Interesting γ-Rays of Point Sources Measured at Different Distances from the LEPD at the INETI, Sacavém

Nuclide	E_γ, keV	2.09 cm	4.05 cm	7.17 cm	10.23 cm	13.15 cm	16.28 cm
Mo-99/ Tc-99m	140.5	0.995	1.000	1.000	1.000	1.000	1.000
La-140	328.8	0.999	1.000	1.000	1.000	1.000	1.000
Ce-141	145.4	1.000	1.000	1.000	1.000	1.000	1.000
Nd-147	91.1	0.999	1.000	1.000	1.000	1.000	1.000
Sm-153	69.7	0.926	0.976	0.992	0.996	0.998	0.998
	103.2	0.967	0.989	0.996	0.998	0.999	1.000
Eu-152	121.8	0.831	0.952	0.980	0.995	0.997	0.999
	244.7	0.767	0.918	0.973	0.988	0.993	0.995
	344.3	0.999	1.000	1.000	1.000	1.000	1.000
Gd-153	97.4	0.853	0.953	0.984	0.992	0.995	0.997
	103.2	0.914	0.973	0.991	0.995	0.997	0.998
Tb-160	86.8	0.974	0.994	0.998	0.999	0.999	0.999
	298.6	0.976	0.992	0.997	0.999	0.999	0.999
Dy-165	94.7	1.000	1.000	1.000	1.000	1.000	1.000
Tm-170	84.3	1.000	1.000	1.000	1.000	1.000	1.000
Yb-169	63.1	0.870	0.956	0.985	0.993	0.996	0.997
	130.5	0.883	0.967	0.988	0.993	0.996	0.997
	177.2	0.829	0.941	0.980	0.990	0.994	0.996
Lu-177	208.4	0.924	0.975	0.991	0.996	0.997	0.998
Ta-182	67.8	0.933	0.978	0.993	0.996	0.998	0.998
	100.1	0.728	0.902	0.967	0.984	0.991	0.994
	152.4	0.922	0.974	0.991	0.996	0.997	0.998
	222.1	0.953	0.984	0.995	0.997	0.998	0.999
Np-239	106.1	0.927	0.977	0.992	0.996	0.997	0.998

Table 2
Characteristics of the LEPD γ-Ray Spectrometer System Used at the INETI, Sacavém *(2,9)*

Dimensions	Resolution FWHM (keV)	Pre-amplifier	Amplifier	MCA	Dead-time correction
Diameter: 2.55 cm Height: 1.50 cm Active area: 5 cm^2 Active volume: 7.7 cm^3	at 4 μs ampl. time constant 320 eV (5.9 keV) 570 eV (122 keV)	Canberra RC	Canberra 2020 Ampli/Pur	Canberra ACCUSPEC	Live-time circuit of MCB

$$
\begin{aligned}
{}^{239}\mathrm{Np}\ 106.1\ \mathrm{keV} - [\mathrm{COI}_{\mathrm{point},\ 2.09\ \mathrm{cm}}]&: \mathrm{exp.} = 0.921 \pm 0.018\\
&: \mathrm{calc.} = 0.927\\
- [\mathrm{COI}_{\mathrm{cyl.},\ 2.09\ \mathrm{cm}}]&: \mathrm{exp.} = 0.920 \pm 0.024\\
&: \mathrm{calc.} = 0.933_5\\
{}^{177}\mathrm{Lu}\ 208.4\ \mathrm{keV} - [\mathrm{COI}_{\mathrm{point},\ 2.09\ \mathrm{cm}}]&: \mathrm{exp.} = 0.940 \pm 0.038\\
&: \mathrm{calc.} = 0.924 \qquad (2)
\end{aligned}
$$

TABULATION OF CALCULATED COINCIDENCE CORRECTION FACTORS

Table 1 shows the calculated COI factors of point sources measured at distances from ~2 up to ~16 cm from the LEPD (INETI, Sacavém). The 140.5-keV γ-ray is emitted by both ^{99}Mo and ^{99m}Tc. For ^{99m}Tc, there are no summing effects. After reaching mother-daughter equilibrium ($t_d \geq 60$ h), the COI factor is given by *(3)*:

$$COI\ (140.5\ keV) = [0.969\gamma_{140.5}\ (Tc) + \gamma_{140.5}\ (Mo) \cdot COI_{140.5}\ (Mo)] / [0.969\gamma_{140.5}\ (Tc) + \gamma_{140.5}\ (Mo)] \qquad (3)$$

where $0.969 = F_2T_{Mo}/(T_{Mo} - T_{Tc})$ is the ^{99m}Tc/^{99}Mo activity ratio. The values shown in Table 1 are valid for the LEPD, that have the characteristics given in Table 2. For other LEPDs, these values can be considered only as indicative.

As observed in Table 1, the true-coincidence correction to be introduced when counting using the LEPD (INETI, Sacavém) can exceed 10% (even ~30% in a few cases). This confirms the relevancy of the true-coincidence correction in INAA with counting using an LEPD. On the other hand, an interesting conclusion is that there is, fortunately, only a negligible summing effect for the comparator ^{99}Mo/^{99m}Tc.

Calculated COI factors were used to correct the measurement of the lanthanides, tantalum, and uranium in reference materials, plant leaves, and lichens via k_0-based INAA using the LEPD *(9)*.

ACKNOWLEDGMENTS

The contribution and advice of W. Bambynek (CBNM, Geel) concerning the orbital electron capture probabilities are gratefully acknowledged. Thanks are also owed to the Belgian National Fund for Scientific Research.

REFERENCES

1. F. De Corte, A. Simonits, F. Bellemans, M. C. Freitas, S. Jovanovič, B. Smodiš, G. Erdtmann, H. Petri, and A. De Wispelaere, *J. Radioanal. Nucl. Chem., Articles* **169,** 125–158 (1993).
2. F. De Corte and M. C. Freitas, *J. Radioanal. Nucl. Chem. Articles* **160,** 253–267 (1992).
3. F. De Corte, The k_0-standardization Method: a Move to the Optimization to Neutron Activation Analysis, habilitation thesis, University of Gent (1987).
4. F. De Corte, M. C. Freitas, and A. De Wispelaere, *Nucl. Instr. Methods Phys. Res.* **A299,** 335–343 (1990).
5. M. C. Freitas and F. De Corte, Proceed. Int. k_0 Users Workshop-Gent, Sept. 30–Oct. 2, 1992, Gent, Belgium, pp. 63–67.
6. J. Hertogen and R. Gijbels, *Anal. Chim. Acta* **56,** 61–82 (1971).
7. L. Moens, J. De Donder, L. Xilei, F. De Corte, A. De Wispelacre, A. Simonits, and J. Hoste, *Nucl. Instr. Methods* **187,** 451–472 (1981).
8. M. C. Freitas, Ph.D. thesis, University of Gent (1993).
9. F. De Corte, A. Simonits, Lin Xilei, and M. C. Freitas, *Biol. Trace Element Res.* this issue.

The Influence of Scattering on the Total Efficiency for Point Sources in γ-Ray Spectrometry

M. KORUN* AND R. MARTINČIČ

J. Stefan Institute, Ljubljana, Slovenia

ABSTRACT

The contribution of γ-ray scattering to the total efficiency for point-source geometries in a γ spectrometer was measured. Scattering of γ-rays on the shield and in the sample was considered. For a 25% efficiency detector in a small shield, about a 1% contribution to the total efficiency was measured, whereas for samples of a few hundred grams, these contributions reached 1–2%. Simple models for description of the effects are presented.

Index Entries: γ Spectrometry; peak efficiency; total efficiency; γ-ray scattering; voluminous samples.

INTRODUCTION

To measure the activity of a sample by γ-ray spectrometry, the probability for total absorption of a γ-ray in the sensitive volume of the detector has to be known. This probability, called also the peak efficiency, depends on the properties of the detector, on the dimensions and the shape of the sample, on the sample material, and on the γ-ray energy. If the decay scheme of the daughter nucleus is sufficiently simple, only one γ-ray is emitted in the de-excitation process of the nucleus. Then the activity of the sample is proportional to the count rate in the measured peak. The proportionality factor is the product of the probability for the emission of the γ-ray and the peak efficiency. In the case of elaborate decay schemes, several γ-rays may be emitted in a time span shorter than the rise time of the detector. Therefore, the spectrometer detects only the sum of all the γ-ray energies. If any of the detected γ-rays only partially

*Author to whom all correspondence and reprint requests should be addressed.

absorb in the detector's sensitive volume, the cascade will not be registered in any of the peaks in the spectrum. Therefore, to calculate the activity in the presence of coincidences, as well as the peak efficiency, the total efficiency of the detector also has to be known.

The probability for the simultaneous detection of several emitted γ-rays belonging to the decay of a single nucleus is given by the product of the probabilities for detection of the individual γ-rays disregarding angular correlations. To calculate such probabilities for voluminous sources, contributions from separate volume elements must be summed *(1)*. Since the probabilities for volume elements are given by point-source efficiencies, the count rates in the peaks are described by integrals over the source volume of the products of total and peak point-source efficiencies.

The influence of the sample material on the peak efficiency is usually taken into account by calculating the self-absorption in the sample. On the other hand, the total efficiencies are also influenced by γ scattering in the sample and on the shield of the spectrometer. It is the purpose of this article to evaluate the importance of these effects and to show how they may be treated to provide realistic estimates for total point-source efficiencies.

THE MODEL

In the present approach, the calculations are made for cylindrical samples placed on the detector so that the symmetry axis of the detector coincides with the axis of the sample. The following processes that contribute to the total efficiency are considered: the detection of unscattered γ-rays, the detection of γ-rays scattered on the walls of the spectrometer shield, and the detection of γ-rays scattered in the sample. It is assumed that those contributions are independent. Therefore, such effects as the influence of the sample on the γ-rays scattered from the shield are neglected. In this approximation, the total efficiency may be written as a sum of three contributions:

$$\varepsilon_T(\mu,E,\vec{r}) = \varepsilon_{T0}(\mu,E,\vec{r}) + \varepsilon_{SH}(E,\vec{r}) + \varepsilon_{SA}(E,\vec{r}) \tag{1}$$

$\varepsilon_{T0}(\mu,E,\vec{r})$ denotes the total efficiency of a point source placed at the point $\vec{r}$, emitting γ-rays with energy E and embedded in an absorbing medium with attenuation coefficient μ. It is assumed that the whole space between the point source and the surface of the detector is filled with the absorbing medium. The total efficiency of the point source is approximated by the product of the peak efficiency of the point source embedded in the absorbing medium $\varepsilon_P(\mu,E,\vec{r})$ and the total-to-peak ratio, estimated without attenuation

$$\varepsilon_{T0}(\mu,E,\vec{r}) = \varepsilon_P(\mu,E,\vec{r}) \cdot \varepsilon_{T0}(E,\vec{r})/\varepsilon_{P0}(E,\vec{r}) \tag{2}$$

provided that the total-to-peak ratio is not influenced by the absorbing medium. The attenuation causes an alteration of the distribution of γ-rays impinging on the sensitive area. The fact that this alteration influences the total efficiency differently from the peak efficiency is therefore neglected. The peak efficiency in the presence of attenuation is calculated from the peak efficiency without attenuation $\varepsilon_{P0}(E,\vec{r})$ using a suitable detector model *(2)*. This is a numerically demanding operation, and it is easier to do it on just the peak efficiency rather than on the peak and total efficiency. Also, the total-to-peak efficiency exhibits a smoother spatial dependence than the total efficiency and is therefore easier to calculate.

$\varepsilon_{SH}(E,\vec{r})$ is the probability of detecting γ-rays scattered on the shield. This probability is measured for each detector–shield combination.

$\varepsilon_{SA}(E,\vec{r})$ is the probability for detection of γ-rays scattered in the sample. In the present approach, this probability is estimated by decomposing the scattering process into two independent steps, referring to the probability of scattering and the probability for the detection of scattered radiation:

$$\varepsilon_{SA}(E,\vec{r}) = P_{SC}(E,H,R) \cdot P_{DET}\,(E_{SC}(E),r_{eff}(R),h_{eff}(H),\vec{r}) \qquad (3)$$

$P_{SC}(E,H,R)$ represents the probability of scattering of γ-rays emitted in the sample. H and R denote the thickness and radius of the sample. The probability is calculated by

$$P_{SC}(E,H,R) = [\mu_{sc}(E)/\mu_{tot}(E)]\,(1 - e^{-\mu_{tot}(E)(RH/(R\,+\,H))}) \qquad (4)$$

The second factor represents the probability that the γ-ray interacts in the sample. No spatial dependence is supposed for that probability; instead the quantity $RH/(R + H)$ is assumed as the characteristic length, transversed by the γ-ray in the sample. The first factor is the probability that the γ-ray is scattered and not absorbed. $P_{DET}(E_{SC},r_{eff},h_{eff},\vec{r})$ represents the probability of detecting the scattered radiation. r_{eff} and h_{eff} denote the average radial and axial coordinates weighted by the peak efficiency. They are calculated as

$$h_{eff} = \int_{h_0}^{h_1} h\varepsilon_{P0}(E,0,h)dh \Big/ \int_{h_0}^{h_1} \varepsilon_{P0}(E,0,h)dh \qquad (5)$$

and

$$r_{eff} = \int_{0}^{R_s} r\varepsilon_{P0}(E,r,h_{eff})rdr \Big/ \int_{0}^{R_s} \varepsilon_{P0}(E,r,h_{eff})rdr \qquad (6)$$

Here h_0 denotes the distance between the detector and the near side of the sample, h_1 the distance between the detector and the opposite side of the sample, and R_s its radius. Those averages are used, since $\varepsilon_{P0}(E,r,h)$ is approximated by analytical expressions *(2)*, so the integrations are done analytically. It should be noted that the average coordinates were evaluated at the incident energy E and not at the energy of the scattered γ-ray E_{SC}. This is because the energy of the scattered γ-ray may be calculated after the coordinates of interactions have been estimated. The energy of the scattered γ-ray is calculated assuming Compton scattering. The probability for the detection of the scattered radiation is then the total efficiency evaluated at the point $\vec{r}_{\text{eff}}$ for the energy E_{SC} for γ-rays emitted in the region $r^2 + h^2 < r_{\text{eff}}^2 + h_{\text{eff}}^2$

$$P_{DET}(E_{SC},r_{eff},h_{eff},r) = \varepsilon_{T0}(E_{SC},r_{eff},h_{eff}) \quad (7)$$

and for $r^2 + h^2 > r_{\text{eff}}^2 + h_{\text{eff}}^2$

$$P_{DET}(E_{SC},r_{eff},h_{eff},r) = \varepsilon_{T0}(E_{SC},r_{eff},h_{eff}) \exp[-\mu_{tot}(E) [(r_{eff} - r)^2 + (h_{eff} - h)^2]^{1/2}] (r_{eff}^2 + h_{eff}^2)/(r^2 + h^2) \quad (8)$$

The last two factors represent the main spatial dependence of the probability for detecting the scattered radiation.

EXPERIMENT

For the measurements of point-source efficiencies, calibrated sources from LMRI sets ECGFA and EGA have been used. Count rates in the spectra varied between several hundred and several thousand counts per second. The effects of dead time and pile-up were controlled by the pulser method *(3,4)*. In all calculations, the contribution of the background has been accounted for.

The spatial dependence of the point-source efficiencies $\varepsilon_{T0}(E,\vec{r})$ and $\varepsilon_{P0}(E,\vec{r})$ were measured in a grid of points covering the space where the sample is placed *(2)*. The measurements were repeated with several sources emitting γ-rays at different energies to measure the energy dependencies. From the ratio of these efficiencies, the total-to-peak ratio used for the total efficiency calculations is obtained.

The probability for detection of γ-rays scattered on the detector shield is presented for an intrinsic Ge detector with relative efficiency of 25%. The measurements were done with three shields. All shields were rather small to emphasize the effect of the scattering. The detector crystal was situated approximately in the center of the shield cavity. Within each shield, the total efficiency was measured with the source in three positions: on the surface of the detector on its geometrical axis, on the surface 3 cm from the axis, and finally 4 cm from the detector on its axis. The first shield was an iron shield with a cavity volume of 2.2 dm^3; the second and the third were lead shields with cavities of 1.8 dm^3 and 1.3 dm^3. To determine the energy dependence, the total efficiency was measured at the seven energies of 60, 88, 168, 392, 662, 835, and 1115 keV. The probability for a detection of the radiation scattered on the shield was calculated as the difference in the total efficiencies measured with and without the shield. Since the result is obtained as a difference, it is rather uncertain. At the achieved accuracy, no systematic relationship could be extracted in order to determine the dependence of the probability on the source position. The energy dependence for all three shields is presented in Fig. 1. The points in the figure are obtained as averages of probabilities in all three positions.

The probability for detection of γ-rays scattered in the sample was measured with aqueous samples in two geometries. In all measurements,

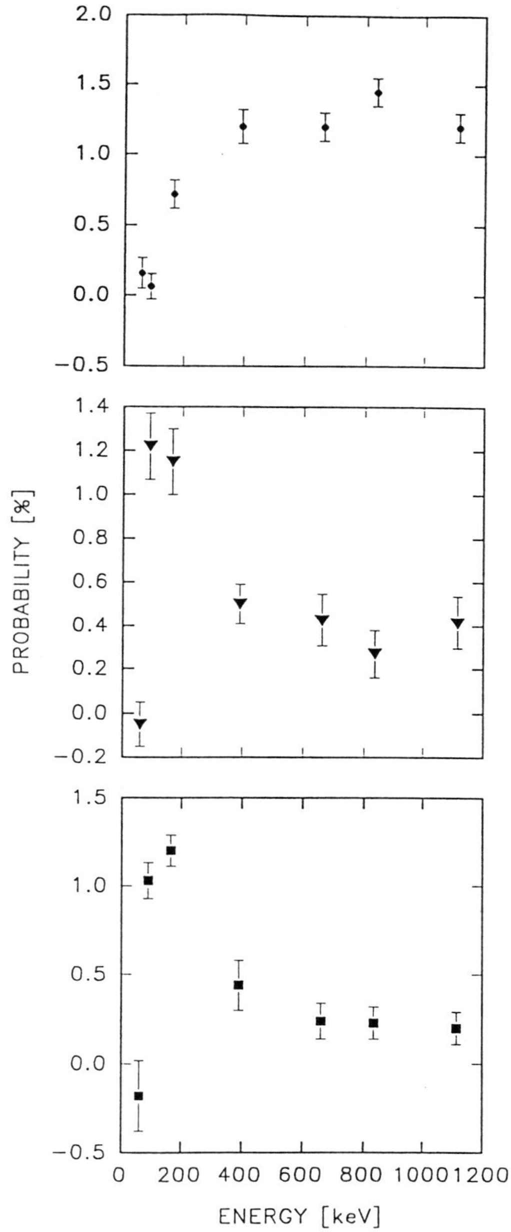

Fig. 1. The measured energy dependence of the probability for radiation scattered on the shield. Full circles designate the data for an iron shield, and triangles and squares the data for the lead shields with 1.3 and 1.8 dm^3 vol, respectively.

the peak and total efficiencies were measured. The diameter of the samples in both geometries was 9 cm, and the thicknesses were 2 and 4 cm. Measurements were done with a ^{137}Cs source in six positions: with the source on the detector axis and at 3 cm from the axis at the distances 0, 1 and 2 cm from the detector. The probability for the detection of scattered γ-rays is obtained according to Eq. 1 and Eq. 2 by

$$\varepsilon_{SA}(E,\vec{r}) = \varepsilon_T(\mu,E,\vec{r}) - \varepsilon_p(\mu,E,\vec{r})\,[\varepsilon_{T0}(E,\vec{r})/\varepsilon_{P0}(E,\vec{r})] \quad (9)$$

Similarly to the case of the probability for the detection of γ-rays scattered on the shield, the probability is obtained as a difference between two larger quantities. Consequently, also in this case, the uncertainties of the results are amplified. The measured probabilities are presented in Table 1.

DISCUSSION

The measurements of the probability for detecting scattered radiation were performed with an iron and two lead shields. Usually, shields are made of lead with inner layers made of lighter materials. So it is expected that in small shields, this probability lies between the values measured and is roughly 1%. Based on geometrical considerations, it is anticipated that this probability is proportional to $(V_{det}/V_{sh})^{2/3}$ where V_{det} and V_{sh} denote the volumes of the sensitive area of the detector and shield cavity. Consequently, in shields with cavities of 10 L, this probability should be on the order of some tenths of a percent.

The measured energy dependence may be qualitatively explained by back scattering and, in the case of the lead shield, also by fluorescence. Assuming a spherical cavity with an isotropic detector and the source placed at its center, the probability of detecting a back scattered photon with incident energy E is

$$P_{back}(E) = [1/\sigma_{tot}(E)]\,[d\sigma(E,180°)/d\Omega]\,\Delta\Omega\,P(E_b)\,[\mu(E)/(\mu(E) + \mu(E_b))] \quad (10)$$

σ_{tot} denotes the total scattering cross-section, $d\sigma/d\Omega$ (180°) the differential cross-section for back scattering of photons, $\Delta\Omega$ the solid angle subtended by the active volume from a point on the cavity surface, and $P(E_b)$ the total probability for detecting the photon if it impinges on the active volume. The product $P(E_b)\Delta\Omega$ resembles the total efficiency for photons emitted from the cavity surface multiplied by 4π. The probability for detecting fluorescent radiation excited by photons with energy E may be expressed as

$$P_{fluo}(E) = \omega\,[\sigma_{k-\mathrm{shell}}(E)/\sigma_{tot}(E)]\,[\mu(E)/(\mu(E) + \mu(E_x))]\,\varepsilon_r(E_T) \quad (11)$$

where ω denotes the fluorescent yield, $\sigma_{k-\mathrm{shell}}(E)$ the cross-section for the ionization of the K shell with photons of energy E, and $\varepsilon_T(E_x)$ the total efficiency for the detection of X-rays emitted from the surface. In Fig. 2, the values for P_{back} for iron are presented with the sum P_{back} + P_{fluo} for lead at the energies where the measurements were performed. The attenuation coefficients were obtained from *(5)*, the fluorescence

Table 1
The Measured Probabilities (in %) for the Detection of Radiation Scattered in a Sample as a Function of Sample Thickness and the Position of the Point Source Within the Sample[a]

Sample thickness	2 cm		4 cm	
Distance from the axis (cm)	0	3	0	3
Distance from the detector (cm)				
0	1.6±0.3	0.7±0.2	1.6±0.3	0.9±0.2
1	1.6±0.2	1.1±0.1	1.8±0.1	1.3±0.1
2	1.3±0.1	0.9±0.1	1.6±0.1	1.4±0.1

[a]The measurements were done with a ^{137}Cs source.

yield from *(6)*, and the cross-section for the K shell photo-ionization from *(7)*. The efficiencies used are the values at 5 cm from the detector on its axis. It may be seen from Fig. 2 that the described model characterizes well the behavior of the scattered radiation, except the decrease of the probability for back scattering on iron at high energies. In the experiment, this decrease with energy is probably compensated for by the effects of multiple scattering. The high value at the energy 88 keV in the probability for scattering on lead is owing to the efficiency for X-rays. Namely, at low energies, the efficiency for γ-rays emitted from points on the detector axis is larger than the efficiency for γ-rays emitted from other points.

The calculated probabilities for detecting radiation scattered on the sample are presented in Table 2. It may be seen that the measured values agree with the calculated ones within the estimated uncertainties, although on the average, the calculation seems to underestimate the effect. It may also be observed that the axial dependence of the probabilities is much smoother than the axial dependence of the total efficiency for the point sources. On the other hand, the radial dependence resembles much more the radial dependence of the total efficiency.

CONCLUSION

The influence of the scattering of γ-rays on the total efficiency of γ spectrometers was assessed. For scattering on the shield, no spatial dependence could be extracted from the measurements. The probability may reach values around 1% in small shields; for larger shields, the value

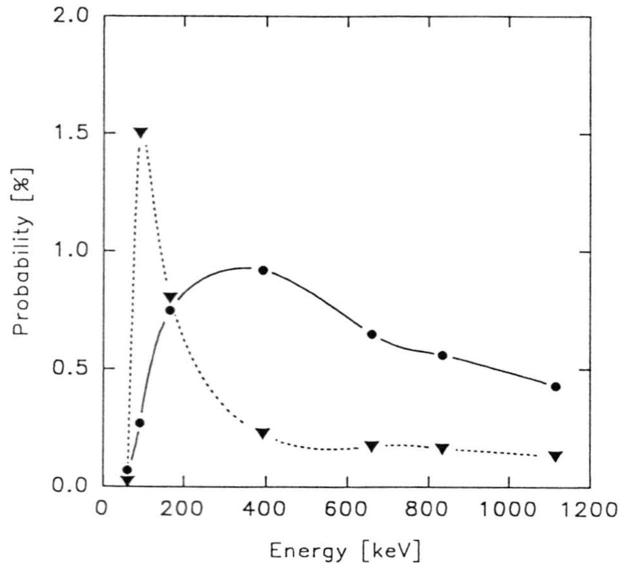

Fig. 2. The calculated energy dependence of the probability for radiation scattered on the shield. Circles designate the data for the iron shield, and triangles the data for the lead shield. The solid lines are drawn only to emphasize the energy dependence.

Table 2
The Calculated Probabilities (in %) for the Detection of Radiation Scattered in a Sample as a Function of Sample Thickness and the Position of the Point Source Within the Sample[a]

Sample thickness	2 cm		4 cm	
Distance from the axis (cm)	0	3	0	3
Distance from the detector (cm)				
0	1.4	1.0	1.6	1.2
1	1.3	0.8	1.5	1.2
2	1.3	0.6	1.3	0.9

[a]The calculations were done for the energy 662 keV.

is lower. For samples of some hundred grams, the contribution from scattering in the sample may amount to 1–2%. The total efficiency for point sources positioned close to a detector with 25% relative efficiency may reach 15–30%, but it falls rapidly and reaches values of some percent at distances of 4 cm. At these distances, scattering presents a considerable contribution to the total efficiency. It may be concluded that coincidence summing corrections for voluminous samples do not fall as much with the sample volume as one would expect, just from point-source efficiency considerations. All findings can be applied for γ-ray spectrometry of large samples of biological or environmental origin.

REFERENCES

1. K. Debertin and U. Schötzig, *Nucl. Instr. and Methods* **185,** 471–477 (1979).
2. M. Korun and R. Martinčič, *Appl. Radiat. Isot.* **43,** 29–35 (1992).
3. K. Debertin, *Atomkernenergie* **17,** 97–102 (1971).
4. F. H. Tenney, *Nucl. Instr. and Methods* **219,** 165–172 (1984).
5. J. H. Hubbell, *Photon Cross Sections, Attenuation Coefficients and Energy Absorption Coefficients from 10 keV to 100GeV*, NSRDS-NBS **29,** (1969).
6. E. Browne and R. B. Firestone, *Table of Radioactive Isotopes*, Wiley, New York, 1986.
7. J. H. Scofield, *Theoretical Photoionization Cross Sections from 1 to 1500 keV*, Lawrence Livermore Laboratory, UCRL-51325, 1973.

Distribution of Na, K, Rb, Cs, and ^{137}Cs in Some Austrian Higher Fungi

S. S. ISMAIL

Austrian Federal Food Control Institute (Linz) and Atom Institute of the Austrian Universities (Vienna)

ABSTRACT

The level of ^{137}Cs in 28 species of mushroom collected and analyzed between 1986 and 1992, after the Chernobyl accident, are reported. *Xerocomus badius* was found to accumulate ^{137}Cs (the median value of 58 samples is 3667 Bq/kg). The ^{137}Cs content of *Rozites caperata, Scleroderma vulgare,* and *Tylopilus felleus* exceeded 600 Bq/kg. Some of these samples collected in August 1992 were investigated for their ^{134}Cs and ^{137}Cs content as well as their Na, K, Rb, and Cs levels by applying instrumental neutron activation analysis (INAA). The results show a high correlation coefficient (0.80) between the contents of ^{137}Cs and Na, but none between ^{137}Cs and K.

Index Entries: ^{137}Cs; ^{134}Cs; mushroom; instrumental neutron activation analysis; sodium; potassium; rubidium; radioactive contamination; cesium.

INTRODUCTION

After the Chernobyl disaster in April 1986, large amounts of radioactive isotopes were distributed in the atmosphere and contaminated a large part of the European environment. In Austria, the ^{137}Cs level of aerosols was 0.011 mBq/m^3 in 1986 preceding the accident, and increased dramatically, about 800,000-fold to 9481 mBq/m^3 after the accident (April 30) *(1)*. The main activities found in Austria as reported *(2)* were because of ^{131}I (23%), ^{132}Te (31%), ^{132}I (26%), ^{103}Ru (5%), and ^{137}CS (about 4%). The ^{90}Sr values were about 1/20 of those of Cs. The main radioactive

*Author to whom all correspondence and reprint requests should be addressed.

Biological Trace Element Research Editor: G. N. Schrauzer © 1994 by Humana Press Inc.

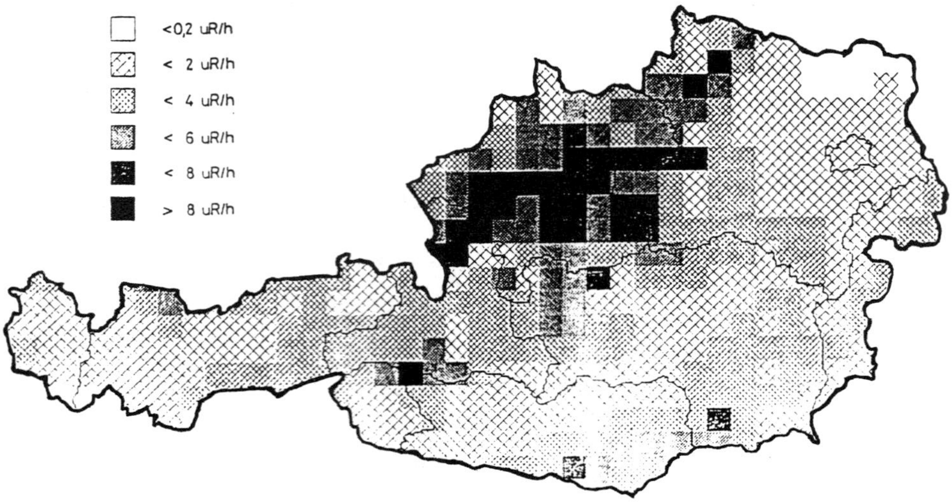

Fig. 1. Austrian external gamma dose rate 6 mo after Chernobyl.

nuclides, which owing to their long half-lives ($t_{1/2}$) cause an environmental impact, are ^{90}Sr ($t_{1/2}$ = 28.7 yr), ^{134}Cs ($t_{1/2}$ = 2.07 yr), and ^{137}Cs ($t_{1/2}$ = 30.2 yr). These nuclides contaminated Austrian soil according to the weather and rain conditions at that time (Fig. 1). One of the most contaminated areas was Upper Austria. Mushrooms are well known for their accumulation *(3)* of toxic and other elements (Cs) and could be very useful as bioindicators for radioactive contamination in certain environmental system.

EXPERIMENT

The samples were collected randomly from different areas in Upper Austria during the time between 1986 and 1991 for investigating the level of ^{137}Cs after the Chernobyl accident. The samples were directly cleaned from soil, washed, and prepared for low-level γ spectroscopy by homogenizing the total sample. About 250 g of each sample were weighed into a suitable precalibrated container. The samples were measured by means of a 30% HPGe detector for about 20,000 s. For the results of ^{137}Cs and ^{134}Cs, the counting statistics give an error of about 5%. In August, some samples were collected on the same day and from the same area. These samples were investigated for their values of ^{134}Cs and ^{137}Cs values, as well as for their content of Na, K, Rb, and Cs by means of INAA.

For INAA, the samples were dried at 70°C for 48 h. One hundred to 150 mg of each were sealed into a quartz tube and irradiated at a flux of 6×10^{13} n/s·cm^2 for 74 h. The samples were cooled for 24 h, and then measured by means of a 20% HPGe having a resolution of 1.5 keV for ^{60}Co (1332.5 keV). ^{24}Na and ^{42}K were measured for 20 min. For measuring ^{46}Sc, ^{57}Cr, ^{59}Fe, ^{60}Co, ^{65}Zn, ^{86}Rb, and ^{124}Sb, the samples were counted second time for 60 min after 14 d of decay. The data were cal-

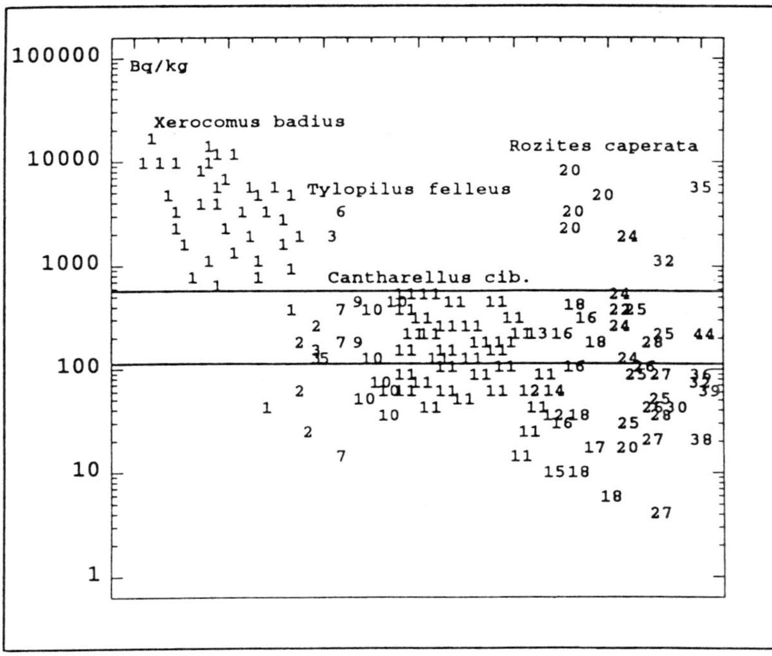

Fig. 2. ^{137}Cs in mushrooms (1986–1992).

culated using self-prepared standards as well as reference materials (i.e., SPINACH NBS-1570, TOMATO LEAVES NBS-1573, PINE-NEEDLES NBS-1575).

RESULTS AND DISCUSSION

Figure 2 shows the distribution of ^{137}Cs in 178 samples of 28 species. Clearly, *Xerocomus badius* (no. 1 in Fig. 2) is one of the most contaminated species with a median of 3667 Bq/kg (wet wt) for 58 samples. Other mushrooms containing a median of 600 Bq/kg up to contamination levels on the order of that of *Xerocomus badius* are *Rozites caperata* (3019 Bq/kg), *Scleroderma vulgare* (6059 Bq/kg), and *Tylopilus felleus* (2963 Bq/Kg), whereas *Cantharellus cibarius*, one of the most commonly consumed mushrooms, has a median value of 119 Bq/kg with a minimum of 14 and a maximum of 521 Bq/kg for 46 samples. Table 1 classifies these species according to their content of ^{137}Cs as highly contaminated species (600 Bq/kg), medium (110–600 Bq/kg), and slightly contaminated species (< 110 Bq/kg wet wt).

Figure 3 shows the distribution (mean and standard error) of ^{137}Cs of *Xerocomus badius* for the specimens collected in Upper Austria between 1986 and 1992. These data indicate that in the last 3 yr, the mean content of ^{137}Cs in *Xerocomus badius* was significantly lower (by several orders of magnitude) than the values obtained in 1986 (the samples are collected in different places). The decrease in ^{137}Cs could be the result not only of radioactive decay, and the transfer of ^{137}Cs from the surface to the deeper

Table 1
Classification of Mushrooms According to Their ^{137}Cs Content

Scientific name	No.[1]	Median (Bq/kg)	S.N[2]	Level
Agaricus arvensis	15	8	1	low
Agaricus bisporus	14	52	1	low
Agaricus campestris	13	25	1	low
Albatrellus ovinus	28	86	1	low
Amanita rubescens	18	28	6	low
Armillariella mellea	25	121	6	med.
Boletus edulis	9	262	6	med.
Boletus pinicola	10	69	10	low
Boletus erytropus	5	96	1	low
Cantharellus cibarius	11	119	46	med.
Clitocybe nebularis	39	50	1	low
Lactarius deterrimus	22	293	1	med.
Lactarius volemus	26	55	2	low
Leccinum scabrum	2	89	6	low
Leccinum testaceo-scabrum	16	129	4	med.
Leccinum versipelle	17	15	1	low
Macrolepiota rhacodes	27	25	3	low
Megacollybia platyphlla	44	333	1	med.
Rozites caperata	20	3019	6	high
Russula cyanoxantha	30	45	1	low
Russula ochroleuca	32	358	2	med.
Scleroderma vulgare	35	6059	1	high
Sparassis crispa	12	40	2	low
Suillus grevillei	7	208	1	med.
Tylopilus felleus	6	2963	1	high
Xerocomus badius	1	3667	58	high
Xerocomus chrysenteron	3	121	4	med.
Xerocomus subtomentosus	24	310	4	med.

1—Number of the mushroom in Fig. 2.
2—Number of analyzed samples.

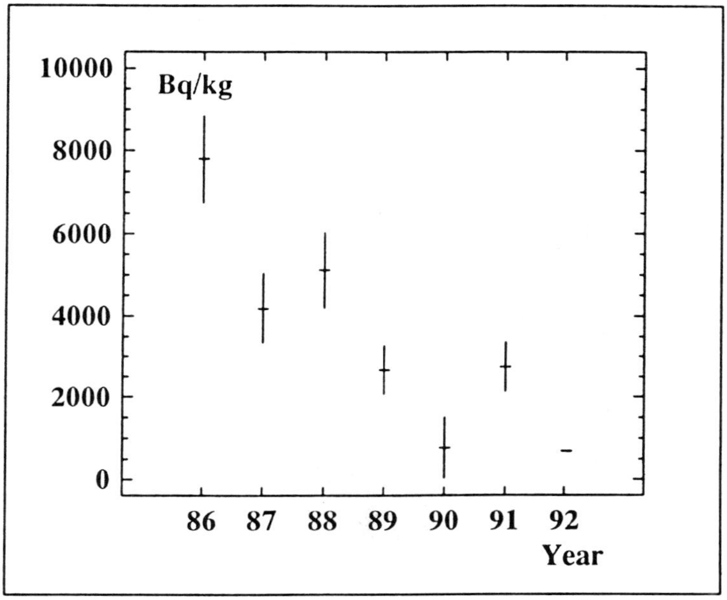

Fig. 3. Mean and standard error of ^{137}Cs in *Xerocomus badius*.

lying soil, thus decreasing the ^{137}Cs available to mushrooms, but also of the likelihood of mushrooms growing in the same place, which would decrease the rate of ^{137}Cs transfer to consecutive mushroom generations, provided the environmental conditions did not significantly change.

The higher levels 2 of ^{137}Cs for Cantharellus cibarius were found in the areas 7 and 14 (Ried and Wels), whereas in area 15 (Salzburg), some samples show higher than median values (119 Bq/kg). The frequency histogram for the distribution of ^{137}Cs in *Xerocomus badius* is shown in Fig. 4.

Table 2 shows the values of ^{134}Cs and ^{137}Cs for the samples collected in August 1992 only. The large variation in ^{137}Cs content of the fruit bodies of the different species collected from one area on the same day could be the result of the different ages of the analyzed species or the specific growing condition of each. On the other hand, the available ^{137}Cs depends on the pH, and on the amount and type of clays and organic materials in the soil, which could differ largely from one region to another. The $^{137}Cs/^{134}Cs$ ratio of the samples ranges from 13.15 up to 17.33. The higher values were found in *Boletus erythropus* (17.33), *Amanita rubescens* (16.54), and *Boletus edulis* (16.61). These higher values indicate contamination from old ^{137}Cs deposition from the nuclear weapons tests *(4)*. According to the reactor conditions at the time of the Chernobyl accident, the ratio of ^{137}Cs to ^{134}Cs was then 2, so this ratio would be about 13 in 1992.

Table 3 shows the data obtained by INAA for Na, K, Rb, and Cs (in dry wt samples). Na has values between 96 and 550 ppm, peak values

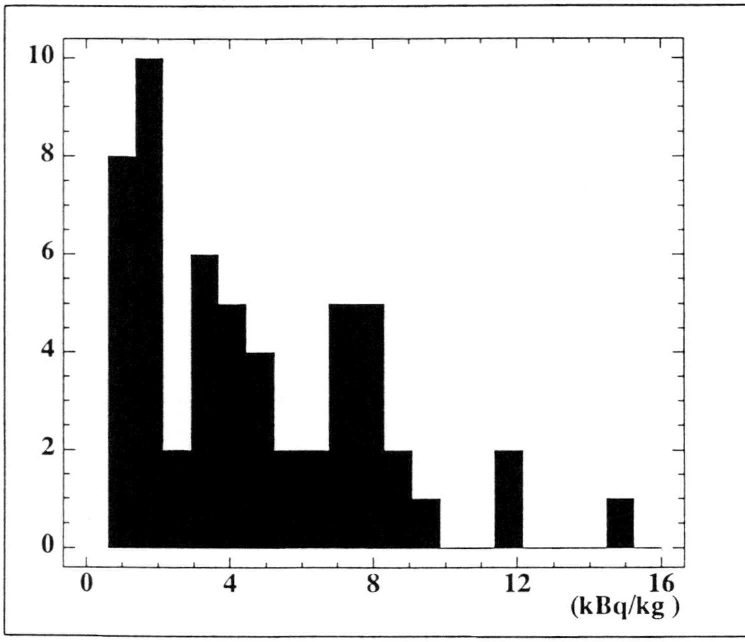

Fig. 4. Frequency histogram of ^{137}Cs in *Xerocomus badius.*

being found in *Xerocomus badius* and *Boletus edulis.* A high correlation coefficient in mushrooms was found for Na and ^{137}Cs (R = 0.8).

A similarly high correlation was found *(5)* between the available Na in soil and ^{137}Cs in grass (0.9). This could indicate that the available Na in soil has an effective role in the transfer of Cs to plants and mushrooms by decreasing the amount of ^{137}Cs absorbed to some clays in soil and thus, increasing the ^{137}Cs available to mushrooms. Maguire et al *(6)* found that less cesium was sorbed by bentonite and kaolinite, but more by illite when soaking the minerals for 7 d with 0.05 *M* NaCl.

K values ranged from 0.18 in *Amantia rubescens* up to 5% in *Xerocomus subtomentosus, Xerocomus badius,* and *Armillariella mellea*. The correlation coefficients for K and Rb with ^{137}Cs were 0.34 and 0.54, respectively. The values of Na, Rb, K, and ^{137}Cs (in dry wt) are represented in Fig. 5.

The levels of Cr, Sc, Fe, Co, Zn, and Sb in *Megacollybia platyphlla* were 134.1 ng/g, 0.57 ng/g, 1.22 µg/g, 41.3 µg/g, 11.3 ng/g, and 1.35 µg/g, respectively, whereas the corresponding values in *Russula ochroleuca* were 52.8 ng/g, 0.89 ng/g, 0.68 µg/g, 367 µg/g, 88.6 ng/g, and 0.03 µg/g. Thus, *Russula ochroleuca* contained eight times more Zn, but 45 times less Sb than *Megacollybia platyphlla*.

CONCLUSION

This study indicates that some mushroom species are contaminated, especially *Xerocomus badius*. The high correlation coefficient between Na

Table 2
Distribution of ^{134}Cs and ^{137}Cs in Mushrooms (wet wt)

No.	Scientific name	^{134}Cs (Bq/kg)	^{137}Cs (Bq/kg)	R
1	Russula ochroleuca	4.4	58.7	13.34
2	Megacollybia platyphlla	22.9	332.6	14.52
3	Xerocomus subtomentosus	8.7	118.0	13.56
4	Xerocomus badius	56.2	797.7	14.19
5	Boletus edulis	20.2	335.5	16.61
6	Armillariella mellea	4.5	68.3	15.17
7	Xerocomus chrysenteron	3.3	43.4	13.15
8	Amanita rubescens	9.5	157.2	16.54
9	Lactarius deliciosus	5.1	73.4	14.39
10	Boletus erythropus	8.0	138.7	17.33

Table 3
Results of INAA for Na, K, Rb and Cs in mushrooms (dry wt)

No	Scientific name	Na (μg/g)	K %	Rb (μg/g)	Cs (μg/g)	^{137}C Bq/kg
1	Russula ochroleuca	246	3.38	178	0.70	734
2	Megacollybia platyphlla	96	2.05	152	5.24	5234
3	Xerocomus subtomentosus	120	5.22	470	1.59	1895
4	Xerocomus badius	550	5.32	713	2.58	12784
5	Boletus edulis	522	3.60	568	9.90	9586
6	Armillariella mellea	97	5.33	420	1.63	813
7	Xerocomus chrysenteron	300	0.88	421	2.03	983
8	Amanita rubescens	226	0.18	74	3.60	3420
9	Lactarius deliciosus	194	0.33	422	4.65	953
10	Boletus erythropus	191	2.00	330	0.33	1928

and ^{137}Cs could indicate a certain role of Na in ^{137}Cs variability in different plants and mushrooms, an assumption that would need confirmation by investigation of a large number of samples from different types of soil. This work also shows that the combination of ^{137}Cs measurement by means of low-level γ spectroscopy and INAA gives important information about the bioavailability of radioactive elements in the environment.

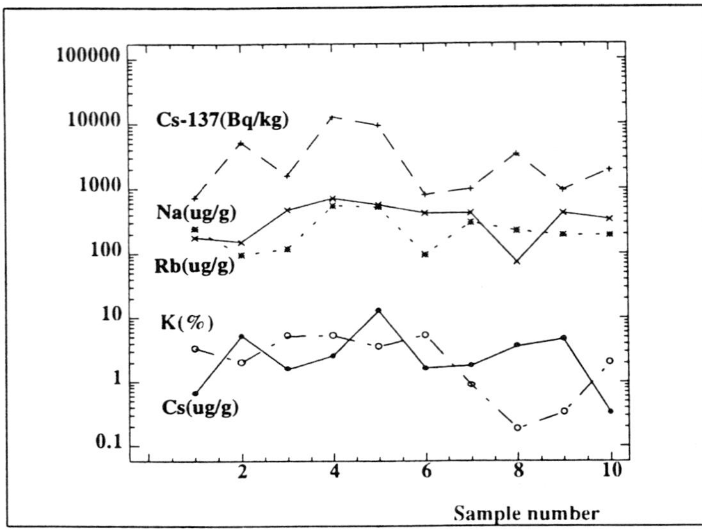

Fig. 5. Level of Na, K, Rb, Cs, and ^{137}Cs in mushrooms (1992).

ACKNOWLEDGMENT

I would like to thank Hofrat D I. Thalhamer, Director of Austrian Federal Food Control Institute Linz, Professor Dr. F. Grass, Atom Institute, Vienna, and my colleagues Dr. R. Konraff, J. Hemetzberger, and M. Bernreiter for their great help during this work.

REFERENCES

1. G. Heinrich., *Mitt. naturwiss. Ver. Steiermark* **117,** 7 (1987).
2. Österreichisches Bundeskanzleramt, Sektion VII, Report 2/88, 1988.
3. J. Horyna, Z. Řanda, and J. Benada, *Zeitschrift für Mykologie* **54,** 2, 179(1988).
4. G. Rückert and J. F. Diehl, *Lebensm. Unters. Forsch.* **185,** 91 (1987).
5. Österreichisches Bundesministerium für Gesundheit, Report "Beweissicherung Kernkraftwerk Temelin," 1993.
6. S. Maguire, I. D. Pulford, G. T. Cook, and A. B. Mackenzie, *J. Soil Sci.* **43,** 689 (1992).

Radioactive Contamination of Forests in Poland

J. W. MIETELSKI,* P. MACHARSKI, M. JASIŃSKA, AND R. BRODA

Henryk Niewodniczanski Institute of Nuclear Physics, PL -31-342 Krakow, Poland, Radzikowskiego 152 st.

ABSTRACT

Partial results of a large survey of radioactive contamination in Polish forests are presented. All measurements were performed with low-background γ spectrometers. The measurement method is briefly described. Activities of ^{137}Cs, ^{134}Cs, ^{125}Sb, ^{106}Ru, ^{144}Ce, ^{154}Eu, ^{155}Eu, and ^{60}Co in A_0 and A_1 layers of forest litter from all over Poland are determined. The geographical distribution of contamination as well as its origin is discussed. Nonuniform composition of the Chernobyl fallout over Poland is confirmed and documented.

Index Entries: Chernobyl; forest contaminations; geographical distribution; radiocesium; hot particles.

INTRODUCTION

Our study was within the Polish government program investigating the forest radioactive contamination, including the natural and artificial radioactivity, with particular emphasis on the Chernobyl component. The samples were collected in the autumn of 1991 by forest inspectors who followed the attached detailed instruction and provided us with 350 sets of samples containing two layers of forest litter: leaf (A_0) and humus (A_1), as well as leaves of blueberry, fruitbodies of *Xerocomus badius*, and other mushroom samples. The sampling covered evenly predominantly coniferous forests in Poland. The first γ-spectroscopic analysis was performed for *Xerocomus badius* samples contained in 280 sets. The results, presented in ref. *(1)*, allowed us to choose 127 "regions of interest" for

*Author to whom all correspondence and reprint requests should be addressed.

Biological Trace Element Research Editor: G. N. Schrauzer © 1994 by Humana Press Inc.

which the whole sets of samples were examined to detect radioactive γ emitters. The specific aims were:

1. Determination of the cesium isotope (^{137}Cs and ^{134}Cs) contamination levels, where the amount of ^{134}Cs enables us to possibly deduce the Chernobyl component;
2. Search for other artificial long-lived γ emitters;
3. Determination of natural radioactivity (mainly ^{40}K) in all samples examined;
4. Determination of cesium isotope transfer factors (T_f) from litter to blueberry and various species of mushrooms;
5. Testing some models of T_f dependence; and
6. Determination of average (relative to *Xerocomus badius*) ability to concentrate radiocesium by various mushroom species.

Since the work is still in progress, this article presents only results obtained for litter samples and consequently deals with problems quoted above as numbers 1 and 2. Another article in this volume by us concerns the transfer factor problems *(2)*.

METHOD

Measurements were performed through 1992 and in the first half of 1993. Two low-background γ spectrometers with HPGe or Ge(Li) detectors, shielded by 10- or 5-cm thick walls of lead with thin cadmium and copper additional shielding, were used. Measurement time of one sample was at least 16 h. Spectra obtained were evaluated using the P.I.M.P computer code *(3)*. Before measurements, each sample was dried for 24 h, homogenized, its density determined, and then placed into a polyethylene vial. Calibration of the spectrometers for various sample volumes and for a given peak energy (186, 605, 662, 796, and 1460 keV) were done using IAEA 154 Whey Powder *(4)* and IAEA Soil 6 *(5)* reference materials. Calibration for peaks that lie between those energies was done by linear interpolation in the space of $\ln(E) - \ln[\varepsilon(E, V)]$ (here E is peak energy and $\varepsilon(E, V)$ is efficiency for given sample volume). The difference in self-absorption between the samples and the standard was taken into account. The method of calibration was described in detail in another paper *(6)*. Additional calibration for various volumes and for low peak energies (105, 123, and 133 keV) was done using an especially prepared standard made of cellulose powder contaminated with known amounts of ^{154}Eu, ^{155}Eu, and ^{144}Ce.

All activities were decay corrected to September 1, 1991. The formula for the activity calculations was as follows:

$$A = [n \exp(\lambda d)\ \mu(\rho,x,E)]/[Y(E)\ \varepsilon(E,V)\ m] \quad (1)$$

where A = activity of sample in Bq/kg, n = number of counts/1 s in given peak area corrected for background peaks, λ = decay constant for

Table 1
Observed Artificial Radionuclides and Properties Used in the Analysis

radio-nuclide	$T_{1/2}$	Energy of γ line [keV]	intensity per decay [%]	Remarks
^{60}Co	5.27 a	1173.2 1332.5	99.9 100.0	resolved from 1167.9 keV (^{134}Cs)
^{106}Ru	368 d	621.9	10.6	intensity with weak 616.2 line
^{125}Sb	2.73 a	427.9 635.9	29.4 11.3	176 keV line also visible in spectra
^{134}Cs	2.02 a	604.7 795.8	97.6 94.1	resolved from 609.3 keV (^{214}Bi) corrected for 794.9 keV (^{228}Ac) intensity with 801.9 keV line
^{137}Cs	30.04a	661.6	85.2	
^{144}Ce	284 d	133.5	11.1	
^{154}Eu	8.59a	123.1	40.4	other lines (1275 & 722 kev) visible, but too large errors
^{155}Eu	4.68a	105.3	21.8	

a given isotope, d = number of days from September 1, 1991 to the day of analysis, $\mu(\rho,x,E)$ = experimentaly obtained function for self-absorption correction for different densities ρ, thickness of sample x, and energy of photon E, $Y(E)$ = intensity of a given peak in % (photons/100 decays), $\varepsilon(E, V)$ = spectrometer efficiency function value for given photon energy E and volume of sample V, and m = sample mass (kg). The list of observed artificial radionuclides is given in Table 1 along with some basic information used in the analysis.

RESULTS

In Table 2, ranges of activity for each artificial radionuclide observed in litter samples are presented. In all litter samples, ^{134}Cs and ^{137}Cs were observed. The geographical distribution of radiocesium contamination in soil samples from Poland has been published by a Warsaw group *(7)*. Our initial study *(1)*, based on measurements of mushrooms (*Xerocomus badius*), basically confirmed these results, although showing relatively higher cesium level in northeast Poland and a strong maximum of contamination in the Opole region (Silesia). The map of ^{137}Cs in litter upper layer, A_0 (Fig. 1A), obtained in this study looks similar to these results,

Table 2
Observed Artificial Radionuclides Activity Ranges in litter Samples from 127 Sites Corrected to September 1, 1991[a]

radio-nuclide	specific activity [Bq/kg d.m.]			
	upper layer - A_0		deeper layer - A_1	
	minimum	maximum	minimum	maximum
^{60}Co	< DL (83%)	4.3 ± 1.0	< DL (87%)	4.8 ± 2.0
^{106}Ru	< DL (67%)	59 ± 16	< DL (69%)	73 ± 17
^{125}Sb	< DL (23%)	73 ± 18	< DL (29%)	44 ± 18
^{134}Cs	4 ± 1	346 ± 13	0.8 ± 0.5	308 ± 18
^{137}Cs	30 ± 2	3320 ± 145	32 ± 4	2880 ± 185
^{144}Ce	< DL (86%)	89 ± 16	< DL (87%)	93 ± 15
^{154}Eu	< DL (90%)	7.9 ± 1.4	< DL (90%)	6.3 ± 1.2
^{155}Eu	< DL (92%)	11 ± 2.3	< DL (95%)	5.4 ± 1.7

[a]A percentage of sites with the activity of given isotope below our detection limit (DL) is presented in parentheses.

but the map of ^{137}Cs in the deeper layer, A_1 (Fig. 1B), does not show the characteristic maxima in Opole or in the Mazurian Lakes regions. Apparently in those regions, the main part of Chernobyl cesium is still present in the upper layer (A_0) of litter. Such conclusion is strongly supported by the fact that for all samples, the ^{134}Cs-to-^{137}Cs ratio was observed to be higher in the A_0 than in the A_1 layers, even if activities in some cases were higher in A_1 than in A_0.

In many samples of litter (both A_0 and A_1 layers), the ^{125}Sb and ^{106}Ru isotopes and traces of ^{60}Co were found to show some increase of activity values in the eastern regions of Poland. Even more significant was the observed inhomogenity in the distribution of the ^{144}Ce and accompanying traces of ^{154}Eu and ^{155}Eu radioactivity, which were found almost solely in northeast Poland.

CONCLUDING REMARKS

The predominant ^{137}Cs and ^{134}Cs activity of forest radioactive contamination could be resolved into the "old" and the Chernobyl origin component and geographically localized. Final conclusions concerning the behavior of both components and transfer factors into forest vegetation await the completion of our analysis. Here we shall present a few remarks regarding the other much less abundant radioisotopes. The detection of such relatively short-lived isotopes as ^{106}Ru and ^{144}Ce indicates very high contamination levels of these isotopes in 1986. In fact, an enormous (relative to cesium) activity for ^{144}Ce in a sample from very northeastern Poland has been reported *(8)*. All isotopes in Table 1 have been established as constituents of hot particles *(9–11)*, but now it has been found by appropriate checks that they are evenly distributed throughout the sample. This can be explained by disintegration of an original hot particle or by the presence in a sample of many small hot particles, i.e., the continous fallout in the limit. Since the radionuclides ^{125}Sb, ^{106}Ru, and ^{144}Ce were typical for the Chernobyl cloud and even *(12)* ^{60}Co was observed in the "continuous" (i.e., small aerosols) fraction of it, one can expect the latter explanation to be the true one. This may also hold for europium isotopes, which have never been reported as a constituent of the "continuous" fraction, but could earlier have been easily omitted because of their low activity and the low energy of their main γ lines, which would be hard to detect in the presence of many short-lived isotopes. In order to clarify this question, it is useful to compare the activity ratios in analyzed samples to those observed in the continuous fraction of the Chernobyl fallout, neglecting possible secondary processes in the litter, which could have affected the ratios during the last 5 yr. The ^{125}Sb, ^{106}Ru, and ^{144}Ce activity ratios to ^{137}Cs in litter samples are presented in Fig. 2 (A, B,and C, respectively). They include only samples with measured values above the detection limit, and all of them are decay corrected to May, 1, 1986. Those ratios in the Chernobyl cloud at Cracow *(13)* were roughly equal to 0.04, 0.5, and 0.02, respectively. The ^{125}Sb and ^{106}Ru to ^{137}Cs activity ratios in litter show values within the order of those mentioned above. However, some samples seem to show a higher amount of ^{106}Ru. In contrast, the observed ^{144}Ce-to-^{137}Cs activity ratio is much higher than in the "continuous" fraction in Cracow. Such a high ^{144}Ce-to-^{137}Cs ratio was observed in areas close to Chernobyl *(14)* and in many other places where the large amount of "fuel-like" hot particles was noticed. Thus, we conclude that ^{125}Sb found in samples originates mainly from "continuous" Chernobyl fallout, the ^{106}Ru seems to be partially from hot particles, and the ^{144}Ce comes mostly from small hot particles of the fuel-like type. The latest information suggests the presence of actinide isotopes of Chernobyl origin in samples in which ^{144}Ce was observed *(15)*. The search for plutonium will be the subject of our next investigation.

ACKNOWLEDGMENTS

The great support of our work from Professor Jan Mikulski is warmly acknowledged. We would like also to thank Dr. Ewa Ochab very much for preparation of the europium-cerium standard.

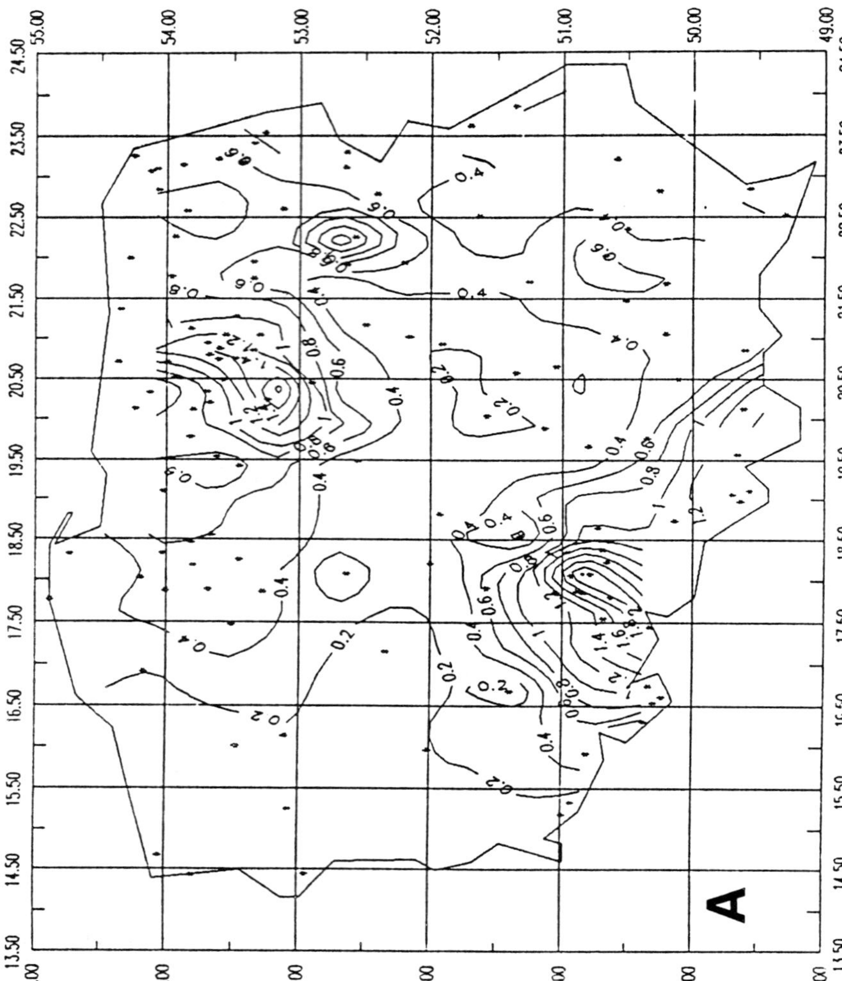
A

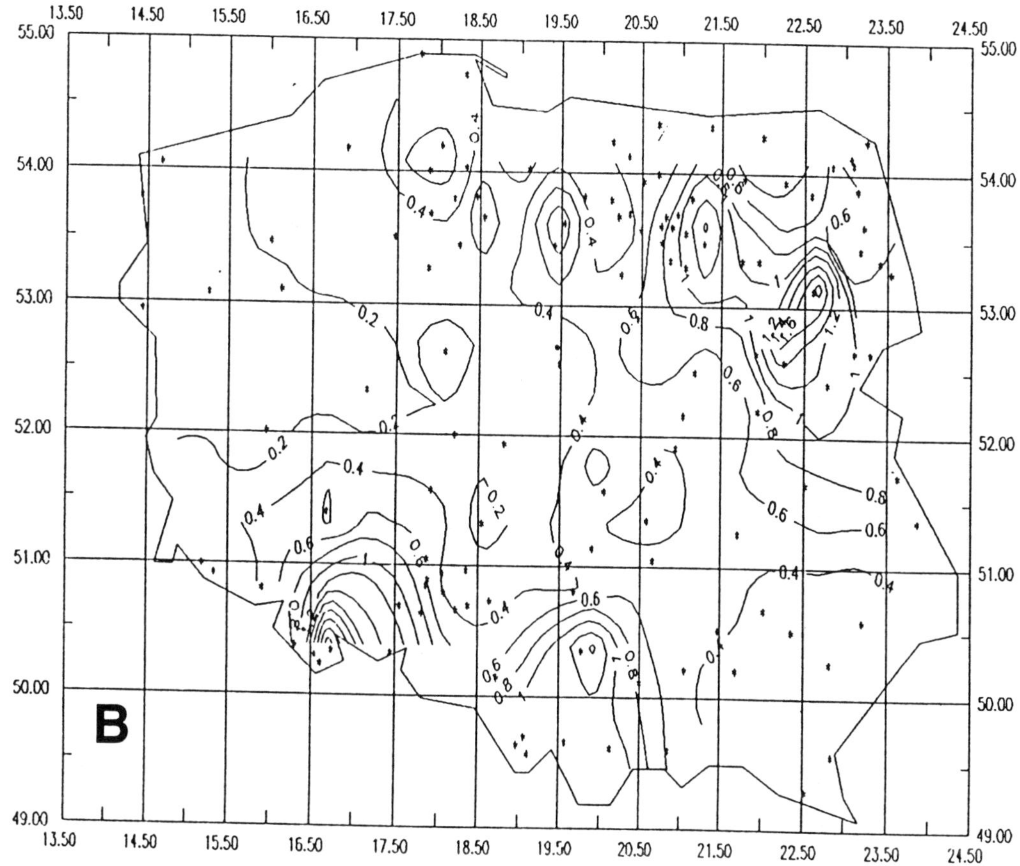

Fig. 1. The geographical distribution of ^{137}Cs activity in forest litter of Poland. A. In A_0 layer. B. In A_1 layer. All activities are given in kBq/kg d.m. and decay corrected to September 1, 1991.

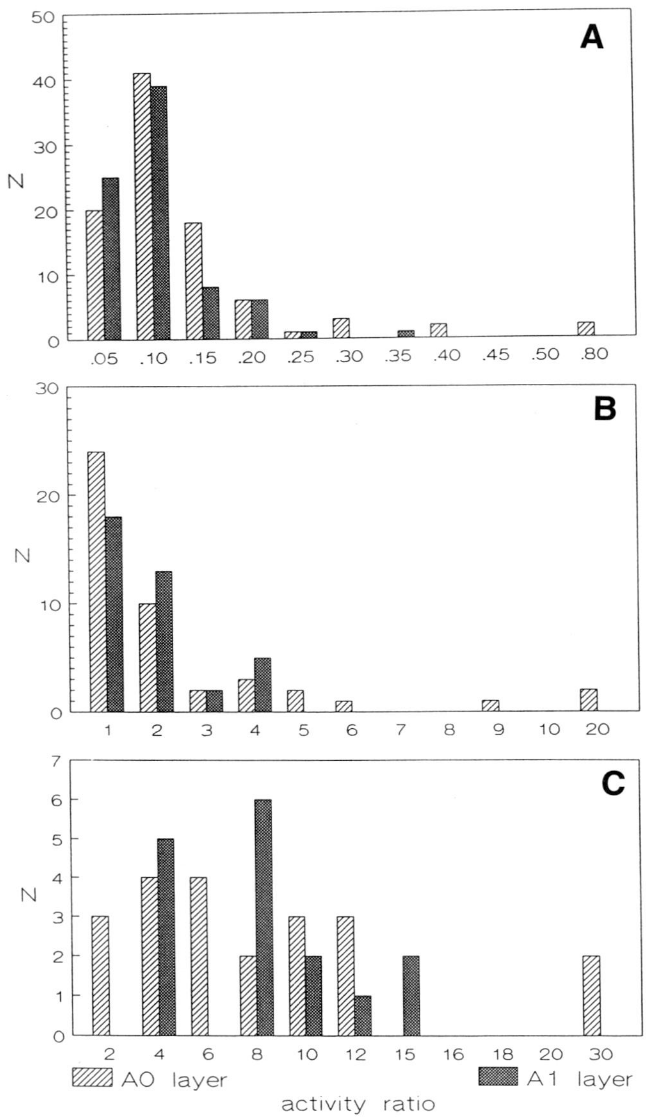

Fig. 2. The distribution of activity ratio in litter samples: A. ^{125}Sb to ^{137}Cs, B. ^{106}Ru to ^{137}Cs, and C. ^{144}Ce to ^{137}Cs. The ratios are decay corrected to May 1, 1986. Data do not include samples where the given activity was below the detection limit.

REFERENCES

1. J. W. Mietelski, M. Jasińska, B. Kubica, K. Kozak, and P. Macharski, *Proceedings of the Seminar on the Dynamic Behaviour of Radionuclides in Forests, Stockholm, May 18–24, 1992* to be published in *Sci. Total Environ.*
2. P. Zagrodzki, J. W. Mietelski, and M. Krósniak, *Biol. Trace Element Res.* this issue.

3. J. W. Mietelski, *Report IFJ* No 1435/s Kraków (1989).
4. IAEA 154 Whey Powder, Reference Sheet, IAEA, Vienna, 1988.
5. IAEA Soil-6, Reference Sheet, IAEA, Vienna, 1984.
6. J. W. Mietelski, M. Jasi´nska, B. Kubica, K. Kozak, and P. Macharski, Proceedings of the Third International Conference on Nuclear and Radiochemistry, Sept. 7–11, 1992, Vienna, to be published in *Radioanal. Nucl. Chem., Articles.*
7. M. Biernacka, J. Henschke, J. Jagielak, and A. Korczy´nski, *Proceedings of the International Symposium on Post-Charnybyl Environmental Radioactivity Studies in East European Countries, Kazimierz, Sept. 17–19, 1990,* UMCS Press, Lublin, 1991, pp. 26–33.
8. L. Pienkowski, J. Jastrzebski, J. Tys, T. Batsch, P. Jaracz, W. Kurcewicz, S. Mirowski, G. Szeflinska, Z. Szeflinski, B. Szweryn, Z. Wilhelmi, and E. T. Józefowicz, *J. Radional. Nucl. Chem. Lett.* **117,** 379 (1987).
9. R. Broda, *Acta Phys. Pol.* **B18,** 935 (1987).
10. I. Osvath, *Proceedings of the International Symposium on Post-Chernobyl Environmental Radioactivity Studies in East European Countries, Kazimierz, Sept. 17–19 1990,* UMCS Press, Lublin, 1991, pp. 113–117.
11. R. Broda, J. W. Mietelski, and J. Sieniawski, *J. Radioanal. Nucl. Chem., Lett.* **166,** 173 (1992).
12. S. H. Faller, D. J. Krask, and P. K. Kuroda, *Radichimica Acta* **54,** 43 (1991).
13. J. W. Mietelski, R. Broda, and J. Sieniawski, *J. Radioanal. Nucl. Chem., Lett.* **127,** 367 (1988).
14. Yu. A. Izrael, V. G. Sokolovskij, V. E. Sokolov, V. A. Vetrov, I. K. Dibobes, D. F. Trusov, I. N. Ryabov, A. P. Povalyaev, L. A. Buldakov, and V. A. Borzilov, *Atomnaja energija* **64,** 28 (1988).
15. R. Broda, B. Kubica, Z. Szeglowski, and K. Zuber, *Radiochimica Acta* **48,** 89 (1989).

The Measurement of Low Concentrations of Kr-85 in Atmospheric Air Samples

LUDMILA WILHELMOVÁ,* MILAN TOMÁŠEK, AND KAREL ŠTUKHEIL

Institute of Radiation Dosimetry, Czech Academy of Sciences, Na Truhlářce 39/64, 180 86 Prague 8, Czech Republic

ABSTRACT

Krypton-85 atmospheric contamination produced by the nuclear industry has been investigated on the basis of regular measurements. ^{85}Kr concentrations observed in ground-level air of Prague between 1983 and 1993 are presented and their time-related changes analyzed. The long-term trend of ^{85}Kr air activity increased steadily over the monitored period with the growth rate of 38 mBq·m^{-3} (STP)/yr and reached about 1.2 Bq·m^{-3} (STP) in 1992. Short-term variations of the baseline level with the average amplitude of 12% show a seasonal periodicity. Occasionally detected extreme values of up to 2.5 Bq·m^{-3} (STP) indicate undispersed emissions from distant nuclear fuel reprocessing plants.

Index Entries: krypton-85; radiation monitoring; annual variations; radioactivity; atmosphere.

INTRODUCTION

Pollution of the atmosphere by the long-lived fission product ^{85}Kr ($t_{1/2}$ = 10.76 yr) results from human activity in exploitation of nuclear energy for peaceful and military purposes. The concentration of this radionuclide in the atmospheric air has been increasing with growth of the nuclear industry during the past several decades *(1,2)*. The main sources of present ^{85}Kr contamination are the reprocessing plants of spent nuclear fuel, which discharge the radioactive gases without applying technology for their retention.

*Author to whom all correspondence and reprint requests should be addressed.

Biological Trace Element Research Editor: G. N. Schrauzer © 1994 by Humana Press Inc.

As an inert gas with the only significant sink by radioactive decay, ^{85}Kr has been accumulating in the earth's atmosphere, enhancing its radiation background. The growth of ^{85}Kr activity in the global environment can have negative effects not only in terms of radiological impact, but also in terms of changes of the ionization conditions of the atmosphere and related geophysical processes *(3)*.

Activity measurement of the globally dispersed radionuclides requires the use of analytical methods with sensitivities significantly higher than those of the commonly used methods for radiation protection. The ^{85}Kr air activity cannot be determined by a direct measurement. For its quantification, highly concentrated and purified samples of natural krypton separated from relatively large volumes of air must be employed. Methods developed and applied for this purpose take advantage of gas chromatography for the krypton sample preparation and internal gas counter or scintillation counting technique for the ^{85}Kr activity measurement *(4)*.

Basic information on ^{85}Kr atmospheric contamination is derived from the measurement of its activity in near-ground air. The published data are mostly related to the source region at the middle latitudes of the Northern Hemisphere, where the highest concentrations are also observed *(2–7)*. This article presents the data on ^{85}Kr air activity in Prague collected during the period 1983–1992. The long-term trend of ^{85}Kr governed by its global occurrence was investigated to provide the basis for the detection of possible local baseline irregularities and their subsequent radiological assessment.

EXPERIMENTAL

The technique used in our laboratory for preparation of atmospheric krypton samples is based on a selective adsorption of air components in beds of solid adsorbents *(8)*. The apparatus consists of molecular sieve traps for removal of H_2O and CO_2, charcoal traps for the krypton retention and concentration, and a gas chromatographic unit enabling the final sample purification and its mass analysis. Since the atmospheric ratio of natural krypton is relatively constant (1.14 $cm^3 \cdot m^{-3}$ STP), the last step allows us to determine the yield of its separation from the treated air in each experiment.

The activity of ^{85}Kr (99.6% β^-, E_{max} = 695 keV, 0.44% γ, E = 514 keV) in the krypton fraction is measured by the counting chamber equipped with a scintillation detector $CaF_2(Eu)$ *(9)*. With the conventional ^{85}Kr activity measurements (efficiency 13%, background 0.5 cps, counting time 5.10^3 s both in the case of background and sample ^{85}Kr activity 5–10 Bq), the relative error of $<(3\%)_{0.95}$ has been achieved.

In routine experiments, an air sample of about 10 m^3 is processed. The collection is performed in the locality of the Institute in Prague (50°04 N, 14°26 E). The results obtained between 1983 and 1988 corre-

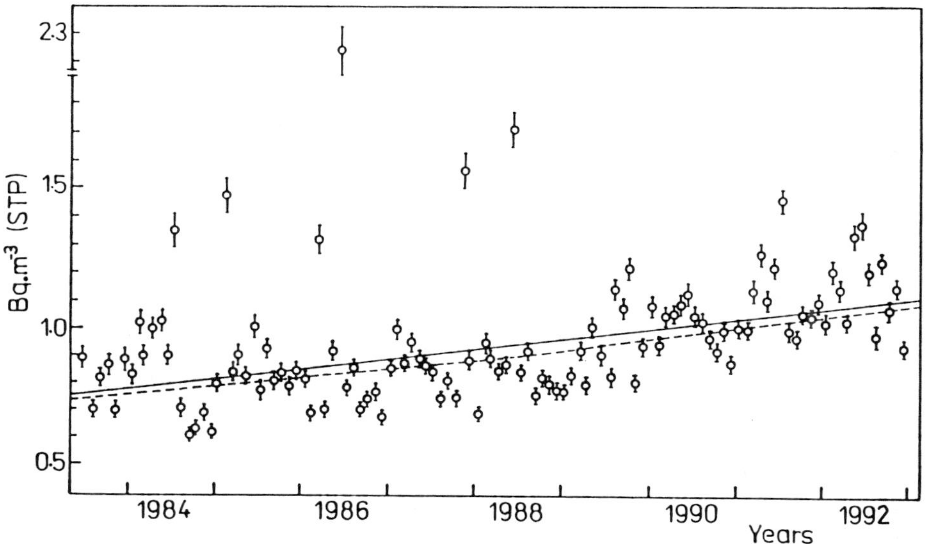

Fig. 1. Concentration of ^{85}Kr in the ground-layer atmosphere of Prague. Full line denotes the linear fit, and dotted line the exponential fit.

spond to the air samples collected regularly for 1 d at monthly intervals. The results since 1989 represent the ^{85}Kr activity of the air samples averaged continuously in monthly intervals. The compatibility of these two sets of data was checked by experiments carried out simultaneously during 1989. Statistical treatment of the data revealed a slightly higher scatter of single-day (spot) observations, but it has proven that the trend and the yearly mean values are identical. The reason for the change of the measurement methodology was the radiological relevancy of the obtained data.

RESULTS AND DISCUSSION

The concentrations of ^{85}Kr observed in the ground-layer atmosphere of Prague between 1983 and 1992 are presented in Fig. 1. It can be seen that the air contamination by this radionuclide has a rising tendency over the monitored period and reached the average level of about 1.15 Bq/m^3 (STP) in 1992. During most of the measurements, the ^{85}Kr activity varied within 20% of typical baseline range and only in a limited number of cases achieved distinctly higher values.

Extreme values of ^{85}Kr activity recorded on 84/7/11, 85/2/13, 86/6/8, 87/11/11, and 88/6/8 were identified in the corresponding sets of yearly data as outlying observations. Because of the absence of significant local sources, their occurrence is probably the result of the passage of the air masses contaminated in the regions with major ^{85}Kr sources. This assumption is to a large extent confirmed by the trajectories of the

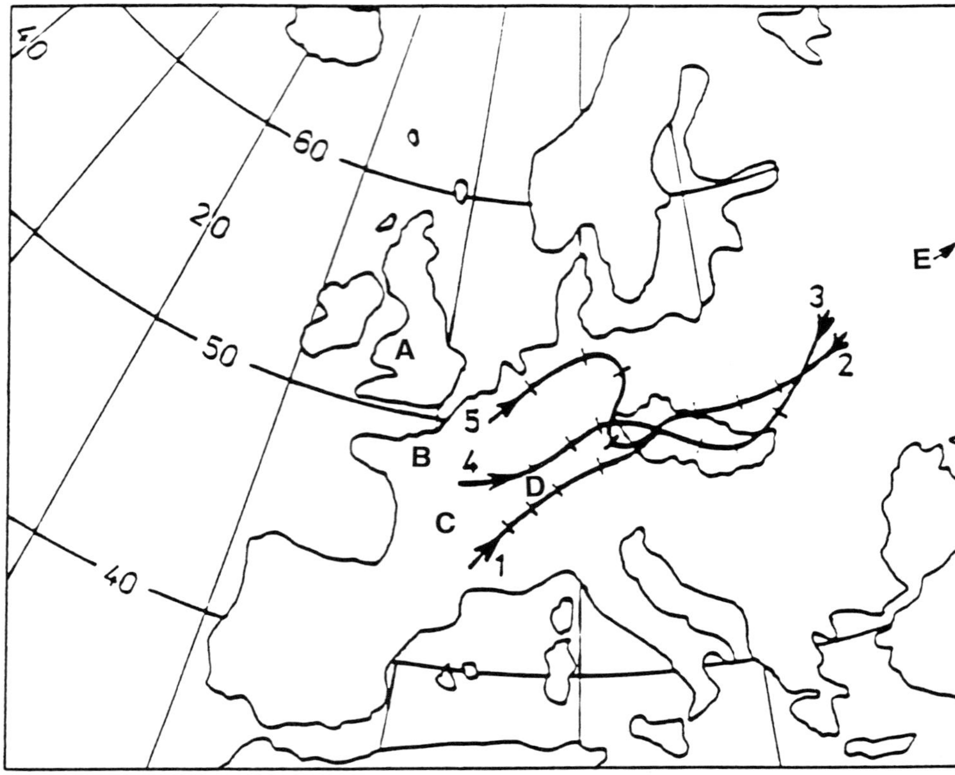

Fig. 2. Backwards trajectories of air masses' movement for the case of outlying observations on days: 1—84/7/11, 2—85/2/13, 3—86/6/8, 4—87/11/11, 5—88/6/8. Locations of the nuclear fuel reprocessing plants: A—Selafield, B—La Hague, C—Marcoule, D—Karlsruhe, E—Middle Ural.

air masses' movement investigated on the 850 hPa level within 96 h before start of the sampling *(10)*. The trajectories together with the locations of the nuclear fuel reprocessing plants are presented in Fig. 2. However, the direct influence of the emissions has a short time character that has been proven by the fact that it had been occasionally recorded only in the spot series of data and not in the air samples integrated over a month.

Further analysis of the obtained data on ^{85}Kr concentration presented in Fig. 1 has shown that the short time variations recorded in the set of typical values are dependent on the season of the year. Variability of this kind has been recorded during all the measurements without any coincidence with the method of air sample collection. A seasonal concentration maximum with an amplitude of approx 12% of the mean baseline value appears in the spring. The reasons for such a behavior of ^{85}Kr have not yet been explained. Referring to several studies, those reasons can originate in the seasonal regime of operation of some reprocessing

Table 1
Characteristics of Long-Term Trend of ^{85}Kr Air Activity

Type of regression	X^2	A_{2000} $Bq.m^{-3}$
$A_t = 0.623 + 0.038(t-1980)$	7.82	1.31
$A_t = 0.662.\exp(0.0405(t-1980))$	8.18	1.50

$X^2_{0.99} = 9.2$ *t*-year AD.

plants *(2)*. Nevertheless, the possible influence of the natural factors cannot be excluded, and the topic therefore requires further investigation.

The long-term trend of the ^{85}Kr activity increase in the atmosphere of Prague recorded during the period 1983–1992 can be approximated by both a linear and an exponential form. Parameters of both fit types together with their closeness (chi-square test) are presented in Table 1. Linear regression yields the increase rate of 38 mBq/m^3 (STP)/yr and leads to a prediction that the average concentration of ^{85}Kr in atmosphere will be 1.39 Bq/m^3 (STP) in the year 2000. The exponential regression results in a 4% yearly increase of ^{85}Kr content in the atmosphere reaching 1.5 Bq/m^3 of ^{85}Kr (STP) in the year 2000.

Mean yearly concentrations of ^{85}Kr in Prague's atmosphere are plotted in Fig. 3 together with the estimations of mean values derived by a statistical processing of the available data recorded at the middle latitudes of the Northern Hemisphere *(11)*. The pattern displayed in Fig. 3 can be considered to be typical for the evolution of environmental contamination by ^{85}Kr in the given geographical zone. Our observations follow this trend well and indicate that the ^{85}Kr concentration above the territory of the Czech Republic reflects in general the changes of its zonally dispersed levels.

The individual dose equivalent for an inhabitant owing due to irradiation by atmospheric ^{85}Kr at its present level (4.7×10^{-7} Sv/yr on skin and 3.5×10^{-9} Sv/yr on whole body) represents about 0.01% of the value given by the hygienic standard. The contribution of this radioactive gas to the atmospheric ionisation is estimated to be 0.5% at present. However, the still existing uncertainty about the effects of the low-radiation doses together with the reasonably predictable changes in climate as a consequence of the excessive ionization of the atmosphere *(3)* evoke fears of its further radioactive contamination. As far as ^{85}Kr is concerned, the situation could stabilize under the condition that provisions for its capture and storage during fuel reprocessing will be introduced.

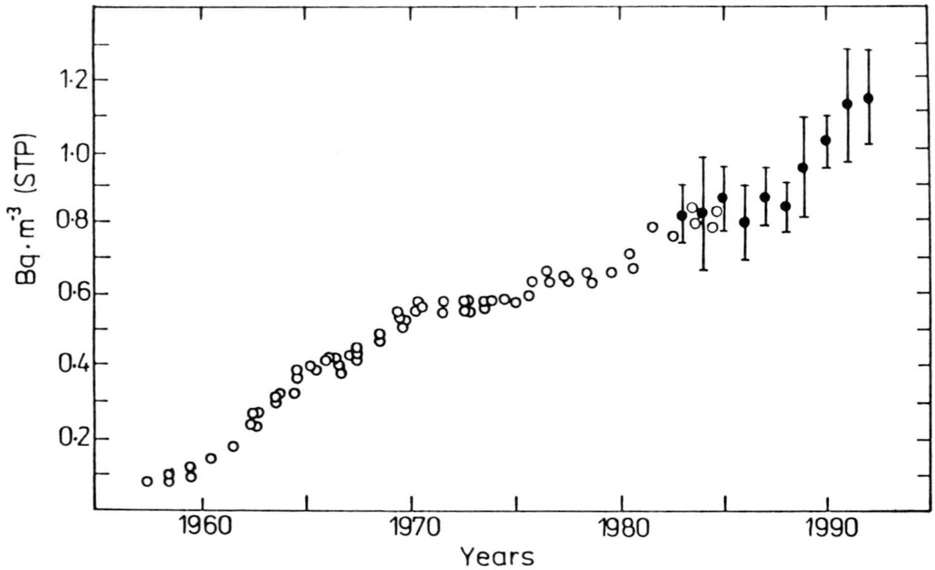

Fig. 3. Evolution of mean ^{85}Kr activity levels in the ground-layer atmosphere of middle latitudes of the Northern Hemisphere. ○—mean annual values derived from literary data; ●—mean annual values derived from our measurements.

REFERENCES

1. Z. Dvořák, A study of atmospheric contamination by Krypton-85 released from nuclear facilities, thesis, Institute of Chemical Technology, Prague, 1987.
2. W. Weiss, H. Sartorius, and H. Stockburger, in Proceedings of Consultants Meeting on Isotopes of Noble Gases as Tracers in Environmental Studies, Vienna 1989, IAEA Vienna, 1992, p. 29.
3. B. Styro and D. Butkus, *Geophysical Problems of Krypton-85 in the Atmosphere,* Moskals, Vilnius, 1988, pp. 112–130.
4. R. J. Jaquish and A. A. Moghissi, in *Noble Gases,* R. E. Stanley and A. A. Moghissi, eds., Government Printing Office, Washington, DC, 1973, p. 169.
5. E. Csongor, in *Proceedings of the 2nd International Conference on Low Level Counting, The High Tatras 1980,* P. Povinec ed., VEDA, Bratislava, 1982, p. 357.
6. Š. Cimbák and P. Povinec, *Environ. Int.* **11,** 69 (1985).
7. E. G. Tertyshnik, A. A. Severin, V. G. Baranov, and S. M. Vakulovskij, *At. Energia* **53,** 114 (1982).
8. L. Wilhelmová, M. Tomášek, and Z. Dvořák, *J. Radional. Nucl. Chem., Lett.* **95,** 45 (1985).
9. L. Wilhelmová, Z. Dvořák, M. Tomášek, and K. Štukheil, *Apppl. Radiation Isotopes* **Part A, 37,** 429 (1986).
10. Czech Hydrometeorological Institute, Prague, personal communication (1989).
11. Z. Dvořák, L. Wilhelmová, and F. Cejnar, *Jaderná energie* **27,** 182 (1981).

Extraction of Uranium from Acidic Solutions by TBP Impregnated Polyurethane Foam

Ş. ÖLMEZ* AND M. ERAL

Ege University, Institute of Nuclear Sciences, 35100 Bornova, İzmir, Turkey

ABSTRACT

Polyurethane foams can be used as a selective absorbent for a number of substances from dilute aqueous solutions. Absorption of uranium (VI) has been investigated using an open-cell polyurethane foam impregnated with tri-butyl phosphate (TBP). The extraction efficiency was observed to depend on the pH of the uranium solution and the concentration of uranyl ions. Also, the effects of TBP concentration, temperature, and squeezing time on the extraction of uranium were examined.

Index Entries: Uranium (separation, absorption, sorption, extraction); tri-butyl phosphate (TBP); polyurethane foam; urethane polymers.

INTRODUCTION

The content of the radionuclides in uranium processing plant-waste solutions must be kept as low as possible. A number of methods have been employed for removing small amounts of radionuclides *(1)*. The methods of ion exchange (IX), solvent extraction (SX), and selective precipitation can be used for processing of waste solutions. Recently, polyurethane foams have been of interest as efficient sorbents for the separation, collection, and concentration of trace amounts of metal ions. Polyurethane foams can be used as an inert support of an extractant for the extraction and recovery of metal ions in analytical and waste-water treatment processes *(1–4)*.

*Author to whom all correspondence and reprint requests should be addressed.

Polyurethane foams are polyesters or polyethers crosslinked by amido linkages and foamed with carbon dioxide during the crosslinking *(5)*. It was explained that the chemical nature of the foam in the previous work *(3)* might be similar to a weak anion exchanger with low capacity, since these foams contains amido and perhaps amino groups.

The use of certain extractants for uranium extraction has been reviewed by several investigators *(1–5)*. The present article deals with the extraction of uranium (VI) on a polyurethane foam impregnated with TBP.

EXPERIMENTAL

Uranium stock solution was prepared by dissolving $UO_2(NO_3)_2 \cdot 6H_2O$ (Merck) to make 1000 mg/L of solution, which was diluted as required with HNO_3. The commercial polyurethane foam was obtained from the Pinar Co. in İzmir-Turkey. The TBP was supplied from Merck. All other chemicals were reagent-grade.

Uranium in aqueous solution was determined by the 1-(2-pyridylazo)-resorcinol (PAR) method (λ = 510 nm, $\varepsilon = 3.87 \times 10^4$ L/mol.cm) *(6)* by using Shimadzu UV-Vis 260 Recording Spectrophotometer.

Polyurethane foam plugs, 4.5 cm in diameter and 2.2 cm long (average weight = 0.5500 ± 0.0020 g), were cut from foam sheet.

Pretreatment was applied by the procedure suggested by Gesser and Gupta *(7)*. Each foam plug was squeezed in 2 mol/L HCl in a batch extractor for 1 h prior to use, washed with distilled water until free of HCl, and again squeezed dry. The foam plug was rinsed with acetone, squeezed, and air-dried over night before using.

The cleaned and dried foam plugs were impregnated in a cyclohexane solution of TBP, and then cyclohexane was removed in air. The extraction was carried out by a dynamic method in an automatic squeezing machine. The amount of uranium extracted (absorbed) by the impregnated foam was determined from the difference between the initial and final concentrations of uranium in aqueous solution.

RESULTS AND DISCUSSION

Effect of pH

The pH of test solution was adjusted to a desired value by addition of HNO_3 or NH_4OH solution. The effect of pH on the extraction of uranium by TBP impregnated foam is shown in Fig. 1.

The low percentage of extraction is probably the result of the hydrolysis of the uranyl ion *(8)* into the nonextractable species or the possibility of desorption in acidic medium. This sorption behavior of the TBP-impregnated foam is essentially similar to that of the foam impregnated with LIX 63 *(1)*.

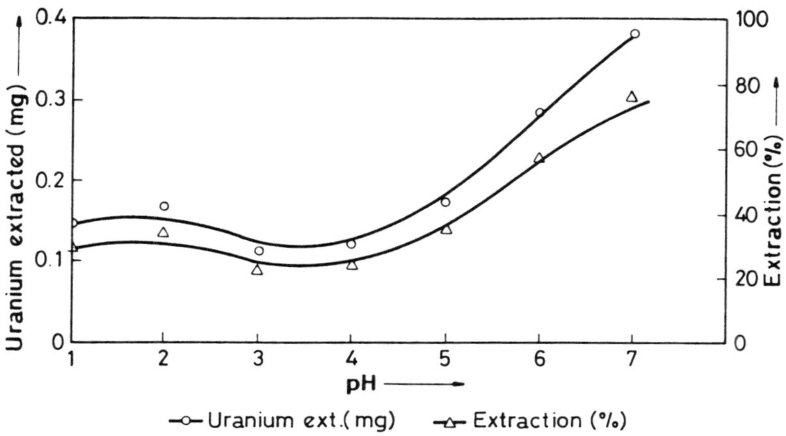

Fig. 1. Effect of pH on the extraction of uranium (VI) onto polyurethane foam impregnated 20% TBP/cyclohexane, [U] = 5 mg/L (in HNO_3), V = 50 mL, 120 min, 25°C, 40–50 stroke/min, average weight of TBP on foam = 1.94 g.

Effect of Uranium Concentration

Extraction behavior of uranium was examined by varying initial concentrations from 5.00 mg/L to 100 mg/L. Figure 2 shows that uranium is quantitatively extracted at low concentrations. At high uranium concentrations, the percentage of extraction falls with increasing owing to deficiency of extractant. An excess of TBP impregnating the foam is required for the quantitative extraction of uranium.

Also, the effect of equilibrium uranyl concentration in the aqueous phase on the distribution coefficient, *D*, is shown as

$$D = \Sigma[U]_{org}/\Sigma[U]_{aq} \qquad (1)$$

Effect of TBP Concentration

Figure 3 presents the extraction of uranium from aqueous solution on the foam impregnated with TBP solution ranging from 5 to 100 vol% in cyclohexane. The uranium absorption capacity of foam increases with impregnating TBP concentration up to 30% TBP/cyclohexane, but decreases in the high TBP concentration region. Jangg and Burker have reported that the adsorbtion of uranium on anion-exchange resins decreases with increasing TBP concentration *(9)*. TBP can be used in concentrations up to 100 vol%, but more usually as a 25–30 vol% solution in a diluent. In some systems, extraction is adversely affected by increasing the concentration of TBP in the solvent *(8)*.

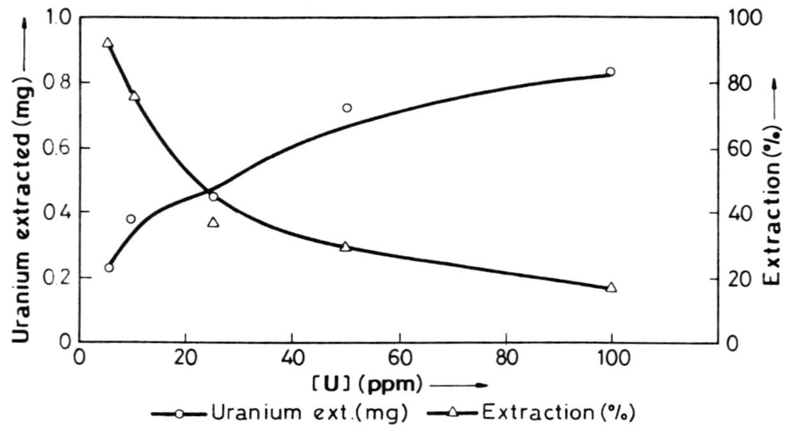

Fig. 2. Effect of uranium concentration on uranium extraction, pH = 7.00, 25°C.

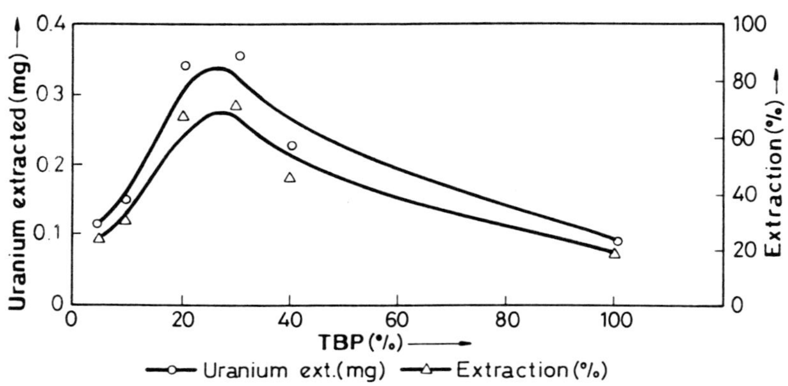

Fig. 3. Effect of TBP concentration [U] = 10 mg/L, pH = 7.00, V = 50 mL.

Rate of Extraction

The time dependence of uranium extraction was examined at different temperatures between 10°C and 50°C. The results are shown in Fig. 4.

Solvent extraction is an equilibrium process. The rate is also an important factor affecting extraction. The extraction (absorption) of uranium increases gradually with contact time at 25°C and 50°C, but decreases after 60 min. The extraction of uranium decreases by increasing time at 10°C.

CONCLUSION

It has been shown that TBP-impregnated polyurethane foam can be used for absorption and extraction of uranium from dilute aqueous solution. Polyurethane foam can act as a cheap and readily available support

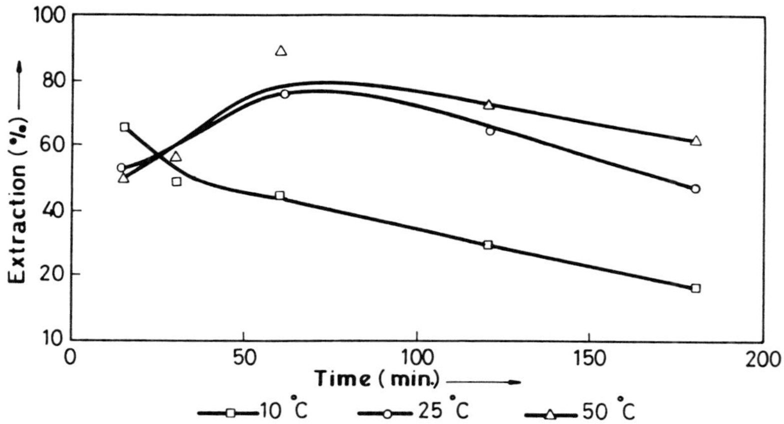

Fig. 4. Effect of temperature on the rate of uranium extraction, pH = 7.00, 30% TBP/cyclohexane on 0.5510 g foam.

material for the neutral organic extractant (TBP). Quantitative collection of uranium in the relatively wide range of concentrations is achieved.

The uranium extraction under determined optimum experimental conditions was obtained as 86.2 ± 5.9%. For the investigation of the desorption behavior and the complete recovery of uranium, the effect of some interfering impurities should be studied.

ACKNOWLEDGMENT

Financial support from the Ege University Research Foundation is gratefully acknowledged.

REFERENCES

1. K. Akiba and H. Hashimoto, *J. Radional. Nucl. Chem., Articles* **130,** 13–20 (1989).
2. H. D. Gesser and S. Ahmed, *J. Radional. Nucl. Chem., Articles* **140,** 395–408 (1990).
3. H. J. M. Bowen, *J. Chem. Soc.* **(A),** 1082–1085 (1970).
4. J. J. Oren, K. M. Gough, and H. D. Gesser, *Can. J. Chem.* **57,** 2032–2036 (1979).
5. F. H. Mark, M. N. Bikales, C. G. Guerberger, and G. Menges, *Encyclopedia of Polymer Science and Engineering, Polyurethanes Wiley,* New York, 1981, p. 243–303.
6. A. I. Busev, V. G. Tipstova, and V. M. Ivanov, *Handbook of The Analytical Chemistry of Rare Elements,* Mir, Moscow, 1981.
7. H. D. Gesser and B. M. Gupta, in *Proceedings of International Solvent Extraction Conference 80,* Pub. Soc. Chem. and Ind., London, Liege-Belgium, Part I, 1980, pp. 80–122.
8. G. M. Ritcey and A. W. Ashbrook, *Solvent Extraction,* Part I, Elsevier, New York, 1984.
9. G. Jangg and A. Burker, *Österr. Chemiker. Z.* **61,** 7 (1960).

Investigation of Uranium Recovery from Dilute Aqueous Solutions Using Silk Fibroin

MAHMOUD A. A. ASLANI* AND MERAL ERAL*

Ege University, Institute of Nuclear Sciences, 35100 Bornova, Izmir, Turkey

ABSTRACT

The uranium uptake ability of silk fibroin was investigated. High ability to uptake uranium from nonsaline water containing 2.500 mg of uranium was observed with the silk fibroin tested. The uranium uptake was very rapid and was dependent on pH, uranium concentration, temperature, and retention time. Almost all uranium taken up is easily eluted with 1 mol/L CH_3COONH_4. This biomatrix, therefore, appears to have potential for use in a commercial process for uranium recovery from uranium-containing waste water.

Index Entries: Silk fibroin; uranium uptake; *Bombyx mori*; separation; waste water; uptake of heavy metals, biomatrix.

INTRODUCTION

It has been reported that the amino acid units of fibroin protein that forms silk are linearly arranged *(1)*. The amino acid and the peptide chain of the *Bombyx mori* silk fibroin has been shown as Gly-Ala-Gly-Ala-Gly-Ser-Gly-Ala-Ala-Gly-(Ser-Gly-(Ala-Gly)n)$_8$-Try sequence (n = 2) *(2)*. There have been some investigations about the uptake of heavy metals, like Cu (II), Ni (II), Hg (II), Pb (II), Sn (II), (IV), and Fe (III) from aqueous solutions using the silk mentioned above *(3,4)*. This article describes the basic features of uranium uptake from aqueous systems by silk fibroin. The elution of uranium from the loaded fibroin is also discussed.

*Authors to whom all correspondence and reprint requests should be addressed.

EXPERIMENTAL

MATERIALS

Native silk was obtained from the Silk Industry Research Center, Bursa, Turkey. 4-(2-Pyridylazo)resorcinol (PAR), $UO_2(NO_3)_2{\cdot}6H_2O$ Merck company and all other chemicals were reagent-grade.

Preparation of Solutions

Aqueous reagents were prepared in deionized water. The standard uranium feed solution was prepared by dissolving pure $UO_2(NO_3)_2{\cdot}6H_2O$ in 0.5 mL of concentrated HNO_3 and diluting with deionized water. The degumming solution was prepared by dissolving 0.5 g of powder soap, and 0.05 g of Na_2CO_3 in 100 mL deionized water. Then, this solution was used for degummation of silk.

Degummation of Silk

The native silk was cut with scissors into lengths about 0.5 cm. Then, 2.00 g of native silk was degummed with soap mixed for 2 h at 99–100°C. The degummed silk was filtered, washed thoroughly with deionized water, and air-dried. Thus, fibroin, which was enclosed with sericin, is saved.

Analysis of Fibroin

Microscopic investigation of the longitudinal view of the fibroin showed that the surface of the fibroin layer was removed almost completely from the sericin layer. Also, the yield of degummation, which was found to be 76.31%, conformed to the literature values *(5)*. The fibroin was then used for uranium uptake tests.

Uptake Experiments

Uptake of Uranium

The fibroin was suspended in 25 mL of the solution (pH = 5) containing 2.500 mg of uranium. The pH of the solution was adjusted to the desired value with 0.1–0.01 mol/L $NaHCO_3$ or 0.1 mol/L HNO_3 solution. The suspension was stirred continuously for 120 min at 25°C. After uptake, the loaded fibroin was collected by filtration. Then, the residual uranium in the filtrate was determined by spectrophotometry using the PAR method *(6)*.

Determination of Uptake Capacity

The determination was carried out by batch techniques. 0.125 g fibroin was equilibrated with a solution (50 mL) containing 0.01 mol/L of UO_2^{2+} as its nitrate at 30°C for 24 h *(7)*. The uranium uptake capacity of fibroin was determined to be 1.17 UO_2^{2+} mmol/g.

Elution of Uranium from the Loaded Fibroin

For this purpose, different elution solutions were being tested. After 2.43 mg of uranium had been taken up on 0.1 g of fibroin, the loaded fibroin was collected by filtration, washed thoroughly with deionized water, and then eluted using 20 mL of 1 mol/L CH_3COONH_4 (pH = 5) as elution solution. The elution tests were performed in a closed Erlenmayer flask and stirred 180 min at 25°C. After elution, fibers were collected by filtration. Then, the residual uranium in the filtrate was determined by the spectrophotometric PAR method.

RESULTS AND DISCUSSION

Parameters Effecting Uranium Uptake

The uptake of uranium by fibroin was very rapid during the first 5 min following the addition of uranium. The analysis of some factors, such as pH, uranium concentration, temperature, retention time, and mixing rate, that influence the uptake of uranium is necessary for the purpose of understanding the recovery of uranium from aqueous systems.

Effect of Temperature

Proteins are known to be very sensitive to temperature. Most of the proteins lose their biological activities by irreversible denaturation above 40°C. The protein separation processes should be carried out carefully at temperatures below the temperature value in which denaturation of protein begins. On the other hand, the temperature in the separation medium should be high enough to prevent protein precipitation *(8)*. In this group of experiments, as seen in Fig. 1, where the temperature ranged from 18 to 50°C, the amounts of uranium taken up by fibroin decreased with increasing temperature, because the uranyl nitrate complex, which is formed in this medium, is rarely influenced by increasing temperature *(9)*. The results indicated that we can make the uptake experiments at 25°C.

Effect of Mixing Rate

It was found that the yield of uranium uptake increased by a factor of two when the mixing rate was 60–120 rpm (Table 1).

Effect of pH on the Uptake of Uranium

The effect of pH on uranium adsorption by fibroin in the pH range 1–10 was investigated. As shown in Fig. 2, maximum uranium uptake was observed at pH = 5, whereas below and above this pH, the uptake of uranium fell off rapidly. This pH value is the stability pH for UO_2CO_3,

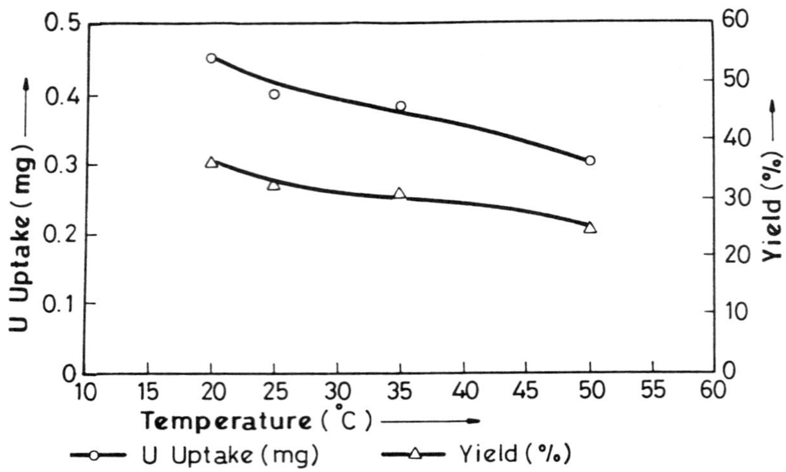

Fig. 1. Effect of temperature on the uptake of uranium (0.1 g fibroin was suspended in 25 mL of the solution containing 1.250 mg of uranium at pH = 2.78 for 360 min).

Table 1
Effect of Mixing Rate

Mixing Rate (rpm)	Amount of Loaded U onto Fibroin (mg)	Yield (%)
60	0.38	30.70
120	0.70	56.09

0.1 g of Fibroin was suspended in 25 mL of 1.25 mg containing standard uranium solution at pH = 2.78 for 360 min at 25°C.

which was selectively absorbed by fibroin *(10)*. From these results, it may be concluded that the decrease in uranium uptake above pH = 6 is the result of the stable $[UO_2(CO_3)_2]^{-2}$ and/or $[UO_2(CO_3)_3]^{-4}$ ions. Below pH = 5, the adsorbent exchanges its protons with uranium in analogy with the functional groups according to the following equation:

$$LH_2 + UO_2^{+2} \Longleftrightarrow LUO_2 + 2H^+ \quad (1)$$

where L represents the functional group. When the equilibrium is shifted to the left at low pH, owing to the increase of H^+, the amount of uranium adsorbed decreases *(11)*.

Effect of Uranium Concentration

The yield of uptake rapidly increased with increasing of concentration and remained at this high value. The yield was determined as approx 90% at 300 mg/L concentration (Fig. 3). The original white color of fibroin silk was gradually darkened to orange color with increasing uranium concentration.

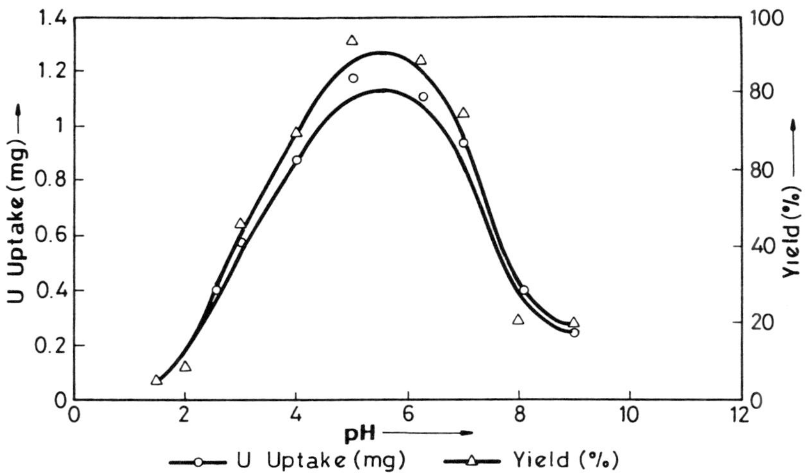

Fig. 2. Effect of pH (0.1 g fibroin was suspended in 25 mL of the solution containing 1.250 mg of uranium for 240 min at 25°C).

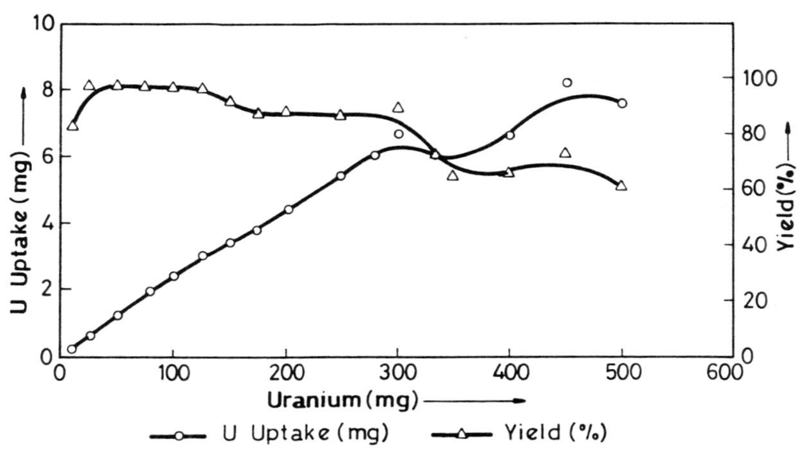

Fig. 3. Effect of the uranium concentration (0.1 g fibroin was suspended in solutions, at pH = 5 for 240 min at 25°C).

Effect of the Retention Time

As shown in Fig. 4, the equilibrium uptake of fibroin was reached within a period of 90 min, which indicates that uranium uptake is rapid and efficient. At the first 5 min, the yield of uranium uptake is 58%. At this condition, we assume that the rate of uptake is dependent on diffusion in fibroin and diffusion in the solution *(12)*. However, the optimum retention time was selected as 120 min. At the determined optimum conditions, the yield of uranium uptake is calculated as 96.95 ± 1.51%.

Selection of Elution Reagent

Different kinds of elution solutions, i.e., CH_3COOH, CH_3COONa, CH_3COONH_4, NH_2CONH_2/MeOH, Na_2CO_3, K_2CO_3, HCl, HNO_3,

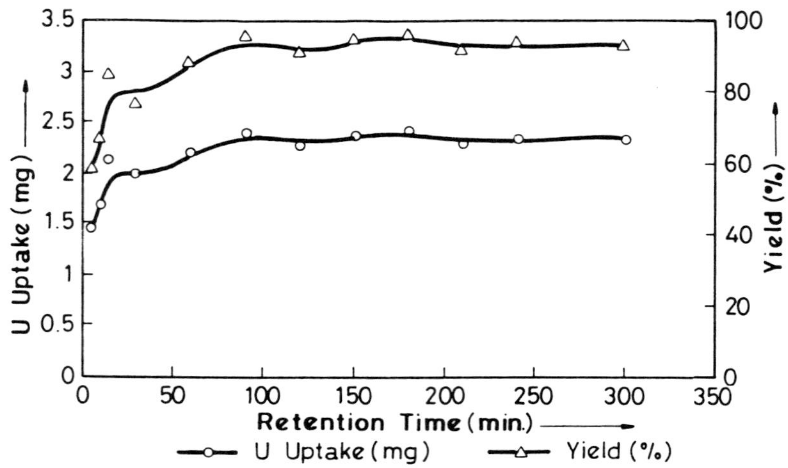

Fig. 4. Effect of the retention time (0.1 g fibroin was suspended in 25 mL of the solution containing 2.500 mg of uranium (pH = 5) at 25°C).

H_2SO_4, $KHSO_4$, $C_4H_6O_6$, $(NH_4)_2SO_4$, and H_2O, were tested. One mole per liter of CH_3COONH_4 solution was found to be the most suitable elution solution.

Parameters Effecting Uranium Elution

The pH of elution solution was adjusted to the desired value of 5 with concentrated CH_3COOH solution. At pHs 5.00 and 7.16, we calculated the elution yields of uranium to be 69.20 and 30.66%, respectively. From these results, the most convenient value of pH was obtained as 5.00.

The effect of retention time was investigated, and the optimum retention time was determined as 180 min. At these optimum conditions, the yield of uranium elution is determined as 85.57 ± 4.07%.

To obtain basic information on the recovery of uranium by the fibroin from aqueous systems, a batch system was studied through three uptake-elution cycles. As described above, uranium retained on fibroin can be quantitatively and easily eluted with 1 mol/L CH_3COONH_4. As shown in Table 2, almost quantitative recovery of uranium from aqueous system can be achieved by the fibroin, and almost all loaded uranium was eluted with 1 mol/L CH_3COONH_4.

ACKNOWLEDGMENTS

We are grateful to Professor Dr. D. Balköse for her encouragement and useful suggestions, and Bursa Silk Research Center for their help.

Table 2
Repetition Test of Uranium Uptake-Elution by Fibroin[a]

Cycle No	U Uptake onto Fibroin (mg)	U Uptake Yield (%)	U Eluted in eluate (mg)	U Eluted Yield (%)
I	2.48	99.20	2.09	84.26
II	2.27	90.84	1.32	58.20
III	2.07	82.87	0.66	31.84

[a]0.1 g of Fibroin was suspended in 25 mL of 2.500 mg containing standard uranium solution at pH = 5 for 90 min at 25°C. After washing with water, the uranium taken up on the fibroin was eluted with 20 mL of 1 mol/L CH_3COONH_4 solution at pH = 5 for 180 min at 25°C.

REFERENCES

1. M. Findik, M. Sc. Thesis, Ege University, Department of Biology, Faculty of Sciences (1991) (in Turkish).
2. T. Asakura, *Macromoleculers*, **17,** 2421–2426 (1984).
3. M. S. Masri and M. Friedman, *Protein Metal Interaction*, Plenum, New York, 1974.
4. D. Balköse, Assoc. Professorship Thesis, Ege University, Chemistry Faculty, June (1982) (in Turkish).
5. P. Carboni, *Silk Biology, Chemistry Technology*, Chapman and Hall Ltd., London, **494/4,** 39, 1952.
6. A. I. Busev, V. G. Tiptsova and V. M. Ironov, *Handbook of the Analytical Chemistry of Rare Elements*, Uranium Mir, Moscow 1981.
7. H. Egawa, N. Kabay, T. Nonaka and T. Shuto, *Bull. Soc. Sea Water Sci. Jpn.* **45,** 87–94 (1991).
8. A. Denizli, M. Olcay, A. Tuncel and E. Pişkin, *Doğa–Turkey J. Chem.* **16,** 135–141 (1992).
9. S. Akyil, M. Sc. Thesis, Ege University, Institute of Nuclear Sciences (1991) (in Turkish).
10. J. Bitte, proceedings of Advances in U Ore Proc. and Rec. from Noncon. Res. IAEA. Vienna 1985, pp. 299–311.
11. T. Sakaguchi and A. Nakajima, *J. Chem. Tech. Biotech.* **40,** 131–141 (1987).
12. D. Balköse and B. Pekin. *J. Sci. Fac.* **A, II,** 141–154 (1978) (in Turkish).

Author Index

Subject Index